Acetylene Chemistry

Edited by
F. Diederich, P. J. Stang,
R. R. Tykwinski

Further Reading from Wiley-VCH

A. de Meijere, F. Diederich (Eds.)

Metal-Catalyzed Cross-Coupling Reactions, 2nd Ed., 2 Vols.

2004, ISBN 3-527-30518-1

R. Mahrwald (Ed.)

Modern Aldol Reactions, 2 Vols.

2004, ISBN 3-527-30714-1

N. Krause, A. S. K. Hashmi (Eds.)

Modern Allene Chemistry, 2 Vols.

2004, ISBN 3-527-30671-4

R. Gleiter, H. Hopf (Eds.)

Modern Cyclophane Chemistry

2004, ISBN 3-527-30713-3

Acetylene Chemistry

Chemistry, Biology and Material Science

Edited by
F. Diederich, P. J. Stang, R. R. Tykwinski

WILEY-
VCH

WILEY-VCH Verlag GmbH & Co. KGaA

Editors:

Prof. Dr. François Diederich
Laboratorium für Organische Chemie
ETH Hönggerberg, HCI
CH-8093 Zürich
Switzerland

Prof. Dr. Peter J. Stang
Department of Chemistry
University of Utah
315S 1400E, Rm. 2020
84112 Salt Lake City
USA

Prof. Dr. Rik R. Tykwinski
Department of Chemistry
University of Alberta
T6G 2G2 Edmonton
Canada

Library of Congress Card No.: applied for
British Library Cataloguing-in-Publication Data:
A catalogue record for this book is available from the British Library.

Bibliographic information published by Die Deutsche Bibliothek
Die Deutsche Bibliothek lists this publication in the Deutsche Nationalbibliografie; detailed bibliographic data is available in the Internet at <http://dnb.ddb.de>.

Printed in the Federal Republic of Germany.
Printed on acid-free paper.

Typesetting hagedorn kommunikation, Viernheim
Printing betz-druck gmbh, Darmstadt
Bookbinding Litges & Dopf Buchbinderei GmbH, Heppenheim

ISBN 3-527-30781-8

Preface

The carbon-carbon triple bond is a common structural motif in organic chemistry. Only during the past two decades, however, has it become a mainstay in the toolbox of synthetic organic chemists, biochemists, and materials scientists. Both as a building block and as a versatile synthon, the fascinating and sometimes unpredictable chemistry associated with the alkyne moiety has fueled many of the most recent advances. A decade ago, the monograph *Modern Acetylene Chemistry* documented an emerging renaissance in the chemistry of the carbon-carbon triple bond. Over the past ten years, this renaissance has evolved at an astounding rate and acetylenes now constitute a principal class of compounds in nearly all areas of chemistry and materials science.

The explosive growth of acetylene chemistry has particularly benefited from the development of new synthetic methodology based on transition metal catalysts and metal acetylides. An acetylene unit can now be introduced into nearly any desired molecule, large or small, often with surprising ease. Metal acetylides are key components for generating new nucleophilic reagents suitable for asymmetric addition reactions into electrophilic multiple bonds. The chemistry of early transition metal acetylides is especially rich, providing a fascinating class of compounds with remarkable synthetic potential. Acetylene chemistry has also driven the development of new methodology such as electrophilic addition reactions that allow the derivatization of this high-energy functional group into hetero- and carbocycles of significant interest to both synthetic and medicinal chemists. Alkynes are also versatile synthetic building blocks for the formation of natural product analogues and hybrid structures. For example, combining the hydrophobic, rigid, and linear attributes of acetylenes with the hydrophilic and chiral framework of carbohydrates affords derivatives with interesting structural properties and biological activity.

The advent of fullerene and nanotube chemistry in the 1990s inspired the search for other molecular carbon allotropes, and the hunt for both linear and cyclic allotropes consisting of sp–hybridized carbon continues in earnest. The study of carbon clusters, both linear and cyclic, spans from astrophysics to fullerene formation and advances in synthetic methodology are propelling efforts on all fronts. Acetylenic carbon rings can now be generated in the gas phase by a number of routes, while acyclic polyynes with lengths of several nanometers can be produced in unprecedented quantities with a range of terminal appendages. Cleverly designed acetylenic scaffolds can also serve as precursors to carbon-rich structures such as fullerenes and bucky onions. These recent discoveries provide for the tantalizing possibility of designer fullerenes with engineered shape and size.

Acetylene Chemistry. Chemistry, Biology and Material Science. Edited by F. Diederich, P. J. Stang, R. R. Tykwinski

ISBN 3-527-30781-8

Structural rigidity and electronic communication which is essentially unperturbed by conformational effects are the hallmarks of the acetylene moiety. These attributes make it a highly versatile component for conjugated scaffolds, especially when coupled with the extraordinary advances that have been achieved in metal catalyzed cross-coupling reactions. Nowhere is this more evident than in the spectacular array of molecules that have been assembled based on an arylene ethynylene framework. The synthesis of macrocycles based on arylacetylenes has now evolved to the point where constitution, physical properties, and chemical reactivity can be controlled in exquisite detail. Many of these molecules have been structurally tailored to exhibit specific properties of use in semiconductors, nonlinear optical media, liquid crystals, and sensors. Shape persistent acetylenic macrocycles can provide ordered systems based both on super- and supramolecular chemistry, resulting in tubular superstructures, two-dimensional networks, hosts for molecular recognition, and even adaptable systems that conform to external stimuli. Oligomeric, dendrimeric, and polymeric systems based on arylene ethynylene subunits are now widely viewed as some of the most important semiconducting organic materials. Vital to the successful application of these materials is the development of a much more thorough understanding of the key aspects of their synthesis, electronic structure, and organization into well-ordered films. The fruits of these efforts include synthetic polymers with programmed solid-state organization and ultra-sensitive molecular sensors for TNT. Arylene ethynylene structures incorporating chiral 1,1'-binaphthyl subunits provide yet another appealing dimension to these materials, and optically active acetylenic scaffolds suitable for asymmetric catalysis, nonlinear optics, and polarized emission have all been realized.

Complementing the efforts of synthetic and experimental chemists are the substantial achievements of theoretical chemists in their ability to both model and predict the properties of acetylene-rich molecules. The ever-increasing power of computational hardware, coupled with the development of new and improved numerical methods, have made theoretical modeling a vital tool in the evolution of modern acetylene chemistry. These efforts have shed light on topics ranging from the fundamentals of homoconjugation to the prospect of utilizing acetylene-based molecular wires as components in molecular electronics.

The eleven expert authors that have contributed to this monograph collectively offer a rich overview of the modern face of acetylene chemistry, as well as a detailed analysis of more subtle aspects that can dictate the success or failure of a particular experiment. Considerable emphasis has been placed on outlining the most recent advances in key areas of this discipline. We hope that this monograph offers to the novice a taste of the fundamental issues that motivate this exciting field of science, and to the expert the specific details on synthesis and applications necessary to stimulate the future of acetylene chemistry.

The editors wish to thank Dr. Elke Maase at Wiley-VCH for an enjoyable collaboration in the preparation of this book and Ms. Annie Tykwinski for designing the cover art.

November, 2004 F. Diederich, P. J. Stang, R. R. Tykwinski

Contents

Acetylene Chemistry. Chemistry, Biology and Material Science. Edited by F. Diederich, P. J. Stang, R. R. Tykwinski

ISBN 3-527-30781-8

Symbols and Abbreviations

A	adenine
ABLA	absolute bond length alternation parameter
Ac	acetyl
ADF	Amsterdam density functional
AFM	atomic force microscopy
AIBN	azobis(isobutyronitrile)
AIM	atoms in molecules
All	allyl
AM1	Austin model 1
APCI	atmospheric pressure chemical ionization
ARCS	aromatic ring current shielding
ASE	aromatic stabilization energy
ATRP	atom transfer radical polymerization
B3LYP	hybrid funtional by Becke and Lee-Yang-Parr
B3PW91	hybrid functional by Becke and Perdew-Wang
BHT	3,5-di-*tert*-butyl-4-hydroxytoluene
BINOL	1,1′-binaphthalene-2,2′-diol
BLYP	Bekce/Lee-Yang-Parr functional
bmim	1-butyl-3-methylimidazolium
Bn	benzyl
Boc	*tert*-butoxycarbonyl
BPO	benzoyl peroxide
bpy	2,2′-bipyridyl
BRE	Breslow resonance energy
BTEACl	benzyltriethylammonium chloride
$BTEAICl_2$	benzyltriethylammonium dichloroiodate
$BTEAICl_3$	benzyltrimethylammonium tribromide
BTMSA	bis(trimethylsilyl)acetylene
$BTMABr_3$	benzyltrimethylammonium tribromide
Bu	butyl
BuLi	butyllithium
Bz	benzoyl

Acetylene Chemistry. Chemistry, Biology and Material Science. Edited by F. Diederich, P. J. Stang, R. R. Tykwinski

ISBN 3-527-30781-8

CAN	ceric ammonium nitrate
CASSCF	complete active-space self consistent field
CC	coupled cluster
CCSDT/ CCSD(T)	coupled cluster singles doubles triples
CD	circular dichroism
CNDO/S-CI	complete neglect differential orbital/singles configuration interaction
COD	1,5-cyclooctadiene
Cp	cyclopentadienyl
Cp′	differently substituted cyclopentadienyl
Cp^*	pentamethylcyclopentadienyl
CPDMS	3-cyanopropyldimethylsilyl
CRD	cavity ring-down (spectroscopy)
CSA	camphorsulfonic acid
$<d>$	average diameter
DABCO	1,4-diazabicyclo[2.2.2]octane
dba	dibenzylideneacetone
DBU	1,8-diazabicyclo[5.4.0]undec-7-ene
DCC	dicyclohexylcarbodiimide
DCE	1,2-dichloroethane
DE	dissociation energy
DEAD	diethyl azodicarboxylate
Dec	decyl
DET	diethyl tartrate
DFT	density functional theory
"D+G" band	"disorder-induced" + "graphite" band
DIB	diffuse interstellar band
DIBAH	diisobutylaluminium hydride
DIC	*N,N′*-diisopropylcarbodiimide
DL	diode laser
DLS	dynamic light scattering
DMAP	4-dimethylaminopyridine
DME	1,2-dimethoxyethane
DMF	*N,N*-dimethylformamide
DMI	1,3-dimethyl-2-imidazolidinone
DMSO	dimethylsulfoxide
Dod	dodecyl
DOKE	differential optical Kerr effect
DOPS-TIPS	$SiMe_2CMe_2CH_2CH_2OSi(CHMe_2)_3$
DP	degree of polymerization
dppe	1,2-bis(diphenylphosphino)ethane
dppp	1,3-bis(diphenylphosphino)propane
DSC	differential scanning calorimetry

EA	electron affinity
ebthi	ethylene-bis-tetrahydroindenyl
ED	electron diffraction
EDX	energy dispersed X-ray emission
ee	enantiomeric excess
EELS	electron energy loss spectroscopy
EEM	electronegativity equalization method
ELF	electron localization function
E_{max}	energy of maximum absorption
EP	ethynylphenylene
EPR	electron paramagnetic resonance
equiv	equivalents
esu	electrostatic unit
Et	ethyl
FBW	Fritsch-Buttenberg-Wiechell
FC	flash chromatography
fs	femtosecond
fur	furyl
FVP	flash vacuum pyrolysis
FWHM	full width at half maximum
γ	molecular second hyperpolarizability
Gal	galactose
GGA	gradient generalized affroximation
Glc	glucose
GPC	gel permeation chromatography
$h\nu$	light
Hep	heptyl
Hex	hexyl
HF	Hartree-Fock
HMPA	hexamethylphosphoramide
HOMO	highest occupied molecular orbital
HOPG	highly oriented pyrolytic graphite
ICR	ion cyclotron resonance
INDO	intermediate neglect differential orbital
IP	ionization potential
K_{assoc}	association constant
l	contour length
λ_{max}	wavelength of maximum absorption
LB	Langmuir–Blodgett

LC	liquid crystal
LDA	lithium diisopropylamide
LD-TOF	laser desorption time of flight
LE	locally excited
LLS	laser light scattering
LiHMDS	lithium hexamethyldisilazide
LUMO	lowest unoccupied molecular orbital
MALDI-TOF	matrix-assisted laser desorption/ionization time-of-flight
Me	methyl
MINDO	modified intermediate neglect differntial orbital
min	minutes
MMC	molecular mechanics for clusters
MM3	molecular mechanics 3
M_n	number average molecular weight
MNDO/3	modified neglect differential orbital number 3
MOM	methoxymethyl
MP2, MP4	Moller Plesset 2, Moller Plesset 4
MRD-CI	multi-reference single and doubly excitation configuration
MRTD	molecular resonant tunneling diode
Ms	mesyl
M_w	weight average molecular weight
NBO	natural bond order
NBS	*N*-bromosuccinimide
NCS	*N*-chlorosuccinimide
NDR	negative differential resistance
NICS	nucleus independant chemical shift
NIS	*N*-iodosuccinimide
NLO	nonlinear optics
NLP	nonlinearity parameter
NME	*N*-methylephedrine
NMO	*N*-methylmorpholine *N*-oxide
NMP	*N*-methyl-2-pyrrolidone
ns	nanosecond
Oct	octyl
ODCB	*ortho*-dichlorobenzene
ODf	difluoromethanesulfonate
OEP	oligoethynylphenylene
OITB	orbital interaction through bonds
OITS	orbital interaction through space
ORTEP	Oak Ridge thermal ellipsoid plot
OTf	trifluoromethanesulfonate

PArE	poly(arylene ethynylene)
PAM	phenylacetylene macrocycle
PCC	pyridinium chlorochromate
PDI	polydispersity index
PDM	phenyldiacetylene macrocycle
PM3	parametric method number 3
PDA	poly(diacetylene)
Ph	phenyl
PhLi	phenyllithium
PHT	phenylheptatriyne
Pic	picrate
Piv	pivaloyl (*tert*-butylcarbonyl)
PmB	*para*-methoxybenzyl
PPE	poly(phenylene ethynylene)
PPV	poly(phenylene vinylene)
Pr	propyl
PS	polystyrene
PTA	poly(triacetylene)
PTFE	polytetrafluoroethylene
PTeM	phenyltetraacetylene macrocycle
PTrM	phenyltriacetylene macrocycle
py	pyridine
QUINAP	1-(2-diphenylphosphanyl-naphthalen-1-yl)-isoquinoline
R2CPD	resonance two-color photodetachment
REMPED	resonance-enhanced multi-photon detachment
RHF	restricted Hartree-Fock
rt	room temperature
SAM	self-assembled monolayer
SCF	self consistent field
SERS	surface plasmon polariton-enhanced Raman spectra
SHG	second harmonic generation
SSH	Su-Schrieffer-Heeger
STM	scanning tunneling microscopy
STO-3G	minimal Pole basis set
T	thymine
TBAF	tetrabutylammonium fluoride
TBDMS	*tert*-butyldimethylsilyl
TBS	*tert*-butyldimethylsilyl
TCNQ	7,7,8,8-tetracyanoquinodimethane
TD	Time dependent
TDDFT	time-dependent DFT

TEBAC	triethylbenzylammonium chloride
TEE	tetraethynyl-ethylene
TEM	transmission electron microscopy
Tf	trifluoromethanesulfonyl
TFA	2,2,2-trifluoroacetic acid
TGA	thermal gravimetric analysis
Th	Thexyl, 1,1,2-trimethylpropyl
THF	tetrahydrofuran
THP	tetrahydropyranyl
TIPS	triisopropylsilyl
TIPSA	triisopropylsilylacetylene
TLC	thin layer chromatography
TMEDA	*N,N,N',N'*-tetramethylethylenediamine
TMS	trimethylsilyl
TMSA	trimethylsilylacetylene
TMU	tetramethylurea
TNT	trinitrotoluene
TOF-MS	time-of-flight mass spectrometry
Tol	4-tolyl (4-methylphenyl)
Ts	*p*-toluenesulfonyl
Tr	trityl (triphenylmethyl)
U	uracil
UPS	ultraviolet photoelectron spectroscopy
VEH/SOS	valence effective Hamiltonien/sum over states
VSEPR	valence shell electron pair repulsion
VPO	vapor pressure osmometry
VUV	vacuum UV
ZINDO	Zerner intermediate neglect differential overlap

List of Contributors

Patrick Aschwanden
Laboratorium für Organische Chemie
ETH Zürich
8093 Zürich
Switzerland

Bruno Bernet
Laboratorium für Organische Chemie
ETH Zürich
8093 Zürich
Switzerland

Erick M. Carreira
Laboratory of Organic Chemistry
ETH Hoenggerberg, HCI
8093 Zürich
Switzerland

Rémi Chauvin
CNRS
Laboraoire de Chimie de Coordination
205 Route Narbonne
31077 Toulouse, Cedex 4
France

Sara Eisler
Department of Chemistry
University of Alberta
Edmonton, Alberta
T6G 2G2
Canada

Michael M. Haley
Department of Chemistry
1253 University of Oregon
Eugene, Oregon 97403-1253
USA

Carissa S. Jones
Department of Chemistry
and the Materials Science Institute
1253 University of Oregon
Eugene, Oregon 97403-1253
USA

Sigurd Höger
Polymer Institut
Universität Karlsruhe
Hertzstraße 16
76187 Karlsruhe
Germany

Richard C. Larock
Department of Chemistry
1605 Gilman Hall
Iowa State University
Ames, IA 50011-3111
USA

Christine Lepetit
CNRS
Laboraoire de Chimie de Coordination
205 Route Narbonne
31077 Toulouse, Cedex 4
France

Matthew J. O'Connor
Department of Chemistry
and the Materials Science Institut
1253 University of Oregon
Eugene, Oregon 97403-1253
USA

Lin Pu
Department of Chemistry
Universtiy of Virginia
PO Box 400319
Charlottesville, VA 22904-4319
USA

Uwe Rosenthal
Institut für Organische
Katalyseforschung
Universität Rostock e.V.
Buchbinderstraße 5–6
18055 Rostock
Germany

Timothy M. Swager
Department of Chemistry
Massachusetts Institute of Technolgoy
77 Massachusetts Ave.
Cambridge, MA 02139
USA

Yoshito Tobe
Department of Chemistry
Faculty of Engineering Science
Osaka University
Toyonaka, Osaka 560-8531
Japan

Rik R. Tykwinski
Department of Chemistry
University of Alberta
Edmonton, Alberta
T6G 2G2
Canada

Andrea Vasella
Laboratory of Organic Chemistry
ETH Hoenggerberg, HCI
8093 Zürich
Switzerland

Tomonari Wakabayashi
Department of Chemistry,
School of Science and Engineering
Kinki University
Higashi-Osaka 557-8502
Japan

1
Theoretical Studies on Acetylenic Scaffolds

Remi Chauvin and Christine Lepetit

1.1
Introduction

Ten years ago, Houk demonstrated that along with experimental analysis, theoretical analysis could not be ignored for advanced understanding of the specific properties of acetylenic compounds [1]. In recent years, the growing power of computational facilities, the optimized implementation of performant methods in various software packages, and the development of new theoretical tools (faster and extended coupled cluster (CC) methods, new density functional theory (DFT) functionals, time-dependent DFT (TDDFT), advanced topological treatments of the information contained in the electron density through the atoms in molecules (AIM) [2], and electron localization function (ELF) [3] methods,...) prompted more and more chemists to consider theoretical modeling as a key tool. This tool is used at either of two levels: (i) a posteriori, for analysis of the properties of known compounds, or (ii) a priori, for selection of synthetic targets with specific properties among potential molecular structures suggested by empirical laws governing the chemist's intuition. This intuition is formally sustained by more systematic suggestion tools, such as the "analogy principle" (variation of some structural unit through structural units with the same valence, such as a terminal atom through all the halogens), the "vinylogy", "cumulogy", or "ethynylogy" principles (insertion of CH=CH, C=C, or C≡C units between a few pairs of conjugated atoms), or the "*carbo*-mer principle" (insertion of C_{sp}-C_{sp} units into at least all symmetry-related bonds of a Lewis structure (Scheme 1.1)) [4]. The last of these principles allows for fine-tuning of the notion of carbon-*rich* molecules [5] to a notion of carbon-

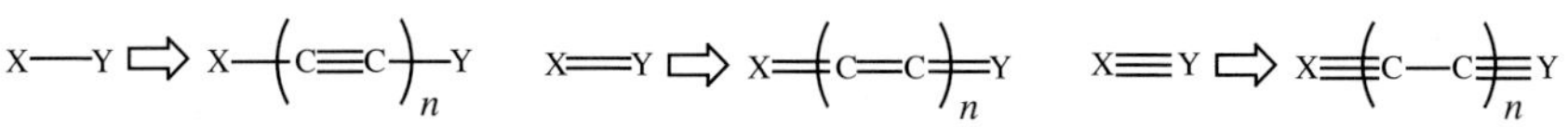

Scheme 1.1 Acetylenic expansions, becoming *carbo*-meric expansions if applied to complete sets of symmetry-related bonds of a parent molecule [4].

Acetylene Chemistry. Chemistry, Biology and Material Science. Edited by F. Diederich, P. J. Stang, R. R. Tykwinski

ISBN 3-527-30781-8

enriched molecules: indeed, *carbo*-merization does not corrupt the personality of a parent molecule "too much", as it preserves many of its basic features (connectivity, symmetry, shape, π-electron resonance, CIP configuration of asymmetric centers, ...), while just expanding its size through its acetylenic content.

A natural question concerns the variation of pre-existing or emerging properties as a function of the acetylenic content. Theoretical modeling allows for systematic screening and targeting prior to the undertaking of possibly tedious experimental syntheses. It should be emphasized, however, that many constructed expanded acetylenic scaffolds intentionally contain butadiynyl units (C_4) and are quite symmetrical. This holds to the efficiency and variety of oxidative C_{sp}-C_{sp} coupling procedures from terminal alkynes, which also motivated theoretical investigations of their mechanism [6].

This chapter presents selected advances (with special emphasis on results published since 1995) in the theoretical analysis or prediction of the properties of acetylenic scaffolds in a sequence based on dimensionality [7]:

1.2 Linear acetylenic scaffolds
1.3 Cyclic acetylenic scaffolds
1.4 Star-shaped acetylenic scaffolds
1.5 Cage acetylenic scaffolds

1.2 Linear Acetylenic Scaffolds

1.2.1 The Dicarbon Molecule and Acetylene

The empirically observed statistical even/odd carbon atom disparity in organic molecules [8] can be arithmetically interpreted by saying that the latter are made of "carbon pairs" rather than of carbon atoms. The bonded carbon pair, C_2, is the unitary "brick" of acetylenic scaffolds. This molecule can be observed in the form of the blue light of hot hydrocarbon flames [9] (e. g., during the pyrolysis of acetylene itself [10]), but also occurs under various conditions (diamond vapor deposition, laser ablation of graphite, in carbon-rich stars and comets...), and it was quickly proposed that it might assemble into the C_{60} fullerene [11]. The MO diagram of the ${}^1\Sigma_g^+$ ground state corresponds to the $(2\sigma_g)^2(2\sigma_u)^2(1\pi_u)^4$ configuration, and is nearly degenerate with the first triplet state (${}^3\Pi_u$). For the ${}^1\Sigma_g^+$ state, the simplest orbital criterion (six bonding and two anti-bonding electrons) corresponds to the :C=C: description. Higher bond orders (approximately 3.7) were, however, assigned by Wiberg or natural population analysis (NPA) [12], in accordance with the experimentally determined bond length (1.242 Å) and harmonic frequency (1855.0 cm^{-1}) [13], which are well reproduced by CCSDT methods and non-iterative approximation thereof [14]. The bond is indeed shorter than more stable C_{sp}=C_{sp} double bonds (e. g., approximately 1.280 Å for butatriene at various experimental and theoretical levels [15]). The ${}^1\Sigma_g^+$ state was thus depicted as a resonance hybrid

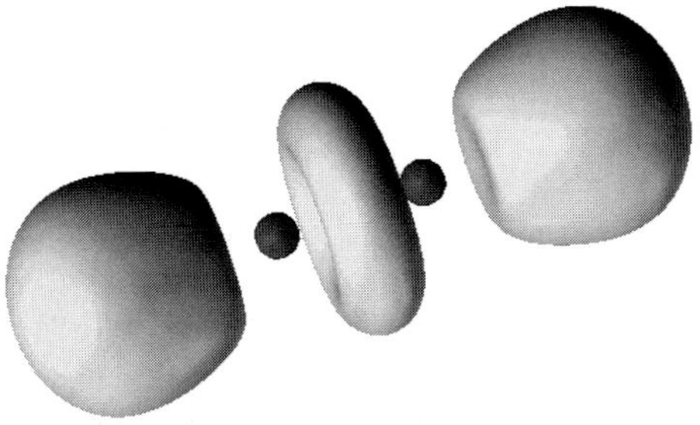

Figure 1.1 Localization domains of the acetylene molecule (ELF = 0.8) [3a]. The two red basins represent the carbon inner shells; the two blue ones the C−H bonds, and the green torus the C≡C bond.

:C=C:↔:C⌢·⌣C·↔·C⌢·⌣C:, as derived from Feynman's diagrams [16]. If the formal C≡C octet form is excluded [12], the classical Lewis resonance accounting both for the spin state and for the bond length of the $^1\Sigma_g^+$ state would be: :C=C:↔$^+$C≡C:$^-$↔$^-$:C≡C$^+$. Similarly, a consistent picture for the $^3\Pi_u$ state would be ˙:C–C:˙↔·C≡C·. These pictures suggest that the C_2 units might efficiently transmit π electron delocalization (vide infra).

In unstrained bonding situations, the biscarbenic character of the C_2 unit disappears. In acetylene, for example, the chemical shape of the three-bond rod can be visualized through the ELF localization domains (Figure 1.1) [3a]. According to the axial symmetry of the molecule, the three bond domains expected for the triple C-C bond are degenerate and merged into a torus: there is no σ/π separation, in variance with the classical view of organic chemists.

1.2.2
Uncapped pure *sp* Carbon Chains [1]

Many theoretical studies of all-*sp* carbon clusters C_n show that the open linear form is preferred over the cyclic form for $n < 10$, though cyclic forms of short *n*-even chains (C_4, C_6, C_8) do compete [17]. The converse holds for $n \geq 10$ (see Section 1.3.2.4), but open chains are known for odd $n = 3$–29 [18].

Optimized geometries at various level of theory indicate that the octet deficiency of carbon chains is partly made up by a strong cumulenic character both for C_{2m+1} chains ($^1\Sigma_g^+$ ground state with equal bond lengths of approximately 1.30 Å) and for C_{2m} chains ($^3\Sigma_g^-$ ground state with less than 2% bond length alternation). Although DFT calculations did not accurately reproduce experimentally determined or CCSD(T) singlet-triplet separation, both the local density approximation (LDA) [19] and the use of gradient-corrected (BLYP) [20] or hybrid (B3LYP) functionals gave relevant results in terms of geometries and harmonic frequencies [21]. Hund rule-reliable Lewis representations of small carbon chains are given in Scheme 1.2. With regard to the contribution of zwitterionic forms, it should be mentioned that a TD-LDA treatment showed that the longitudinal polarizability increases from approximately 7 Å^3 ($n = 3$) to approximately 490 Å^3 ($n = 21$), namely more rapidly than *n*, and more rapidly than the polarizability of the corresponding polyenes with the same number of valence electrons [18].

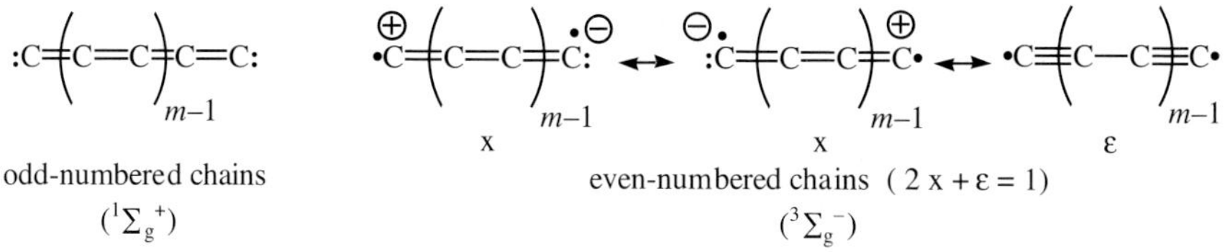

Scheme 1.2 Main Lewis forms for ground state C_{2m+1} and C_{2m} (m = 1–8) accounting for both the bond orders and the spin state.

Calculations of various excited or charged (C_n^-, C_n^+) states of linear carbon chains have been carried out at high levels of theory, especially by coupled cluster methods, which proved to be useful for reproduction or prediction of experimentally measured electronic spectra [22], electron affinities (EAs), ionization potentials (IPs), or dissociation energies (DEs) [17].

Efforts devoted to theoretical studies of linear carbon chains are motivated by their uniques properties, such as their role in fullerene growth or, more recently, in the field emission of electrons from carbon nanotubes or in the origin of the diffuse interstellar bands (DIBs) [18].

- C_n chains of nanometer sizes (n = 10–100) can be generated by electric field-induced unraveling of the inner graphene wall layers of carbon nanotubes, and are then responsible for emission currents of 0.1–1 μA under bias voltages of less than 80 V [23]. Indeed, linear C_n chains (n = 3–11) were calculated to pass relevant tests to qualify as metallic atomic wires with ease, but open-shell n-even chains give currents 100 times as intense as their closed-shell n-odd neighbors [19].
- A challenging goal in radio-astronomy is understanding of the origin of the DIBs, silent bands of frequency in the visible-IR spectra of interstellar clouds [22]. DIBs were previously proposed to be caused by linear carbon chains [24], and some of them could recently be identified with optical gas-phase electronic transitions of anionic representatives [25]. An alternative to the difficult experimental determination of the characteristics of such transitions is their calculation. A partial attempt using experimental energies and theoretical (TD-LDA) oscillator strengths of C_5–C_{21} carbon chains provided a few qualitative coincidences with observed DIBs [18].

The octet deficiency of carbon chains can be completed by closure to cyclo[n]carbons (see Section 1.3.2.4), by simple doping, by end-capping, or by passing from the molecular status to a material status through infinite expansion [26]. Whereas end-capping is the topic of the next section, doping can be exemplified by neutral and charged carbon clusters $[C_{2n}X]^q$, which have been studied at the DFT level when X is a main group metal atom [27]: the cyclic ω-diacetylide-type structures C≡C---X---C≡C are favored over the open-chain isomers for q = 0 and X divalent (e.g., X = Mg) [28], or for q = –1 and X monovalent (e.g., X = Na) [27]. Similar studies have been conducted with transition metal atoms, including with X = Pt at the B3LYP level [29].

Finally, infinite expansion affords carbyne (1D *sp* carbon allotrope), the missing link between atomic carbon and graphite (2D sp^2 carbon allotrope), which may be hybridized with the latter to design graphyne and related allotropic structures [30]. Although the carbon allotropes are beyond the scope of this chapter [31], it should be mentioned that several theoretical studies have focused on the possible structures and promising properties of these challenging materials [32].

1.2.3
Capped all-*sp* Oligoacetylenic Chains

1.2.3.1 all-*sp* Oligoacetylenic Chains between Redox-Silent Organic Groups

The recently surveyed state of the art of the characterization of doubly capped acetylenic chains $XC_{2n}Y$ prompted a comprehensive study of lower representatives (Scheme 1.3, $n \leq 6$) at various level of theory (HF, MP2, MP4, CCSD(T), B3P86, B3LYP, with various basis sets) [33a].

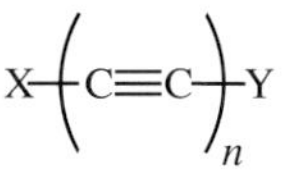

X, Y = H, CH_3, Ph, Na, F, I, CN, O, S, Se,...

Scheme 1.3 Capped polyacetylenics.

For X = Y = H, let us first remind ourselves that the possible structures of small C_mH_2 molecules (m = 3, 5, 6, 7) have been investigated by various ab initio calculations (see, for example, refs. [33b–f]). For larger polyacetylenic representatives ($m = 2n$, $n \leq 6$), geometrical, vibrational, and energetic (heats of formation, ionization potentials, electronic affinities (EAs)) properties display regular variations with respect to n [33a]. At the HF level, the C≡C and C–C bond lengths were found to converge to different values (1.190 Å and 1.366 Å, respectively), while the EAs decrease from 4.15 eV for n = 1 to 1.19 eV for n = 5. These results were extended by DFT studies of higher representatives (n = 6–12) [33]. The B3LYP level was argued to be relevant not only for the ${}^1\Sigma_g{}^+$ ground states of the neutral species, but also for the ${}^2\Pi_u$ or ${}^2\Pi_g$ ground states of the radical anions $[HC_{2n}H]^{\cdot -}$. In the latter, the bond length equalization increases with increasing n, while the adiabatic EAs of the neutral species increase continuously from 1.78 eV (n = 6) to 2.95 eV (n = 12). The number of calculated stretches in the 1900–2200 cm^{-1} region support the hypothesis that long $[HC_{2n}H\ ]^q$ species ($q = 0, -1$, $n \geq 6$) could contribute to the DIBs [34]. It may be commented that the results from references [33a] and [34], though referring to different methods, could suggest that the EA is at its minimum around $n \approx 5$.

Chaquin et al. recently reported a comprehensive structural and vibrational study of the $HC_{2n}H$ molecules up to n = 20 at the B3LYP level [35]. The results confirm the persistency of bond length alternation with increasing n, and a systematic correlation scheme afforded asymptotic values of 1.229 Å and 1.329 Å for triple and single bonds, respectively. As would be expected, the asymptotic bond length alternation (0.10 Å) is smaller than that obtained at the HF level (0.17 Å) [33a], but the DFT estimate is supported by excellent fits with experimentally obtained data. The alternation was also visualized through the ELF localization domains in $C_{30}H_2$ [35].

Polyynes have also received much attention, as they are predicted to exhibit large second hyperpolarizabilities [36]. Beyond ab initio calculations on simple diacetylene derivatives [36d–e], high level ab initio studies (SCF, MP, CCSD(T)) with large basis sets show that the static hyperpolarizability increases with increasing C≡C bond lengths [37]. The bond length alternation (BLA) parameter, which is a determining factor of nonlinear optical properties, was compared for polyyne chains at various calculation levels (PM3, ab initio, DFT) [38, 39]. The static second hyperpolarizabilities (γ of linear polyynes up to $C_{160}H_2$) was calculated [36b, 39] and, as in the case of polyenes, γ/n_C versus the number of carbon atoms n_C saturates beyond $n_C \approx 50$. Similarly, deprotonation and double deprotonation energies of polyynes containing up to 40 carbon atoms were shown to decrease towards an asymptotic limit of 370 kcal mol^{-1} [38b].

For X = Y = Ph (Scheme 1.3, $n \leq 6$), DFT studies of α,ω-diphenyl polyynes of D_{2h} symmetry accounted for the effects both of the chain length and of the phenyl end-caps on the fluorescence properties [40]. The fluorescent excited state switches from $1^1B_{2u}(\pi_x\pi_x^*)$ for $n \leq 2$ to 1^1A_u ($\pi_x\pi_y^*$ or $\pi_y\pi_x^*$) for $n \geq 3$.

For X = Y = O, S, Se, the acetylenic character is expected a priori for the triplet states (Scheme 1.3). The $OC_{2n}O$ molecules are much less stable than the cumulenic "odd" carbon suboxides $OC_{2n+1}O$. The first member of this series is ethylenedione C_2O_2 (i. e., the dimer of CO or the *carbo*-mer of O_2), which has hitherto evaded unambiguous experimental observation. On the basis of CCSD(T) calculations, it was recently suggested to be an intrinsically short-lived molecule [41]. The monosulfur analogue C_2OS [42] and the higher cumulogue C_4O_2 could, however, be identified by comparison between calculated and experimentally measured IR or UV spectra [43]. Recent calculations at the UB3LYP/6-311G* level support the view that all triplet structures $XC_{2n}X$ (X = O, S, Se; $n \leq 4$) possess a $^3\Sigma_g$ ground state with trivial spin contamination [44]. Both the C-C bond length alternation and the C-X bond lengths confirm the contribution of the acetylenic form $^\bullet$X–(C≡C)$_n$–X$^\bullet$.

For X, Y = H, CN (Scheme 1.3, $n \leq 9$), the corresponding cyanopolyynes H-$(C_2)_n$-CN and dicyanopolyynes NC-$(C_2)_n$-CN are regarded as isolable model substances of the carbyne allotrope [45]. The variation of the HOMO-LUMO gap (AM1, HF or MP2) and of the ZINDO-calculated UV spectra accounts for the experimentally observed Lewis–Calvin bathochromic shift with increasing n ($\lambda^2 = kn$). The observed regioselectivity of dienophile cycloadditions at the $C_\beta≡C_\gamma$ bonds is consistent with the highest positive Mulliken charges at the C_β and C_γ atoms (Scheme 1.4). Since the positive charge is greater at C_β than at C_γ, the weight of the long-range charge separation (y) is greater than that of the shorter-range one (x) (Scheme 1.4). The increase in cumulenic character with increasing n, also systematically evidenced at the B3LYP level [35], would thus be consistent with a parallel increase in the y/x ratio. The occurrence of the zwitterionic resonance forms would also account for the σ-coordination ability of the nitrogen atoms toward main group metals as evidenced by DFT calculations [46]. From a more applied

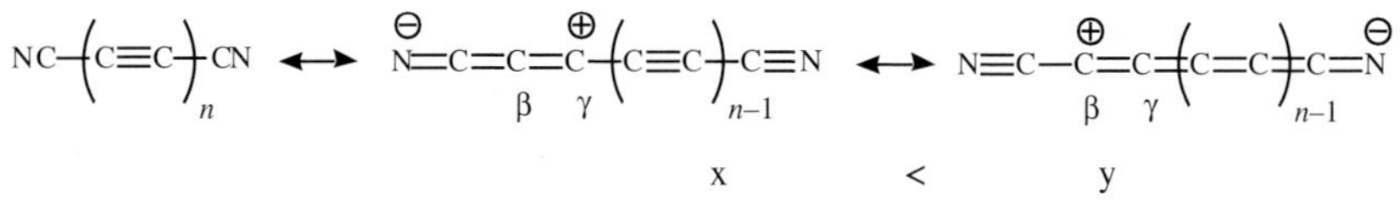

Scheme 1.4 Schematic charge separations in dicyanopolyynes.

standpoint, DFT-calculated low vibration frequencies of cyanopolyynes have also been claimed to be relevant for possible assignments of spectroscopic absorptions in interstellar media [35].

For X, Y = H, F, or Na (Scheme 1.3, $n \leq 5$), the electron distribution in the corresponding polyynes has been studied at the MP2/6-31G** level within the formalism of the electronegativity equalization method (EEM) [47]. In contrast to the atomic charges, the charges of the X-C≡C- and -C≡C-Y termini and those of the (n–2) intermediate -C≡C- units were found to be additive. For example, given n = 5 and X, Y = Na, F, the charges of the ith unit in the heteronuclear and homonuclear molecules are bound through: $[Q_i(FC_{10}F) + Q_i(NaC_{10}Na)] \approx [Q_i(NaC_{10}F) + Q_i(FC_{10}Na)]$. This gives support to the view of the C_2 units as "compact" building blocks (see Section 1.2.1).

Much weaker than conventional hydrogen bonds, intermolecular $\pi \ldots \pi$ or C–H…π interactions have been attracting increasing attention [48]. In particular, several ab initio and DFT studies have focused on the hydrogen bond-accepting ability of triple bonds [49]. The ab initio intermolecular potential for acetylene sliding perpendicularly along a polyyne (Figure 1.2) was calculated at the MP2 level and was used to refine an MMC (molecular mechanics for clusters) model potential for polyynes [50]. The T-shaped interaction energy of acetylene with the triple bond at the end of the chain reaches an asymptote of $-372\ \mathrm{cm}^{-1}$. As the acetylene molecule slides along the chain, the intermolecular potential is essentially flat with only small bumps (Figure 1.2) (acetylene interacts slightly more with the triple C≡C bonds than with the single C-C bonds). This extended smooth intermolecular potential zone produces weak bonding slipperiness of acetylene along the polyyne chain.

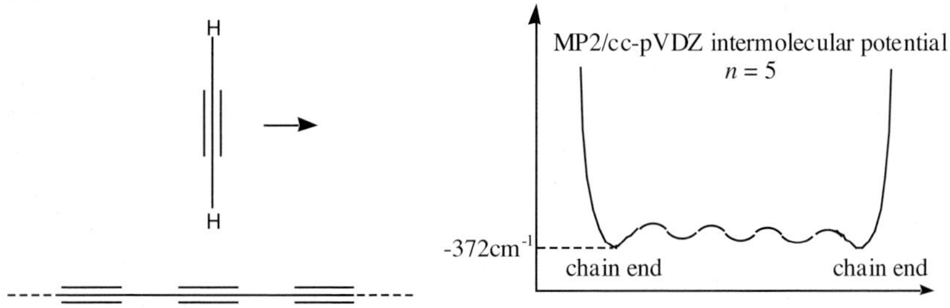

Figure 1.2 Perpendicular sliding of acetylene along a polyyne $H(C{\equiv}C)_nH$ (n = 2–5) at fixed distance between the center of mass of acetylene and the polyyne molecular axis (4.5 Å) [50].

1.2.3.2 All-*sp* Oligoacetylenic Chains between Organometallic Termini

Besides their potential applications in homogeneous and heterogeneous catalysis, alkynylmetal complexes are thermally robust, available in high yield, and attractive for their conducting and nonlinear optical properties. Among a very large number of theoretical studies on such complexes, a few recent results are highlighted below.

In monometallic complexes, theoretical studies have aimed at better understanding of metal-alkynyl bonding [51]. Through the use of the ADF program and the bond energy decomposition scheme designed by Ziegler and Rauk [52], metal-to-ligand backbonding energies were calculated for a set of σ-acetylide complexes of electron-rich metal centers such as $M(X)(PH_3)_n$, with M = Fe, Ru, Os (X = Cl and n = 4, or X = C_5H_5 and n = 2) [53, 54]. As expected, the π backbonding increases with the electron-withdrawing ability of the η^1-acetylide substituent and increases from Fe to Os. However, π bonding effects are relatively small in relation to the dominant metal-alkyne σ bonding. The backbonding energy is accurately correlated with the experimentally determined or calculated quadratic hyperpolarizability [55], but not with the υ(MC≡C) IR stretching frequency, which is indeed lowered by coupling with the stretching of the adjacent bonds. A similar DFT study of the influence of terminal substituents on the π backbonding in metallacumulene complexes has also been reported [56].

Studies of the homo- and heterobimetallic complexes have focused on the electronic communication between the metal centers through the polyyne bridge (Scheme 1.5) [57, 58, 59].

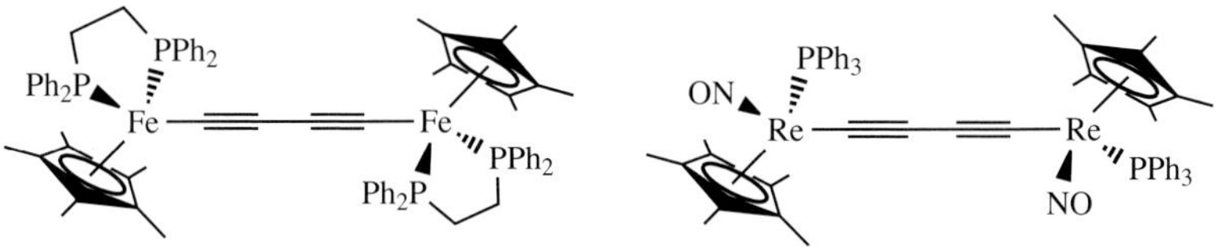

Scheme 1.5 Examples of homobimetallic complexes where electronic communication is assumed by polyyne bridges [64b].

In $L_mM{\cdots}C{\cdots}C{\cdots}ML_m$ complexes, the C_2 molecule (Section 1.2.1) is stabilized as a μ-diacetylide ligand [31, 60]. DFT calculations allow two classes of dicarbido complexes to be identified, the metals lying in pseudotetrahedral coordination spheres [61]. For early metals of the Ti, V, and Cr triads in high oxidation states with π-donor ligands, the coordination mode (M−C≡C−M, M=C=C=M, or M≡C–C≡M) depends on the d^n configuration. In contrast, for metals of the late triads (from Mn onwards) in low oxidation states with π-acceptor ligands, only the M−C≡C−M coordination mode is observed whatever the metal d^n configuration.

The generalized valence flexibility (M–C≡↔M=C=↔M≡C–) has been elegantly studied by natural bond order (NBO) and AIM analyses for even- [62] or odd-numbered [63] carbon chains, depending on the metallic end-caps. In the case of C_4 chains bonded to electron-rich metal fragments (M = Fe, Re, Ru,

Scheme 1.5) [64], the neutral complexes are best described by a M-C≡C-C≡C-M Lewis form, while the doubly oxidized ground state is best described by a cumulenic $^{+}$M=C=C=C=C=M^{+} form. However, the triplet spin state with a dominant $^{+\cdot}$M-C≡C-C≡C-M$^{\cdot+}$ character is close in energy. The diversity of the low-lying electronic states of these complexes would be expected to be associated with switchable physical or chemical properties.

The localization properties of sequential triple bonds bridging a mononuclear Ru fragment to a Co-carbon cluster were compared by ELF analysis. The shape of the ELF localization domains is related to the nature of the metal-ligand bonding (i. e., σ-donation versus π backbonding) [65]. The electronic structures of complexes consisting of two {FeL_n}-C≡CC≡C- chains linked by a third metal center (Hg) have recently been described [66].

For long metal-capped polyynes asymptotically approaching the *sp*-carbon allotrope, Gladysz and Szafert proposed a classification of the carbon chain conformation based on a nonlinearity parameter (NLP) calculated from X-ray diffraction data [67].

1.2.4
Hybrid *sp*-*sp*^{2} Oligoacetylenic Molecules

1.2.4.1 **Oligoenynes**

The simplest representatives of this class are *cis*-α,β-enediynes. These have attracted considerable experimental and theoretical interest owing to their behavior in the Bergman cyclization, which is responsible for the anti-cancer activity of several derivatives [68]. Their gas-phase structures were thus studied at a high level of theory [69] in the context of accurate studies of the cyclization mechanism, which has been reviewed recently and so is not repeated here [68].

Higher oligoenynes capped with suitable termini to ensure stability and solubility are model systems of poly(diacetylene) (PDA) or poly(triacetylene) (PTA) (Scheme 1.6), which may exist in *trans* or *cis* forms. These conjugated organic oligomers and polymers can be easily processed, and have potential applications in conductivity, magnetism, photoconductivity, electronically stimulated light emission, nonlinear optics, and photorefractivity. The structural features governing the conducting properties of PDA (as compared with those of polyacetylene or polythiophenes) have thus long attracted theoretical interest [70]. The interconversion

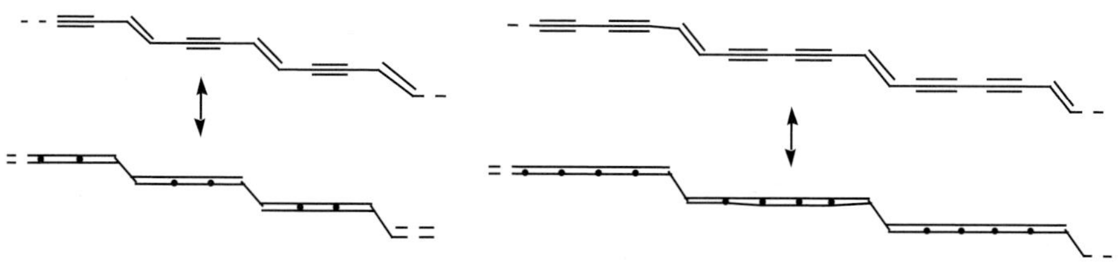

Scheme 1.6 Acetylenic and cumulenic forms of poly(diacetylene) (left) and poly(triacetylene) (right).

between the polyenyne and polybutatriene forms (Scheme 1.6) has been studied at the CASSCF level, through the characterization of the low-lying excited states of both forms [71].

The dependence of electronic and optical properties of PTA oligomers on the chain length have been studied up to the hexadecamer [72]. AM1 geometries, INDO gas-phase ionization potentials and electron affinities (EAs), and VEH/SOS second-order hyperpolarizabilities (γ) have been calculated: EA and γ values increase up to the octamer, and then saturate at around ten repeating units, in agreement with experimentally obtained results.

The third-order optical nonlinearity of PDAs and PTAs would be expected to be enhanced in the *trans*-enyne forms. Phenyl-terminated *cis*-oligodiacetylenes were expected to be locked towards *cis/trans* isomerization through incorporation of the ene moieties into ring systems (Scheme 1.7). Whereas the crystallographic structure is found to be planar, gas-phase PM3 calculations show that the helical conformation is more stable, consistently with NMR measurements suggesting that the *cisoid* conformation of the dienyne units is preferred in solution [73].

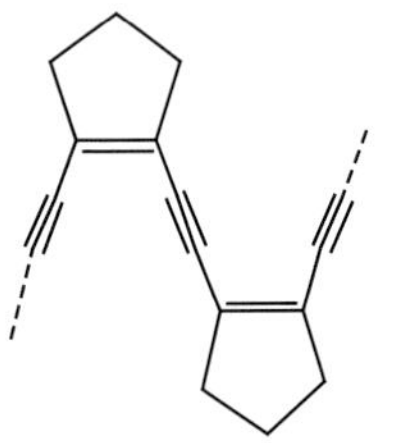

Scheme 1.7 *cis*-Oligodiacetylenes with the ene moieties parts of cyclopentene rings.

Backbone-substituted oligo(triacetylene) chromophores have also been designed, for tuning of their third-order nonlinear optical properties [74]. Insertion of a π chromophore heterospacer into a central butadiyne unit (Figure 1.3) expands the conjugation in two dimensions, and was expected to enhance the second hyperpolarizability [75]. Moreover, transition metals coordinated to these hybrid oligomers may act as electronic bridges in star-shaped associations (Figure 1.3). The calculated variation of the second hyperpolarizabilities with the nature of the spacer is in good agreement with experimentally obtained results.

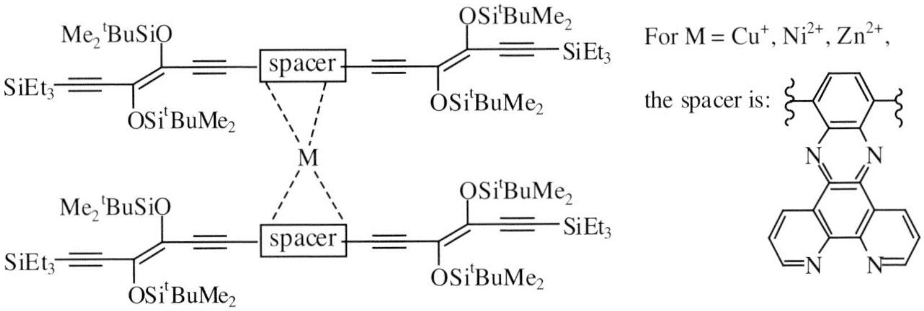

Figure 1.3 Oligo(triacetylene) chromophores with a central heterospacer between two diethynylene units [74].

Let us finally mention that the poly(triacetylene) hexamer served as an illustrative example of the applicability of the new program SIXW. C. This program is based on the CNDO/S-CI method with inclusion of double excitations, and allows UV/visible spectra of fairly large molecules to be computed [76].

1.2.4.2 Oligo(ethynylphenylene)s

Oligomers of ethynylphenylene (for which the acronym "EP" is adopted here) (OEPs) have been extensively studied for molecular electronics and photonics purposes. While the *m*- and *o*-EP oligomers may adopt helical conformations [77], *p*-EP oligomers (Scheme 1.8) are considered to be rigid rods, suited for the design of molecular wires with various properties (reversible one-electron oxidation or reduction, electroluminescence, nonlinear optics,...).

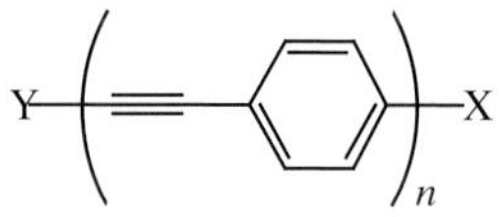

Scheme 1.8 Generic structure of *p*-EP oligomers (hydrocarbon *p*-EP oligomers correspond to X = H, Y = Ph).

The geometries of linear and cyclic oligomers made of $n = 3–10$ *p*-PE repeating units were calculated at the HF/6-31G level, and the corresponding electronic spectra and second hyperpolarizabilities (γ) were calculated at the semiempirical level [78]. It was found that the asymptotic γ/n values extrapolate to 5.15 and 3.93×10^5 a. u. for chain and ring compounds respectively, indicating they are promising third-order NLO materials with good transparency to visible light. Mixed vinylphenylene/ethynylphenylene oligomers were also investigated at the semiempirical level [79]. Their simulated spectra are in good agreement with experimentally obtained data, and highlight the effect of the central aromatic ring on the nature of the lowest excited state.

The conformational behaviors of OEPs are closely related to the effective extents of conjugation. DFT torsional barriers of simple tolan analogues (possibly with aromatic nitrogen heterocycles instead of the phenylene rings in Scheme 1.8) have been calculated at the B3LYP/6-31+G* level. The molecular flexibility of these compounds is different in the gas phase and in aqueous solution, as evidenced by classical molecular dynamics (MM3 force-field) [80]. Even at low temperature, free ring twisting dominates in the gas phase, becoming almost periodic after a few picoseconds of equilibration, while random conformations are predicted in solution. This random twisting is expected to affect the fluorescence properties of these compounds significantly. Indeed, rigidity favors fluorescence, making intramolecular dissipation difficult. Molecular dynamics was also used to describe the arrangement of the chains of block copolymers containing one conjugated *p*-EP segment and one or two non-conjugated segments (e. g., -$(OCH_2CH_2)_m$-) [81].

EP polymers also possess light-emitting diode properties. The detailed mechanism of this phenomenon, and especially the role of the triplet excitations, is not yet established. Configuration interaction calculations have thus been performed in

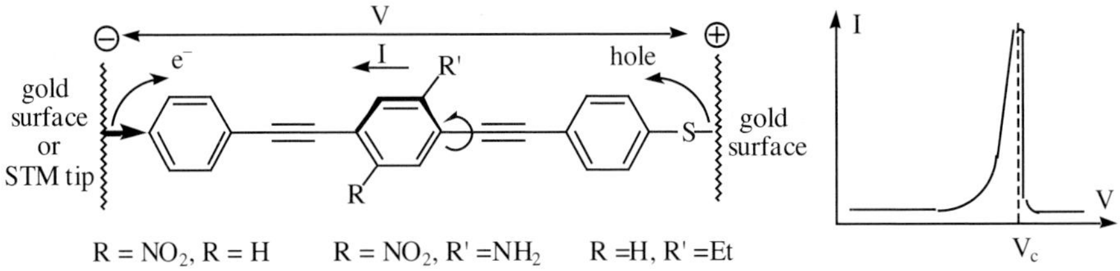

Figure 1.4 Schematic representation of the MRTD and NDR properties of OEPs.

order to describe the singlet and triplet excited states of OEPs [82]. Intersystem crossing from the singlet to the triplet manifold by spin-orbit coupling is possible, and results in a mixing of the singlet and triplet wavefunctions. The role of chain length and of ring twisting on the intersystem crossing process has been studied.

Another promising property of OEPs in molecular electronics is their behavior as molecular resonant tunneling diodes (MRTDs) with negative differential resistance (NDR) when disymmetrically grafted between two metallic contacts (Figure 1.4). The structure-activity relationship and the mechanism of this phenomenon are currently attracting a lot of experimental and theoretical efforts [83].

In the case of R = NO_2, R′ = NH_2 at the DFT (B3PW91) level, it was proposed that electron conduction through the LUMO of the planar conformation was triggered as soon as the molecule was charged by a single electron. The temperature dependence of the phenomenon was attributed to the flexibility of the chain, which may adopt all-planar or twisted conformations [84].

In the case of R = H, R′ = Et, at the INDO semiempirical level, Brédas et al. proposed a resonant tunneling process for both electrons and holes in the twisted conformation (Figure 1.4), based on the variation of MO levels with the magnitude of an electric field applied along the molecular axis (Figure 1.5) [85, 86]. Very recently,

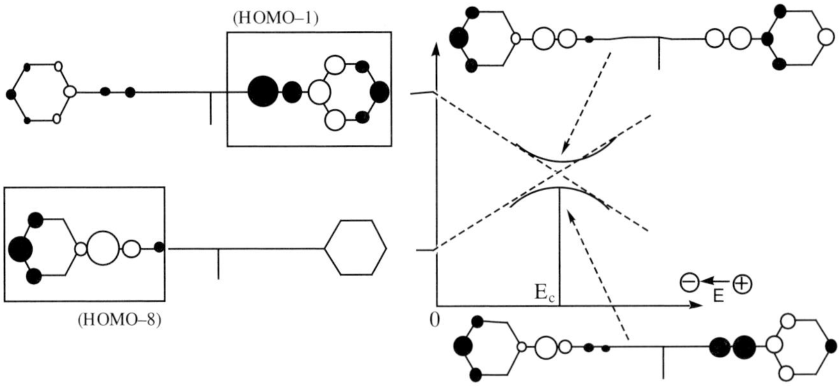

Figure 1.5 Brédas's resonant tunneling mechanism for hole conduction and NDR in a twisted OEP (R = H, R′ = Et, Figure 1.4). A similar scheme for antibonding MOs accounts for NDR in electron conduction. The electronic splitting and the broadness of the curves near resonance (at Ec = 3.3 × 107 V cm^{-1}) are here graphically much exaggerated with respect to the initial HOMO–1/HOMO–8 gap.

this mechanism was shown also to be consistent in the case of R = NO_2, R′ = NH_2 [87].

OEPs capped by push-pull substituents resembling merocyanines (Y = 4-$C_5H_4NMe^+$, X = O^-, Scheme 1.8) were predicted on the basis of semiempirical ZINDO calculations to possess very large gas-phase static quadratic hyperpolarizabilities (β_0) [88]. The use of this method allowed a broad range of structures to be covered for systematic comparison, and the results were considered to have at least relative significance. The hyperpolarizabilities of either of the isomers $^-OC_6H_4$-C≡C-C_6H_4-$C_5H_4N^+Me$ and $^-OC_6H_4$-C_6H_4-C≡C-$C_5H_4N^+Me$ were first calculated to be as high as $\beta_0 \approx 800 \times 10^{-30}$ cm^5 esu^{-1}. The double ethynylogue $^-$O-C_6H_4–C≡C–C_6H_4–C≡C–$C_5H_4N^+$–Me affords a double β_0 value (1982×10^{-30} cm^5 esu^{-1}). The "*p*-PE effect" becomes dramatic in $^-$O-C_6H_4–C≡C–(C_6H_4–C≡C)$_2$–$C_5H_4N^+$–Me, for which the calculated value is unprecedented in small organic chromophores: $\beta_0 = 33856 \times 10^{-30}$ cm^5 esu^{-1}. The definition of a generalized absolute bond length alternation parameter (ABLA) over the OPE topology affords a regular variation of β_0 with regard to ABLA for a homogeneous set of push-pull merocyanine-like chromophores (Figure 1.6).

A Lorentzian-based fit was heuristically derived from simple approximations in a two-level model, and the relevance of the results is supported by a strong structure-activity relationship. Indeed, one chemical way to draw nearer to the critical ABLA value (ABLA° ≈ 0.04 Å. Figure 1.6) is a sequential insertion of *p*-EP units between pyridinium and phenolate termini. Although experimentally determined β coefficients of the recently synthesized molecules are not yet available [89], it should be stressed that the solvent and temperature effects were not taken into account in the calculations.

Much lower quadratic hyperpolarizabilities have been calculated ($\beta_{1907(ZINDO)} < 45 \times 10^{-30}$ cm^5 esu^{-1}) or measured ($\beta_{1604(corr.)} < 232 \times 10^{-30}$ cm^5 esu^{-1}) for inorganic analogues in which the donor moiety is Y = (C_5H_5)(PPh_3)$_2$Ru(II) and the acceptor

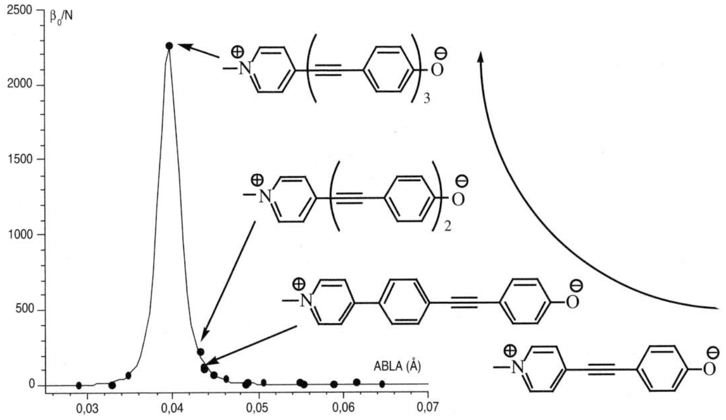

Figure 1.6 Plot of β_0/N vs. ABLA for a set of merocyanine-like chromophores (*N* = bridge length between the donor and the acceptor ends). From reference [88].

is X = NO_2 (Scheme 1.8) [90]. The NLO efficiency was found to vary with the nature of the bridging units, and in particular as: *E*-ene linkage > yne-linkage. The reverse order was observed for the metal-free merocyanine-like chromophores discussed above [88]. Quite large first-order hyperpolarizabilities (≈ 1000 × 10^{-30} cm^5 esu^{-1}) were also computed for push-pull arylethynyl porphyrins [91] and phthalocyanines [92], in agreement with EFISH measurements.

The hyperpolarizability of *p*-PE-bridged chromophores can also be switched by photon, proton, or electron transfers. Thus, the first hyperpolarizability of electron-rich iron σ-aryl acetylides (Scheme 1.8: Y = [(η^2-dppe)(η^5-C_5Me_5)Fe], dppe=1,2-bis(diphenylphosphino)ethane) may be switched through redox processes [93]. An electrochromic NLO switching has also been reported for a chloro (alkynyl)bis(diphosphine)ruthenium complex [94]. Upon one-electron oxidation, the second order hyperpolarizability is switched on (the γ modulus increases from zero to 2600 × 10^{-36} esu) or, alternatively, the sign of its imaginary or real component is reversed. This has been related to a change in the dominant electronic transition upon oxidation. TD-DFT calculations show that the dominant low-energy transition of the neutral species has a strong metal-to-alkynyl charge-transfer character, which reverts to a chloro/alkynyl-to-metal charge-transfer process in the oxidized species [95].

OEP bridges in mixed-valence organometallics. The photoluminescence of Re(bipyridine)$(CO)_3$(C≡C–C_6H_4–C≡C)Fe(C_5Me_5)(dppe) may be switched on by one-electron oxidation. Theoretical studies have provided evidence that the quenching pathway of the emissive dπ(Re)→π*(bipyridine) state, by a low-lying excited state involving the iron moiety, disappears upon oxidation [96].

Cyclometallated $Ru(bpy)_4(pp)$ units (bpy = 2,2′-bipyridine; pp = 2-phenylpyridine) linked by various spacers including C≡C bonds have been prepared in order to study intramolecular electron transfer between the two metallic sites [97]. The efficiency of the bridge at mediating electron transfer was estimated by the electronic coupling term V_{ab}. The latter may be extracted either experimentally from the experimentally measured UV/Vis spectrum, or theoretically by use of the "dimer splitting" method [98] and the energy splitting of the relevant orbitals (extended Hückel level). Theoretical and experimental results are in good agreement. Insertion of a second triple bond or/and a phenylene unit results in a reduced coupling relative to a linkage made of a single triple bond, although replacement of the phenylene with an anthracenylene enhances the coupling markedly.

1.2.5
Hybrid *sp*-*sp*^{3} Oligoacetylenic Molecules

Several β-diynes (or "skipped" diynes [99]) $X(C{\equiv}CH)_2$ (X = CR'_2, SiR'_2, PR′, S,...) are known, but the elusiveness of the unsubstituted diethynyl ether (X = O) [100] motivated comparative ab initio theoretical comparisons with the analogues X = S [101] or X = Se [102]. A systematic DFT study recently suggested that heteroatom-skipped β-diynes might replace *cis*-β-enediynes substrates in Bergman-like cycloaromatization, leading to open-shell precursors of five-membered heteroaromatics

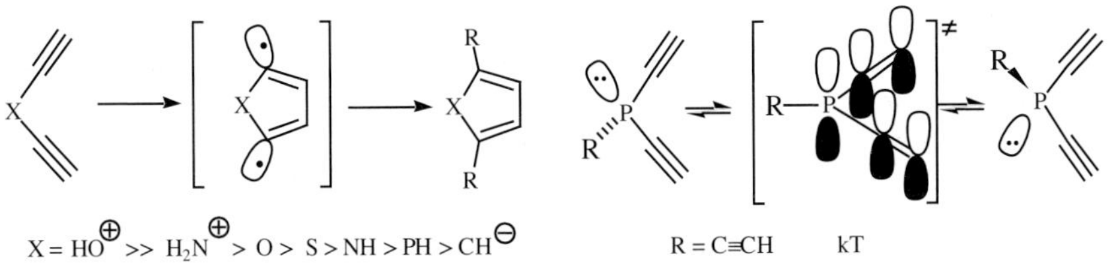

Scheme 1.9 Reactivity of β-diynes: theoretically predicted Bergman-like cycloaromatization [103] and observed inversion of stereogenic propargylic phosphorus centers [105].

(see Section 1.1.4.1) [103]. It was shown that the more electronegative the X group, the easier the reaction (Scheme 1.9).

In skipped polyynes, the doubly propargylic character of the sp^3-CH_2 groups makes them rather unstable [104]. In related phosphorus analogues (R′C≡C–P(R)–C≡C–)$_2$, the doubly propargylic character of the phosphorus lone pairs has a positive effect, namely the thermal coalescence of the stereoisomeric mixture resulting from the expected stereogenicity of the P atoms [105]. Indeed, DFT calculations on model alkynylphosphines $PH_{3-n}(C{\equiv}CH)_n$ ($n = 1$–3) showed that the ΔG^{*} value of the inversion barrier decreases with increasing n: the planar transition state is stabilized by conjugation of the lone pair with the surrounding acetylenic π orbitals.

1.3 Cyclic Acetylenic Scaffolds

1.3.1 Hybrid *sp*-*sp*^{3} Rings

1.3.1.1 Cycloalkadiynes

In cycloalkadiynes, extensively studied by Gleiter et al., the transannular interaction of the in-plane π orbitals of the triple bonds facing each other is relayed by the σ system [106]. Depending on the length of the σ bridge, MO calculations allowed for qualitative and quantitative interpretation of the differences between observed photoelectron (PE) spectra. In related systems in which the σ bridge is interrupted by heteroatoms or double bonds, thus “insulating” carbon sp^3 orbitals between π orbitals, the question of extended homoconjugation was addressed by similar PE/MO studies [107a]. When generalized to cyclic skipped polyynes (“pericyclyne”) [107b], the question meets the challenging debate on the possibility of neutral homoaromaticity, theoretical aspects of which are surveyed in the following section.

1.3.1.2 Pericyclynes

The state of the art in the field of macrocyclic homoconjugated oligoacetylenic molecules was reviewed in 1999 [108]. The subclass of [*n*]pericyclynes (**1–4a**, Figure 1.7, $3 \leq n \leq 6$) was defined and demonstrated by Scott et al. in 1983

[109]. On the very outset, the constrained vicinity of several triple bonds could favor rearrangements (e. g., to n-cyclopropa[2n]annulenes **1–4b**, Figure 1.7, $3 \leq n \leq 6$), and their “astonishing” stability was tentatively attributed to homoaromaticity early in the course of these investigations [110]. Recent studies at the B3PW91/6-31G** level (Figure 1.7) [111] reproduced the geometries obtained earlier at the MNDO level [112] or even at the MM2 level [110], thus suggesting the absence of non-additive effects from the cyclic proximity of the n triple bonds. In planar pericyclynes, however, analysis of the energy and symmetry of the STO-3G MOs showed that in-plane π_{xy} MOs of the triple bonds interact through space (OITS, $2n$ electrons), while out-of-plane π_z MOs interact through adjacent CH_2 π_σ orbitals (OITB, $4n$ electrons) [110]. As a whole, however, these interactions result in no structural effect.

With increasing n, the regular decrease in the out bending of the C≡C–C units (from 20.7° in [3]pericyclyne **1a** to 0.6° in [6]pericyclyne **4a**: Figure 1.7), is just as it would be expected on a simple VSEPR basis. On the other hand, the NICS values calculated at the center of the rings remain very small. Finally, the homoaromaticity of pericyclynes was studied with the aid of a novel tool: the ELF weighting of

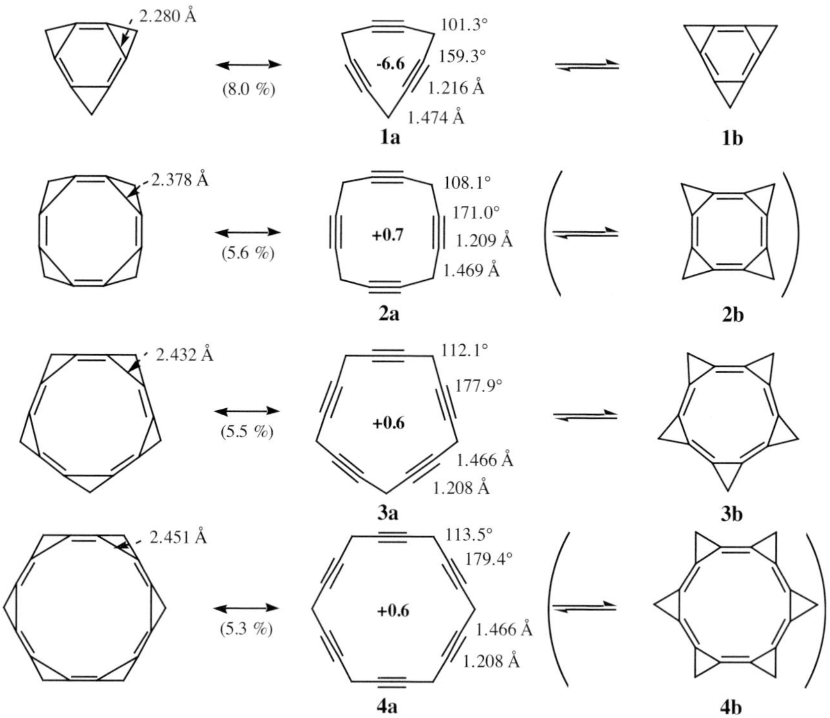

Figure 1.7 Main resonance forms and equilibrium geometrical parameters of [n]pericylynes (**1–4a**, n = 3–6) at the B3PW91/6-31G** level. All structures are D_{nh}-symmetric, except [6]pericyclyne, which is D_{3d}. In bold: central NICS values at the HF/6-31+G* level. Percentages in brackets correspond to the ELF-derived relative weight of resonance forms featuring cyclic homodelocalization [111]. The putative n-cyclopropa[2n]annulene isomers (**1–4b**) formally resulting from thermally forbidden [$n \times {}_{\pi}2_{\sigma}$] processes are indicated in brackets.

resonance forms [111]. The relative weight of the resonance forms featuring cyclic homodelocalization was found to be less than 8 %, thus confirming that [3]pericyclyne is at most poorly homoaromatic, and that higher pericyclynes are not (Figure 1.7).

- [3]Pericyclyne **1a**. Dewar was quick to consider the possible isomerization of **1a** to the tricyclopropabenzene valence isomer **1b** (e. g., through a thermally allowed $[_{\pi}2_{\sigma}+_{\pi}2_{\sigma}+_{\pi}2_{\sigma}]$ process [112, 113]). Though more strained at the level of sp^3 and sp^2 carbons, this isomer could be stabilized by aromaticity. Geometry optimization indeed reveals a negligible Mills–Nixon effect with (*endo*-C-C, *exo*-C-C) bond lengths of (1.359 Å, 1.356 Å) and (1.371 Å, 1.361 Å) at the RHF-SCF/6-31G* [114] and CASSCF/6-31G* [115] levels, respectively. From a magnetic standpoint at the CTOCD-DZ/6-31G** level, the maximum of the π ring-current density mapped 1 a_0 above the central ring of **1b** is equal to the corresponding maximum for the parent benzene [116]. At the B3PW91/6-31G** level, however, [3]pericyclyne **1a** is more stable than its **1b** isomer (D_{3h}, *endo*-C-C = 1.385 Å, *exo*-C-C = 1.357 Å) by 29.2 kcal mol^{-1} [117]. Benzenoid aromaticity is therefore not sufficient for favoring intramolecular cyclotrimerization.
  Whereas triphospha-[118] and trisila-[119] [3]pericyclynes proved to be stable, monocyclic all-carbon analogues have so far remained unknown. Let us remind ourselves here that the effect of the silicon vertices on homoconjugation was discussed on the basis of a comparison between photoelectron spectra and MNDO/3 or MINDO orbital calculations [120].
- [4]Pericyclynes. Few representatives of [4]pericyclynes are known [109]. Although planar, the unsubstituted model **2a** is not homoaromatic according to both NICS and ELF-resonance criteria (Figure 1.7). The optimized geometry was recently reproduced at the B3LYP/6-31+G* level [121]: the ring strain was estimated to be quite weak (9 kcal mol^{-1}) and the coordinating ability toward Li^+ and Na^+ was studied. As can be seen in Figure 1.7, the angular strain is taken over by the *sp* carbon atoms and so might be further relieved by insertion of additional triple bonds. Partially expanded [4]pericyclynes containing butadiynyl units on one of the edges were thus synthesized: with respect to the octamethyl version of this pentayne, the tetracyclopropylidene version **5** (Scheme 1.10) was shown to experience reduced OITS and enhanced OITB, as simply predicted from the π donating properties and the Walsh orbitals of the cyclopropane rings [122]. Several square totally expanded [4]pericyclynes [123] and heteroatomic analogues [105] are known. A model of an expanded metalla[4]pericyclyne **6** (Scheme 1.10) was recently studied at the DFT level [124]. On the basis of experimentally determined and optimized geometries, the Pt(II) vertices exert no homoaromatic geometrical effect. The extended Hückel MO diagram, however, indicates that several orbitals in the HOMO region, resulting from the interaction of *d*-type metallic orbitals with C-C π orbitals, are highly delocalized over the Pt_4C_{16} ring.
- [5]Pericyclynes. Although decamethyl[5]pericyclyne was shown to exhibit an envelope conformation in the crystal state [110], [5]pericyclyne itself (**3a**) was calculated to have a planar D_{5h} geometry, at the MM2 [110], RHF/3-21G [125], B3LYP/

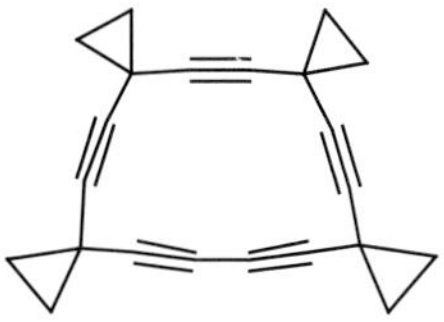
5

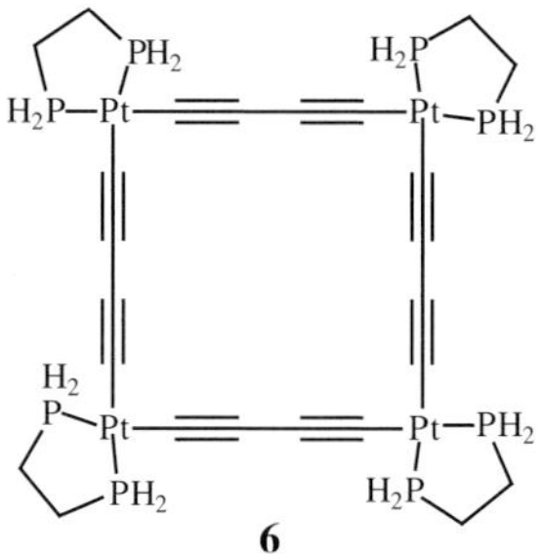

6

Scheme 1.10 Expanded [4] pericyclynic rings.

6-31G* [126], or B3PW91/6-31G** [127] levels of theory. This observation, as well as electronic spectroscopy and MO calculations [128], stimulated several studies on the possible homoaromaticity of [5]pericyclyne according to classical aromaticity criteria (see Section 1.3.2.3).

Structural criterion. Despite a planar equilibrium structure, the bond lengths and angles are very close to those calculated in 1,4-pentadiynes, ruling out any structural homoaromaticity.

Energetic criterion. The energetic homoaromaticity of a cyclic molecule is measured by its stability with respect to acyclic reference molecules. The Dewar-type resonance energy of **3a**, calculated from the averaged contribution of -C≡C–CH_2- units to the enthalpy of skipped polyynes of various lengths, corresponds to a slight destabilization ($\Delta H°$(298 K) = +0.9 kcal mol^{-1}, $\Delta H°$(0 K) = +2.2 kcal mol^{-1}) [125]. Similarly, the homodesmotic equation 5 $(MeC{\equiv}C)_2CH_2 \rightarrow$ **3a** + 5 MeC≡CMe (corrected for angle strain) indicates a very weak destabilization of **3a** by +1.8 kcal mol^{-1} [126]. [5]Pericyclyne is therefore not energetically homoaromatic.

The stability of [5]pericyclynes with respect to cyclic isomers has also been considered. Although [10]annulene itself was definitely shown to be non-planar, pentacyclopropa[10]annulene **3b** was calculated to exists in a D_{5h} form (B3LYP) with a weak Mills-Nixon effect (*endo*-C-C = 1.379 Å > *exo*-C-C = 1.358 Å) [129]: this putative isomer of **3a** is therefore aromatic in the structural sense. Nevertheless, as in the case of tricyclopropabenzene **1b**, the extreme strain of **3b** prevents its formation from the non-aromatic, but weakly strained, **3a**.

A cyclic isomer of the trihydroxy[5]pericyclyne **8** is the cyclotetrayne **7** (Figure 1.8), which has been demonstrated experimentally [130]. DFT calculations on model compounds of isoconfigurational isomers showed that the functional pericyclyne **8** is slightly less stable than its allenylidene isomer **7** (Figure 1.8): the local conjugation of the allenyne sequence is probably sufficient to overcome the excess ring strain of the C13 cyclotetrayne **7** with respect to the C15 cyclopentayne **8**. It must, however, be emphasized that several pentaoxy[5]pericyclynes have been prepared recently [131].

Magnetic criterion. [5]Pericyclyne **3a** is associated with a negligible magnetic susceptibility exaltation (Λ = –1.2 ppm cgs), which is also consistent with the very weak shielding experienced by either a $^7Li^+$ cation or a ghost atom located

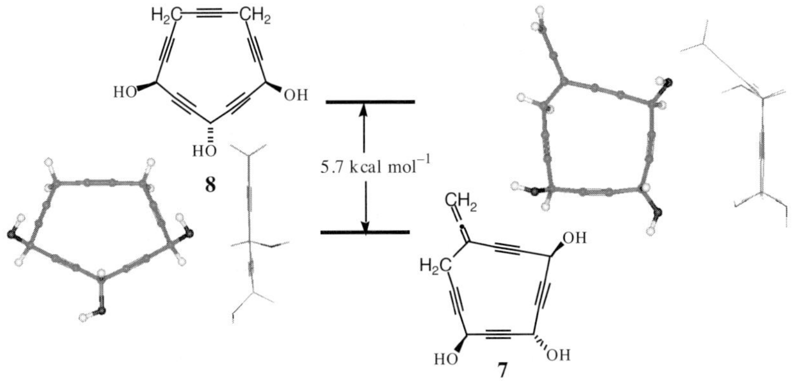

Figure 1.8 Geometry optimization of (3*S*,6*R*,9*R*)-trihydroxy-[5]pericyclyne **8**, and (3*S*,6*R*,9*R*)-2-ethenylidene-trihydroxy-cyclotrideca-1,4,7,10-tetrayne 7 (B3PW91/6-31G**) [130].

at the center of the ring ($\delta_{7Li}{}^{+} = -1.7$ ppm [126]; NICS = +0.6 ppm [127]). Finally, the ELF analysis confirms the weakness of the cyclic homodelocalization. The integration of the electron density in the C_{sp3}–C_{sp} and C_{sp}–C_{sp} valence basins is indeed consistent with pure single and triple bonds, respectively (Figure 1.9). More precisely, the ELF weight of resonance forms featuring homodelocalization is not significant (5.5 %, Figure 1.7) [111].

All the above results consistently imply that [5]pericyclyne is definitely not homoaromatic. Although the π character of cyclopropane rings might a priori be expected to favor extended interaction with the triple bond MOs, penta(cyclopropylidene)[5]pericyclyne was calculated to be as poorly homoaromatic as unsubstituted [5]pericyclyne [126].

- [6]Pericyclynes. The [6]pericyclyne structure has been demonstrated in the form of peralkyl [109, 132] and hexaoxy [133] derivatives. According to MM2 calculations, [6]pericyclyne **4a** and its dodecamethyl derivative should possess several isoenergetic conformations (chair, boat, twist-boat). At the B3PW91/6-31G** level, however, the chair conformation (D_{3d}) [111] directly shows the absence of homoaromaticity. This is confirmed by the central NICS value and the ELF analysis (Figure 1.7).

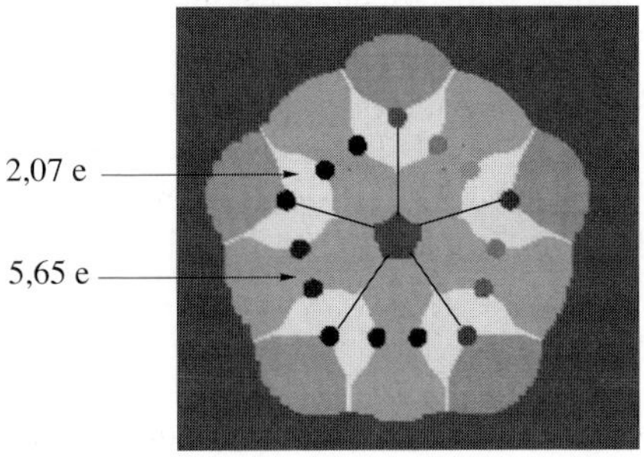

Figure 1.9 ELF partition map within the plane of [5]pericyclyne **3a** (B3PW91/6-31G**) [111].

1.3.2
Hybrid *sp*-*sp*^{2} Rings (Dehydroannulenes)

The recent experimental chemistry of annulenes and dehydroannulenes has been reviewed [134]. The main quality assigned to both annulenes and dehydroannulenes is their relative aromaticity, for which three criteria are usually distinguished [135]: structural, magnetic, and energetic. Dehydroannulenes can also be regarded as derived from annulenes of smaller ring size through insertions of triple bonds. This view dates back to 1948, when Sworski suggested that two or three C-C bonds of benzene could be elongated by insertion of C_2 groups [136]. When the C_2 insertion is applied to all the C-C bonds, the resulting structure has the same maximal symmetry as benzene and so was named [C,C]$_6$*carbo*-benzene or, more simply here, *carbo*-benzene (see Section 1.3.2.3) [4].

1.3.2.1 Dehydroannulenes of "Low" Symmetry

If Sworski's insertion is applied twice on opposite edges of benzene, the resulting tetradehydro[10]annulene **9** (Scheme 1.11) still satisfies the Hückel rule and is a known, but fragile, molecule [137]. If Sworski's insertion is applied three times, the resulting hexadehydro[12]annulene **10** (Scheme 1.11) now satisfies the Hückel criterion of antiaromaticity. According to bond distances, a "delocalized" butatrienic structure **10b** was obtained at the B3LYP/6-31G* level [138a], while, in accordance with early NMR data [138b], a slightly more localized acetylenic equilibrium structure **10a** is obtained at the MP2 or DFT-GGA levels [71, 139] (Scheme 1.11). The central NICS value of **10b** amounts to +106.1 ppm at the B3LYP level, and reduces but remains highly positive at the RHF level (+23.3 ppm) [138a]. The central NICS value of **10a** is equal to +10.6 ppm at the HF level [139]. Several alternative magnetic criteria (ring current susceptibilities, aromatic ring current shielding (ARCS) plots) show that the acetylenic form **10a** sustains a paratropic ring current (RC). These results demonstrate the antiaromatic character of **10**. It should be mentioned that the η^6 coordinating properties of **10** (and tribenzocyclyne **14**, Section 1.3.2.2) are well established [140]. B3LYP calculations showed that their binding energy towards transition metal cations is very high, and the symmetry of the nickel and copper complexes was optimized to D_{3h}, with the metal cations inserted at the center of the ligand plane [141].

Sworski's insertion of cumulated triple bonds gives rise to larger dehydroannulenes. As would be expected from the Hückel rule, hexa- and octadehydro[14]annulenes (NICS $\approx$ −8.5 ppm) [142] and dodecadehydro[18]annulene **12** (NICS = −5.6 ppm) are magnetically aromatic according to the same criteria as above, and in agreement with experimental findings [139].

Sworski-type dehydroannulenes derived from cyclooctatetraene, such as **11** (Scheme 1.11), were also investigated with AM1, PM3, HF, and MP2 calculations. In this series, the relative energy of the puckered and planar conformations correlates well with the Hückel rule for the π_z electrons [143].

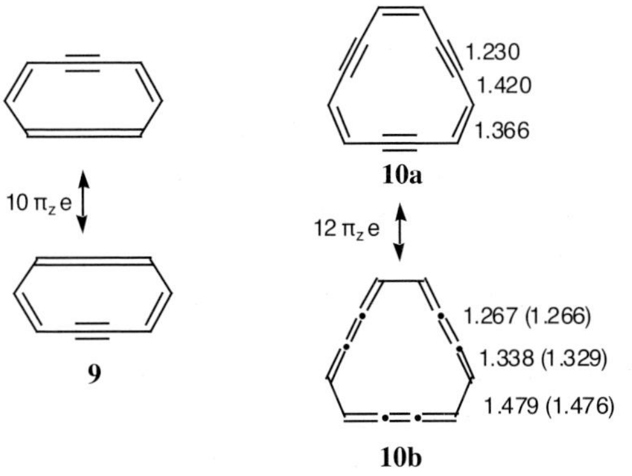

Scheme 1.11 Sworski-type dehydroannulenes of D_{2h} and D_{3h} symmetry derived from benzene (**9**, **10**, **12**) and cyclooctatetraene (**11**). Bond lengths of the two forms of **10** are in Å at the MP2/6-311G(d,p) level (the butatrienic structure **10b** is obtained by proportionality constraints on the bond lengths) [71]. Values in brackets correspond to the B3LYP/6-31G* level [138a].

1.3.2.2 Benzocyclynes and Related Systems

Dehydro[2*n*]annulenes, in particular for *n* = 2*m*, can be stabilized by embedding the remaining double bonds in *o*-, *m*- or *p*-phenylene units. Here we consider only the structures in which all the double bonds are benzannelated. Fully alternate benzannelated *n*-dehydro[2*n*]annulenes thus correspond to cyclic *o*-, *m*-, or *p*-EP oligomers (Section 1.2.4.2). The synthesis of large (C_{15}, C_{20}) planar *m*-EP rings and very large (C_{36}, C_{48}) belt-shaped *p*-EP rings was precedented by AM1 calculations, which turned out to be quite accurate, at least regarding comparison of the *sp* bond angles with those in the X-ray crystal structure [144]. The chemistry of *o*-EP rings, and more generally of benzannelated dehydro[2*n*]annulene, commonly known as "benzocyclynes", is more thoroughly documented [145]. The antiaromatic *n*-even case is surveyed first (the case *n* = 2 is degenerate with Vollhardt's oligophenylenes [146]).

For n = 4, two regioisomeric doubly benzannelated molecules are possible, one of them – **15a** (Scheme 1.12) – being quite stable [147]. The paratropic ring current of the C_8 ring and the lowered diatropic ring currents of the phenylene rings are confirmed by NICS calculation (+4.04 and −4.59 ppm, respectively) on the fully optimized structure at the B3LYP/6-31G* level [148].

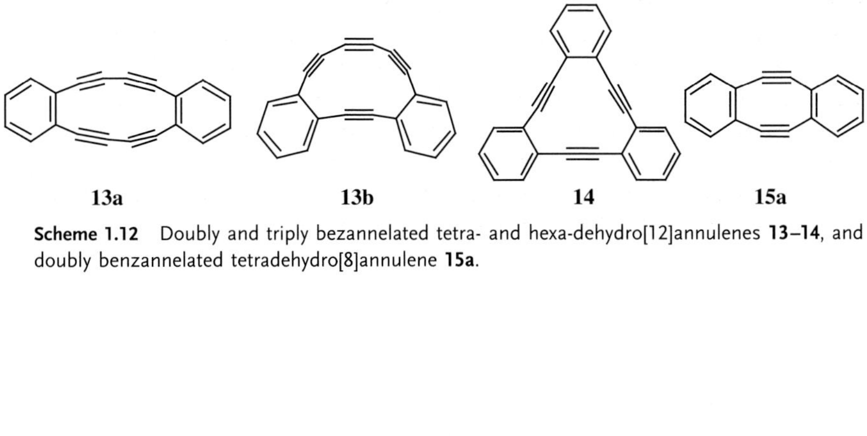

Scheme 1.12 Doubly and triply bezannelated tetra- and hexa-dehydro[12]annulenes **13–14**, and doubly benzannelated tetradehydro[8]annulene **15a**.

For $n = 6$, double benzannelation gives three possible isomers; two of them, **13a** and **13b**, are depicted in Scheme 1.12. The cyclotetrayne **13a** and substituted derivatives thereof, though quite reactive, have long been known [149]. The paratropicity of the C_{12} ring is confirmed by a positive NICS value (+5.75 ppm) [148]. Mixed experimental and theoretical evidence for the isomer **13b** was recently reported by Tobe et al. [150]. Geometry optimization at the B3LYP/6-31G* level indicates a C_{2v}-symmetric structure in which the tolan triple bond bows "in" by 7.8° about the *sp* carbon atoms, while the opposite $C_{sp}{\equiv}C_{sp}{-}C_{sp}$ bond angles are bent to 146.4°. Despite this intrinsic strain, **13b** was generated by photolysis in an argon matrix at 20 K and could be spectroscopically characterized in admixture with its precursor: cross-comparison of the experimentally measured and the calculated IR spectra is indeed convincing.

In the tris-*o*-benzannelated series, the ring current of the very stable D_{3h} tribenzocyclyne **14** (Scheme 1.12) [151] is less paratropic than that of the D_{2h} homologue **13a** discussed above, as indicated by a less strongly positive central NICS value (+4.4 and +2.9 ppm at the B3LYP [148] and HF [139] levels, respectively). This moderate value is due to the phenylene rings [148], since the NICS value of the non-benzannelated version **10** is much higher (see Section 1.3.2.1) [138a].

Tribenzocyclyne **14** (Scheme 1.12) is the elementary unit of graphyne and related carbon allotropes, which were first proposed and studied at the theoretical level [152] before becoming the target of active experimental investigations [145]. Several derivatives were thus investigated. Unlike the redox properties, the chromophoric properties of such molecules were shown to be poorly influenced by fluorine substituents at the phenylene rings: while both the experimentally determined reduction potential and the LUMO energy at the B3LYP level are significantly lowered, the experimentally determined λ_{max} value and the HOMO-LUMO gap remain unchanged (approximately 3 eV) [153].

The series of benzannelated *n*-odd dehydro[2*n*]annulenes starts with phenanthryne ($n = 3$) [154]. For $n = 5$, the unknown dibenzannelated triyne **15b** (Scheme 1.13) was predicted to be significantly diatropic [148]. For $n = 7$ and 9, the diatropicity decreases with the ring size and the benzannelation level [142, 148]. Thus, for the tribenzannelated representatives **16** and **17**, the central NICS almost vanishes at either the B3LYP [148] or the HF [139, 142] levels (Scheme 1.13).

Annelations of dehydroannulenes by other unsaturated rings, such as cyclobutadiene [139], cyclobutene [155a–b], dimethyldihydropyrene, thiophene [156], or tetrathiafulvalenes [155c], at various theoretical levels have also been envisioned. For the cyclobutadiene-annelated homologues of **14** and **17**, in which the external macrocycles host 18 and 24 π_z electrons, respectively [139], the bond lengths reveal a marked cumulenic character. Their overall non-aromaticity is confirmed by the NICS and ring current susceptibility values [139]. The cyclobutadiene-annelation thus convert aromatic and antiaromatic structures into non-aromatic ones, as does benzannelation.

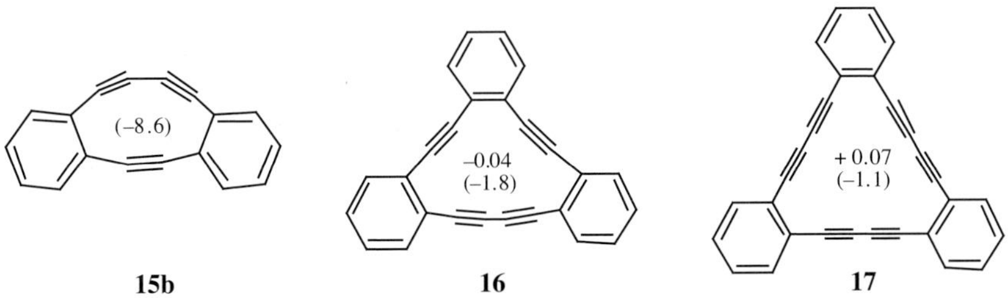

15b **16** **17**

Scheme 1.13 Di- and Tribenzannelated *n*-odd ("aromatic") dehydro[2*n*]annulenes. The central NICS values are given in ppm at the HF level [139, 142]. Values in brackets correspond to the B3LYP level [148].

1.3.2.3 Dehydroannulenes of Maximal Symmetry: *carbo*-[*N*]Annulenes

A *carbo*-[*N*]annulene is a Sworski-type 2*N*-dehydro[3*N*]annulene with the same ideal symmetry as the parent [*N*]annulene [4]. Their respective aromaticities were first compared by the three classical criteria, relating to structural, magnetic, and energetic observables, respectively [135]. A fourth criterion was then designed by considering that the electron density $\rho(\mathbf{r})$, long calculable from the (non-observable) polyelectronic wavefunction, is now gaining the status of an observable through recent developments in X-ray diffraction techniques [157]. The "electronic" aromaticity criterion is based on the ELF analysis of $\rho(\mathbf{r})$ [3]. In particular, in contrast to annulenes and analogously with pericyclynes (Section 1.3.1.2), *carbo*-annulenes possess a cyclically homoconjugated in-plane π_{xy} MO system, and the ELF tool allows for a relative weighting of resonance forms associated with in-plane and out-of-plane cyclic delocalization [111]. The aromaticity of the ring *carbo*-mers of the first annulenes was theoretically investigated according to these four criteria.

The process is detailed first for *carbo*-benzene ($C_{18}H_6$), which was proposed as a synthetic challenge in 1995 [158]. Although the dehydroannulene isomer **12** had been known for some time (Scheme 1.11) [159], the first examples of substituted *carbo*-benzenes appeared the same year [133, 160]. Very recently, a tetraphenyl *carbo*-benzene with two hydrogen atoms in *ortho* positions (Figure 1.10) has been synthesized [161a]. Both hexaphenyl- and tetraphenyl-*carbo*-benzenes fulfill the criteria of aromatic structures: they are stable, diatropic (deshielded ^{1}H NMR chemical shifts), with a quasi D_{6h} central ring (according to X-ray crystal structures). These experimental data served as references for following theoretical investigations.

- **Structural criterion.** At the B3PW91/6-31G** level, the very good agreement between the experimentally determined [161a] and the calculated [161b] structures of tetraphenyl-*carbo*-benzene indicates that hybrid DFT is well suited for this family of molecules (Figure 1.10). At the same level, the calculated structure of unsubstituted *carbo*-benzene is perfectly D_{6h}-symmetric with similar bond lengths (1.239 Å and 1.369 Å) [127,162]. In contrast, HF [162] or AM1 calculations [163] failed to reproduce the D_{6h} symmetry.

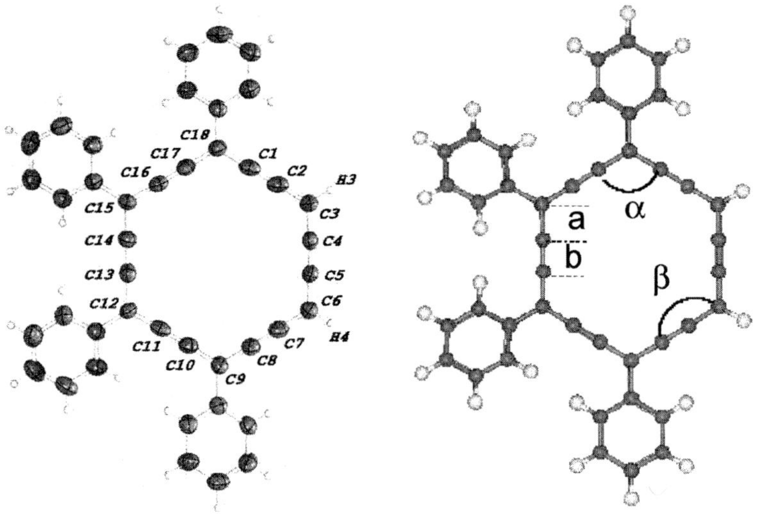

Figure 1.10 Experimentally determined (X-ray) and calculated (B3PW91/6-31G*) structures of tetraphenyl *carbo*-benzene. Exp. (calc.): a = 1.372 (1.374) Å, b = 1.233 (1.237) Å, α = 119.9° (118.5°), β = 178.3° (178.6°) [161].

- **Magnetic criterion.** The NICS value calculated at the center of the ring is equal to −17.9 ppm (B3LYP/6-31+G*) [127], namely more than twice the value of benzene (−8.0 ppm) and three times the value of the D_{3h} dehydroannulene isomer **12** (−5.6 ppm, Section 1.3.2.1) [139]. The total ring current density map 1 a_0 above the ring plane was established within the framework of the CTOCD-DZ formalism [164]: a strong diamagnetic circulation follows the line of the carbon ring. Orbital analyses indicate that the HOMO levels are of π_z character (E_{2u}) [162], and that their four electrons are responsible for the main part of the ring current [164].
- **Energetic criterion.** Several aromatic stabilization energies (ASEs) [135, 165] and resonance energies (RE) have been calculated for *carbo*-benzene and compared with those of benzene [166]. A few of them are reviewed below.
  - ***Isogyric ASE and Breslow resonance energy.*** The simple equations of Scheme 1.14 directly compare a single C_6 (or C_{18}) conjugated ring with the same C_6 (or C_{18}) conjugated chain in the open state. Since a ring closure turns on the possibility of a ring current [167], this isogyric equation is *fundamentally* related to the magnetic criterion of aromaticity. At the B3PW91/6-31G** level, the corresponding ASEs are equal to −17.9 kcal mol^{-1} for benzene, and to −11.2 kcal mol^{-1} for *carbo*-benzene. In accordance with Shaik's theory [168], these ASEs explicitly refer to both the π and the σ systems. However, since the geometry of the acyclic structure is considered in its most stable all-*trans* conformation, these ASEs underestimate the pure topological cyclicity effect.

    Fictitious linear reference molecules can, however, be partly described by their π system within the framework of the topological Hückel (HMO) theory.

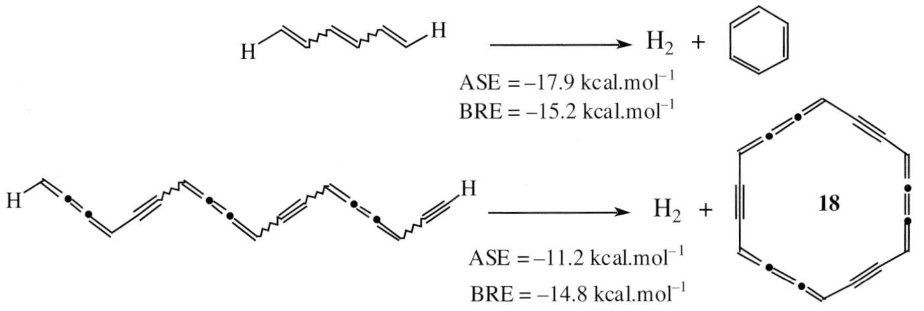

Scheme 1.14 "Ring-closing" isogyric equations defining an aromatic stabilization energy (ASE) and the Breslow resonance energy (BRE) of benzene and *carbo*-benzene **18** [166].

Within the HMO framework, Breslow's definition of the resonance energy of benzene (E_π(benzene) − E_π(1,3,5-hexadiene)) [169] can be generalized for any [*N*]annulene and reads (Equation 1.1):

$$\mathrm{BRE} = 2\,\beta_N\,[\Sigma\, n_k \cos(2k\pi/\mathrm{N}) - \Sigma\, n'_k \cos(k\pi/(\mathrm{N}+1))] \qquad (1.1)$$

where β_N is an average C-C resonance integral, and n_k and n'_k are the numbers of electrons occupying the kth π MO in the ground states of the [*N*]annulene and the corresponding *n*-polyene, respectively ($\Sigma n_k = \Sigma n'_k$ = N). This formula also holds for any dehydro[*N*]annulene as soon as the *sp-sp* and *sp-sp*^{2} resonance integrals can be averaged. This was found to be possible for *carbo*-annulenes [162], since DFT and HF π MO levels vary accurately as cos($2k\pi$/N). The "spectroscopic" proportionality factors β_N can be renormalized to "thermochemical" values (by assuming β_∞ = −20 kcal mol^{-1}), thus allowing direct comparison of the BRE with the ASE. The cases of N = 6 and N = 18 ultimately give BRE(C_6H_6) = −15.2 kcal mol^{-1} and BRE($C_{18}H_6$) = −14.8 kcal mol^{-1}, respectively. These results confirm that *carbo*-benzene is only slightly less energetically aromatic than benzene.

The preceding Hückel MO analysis was limited to out-of-plane π_z orbitals. A similar treatment can be applied to the homoconjugated in-plane p_{xy} orbitals [162], where the fictitious linear references are now "skipped" polyynes $H(C{\equiv}CCH_2)_NH$. For *carbo*-benzene, more tedious calculation afforded a Breslow-type resonance energy of BRE′ = +0.01 kcal mol^{-1}. Though formally positive, this value, and the corresponding energetic in-plane homoaromaticity, are negligible.

– ***Homodesmotic ASE.*** Cyclizing homodesmotic reactions [170] are commonly employed to evaluate the energetic aromaticity of highly symmetrical annulene-type rings made up of alternating single and double bonds only [165, 166]. Thus, by use of butadiene and ethylene as acyclic reference compounds for benzene (Scheme 1.15), calculations at the B3LYP/6-31G* level afford $\Delta E(C_6H_6)$ = −19.5 kcal mol^{-1} [171]. The corresponding equation for *carbo*-ben-

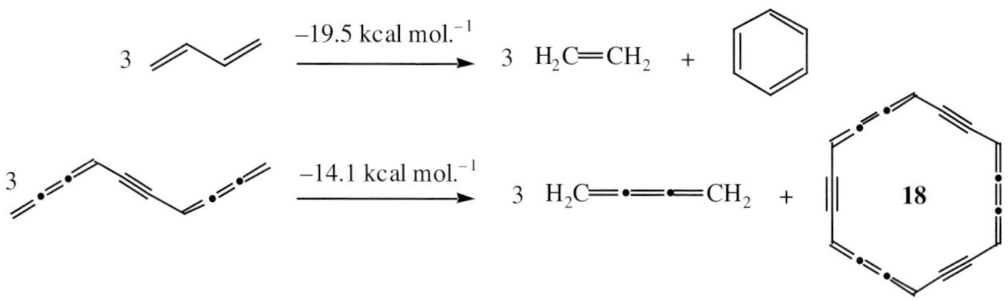

Scheme 1.15 Cyclizing homodesmotic reactions and ASEs for benzene and carbo-benzene [166].

zene is defined by replacing each reactant and product for its skeleton *carbo*-mer (Scheme 1.15): at a similar level of theory (B3PW91/6-31G**), the ASE is found to be only 25 % less negative: $\Delta E(C_{18}H_6) = -14.1$ kcal/mol^{-1}.
Other ASEs of the homodesmotic type and the resonance energy of the Mulliken–Parr type [172] have also been calculated [166]. All the results confirm that *carbo*-benzene is slightly less energetically aromatic than benzene.

- **Electronic criterion.** The populations of the ELF bonding valence basins directly show that the Kékulè structures of *carbo*-benzene are sound. The ELF-derived relative weights of the resonance forms featuring cyclic homodelocalization are negligible (approximately 3.5 %). Finally, the bifurcation-tree diagram (Figure 1.11) shows that the most nucleophilic basins are those of triple bonds: the C_{sp}-C_{sp} domain indeed reduces to two irreducible domains at a very high ELF value (0.87) [111]. Therefore, electrophilic attack is predicted to occur at the *sp* carbon atoms prior to the sp^2 ones, and should not preserve the *carbo*-benzenic structure.

Finally, all the aromaticity criteria reach their conclusions as to the aromaticity of *carbo*-benzene in a parallel manner (none is "orthogonal" to the others [173]). By use of the same methods, the aromaticity of ring *carbo*-mers of other annulenes has also been addressed for a complete family of *carbo*-annulenes $[(C_3H)_N]^q$ satisfying the orbital Hückel rule for their out-of-plane π_z electrons (Figure 1.12) [174].

The *carbo*-cyclopentadienyl cation **19** (Figure 1.12) remains experimentally unknown [131], but possesses virtually the same NICS and BRE values as *carbo*-benzene [162]. The ELF-derived weight of the resonance forms featuring cyclic homodelocalization in the free cation is much higher (13 %) than in the neutral *carbo*-benzene **18** [111] and the zwitterionic *carbo*-cyclopentadienone **26b** (see Section 1.3.3). It should be mentioned that the neutral *carbo*-cyclopentadienylene substructure was considered theoretically in *carbo*-pentalene **23** (Figure 1.14) [175]. Within the framework of Vogler's early theory for the calculation of ^{1}H shielding in hydrocarbons containing conjugated triple bonds [176] the protons were calculated to experience a strong paramagnetic RC effect. The antiaromaticity of pentalene could thus be preserved in its *carbo*-mer.

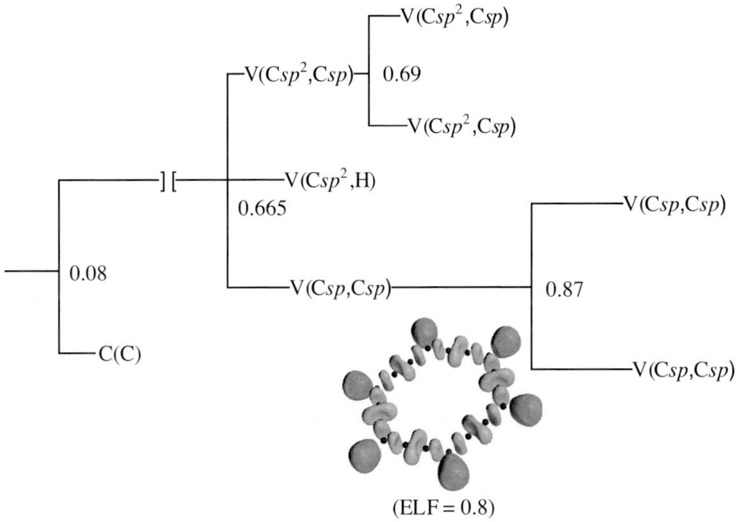

Figure 1.11 ELF bifurcation-tree diagram of *carbo*-benzene. At ELF = 0.8, the two disynaptic valence basins of each C_{sp}-C_{sp} bond are still connected by a lens-like domain [111].

DFT calculations accurately reproduce the equilibrium geometry (D_{2h}) of singlet *carbo*-cyclobutadiene **21** obtained by AM1 calculations [163]. The ELF weight of the resonance forms featuring cyclic in-plane homodelocalization is negligible, in agreement with the non-aromaticity of the corresponding valence isomer as predicted at the PM3 level [177]. The BRE, though still negative, is quite small (–5.8 kcal mol^{-1}), and its NICS value (+53.8 ppm) indicates a strong paratropic character. In contrast, triplet *carbo*-cyclobutadiene **20** (Figure 1.12) possesses all the features of an aromatic molecule.

Finally, the *carbo*-cyclopropenyl anion **22** appears strongly aromatic, with the same NICS and BRE value as its parent cyclopropenyl cation. Its formal analogy with the cyclopentadienyl anion suggests that this anion might behave as a strong π ligand for transition metals. Geometry optimization of the $[Fe(C_9H_3)]^+$ complex at the B3PW91 levels indicates a C_{3v} η^9-coordination, with a formal count of 18 valence electrons on iron (Figure 1.13) [178].

Ring *carbo*-mers of higher [*N*]annulenes and [*N*]annulenic ions were also considered. For $N = 7$, the same level as used above indicates that the Hückel rule for the π_z electrons remains valid for the *carbo*-tropylium anion ($[C_{21}H_7]^-$, 22 π_z electrons): the equilibrium symmetry is D_{7h} and the central NICS is equal to –20.4 ppm [179]. For $N = 8$, AM1 calculations afforded a non-planar tub geometry for *carbo*-cyclooctatetraene, nearly degenerate with the planar transition state of inversion ($\Delta E^* =$ 1.2 kcal mol^{-1}) [163].

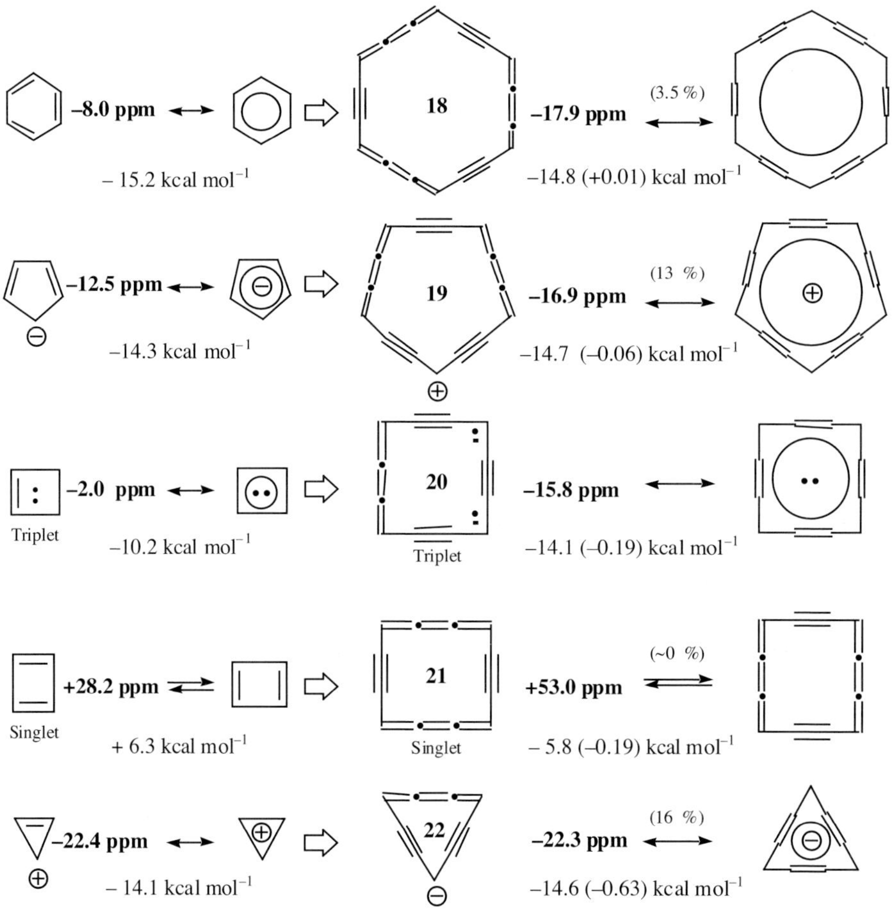

Figure 1.12 Compared aromaticity measures of annulenic species and their ring *carbo*-mers **18–22**: central NICS (in bold), BREs for the out-of-plane aromaticity (in brackets: BREs for the in-plane homoaromaticity), ELF-weights of resonance forms featuring cyclic homodelocalization (percentages in brackets above the arrows) [111, 127, 162].

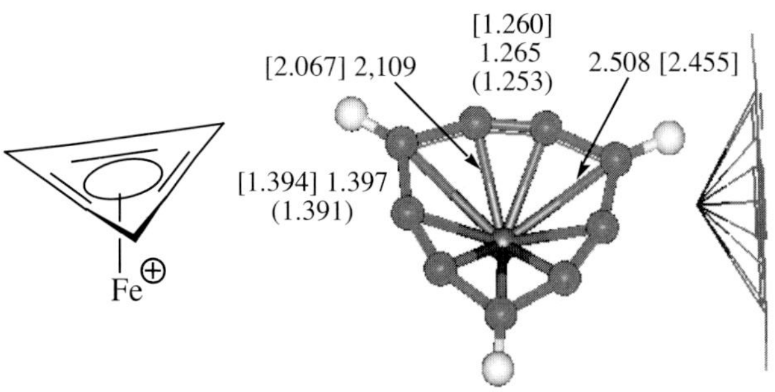

Figure 1.13 Geometry of the $[Fe(C_9H_3)]+$ complex at the B3PW91/6-31G**/LANL2DZ(Fe) level. Bond lengths are in Å. Values in square brackets are bond lengths at the B3PW91/6-311+G* level. Values in round brackets, are those of the free anion at the B3PW91/6-31G** level [178].

Further speculative synthetic targets related to the *carbo*-benzene structure were identified on the basis of DFT calculations. Soncini, Fowler, et al. studied a hexa-azahexaborine analogue of *carbo*-benzene in which BN (in place of C_2) units are inserted in a head-to-tail manner into the bonds of a benzene ring [164]. The molecule was found to be aromatic in the structural and magnetic senses (C_{6h} symmetry, homogeneous diatropic ring current).

Push-pull chromophores involving *carbo*-phenylene units were investigated for possible first-order NLO properties. At the ZINDO level, the static quadratic hyperpolarizability of the full *carbo*-mer of *p*-nitroaniline ($\beta_0 = 841 \times 10^{-30}$ cm^5 esu^{-1}) is 80 times greater than that of *p*-nitroaniline ($\beta_0 = 10 \times 10^{-30}$ cm^5 esu^{-1}) [180]. A systematic study of 32 partial *carbo*-mers, optimized at the MM level, showed that much simpler derivatives such as **24** (Scheme 1.16) still exhibit a promising β_0 exaltation: $\beta_0(X = NH_2) = 810 \times 10^{-30}$ cm^5 esu^{-1}, $\beta_0(X = H) = 633 \times 10^{-30}$ cm^5 esu^{-1}. However, *carbo*-merization of the acceptor NO_2 group remains essential. The properties of the NC_4O_2 group were thus investigated at higher levels of theory (DFT and CASSCF). Several nearly degenerate conformational minima were found, the *sp* carbon atoms exhibiting strong carbenic character. This suggests a high flexibility and instability of the NC_4O_2 group, thus discouraging attempts at experimental synthesis.

Chromophores made up of ethynyl-*carbo*-phenylene units bridging pyridinium and phenolate end groups have also been envisioned [88], but no β_0 exaltation was predicted. These structures were, however, included in the β_0 vs. ABLA correlation (Figure 1.6) discussed in Section 1.2.4.2.

The "second" ring *carbo*-mer of benzene, **25** (Scheme 1.16), was also considered [179]. Its optimized structure is D_{6h} and its central NICS value, of −18.9 ppm (B3LYP/6-31+G*), is virtually identical to that of the "first" ring *carbo*-mer **18** (Figure 1.12). It is noteworthy that the recognized efficiency of procedures for C_{sp}-C_{sp} coupling (with respect to C_{sp}-C_{sp3} coupling) might facilitate the synthesis of *carbo*^{2}-benzene derivatives (with respect to *carbo*-benzene) [181].

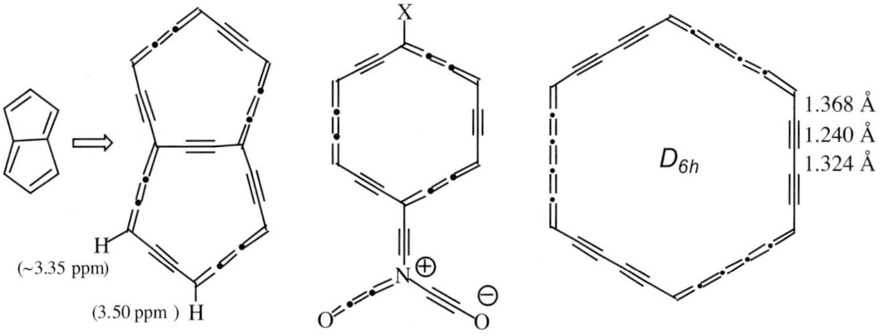

23 **24** **25**

Scheme 1.16 *Carbo*-pentalene **23** [175], partial *carbo*-mer of *p*-nitroaniline **24** [180], and *carbo*^{2}-benzene **25** (bond lengths calculated at the B3PW91/6-31G* level) [179].

1.3.2.4 Cyclo[*n*]carbons

As emphasized in Section 1.2, *sp* carbon chains stabilize by ring-closing for $n \geq 10$. Despite their higher theoretical stability, the carbon rings remain experimentally elusive under the high-temperature conditions (> 2000°C) at which gaseous carbon clusters are produced. The open chains are indeed entropically favored through the possibility of large-amplitude motions [17]. The temperature-dependent rate constants for the cyclization and the ring snapping of C_8, C_{10}, and C_{12} were thus calculated within the rotor-harmonic oscillator approximation at the B3LYP/6-311G* level [182]. Above 1000 K, it was shown that both processes, in either the singlet or triplet series, are already very fast on the timescale of cluster production, and that above 3000 K, cyclization is slower than ring snapping (Scheme 1.17). It was proposed that trapping of carbon rings by fast cooling to ~1000 K could be envisioned [182].

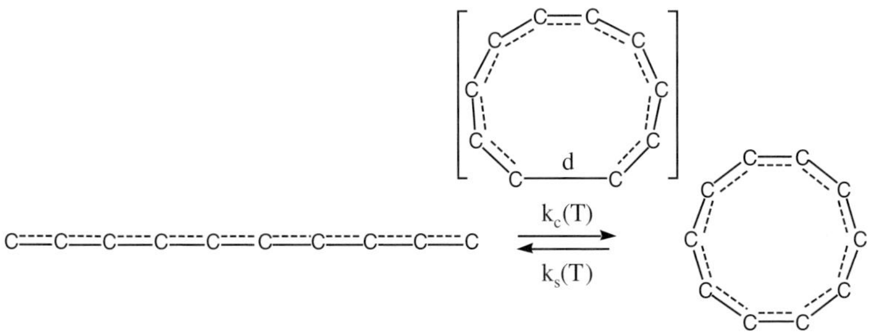

Scheme 1.17 Interconversion between linear and cyclic forms of C_{10} through cyclisation and ring snapping processes. The incipient bond of the transition state is longer in the singlet series (d = 2.94 Å) than in the triplet series (d = 2.53 Å) (B3LYP/6-311G*) [182].

Thanks to their possible stabilization by double aromaticity [183], cyclo[4*n*+2]carbons have attracted specific interest [1]. Since the C_{18} member of this series was detected by Diederich et al. [184], several efficient molecular precursors for their generation have been designed [185]. In a critical review, Houk et al. considered four types of geometry for potentially aromatic cyclo[4*n*+2]carbons (Figure 1.14), and reported that calculations at the RHF/6-31G* level predict C_{18} to possess a C_{9h} polyacetylenic structure [1, 186], the cumulenic and polyacetylenic D_{9h} structures being less stable by 32 and 0.3 kcal mol^{-1}, respectively. Introducing electron correlation through the nonlocal BLYP functional and applying an empirical correction for the recognized overestimate of the energy of allenylene units by DFT methods (approximately 6 kcal mol^{-1} with respect to propynylene isomers), the authors confirmed that the C_{9h} polyyne is approximately 36 kcal mol^{-1} more stable than the D_{9h} cumulene. The polyacetylenic D_{9h} structure is actually the lowest transition state for valence interconversion between two C_{9h} minima, just as the D_{4h} structure of singlet cyclobutadiene is the transition state between two D_{2h} minima. This behavior is not anticipated from the "aromatic" Hückel electron counts of C_{18} for both π systems. The low symmetry of the equilibrium geometry

Bond length alternation \ Bond angle alternation	No (rounded)	Yes (flattened)
No (cumulene)	=C=C= D_{2nh}	=C=C= D_{nh}
Yes (polyyne)	−C≡C− D_{nh}	−C≡C− C_{nh}

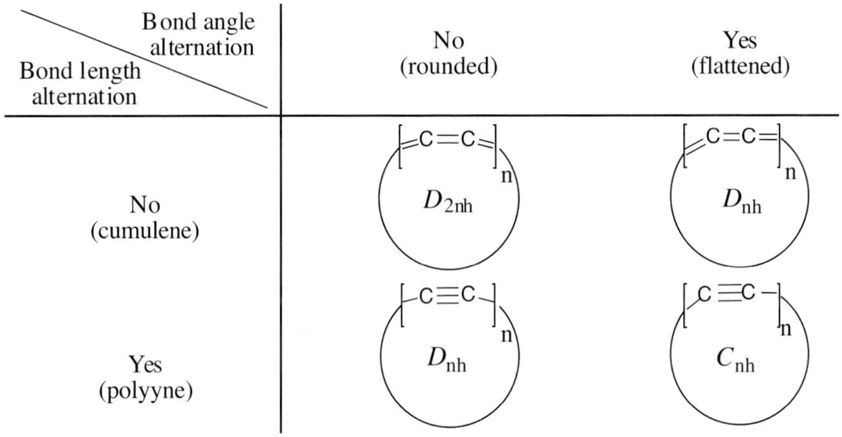

Figure 1.14 Four types of planar cyclo[2n]carbons [186].

for cyclo[4n+2]carbons, however, is specific to the large size of C_{18}. The "normal" tendency is restored for the smaller cyclo[10]carbon: the cumulenic D_{5h} structure (more symmetrical in terms of bond lengths) is a minimum, 25 kcal mol^{-1} below the polyacetylenic D_{5h} structure. At the same level of theory, the validity threshold of the Hückel rule for cyclo[4n+2]carbons is claimed to occur just below $n = 3$: for C_{14}, the D_{7h} and C_{7h} polyacetylenic structures already lie a few kcal mol^{-1} below the D_{7h} cumulenic structure. The polyacetylenic character of C_{14}, C_{18}, and higher homologues was later confirmed at the RHF level of theory [187]. Finally, limited MCSCF or CCSD calculations from HF geometries, and quantum Monte Carlo calculations on C_{14} and C_{18}, showed that electron correlation results in decreased bond length alternation but still favors polyacetylenic structures [188]. For all C_{4n+2} molecules, the maximum $D_{(4n+2)h}$ symmetry is thus at least reduced to $D_{(2n+1)h}$ through either kinds of structure, cumulenic (bond angle alternation) or polyacetylenic (bond length alternation) (Figure 1.14).

- For small C_{4n+2} molecules such as C_6 (or C_{10}), most calculation methods, including post-HF [188], indicate that they undergo bond angle alternation, but preserve bond length equalization. The σ-π separation is a prerequisite of the Hückel model. In small planar rings, however, it does not hold for the π_{xy} MOs, which mix with σ MOs: the six (or ten) electrons in the pure π_z MOs obey the Hückel rule and tend to induce D_{6h} (or D_{10h}) symmetry through equalization both of bond lengths and of bond angles. The σ-π_{xy} mixing induces a second-order Jahn–Teller distortion, which partly reduces the symmetry to D_{3h} (D_{5h}) at the angular level only [187].
- For larger C_{4n+2} molecules, the σ-π_{xy} mixing tends to disappear. In spite of the formal validity of the Hückel model for both π systems, reduced symmetries of the polyacetylenic kind are calculated. This discrepancy is ascribed here to a Peierls distortion [187]. Cyclo[4n+2]carbons of increasing size indeed become semi-me-

tallic, and a simple modified Su–Schrieffer–Heeger (SSH [189]) treatment (based on the σ-π separation) shows that the Peierls transition must take place for some n value depending on the electron-phonon coupling parameter $\lambda_{ep} = 8\gamma^2/(\pi K ß)$. Here, γ is a linear coefficient describing the variation of the hopping integral $ß$ with the bond length alternation parameter δ_d ($\Delta ß = \gamma\ \delta_d$; $E_\pi = f_{\gamma,ß}(\delta_d)$) and K is the coefficient of the harmonic potential describing the σ bonds/electrostatic framework ($E_\sigma = n\ K\ \delta_d^{\ 2}/2$). A fit of the HF results for cyclocarbons from C_4 to C_{42} gives $\lambda_{ep} \approx 0.44$, which corresponds to medium electron-phonon coupling. An analogous analysis at the DFT (LDA/plane-waves) level afforded a stronger coupling parameter $\lambda_{ep} \approx 0.24$ (at this level, cyclo[4n+2]carbons up to C_{42} are predicted to be cumulenic). In accordance with spectroscopic evidence, the Peierls transition should occur beyond C_{22} [190].

The Peierls transition can be regarded as a breakdown of the formal Hückel rule for predicting *structural* aromaticity. On the other hand, the variation of *energetic* aromaticity of the C_{2n} rings can be roughly visualized from the plot of atomization energies against n [187]. Their relative stability with respect to isomeric chains, cages, plates, and bowls ($2 \geq n \geq 16$) has also been discussed in more detail on the basis of various DFT treatments [191]. Relevant energetic information was also gained from comparison of experimentally obtained electronic spectra of the C_{10}, C_{12}, and C_{14} rings with results of ab initio calculations [192].

1.3.3
carbo-Heteroannulenes

At the B3PW91/6-31G** level, the structures of *carbo*-pyridine **25a** and *carbo*-phosphinine **25b** (Scheme 1.18) exhibit C_{2v} symmetry with C-C bond lengths and central NICS values similar to those of *carbo*-benzene [193]. These data support strong aromatic character in **25a** and **25b**.

The ELF bifurcation-trees indicate that the P lone pair domain of **25b** separates at ELF = 0.68, just after the C–H basins (occurring at ELF = 0.67). In accordance with

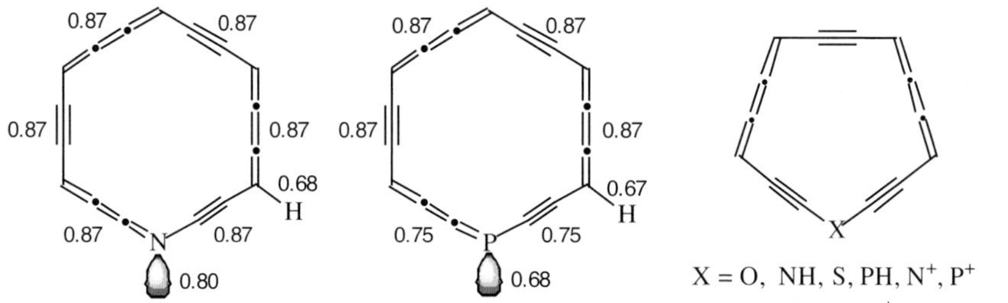

25a **25b**

X = O, NH, S, PH, N^+, P^+
X = PH=O ↔ PH^+– O^-: **26a**
X = C=O ↔ C^+– O^-: **26b**

Scheme 1.18 Ring *carbo*-mers of cyclically conjugated heterocycles. The values attached to the lone pairs and C-H and C_{sp}-C_{sp} bonds of **25a** and **25b** correspond to the latest ELF bifurcation values of the corresponding domains (ELF-nucleophilicity indices) [193].

the stronger nucleophilicity of nitrogen, the N lone pair domain of **25a** separates at a much higher value (0.80). Nevertheless, as in *carbo*-benzene (Figure 1.12), the most nucleophilic sites remain the $C_{sp}\equiv C_{sp}$ edge units, which separate at ELF = 0.87, while the $C_{sp2}\equiv C_{sp}$ vertex units separate earlier (at ELF ≈ 0.70): as in *carbo*-benzene, electrophilic attack is not directed towards the vertices, and should break the aromatic C_{2v} nucleus.

As expected from the Hückel rule (16 p/π_z electron counts), the ring *carbo*-mers of furan, pyrrole, thiophene, and phosphole are strongly antiaromatic by structural and magnetic criteria. Formal oxidation of *carbo*-phosphole through oxygenation or hydride abstraction reduces the p/π_z electron counts to 14, and so restores aromaticity. On going from *carbo*-phosphole to *carbo*-phospholylium, the symmetry thus increases from highly distorted C_1 to C_{2v}, while the NICS value decreases from the "abnormal" value of +174.3 ppm to −15.5 ppm. *Carbo*-phosphole oxide **26a** (C_s, NICS = −4.9 ppm) has been compared with the ring *carbo*-mer of *carbo*-cyclopentadienone **26b** (C_{2v}, NICS = − 8.1 ppm). ELF-weighting of resonance forms shows that they both exhibit negligible in-plane cyclic homodelocalization, and 44 % out-of-plane cyclic (homo)delocalization (Figure 1.15). The ELF basin of the P-O bond of **26a** contains less than two electrons, enforcing the occurrence of the no-bond resonance forms. This picture suggests that the delocalization of the 14 π_z electrons through the C(O) (or PH(O)) vertex takes place through conjugation (or homo-conjugation) of p_z AOs. In the case of **26a**, the homoconjugation is based on a negative hyperconjugative effect of the σ P-O bond.

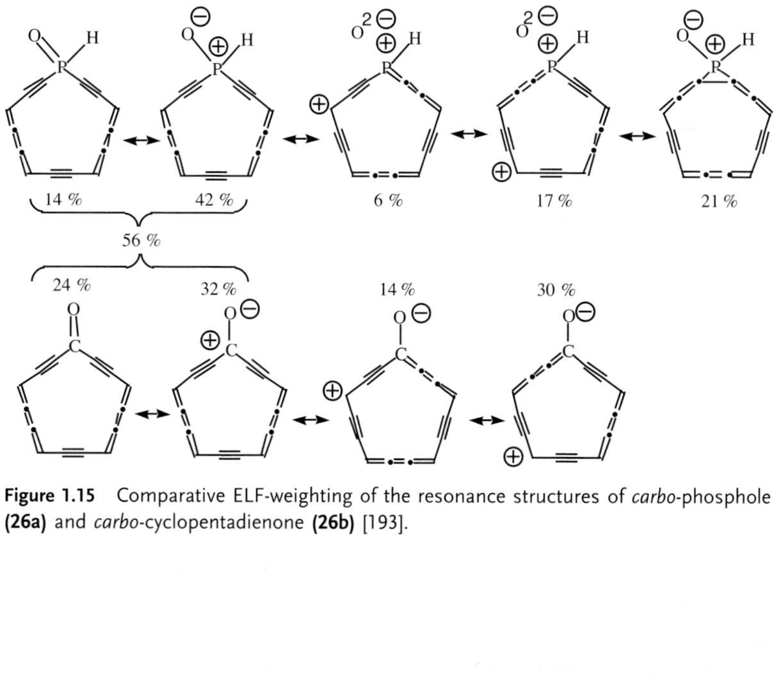

Figure 1.15 Comparative ELF-weighting of the resonance structures of *carbo*-phosphole oxide **(26a)** and *carbo*-cyclopentadienone **(26b)** [193].

1.4 Star-Shaped Acetylenic Scaffolds

Hexaethynylbenzene is a paradigm of star-shaped acetylenic scaffolds [151]. This class is extended here to any molecule with cross-conjugated or radially "homo"-conjugated triple bonds in at least three directions from a central core. The central core can be variously an atom, a rod, a ring, or even a cage (expanded adamantane [31] or cubane [194]).

1.4.1 Atomic Cores

The synthesis and the X-ray crystal structure of tetraethynylmethane (Scheme 1.19) (or "*carbo*-methane" [4]) have been reported by Feldman [195]. On the basis of Mulliken charges and substituent electronegativity effect analyses, Schaefer III et al. explained the discrepancy between experimentally determined (X-ray) and calculated (HF, MP2) structures in terms of a π electron compression effect [196], elongating the C_{sp3}–C_{sp} bonds and shifting the electron density of the corresponding internal C_{sp} atoms towards the external C_{sp} atoms: the electron density-based X-ray diffraction analysis would thus locate the internal C_{sp} atom closer to the other C_{sp} atom (1.14 Å) than actually seen for the nucleus (approximately 1.20 Å), thus giving an apparent contracted $C_{sp}{\equiv}C_{sp}$ bond length. This experimentally determined shortening is also not reproduced by semiempirical calculations (PM3, AM1, MNDO) [197]. The weak strain of the molecule was also estimated through the homodesmotic reaction: 4 CH_3-C≡CH → C(C≡CH)$_4$ + 3 CH_4, $\Delta H°$(MP2) = 5 kcal mol^{-1}. The strain of tetrabutadiynylmethane (R = C≡CH, Scheme 1.19) is even lower.

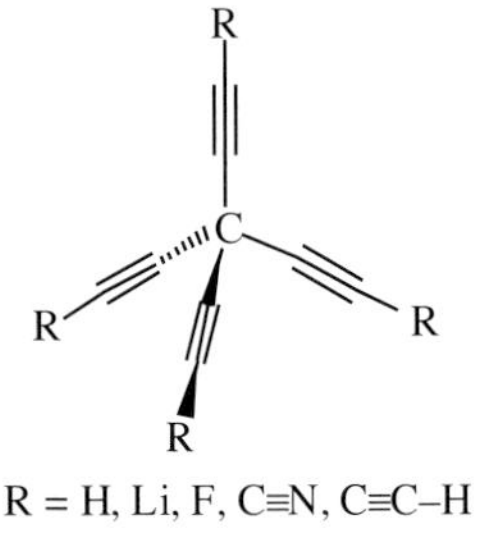

Scheme 1.19 Tetraethynylmethane and substituted derivatives studied at the SCF level [196].

1.4.2 Rod Cores

Tetraethynyl-ethylene (TEE) **27** [198], -allene **28** [199], and -butatriene **29** [200] (or *carbo*-ethylene [4]) (Scheme 1.20) were studied at the SCF level. The equilibrium geometry and vibrational frequencies of **27** and **29** are in good agreement with experimental findings [196]. The remote (cross-) conjugation energy of these structures was estimated through the enthalpy of isodesmic formation from vinylacetylene. At the MP2 level it is close to zero, except for **28** (+9 kcal mol^{-1}), in which the two π systems are indeed orthogonal. The heats of formation of these molecules are very high ($\Delta H°_f$(**27**, MP2)= +317 kcal mol^{-1}). As a complementary result, the stabilizing effect of the $SiMe_3$ substituent could be partly attributed to a hyperconjugation effect. Finally, the polarizability $<\alpha>$ and second hyperpolarizability $<\gamma>$ were calculated with an augmented STO-3G basis set: the $<\alpha>$ and $<\gamma>$ values consistently increase with the extension of π conjugation. In particular, **28** exhibits a lower $<\gamma>$ value (39 980 a. u.) than **27** (46 030 a. u.) and **29** (72 210 a. u.).

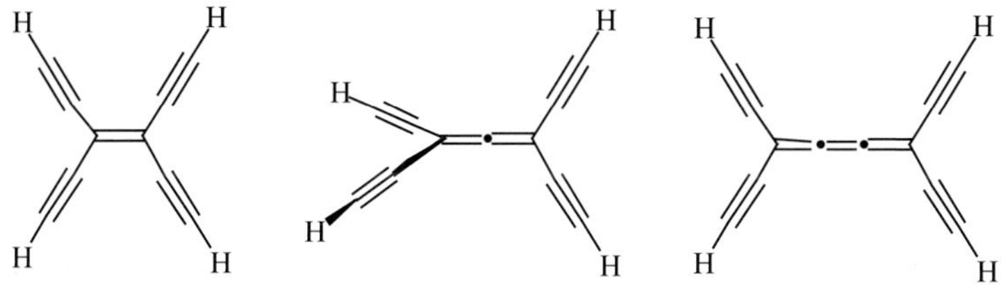

27 **28** **29**

Scheme 1.20 Star-shaped acetylenic scaffolds with rod-shaped cores.

Both linear and cross-conjugative pathways for electron delocalization are operative within the TEE framework of **27** (Scheme 1.21) [7].

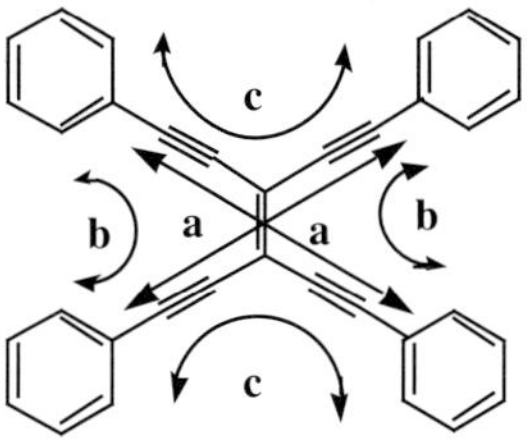

Scheme 1.21 Planar phenyl-substituted TEE exhibiting full two-dimensional conjugation. Paths a and b depict *trans*- and *cis*-linear conjugation, whereas path c depicts geminal cross-conjugation. From reference [7].

Physical characteristics of TEEs may be tuned by suitable choice of donor and/or acceptor substituents. Geometries of mono- and bis(*p*-nitrophenyl) TEE derivatives were calculated at the HF/6-31G** level [201]. The enediyne character is dominant in the neutral and the one-electron reduced compounds, and is similar in the parent TEE and in its substituted derivatives. Upon bielectronic reduction, however, the molecule adopts a cumulenic/quinoid form, whereas the bonds located outside the main conjugation path are affected neither by the substitution nor by the reduction process. In contrast to tetraphenylethylene and tetracyanoethylene, which exhibit twisted equilibrium structures in their reduced states, the disubstituted TEEs retain planarity in all reduced states.

The reversible photochemical *trans-cis* isomerization of TEEs substituted with electron-donating (*p*-dialkylaminophenyl) and/or electron-withdrawing (*p*-nitrophenyl) groups was investigated at the semiempirical AM1 level [202]. As the frontier orbitals are concentrated about the central olefinic bond, the (photochemical) promotion of an electron from the HOMO to the LUMO results in an elongation of this bond. The barrier to rotation is therefore strongly reduced, enabling facile isomerization.

Substitution of *p*-nitrophenyl acceptors by 5-nitro-2-thienyl moieties allows the third-order NLO efficiency of donor-acceptor functionalized TEEs to be improved [203]. Semiempirical studies of these compounds have attested to the dual role of the nitrothienyl group. This acts as a strong electron acceptor, but its electron-rich thiophene ring is also involved as a donor in a few electronic transitions. The TEE backbone is also able, depending on the substituents and the substitution pattern (*cis, trans* or geminal), to act either as a donor or as a weak acceptor.

N,N-Dialkylaniline-substituted TEEs exhibit dual fluorescence [204]. This may be described in terms of a first higher-energy emission, due to the initial excitation to the locally excited (LE) state, and a second lower-energy band (A-band), due to emission from an internal charge-transfer (CT) state. The twisted intramolecular charge transfer (TICT) model assumes that the molecule relaxes from its LE state to a minimum on the excited state surface through twisting of the donor group into a plane perpendicular to the acceptor group. Along this twisting coordinate, there is an increase in the CT, which gives rise to the highly polar structure responsible for the A band. TDDFT computations allow such a TICT to be inferred for the above *N,N*-dialkylaniline-substituted TEEs [204].

Expanded dendralenes with *iso*-PTA [205] or *iso*-PDA [206] backbones have been reported (Scheme 1.22). Semiempirical PM3 calculations indicate that the planar all-*s-trans* conformation is not the most stable, and that slightly twisted *s-cis* and "U-shaped" geometries are preferred. This illustrates that effective π conjugation between adjacent C≡C and C=C bonds does not require full planarity of the C_4 fragment [207].

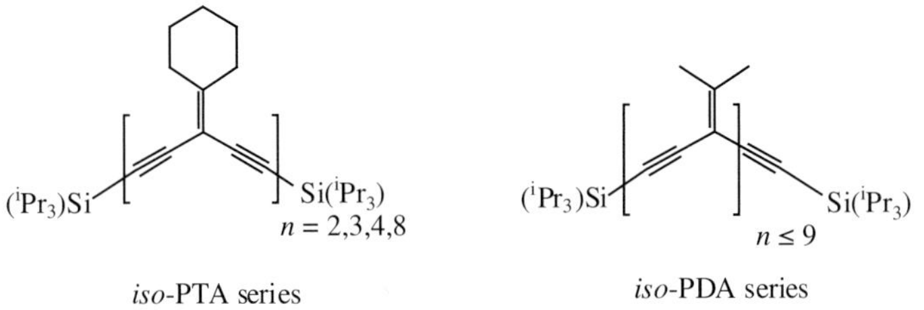

Scheme 1.22 Expanded dendralenes from reference [207].

1.4.3
Cyclic Cores

1.4.3.1 Annulene and Dehydroannulene Cores

Hexaethynylbenzene (HEB) **31** [208], tetraethynylcyclobutadiene **30** (known as η^4 ligand in metal complexes [151]), and octaethynylcyclotetraene **32** (Scheme 1.23), have been studied at the AM1 level and compared with their ring *carbo*-annulene isomers [163]. Structures **30** and **32** were found to be less stable than their "anti-aromatic" *carbo*-cyclobutadiene (**21**) and *carbo*-cyclotetraene isomers, by 40.3 and 12.2 kcal mol^{-1}, respectively.

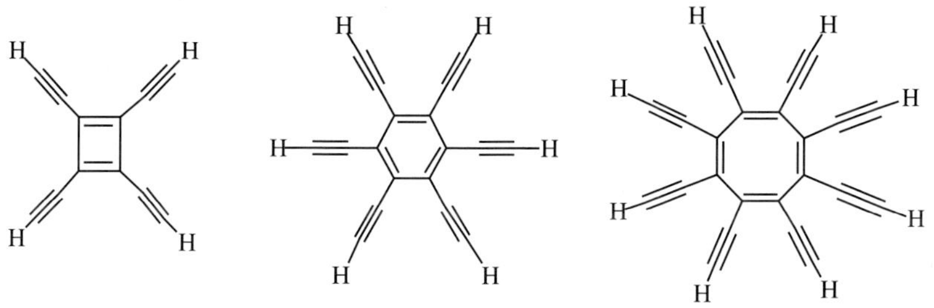

30 **31** **32**

Scheme 1.23 Neutral perethynyl annulenes.

After early MNDO and HF calculations reproducing the experimental geometry of HEB **31** [209], excited singlet electronic states and chromophoric properties of phenyl derivatives thereof were characterized from single (limited to π MOs) and double configuration interactions [210]. The electric and magnetic properties of hexaethynylbenzene **31** were also investigated at the coupled HF level [211]. HEB was thus calculated to be three times as polarizable as benzene. By use of distributed-origin methods for the vector potential (CTOCD-DZ or -PZ), ring current density maps in and 1 a_0 above the molecular plane indicate that no current links the ethynyl arms, either through space or through bonds.

Dissymetrical HEBs could be obtained by decarbonylative Diels–Alder reactions of alkynes with tetraethynylcyclopentadienone [212]. Regioselectivity features could be analyzed from the relative localization of PM3-calculated frontier orbitals of the star-shaped substrate. On this occasion, the octaethynyldibenzo[8]cyclyne **33** (Scheme 1.24) was obtained, and the determination of its X-ray crystal structure was assisted by comparison with its PM3 or ab initio optimized geometry [212]. Recourse to PM3 calculation was also useful for interpretation of the steric strain in the X-ray crystal structure of a tetraethynylcyclobutadiene cobalt complex [213].

Hexaethynyl substitution of dehydroannulenes has also been envisioned. The magnetic criteria indicate that hexaethynyl substitution slightly reduces the antiaromaticity or aromaticity of the dehydroannulene rings: NICS(**34**) = +8.8 < NICS(**10**) = +10.6 ppm, and NICS(**12**) = −5.6 < NICS(**35**) = −4.9 ppm

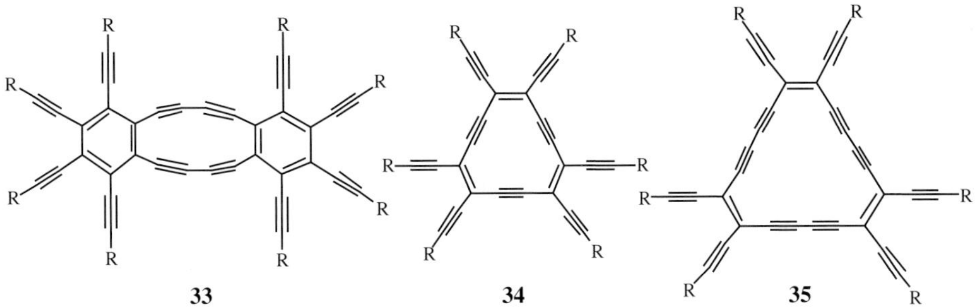

33 34 35

Scheme 1.24 Star-shaped acetylenic scaffolds from dehydroannulene cores.

(Schemes 1.11, 1.24) [139]. The effect of hexaethynyl substitution is, however, much less pronounced than in the case of *carbo*-radialenes (Section 1.4.3.2).

HEB and lower expanded versions are currently used as electronic transmitters [214]. Thus, a ferromagnetic metal-metal exchange interaction was detected for $[(C_5Me_5)(dppe)Fe(III)-]^+$ units bridged through 1,3-diethynylbenzene or 1,3,5-triethynylbenzene spacers [58, 215]. Theoretical studies support the experimentally obtained results [216].

Of the star-shaped structures, dendrimers should not be omitted. Second hyperpolarizabilities (γ) were calculated at the semiempirical level for ethynylphenylene (EP) dendrimers. In the case of 24 phenylacetylene units, the γ value was found to be six times smaller than that of the linear *p*-EP oligomer of the same size [217].

1.4.3.2 Radialene and Expanded Radialene Cores

The longstanding discussion relating to the weak aromaticity of radialenes [218] has been theoretically resumed within a more general framework for [3]radialene and acetylenic expansions thereof [219]. The starting point was the observation that perethynyl derivatives of doubly expanded [3]- and [4]radialenes (e. g., the hexakistrimethylsilyl derivative of **36b**, with R = $C{\equiv}C\text{-}SiMe_3$ in Scheme 1.25 [220]) possess low reduction potentials, and the question at stake was whether enhanced aromaticity of the anions could be the driving force.

The geometry and the central NICS of a complete set of eighteen *carbo*-[3]radialenic species in the singlet or doublet spin states were calculated at the B3PW91/6-31G** and B3PW91/6-31+G** levels, respectively (Scheme 1.25). The central rings in all the structures were found to be planar with a symmetry close to D_{3h}. For hexaethynylradialene **36b** ($C_{18}H_6$, an isomer of *carbo*-benzene; see Section 1.3.2.3) the bond lengths show a stronger delocalization than those previously obtained at the SCF level [196] (± 0.02 Å), as would be expected from the well known relative tendency of the DFT methods in relation to HF methods. The central NICS values are all negative, except for the singly expanded monoanion $[\mathbf{37a}]^-$ (C_{2v}, NICS = +4.0 ppm). Interestingly, hexaethynyl substitution has a dramatic effect, since the NICS value drops to −16.4 ppm in $[\mathbf{37b}]^-$. Over all species with a non-expanded

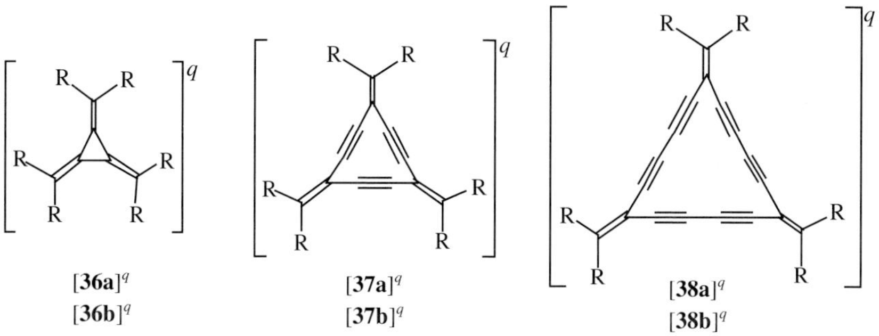

$[36a]^q$ $[36b]^q$ $[37a]^q$ $[37b]^q$ $[38a]^q$ $[38b]^q$

$q = 0, -1, -2$ { **a**: R = H, **b**: R = C≡C–H;

Scheme 1.25 A complete set of *carbo*-[3]radialene species studied at the B3PW91 level [219].

(**36**) or doubly expanded (**38**) ring, a remarkable linear correlation is obtained between NICS and σ(d), the root mean square deviation from the average bond length of the radiacycle, showing that structural and magnetic aromaticities vary in a highly parallel manner (NICS ≈ −34.15 + 405.5σ(d); R = 0.9986). Furthermore, a correlated exaltation of structural and magnetic aromaticities is observed through all the reduction processes, which all result in an average bond length equalization over the radiacycle and in a decrease in the NICS (but the **37a** → $[\mathbf{37a}]^-$ reduction). According to the ELF basin populations, the neutral structures **36a–38a** exhibit some electron delocalization. The additional electrons of the monoanions are then homogeneously absorbed by the ring valence basins in $[\mathbf{36a}]^-$ and $[\mathbf{38a}]^-$, and repelled to the peripheral double bond in $[\mathbf{37a}]^-$ (Figure 1.16). Finally, the additional electrons of the dianion $[\mathbf{38a}]^{2-}$ partly localize in monosynaptic valence basins corresponding to CH_2^- lone pairs.

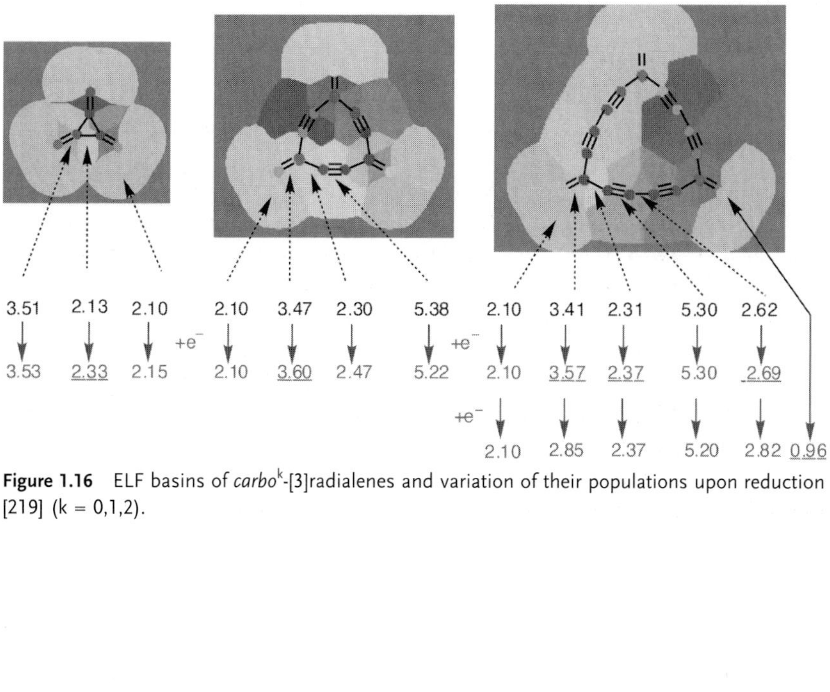

Figure 1.16 ELF basins of *carbo*^{k}-[3]radialenes and variation of their populations upon reduction [219] (k = 0,1,2).

The special status of the mono-expanded series (**37**) is also apparent in the MO diagrams. Indeed, the HOMO and LUMO of **37a** are higher in energy than the corresponding MOs of **36a** and **38a**, although their gaps rank, as expected, in the order **36a** > **37a** > **38a**.

The ELF-weighting of the resonance forms of [**36a**]^q, [**37a**]^q, and [**38a**]^q ($q = 0, -1$) is consistent with AIM atomic charges and spin densities, and allows for the definition of a formal average number of *endocyclic paired* π_z *electrons, n*. It was incidentally observed that the closer to a $4m+2$ integer the n value, the more negative the NICS value, and a general quantitative correlation (including the monoexpanded series **37**) between the magnetic (NICS), structural (σ(d)), and formal (n, Hückel-like) aromaticity measures was heuristically proposed.

Vertical and adiabatic first and second electron affinities (EAs) of unsubstituted (**a**) and perethynylated (**b**) structures were calculated. Experimentally measured EAs or reduction potentials for a systematic calibration are still lacking, but regardless of their absolute values, the calculated EAs systematically increase by 1–2 eV upon hexaethynyl substitution. This observation is consistent with the surprisingly low experimentally determined reduction potentials of hexaethynylradialenes. This phenomenon is basically explained by detailed analysis of aromaticity enhancement upon reduction and ethynyl substitution.

In the realm of still fictitious but beautiful molecules, radialene-like pure carbon dianions $(CC_2)_n^{2-}$ ($n = 3$–6) deserve to be evoked [221]. They have been investigated at the RHF, MP2, and CCSD levels, and their representation is reminiscent of that of oxocarbon dianions $(CO)_2^{2-}$, in which the oxygen atoms would be replaced with C_2 units. Their aromaticity has been studied with classical tools traditionally used for more stable organic molecules (resonance forms, NICS,...).

1.5 Cage Acetylenic Scaffolds

Whereas oligoacetylenic cage molecules with sp^2 vertices are studied because of their close resemblance to fullerenes [151], those with sp^3 vertices remained unknown until only recently. At the theoretical level, we should just mention SCF and DFT calculations on a highly putative strained cage (dehydropaddlane) structure containing two coupled tetraethynylmethane units [222]. In 2002, Diederich et al. reported the characterization of an expanded cubane and its conversion into fullerene cations [223]. Their exceptional strain is revealed by their heats of formation at the AM1 and PM3 levels (approximately 1000 kcal mol^{-1}). Computational studies of a related structures at the B3LYP level appeared shortly afterwards [121].

The lower homologue expanded by simple C_2 units (which could be called "[C,C]$_{12}$-*carbo*-cubane " [4]) exhibits a O_h geometry. In comparison with [4]pericyclyne (Section 1.3.1.2), the C-C bond lengths are essentially identical, but the C–C≡C angles close from 171° to 166°. This allows almost normal bond angles to be preserved at the sp^3 vertices (107°). Through the use of an isodesmic scheme, the ring strain was estimated as 51 kcal mol^{-1}, hence quite high but much lower

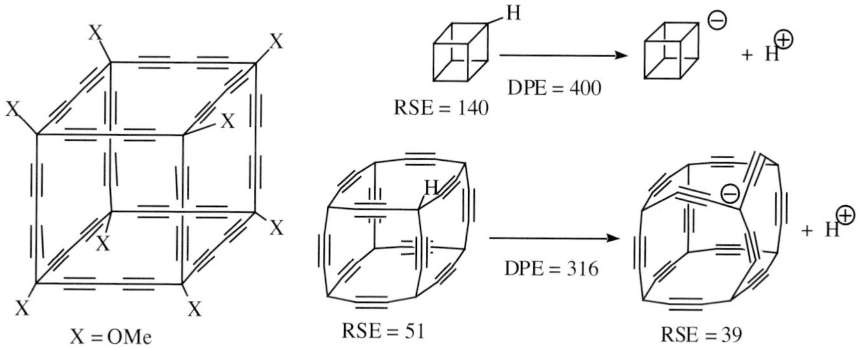

Scheme 1.26 Expanded cubanes. The doubly expanded structure is experimental [223]. Deprotonation energies (DPE) and rings strain energies (RSE) are in kcal mol^{-1} (B3LYP/6-31+G*) [121].

than in the parent cubane (140 kcal mol^{-1}). The expanded cubane is predicted to be quite acidic because of trigonal conjugation of the negative charge with the triple bonds and a lowering of ring strain in the conjugated base (Scheme 1.26): the deprotonation energy at the B3LYP/6-31+G* is 316 kcal mol^{-1}, while it is 400 kcal mol^{-1} for cubane. The complexing properties of the neutral expanded cubane was also investigated: the Li^+ complex is quasi face-centered (C_{4v}), while the Na^+ complex is cage-centered (O_h).

1.6 Conclusion

As a synthetic epilogue, the valence flexibility of the acetylenic linkage –C≡C– might be partly compared with that of a low-valent metal center, and indeed, several specific properties provided by either unit are of the same nature (e. g., redox, conducting, NLO,…). Nevertheless, any analogy between acetylenic chemistry and organometallic chemistry remains limited. Indeed, while organometallic complexes are prepared by anchoring ligands to a metallic center, acetylenic scaffolds cannot be generally obtained by direct insertion of C_2 units into parent molecules. It may be mentioned, however, that acetylene has been observed as a source of C_2 in a model reaction for the growth of polycyclic aromatic hydrocarbons (PAH) by C_2 accretion [224]. From a more fundamental standpoint, crossed beam experiments under very low pressure conditions (approximately 10^{-10} bar) coupled with ab initio calculations showed that insertion of dicarbon into one (central) bond of simple molecules such as dihydrogen [225], methane [226], ethylene [227], or water [228] is chemically feasible. The case of acetylene was detailed recently [229]. The generation of butadiynyl and hydrogen radicals from dicarbon and acetylene (e. g., at a collision energy of 24 kJ mol^{-1}) was shown to take place via the D_{∞} butadiyne intermediate at the CCSD(T)//B3LYP level (Scheme 1.27). But let us stop dreaming of any generalization...

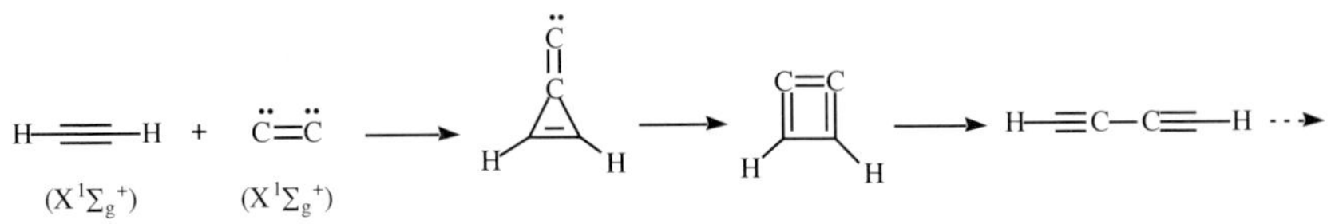

Scheme 1.27 Schematic generation of diacetylene by insertion of singlet (ground state) dicarbon into the triple bond of acetylene (CCSD(T)/6-311+G(3df,2p)//B3LYP/6-311G(d,p)) [229].

Acknowledgements

The authors would like to thank F. Diederich, N. Mogens, P. Lacroix, and B. Silvi for fruitful collaborations and discussions. R. C. also thanks the Ministère de l'Enseignement de la Recherche et de la Technologie for ACIJ funding.

References

[1] D. A. Plattner, Y. Li, K. N. Houk in *Modern Acetylene Chemistry* (P. J. Stang, F. Diederich Eds.), VCH, Weinheim, **1995**, pp. 1–32.
[2] R. F. W. Bader, *Chem. Rev.* **1991**, *91*, 893–928.
[3] a) B. Silvi, A. Savin, *Nature* **1994**, *371*, 683–686. b) A. Savin, R. Nesper, S. Wengert, T. F. Fässler, *Angew. Chem. Int. Ed. Engl.* **1997**, *36*, 1808–1832. *Angew. Chem.* **1997**, *109*, 1892–1918.
[4] R. Chauvin, *Tetrahedron Lett.* **1995**, *36*, 397–400.
[5] a) "Carbon Rich Compounds I", *Top. Curr. Chem.* (A. de Meijere, Ed.), **1998**, *196*, 1–220. b) "Carbon Rich Compounds II", *Top. Curr. Chem.* (A. de Meijere, Ed.), **1999**, *201*, 1–223.
[6] L. Fomina, B. Vazquez, E. Tkatchouk, S. Fomine, *Tetrahedron* **2002**, *58*, 6741–6747.
[7] M. B. Nielsen, F. Diederich, *Chem. Rec.* **2002**, 2, 189–198.
[8] J. A. R. P. Sarma, A. Nangia, G. R. Desiraju, E. Zass, J. D. Dunitz, *Nature* **1996**, *384*, 320–320.
[9] R. Hoffmann, *American Scientist* **1995**, *83*, 309–311.
[10] W. H. Beck, J. C. Mackie, *J. Chem. Soc. Faraday Trans. 1* **1975**, *71*, 1363–1371.
[11] R. M. Baum, *Chem. Eng. News* **1993**, *71*, 32–34.
[12] P. v. R. Schleyer, P. Maslak, J. Chandrasekhar, R. S. Grev, *Tetrahedron Lett.* **1993**, *34*, 6387–6390.
[13] M. Douay, R. Nietmann, P. F. Bernath, *J. Mol. Spectrosc.* **1988**, *131*, 250–260.
[14] a) J. D. Watts, R. J. Bartlett, *J. Chem. Phys.* **1992**, *96*, 6073–6084. b) S. A. Kucharski, M. Kolaski, R. J. Bartlett, *J. Chem. Phys.* **2001**, *114*, 692–700.
[15] C. Liang, Y. Xie, H. F. Schaefer III, K. S. Sim, H. S. Kim, *J. Am. Chem. Soc.* **1991**, *113*, 2452–2458.
[16] T. E. Sorensen, W. B. England, *J. Chem. Phys.* **1998**, *108*, 5205–5215.
[17] A. van Orden, R. J. Saykally, *Chem. Rev.* **1998**, *98*, 2313–2358.
[18] M. Bianchetti, P. F. Bonsuante, F. Ginelli, H. E. Roman, R. A. Broglia, F. Alasia, *Phys. Rep.* **2002**, *357*, 459–513.
[19] A. Lorenzoni, H. E. Roman, F. Alasia, R. A. Broglia, *Chem. Phys. Lett.* **1997**, *276*, 237–241.
[20] J. Hutter, H. P. Lüthi, F. Diederich, *J. Am. Chem. Soc.* **1994**, *116*, 750–756.
[21] J. M. L. Martin, J. El–Yazal, J.-P. François, *Chem. Phys. Lett.* **1995**, *242*, 570–579.

[22] J. P. Maier, *J. Phys. Chem. A* **1998**, *102*, 3462–3469.
[23] A. G. Rinzler, J. H. Hafner, P. Nikolaev, L. Lou, S. G. Kim, D. Tomanek, P. Nordlander, D. T. Colbert, R. E. Smalley, *Science* **1995**, *269*, 1550–1553.
[24] A. E. Douglas, *Nature* **1977**, *269*, 130–132.
[25] M. Tulej, D. A. Kirkwood, M. Pachkov, J. P. Maier, *Astrophys. J. Lett.* **1998**, *506*, L69–L73.
[26] R. J. Lagow, J. J. Kampa, H.-C. Wei, S. L. Battle, J. W. Genge, D. A. Laude, C. J. Harper, R. Bau, R. C. Stevens, J. F. Haw, E. Munson, *Science* **1995**, *267*, 362–367.
[27] For a recent paper, see: P. Redondo, C. Barrientos, A. Cimas, A. Largo, *J. Phys. Chem. A* **2004**, *108*, 212–224, and references therein.
[28] P. Redondo, C. Barrientos, A. Cimas, A. Largo, *J. Phys. Chem. A* **2003**, *107*, 6317–6325.
[29] T. F. Miller III, M. B. Hall, *J. Am. Chem. Soc.* **1999**, *121*, 7389–7396.
[30] W. B. Wan, S. C. Brand, J. J. Pak, M. M. Haley, *Chem. Eur. J.* **2000**, *6*, 2044–2052.
[31] F. Diederich, Y. Rubin, *Angew. Chem. Int. Ed. Engl.* **1992**, *31*, 1101–1123. *Angew. Chem.* **1992**, *104*, 1123.
[32] See, for example: a) N. Narita, S. Nagai, S. Suzuki, K. Nakao, *Phys. Rev. B: Cond. Matter Mater. Phys.* **1998**, *58*, 11009–11014. b) N. Narita, S. Nagai, S. Suzuki, K. Nakao, *Phys. Rev. B: Cond. Matter Mater. Phys.* **2000**, *62*, 11146–11151.
[33] a) R. Zahradnik, L. Sroubkova, *Helv. Chim. Acta* **2003**, *86*, 979–1000. b) R. Herges, A. Mebel, *J. Am. Chem. Soc.* **1994**, *116*, 8229–8237. c) B. S. Jursic, *J. Mol. Struct. (THEOCHEM)* **1999**, *491*, 193–203. d) S. J. Blanksby, S. Dua, J. H. Bowie, D. Schröder, H. Schwarz, *J. Phys. Chem. A* **1998**, *102*, 9949–9956. e) S. Dua, S. J. Blanksby, J. H. Bowie, *J. Phys. Chem. A* **2000**, *104*, 77–85. f) K. W. Sattelmeyer, J. F. Stanton, *J. Am. Chem. Soc.* **2000**, *122*, 8220–8227.
[34] L. Horny, N. D. K. Petraco, C. Pak, H. F. Schaefer III, *J. Am. Chem. Soc.* **2002**, *124*, 5861–5864.
[35] A. Scemama, P. Chaquin, M.-C. Gazeau, Y. Bénilan, *J. Phys. Chem. A* **2002**, *106*, 3828–3837.
[36] a) J. O. Morley, *Int. J. Quantum Chem.* **1993**, *46*, 19–26. b) W. J. Buma, M. Fanti, F. Zerbetto, *Chem. Phys. Lett.* **1999**, *313*, 426–430. c) J. Y. Lee, S. B. Suh, K. S. Kim, *J. Chem. Phys.* **2000**, *112*, 344–348. d) W. J. Lauderdale, M. B. Coolidge, *J. Phys. Chem.* **1995**, *99*, 9368–9373. e) P. Karamanis, G. Maroulis, *J. Mol. Struct. (THEOCHEM)* **2003**, *621*, 157–162.
[37] P. Karamanis, G. Maroulis, *Chem. Phys. Lett.* **2003**, *376*, 403–410.
[38] a) E. J. Bylaska, J. H. Weare, R. Kawai, *Phys. Rev. B* **1998**, *58*, R7488–R7491. b) U. Mölder, P. Burk, I. A. Koppel, *Int. J. Quant. Chem.* **2001**, *82*, 73–85.
[39] J. L. Toto, T. T. Toto, C. P. de Melo, B. Kirtman, K. Robins, *J. Chem. Phys.* **1996**, *104*, 8586–8592.
[40] Y. Nagano, T. Ikoma, K. Akiyama, S. Tero–Kubota, *J. Am. Chem. Soc.* **2003**, *125*, 14103–14112.
[41] D. Schröder, C. Heinemann, H. Schwartz, J. N. Harvey, S. Dua, R. J. Blanksby, J. H. Bowie, *Chem. Eur. J.* **1998**, *4*, 2550–2557.
[42] G. Maier, H. P. Reisenhauer, R. Ruppel, *Angew. Chem. Int. Ed. Engl.* **1997**, *36*, 1862–1864. *Angew. Chem.* **1997**, *109*, 1964–1966.
[43] G. Maier, H. P. Reisenauer, H. Balli, W. Brandt, R. Janoschek, *Angew. Chem. Int. Ed. Engl.* **1990**, *29*, 905–908. *Angew. Chem.* **1990**, *102*, 920.
[44] a) H.-Y. Wang, X. Lu, R. B. Huang, L.-S. Zheng, *J. Mol. Struct. (THEOCHEM)* **2002**, *593*, 187–197. b) K.-H. Kim, B. Lee, S. Lee, *Bull. Korean Chem. Soc.* **1998**, *19*, 553–557.
[45] G. Schermann, T. Grösser, F. Hampel, A. Hirsch, *Chem. Eur. J.* **1997**, *3*, 1105–1112.
[46] S. Petrie, *J. Phys. Chem. A* **2003**, *107*, 10441–10449.
[47] J. Cioslowski, M. Martinov, *J. Phys. Chem.* **1996**, *100*, 6156–6160.
[48] E. A. Meyer, R. K. Castellano, F. Diederich, *Angew. Chem. Int. Ed.* **2003**,

42, 1210–1250. *Angew. Chem.* **2003**, *115*, 1244–1287.

[49] a) M.-F. Fan, Z. Lin, J. E. McGrady, D. M. P. Mingos, *J. Chem. Soc. Perkin Trans. 2*, **1996**, 563–568. b) T.-H. Tang, Y.-P. Cui, *Can. J. Chem.* **1996**, *74*, 1162–1170. c) D. Philp, J. M. A. Robinson, *J. Chem. Soc. Perkin Trans. 2* **1998**, 1643–1650. d) S. Scheiner, S. J. Grabowski, *J. Mol. Struct. (THEOCHEM)* **2002**, *615*, 209–218. e) A. Karpfen, *J. Phys. Chem. A*, **1999**, *103*, 11431–11441.

[50] K. Chenoweth , C. E. Dykstra, *J. Phys. Chem. A* **2002**, *106*, 8117–8123.

[51] a) R. F. Winter, K.-W. Klinkhammer, S. Zalis, *Organometallics* **2001**, *20*, 1317–1333. b) O. F. Koentjoro, R. Rousseau, P. J. Low, *Organometallics* **2001**, *20*, 4502–4509. c) N. Auger, D. Touchard, S. Rigaut, J.-F. Halet, J.-Y. Saillard, *Organometallics* **2003**, *22*, 1638–1644.

[52] a) T. Ziegler, A. Rauk, *Theor. Chim. Act.* **1977**, *46*, 1–10. b) T. Ziegler, A. Rauk, *Inorg. Chem.* **1979**, *18*, 1558–1565. c) T. Ziegler, A. Rauk, *Inorg. Chem.* **1979**, *18*, 1755–1759.

[53] J. E. McGrady, T. Lovell, R. Stranger, M. G. Humphrey, *Organometallics* **1997**, *16*, 4004–4011.

[54] C. D. Delfs, R. Stranger, M. G. Humphrey, A. M. McDonagh, *J. Organomet. Chem.* **2000**, *607*, 208–212.

[55] C. E. Powell, M. P. Cifuentes, A. M. McDonagh, S. K. Hurst, N. T. Lucas, C. D. Delfs, R. Stranger, M. G. Humphrey, S. Houbrechts, I. Asselberghs, A. Persoons, D. C. R. Hockless, *Inorg. Chim. Act.* **2003**, *352*, 9–18.

[56] a) N. Re, A. Sgamellotti, C. Floriani, *Organometallics* **2000**, *19*, 1115–1122. b) A. Marrone, N. Re, *Organometallics* **2002**, *21*, 3562–3571.

[57] D. Beljonne, M. C. B. Colbert, P. R. Raithby, R. H. Friend, J.-L. Brédas, *Synth. Met.* **1996**, *81*, 179–183.

[58] T. Weyland, C. Lapinte, G. Frapper, M. J. Calhorda, J.-F. Halet, L. Toupet, *Organometallics* **1997**, *16*, 2024–2031.

[59] N. Re, A. Sgamellotti, C. Floriani, *J. Chem. Soc. Dalton Trans.* **1998**, 2521–2529.

[60] H. Lang, *Angew. Chem. Int. Ed. Engl.* **1994**, *33*, 547–550. *Angew. Chem.* **1994**, *106*, 569.

[61] a) P. Belanzoni, N. Re, M. Rosi, A. Sgamellotti, C. Floriani, *Organometallics* **1996**, *15*, 4264–4273. b) P. Belanzoni, N. Re, A. Sgamellotti, C. Floriani, *J. Chem. Soc. Dalton Trans.* **1997**, 4773–4782.

[62] P. Belanzoni, N. Re, A. Sgamellotti, C. Floriani, *J. Chem. Soc. Dalton Trans.* **1998**, 1825–1835.

[63] a) M. Brady, W. Weng, Y. Zhou, J. W. Seyler, A. J. Amoroso, A. M. Arif, M. Böhme, G. Frenking, J. A. Gladysz, *J. Am. Chem. Soc.* **1997**, *119*, 775–788. b) H. Jiao, J. A. Gladysz, *New J. Chem.* **2001**, *25*, 551–562.

[64] a) M. I. Bruce, P. J. Low, K. Costuas, J.-F. Halet, S. P. Best, G. A. Heath, *J. Am. Chem. Soc.* **2000**, *122*, 1949–1962. b) H. Jiao, K. Costuas, J. A. Gladysz, J.-F. Halet, M. Guillemot, L. Toupet, F. Paul, C. Lapinte, *J. Am. Chem. Soc.* **2003**, *125*, 9511–9522. c) F. Coat, F. Paul, C. Lapinte, L. Toupet, K. Costuas, J.-F. Halet, *J. Organomet. Chem.* **2003**, *683*, 368–378.

[65] P. J. Low, R. Rousseau, P. Lam, K. A. Udachin, G. D. Enright, J. S. Tse, D. D. M. Wayner, A. J. Carty, *Organometallics* **1999**, *18*, 3885–3897.

[66] M. I. Bruce, J.-F. Halet, B. Le Guennic, B. W. Skelton, M. E. Smith, A. H. White, *Inorg. Chim. Act.* **2003**, *350*, 175–181.

[67] S. Szafert, J. A. Gladysz, *Chem. Rev.* **2003**, *103*, 4175–4206.

[68] H. H. Wenk, M. Winkler, W. Sander, *Angew. Chem. Int. Ed.* **2003**, *42*, 502–528. *Angew. Chem.* **2003**, *115*, 518–546.

[69] R. J. McMahon, R. J. Halter, R. L. Fimmen, R. J. Wilson, S. A. Peebles, R. L. Kuczkowski, J. F. Stanton, *J. Am. Chem. Soc.* **2000**, *122*, 939–949.

[70] See, for example: J.-L. Brédas, A. J. Heeger, *Macromolecules* **1990**, *23*, 1150–1156.

[71] M. Turki, T. Barisien, J.-Y. Bigot, C. Daniel, *J. Chem. Phys.* **2000**, *112*, 10526–10537.

[72] R. E. Martin, U. Gubler, J. Cornil, M. Balakina, C. Boudon, C. Bosshard,

J. P. Gisselbrecht, F. Diederich, P. Günter, M. Gross, J.-L. Brédas, *Chem. Eur. J.* **2000**, *6*, 3622–3635.

[73] C. Kosinski, A. Hirsch, F. W. Heinemann, F. Hampel, *Eur. J. Org. Chem.* **2001**, 3879–3890.

[74] S. Concilio, I. Biaggio, P. Günter, S. P. Piotto, M. J. Edelmann, J.-M. Raimundo, F. Diederich *J. Opt. Soc. Am. B* **2003**, *20*, 1656–1660.

[75] a) R. R. Tykwinski, U. Gubler, R. E. Martin, F. Diederich, C. Bosshard, P. Günter, *J. Phys. Chem. B* **1998**, *102*, 4451–4465. b) U. Gubler, R. Spreiter, C. Bosshard, P. Günter, R. R. Tykwinski, F. Diederich, *Appl. Phys. Lett.* **1998**, *73*, 2396–2398.

[76] H. Baumann, R. E. Martin, F. Diederich, *J. Comput. Chem.* **1999**, *20*, 396–411.

[77] T. V. Jones, R. A. Blatchly, G. N. Tew, *Org. Lett.* **2003**, *5*, 3297–3299.

[78] Y.-Z. Zhou, *Mater. Sci. Engineer. B* **2003**, *99*, 593–596.

[79] A. P. H. J. Schenning, A. C. Tsipis, S. C. J. Meskers, D. Beljonne, E. W. Meijer, J.-L. Brédas, *Chem. Mater.* **2002**, *14*, 1362–1368.

[80] A. Göller, E. Klemm, D. A. M. Egbe, *Int. J. Quantum Chem.* **2001**, *84*, 86–98.

[81] Ph. Leclère, A. Calderone, K. Müllen, J.-L. Brédas, R. Lazzaroni, *Mater. Sci. Technol.* **2002**, *18*, 749–754.

[82] D. Beljonne, Z. Shuai, G. Pourtois, J.-L. Brédas, *J. Phys. Chem. A* **2001**, *105*, 3899–3907.

[83] See for example: a) Z. J. Donhauser, B. A. Mantooth, K. F. Kelly, L. A. Bumm, J. D. Monnell, J. J. Stapleton, D. W. Price Jr., A. M. Rawlett, D. L. Allara, J. M. Tour, P. S. Weiss, *Science* **2001**, *292*, 2303–2307. b) D. W. Price Jr., J. M. Tour, *Tetrahedron* **2003**, *59*, 3131–3156. c) F.-R. F. Fan, R. Y. Lai, J. Cornil, Y. Karzazi, J.-L. Brédas, L. Cai, L. Cheng, Y. Yao, D. W. Price Jr., S. M. Dirk, J. M. Tour, A. J. Bard, *J. Am. Chem. Soc.* **2004**, *126*, 2568–2573.

[84] J. M. Seminario, A. G. Zacarias, J. M. Tour, *J. Am. Chem. Soc.* **2000**, *122*, 3015–3020.

[84] J. Cornil, Y. Karzazi, J.-L. Brédas, *J. Am. Chem. Soc.* **2002**, *124*, 3516–3517.

[86] Y. Karzazi, J. Cornil, J.-L. Brédas, *Adv. Funct. Mater.* **2002**, *12*, 787–794.

[87] Y. Karzazi, J. Cornil, J.-L. Brédas, *Nanotechnology* **2003**, *14*, 165–171.

[88] a) C. Lepetit, P. Lacroix, V. Peyrou, C. Saccavini, R. Chauvin *J. Comp. Methods in Sci. and Engineer.*, in press. b) A study on related push-pull but non zwitterionic systems has just appeared: H. Meier, B. Mühling, H. Kolshorn, *Eur. J. Org. Chem.* **2004**, 1033–1042.

[89] S. Kolotilov, C. Saccavini, Z. Voitenko, R. Chauvin, unpublished results.

[90] I. R. Whittall, M. P. Cifuentes, M. G. Humphrey, B. Luther-Davies, M. Samoc, S. Houbrechts, A. Persoons, G. A. Heath, D. C. R. Hockless, *J. Organomet. Chem.* **1997**, *549*, 127–137.

[91] a) S. M. LeCours, H.-W. Guan, S. G. DiMagno, C. H. Wang, M. J. Therien, *J. Am. Chem. Soc.* **1996**, *118*, 1497–1503. b) S. Priyadarshy, M. J. Therien, D. N. Beratan, *J. Am. Chem. Soc.* **1996**, *118*, 1504–1510. c) M. Yeung, A. C. H. Ng, M. G. B. Drew, E. Vorpagel, E. M. Breitung, R. J. McMahon, D. K. P. Ng, *J. Org. Chem.* **1998**, *63*, 7143–7150.

[92] E. M. Maya, E. M. Garcia-Frutos, P. Vazquez, T. Torres, G. Martin, G. Rojo, F. Agullo-Lopez, R. Gonzalez-Jonte, V. R. Ferro, J. M. Garcia de la Vega, I. Ledoux, J. Zyss, *J. Phys. Chem. A* **2003**, *107*, 2110–2117.

[93] F. Paul, K. Costuas, I. Ledoux, S. Deveau, J. Zyss, J.-F. Halet, C. Lapinte, *Organometallics* **2002**, 21, 5229–5235.

[94] C. E. Powell, M. P. Cifuentes, J. P. Morrall, R. Stranger, M. G. Humphrey, M. Samoc, B. Luther-Davies, G. A. Heath, *J. Am. Chem. Soc.* **2003**, *125*, 602–610.

[95] J. P. Morrall, C. E. Powell, R. Stranger, M. P. Cifuentes, M. G. Humphrey, G. A. Heath, *J. Organomet. Chem.* **2003**, *670*, 248–255.

[96] K. M.-C. Wong, S. C.-F. Lam, C.-C. Ko, N. Zhu, V. W.-W. Yam, S. Roué, C. Lapinte, S. Fathallah, K. Costuas, S. Kahlal, J.-F. Halet, *Inorg. Chem.* **2003**, *42*, 7086–7097.

[97] a) S. Fraysse, C. Coudret, J.-P. Launay, *Eur. J. Inorg. Chem.* **2000**, 1581–1590. b) S. Fraysse, C. Coudret, J.-P. Launay, *J. Am. Chem. Soc.* **2003**, *125*, 5880–5888.

[98] C. Patoux, J.-P. Launay, M. Beley, S. Chodorowski-Kimmes, J.-P. Collin, S. James, J.-P. Sauvage, *J. Am. Chem. Soc.* **1998**, *120*, 3717–3725.

[99] K. G. Migliorese, Y. Tanaka, S. I. Miller, *J. Org. Chem.* **1974**, *39*, 739–747.

[100] I. H. Ooi, R. H. Smithers, *J. Org. Chem.* **1989**, *54*, 1479–1480.

[101] A. J. Matzger, K. D. Lewis, C. E. Nathan, S. A. Peebles, R. A. Peebles, R. L. Kuczkowski, J. F. Stanton, J. J. Oh, *J. Phys. Chem. A*, **2002**, *106*, 12110–12116.

[102] T. Murai, A. Shimizu, S. Tatematsu, K. Ono, T. Kanda, S. Kato, *Heteroatom Chem.* **1994**, *5*, 31–35.

[103] S. P. Kawatkar, P. R. Schreiner, *Org. Lett.* **2002**, *4*, 3643–3646.

[104] See, for example: M. Brudermüller, H. Musso, A. Wagner, *Chem. Ber.* **1988**, *121*, 2239–2244.

[105] G. Märkl, T. Zollitsch, P. Kreitmeier, M. Prinzhorn, S. Reithinger, E. Eibler, *Chem. Eur. J.* **2000**, *6*, 3806–3820.

[106] R. Gleiter, *Angew. Chem. Int. Ed. Engl.* **1992**, *31*, 27–44. *Angew. Chem.* **1992**, *104*, 29.

[107] a) R. Gleiter, R. Merger, H. Irngartinger, *J. Am. Chem. Soc.* **1992**, *114*, 8927–8932.
b) See a just appeared reference: D. B. Wertz, R. Gleiter, *Organic Lett.* **2004**, 6, 589–592.

[108] A. de Meijere, S. I. Kozhushkov, *Top. Curr. Chem.* **1999**, *201*, 1–42.

[109] L. T. Scott, M. J. Cooney in *Modern Acetylene Chemistry* (P. J. Stang, F. Diederich Eds.), VCH, Weinheim, **1995**, p. 321–351.

[110] K. N. Houk, L. T. Scott, N. G. Rondan, D. C. Spellmeyer, G. Reinhardt, J. L. Hyun, G. J. DeCicco, R. Weiss, M. H. M. Chen, L. S. Bass, J. Clardy, F. S. Jorgensen, T. A. Eaton, V. Sarkozi, C. M. Petit, L. Ng, K. D. Jordan, *J. Am. Chem. Soc.* **1985**, *107*, 6556–6562.

[111] C. Lepetit, B. Silvi, R. Chauvin, *J. Phys. Chem. A* **2003**, *107*, 464–473.

[112] M. J. S. Dewar, M. K. Holloway, *J. Chem. Soc. Chem. Comm.* **1984**, *17*, 1188–1191.

[113] R. Hoffmann, R. B. Woodward, *J. Am. Chem. Soc.* **1965**, *87*, 2046–2048.

[114] K. K. Baldridge, J. S. Siegel, *J. Am. Chem. Soc.* **1992**, *114*, 9583–9587.

[115] S. Sakai, *J. Phys. Chem. A* **2002**, *106*, 11526–11532.

[116] A. Soncini, R. W. A. Havenith, P. W. Fowler, W. Jenneskens, E. Steiner, *J. Org. Chem.* **2002**, *67*, 4753–4758.

[117] C. Lepetit, R. Chauvin, unpublished result.

[118] L. T. Scott, M. Unno, *J. Am. Chem. Soc.* **1990**, *112*, 7823–7825.

[119] H. Sakurai, Y. Eriyama, A. Hosomi, Y. Nakadaira, C. Kabuto, *Chem. Lett.* **1984**, 595–598.

[120] R. Gleiter, W. Schäfer, H. Sakurai, *J. Am. Chem. Soc.* **1985**, *107*, 3046–3050.

[121] S. M. Bachrach, *J. Phys. Chem. A* **2003**, *107*, 4957–4961.

[122] L. T. Scott, M. J. Cooney, C. Otte, C. Puls, T. Haumann, R. Boese, P. J. Carroll, A. B. Smith III, A. de Meijere, *J. Am. Chem. Soc.* **1994**, *116*, 10275–10283.

[123] For a recent reference, see: B. Leibrock, O. Vostrowsky, A. Hirsch, *Eur. J. Org. Chem*, **2001**, 4401–4409, and references therein.

[124] M. I. Bruce, K. Costuas, J.-F. Haley, B. C. Hall, P. J. Low, B. K. Nicholson, B. W. Skelton, A. H. White, *J. Chem. Soc., Dalton Trans.* **2002**, 383–398.

[125] L. J. Schaad, B. A. Hess, Jr., L. T. Scott, *J. Phys. Org. Chem.* **1993**, *6*, 316–318.

[126] H. Jiao, N. J. R. v. E. Hommes, P. v. R. Schleyer, A. de Meijere, *J. Org. Chem.* **1996**, *61*, 2826–2828.

[127] C. Godard, C. Lepetit, R. Chauvin, *Chem. Commun.* **2000**, 1833–1834.

[128] a) L. T. Scott, M. J. Cooney, D. W. Rogers, K. Dejroongruang, *J. Am. Chem. Soc.* **1988**, *110*, 7244–7245.
b) R. V. Williams, H. A. Kurtz, *Adv. Phys. Org. Chem.* **1994**, *29*, 273–331.

[129] P. v. R. Schleyer, H. Jiao, H. M. Sulzbach, H. F. Schaefer III, *J. Am. Chem. Soc.* **1996**, *118*, 2093–2094.

[130] L. Maurette, C. Godard, S. Frau, C. Lepetit, M. Soleilhavoup, R. Chauvin, *Chem. Eur. J.* **2001**, *7*, 1165–1170.
[131] L. Maurette, C. Tedeschi, E. Sermot, F. Hussain, M. Soleilhavoup, B. Donnadieu, R. Chauvin, *Tetrahedron* **2004**, *60*, 10077–10098.
[132] L. T. Scott, G. J. DeCicco, J. L. Hyun, G. Reinhardt, *J. Am. Chem. Soc.* **1985**, *107*, 6546–6555.
[133] R. Suzuki, H. Tsukuda, N. Watanabe, Y. Kuwatani, I. Ueda, *Tetrahedron* **1998**, *54*, 2477–2496.
[134] R. D. Kennedy, D. Lloyd, H. McNab, *J. Chem. Soc. Perkin Trans 1* **2002**, 14,1601–1621.
[135] a) V. I. Minkin, M. N. Glukhovtsev, B. Y. Simkin, *Aromaticity and Antiaromaticity. Electronic and Structural Aspects*, Wiley, New York, **1994**. b) Whole issue of *Chem. Rev.* **2001**, *101* (P. v. R. Schleyer Ed.), 1115–1566.
[136] T. J. Sworski, *J. Chem. Phys.* **1948**, *16*, 550.
[137] A. G. Myers, N. S. Finney, *J. Am. Chem. Soc.* **1992**, *114*, 10986–10987.
[138] a) I. Alkorta, I. Rozas, J. Elguero, *Tetrahedron* **2001**, *57*, 6043–6049. b) K. G. Untch, D. C. Wysoki, *J. Am. Chem. Soc.* **1966**, *88*, 2608–2610.
[139] J. Jusélius, D. Sundholm, *Phys. Chem. Chem. Phys.* **2001**, *3*, 2433–2437.
[140] W. J. Youngs, C. A. Tessier, J. D. Bradshaw, *Chem. Rev* **1999**, *99*, 3153–3180.
[141] S. J. Klippenstein, C.-N. Yang, *Int. J. Mass Spectrom.* **2000**, *201*, 253–267.
[142] A. J. Boydston, M. M. Haley, R. V. Williams, J. R. Armantrout, *J. Org. Chem.* **2002**, *67*, 8812–8819.
[143] I. Yavari, H. Norouzi-Arasi, *J. Mol. Struct. (THEOCHEM)* **2002**, *593*, 199–207.
[144] T. Kawase, N. Ueda, M. Oda, *Tetrahedron Lett.* **1997**, *38*, 6681–6684.
[145] M. M. Haley, *Synlett* **1998**, 557–565.
[146] C. Eickmeier, H. Junga, A. J. Matzger, F. Scherhag, M. Shim, K. P. C. Vollhardt, *Angew. Chem. Int. Ed. Engl.* **1997**, *36*, 2103–2107. *Angew. Chem.* **1997**, *109*, 2194–2197.
[147] H. N. C. Wong, P. J. Garratt, F. Sondheimer, *J. Am. Chem. Soc.* **1974**, 96, 5604–5605.
[148] A. J. Matzger, K. P. C. Vollhardt, *Tetrahedron Lett.* **1998**, *39*, 6791–6794.
[149] Q. Zhou, P. J. Carroll, T. M. Swager, *J. Org. Chem.* **1994**, *59*, 1294–1301.
[150] Y. Tobe, I. Ohki, M. Sonoda, H. Niino, T. Sato, T. Wakabayashi, *J. Am. Chem. Soc.* **2003**, *125*, 5614–5615, and Supporting Information.
[151] U. H. F. Bunz, Y. Rubin, Y. Tobe, *Chem. Soc. Rev.* **1999**, *28*, 107–119.
[152] R. H. Baughman, H. Eckhardt, M. Kertesz, *J. Chem. Phys.* **1987**, *87*, 6687–6699.
[153] T. Nishinaga, N. Nodera, Y. Miyata, K. Komatsu, *J. Org. Chem.* **2002**, *67*, 6091–6096.
[154] H. Tomioka, A. Okuno, T. Sugiyama, S. Murata, *J. Org. Chem.* **1995**, *60*, 2344–2352.
[155] a) Y. Li, Y. Rubin, F. Diederich, K. N. Houk, *J. Am. Chem. Soc.* **1990**, *112*, 1618–1623. b) Y. Tobe, T. Fujii, H. Matsumoto, K. Tsumuraya, D. Noguchi, N. Nakagawa, M. Sonoda, K. Naemura, Y. Achiba, T. Wakabayashi, *J. Am. Chem. Soc.* **2000**, *122*, 1762–1775. c) D. Solooki, T. C. Parker, S. I. Khan, Y. Rubin, *Tetrahedron Lett.* **1998**, *39*, 1327–1330.
[156] D. B. Kimball, M. M. Haley, R. H. Mitchell, T. R. Ward, S. Bandyopadhyay, R. V. Williams, J. R. Armantrout, *J. Org. Chem.* **2002**, *67*, 8798–8811.
[157] a) P. Coppens, *X-ray Charge Densities and Chemical Bonding*, IUCR. Ed., Oxford University Press, **1997**. b) C. Lecomte, *Adv. Mol. Struct. Res.* **1995**, *1*, 261–302.
[158] R. Chauvin, *Tetrahedron Lett.* **1995**, *36*, 401–404.
[159] W. H. Okamura, F. Sondheimer *J. Am. Chem. Soc.* **1967**, *89*, 5991–5992.
[160] Y. Kuwatani, N. Watanabe, I. Ueda, *Tetrahedron Lett.* **1995**, *36*, 119–122.
[161] a) L. Maurette, C. Sui-Seng, M. Soleilhavoup, B. Donnadieu, R. Chauvin, unpublished result. b) C. Lepetit, R. Chauvin, unpublished result.
[162] C. Lepetit, C. Godard, R. Chauvin, *New. J. Chem.* **2001**, *25*, 572–580.
[163] I. Yavari, A. Jabbari, M. Samadizadeh, *J. Chem. Res. (S)* **1999**, 152–153.

[164] A. Soncini, P. W. Fowler, I. Cernusak, E. Steiner, *Phys. Chem. Chem. Phys.* **2001**, *3*, 3920–3923.

[165] For a recent reference, see for example: I. Fishtik, R. Datta, *J. Phys. Chem.* A **2003**, *107*, 10471–10476.

[166] C. Lepetit, R. Chauvin, manuscript in preparation.

[167] The ring current is generally considered as being induced by an external magnetic field and bears analogy with a superconducting current (see, for example: R. C. Haddon, *J. Am. Chem. Soc.* **1979**, *101*, 1722–1728). Formal intrinsic ring currents can be also be defined (C. Lepetit, R. Chauvin, manuscript in preparation).

[168] S. Shaik, P. C. Hiberty, J.-M. Lefour, G. Ohanessian, *J. Am. Chem. Soc.* **1987**, *109*, 363–374.

[169] a) R. Breslow, E. Mohacsi, *J. Am. Chem. Soc.* **1963**, *85*, 431–434. b) J. Aihara, H. Ichikawa, *Bull. Chem. Soc. Jpn.* **1988**, *61*, 223–228.

[170] P. Georges, M. Trachtman, C. W. Bock, A. M. Brett, *Tetrahedron* **1976**, *32*, 317–323.

[171] H. Jiao, P. v. R. Schleyer, *Angew. Chem. Int. Ed. Engl.* **1996**, *35*, 2383–2386. *Angew. Chem.* **1996**, *108*, 2634–2637.

[172] R. S. Mulliken, R. G. Parr, *J. Chem. Phys.* **1951**, *19*, 1271–1278.

[173] a) A. R. Katritsky, P. Barczynski, G. Musumarra, D. Pisano, M. Szafran, *J. Am. Chem. Soc.* **1989**, *111*, 7–15. b) P. v. R. Schleyer, P. K. Freeman, H. Jiao, B. Goldfuss, *Angew. Chem. Int. Ed. Engl.* **1995**, *34*, 337–340. *Angew. Chem.* **1995**, *107*, 332.

[174] E. Hückel, *Zeitsch. Phys.* **1931**, *70*, 204–286.

[175] H. Vogler, *Org. Magn. Resonance* **1979**, *12*, 306–312.

[176] a) H. Vogler, *J. Am. Chem. Soc.* **1978**, *100*, 7464–7471. b) H. Vogler, *Tetrahedron* **1979**, *35*, 657–661.

[177] R. F. Langler, A. M. McBain, *Aust. J. Chem.* **2002**, *55*, 727–731.

[178] C. Lepetit, J.-M. Ducéré, R. Chauvin, abstract, 7ème Université d'été de Physico-Chimie Théorique, 30 septembre 2001, Aussois-France.

[179] C. Lepetit, R. Chauvin, unpublished result.

[180] J.-M. Ducéré, C. Lepetit, P. G. Lacroix, J.-L. Heully, R. Chauvin, *Chem. Mater.* **2002**, *14*, 3332–3338.

[181] a) C. Tedeschi, C. Saccavini, L. Maurette, M. Soleilhavoup, R. Chauvin, *J. Organomet. Chem.* **2003**, *670*, 151–169. b) C. Sui-Seng, M. Soleilhavoup, L. Maurette, C. Tedeschi, B. Donnadieu, R. Chauvin, *Eur. J. Org. Chem.* **2003**, *9*, 1641–1651.

[182] F. Zerbetto, *J. Am. Chem. Soc.* **1999**, *121*, 10958–10961.

[183] P. v. R. Schleyer, H. Jiao, M. N. Glukhovtsev, J. Chandrasekhar, E. Kraka, *J. Am. Chem. Soc.* **1994**, *116*, 10129–10134.

[184] F. Diederich, Y. Rubin, C. B. Knobler, R. L. Whetten, K. E. Schriver, K. N. Houk, Y. Li, *Science* **1989**, *245*, 1088–1090.

[185] See, for example: a) G. A. Adamson, C. W. Rees, *J. Chem. Soc. Perkin. Trans. 1* **1996**, 1535–1543. b) Y. Tobe, R. Umeda, N. Iwasa, M. Sonoda, *Chem. Eur. J.* **2003**, *9*, 5549–5559.

[186] D. A. Plattner, K. N. Houk, *J. Am. Chem. Soc.* **1995**, *117*, 4405–4406.

[187] E. J. Bylaska, R. Kawai, J. H. Weare, *J. Chem. Phys.* **2000**, *113*, 6096–6106.

[188] T. Torelli, L. Mitas, *Phys. Rev. Lett.* **2000**, *85*, 1702–1705.

[189] W. P. Su, J. R. Schrieffer, A. J. Heeger, *Phys. Rev. B* **1980**, *22*, 2099–2111.

[190] T. Wakabayashi, M. Kohno, Y. Achiba, H. Shiromaru, T. Momose, T. Shida, K. Naemura, Y. Tobe, *J. Chem. Phys.* **1997**, *107*, 4783–4787.

[191] R. O. Jones, *J. Chem. Phys.* **1999**, *110*, 5189–5200.

[192] M. Grutter, M. Wyss, E. Riaplov, J. P. Maier, S. D. Peyerimhoff, M. Hanrath, *J. Chem. Phys.* **1999**, *111*, 7397–7401.

[193] C. Lepetit, V. Peyrou, R. Chauvin, *Phys. Chem. Chem. Phys.* **2004**, *6*, 303–309.

[194] P. E. Eaton, E. Galoppini, R. Gilardi, *J. Am. Chem. Soc.* **1994**, *116*, 7588–7596.

[195] a) K. S. Feldman, M. Kraebel, M. Parvez, *J. Am. Chem. Soc.* **1993**, *115*, 3846–3847. b) K. S. Feldman, C. K. Weinreb, W. J. Youngs, J. D. Bradshaw, *J. Am. Chem. Soc.* **1994**, *116*, 9016–9026.

[196] B. Ma, H. M. Sulzbach, Xie, F. H. Schaefer III, *J. Am. Chem. Soc.* **1994**, *116*, 3529–3538.

[197] J. F. Capitani, *J. Mol. Struct. (THEOCHEM)* **1995**, *332*, 21–23.

[198] Y. Rubin, C. B. Knobler, F. Diederich, *Angew. Chem. Int. Ed. Engl.* **1991**, *30*, 698–700. *Angew. Chem.* **1991**, *103*, 708.

[199] T. Lange, J.-D. van Loon, R. R. Tykwinski, M. Schreiber, F. Diederich, *Synthesis* **1996**, 537–550.

[200] J.-D. van Loon, P. Seiler, F. Diederich, *Angew. Chem. Int. Ed. Engl.* **1993**, *32*, 1187–1189. *Angew. Chem.* **1993**, *105*, 1235.

[201] A. Hilger, J.-P. Gisselbrecht, R. R. Tykwinski, C. Boudon, M. Schreiber, R. E. Martin, H. P. Lüthi, M. Gross, F. Diederich, *J. Am. Chem. Soc.* **1997**, *119*, 2069–2078.

[202] R. E. Martin, J. Bartek, F. Diederich, R. R. Tykwinski, E. C. Meister, A. Hilger, H. P. Lüthi, *J. Chem. Soc. Perkin Trans. 2* **1998**, 233–241.

[203] R. R. Tykwinski, A. Hilger, F. Diederich, H. P. Lüthi, P. Seiler, V. Gramlich, J.-P. Gisselbrecht, C. Boudon, M. Gross, *Helv. Chim. Act.* **2000**, *83*, 1484–1508.

[204] L. Gobbi, N. Elmaci, H. P. Lüthi, F. Diederich, *ChemPhysChem* **2001**, *2*, 423–433.

[205] A. M. Boldi, J. Anthony, V. Gramlich, C. B. Knobler, C. Boudon, J.-P. Gisselbrecht, M. Gross, F. Diederich, *Helv. Chim. Act.* **1995**, *78*, 779–796.

[206] a) S. Eisler, R. R. Tykwinski, *Angew. Chem. Int. Ed.* **1999**, *38*, 1940–1943. *Angew. Chem.* **1999**, *111*, 2138–2141. b) Y. Zhao, K. Campbell, R. R. Tykwinski, *J. Org. Chem.* **2002**, *67*, 336–344.

[207] E. Burri, F. Diederich, M. B. Nielsen, *Helv. Chim. Act.* **2002**, *85*, 2169–2182.

[208] R. Diercks, J. C. Armstrong, R. Boese, K. P. C. Vollhardt, *Angew. Chem. Int. Ed. Engl.* **1986**, *25*, 268–269. *Angew. Chem.* **1986**, *98*, 270–271.

[209] M. S. El-Shall, K. P. C. Vollhardt, *J. Mol. Struct. (THEOCHEM)* **1989**, *183*, 175–181.

[210] S. Marguet, A. Germain, P. Millie, *Chem. Phys.* **1996**, *208*, 351–373.

[211] P. W. Fowler, E. Steiner, R. Zanasi, B. Cadioli, *Mol. Phys.* **1999**, *96*, 1099–1108.

[212] J. D. Tovar, N. Jux, T. Jarrosson, S. I. Khan, Y. Rubin, *J. Org. Chem.* **1997**, *62*, 3432–3433, and addendum *J. Org. Chem.* **1998**, *63*, 4856–4856.

[213] M. Laskoski, G. Roidl, H. L. Ricks, J. G. M. Morton, M. D. Smith, U. H. F. Bunz, *J. Organomet. Chem.* **2003**, *673*, 13–24. See also: M. Laskoski, W. Steffen, J. G. M. Morton, M. D. Smith, U. H. F. Bunz *J. Organomet. Chem.* **2003**, *673*, 25–39.

[214] See for example: S. Ito, H. Inabe, N. Morita, K. Ohta, T. Kitamura, K. Imafuku, *J. Am. Chem. Soc.* **2003**, *125*, 1669–1680.

[215] a) T. Weyland, K. Costuas, A. Mari, J.-F. Halet, C. Lapinte, *Organometallics* **1998**, *17*, 5569–5579. b) F. Paul, C. Lapinte, *Phys. Organom. Chem.*, **2002**, *3*, Eds: M. Gielen, R. Willem, B. Wrackmeyer, John Wiley, Chichester, 220–295.

[216] T. Weyland, K. Costuas, L. Toupet, J.-F. Halet, C. Lapinte, *Organometallics* **2000**, 19, 4228–4239.

[217] M. Nakano, H. Fujita, M. Takahata, K. Yamaguchi, *J. Am. Chem. Soc.* **2002**, *124*, 9648–9655

[218] H. Hopf, G. Maas, *Angew. Chem. Int. Ed. Engl.* **1992**, *31*, 931–954. *Angew. Chem.* **1992**, *104*, 953.

[219] C. Lepetit, M. B. Nielsen, F. Diederich, R. Chauvin, *Chem. Eur. J.* **2003**, *9*, 5056–5066.

[220] T. Lange, V. Gramlich, W. Amrein, F. Diederich, M. Gross, C. Boudon, J.-P. Gisselbrecht, *Angew. Chem. Int. Ed. Engl.* **1995**, *34*, 805–809. *Angew. Chem.* **1995**, *107*, 898–902.

[221] S. Feuerbacher, A. Dreuw, L. S. Cederbaum, *J. Am. Chem. Soc.* **2002**, *124*, 3163–3168.

[222] H. Dodziuk, J. Leszczynski, K. S. Nowinski, *J. Mol. Struct. (THEOCHEM)* **1997**, *391*, 201–205.

[223] P. Manini, W. Amrein, V. Gramlich, F. Diederich, *Angew. Chem. Int. Ed.* **2002**, *41*, 4339–4343. *Angew. Chem.* **2002**, *114*, 4515–4519.

[224] A. Necula, L. T. Scott, *J. Am. Chem. Soc.* **2000**, *122*, 1548–1549.

[225] X. Zhang, Y.-H. Ding, Z.-S. Li, X.-R. Huang, C.-C. Sun, *Chem. Phys. Lett.* **2000**, *330*, 577–584.

[226] D. A. Horner, L. A. Curtiss, D. M. Gruen, *Chem. Phys. Lett.* **1995**, *233*, 243–248.

[227] R. I. Kaiser, T. N. Le, T. L. Nguyen, A. M. Mebel, N. Balucani, Y. T. Lee, F. Stahl, P. v. R. Schleyer, H. F. Schaefer III, *Faraday Discuss.* **2001**, *119*, 51–66.

[228] J.-H. Wang, K.-L. Han, G.-Z. He, Z. Li, V. R. Morris, *J. Phys. Chem. A* **2003**, *107*, 9825–9833.

[229] R. I. Kaiser, N. Balucani, D. O. Charkin, A. M. Mebel, *Chem. Phys. Lett.* **2003**, *382*, 112–119.

2
Synthesis of Heterocycles and Carbocycles by Electrophilic Cyclization of Alkynes

Richard C. Larock

2.1
Introduction

Electrophilic, nucleophilic, and free radical additions to alkenes have been employed in the synthesis of heterocycles and carbocycles for a long time [1], but analogous reactions of acetylenes have only recently found widespread application. We earlier reviewed the addition of H-X reagents, where X was a halogen or heteroatom, to carbon-carbon double and triple bonds [2]. This chapter will review many synthetically useful electrophilic acetylene addition reactions that generate heterocycles and carbocycles and a few closely related nucleophilic cyclizations. While there are many other very useful reactions of alkynes that lead to heterocycles and carbocycles, they are beyond the scope of this review, so no attempt will be made to cover intramolecular Diels–Alder reactions of alkynes; transition metal-catalyzed alkyne and enyne metathesis chemistry; cycloaddition reactions of alkynes; transition metal carbene-promoted annulations of alkynes; Bergman and related diyne cyclizations; the synthesis of arenes by alkyne cycloaddition chemistry; the Pauson–Khand reaction; intramolecular photochemical, nucleophilic (with a few exceptions), or free radical cyclizations of alkynes; or numerous transition metal-catalyzed reactions of enynes and diynes that produce heterocycles or carbocycles. Many of these topics have recently been reviewed, or leading references can be found in recent publications.

2.2
Cyclization of Oxygen Compounds

2.2.1
Cyclization of Acetylenic Alcohols

A variety of oxygen-containing heterocycles have been obtained by the cyclization of acetylenic alcohols. Furans have been obtained by the base-promoted cyclization of alkenynols (Equations 2.1 and 2.2) [3].

Acetylene Chemistry. Chemistry, Biology and Material Science. Edited by F. Diederich, P. J. Stang, R. R. Tykwinski

ISBN 3-527-30781-8

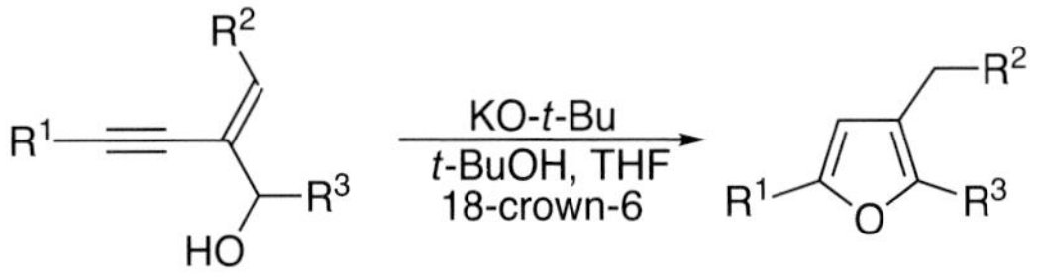

Equation 2-1

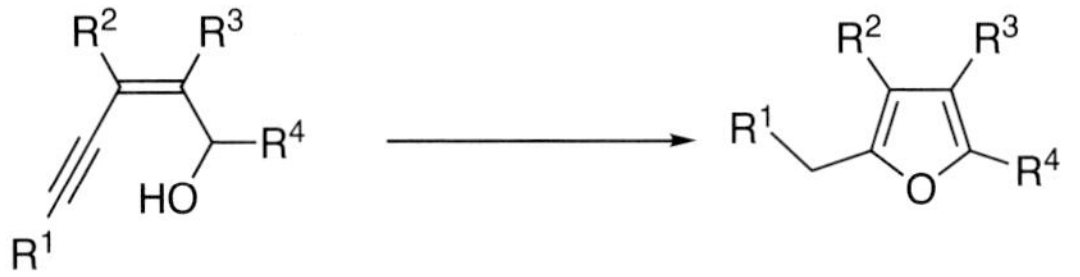

Equation 2-2

Treatment of 3-alkyne-1,2-diols with I_2 in the presence of $NaHCO_3$ provides a very convenient synthesis of 3-iodofurans (Equation 2.3) [4]. This reaction presumably involves electrophilic cyclization, followed by dehydration. The corresponding silyl ethers can also be employed as starting materials [4b].

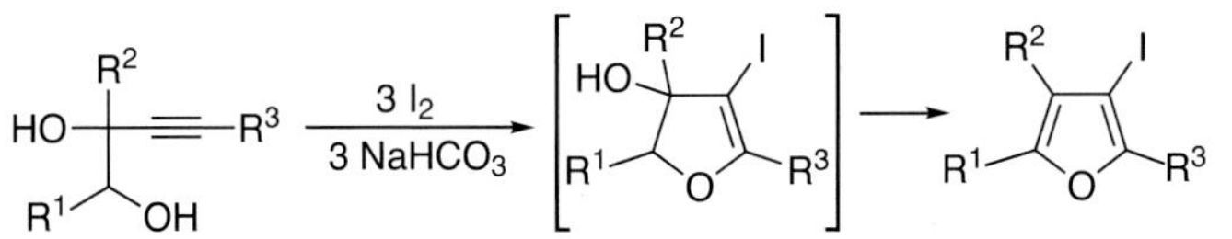

Equation 2-3

Propargylic alcohols react with $HgCl_2$ to generate vinylic mercurials that are readily cyclized to furylmercurials [5] (Equation 2.4) or carbonylated to butenolides (Equation 2.5) [6].

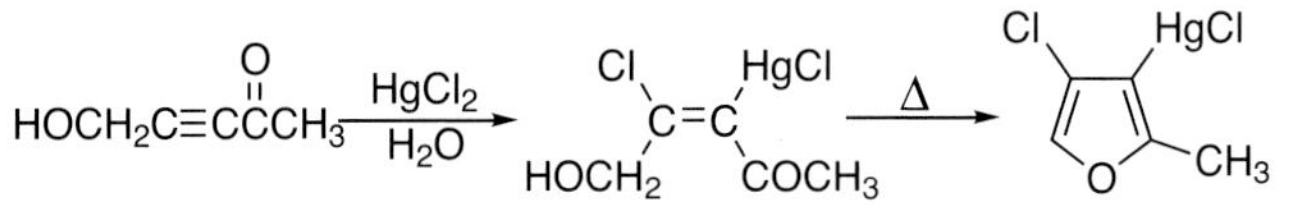

Equation 2-4

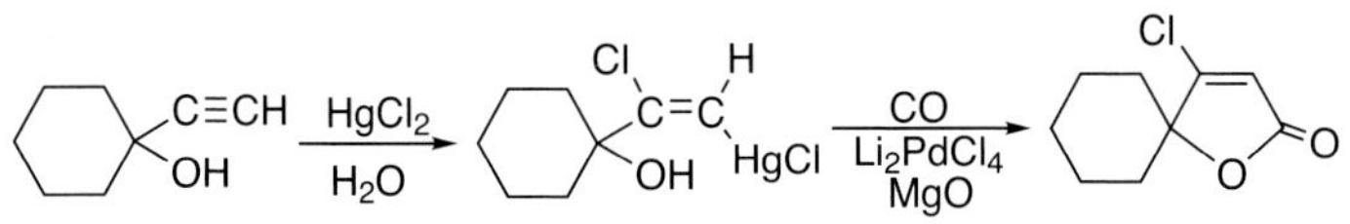

Equation 2-5

4-Alkyn-1-ols have also been cyclized to vinylic ethers by using mercury salts (Equation 2.6) [7]. If halosuccinimides are added to the reaction, the corresponding vinylic halides can be obtained.

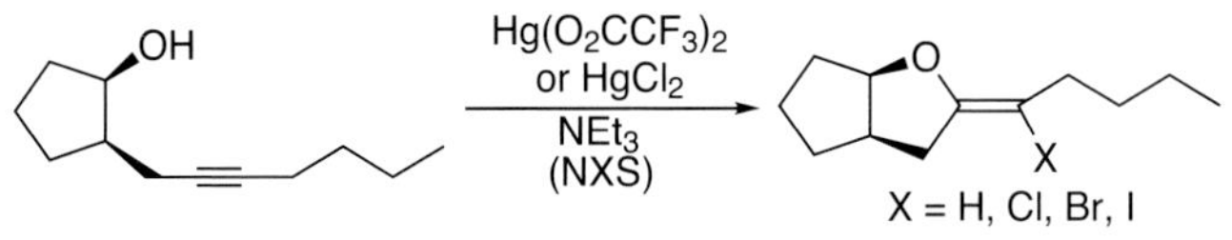

Equation 2-6

The isobenzofuran ring has been prepared from an acetylenic benzyl alcohol by use either of stoichiometric $Hg(OAc)_2$, followed by $NaBH_4$, or of catalytic amounts of HgO plus BF_3 (Equation 2.7) [8].

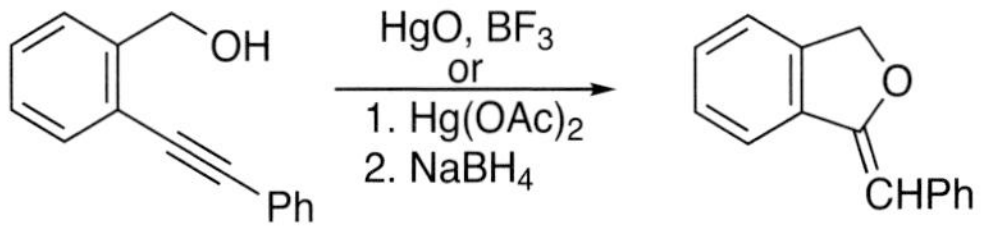

Equation 2-7

Palladium catalysts have been widely employed to cyclize alkynols to heterocycles. Thus, 2-methoxy-3-alkyn-1-ols readily cyclize to furans (Equation 2.8) [9]. Similarly, alkenynols can be cyclized to furans [10] (Equation 2.9) and pyrans [11] (Equation 2.10), and alkynediols have been cyclized to bicyclic acetals [9] (Equation 2.11).

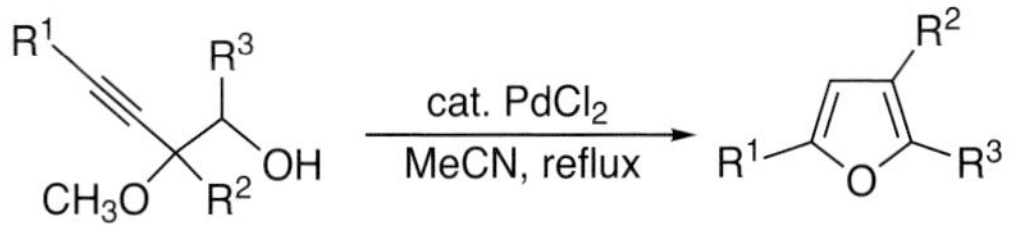

Equation 2-8

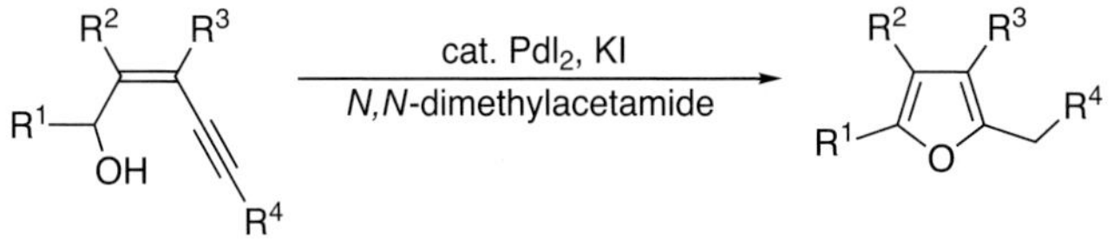

Equation 2-9

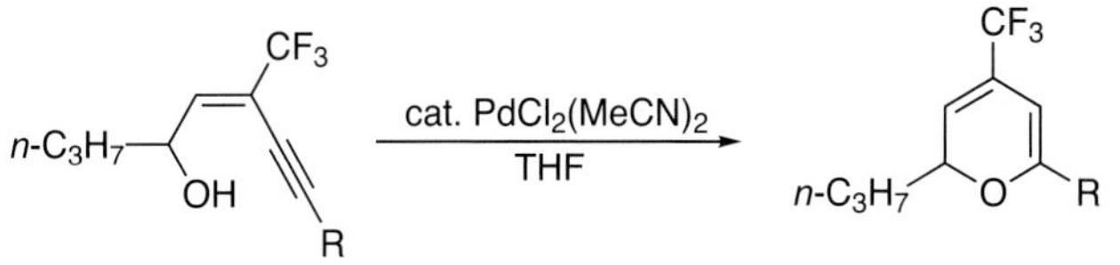

Equation 2-10

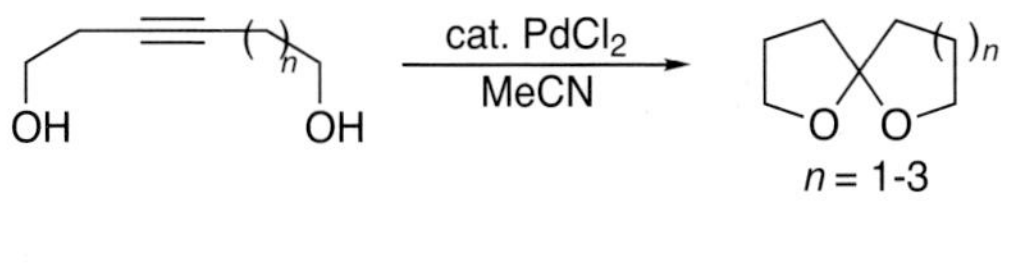

Equation 2-11

Palladium and copper catalyze the cyclization of acetylenic alcohols to a variety of lactones and heterocyclic esters. Propargylic alcohols have been cyclized either to (*Z*)-α-chloromethylene-β-lactones in the presence of CO (Equation 2.12) [12] or to cyclic carbonates in the presence of CO_2 (Equation 2.13) [13]. The carbonylation of 1-alkyn-4-ols affords α-methylene-γ-butyrolactones (Equation 2.14) [14]. Analogous silylalkynols have been converted either into simple saturated lactones [15] or into lactones incorporating two carbonyl groups [16], depending on the reaction conditions (Equation 2.15). Under the latter reaction conditions, simple 3-alkyn-1-ols afford α-methoxyalkylidene-γ-butyrolactones (Equation 2.16) [17]. Similarly, 1-alkyn-5-ols cyclize to the corresponding tetrahydrofuran-containing esters (Equation 2.17) [18]. Modest enantioselectivity has been achieved in this latter process with utilization of a chiral ligand [18]. Finally, furyl esters can be obtained from

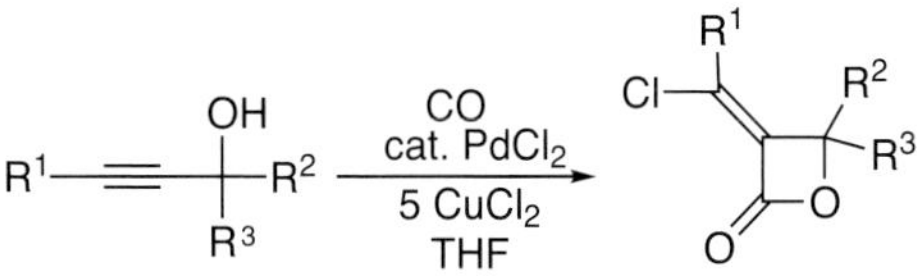

Equation 2-12

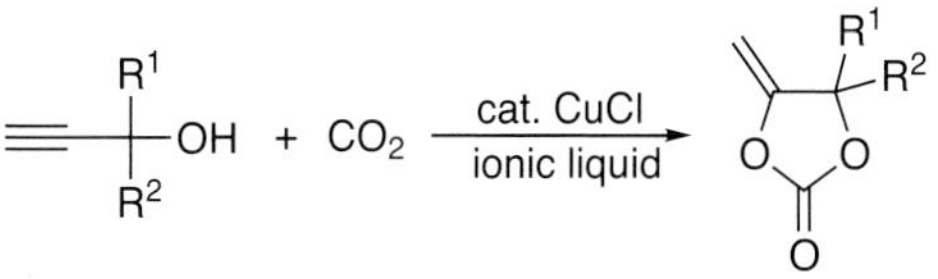

Equation 2-13

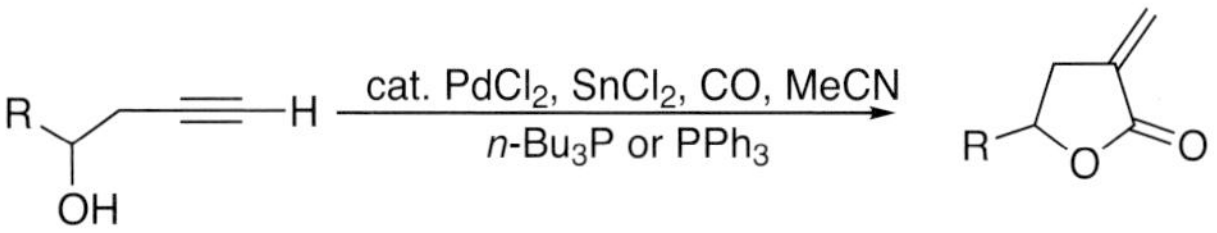

Equation 2-14

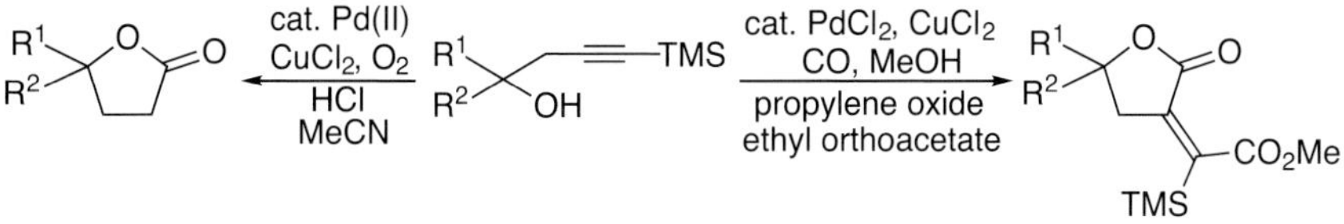

Equation 2-15

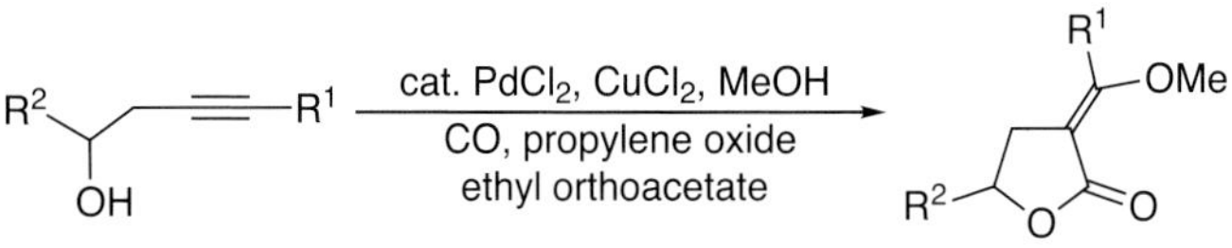

Equation 2-16

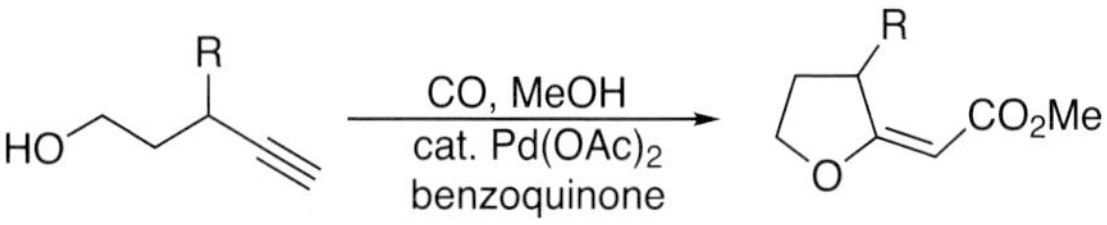

Equation 2-17

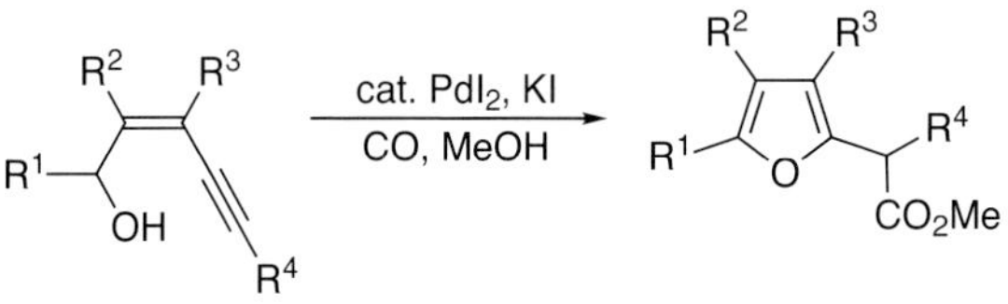

Equation 2-18

alkenynols (Equation 2.18) [19].

Tungsten chemistry has provided a novel approach to cyclic ethers. The photocyclization of 1-alkyn-5-ols in the presence of $W(CO)_6$ plus DABCO affords either five- or six-membered ring unsaturated ethers depending on the substitution pattern of the alkynol (Equation 2.19) [20]. Tungsten can also effect cyclization of 4-alkyn-1-ols with incorporation of an aldehyde moiety (Equation 2.20) [21].

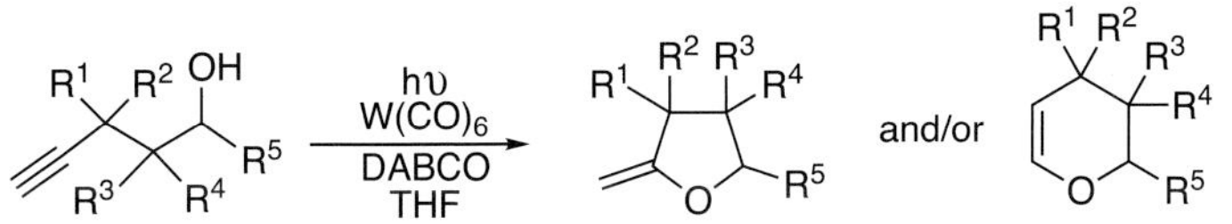

Equation 2-19

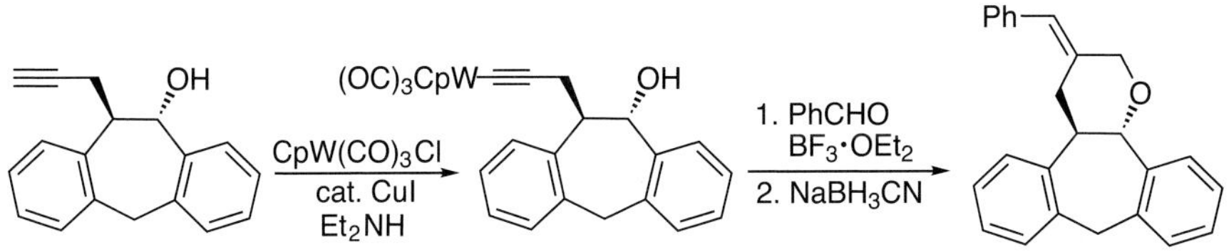

Equation 2-20

2.2.2
Cyclization of Acetylenic Phenols

The cyclization of acetylenic phenols provides a very convenient route to benzofurans. Benzofurans and 4-azabenzofurans have been prepared by reactions between 2-ethynylphenols or -pyridinols and aryl halides or vinylic triflates (Equation 2.21) [22]. Undoubtedly, the reaction initially affords the corresponding substituted alkynylphenol, which is then cyclized by the Pd and/or Cu salts present in the reaction. Analogous silylalkynes undergo a similar substitutive cyclization in the presence of microwave irradiation (Equation 2.22) [23].

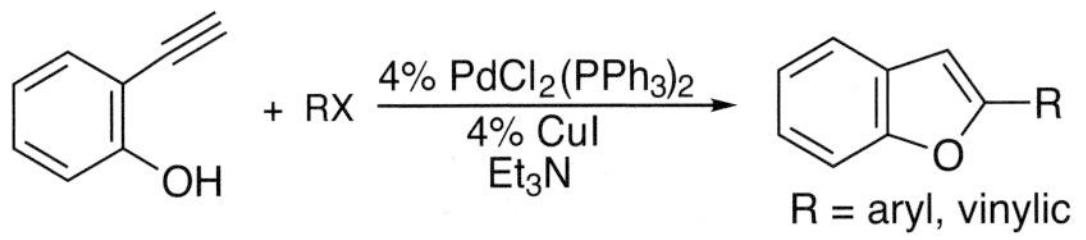

Equation 2-21

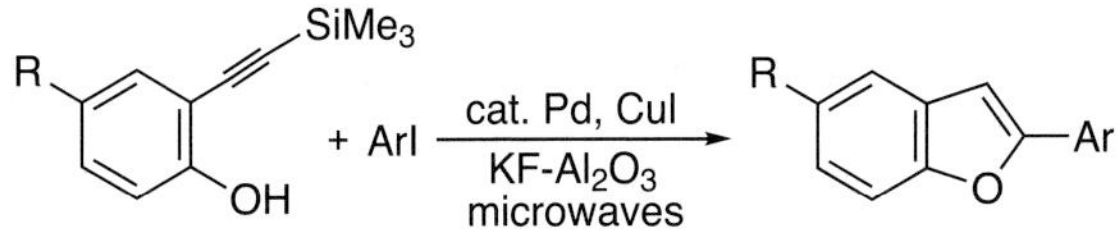

Equation 2-22

2-(1-Alkynyl)phenols are readily cyclized to 3-substituted benzofurans by various electrophiles. For example, I_2 plus $NaHCO_3$ [24] or IPy_2BF_4 plus HBF_4 [25] afford 3-iodobenzofurans (Equation 2.23).

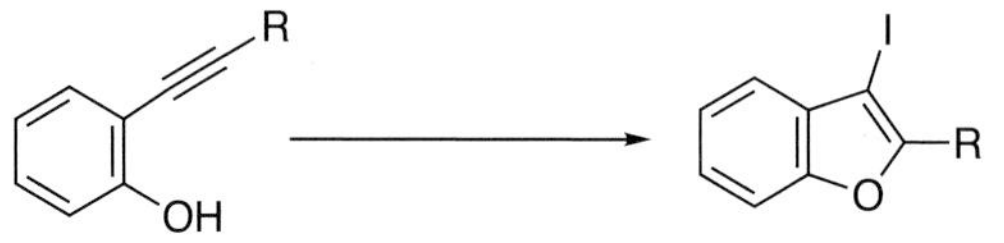

Equation 2-23

Organopalladium compounds react readily with 2-(1-alkynyl)phenols to afford a variety of benzofurans. For example, allylic carbonates react with 2-(1-alkynyl)phenols in the presence of a Pd catalyst to produce 3-allylic benzofurans (Equation 2.24) [26]. Propargylic carbonates afford analogous 3-allenylbenzofurans [27]. Aryl or vinylic triflates react similarly to afford 3-substituted benzofurans (Equation 2.25) [22b]. These reactions proceed by: (1) oxidative addition of the organic carbonate or triflate to Pd(0), (2) electrophilic attack of the organopalladium intermediate on the alkyne to produce the benzofuran with the organopalladium moiety in the 3-position, and (3) reductive elimination of Pd(0), which produces the observed product and regenerates the Pd(0) catalyst.

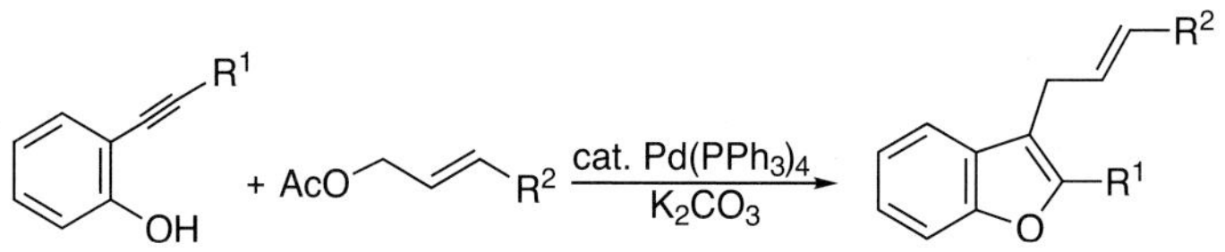

Equation 2-24

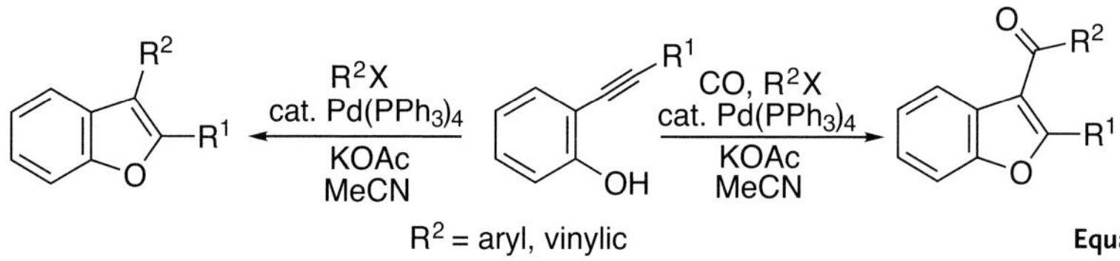

Equation 2-25

In the presence of CO and a palladium catalyst, carbonyl-containing products are produced. When the reaction with aryl or vinylic triflates is run in the presence of CO, the corresponding ketones are generated (Equation 2.25) [22b, 28]. Surprisingly, the reaction of 2-ethynylphenols, CO, and vinylic triflates affords 3-alkylidene-2-coumaranones (Equation 2.26) [29], while in the presence of just CO and an alcohol, internal alkynes generate benzofuran esters [30] (Equation 2.27) and lactones [31] (Equation 2.28).

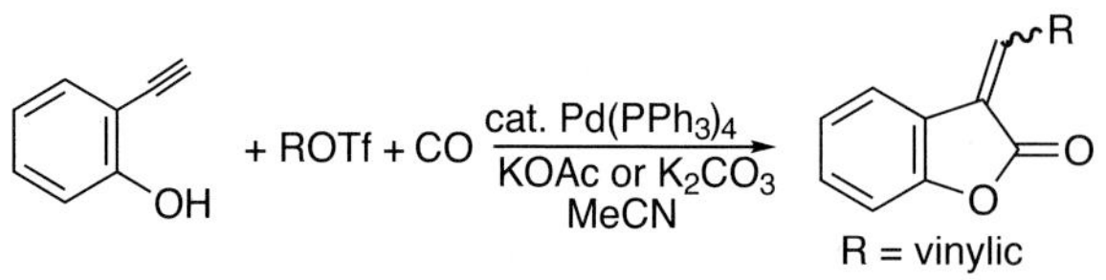

Equation 2-26

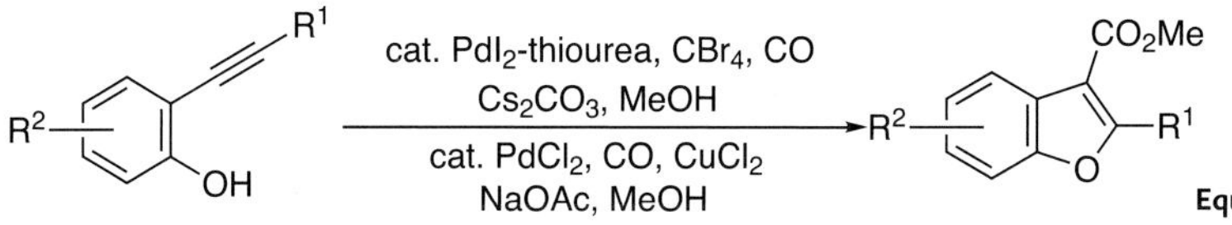

Equation 2-27

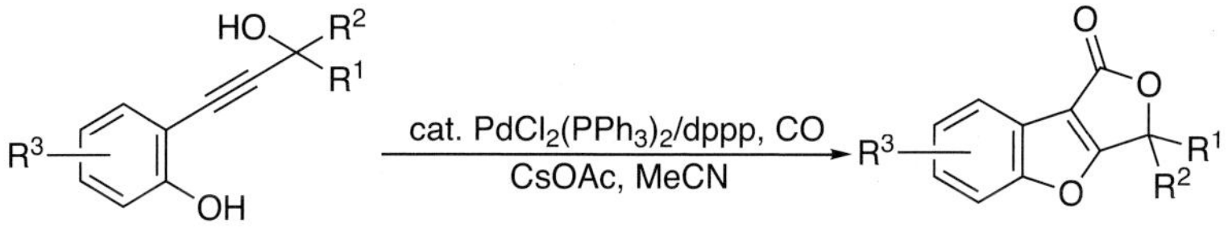

Equation 2-28

When the acetylene unit is further removed from the phenol moiety, palladium-catalyzed cross-coupling with organic halides and cyclization are still feasible (Equations 2.29 and 2.30) [32]. Palladium has also been reported to catalyze the cyclization of *o*-hydroxyphenyl phenylethynyl ketone to aurone (Equation 2.31) [33].

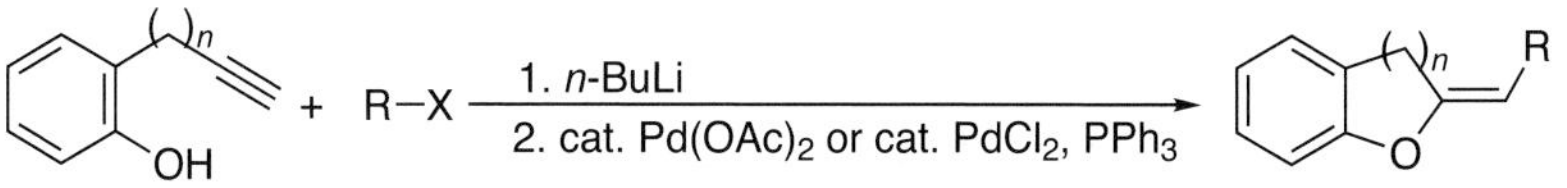

R = Me, Ar, PhCH$_2$; X = I, Br; *n* = 0, 1

Equation 2-29

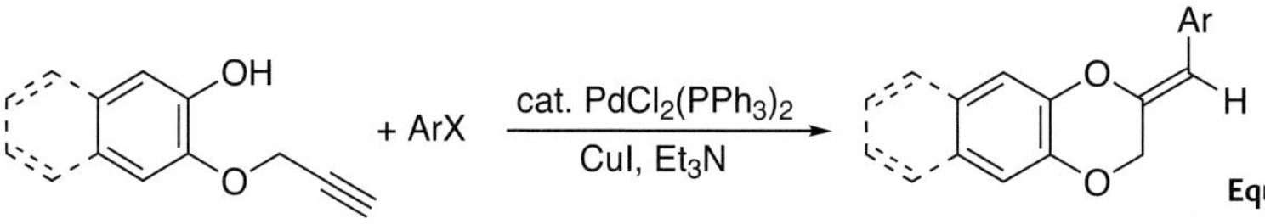

Equation 2-30

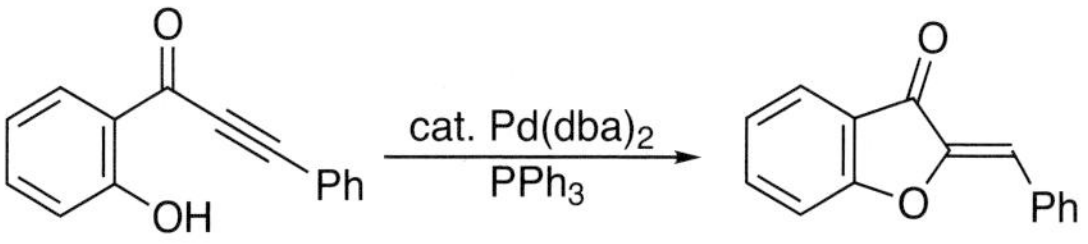

Equation 2-31

2.2.3
Cyclization of Acetylenic Ethers

There are a few recent examples of acetylenic ethers being nicely cyclized to oxygen heterocycles. For example, an alkynyl acetal of diacetone glucose [34] (Equation 2.32) and *o*-benzyloxy- and *o*-acetoxyalkynylpyridines [35] (Equation 2.33) react with NIS and with I_2, respectively, to produce cyclic ethers. The alkynylpyridines can also be readily carbonylated in the presence of a Pd catalyst to produce the corresponding heterocyclic esters (Equation 2.33) [35].

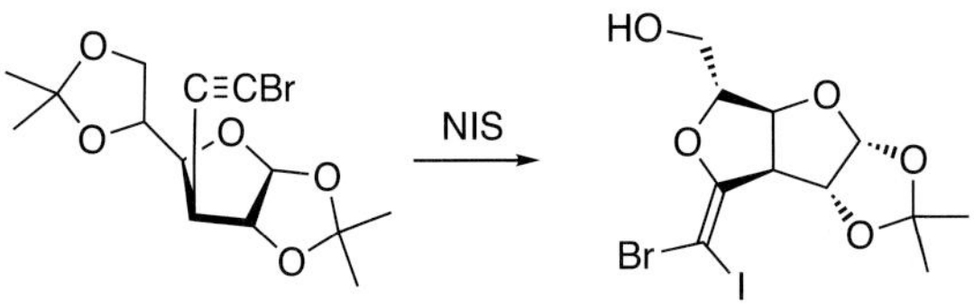

Equation 2-32

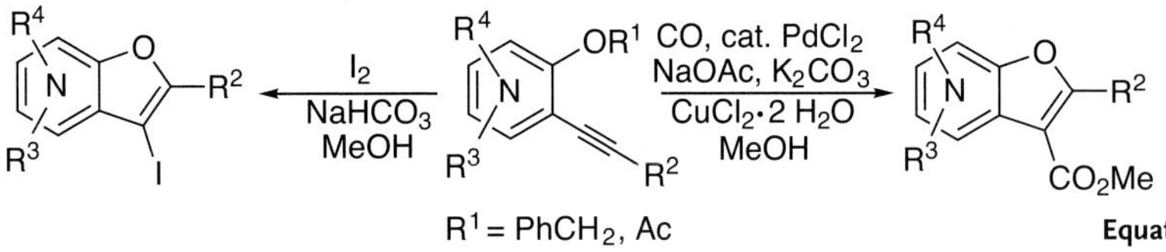

Equation 2-33

We have observed that simple methyl ethers of 2-(1-alkynyl)phenols can be very efficiently cyclized to 3-substituted benzofurans under very mild reaction conditions by the use of halogen, S, Se, and Hg electrophiles (Equation 2.34) [36]. The mercury chemistry also works on other alkyne-containing methyl ethers (Equations 2.35 and 2.36) [36b]. The resulting organomercurials are readily reduced, halogenated, or carbonylated to esters.

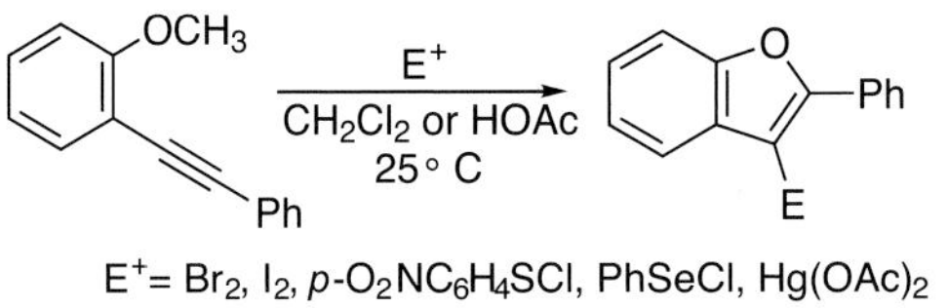

Equation 2-34

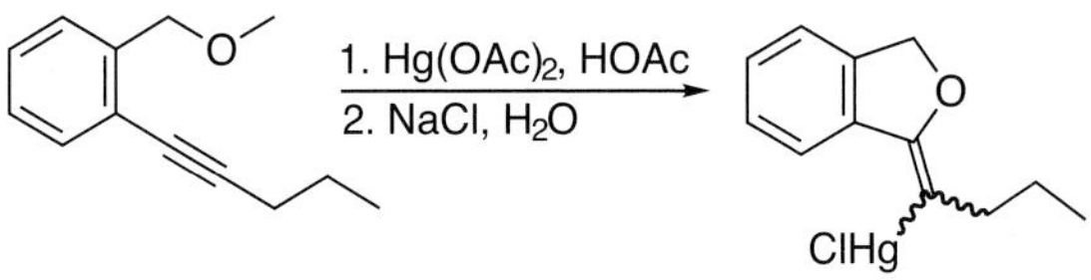

Equation 2-35

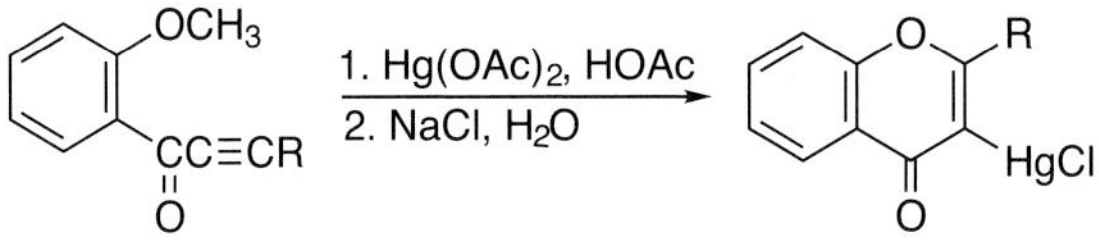

Equation 2-36

Propargylic *o*-(1-alkynyl)phenyl ethers will rearrange in the presence of a Pd catalyst to afford mixtures of 3-allenyl- and 3-propargylbenzofurans (Equation 2.37) [37]. In a similar manner, analogous allylic ethers have been rearranged to 3-allylic benzofurans under similar conditions [26].

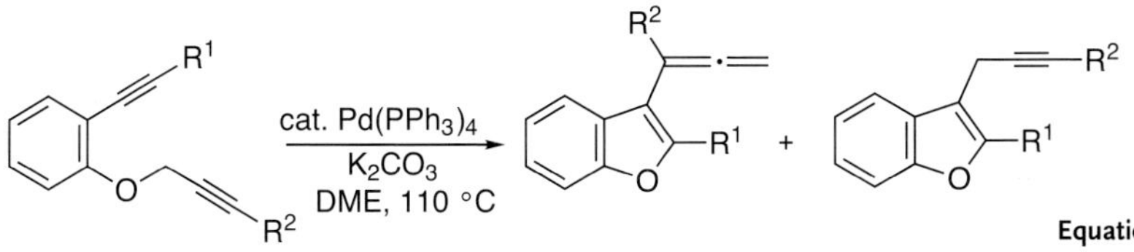

Equation 2-37

Silyl ethers of alkynones have been cyclized by diethylamine to flavones [38] and cross-coupled with aryl halides in the presence of a palladium catalyst to produce aurones [39] (Equation 2.38).

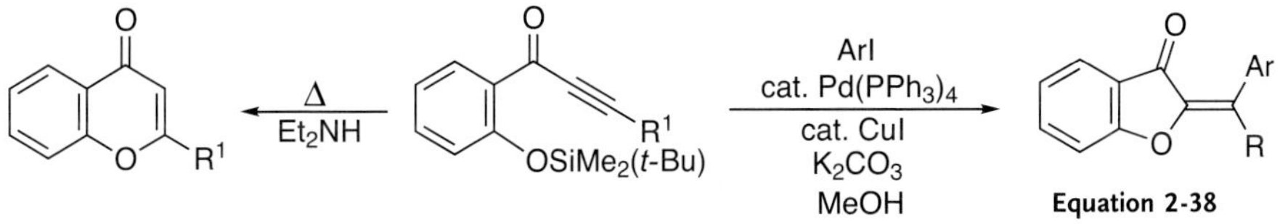

Equation 2-38

Finally, acetylenic epoxides have been rearranged to oxygen heterocycles. Thus, alkynyloxiranes react with KH or KO-*t*-Bu to afford furans [40] (Equation 2.39), and tungsten intermediates have been cyclized to lactones (Equation 2.40) [41].

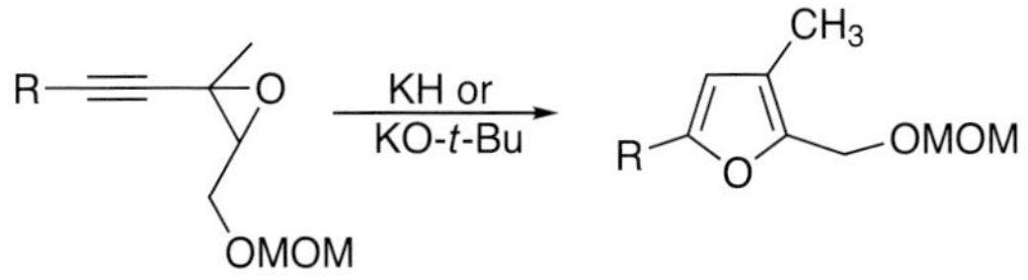

Equation 2-39

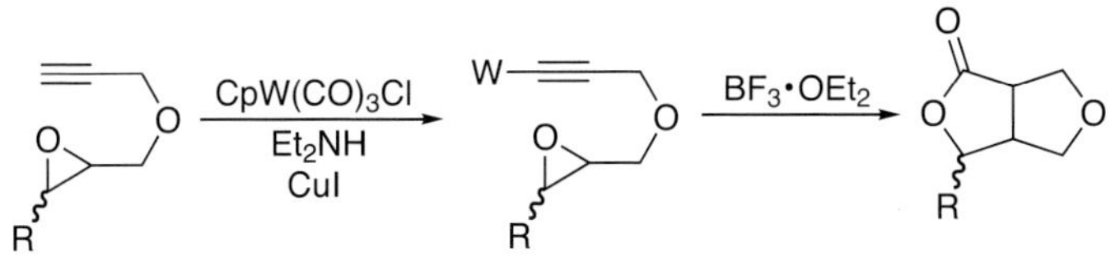

Equation 2-40

2.2.4
Cyclization of Acetylenic Acids and Derivatives

Unsaturated five- and six-membered ring lactones are readily synthesized by the cyclization of 4- and 5-alkynoic acids by bases or catalytic amounts of Rh, Pd, Ag, and Hg salts (Equation 2.41) [42]. 3-Alkynoic acids afford butenolides in the presence of a Pd catalyst (Equation 2.42) [43]. Halogen reagents effect similar cyclizations to halolactones (Equation 2.43) [44].

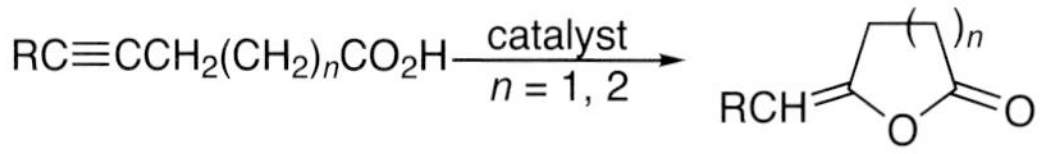

$$RC\equiv CCH_2(CH_2)_nCO_2H \xrightarrow[n=1,2]{\text{catalyst}} \text{[lactone]}$$

Equation 2-41

$$RC\equiv CCH_2CO_2H \xrightarrow[Et_3N]{\text{cat. PdCl}_2(\text{PhCN})_2} \text{[lactone]}$$

Equation 2-42

$$RC\equiv CCH_2(CH_2)_nCO_2H \xrightarrow{X^+} \text{[halolactone]}$$

$X^+ = Br, I_2, NCS, NBS, NIS$

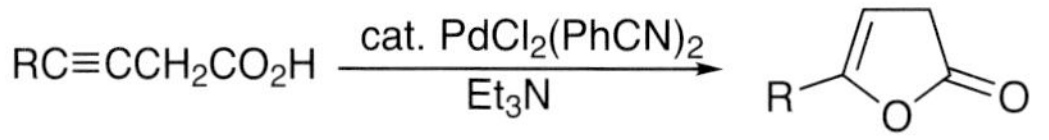

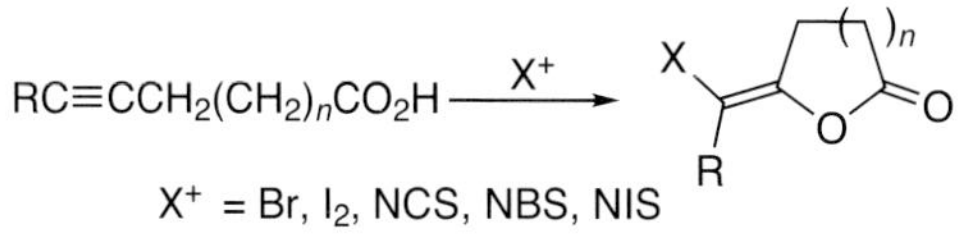

Equation 2-43

Lithium alkynoate salts react with Pd salts and vinylic or allylic halides [45] or allylic acetates [46] to produce similar lactones substituted on the carbon-carbon double bond by a vinylic or an allylic group (Equation 2.44). The allylic products can also be obtained directly from the corresponding allylic alkynoate esters [46]. Propargylic acetates afford the corresponding allenic products [47]. If acrolein is employed as the olefin, aldehyde-containing lactones can be produced (Equation 2.45) [48]. The products are consistent with the formation of a vinylic palladium species by cyclization of the alkynoate salt, which then reacts by adding to the double or triple bond of the olefin or alkyne.

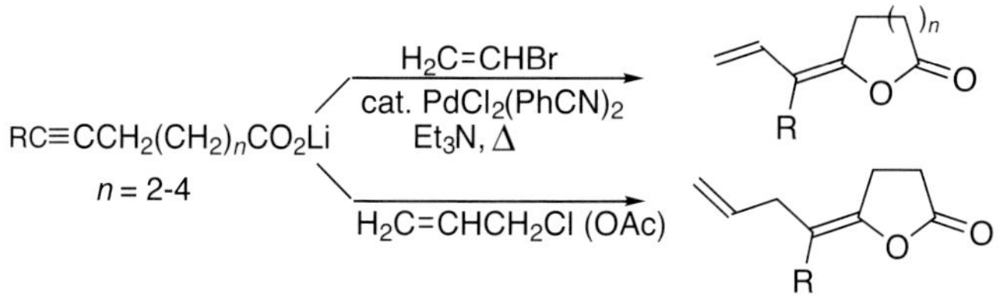

Equation 2-44

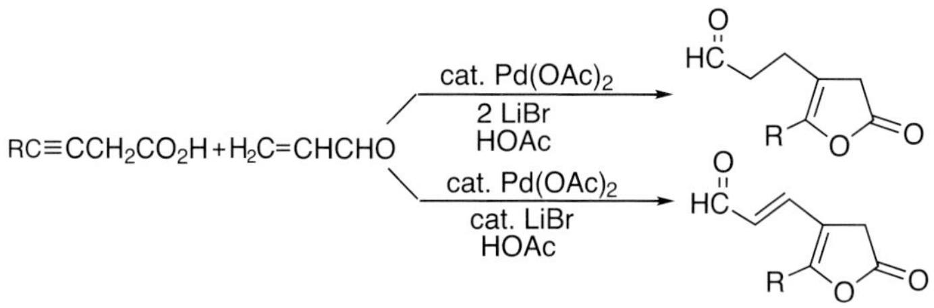

Equation 2-45

The reaction of aryl halides and vinylic triflates or halides [49] or 1-bromo-1-alkynes [50] with 4-alkynoic acids in the presence of a Pd catalyst affords cross-coupled butyrolactones (Equation 2.46). Intramolecular examples of this chemistry have also been reported (Equation 2.47) [51].

$$HC \equiv CCH_2CR^1{}_2CO_2H + R^2X \xrightarrow[\text{Et}_3N, n\text{-Bu}_4NCl]{\text{cat. Pd(OAc)}_2(\text{PPh}_3)_2}$$

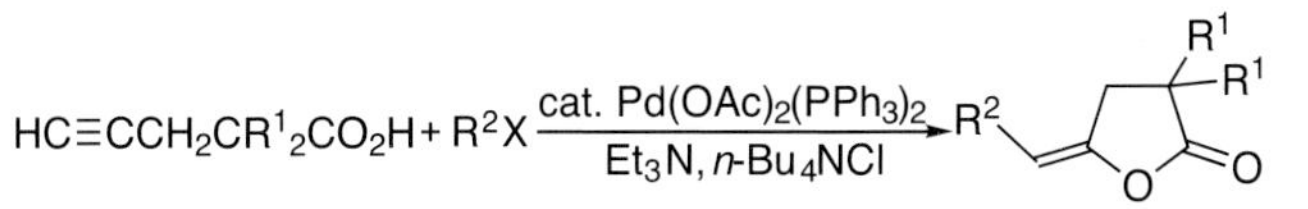

R^2 = aryl, vinylic, alkynyl **Equation 2-46**

$$\xrightarrow[\text{KO-}t\text{-Bu}]{\text{cat. Pd(OAc)}_2 \\ \text{P(2-furyl)}_3}$$

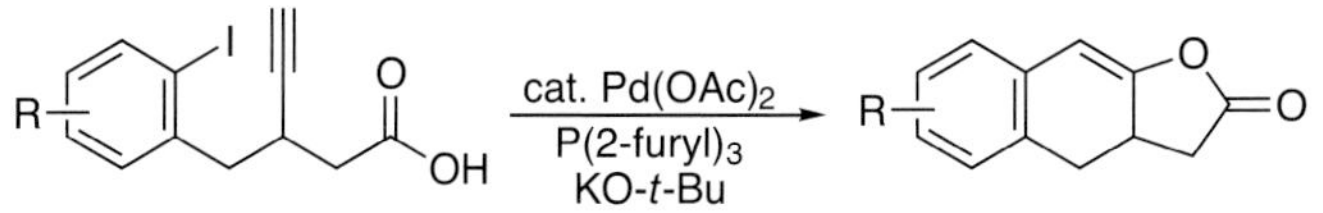

Equation 2-47

(Z)-2-Alken-4-ynoic acids react with catalytic amounts of Ag or AgI to afford primarily the corresponding five-membered ring lactones, while HgO gives mixtures of five- and six-membered ring lactones (Equation 2.48) [42b]. The reagents I_2 and NaHCO$_3$ in MeCN or ICl in CH$_2$Cl$_2$ afford mixtures of iodofuranones and -pyranones, in which the latter predominate (Equation 2.48) [52]. The corresponding methyl esters cyclize with I_2 or ICl with comparable or better regioselectivity [52d, 53]. When aryl halides react with these acids in the presence of catalytic amounts of Pd(PPh$_3$)$_4$ and K$_2$CO$_3$, mixtures of aryl-substituted furanones and pyranones are generated in low yields, with the former predominating (Equation 2.48) [54].

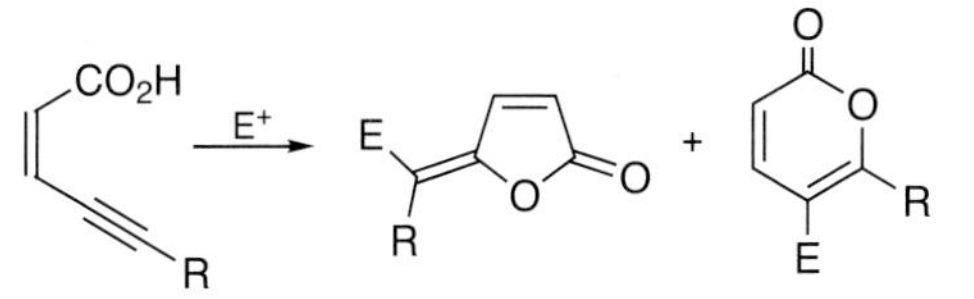

E^+ = H$^+$ [Ag, AgI, or HgO], I_2, ICl, ArX [cat. Pd(PPh$_3$)$_4$, K$_2$CO$_3$] **Equation 2-48**

Similar results have been observed in the cyclization of 2-(1-alkynyl)benzoic acids. Cyclization by use of strong acids has been reported to produce both five- and six-membered ring lactones (Equation 2.49) [8, 55]. Catalytic amounts of AgClO$_4$, AgOTf, or AgNO$_3$ afford predominantly isocoumarins, while Ag powder or AgI affords phthalides as the major products (Equation 2.49) [42b, 55]. Treatment with aryl halides and a Pd catalyst affords predominantly aryl-substituted phthalides [54]. Halolactonization by N-halosuccinimides in the presence of NaHCO$_3$ and Triton-B in CH$_2$Cl$_2$ [56] or I_2 plus NaHCO$_3$ in MeCN [57] is reported to produce good yields of the corresponding 4-haloisocoumarins. The corresponding methyl esters have also been cyclized to isocoumarins in good yields by use of HI [53b], Br$_2$ [58], I_2 [52d, 53b, 57], ICl [53a,b, 57], p-O$_2$NC$_6$H$_4$SCl [53b], PhSeCl [53b], and Hg(OAc)$_2$ [36b, 59].

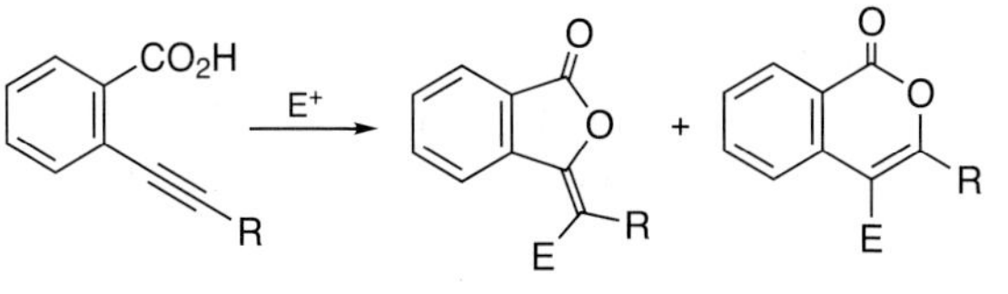

E^+ = H$^+$ [also cat. Ag$^+$ or Ag], NCS, NBS, NIS, I_2, ArX [cat. Pd(PPh$_3$)$_4$, K$_2$CO$_3$] **Equation 2-49**

2-(1-Alkynyl)benzoate esters and thioesters react with vinylic ethers in the presence of 10% $PtCl_2$ to produce naphthyl ketones by a cycloaddition/ring-opening process (Equation 2.50) [60].

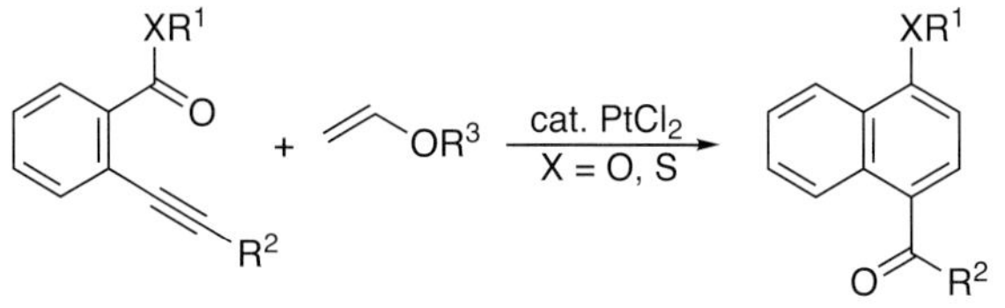

Equation 2-50

Benzofurans have been prepared by the direct cyclization of acetylene-containing phenolic acetates by alkoxide bases (Equation 2.51) [61]. Related silyl derivatives have been directly cross-coupled and cyclized by use of a strong base plus a palladium catalyst (Equation 2.52) [22b].

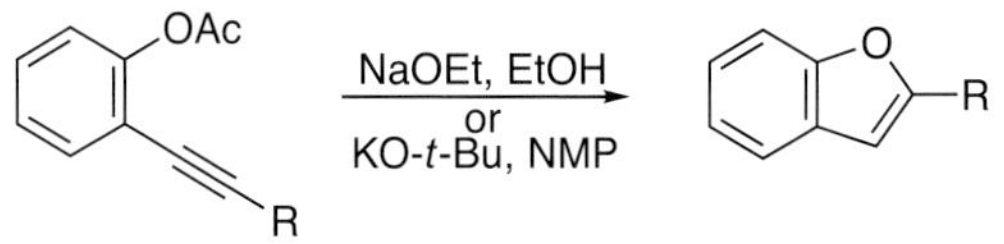

Equation 2-51

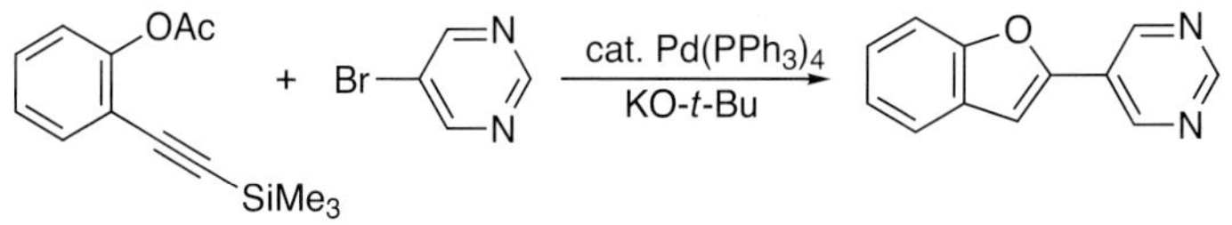

Equation 2-52

A number of different heterocycles have been obtained by cyclization of acetylenic amides. For example, the Pd-catalyzed cyclization of *N*-propargylamides has provided oxazoles [62] and oxazolines [63] (Equation 2.53). In a similar manner, the palladium-catalyzed carbonylative cyclization of acetylenic aniline derivatives affords 4*H*-3,1-benzoxazines, quinazolin-2-ones, and quinolin-4-ones [64].

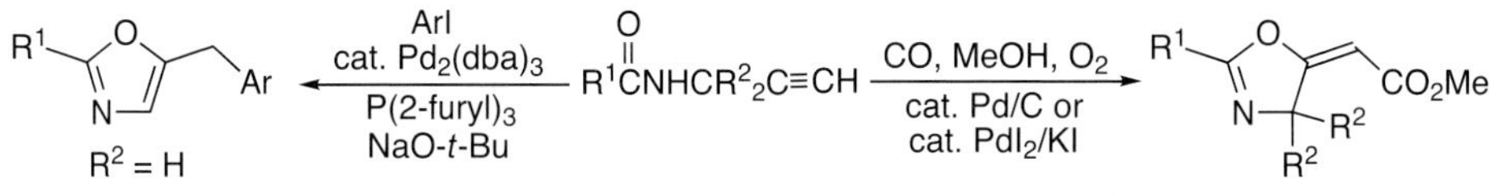

Equation 2-53

The cyclization of acetylenes onto the carbonyl oxygen of nucleosides and related amides has been carried out with cat. CuI [65] and halogen reagents [66] (Equation 2.54). In a similar manner, 3-(1-alkynyl)pyridones react with arylpalladium intermediates to produce furo[2,3-*b*]pyridones (Equation 2.55) [67].

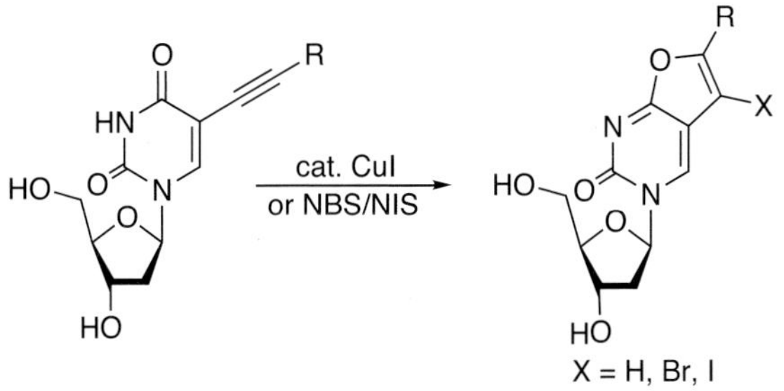

Equation 2-54

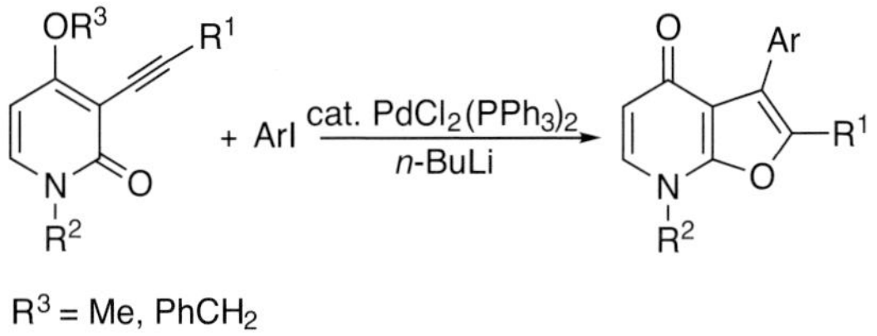

Equation 2-55

Finally, phosphaisocoumarins have been obtained by proto- [68] and iodocyclization [69] of the corresponding phosphorous esters (Equation 2.56).

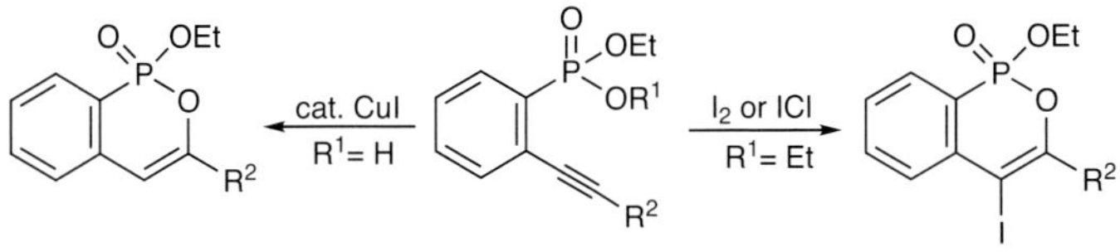

Equation 2-56

2.2.5
Cyclization of Acetylenic Aldehydes and Ketones

Acetylenic aldehydes can be readily cyclized to vinylic ethers and acetals. In methanol, simple 4-alkynals afford modest yields of mixtures of methoxy-substituted five- and six-membered ring ethers, with the five-membered ring products predominating (Equation 2.57) [70]

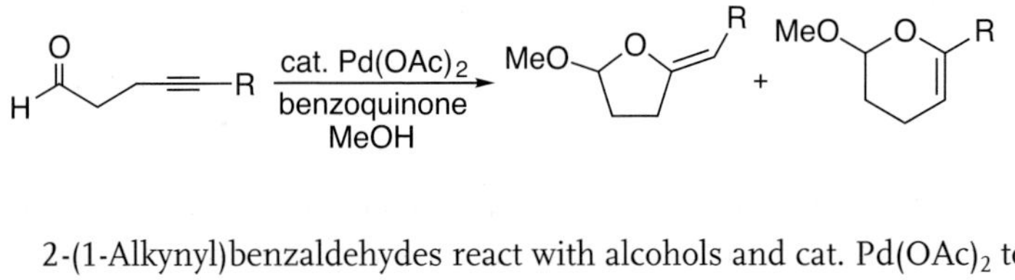

Equation 2-57

2-(1-Alkynyl)benzaldehydes react with alcohols and cat. $Pd(OAc)_2$ to produce cyclic acetals (Equation 2.58) [70, 71]. IPy_2BF_4 plus HBF_4 and various nucleophiles generate substituted iodoisochromenes (Equation 2.59) [72]. Alcohols produce acetals and allylic silanes, and enol silanes and electron-rich arenes generate new carbon-carbon bonds. When the reaction is run in the presence of simple alkenes and alkynes, naphthalene derivatives are produced by a process that involves cycloaddition and cycloreversion (Equation 2.60) [73]. We have recently found that even better yields can be obtained in both of these processes by simply using I_2 instead of the expensive, difficult to handle iodonium salt [74]. These same aldehydes react with alkynes and catalytic amounts of AuX_3 (X = Cl or Br) to produce the same naphthyl ketones, while catalytic $Cu(OTf)_2$ affords simple naphthalenes (Equation 2.61) [75]. One can also start with aryl ketones here.

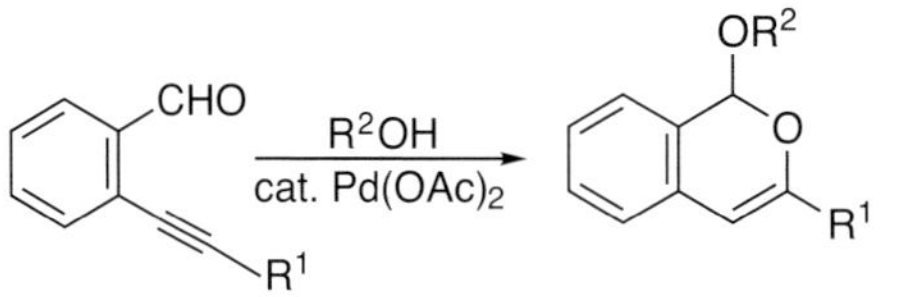

Equation 2-58

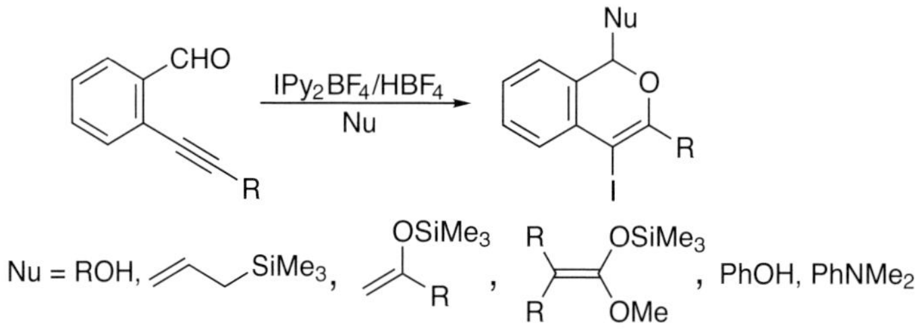

Equation 2-59

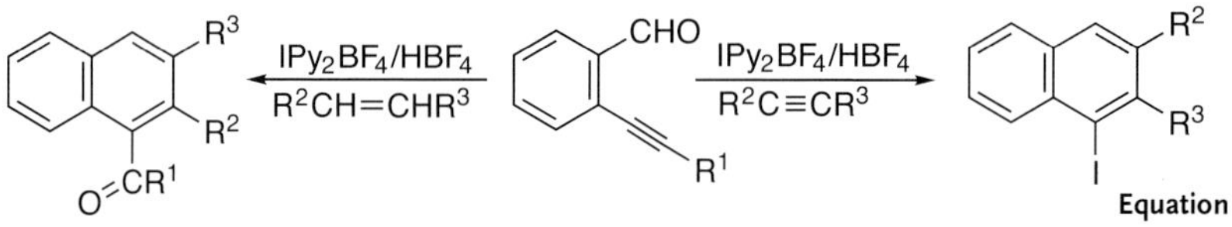

Equation 2-60

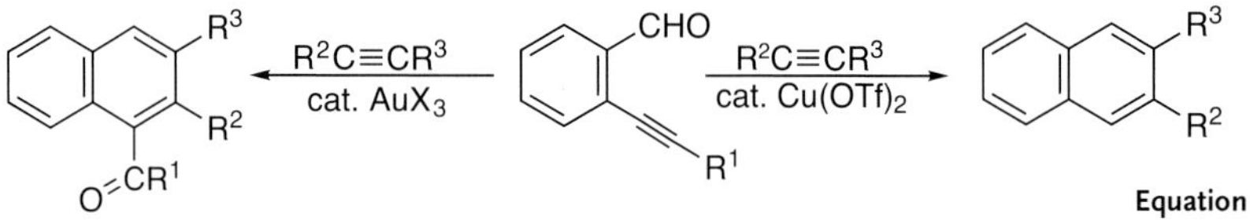

Equation 2-61

A wide variety of alkynones have been cyclized to furans. Thus, *p*-TsOH has been used to cyclize 4-alkyn-1-ones to furans (Equation 2.62) [76]. Similarly, $PdCl_2$ catalyzes the cyclization of 3-alkyn-1-ones to furans (Equation 2.63) [77]. If allylic chlorides are added to this latter reaction, 3-allylic furans can be obtained in good yields [59]. Palladium has also been used to cyclize 2-alkyn-1-ones to furans, but substantial amounts of bifuryls are also formed (Equation 2.64) [78].

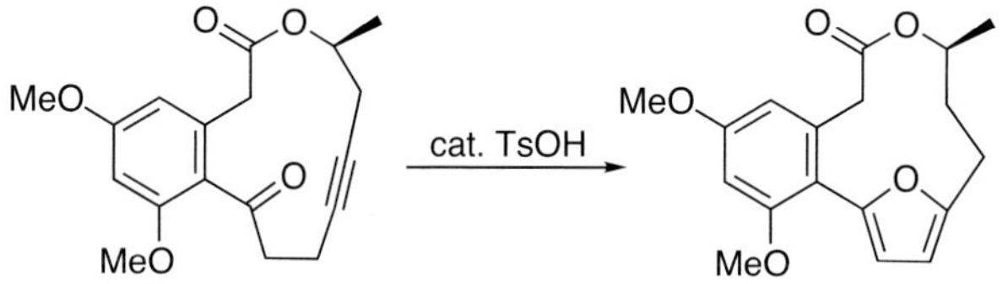

Equation 2-62

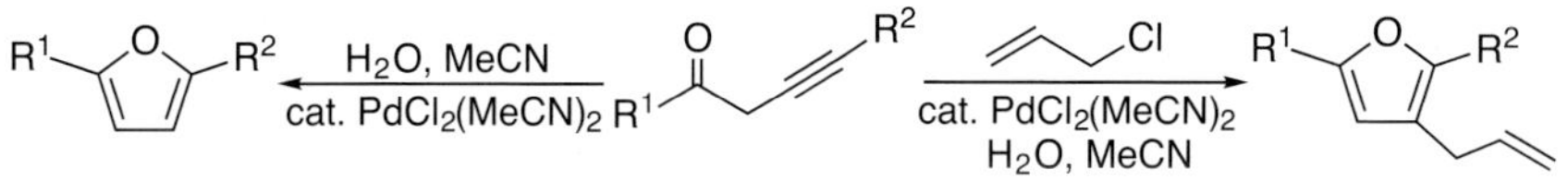

Equation 2-63

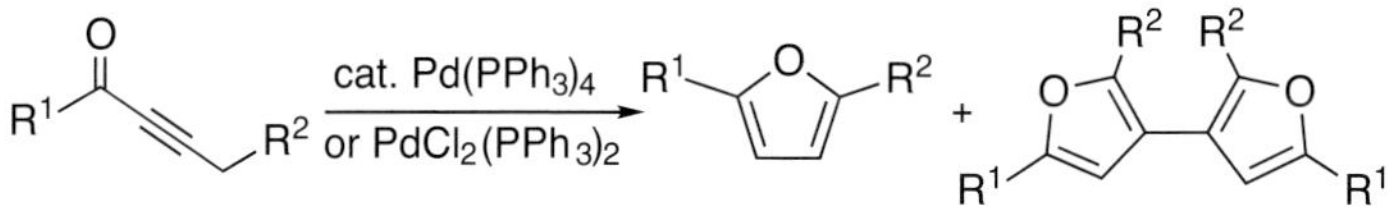

Equation 2-64

A number of acetylenic dicarbonyl compounds have been cyclized by metal catalysts. Thus, catalytic amounts of $Pd(OAc)_2$, $PdCl_2(MeCN)_2$, $PtCl_2$, $W(CO)_5 \cdot THF$, or $CpRuCl(PPh_3)_2$ efficiently cyclize cyclic β-diketones to the corresponding cyclic vinylic ethers (Equation 2.65) [79]. The regiochemistry depends on the ring size of the diketone, the substitution pattern of the acetylene, and the catalyst. Palladium also catalyzes the cross-coupling of vinylic or aryl triflates or halides with related β-diketones or β-keto esters, amides or nitriles to furans (Equation 2.66) [80]. When CO is added to the reaction, keto furans are obtained (Equation 2.67) [80c, 81]. In the absence of the organic halide or triflate, CO and methanol react with these same acetylenic ketones to afford tetrahydrofuran-containing esters (Equation 2.67) [82].

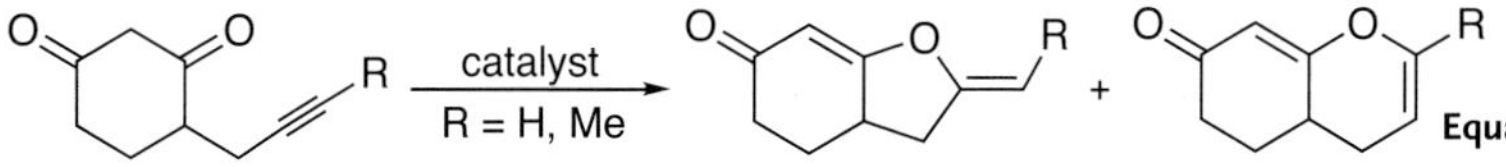

Equation 2-65

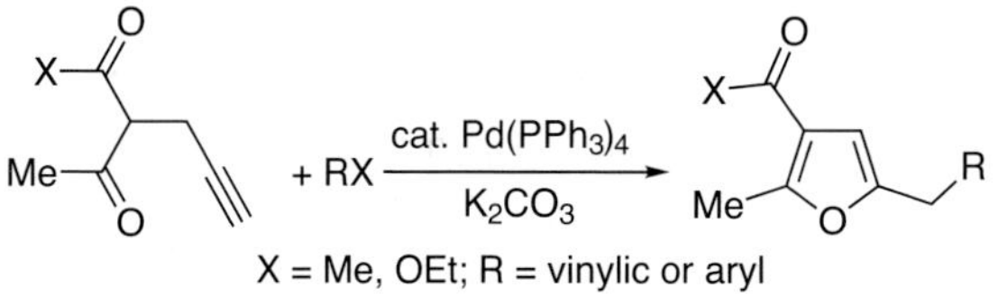

Equation 2-66

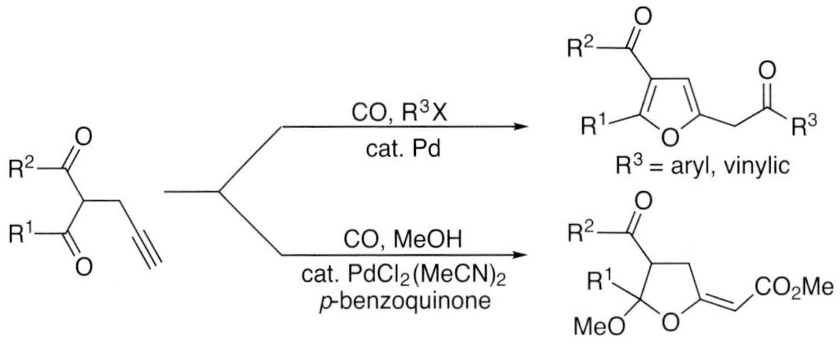

Equation 2-67

2.3 Cyclization of Sulfur and Selenium Compounds

Acetylenic thioanisoles have been cyclized to benzothiophenes by a variety of electrophiles (Equation 2.68) [36b, 83]. The analogous methylselenium-containing compounds can also be very efficiently cyclized by Br_2, I_2, ICl, and PhSeCl to produce 3-substituted selenophenes [84]. Iodine also induces the cyclization of sulfur-containing acetylenic alcohols to either ketone- or vinylic iodide-containing benzothiophenes (Equation 2.69) [85]. Methylsulfur and -selenium compounds with a neighboring alkynone moiety have also been cyclized in the presence of aldehydes and $BF_3 \cdot Et_2O$ to generate sulfur- and selenium-containing chromen-4-ones (Equation 2.70) [86].

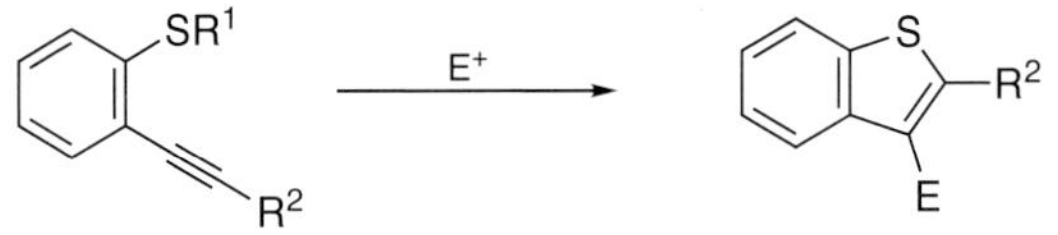

R^1 = Me, SCH_2Ph; E^+ = Br_2, NBS, I_2, p-$O_2NC_6H_4SCl$, PhSeCl, $Hg(OAc)_2$

Equation 2-68

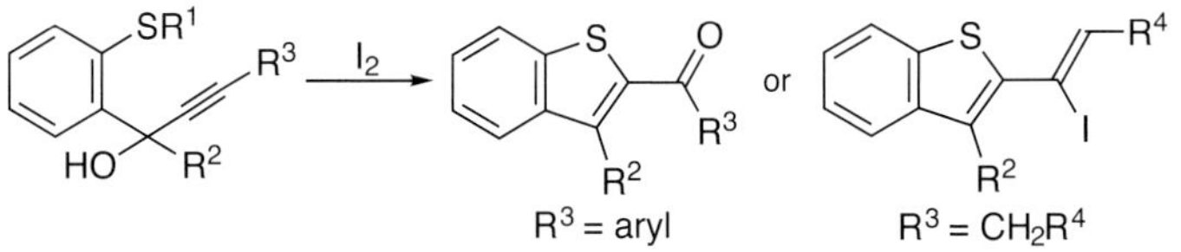

Equation 2-69

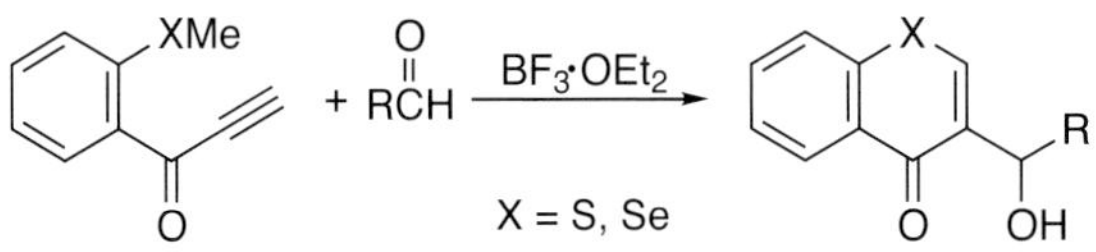

Equation 2-70

Finally, 2-(thioformylamino)diarylacetylenes cyclize at room temperature in the presence of DBU to afford 3,1-benzothiazines (Equation 2.71) [87].

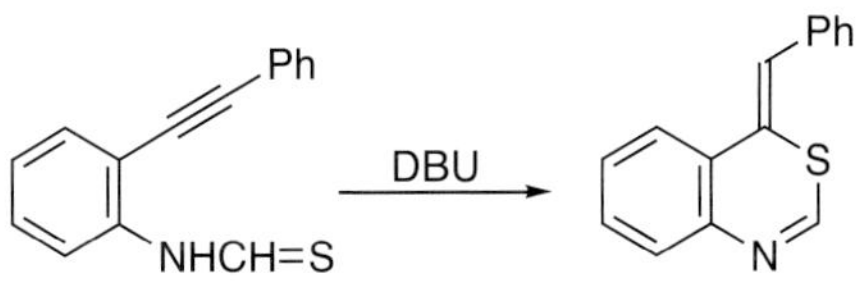

Equation 2-71

2.4 Cyclization of Nitrogen Compounds

2.4.1 Cyclization of Acetylenic Amines

Simple alkynylamines are readily cyclized to five- and six-membered ring imines [88] by use of catalytic amounts of $CpTiCl_3$ [89], $CpTiMe_2Cl$ [89b, 90], $CpTiCl(NEt_2)_2$ [91], $Ti(NMe_2)_4$ [92], Cp_2TiMe_2 [93], $CpZrMe_2Cl$ [89b, 90], $NaAuCl_4$ [94], $[Cu(MeCN)_4]PF_6$ [95], $Zn(OTf)_2$ [95], $Pd(OTf)_2$(triphos) [95], $PdCl_2$ [96], and $(\eta^5\text{-}Me_5C_5)_2LnCH(TMS)_2$ (Ln = La, Nd, Sm, Lu) [97] (Equation 2.72).

Equation 2-72

The intermediates from the Ti chemistry have been trapped by acyl cyanides to give either nitriles or ketones, depending on the substitution pattern of the acetylene (Equation 2.73) [90, 91].

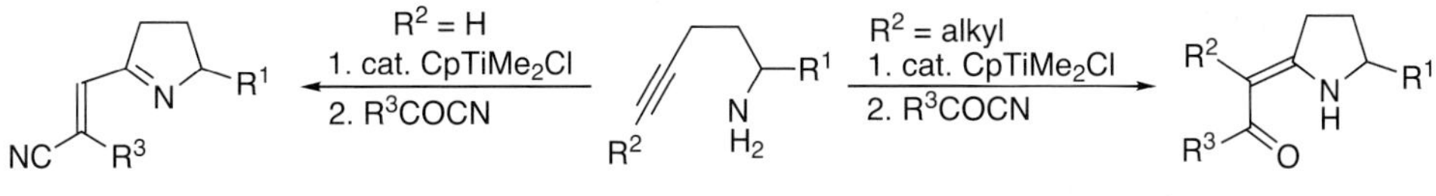

Equation 2-73

Recently, the palladium-catalyzed cyclization of acetylenic amines has provided unsaturated cyclic amines instead [98], and this process can be made enantioselective by employing of a chiral diphosphine [99] (Equation 2.74).

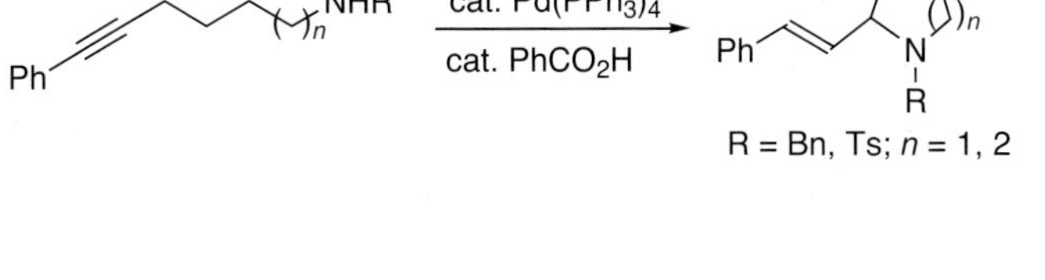

Equation 2-74

Pyrroles are readily prepared by the palladium-catalyzed cyclization of 1-amino-3-alkyn-2-ols [100] (Equation 2.75) or 2-alken-4-ynylamines [101] (Equation 2.76).

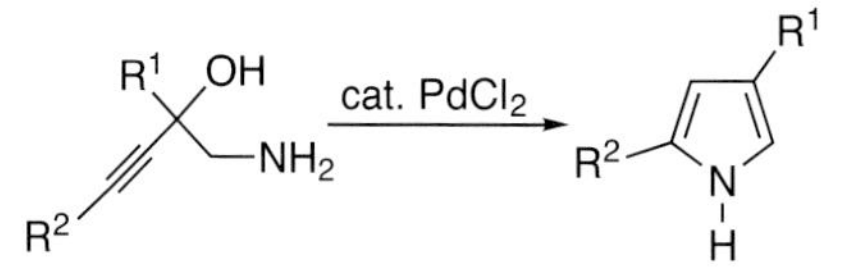

Equation 2-75

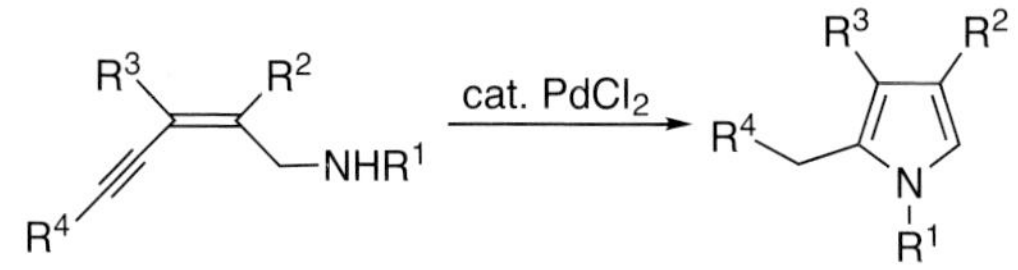

Equation 2-76

Numerous very useful routes from acetylenic amines to indoles have been reported. For example, 2-chlorophenyl acetylenic amines react with Cp_2TiMe_2, followed by a Pd catalyst, to produce annulated indoles (Equation 2.77) [102].

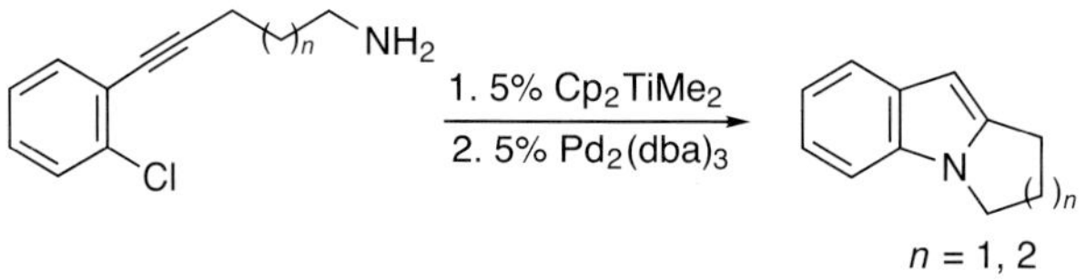

Equation 2-77

Simple 2-(1-alkynyl)anilines can be cyclized to indoles by KO-*t*-Bu [103], $NaNH_2$ [8], cat. $Cu(OAc)_2$ [104], cat. $Cu(O_2CCF_3)_2$ [104], cat. $(Et_3N)Mo(CO)_5$ [105], cat. $PdCl_2$ [106], cat. $NaAuCl_4$ [106c], or cat. $InBr_3$ [107] (Equation 2.78). Analogous silylacetylenes have been cyclized with simultaneous desilylation by heating with 2 equiv. of CuI [108].

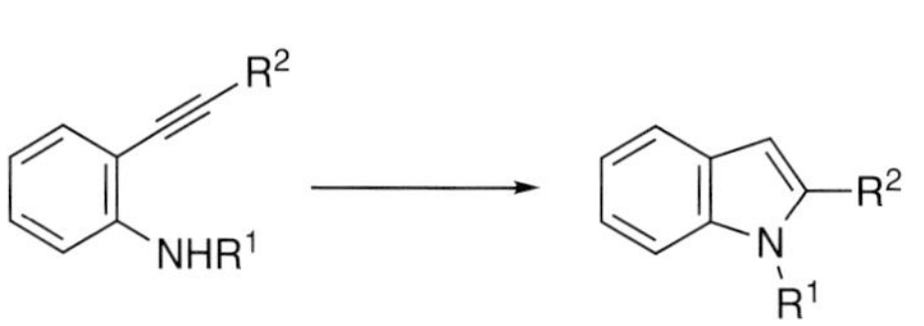

Equation 2-78

The Pd/Cu-catalyzed cross-coupling of 2-ethynylaniline and vinylic triflates, followed by $PdCl_2$-catalyzed cyclization, provides a convenient route to 2-substituted indoles (Equation 2.79) [106b]. In a similar fashion, a 2-ethynylaniline has been cross-coupled and cyclized in a one-pot fashion to generate a 2-dienylindole [109]. When this type of cyclization is carried out in the presence of CO and methanol, either indole esters [30c,d] (Equation 2.80) or indol-2-ones [110] (Equation 2.81) can be obtained.

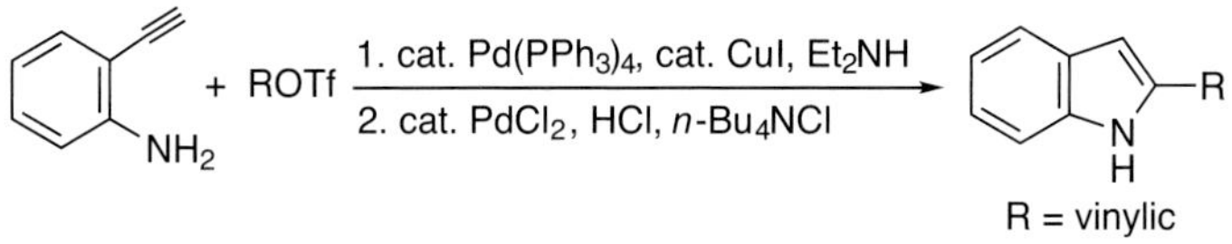

Equation 2-79

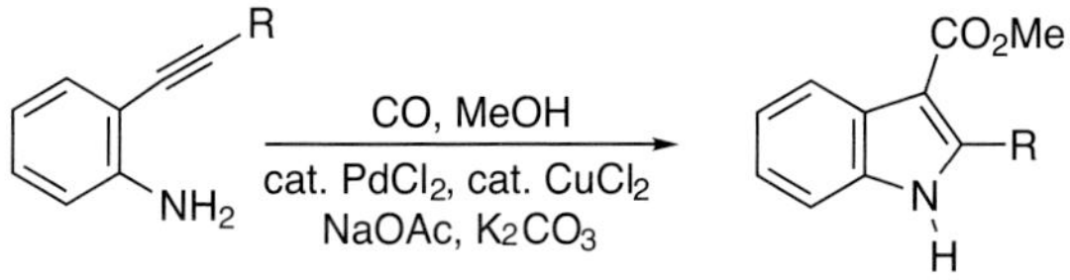

Equation 2-80

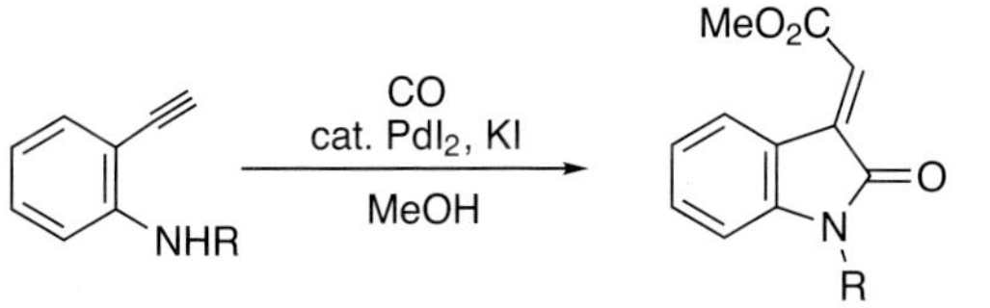

Equation 2-81

Palladium-catalyzed cross-couplings of acetylenic amines have been employed to generate other nitrogen heterocycles as well. When 2-ethynylaniline is cross-coupled with aryl halides, an amine, and CO in the presence of a Pd catalyst, the initially generated alkynones are apparently readily cyclized to afford 4-amino-2-arylquinolines (Equation 2.82) [111]. Quinolines have also been obtained by the Pd-catalyzed cyclization of alcohol-containing acetylenic anilines (Equation 2.83) [112]. The Pd apparently isomerizes the propargylic alcohol to the corresponding enone, which then cyclizes. *N*-Tosyl-*N*-propargylhydrazine has been cross-coupled with aryl halides and vinylic triflates in the presence of a Pd catalyst to generate substituted pyrazoles (Equation 2.84) [113]. Palladium has also been used to catalyze the cyclization of amines containing a silylacetylene to a substituted benzimidazole (Equation 2.85) [114] and a lactam (Equation 2.86) [115].

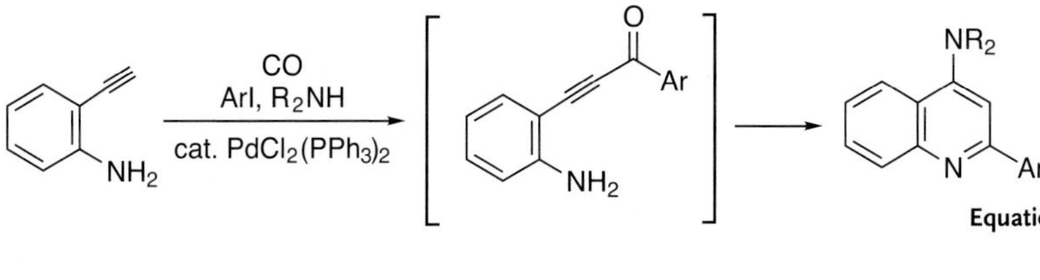

Equation 2-82

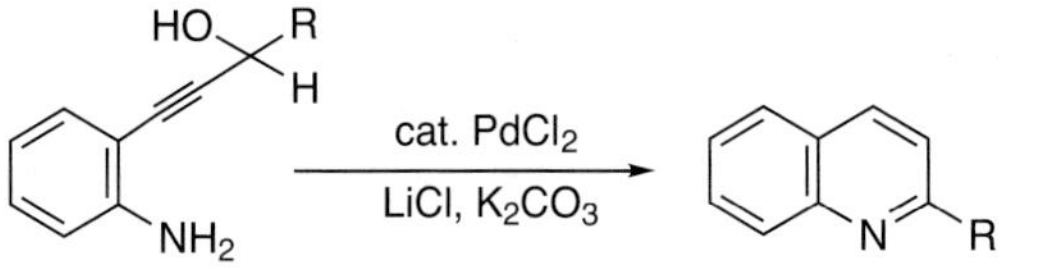

Equation 2-83

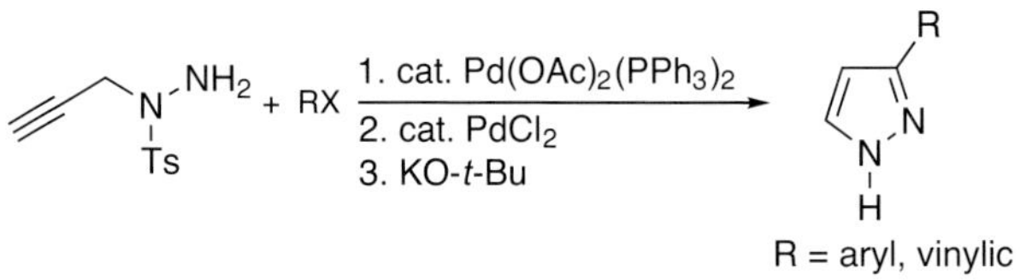

Equation 2-84

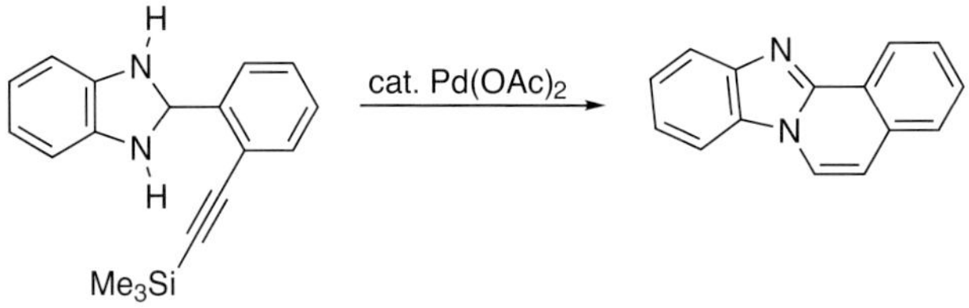

Equation 2-85

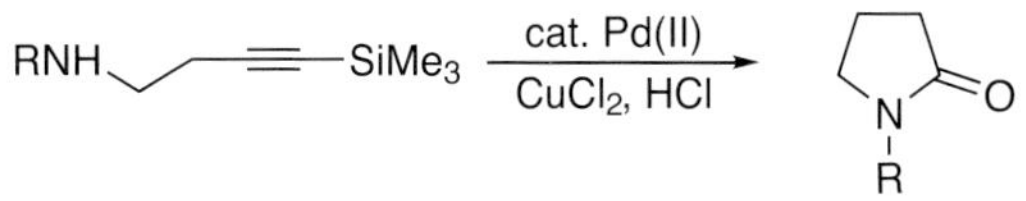

Equation 2-86

Finally, 3-iodoindoles can be readily prepared in high yields by the iodocyclization of simple acetylenic anilines by use of IPy_2BF_4 plus HBF_4 [25] or *N,N*-dimethylanilines by use of I_2 [116] (Equation 2.87).

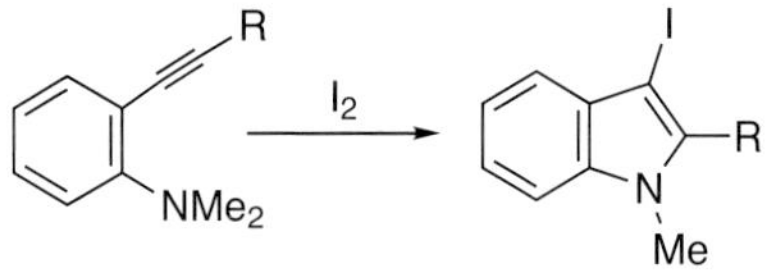

Equation 2-87

2.4.2
Cyclization of Acetylenic Amides

In a few cases, acetylenic amides have been cyclized on oxygen, as discussed in Section 2.2.4, but the majority of acetylenic amides have been cyclized on nitrogen to afford nitrogen heterocycles. Simple carboxamides of 2-(1-alkynyl)anilines are readily cyclized to indoles with loss of the acyl functionality by use of KO-*t*-Bu [103, 117] or cat. $Pd_2(dba)_3$ plus K_2CO_3 (trifluoroacetamides) [118], whereas stoichiometric amounts of *n*-Bu_4NF [119] or catalysis by $NaAuCl_4$ [106c] or $PdCl_2$ [106c, 120] affords the acylindoles (Equation 2.88). The latter reaction in the presence of allylic chlorides produces the corresponding 3-allylic 1-acylindoles [106c].

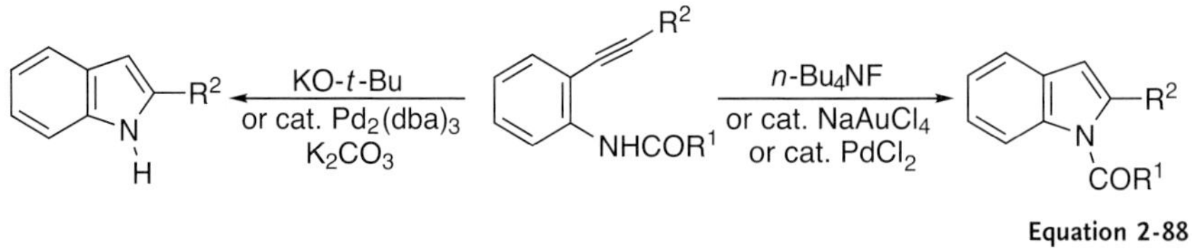

Equation 2-88

Trifluoroacetanilides bearing tertiary or secondary acetylenic alcohols cyclize to indoles or quinolines on treatment with cat. $Pd(OAc)_2$ or NaOEt, respectively (Equation 2.89) [112].

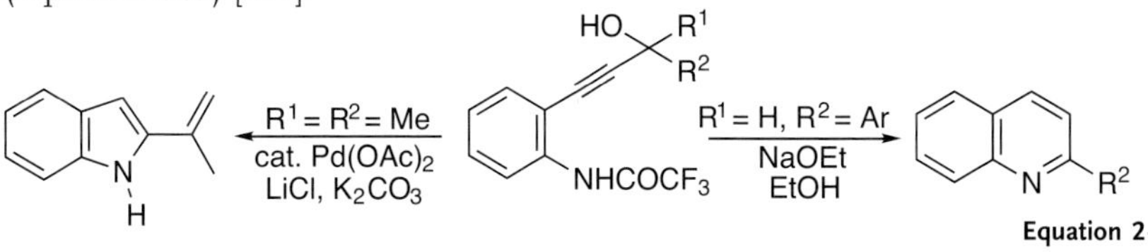

Equation 2-89

2-(1-Alkynyl)arenecarboxamides can be easily prepared and cyclized to 3-aryl(alkyl)idene isoindolin-1-ones by use either of NaOEt in EtOH or of cat. $Pd(OAc)_2$ (Equation 2.90) [121]. When these substrates are allowed to react with cat. $PhCH_2PdCl(PPh_3)_2$ plus Et_3N [122] or cat. $PdCl_2(MeCN)_2$ plus $CuCl_2$ and NaH [123], isoquinolinones are produced (Equation 2.90).

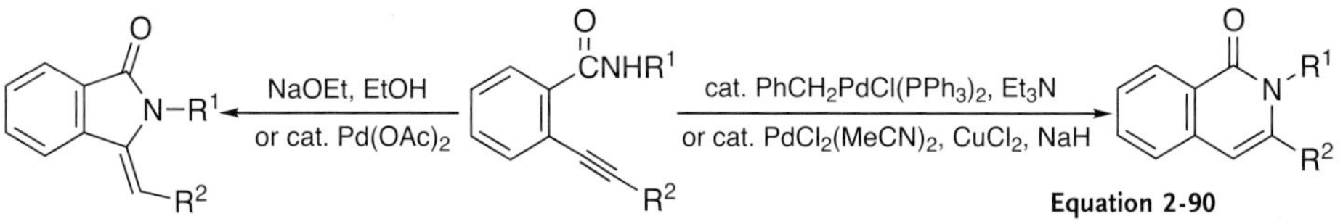

Equation 2-90

Many of these same acetylenic amides will also react with organic substrates plus a Pd catalyst to give more highly substituted nitrogen heterocycles [124]. For example, 2-(1-alkynyl)trifluoroacetanilides react with vinylic triflates [118, 125], aryl [125, 126] or heteroaryl [126] halides, benzyl bromide [127], allylic esters [128], ethyl iodo- or bromoacetate [127, 129], and α-bromo ketones [129] to produce the corresponding 3-substituted indoles minus the trifluoroacetyl group (Equation 2.91). The chemistry with vinylic triflates has been utilized on trifluoroacetanilides attached to a solid support [130]. This approach has recently been extended to the synthesis of pyrrolo[2,3-*b*]quinoxalines [131]. When this reaction is carried out on bis(*o*-trifluoroacetamidophenyl)acetylene and aryl or vinylic halides or triflates, 12-aryl(vinylic)indolo[1,2-*c*]quinazolines are formed (Equation 2.92) [132]. Analogous chemistry on a diyne has provided a unique route to a rebeccamycin-related indolocarbazole (Equation 2.93) [133].

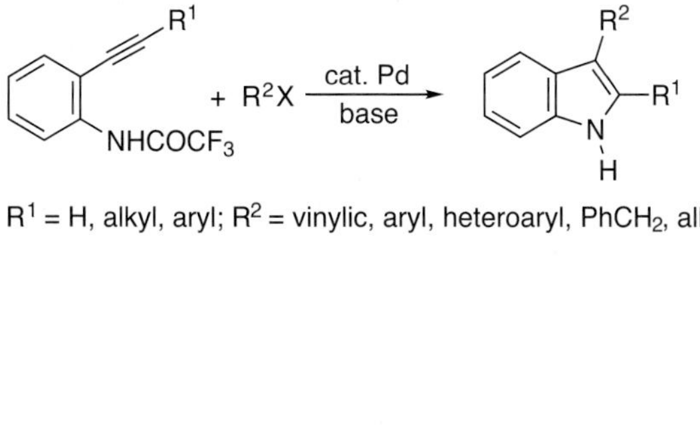

R^1 = H, alkyl, aryl; R^2 = vinylic, aryl, heteroaryl, $PhCH_2$, allylic, CH_2CO_2Et, CH_2COAr

Equation 2-91

Equation 2-92

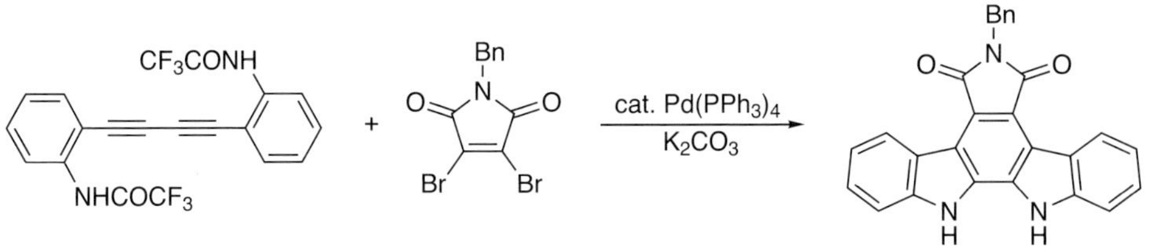

Equation 2-93

When the aryl halide reactions are run in the presence of CO, 3-aroylindoles are formed (Equation 2.94) [134]. The reaction of bis(*o*-trifluoroacetamidophenyl)acetylene, CO, and aryl or vinylic halides or triflates affords the corresponding 12-acylindolo[1,2-*c*]quinazolines (Equation 2.92) [135]. 6-Aryl-11*H*-indolo[3,2-*c*]quinolines are prepared by the Pd-catalyzed carbonylative cyclization of *o*-(*o*′-aminophenylethynyl)trifluoroacetanilide and aryl halides (Equation 2.95) [136].

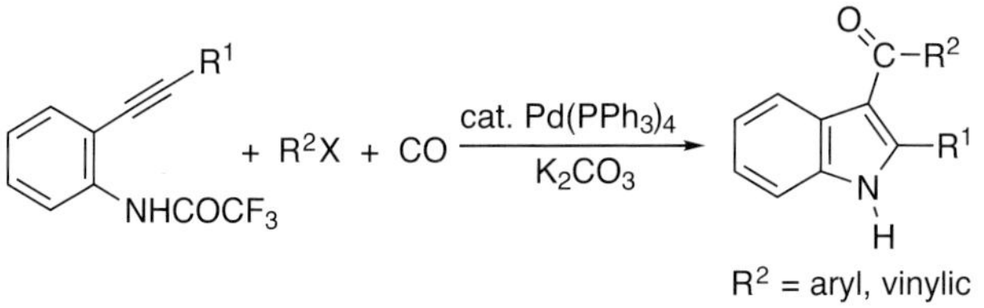

Equation 2-94

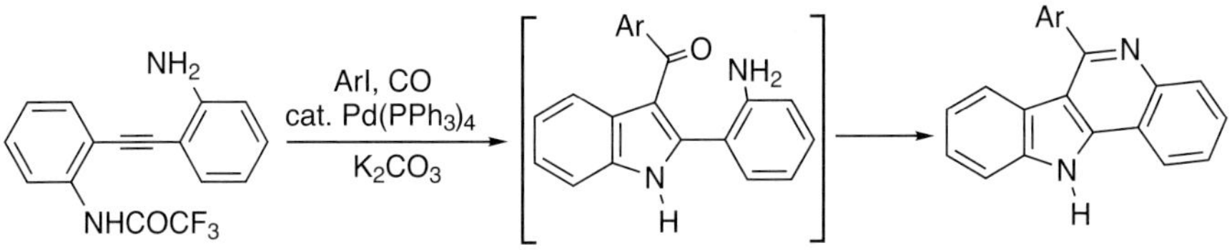

Equation 2-95

Bicyclic lactams have also been prepared by the Pd-catalyzed cross-coupling of simple acetylene-containing lactams and aryl or vinylic halides (Equation 2.96) [137].

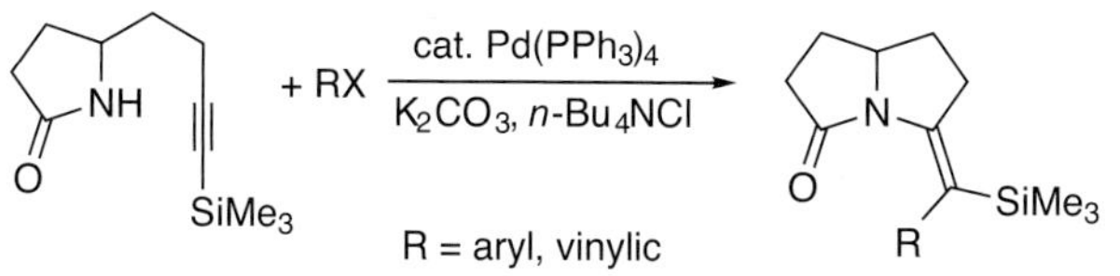

Equation 2-96

Finally, simple alkenynamides react readily with ICl to afford stereoisomeric mixtures of the corresponding five-membered ring iodolactams (Equation 2.97) [138].

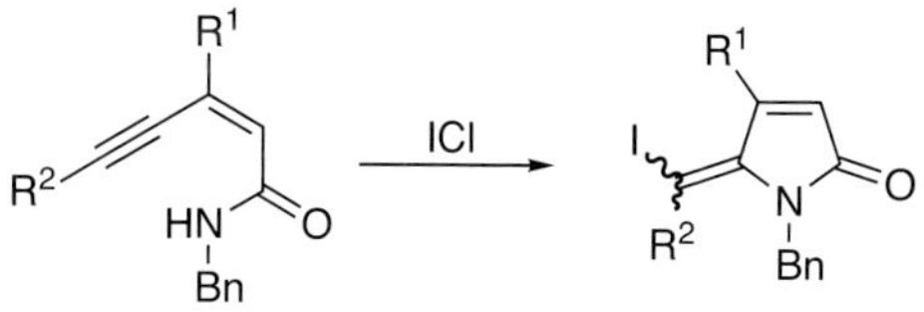

Equation 2-97

2.4.3
Cyclization of Acetylenic Carbamates

Many of the same reactions that have been carried out with amides have also been carried out with acetylenic carbamates. Thus, acetylenic Boc derivatives have been cyclized by Pd(II) to provide cyclic enamides (Equation 2.98) [106d]. Stoichiometric amounts of Mo and W compounds will also effect this same transformation (Equation 2.99) [105]. The Mo reagent has also been used to prepare *N*-Boc-indole [105].

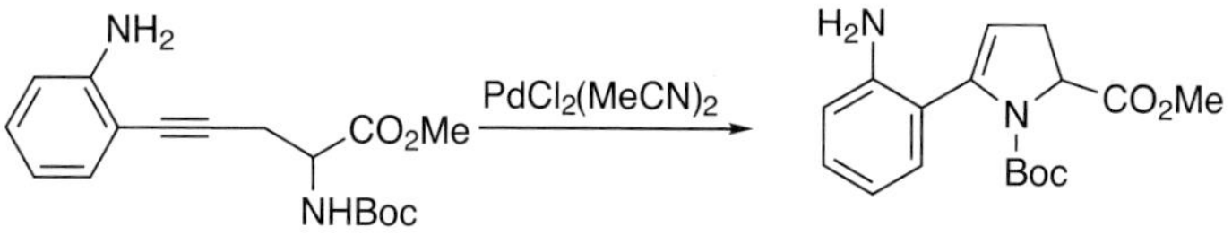

Equation 2-98

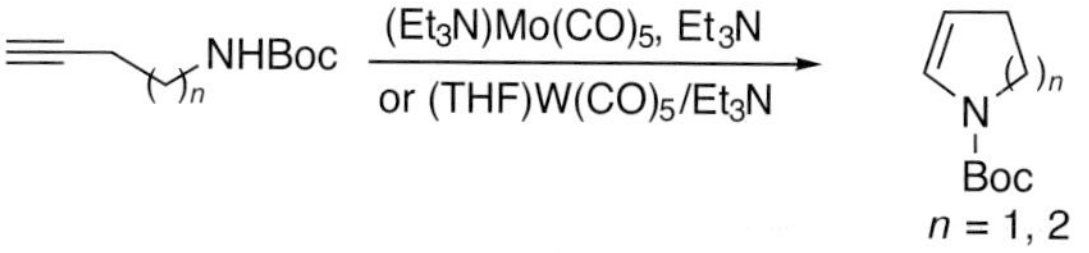

Equation 2-99

A number of *N*-alkoxycarbonyl-protected 2-(1-alkynyl)anilines have been cyclized to the corresponding indoles by use of cat. $PdCl_2$ [106c], $NaAuCl_4$ [106c], $Cu(OAc)_2$ [104, 139], $Cu(OTf)_2$ [104, 139], NaOEt in EtOH [140], and *n*-Bu_4NF [119] (Equation 2.100). 5-Azaindoles have been prepared in a similar manner by use of cat. CuI [141]. Silylalkynes are readily cyclized with desilylation by NaOEt in EtOH [140a, 142], KO-*t*-Bu in *t*-BuOH [142, 143], and *n*-Bu_4NF [119]. When the reactions with $PdCl_2$ are run in the presence of allylic halides [106c] or simple alkenes [144], 3-allylic or -vinylic indoles are obtained (Equation 2.101).

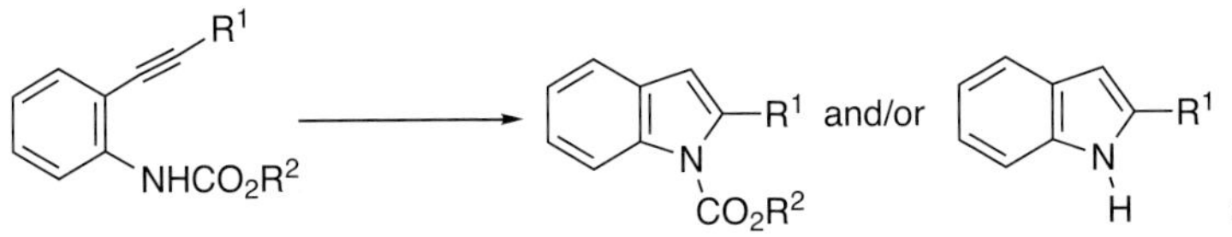

Equation 2-100

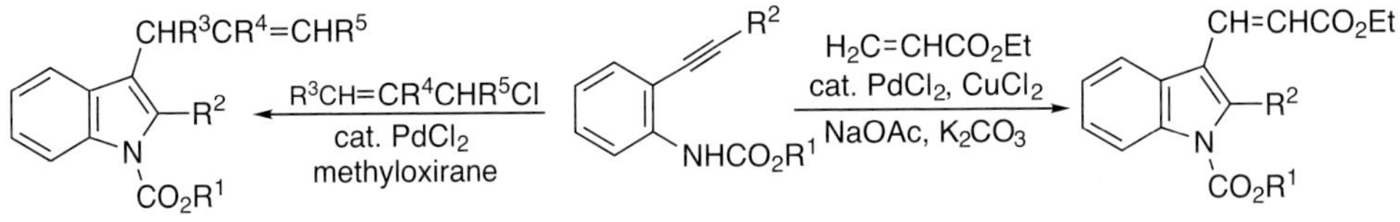

Equation 2-101

Propargylic tosylcarbamates react with cat. CuI or with aryl iodides/vinylic triflates and a palladium catalyst to afford oxazolidinones (Equation 2.102) [145]. These same substrates react with $Pd(OAc)_2$ and an enone to afford cross-coupled products (Equation 2.103) [146].

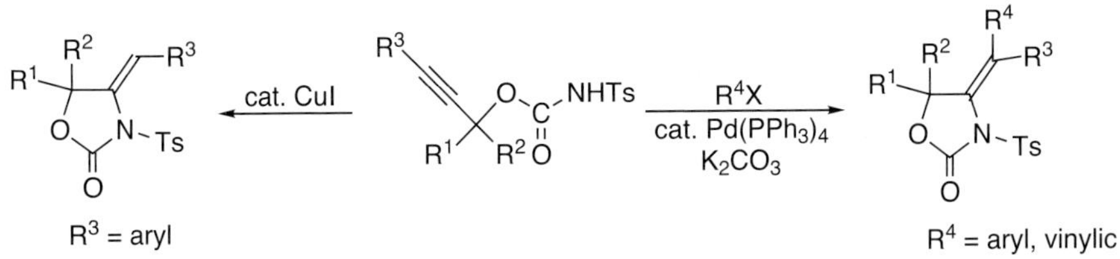

Equation 2-102

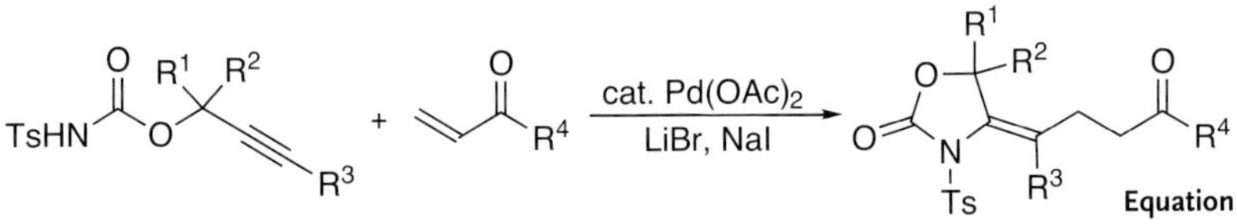

Equation 2-103

Finally, 3-iodoindole derivatives can be obtained in good yields by iodocyclization of the corresponding carbamates with IPy_2BF_4 plus HBF_4 (Equation 2.104) [25].

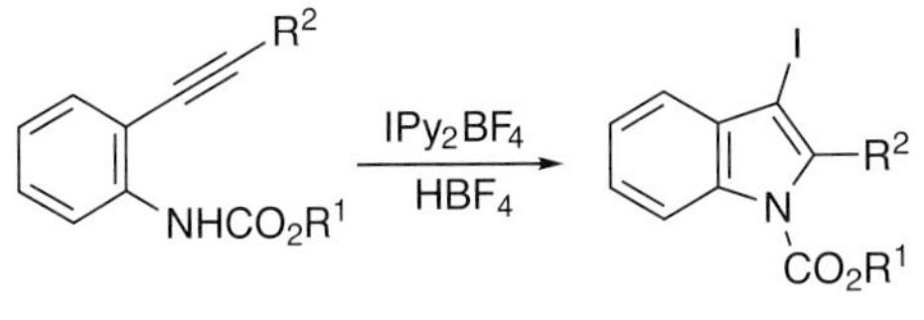

Equation 2-104

2.4.4
Cyclization of Acetylenic Sulfonamides

A number of examples of the cyclization of acetylenic sulfonamides to nitrogen heterocycles have been reported. Tosylamides of aminoesters are readily cyclized to five-membered ring heterocycles by various palladium catalysts (Equation 2.105) [147]. Analogous propargylic alcohols have been cyclized and dehydrated by TsOH to form pyrroles (Equation 2.106) [148]. Recently, acetylenic nonafluorobutanesulfonamides have been enantioselectively cyclized to five- and six-membered ring nitrogen heterocycles by palladium (Equation 2.107) [98, 99], while 2-(1-alkynyl)aniline sulfonamides are readily cyclized to the corresponding indoles by treatment with cat. $Cu(OAc)_2$ [104, 139], CuI [140f], $Cu(OTf)_2$ [104, 139] or 2–3 equiv. of n-Bu_4NF [119] (Equation 2.108).

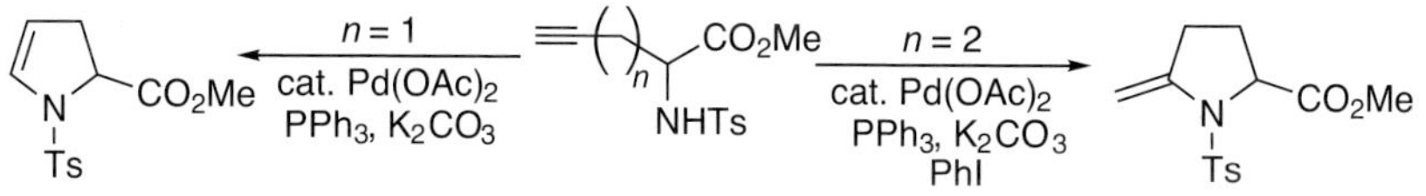

Equation 2-105

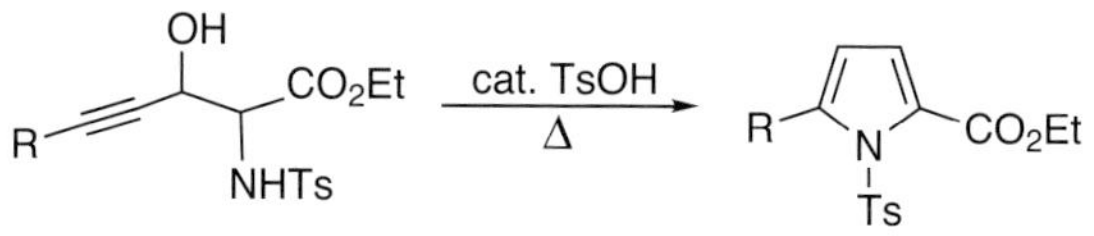

Equation 2-106

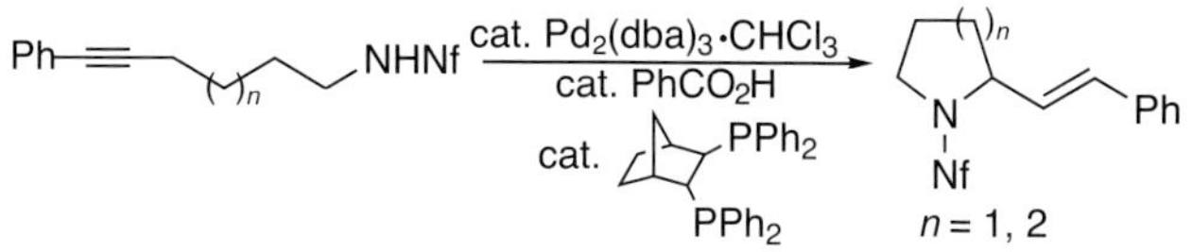

Equation 2-107

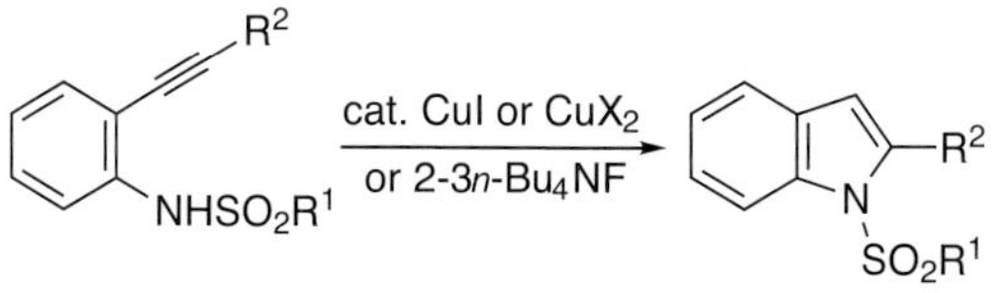

Equation 2-108

The iodocyclization of simple acetylenic tosylamides by I_2 (Equation 2.109) [149] and sulfonamide derivatives of 2-(1-alkynylanilines) by I_2 plus K_2CO_3 [150] or IPy_2BF_4 plus HBF_4 (Equation 2.110) [25] provides easy access to the corresponding five-membered ring heterocycles.

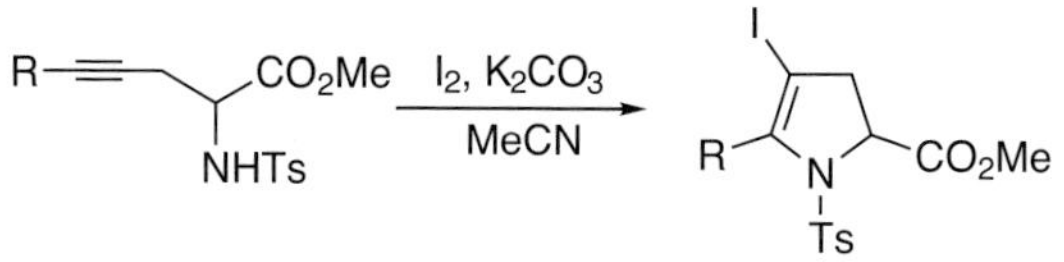

Equation 2-109

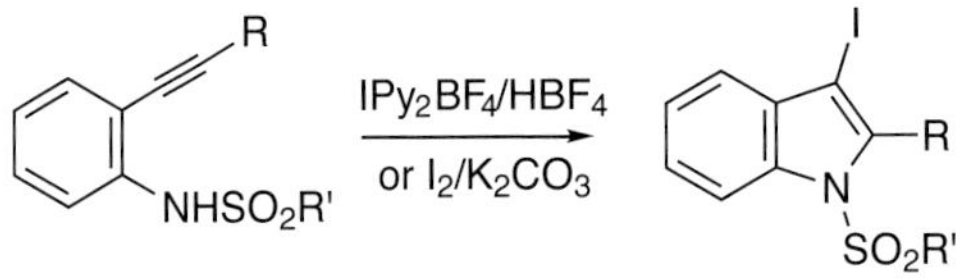

Equation 2-110

Acetylenic tosylamides have been cross-coupled with aryl and vinylic halides or triflates in the presence of a Pd catalyst to produce tetrahydropyrroles (Equation 2.111) [147a,b], alkylidenepyrrolidines and -piperidines (Equation 2.112) [151], and tetrahydroquinoxalines (Equation 2.113) [152].

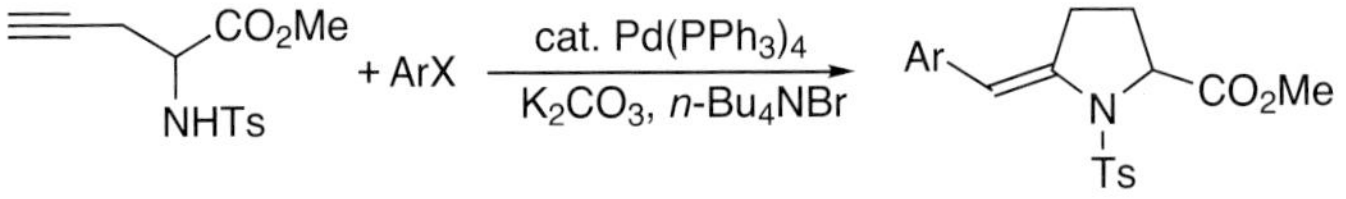

Equation 2-111

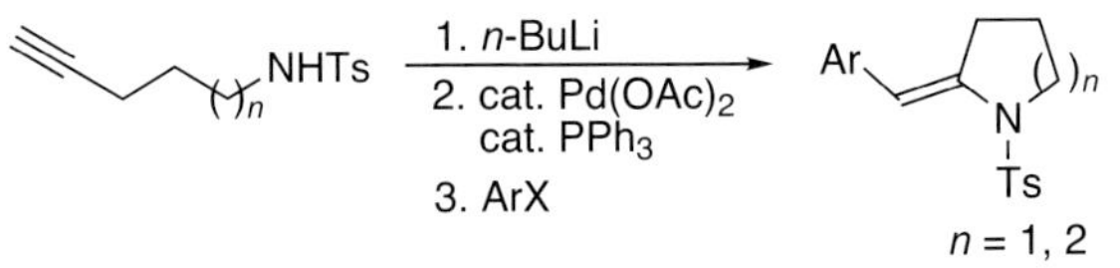

Equation 2-112

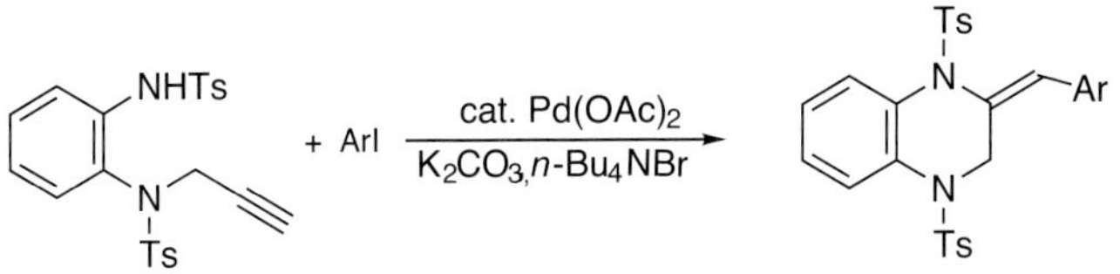

Equation 2-113

The cyclization of methanesulfonamides of 2-(1-alkynyl)anilines by Pd(II) salts in the presence of olefins [144] or CO [30c,d] provides a very useful route to 3-substituted indoles (Equation 2.114).

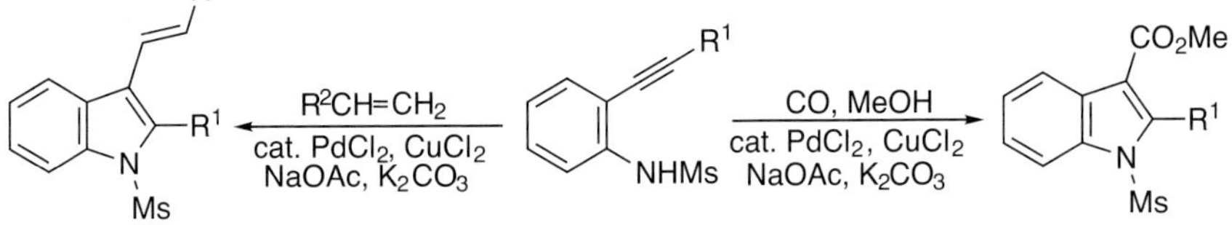

Equation 2-114

2.4.5
Cyclization of Acetylenic Enamines and Imines

A number of different nitrogen heterocycles have been prepared by the cyclization of acetylenic enamines or imines. For example, ketones react with propargylamine to produce enamines/imines that readily cyclizable to pyridines in the presence of a copper or gold catalyst (Equation 2.115) [153]. In a similar manner, acetylene-containing 1,3-dicarbonyl compounds react with primary amines in the presence of a gold catalyst to generate pyrroles (Equation 2.116) [154].

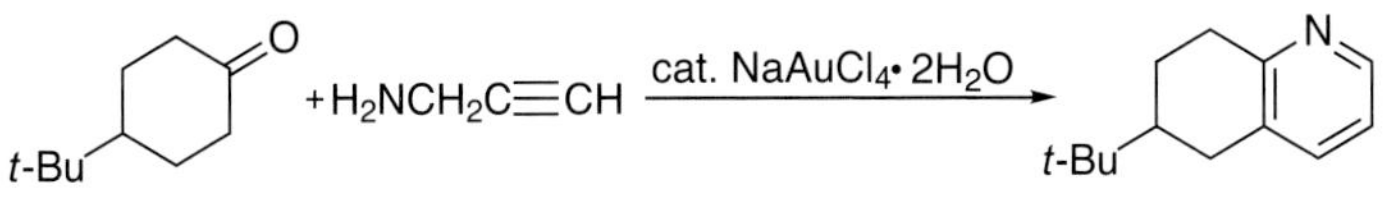

Equation 2-115

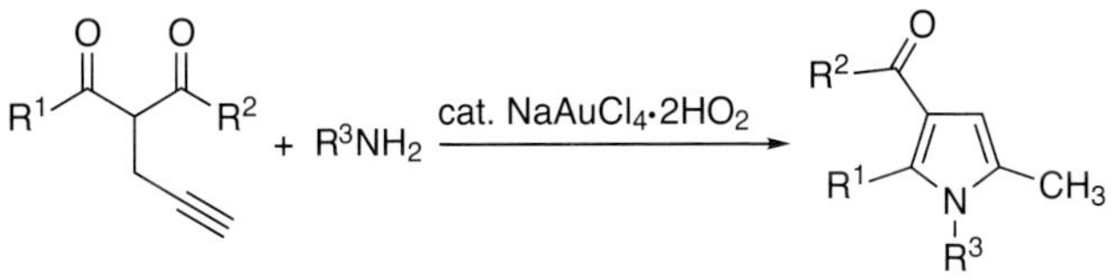

Equation 2-116

Copper catalyzes the cyclization of imines derived from 2-(1-alkynyl)anilines (Equation 2.117) [155], while acetylenic ketones have been cyclized to naphthylamines in the presence of an amine and a palladium/copper catalyst (Equation 2.118) [156].

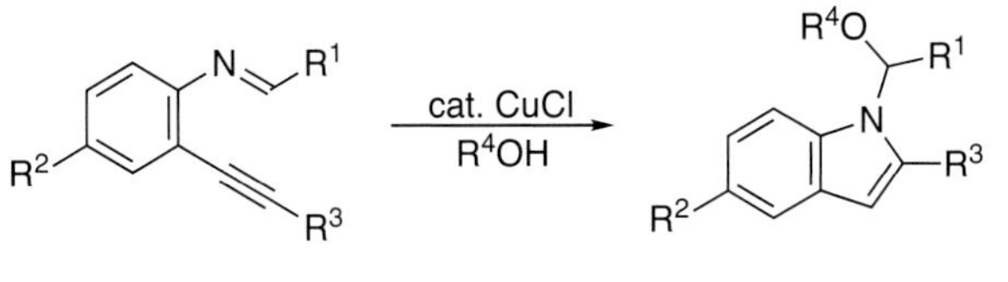

Equation 2-117

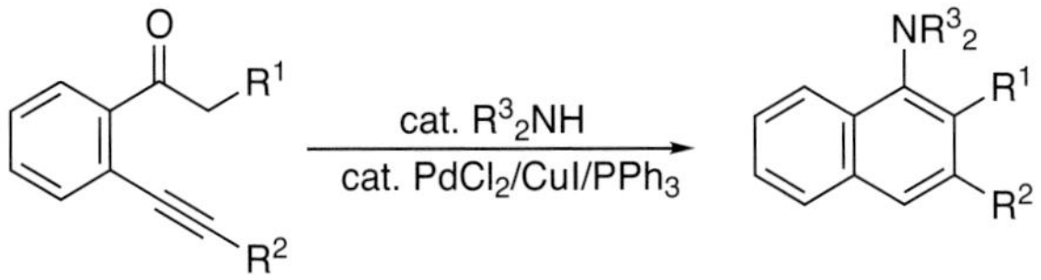

Equation 2-118

Indole-containing acetylenic ketones react with ammonia to afford pyrazinoindoles (Equation 2.119) [157].

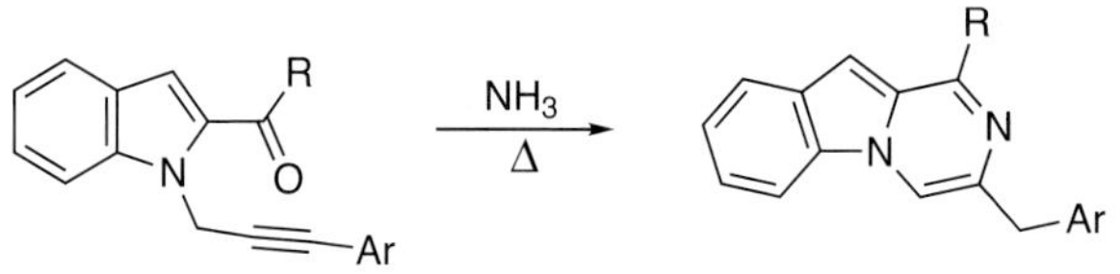

Equation 2-119

A wide variety of 4-substituted isoquinolines have been prepared by the cyclization of the imines of 2-(1-alkynyl)benzaldehydes. NH_3, for example, reacts with these aldehydes to afford simple isoquinolines (Equation 2.120) [158]. We have observed that analogous *tert*-butylimines can be cyclized to isoquinolines with loss of the *tert*-butyl group by use of catalytic amounts of CuI [159] or $AgNO_3$ [159b, 160] (Equation 2.121). Similarly, the thermal or CuI-catalyzed cyclization of *tert*-butylimines of indole-containing acetylenes has been employed to prepare β- and γ-carbolines [161]. These same imines also cyclize under very mild reaction conditions when treated with I, S, and Se electrophiles (Equation 2.121) [159b, 160].

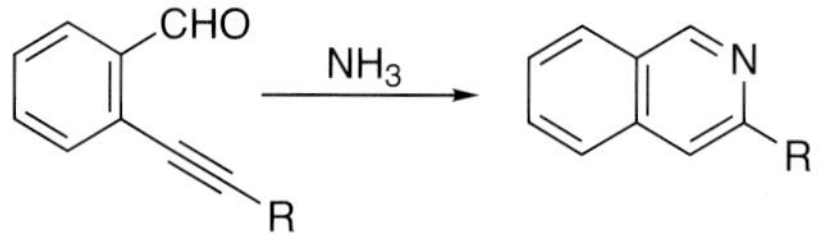

Equation 2-120

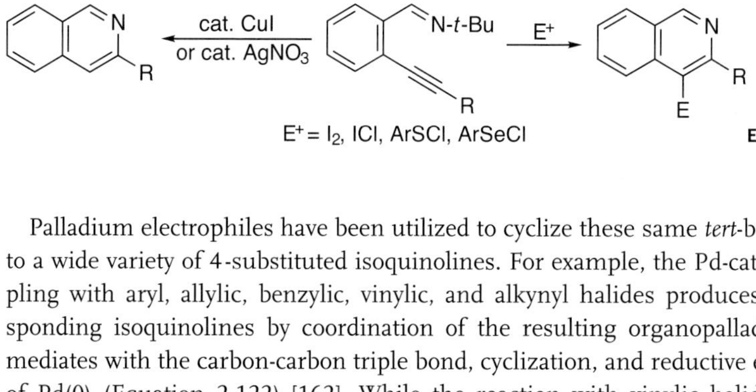

Equation 2-121

Palladium electrophiles have been utilized to cyclize these same *tert*-butyl imines to a wide variety of 4-substituted isoquinolines. For example, the Pd-catalyzed coupling with aryl, allylic, benzylic, vinylic, and alkynyl halides produces the corresponding isoquinolines by coordination of the resulting organopalladium intermediates with the carbon-carbon triple bond, cyclization, and reductive elimination of Pd(0) (Equation 2.122) [162]. While the reaction with vinylic halides did not prove to be very general, the 4-vinylic isoquinolines can be readily prepared by employing of $PdBr_2$ as a catalyst in the presence of olefins (Equation 2.123) [163]. This process proceeds by Pd(II)-induced cyclization, followed by a Heck reaction. Finally, the cross-coupling of these same imines with aryl halides and CO provides the corresponding 4-aroylisoquinolines (Equation 2.124) [164].

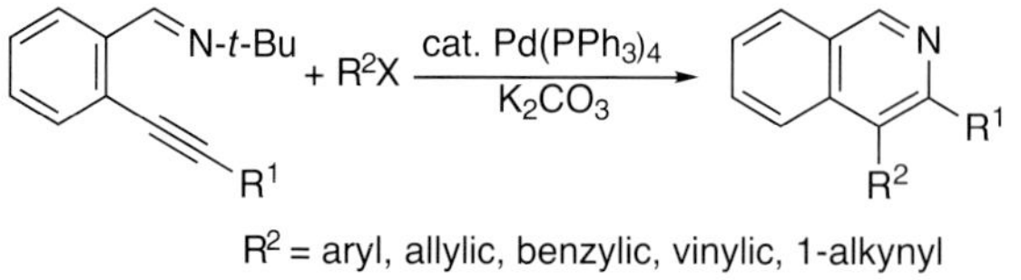

Equation 2-122

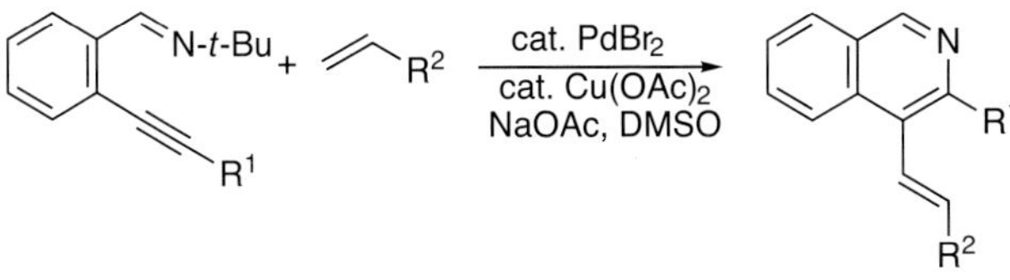

Equation 2-123

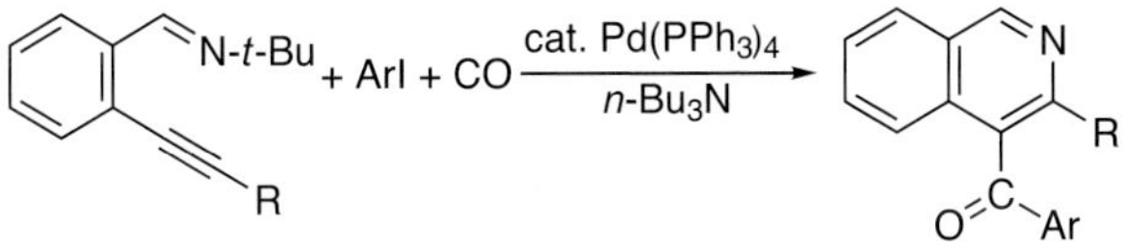

Equation 2-124

2.4.6
Cyclization of Other Acetylenic Nitrogen Functional Groups

A variety of nitrogen heterocycles have been formed by the cyclization of acetylenes containing other nitrogen functional groups. Six-membered ring nitrogen heterocycles with either endocyclic or exocyclic double bonds bearing halogens have been formed by the cyclization of *N,O*-acetals of 3- or 4-alkynylamines, respectively (Equation 2.125) [165]. Analogous products have been formed by starting with the corresponding secondary amine and an aldehyde plus a sulfonic acid and NaI or n-Bu_4NX (X = Br, I) [166]. The X group introduced can also be N_3, SCN, or SPh [167]. This chemistry has been employed in the synthesis of pumiliotoxins [168].

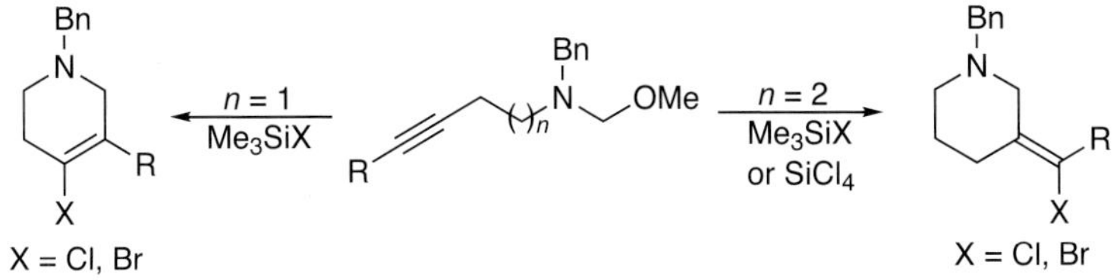

Equation 2-125

Highly substituted pyrroles have been obtained by the palladium-catalyzed cross-coupling of aryl halides and acetylenic tosylhydrazones (Equation 2.126) [169].

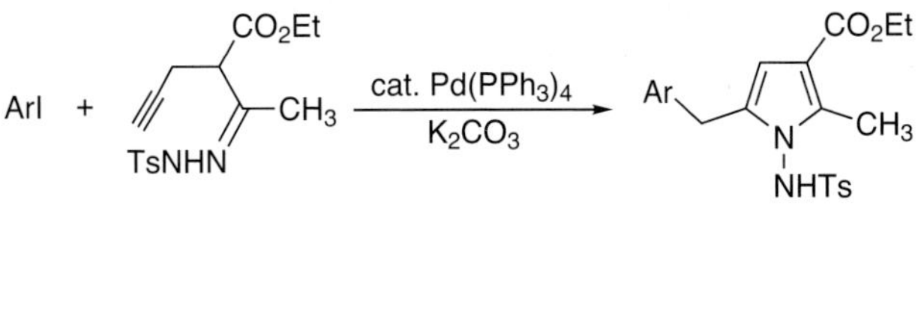

Equation 2-126

N-(Alkoxycarbonyl)indoles have been synthesized by reactions between 2-(1-alkynyl)phenylisocyanates and alcohols in the presence of catalytic amounts of Na_2PdCl_4 or $PtCl_2$ (Equation 2.127) [123]. These same alkynes react with allyl carbonates in the presence of catalytic amounts of $Pd(PPh_3)_4$ and CuCl to afford 3-allylindoles (Equation 2.127) [170, 171]. Related isonitriles react with allyl methyl carbonate to form analogous indoles (Equation 2.128) [172].

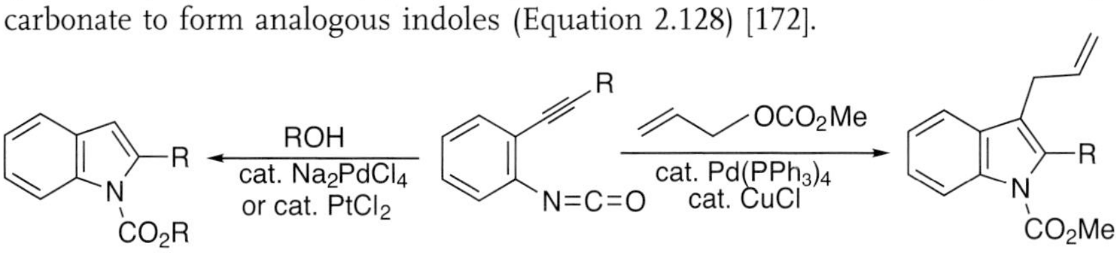

Equation 2-127

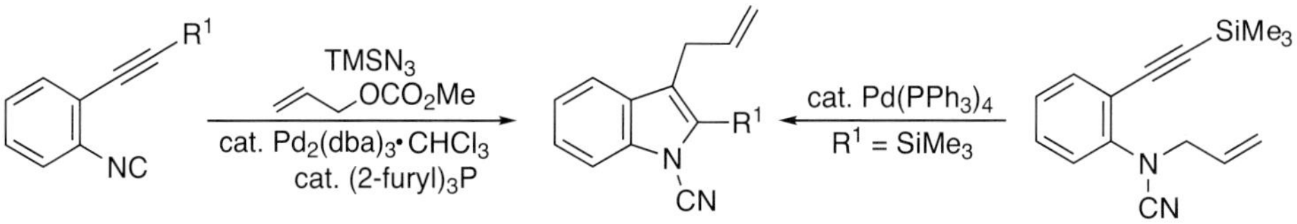

Equation 2-128

Analogous diazonium salts [173] and nitro compounds [174] have been cyclized to five-membered ring nitrogen heterocycles (Equations 2.129 and 2.130), while analogous nitriles have been cross-coupled with aryl halides to produce five- or six-membered ring heterocycles (Equation 2.131) [175].

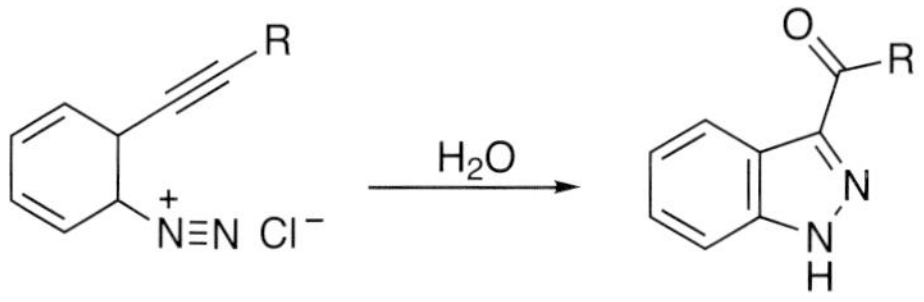

Equation 2-129

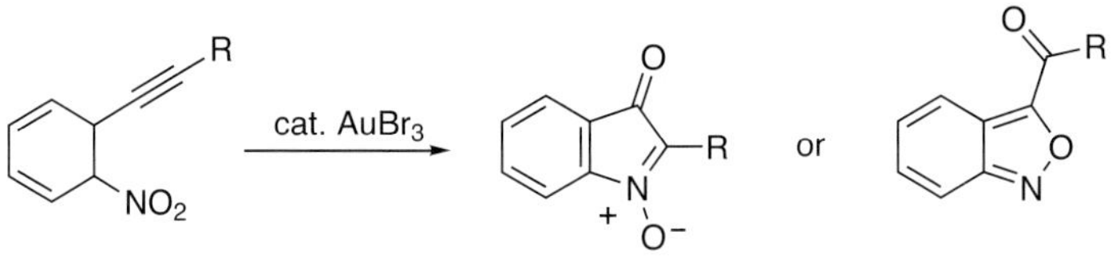

Equation 2-130

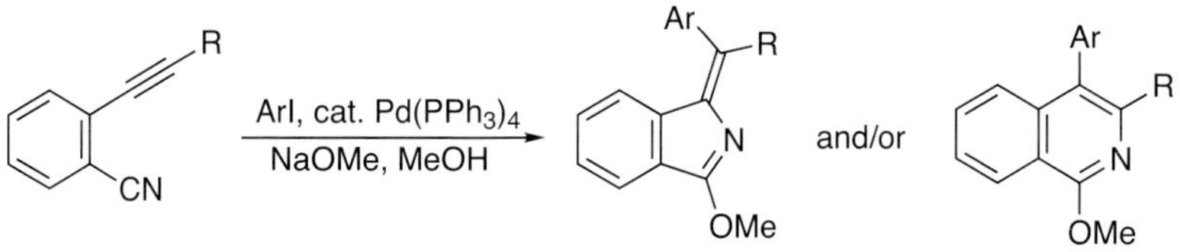

Equation 2-131

Quinoline derivatives have been prepared by the base-promoted cyclization of 2-(1-alkynyl)benzaldehyde oximes [176] and hydrazones [177] (Equation 2.132).

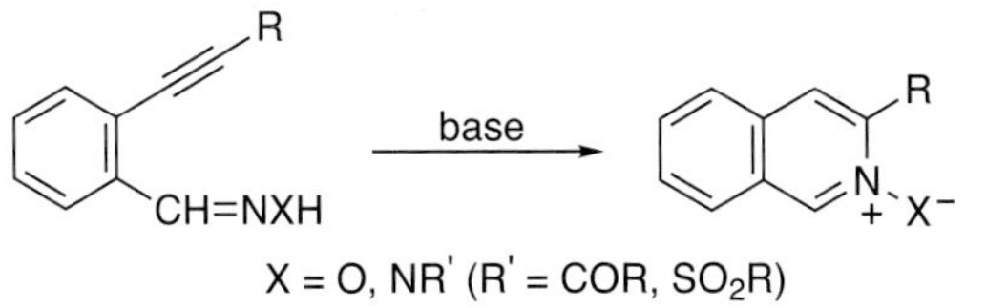

Equation 2-132

Finally, acetylenic triazin-5(2*H*)-ones have been cyclized to isomeric triazinones by treatment with a Pd catalyst (Equation 2.133) [178].

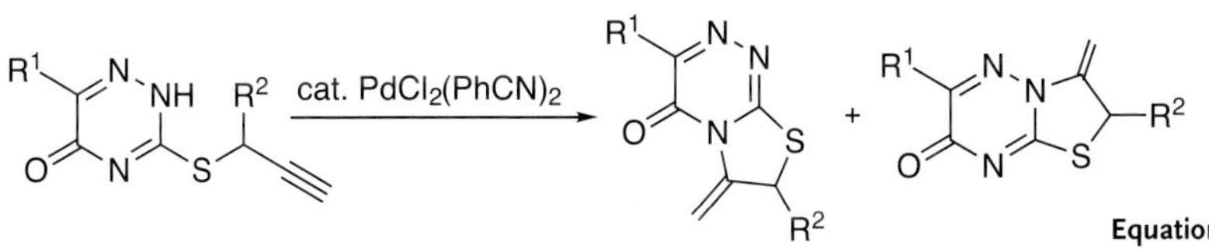

Equation 2-133

2.5
Cyclization of Carbon onto Acetylenes

2.5.1
Cyclization of Acetylenic Carbonyl Compounds and Derivatives.

A wide number of carbocycles have been prepared by the cyclization of carbon onto acetylenes. Unfortunately, because of the large amount of work that has been done and limitations of space, none of the work carried out on intramolecular nucleophilic or free radical additions to alkynes or the metal-catalyzed cycloaddition, cyclization, cycloisomerization, or metathesis of enynes and diynes is covered here. This section focuses primarily on the cyclization of carbonyl compounds and arenes onto acetylenes and a few related palladium-catalyzed cyclizations. Note the earlier examples of carbon cyclization onto acetylenes covered in Sections 2.4.5 and 2.4.6.

The cyclization of carbonyl compounds onto acetylenes can be quite useful synthetically. Although the Conia-ene reaction effects just such a transformation, this process is limited by the high temperatures required (Equation 2.134) [179]. More recently, β-dicarbonyl and related compounds have been cyclized, primarily to five-membered rings, by use of $HgCl_2/HCl$ [180], $ZnCl_2/Et_3N/H_3O^+$ [181], $TiCl_4/Et_3N/HCl$ [181], CuI/KO-*t*-Bu [182], $(Et_3N)Mo(CO)_5/NaH/h\nu$ [183], $CpCo(CO)_2/h\nu$ [184], Pd(dppe)/KO-*t*-Bu [185], and $AuCl(PPh_3)$ [186] (Equation 2.135).

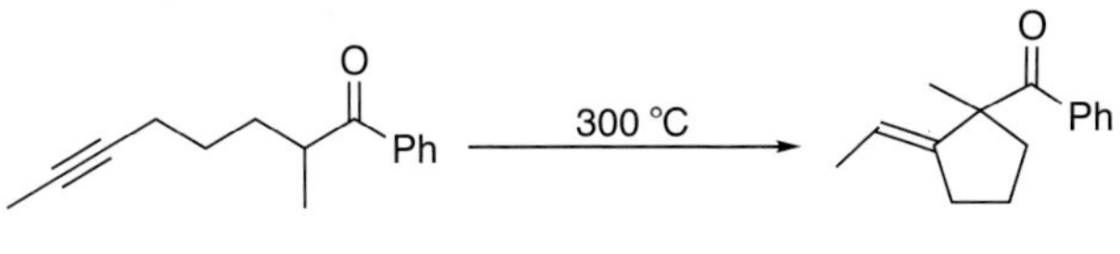

Equation 2-134

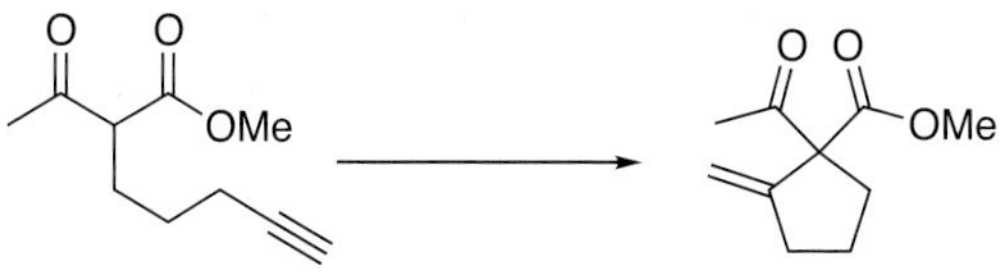

Equation 2-135

One can combine a Michael addition with this cyclization to prepare heterocycles (Equation 2.136) [187]. The analogous palladium-catalyzed cross-coupling of acetylenic dicarbonyl compounds and aryl or vinylic halides affords carbocycles (Equation 2.137) [188]. 1-Halo-1-alkynes can also be employed in this latter reaction if a Pd/Cu catalyst is utilized [189].

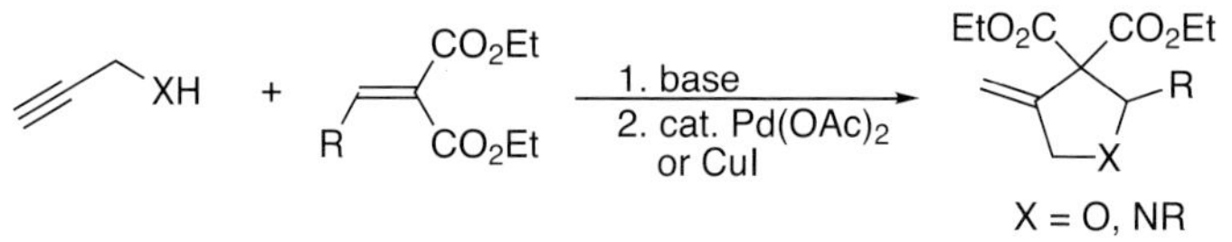

Equation 2-136

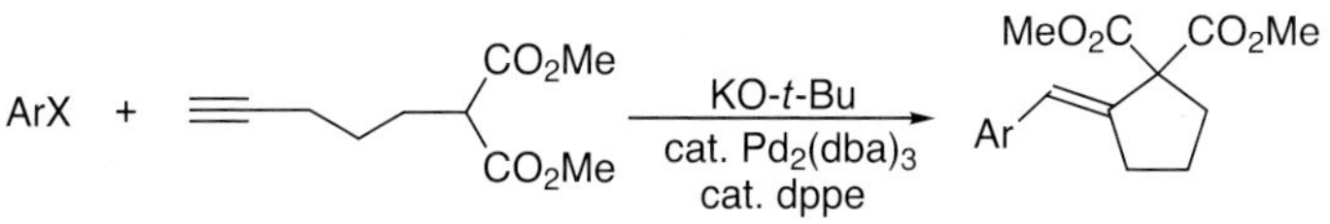

Equation 2-137

The stereoselective iodocyclization of acetylenic diesters can also be effected by use of titanium chemistry (Equation 2.138) [181].

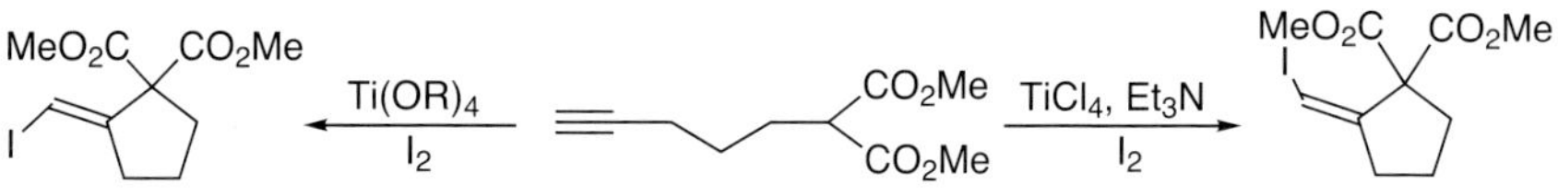

Equation 2-138

Derivatives of carbonyl compounds can also be cyclized onto acetylenes. Metal salts of Pt(II), Pd(II), Au(III), and Cu(I), for example, will cyclize acetylenic vinylic ethers in the presence of methanol to produce five- and six-membered ring carbocyclic acetals (Equation 2.139) [190]. Acetylenic enol silanes have been cyclized by treatment with catalytic amounts of $PtCl_2$ [188] or $W(CO)_5(THF)$ [191] (Equation 2.140). Isomeric naphthalenes can be obtained by the cyclization of enol silanes with various transition metal catalysts (Equation 2.141) [192]. Acetylene-containing enol silanes also smoothly cyclize in the presence of $HgCl_2$ to afford vinylmercurials, which have subsequently been cleaved by protic acids or halogenating agents (NBS, NIS) or carbonylated to esters (CO, MeOH, $PdCl_2$, $CuCl_2$, LiCl) to afford a wide variety of cyclic, bicyclic, and spirocyclic unsaturated ketones and several natural products (Equation 2.142) [193]. The *endo* cyclization of enol silanes has been accomplished by use of $EtAlCl_2$ (Equation 2.143) [194]. Acetylene-containing enamines formed *in situ* from propargylamine and ketones can be readily cyclized to pyridines as noted earlier (see Section 2.4.5).

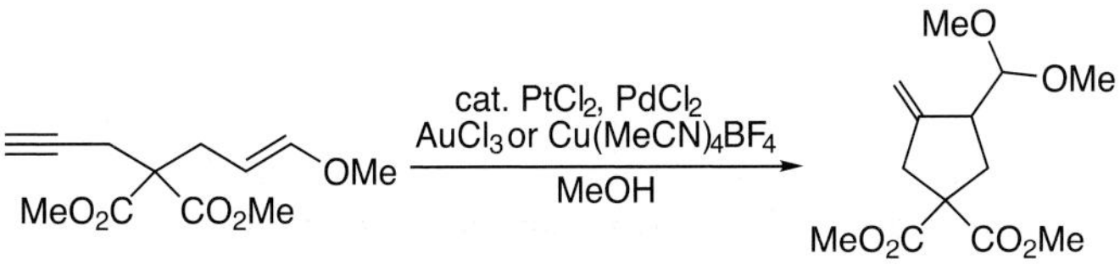

Equation 2-139

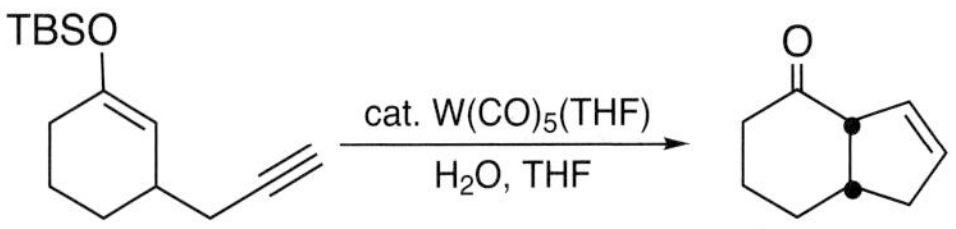

Equation 2-140

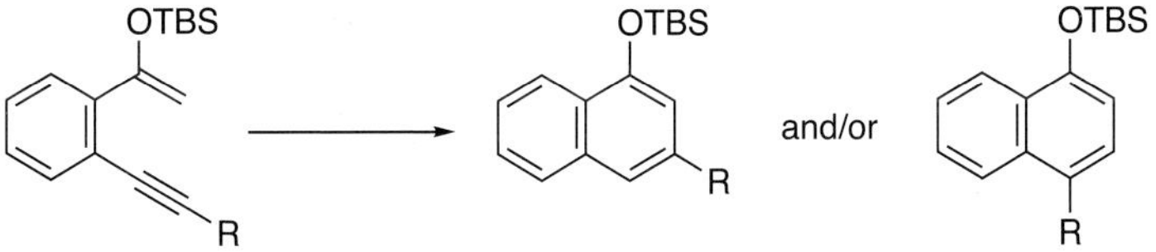

Equation 2-141

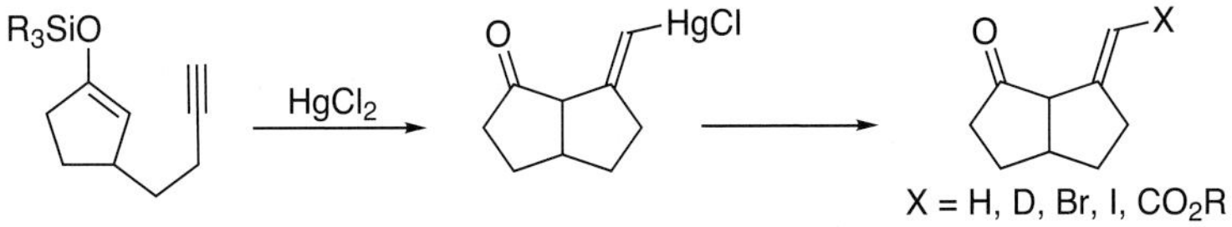

Equation 2-142

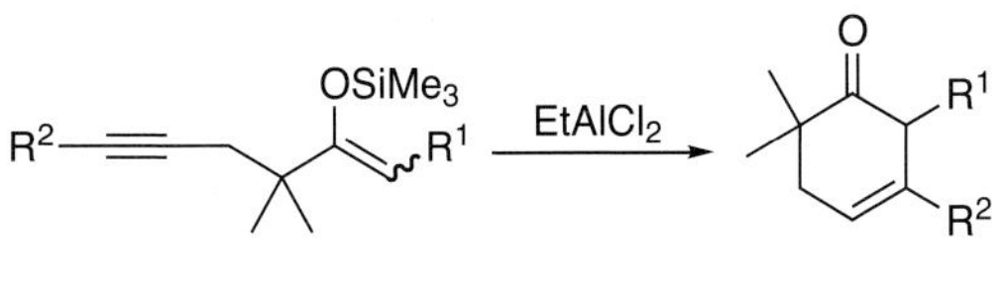

Equation 2-143

2.5.2
Cyclization of Diacetylenes

Although there are many transition metal-catalyzed reactions of diynes, there appears to be only one example of the electrophilic cyclization of an alkyne onto another alkyne, which produces sulfur heterocycles (Equation 2.144) [195].

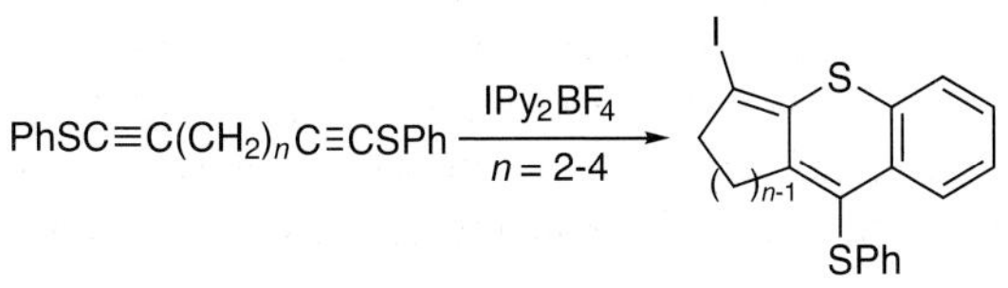

Equation 2-144

2.5.3
Cyclization of Aryl Acetylenes

There are a number of examples of acetylenic arenes being cyclized by various metals and electrophiles to produce aromatic carbocycles and heterocycles. For example, catalytic amounts of $GaCl_3$ [196], $PtCl_2$ [197], $[RuCl_2(CO)_3]_2$ [197], and $[RuCl_2(CO)_3]_2$ plus AgOTf [197] cyclize 5-aryl-1-alkynes to dihydronaphthalenes (Equation 2.145). $Hg(OTf)_2$ plus 3 equiv. of tetramethylurea (TMU) catalyzes the cyclization of acetylenic arenes to 1,2-dihydronaphthalenes, benzopyrans, and 1,2-dihydroquinoline analogues (Equation 2.146) [198]. Several groups have reported the cyclization of aryl propargylic ethers to benzopyrans by other Hg(II) salts in the presence of strong acids (Equation 2.147) [199]. By variation of the Hg salt and reaction conditions, vinylic mercury compounds corresponding to 1,2-dihydronaphthalenes and benzopyrans can be isolated in low yields [36b]. Catalytic amounts of $PtCl_2$ or $PtCl_4$ efficiently produce analogous benzopyrans, 1,2-dihydroquinoline derivatives, and coumarins from acetylenic starting materials [200], while catalytic amounts of $Pd(OAc)_2$ in trifluoroacetic acid also nicely cyclize acetylenic esters or amides to coumarins or 2(1*H*)-quinolinones, respectively (Equation 2.148) [201]. Finally, strong acids will cyclize 1,3-diarylpropynones to 3-arylindenones (Equation 2.149) [202].

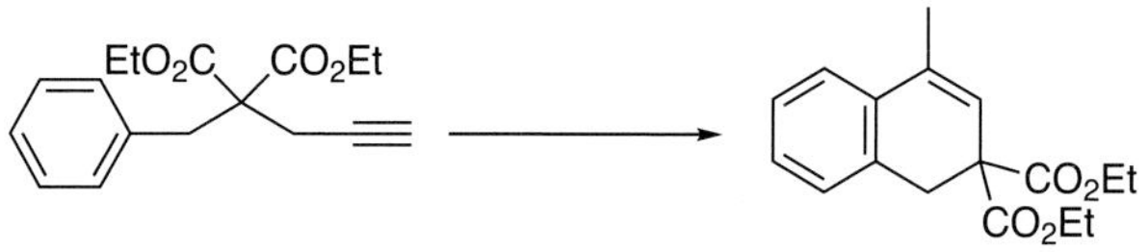

Equation 2-145

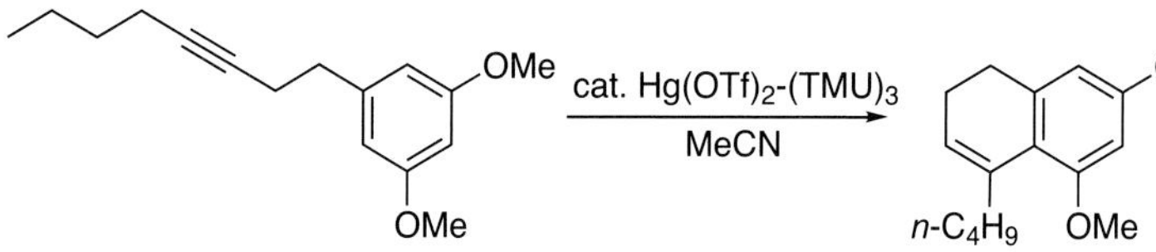

Equation 2-146

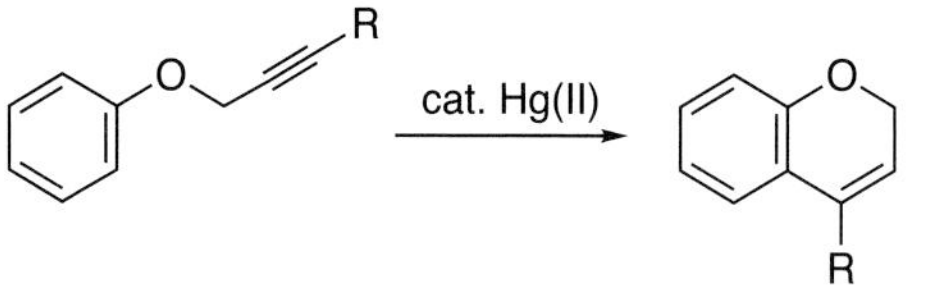

Equation 2-147

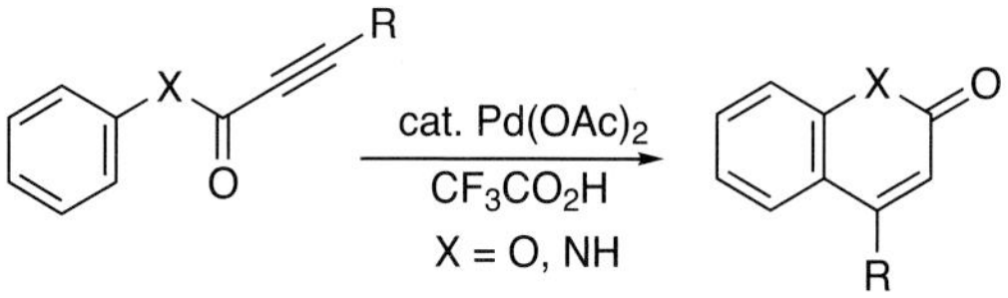

Equation 2-148

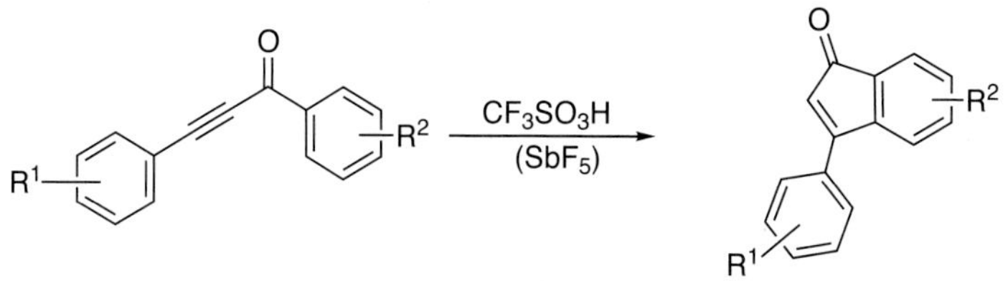

Equation 2-149

Other heterocycles have been prepared in a similar manner. Thus, a perinaphthothioxanthene (Equation 2.150) [203] and quinolines (Equation 2.151) [204] have been prepared by transition metal-catalyzed cyclizations.

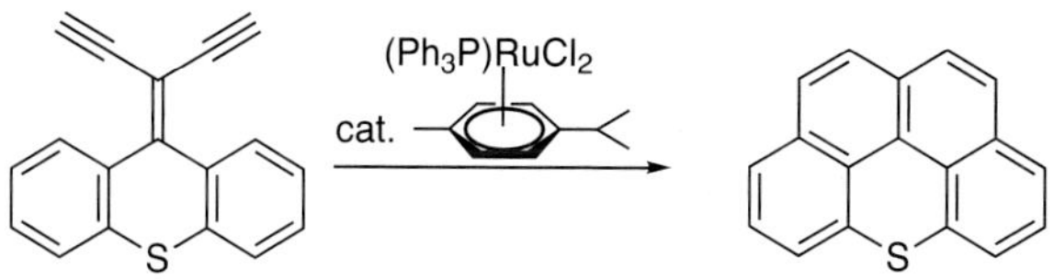

Equation 2-150

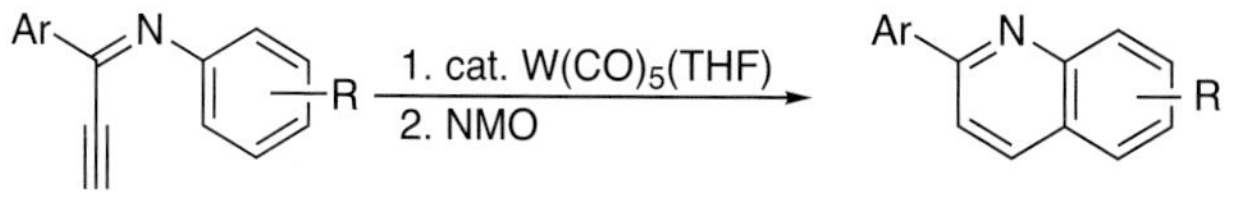

Equation 2-151

Biaryls substituted by an acetylene unit in the *ortho* position can be cyclized to phenanthrenes and heterocyclic analogues by use of catalytic amounts of $PtCl_2$, $AuCl_3$, $GaCl_3$, or $InCl_3$ (Equation 2.152) [205]. Trifluoroacetic acid also effects this same cyclization [206].

E^+ = H(cat. $PtCl_2$, $AuCl_3$, $GaCl_3$, or $InCl_3$, or CF_3CO_2H)
 = I (IPy_2BF_4 or ICl)

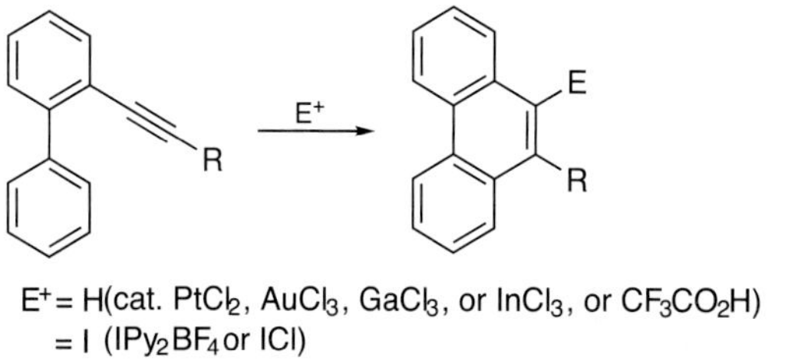

Equation 2-152

The iodocyclization of aryl acetylenes provides a very useful route to various ring systems. Thus, the reaction of acetylenic biaryls and IPy_2BF_4 affords iodophenanthrenes, but apparently requires an alkoxy-substituted arene on the remote end of the acetylene (Equation 2.152) [206a,b]. The same reagent has been reported to cyclize 1,4-diphenyl-1-butyne to 3-iodo-4-phenyl-1,2-dihydronaphthalene [207] and an acetylenic indole derivative [25]. We have found that ICl is a more general reagent for the synthesis of iodophenanthrenes and that it efficiently produces a variety of heterocyclic analogues (Equation 2.152) [208].

A variety of polycyclic iodoarenes can readily be prepared by the room temperature cyclization of arene-containing acetylenic alcohols (Equation 2.153) [209]. When aryl propargylic amines are allowed to react with I_2, 3-iodoquinolines are produced in good yields (Equation 2.154) [209]. When the aromatic ring bears a methoxy group in the *para* position, azaspirodienones are produced instead (Equation 2.155) [209]. Sulfur-substituted acetylenes undergo analogous halocyclizations (Equation 2.156) [210].

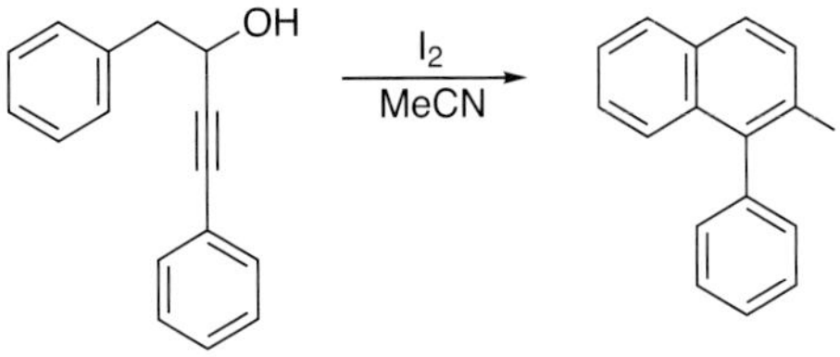

Equation 2-153

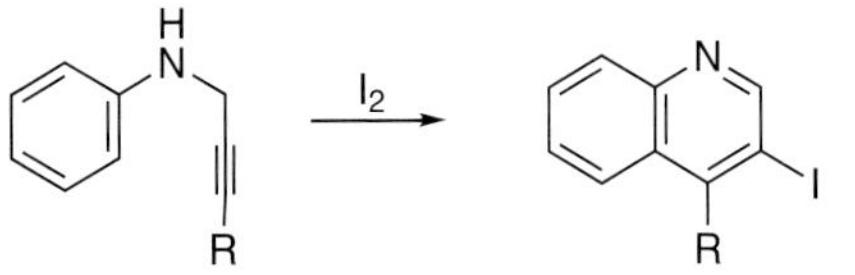

Equation 2-154

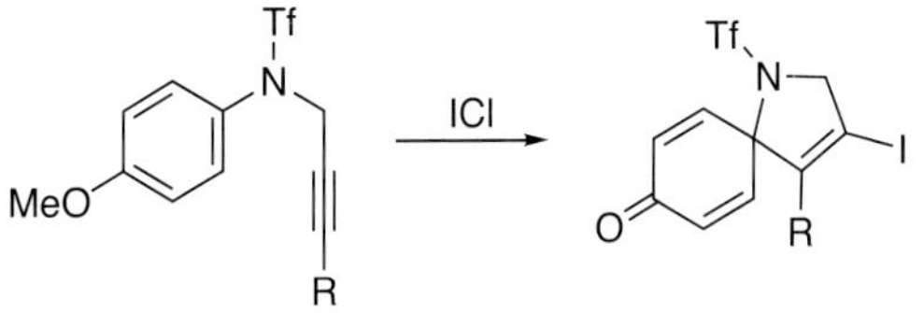

Equation 2-155

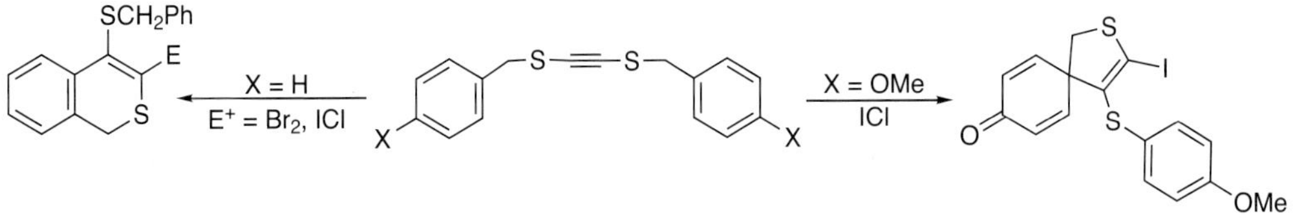

Equation 2-156

o-Phenoxybiaryls react with a variety of electrophiles to produce the corresponding seven-membered ring heterocycles (Equation 2.157) [211].

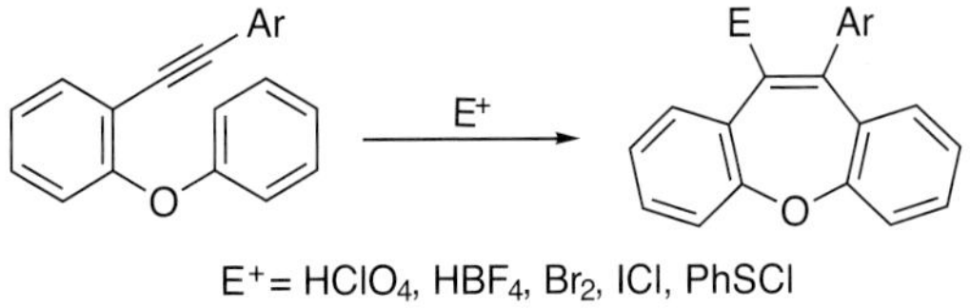

Equation 2-157

Furan-containing acetylenes have been cyclized to a variety of products. For example, 5-(2-furyl)-1-alkynes have been cyclized to phenols in the presence of catalytic amounts of $AuCl_3$ [212] or $PtCl_2$ [213] (Equation 2.158), and an enyne-containing furan has been cyclized to the corresponding benzofuran (Equation 2.159) [214].

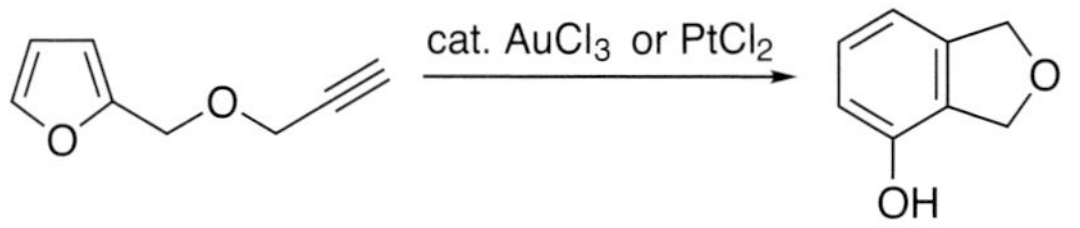

Equation 2-158

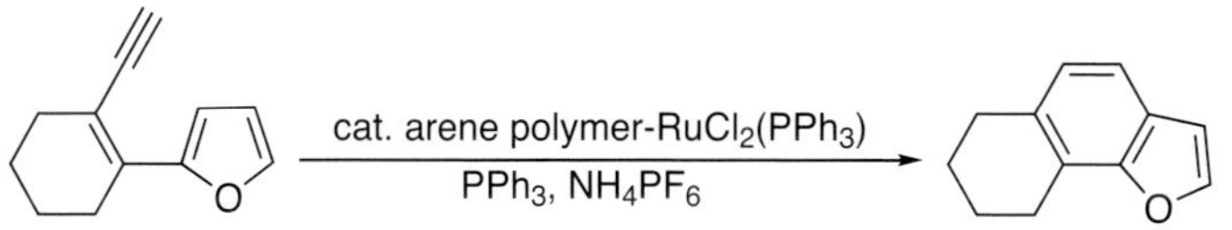

Equation 2-159

Intramolecular acetylene additions of organopalladium compounds generated by the oxidative addition of organic halides to Pd(0) have provided a very useful procedure for the formation of a wide variety of carbocyclic and heterocyclic alkenes [215]. The resulting vinylic palladium intermediates can be trapped by: (1) reducing agents (Equation 2.160) [216]; (2) organoboron [217], -tin [218], and -zinc [217a] reagents (Equation 2.161); (3) internal arenes [219] (Equation 2.162); (4) internal alkenes [220] (Equation 2.163); and (5) internal alkynes [221] (Equation 2.164).

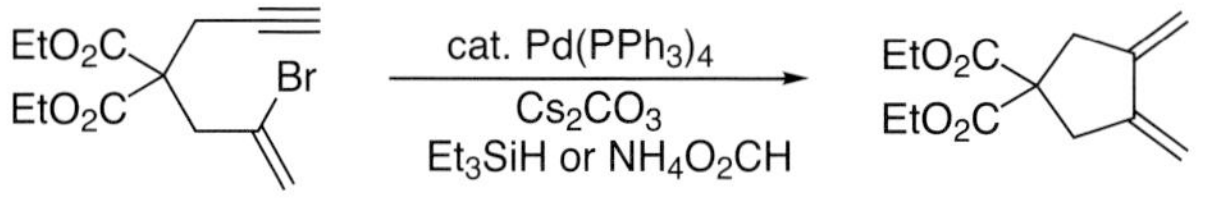

Equation 2-160

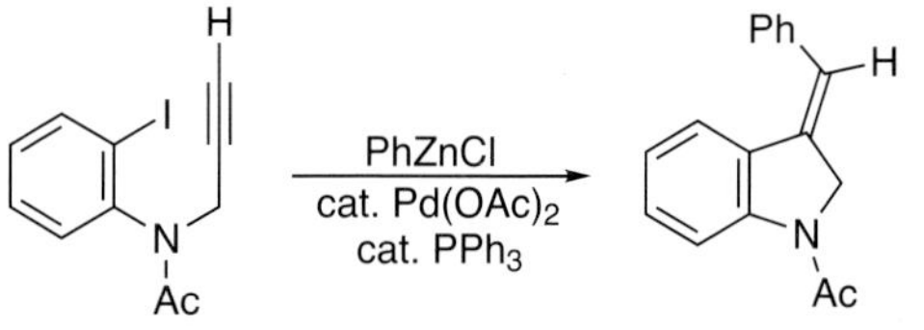

Equation 2-161

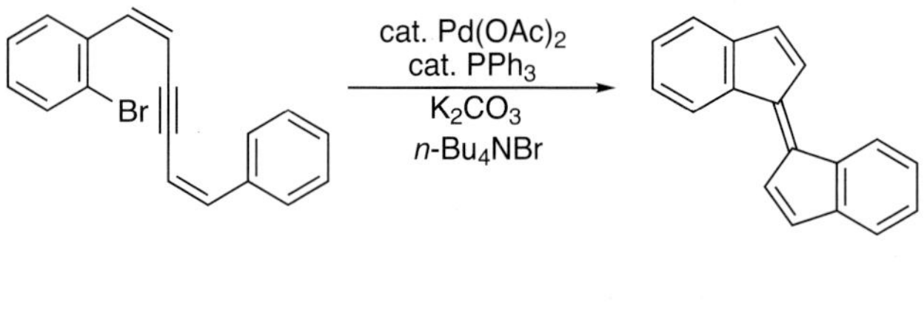

Equation 2-162

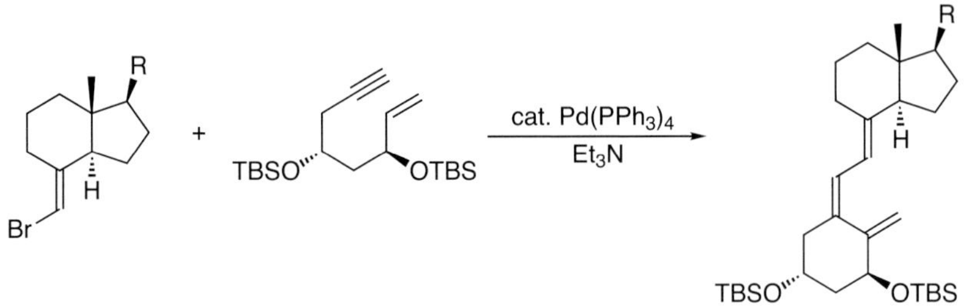

Equation 2-163

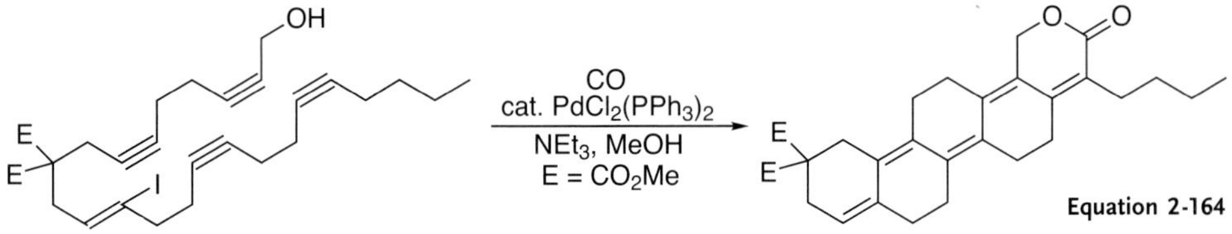

Equation 2-164

Intermolecular organopalladium additions to alkynylcyclobutanols effect ring expansion to 2-alkylidenecyclopentanones (Equation 2.165) [222].

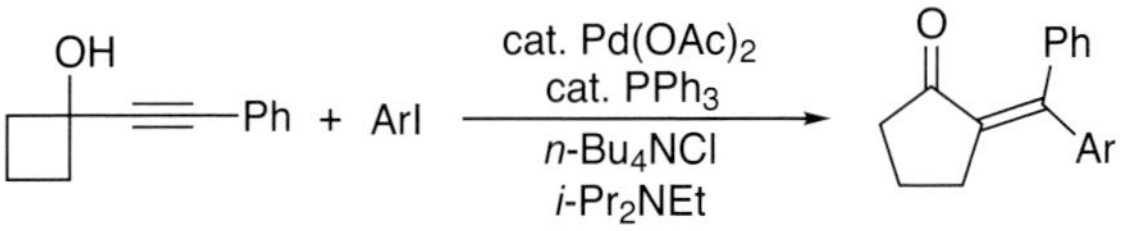

Equation 2-165

Palladium(II) salts have also been employed to cyclize acetylenic esters onto neighboring olefins (Equation 2.166) [223] and dienes (Equations 2.167 [224] and 2.168 [225]).

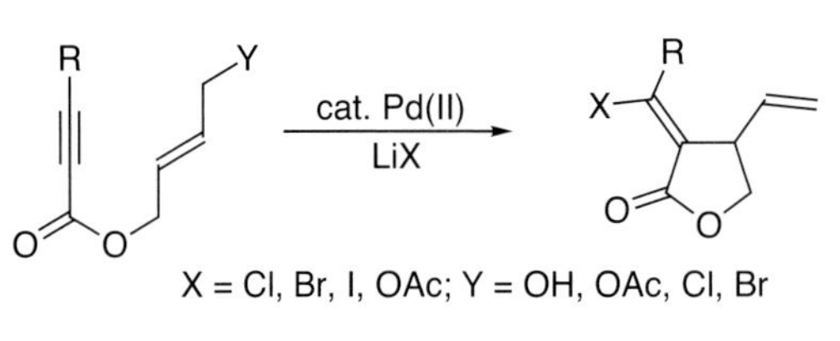

Equation 2-166

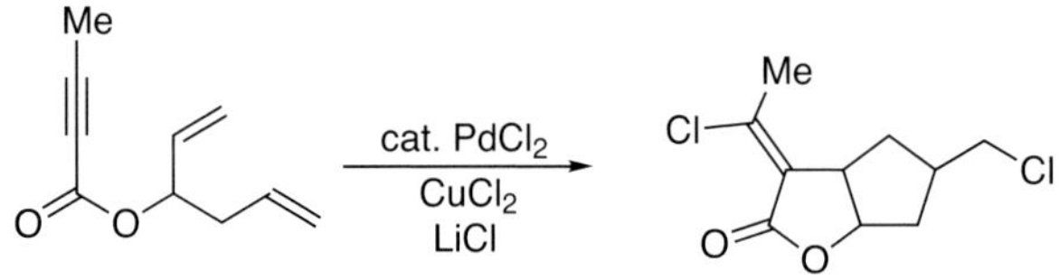

Equation 2-167

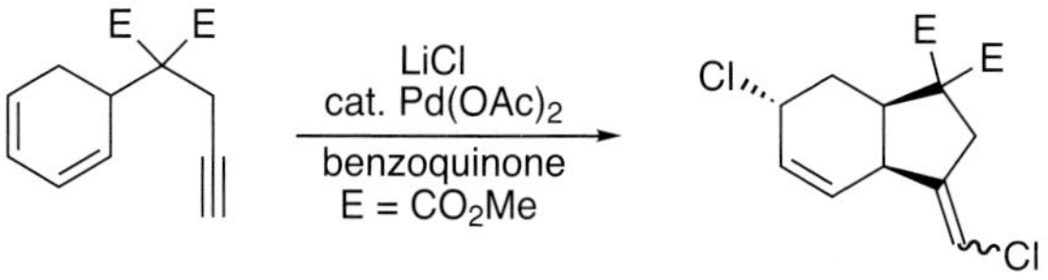

Equation 2-168

2.5.4
Cyclization of Acetylenic Organometallics

Finally, there are a number of other organometallic cyclization reactions that generate carbocycles from acetylenes. Lewis acids will cyclize acetylenic vinylic silanes to carbocyclic dienylsilanes (Equation 2.169) [226]. Analogous allylic silanes afford cyclic vinylic silanes in the presence of cat. $HfCl_4$ (Equation 2.170) [227] and dienylmercurials in the presence of $HgCl_2$ (Equation 2.171) [228]. Various Ru, Pd, Pt, and Ag salts catalyze the cyclization of acetylene-containing allylic silanes and stannanes to simple dienes (Equation 2.172) [229].

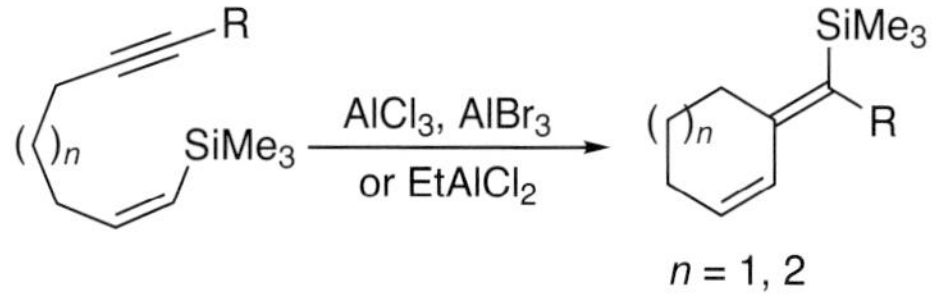

Equation 2-169

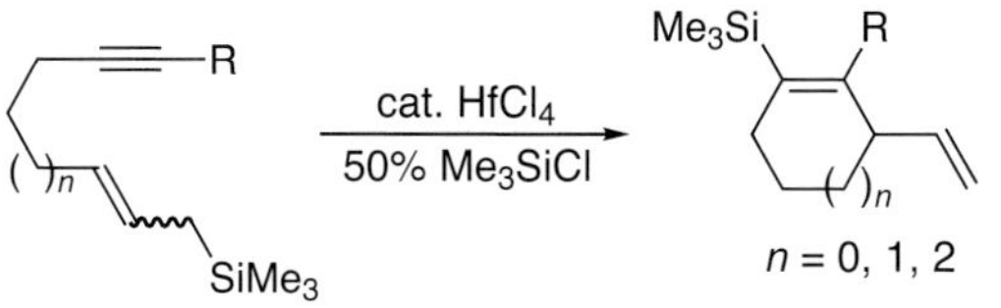

Equation 2-170

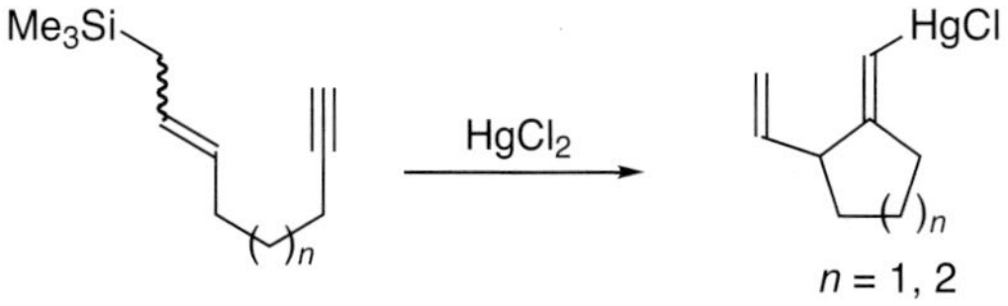

Equation 2-171

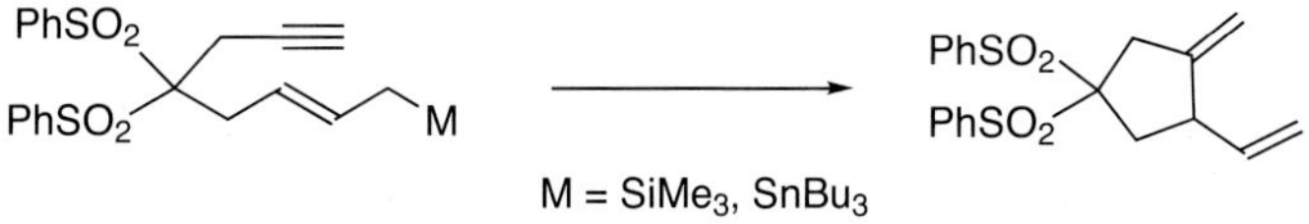

Equation 2-172

2.6
Conclusions

A wide variety of acetylene-containing oxygen, sulfur, selenium, nitrogen, and carbon compounds have been cyclized by treatment with acid, halogen, sulfur, selenium, metallic, organometallic, and other electrophiles to yield a wide array of heterocycles and carbocycles by electrophilic processes. Much of this work is quite recent, and many of the processes provide particularly convenient, high-yielding approaches to the systems in question. One can expect continued efforts in this area to provide very useful new routes to heterocycles and carbocycles of great interest to the synthetic organic and medicinal chemist.

2.7
Representative Experimental Procedures

2.7.1
Synthesis of α-Methylene-γ-butyrolactones by Carbonylation of 1-Alkyn-4-ols (Equation 2.14) [14]

The 1-alkyn-4-ol (10 mmol) was added to a solution of $PdCl_2$ (0.16 mmol), anhydrous $SnCl_2$ (0.16 mmol), and PPh_3 (0.32 mmol) in MeCN (30 mL) at 75 °C, and the mixture was then stirred under CO (100 psi) for 4 h. The resulting solution was reduced in volume on a rotary evaporator, taken up in diethyl ether (300 mL), and then passed through a 4 in. × 1 in. silica gel column to remove Pd complexes and metal. The Et_2O was removed *in vacuo*, and the resulting yellow oil was distilled at reduced pressure to yield the lactone.

2.7.2
Synthesis of 1-Alkoxyisochromenes by Cyclization of 2-(1-Alkynyl)benzaldehydes (Equation 2.58) [70]

MeOH (1.0 mmol) was added at 10 °C under an Ar atmosphere to a stirred mixture of 2-(1-alkynyl)benzaldehyde (0.5 mmol), $Pd(OAc)_2$ (5 mol%), and benzoquinone (0.5 mmol) in 1,4-dioxane (1 mL). After the mixture had been stirred for 30 min, the resulting solution was filtered through a short column of silica gel. The solvent was removed under reduced pressure to give the crude product, which was purified by silica gel column chromatography with 20:1 hexane/Et_2O as eluent.

2.7.3
Synthesis of 3-Aryl(vinylic)indoles by Palladium-catalyzed Cross-coupling of Aryl Halides or Vinylic Triflates and 2-(1-Alkynyl)trifluoroacetanilides (Equation 2.91) [125]

The aryl halide or vinylic triflate (0.59 mmol), K_2CO_3 (2.95 mmol), and $Pd(PPh_3)_4$ (0.03 mmol) were added to a solution of 2-(1-alkynyl)trifluoroacetanilide

(0.59 mmol) in MeCN (2.5 mL). The reaction mixture was stirred at room temperature under an N_2 atmosphere for 1.5 h. The mixture was then diluted with diethyl ether, and HCl (0.1 N) was added. The organic layer was separated, washed with water, dried (Na_2SO_4), and evaporated under vacuum. The residue was purified by chromatography on silica gel with elution with an *n*-hexane/EtOAc mixture.

2.7.4
Synthesis of Pyridines by the Gold-catalyzed Cross-coupling of Ketones and Propargyl Amine (Equation 2.115) [153]

Propargylamine (2.52 mmol) and the $NaAuCl_4 \cdot 2H_2O$ catalyst (0.03 mmol) were added to a 50 mL stainless steel autoclave charged with a solution of the ketone (1.26 mmol) in absolute EtOH (5 mL). The resulting mixture was heated at reflux with stirring or at 100 or 140 °C. The reaction was monitored by TLC and GC-MS. After cooling, the mixture was filtered to remove the catalyst, and the solvent was concentrated under reduced pressure. The residue was purified by flash chromatography (silica gel, *n*-hexane/ethyl acetate mixtures) to give the pyridine.

2.7.5
Synthesis of 4-Iodoisoquinolines by the Cyclization of Iminoalkynes (Equation 2.121) [159b]

I_2 (1.50 mmol), NaOAc (0.75 mmol), and MeCN (5 mL) were placed in a 2 dram vial. The iminoalkyne (0.25 mmol) in MeCN (2 mL) was added dropwise. The vial was flushed with Ar, and the reaction mixture was stirred at room temperature for 0.5 h. The reaction mixture was then diluted with ether (25 mL), washed with satd $Na_2S_2O_3$ (25 mL), dried (Na_2SO_4), and filtered. The solvent was evaporated under reduced pressure, and the product was purified by flash chromatography (3:1 hexane/EtOAc).

2.7.6
Synthesis of Cyclic Amines by Acetylene-Iminium Ion Cyclizations (Equation 2.125) [166b]

A 250 mL one-necked, round-bottomed flask containing a magnetic stirring bar and a reflux condenser topped with a rubber septum inlet was flushed with Ar and charged with alkynylamine (21 mmol), NaI (73 mmol), formaldehyde solution (37% w/w, 35 mL), camphorsulfonic acid monohydrate (22 mmol), and water (80 mL). The resulting mixture was heated at reflux under an Ar atmosphere for 15 min and was then allowed to cool to room temperature. This solution was made basic by addition of aq. KOH solution (5 M) and then poured into a separatory funnel, where it was extracted with CH_2Cl_2 (3 × 50 mL). The combined organic layers were dried over Na_2SO_4, filtered, and concentrated on a rotary evaporator. The resulting residue was purified by flash chromatography (ca. 150 g silica gel, 1:1 hexane/ethyl ether containing 5% Et_3N as eluent).

Acknowledgements

We gratefully acknowledge the Petroleum Research Fund, administered by the American Chemical Society, the National Science Foundation, and the National Institute of General Medical Sciences for generously funding our past and present research on acetylene chemistry. The author is particularly indebted to the many students whose names are cited in the Larock references, whose dedication, hard work, experimental skills, and intellectual contributions have added so much to our own research effort in this area. Finally, the close friendship and invaluable discussions of two close chemistry friends and major contributors to acetylene chemistry, Professors Sandro Cacchi and Yoshinori Yamamoto, are deeply appreciated.

References

[1] (a) G. Cardillo, M. Orena, *Tetrahedron* **1990**, *46*, 3321–3408; (b) O. Kitagawa, T. Inoue, T. Taguchi, *Rev. Heteroatom Chem.* **1996**, *15*, 243–262; (c) M. Frederickson, R. Grigg, *Org. Prep. Proc. Intl.* **1997**, *29*, 33–62.

[2] R. C. Larock, W. W. Leong, *Comp. Org. Synthesis*, Pergamon, New York, **1991**, 269–327.

[3] J. A. Marshall, W. J. DuBay, *J. Org. Chem.* **1993**, *58*, 3435–3443.

[4] (a) S. P. Bew, D. W. Knight, *Chem. Commun.* **1996**, 1007–1008; (b) G. M. M. El-Taeb, A. B. Evans, S. Jones, D. W. Knight, *Tetrahedron Lett.* **2001**, *42*, 5945–5948.

[5] R. C. Larock, C.-L. Liu, *J. Org. Chem.* **1983**, *48*, 2151–2158.

[6] R. C. Larock, B. Riefling, C. A. Fellows, *J. Org. Chem.* **1978**, *43*, 131–137.

[7] (a) M. Riediker, J. Schwartz, *J. Am. Chem. Soc.* **1982**, *104*, 5842–5844; (b) F. Compernolle, H. Mao, A. Tahri, T. Kozlecki, E. Van der Eycken, B. Medaer, G. J. Hoornaert, *Tetrahedron Lett.* **2002**, *43*, 3011–3015.

[8] D. Villemin, D. Goussu, *Heterocycles* **1989**, *29*, 1255–1261.

[9] K. Utimoto, *Pure Appl. Chem.* **1983**, *55*, 1845–1852.

[10] (a) B. Gabriele, G. Salerno, E. Lauria, *J. Org. Chem.* **1999**, *64*, 7687–7692; (b) B. Gabriele, G. Salerno, *Chem. Commun.* **1997**, 1083–1084; (c) F.-L. Qing, W.-Z. Gao, J. Ying, *J. Org. Chem.* **2000**, *65*, 2003–2006.

[11] F.-L. Qing, W.-Z. Gao, *Tetrahedron Lett.* **2000**, *41*, 7727–7730.

[12] S. Ma, B. Wu, S. Zhao, *Org. Lett.* **2003**, *5*, 4429–4432.

[13] Y. Gu, F. Shi, Y. Deng, *J. Org. Chem.* **2004**, *69*, 391–394.

[14] (a) T. F. Murray, E. G. Samsel, V. Varma, J. R. Norton, *J. Am. Chem. Soc.* **1981**, *103*, 7520–7528; (b) T. F. Murray, J. R. Norton, *J. Am. Chem. Soc.* **1979**, *101*, 4107–4119; (c) J. R. Norton, K. E. Shenton, J. Schwartz, *Tetrahedron Lett.* **1975**, *16*, 51–54.

[15] (a) P. Compain, J. Goré, J.-M. Vatèle, *Tetrahedron* **1996**, *52*, 10405–10416; (b) P. Compain, J.-M. Vatèle, J. Goré, *Synlett* **1994**, 943–945.

[16] Y. Tamaru, M. Hojo, Z.-I. Yoshida, *J. Org. Chem.* **1991**, *56*, 1099–1105.

[17] K. Kato, A. Nishimura, Y. Yamamoto, H. Akita, *Tetrahedron Lett.* **2001**, *42*, 4203–4205.

[18] (a) K. Kato, M. Tanaka, Y. Yamamoto, H. Akita, *Tetrahedron Lett.* **2002**, *43*, 1511–1513; (b) K. Kato, A. Nishimura, Y. Yamamoto, H. Akita, *Tetrahedron Lett.* **2002**, *43*, 643–645.

[19] (a) B. Gabriele, G. Salerno, F. De Pascali, M. Costa, G. P. Chiusoli, *J. Org. Chem.* **1999**, *64*, 7693–7699;

(b) B. Gabriele, G. Salerno, F. De Pascali, M. Costa, G. P. Chiusoli, *J. Organomet. Chem.* **2000**, *593*, 409–415.

[20] P. Wipf, T. H. Graham, *J. Org. Chem.* **2003**, *68*, 8798–8807.

[21] H. Mao, M. Koukni, T. Kozlecki, F. Compernolle, G. J. Hoornaert, *Tetrahedron Lett.* **2002**, *43*, 8697–8700.

[22] (a) A. Arcadi, S. Cacchi, S. Di Giuseppe, G. Fabrizi, F. Marinelli, *Synlett* **2002**, 453–457; (b) A. Arcadi, S. Cacchi, M. Del Rosario, G. Fabrizi, F. Marinelli, *J. Org. Chem.* **1996**, *61*, 9280–9288.

[23] G. W. Kabalka, L. Wang, R. M. Pagni, *Tetrahedron* **2001**, *57*, 8017–8028.

[24] A. Arcadi, S. Cacchi, G. Fabrizi, F. Marinelli, L. Moro, *Synlett* **1999**, 1432–1434.

[25] J. Barluenga, M. Trincado, E. Rubio, J. M. González, *Angew. Chem., Int. Ed.* **2003**, *42*, 2406–2409; *Angew. Chem.* **2003**, *115*, 2508–2511.

[26] S. Cacchi, G. Fabrizi, L. Moro, *Synlett* **1998**, 741–745.

[27] N. Monteiro, A. Arnold, G. Balme, *Synlett* **1998**, 1111–1113.

[28] Y. Hu, Y. Zhang, Z. Yang, R. Fathi, *J. Org. Chem.* **2002**, *67*, 2365–2368.

[29] A. Arcadi, S. Cacchi, G. Fabrizi, L. Moro, *Eur. J. Org. Chem.* **1999**, 1137–1141.

[30] (a) H. Lütjens, P. J. Scammells, *Synlett* **1999**, 1079–1081; (b) Y. Nan, H. Miao, Z. Yang, *Org. Lett.* **2000**, *2*, 297–299; (c) Y. Kondo, T. Sakamoto, H. Yamanaka, *Heterocycles* **1989**, *29*, 1013–1016; (d) Y. Kondo, F. Shiga, N. Murata, T. Sakamoto, H. Yamanaka, *Tetrahedron* **1994**, *50*, 11803–11812; (e) C. C. Li, Z. X. Xie, Y. D. Zhang, J. H. Chen, Z. Yang, *J. Org. Chem.* **2003**, *68*, 8500–8504.

[31] Y. Hu, Z. Yang, *Org. Lett.* **2001**, *3*, 1387–1390.

[32] (a) F.-T. Luo, I. Schreuder, R.-T. Wang, *J. Org. Chem.* **1992**, *57*, 2213–2215; (b) C. Chowdhury, G. Chaudhuri, S. Guha, A. K. Mukherjee, N. G. Kundu, *J. Org. Chem.* **1998**, *63*, 1863–1871.

[33] Z. An, M. Catellani, G. P. Chiusoli, *J. Organomet. Chem.* **1990**, *397*, 371–373.

[34] M. Blandino, E. McNelis, *Org. Lett.* **2002**, *4*, 3387–3390.

[35] A. Arcadi, S. Cacchi, S. Di Giuseppe, G. Fabrizi, F. Marinelli, *Org. Lett.* **2002**, *4*, 2409–2412.

[36] (a) D. W. Yue, R. C. Larock, work in progress; (b) R. C. Larock, L. W. Harrison, *J. Am. Chem. Soc.* **1984**, *106*, 4218–4227.

[37] S. Cacchi, G. Fabrizi, L. Moro, *Tetrahedron Lett.* **1998**, *39*, 5101–5104.

[38] A. S. Bhat, J. L. Whetstone, R. W. Brueggemeier, *Tetrahedron Lett.* **1999**, *40*, 2469–2472.

[39] C.-F. Lin, W.-D. Lu, I.-W. Wang, M.-J. Wu, *Synlett* **2003**, 2057–2061.

[40] J. A. Marshall, W. J. DuBay, *J. Am. Chem. Soc.* **1992**, *114*, 1450–1456.

[41] R. J. Madhushaw, C.-L. Li, H.-L. Su, C.-C. Hu, S.-F. Lush, R.-S. Liu, *J. Org. Chem.* **2003**, *68*, 1872–1877.

[42] (a) R. C. Larock, *Comprehensive Organic Transformations*, Wiley-VCH, New York, **1999**, 1895–1896; (b) Y. Ogawa, M. Maruno, T. Wakamatsu, *Heterocycles* **1995**, *41*, 2587–2599.

[43] C. Lambert, K. Utimoto, H. Nozaki, *Tetrahedron Lett.* **1984**, *25*, 5323–5326.

[44] R. C. Larock, *Comprehensive Organic Transformations*, Wiley-VCH, New York, **1999**, 1896.

[45] (a) N. Yanagihara, C. Lambert, K. Iritani, K. Utimoto, H. Nozaki, *J. Am. Chem. Soc.* **1986**, *108*, 2753–2754; (b) K. Iritani, N. Yanagihara, K. Utimoto, *J. Org. Chem.* **1986**, *51*, 5499–5501.

[46] T. Tsuda, Y. Ohashi, N. Nagahama, R. Sumiya, T. Saegusa, *J. Org. Chem.* **1988**, *53*, 2650–2653.

[47] D. Bouyssi, J. Goré, G. Balme, D. Louis, J. Wallach, *Tetrahedron Lett.* **1993**, *34*, 3129–3130.

[48] Z. Wang, X. Lu, *J. Org. Chem.* **1996**, *61*, 2254–2255.

[49] A. Arcadi, A. Burini, S. Cacchi, M. Delmastro, F. Marinelli, B. R. Pietroni, *J. Org. Chem.* **1992**, *57*, 976–982.

[50] D. Bouyssi, J. Goré, G. Balme, *Tetrahedron Lett.* **1992**, *33*, 2811–2814.

[51] M. Cavicchioli, D. Bouyssi, J. Goré, G. Balme, *Tetrahedron Lett.* **1996**, *37*, 1429–1432.
[52] (a) F. Bellina, M. Biagetti, A. Carpita, R. Rossi, *Tetrahedron* **2001**, *57*, 2857–2870; (b) M. Biagetti, F. Bellina, A. Carpita, S. Viel, L. Mannina, R. Rossi, *Eur. J. Org. Chem.* **2002**, 1063–1076; (c) F. Bellina, M. Biagetti, A. Carpita, R. Rossi, *Tetrahedron Lett.* **2001**, *42*, 2859–2863; (d) M. Biagetti, F. Bellina, A. Carpita, P. Stabile, R. Rossi, *Tetrahedron* **2002**, *58*, 5023–5038.
[53] (a) T. Yao, R. C. Larock, *Tetrahedron Lett.* **2002**, *43*, 7401–7404; (b) T. Yao, R. C. Larock, *J. Org. Chem.* **2003**, *68*, 5936–5942.
[54] R. Rossi, F. Bellina, M. Biagetti, A. Catanese, L. Mannina, *Tetrahedron Lett.* **2000**, *41*, 5281–5286.
[55] F. Bellina, D. Ciucci, P. Vergamini, R. Rossi, *Tetrahedron* **2000**, *56*, 2533–2545.
[56] A. Nagarajan, T. R. Balasubramanian, *Indian J. Chem.* **1988**, *27B*, 380.
[57] R. Rossi, A. Carpita, F. Bellina, P. Stabile, L. Mannina, *Tetrahedron* **2003**, *59*, 2067–2081.
[58] M. A. Oliver, R. D. Gandour, *J. Org. Chem.* **1984**, *49*, 558–559.
[59] A. Nagarajan, T. R. Balasubramanian, *Indian J. Chem.* **1987**, *26B*, 917–919.
[60] H. Kusama, H. Funami, J. Takaya, N. Iwasawa, *Org. Lett.* **2004**, *6*, 605–608.
[61] (a) N. G. Kundu, M. Pal, J. S. Mahanty, M. De, *J. Chem. Soc., Perkin Trans. 1* **1997**, 2815–2820; (b) W.-M. Dai, K. W. Lai, *Tetrahedron Lett.* **2002**, *43*, 9377–9380.
[62] A. Arcadi, S. Cacchi, L. Cascia, G. Fabrizi, F. Marinelli, *Org. Lett.* **2001**, *3*, 2501–2504.
[63] A. Bacchi, M. Costa, B. Gabriele, G. Pelizzi, G. Salerno, *J. Org. Chem.* **2002**, *67*, 4450–4457.
[64] M. Costa, N. Della Cà, B. Gabriele, C. Massera, G. Salerno, M. Soliani, *J. Org. Chem.* **2004**, *69*, 2469–2477.
[65] (a) C. McGuigan, C. J. Yarnold, G. Jones, S. Velázquez, H. Barucki, A. Brancale, G. Andrei, R. Snoeck, E. De Clercq, J. Balzarini, *J. Med. Chem.* **1999**, *42*, 4479–4484; (b) I. N. Houpis, W. B. Choi, P. J. Reider, A. Molina, H. Churchill, J. Lynch, R. P. Volante, *Tetrahedron Lett.* **1994**, *35*, 9355–9358.
[66] M. S. Rao, N. Esho, C. Sergeant, R. Dembinski, *J. Org. Chem.* **2003**, *68*, 6788–6790.
[67] E. Bossharth, P. Desbordes, N. Monteiro, G. Balme, *Org. Lett.* **2003**, *5*, 2441–2444.
[68] A.-Y. Peng, Y.-X. Ding, *J. Am. Chem. Soc.* **2003**, *125*, 15006–15007.
[69] A.-Y. Peng, Y.-X. Ding, *Org. Lett.* **2004**, *6*, 1119–1121.
[70] N. Asao, T. Nogami, K. Takahashi, Y. Yamamoto, *J. Am. Chem. Soc.* **2002**, *124*, 764–765.
[71] S. Mondal, T. Nogami, N. Asao, Y. Yamamoto, *J. Org. Chem.* **2003**, *68*, 9496–9498.
[72] J. Barluenga, H. Vázquez-Villa, A. Ballesteros, J. M. González, *J. Am. Chem. Soc.* **2003**, *125*, 9028–9029.
[73] J. Barluenga, H. Vázquez-Villa, A. Ballesteros, J. M. González, *Org. Lett.* **2003**, *5*, 4121–4123.
[74] D. Yue, N. Della Cà, R. C. Larock, *Org. Lett.* **2004**, *6*, 1581–1584.
[75] (a) N. Asao, K. Takahashi, S. Lee, T. Kasahara, Y. Yamamoto, *J. Am. Chem. Soc.* **2002**, *124*, 12650–12651; (b) N. Asao, T. Nogami, S. Lee, Y. Yamamoto, *J. Am. Chem. Soc.* **2003**, *125*, 10921–10925.
[76] A. Fürstner, A.-S. Castanet, K. Radkowski, C. W. Lehmann, *J. Org. Chem.* **2003**, *68*, 1521–1528 and references therein.
[77] Y. Fukuda, H. Shiragami, K. Utimoto, H. Nozaki, *J. Org. Chem.* **1991**, *56*, 5816–5819.
[78] (a) A. Jeevanandam, K. Narkunan, C. Cartwright, Y.-C. Ling, *Tetrahedron Lett.* **1999**, *40*, 4841–4844; (b) A. Jeevanandam, K. Narkunan, Y.-C. Ling, *J. Org. Chem.* **2001**, *66*, 6014–6020.
[79] M. Gulías, J. R. Rodríguez, L. Castedo, J. L. Mascareñas, *Org. Lett.* **2003**, *5*, 1975–1977.
[80] (a) A. Arcadi, S. Cacchi, R. C. Larock, F. Marinelli, *Tetrahedron Lett.* **1993**, *34*, 2813–2816; (b) S. Cacchi, G. Fabrizi, L. Moro, *J. Org. Chem.* **1997**, *62*, 5327–5332; (c) A. Arcadi, S. Cacchi,

G. Fabrizi, F. Marinelli, L. M. Parisi, *Tetrahedron* **2003**, *59*, 4661–4671.

[81] A. Arcadi, E. Rossi, *Tetrahedron Lett.* **1996**, *37*, 6811–6814.

[82] K. Kato, Y. Yamamoto, H. Akita, *Tetrahedron Lett.* **2002**, *43*, 4915–4917.

[83] (a) B. L. Flynn, P. Verdier-Pinard, E. Hamel, *Org. Lett.* **2001**, *5*, 651–654; (b) R. C. Larock, D. Yue, *Tetrahedron Lett.* **2001**, *42*, 6011–6013; (c) D. Yue, R. C. Larock, *J. Org. Chem.* **2002**, *67*, 1905–1909.

[84] T. Kesharwani, R. C. Larock, work in progress.

[85] K. O. Hessian, B. L. Flynn, *Org. Lett.* **2003**, *5*, 4377–4380.

[86] T. Kataoka, H. Kinoshita, S. Kinoshita, T. Iwamura, *Tetrahedron Lett.* **2002**, *43*, 7039–7041.

[87] M. A. Fernandes, D. H. Reid, *Synlett* **2003**, 2231–2233.

[88] For a review, see T. E. Müller, M. Beller, *Chem. Rev.* **1998**, *98*, 675–703.

[89] (a) P. L. McGrane, T. Livinghouse, *J. Org. Chem.* **1992**, *57*, 1323–1324; (b) P. L. McGrane, M. Jensen, T. Livinghouse, *J. Am. Chem. Soc.* **1992**, *114*, 5459–5460.

[90] D. Fairfax, M. Stein, T. Livinghouse, M. Jensen, *Organometallics* **1997**, *16*, 1523–1525.

[91] P. L. McGrane, T. Livinghouse, *J. Am. Chem. Soc.* **1993**, *115*, 11485–11489.

[92] L. Ackermann, R. G. Begman, R. N. Loy, *J. Am. Chem. Soc.* **2003**, *125*, 11956–11963.

[93] I. Bytschkov, S. Doye, *Tetrahedron Lett.* **2002**, *43*, 3715–3718.

[94] (a) Y. Fukuda, K. Utimoto, *Synthesis* **1991**, 975–978; (b) Y. Fukuda, K. Utimoto, H. Nozaki, *Heterocycles* **1987**, *25*, 297–300.

[95] T. E. Müller, M. Grosche, E. Herdtweck, A.-K. Pleier, E. Walter, Y.-K. Yan, *Organometallics* **2000**, *19*, 170–183.

[96] (a) Y. Fukuda, S. Matsubara, K. Utimoto, *J. Org. Chem.* **1991**, *56*, 5812–5816; (b) K. Utimoto, *Pure Appl. Chem.* **1983**, *55*, 1845–1852.

[97] (a) Y. Li, P.-F. Fu, T. J. Marks, *Organometallics* **1994**, *13*, 439–440; (b) Y. Li, T. J. Marks, *J. Am. Chem. Soc.* **1996**, *118*, 9295–9306.

[98] I. Kadota, A. Shibuya, L. M. Lutete, Y. Yamamoto, *J. Org. Chem.* **1999**, *64*, 4570–4571.

[99] L. M. Lutete, I. Kadota, Y. Yamamoto, *J. Am. Chem. Soc.* **2004**, *126*, 1622–1623.

[100] K. Utimoto, H. Miwa, H. Nozaki, *Tetrahedron Lett.* **1981**, *22*, 4277–4278.

[101] (a) B. Gabriele, G. Salerno, A. Fazio, M. R. Bossio, *Tetrahedron Lett.* **2001**, *42*, 1339–1341; (b) B. Gabriele, G. Salerno, A. Fazio, *J. Org. Chem.* **2003**, *68*, 7853–7861.

[102] H. Siebeneicher, I. Bytschkov, S. Doye, *Angew. Chem., Int. Ed.* **2003**, *42*, 3042–3044; *Angew. Chem.* **2003**, *115*, 3151–3153.

[103] W.-M. Dai, L.-P. Sun, D.-S. Guo, *Tetrahedron Lett.* **2002**, *43*, 7699–7702.

[104] K. Hiroya, S. Itoh, T. Sakamoto, *J. Org. Chem.* **2004**, *69*, 1126–1136.

[105] F. E. McDonald, A. K. Chatterjee, *Tetrahedron Lett.* **1997**, *38*, 7687–7690.

[106] (a) A. Arcadi, S. Cacchi, F. Marinelli, *Tetrahedron Lett.* **1989**, *30*, 2581–2584; (b) S. Cacchi, V. Carnicelli, F. Marinelli, *J. Organomet. Chem.* **1994**, *475*, 289–296; (c) K. Iritani, S. Matsubara, K. Utimoto, *Tetrahedron Lett.* **1988**, *29*, 1799–1802; (d) B. C. J. van Esseveldt, F. L. van Delft, R. de Gelder, F. P. J. T. Rutjes, *Org. Lett.* **2003**, *5*, 1717–1720.

[107] N. Sakai, K. Annaka, T. Konakahara, *Org. Lett.* **2004**, *6*, 1527–1530.

[108] (a) J. Ezquerra, C. Pedregal, C. Lamas, J. Barluenga, M. Pérez, M. A. García-Martín, J. M. González, *J. Org. Chem.* **1996**, *61*, 5804–5812; (b) J. Soloducho, *Tetrahedron Lett.* **1999**, *40*, 2429–2430.

[109] M. S. Yu, L. Lopez de Leon, M. A. McGuire, G. Botha, *Tetrahedron Lett.* **1998**, *39*, 9347–9350.

[110] B. Gabriele, G. Salerno, L. Veltri, M. Costa, C. Massera, *Eur. J. Org. Chem.* **2001**, 4607–4613.

[111] S. Torii, L. H. Xu, M. Sadakane, H. Okumoto, *Synlett* **1992**, 513–514.

[112] J. S. Mahanty, M. De, P. Das, N. G. Kundu, *Tetrahedron* **1997**, *53*, 13397–13418.

[113] S. Cacchi, G. Fabrizi, A. Carangio, *Synlett* **1997**, 959–961.

[114] Q. Sun, E. J. LaVoie, *Heterocycles* **1996**, *43*, 737–743.

[115] H. D. Doan, J. Goré, J.-M. Vatéle, *Tetrahedron Lett.* **1999**, *40*, 6765–6768.

[116] (a) R. W. M. ten Hoedt, G. van Koten, J. G. Noltes, *Synth. Commun.* **1977**, *7*, 61–69; (b) D. Yue, R. C. Larock, *Org. Lett.* **2004**, *6*, 1037–1040.

[117] W.-M. Dai, D.-S. Guo, L.-P. Sun, *Tetrahedron Lett.* **2001**, *42*, 5275–5278.

[118] S. Cacchi, G. Fabrizi, F. Marinelli, L. Moro, P. Pace, *Synlett* **1997**, 1363–1366.

[119] A. Yasuhara, Y. Kanamori, M. Kaneko, A. Numata, Y. Kondo, T. Sakamoto, *J. Chem. Soc., Perkin Trans. 1* **1999**, 529–534.

[120] (a) D. E. Rudisill, J. K. Stille, *J. Org. Chem.* **1989**, *54*, 5856–5866; (b) E. C. Taylor, A. H. Katz, H. Salgado-Zamora, A. McKillop, *Tetrahedron Lett.* **1985**, *26*, 5963–5966.

[121] N. G. Kundu, M. W. Khan, *Tetrahedron* **2000**, *56*, 4777–4792.

[122] H. Sashida, A. Kawamukai, *Synthesis* **1999**, 1145–1148.

[123] A. Nagarajan, T. R. Balasubramanian, *Indian J. Chem.* **1989**, *28B*, 67–68.

[124] G. Battistuzzi, S. Cacchi, G. Fabrizi, *Eur. J. Org. Chem.* **2002**, 2671–2681.

[125] A. Arcadi, S. Cacchi, F. Marinelli, *Tetrahedron Lett.* **1992**, *33*, 3915–3918.

[126] S. Cacchi, G. Fabrizi, D. Lamba, F. Marinelli, L. M. Parisi, *Synthesis* **2003**, 728–734.

[127] A. Arcadi, S. Cacchi, G. Fabrizi, F. Marinelli, *Synlett* **2000**, 394–396.

[128] S. Cacchi, G. Fabrizi, P. Pace, *J. Org. Chem.* **1998**, *63*, 1001–1011.

[129] A. Arcadi, S. Cacchi, G. Fabrizi, F. Marinelli, *Synlett* **2000**, 647–649.

[130] M. D. Collini, J. W. Ellingboe, *Tetrahedron Lett.* **1997**, *38*, 7963–7966.

[131] A. Arcadi, S. Cacchi, G. Fabrizi, L. M. Parisi, *Tetrahedron Lett.* **2004**, *45*, 2431–2434.

[132] A. Arcadi, S. Cacchi, A. Cassetta, G. Fabrizi, L. M. Parisi, *Synlett* **2001**, 1605–1607.

[133] M. G. Saulnier, D. B. Frennesson, M. S. Deshpande, D. M. Vyas, *Tetrahedron Lett.* **1995**, *36*, 7841–7844.

[134] A. Arcadi, S. Cacchi, V. Carnicelli, F. Marinelli, *Tetrahedron* **1994**, *50*, 437–452.

[135] G. Battistuzzi, S. Cacchi, G. Fabrizi, F. Marinelli, L. M. Parisi, *Org. Lett.* **2002**, *4*, 1355–1358.

[136] S. Cacchi, G. Fabrizi, P. Pace, F. Marinelli, *Synlett* **1999**, 620–622.

[137] W. F. J. Karstens, M. Stol, F. P. J. T. Rutjes, H. Kooijman, A. L. Spek, H. Hiemstra, *J. Organomet. Chem.* **2001**, *624*, 244–258.

[138] K. Cherry, J. Thibonnet, A. Duchêne, J.-L. Parrain, *Tetrahedron Lett.* **2004**, *45*, 2063–2066.

[139] K. Hiroya, S. Itoh, M. Ozawa, Y. Kanamori, T. Sakamoto, *Tetrahedron Lett.* **2002**, *43*, 1277–1280.

[140] (a) T. Sakamoto, Y. Kondo, H. Yamanaka, *Heterocycles* **1986**, *24*, 31–32; (b) T. Sakamoto, Y. Kondo, H. Yamanaka, *Heterocycles* **1986**, *24*, 1845–1847; (c) K. Shin, K. Ogasawara, *Synlett* **1996**, 922–924; (d) K. Shin, K. Ogasawara, *Synlett* **1995**, 859–860; (e) K. Shin, K. Ogasawara, *Chem. Lett.* **1995**, 289–290; (f) T. Sakamoto, Y. Kondo, S. Iwashita, T. Nagano, H. Yamanaka, *Chem. Pharm. Bull.* **1988**, *36*, 1305–1308; (g) S. Takano, T. Sato, K. Inomata, K. Ogasawara, *J. Chem. Soc., Chem. Commun.* **1991**, 462–464.

[141] L. Xu, I. R. Lewis, S. K. Davidsen, J. B. Summers, *Tetrahedron Lett.* **1998**, *39*, 5159–5162.

[142] Y. Kondo, S. Kojima, T. Sakamoto, *Heterocycles* **1996**, *43*, 2741–2746.

[143] Y. Kondo, S. Kojima, T. Sakamoto, *J. Org. Chem.* **1997**, *62*, 6507–6511.

[144] A. Yasuhara, M. Kaneko, T. Sakamoto, *Heterocycles* **1998**, *48*, 1793–1799.

[145] (a) D. Bouyssi, M. Cavicchioli, G. Balme, *Synlett* **1997**, 944–946; (b) A. Arcadi, *Synlett* **1997**, 941–943.

[146] Q. Sun, E. J. LaVoie, *Heterocycles* **1996**, *43*, 737–743.

[147] (a) L. B. Wolf, K. C. M. F. Tjen, H. T. ten Brink, R. H. Blaauw, H. Hiemstra, H. E. Schoemaker, F. P. J. T. Rutjes, *Adv. Synth. Catal.* **2002**, *344*, 70–83; (b) L. B. Wolf, K. C. M. F. Tjen, F. P. J. T. Rutjes, H. Hiemstra,

H. E. Schoemaker, *Tetrahedron Lett.* **1998**, *39*, 5081–5084; (c) B. C. J. van Esseveldt, F. L. van Delft, J. M. M. Smits, R. de Gelder, F. P. J. T. Rutjes, *Synlett* **2003**, 2354–2358.
[148] D. W. Knight, C. M. Sharland, *Synlett* **2003**, 2258–2260.
[149] (a) D. W. Knight, A. L. Redfern, J. Gilmore, *Chem. Commun.* **1998**, 2207–2208; (b) D. W. Knight, A. L. Redfern, J. Gilmore, *J. Chem. Soc., Perkin Trans. 1* **2002**, 622–628.
[150] M. Amjad, D. W. Knight, *Tetrahedron Lett.* **2004**, *45*, 539–541.
[151] F.-T. Luo, R.-T. Wang, *Tetrahedron Lett.* **1992**, *33*, 6835–6838.
[152] R. Mukhopadhyay, N. G. Kundu, *Synlett* **2001**, 1143–1145.
[153] G. Abbiati, A. Arcadi, G. Bianchi, S. Di Giuseppe, F. Marinelli, E. Rossi, *J. Org. Chem.* **2003**, *68*, 6959–6966.
[154] (a) A. Arcadi, S. Di Guiseppe, F. Marinelli, E. Rossi, *Tetrahedron: Asymmetry* **2001**, *12*, 2715–2720; (b) A. Arcadi, S. Di Guiseppe, F. Marinelli, E. Rossi, *Adv. Synth. Catal.* **2001**, *343*, 443–446.
[155] S. Kamijo, Y. Sasaki, Y. Yamamoto, *Tetrahedron Lett.* **2004**, *45*, 35–38.
[156] J. W. Herndon, Y. Zhang, K. Wang, *J. Organomet. Chem.* **2001**, *634*, 1–4.
[157] G. Abbiati, A. Arcadi, E. Beccalli, E. Rossi, *Tetrahedron Lett.* **2003**, *44*, 5331–5334.
[158] T. Sakamoto, Y. Kondo, H. Yamanaka, *Heterocycles* **1988**, *27*, 2225–2249.
[159] (a) K. R. Roesch, R. C. Larock, *Org. Lett.* **1999**, *1*, 553–556; (b) Q. Huang, J. A. Hunter, R. C. Larock, *J. Org. Chem.* **2002**, *67*, 3437–3444.
[160] Q. Huang, J. A. Hunter, R. C. Larock, *Org. Lett.* **2001**, *3*, 2973–2976.
[161] (a) H. Zhang, R. C. Larock, *Tetrahedron Lett.* **2002**, *43*, 1359–1362; (b) H. Zhang, R. C. Larock, *J. Org. Chem.* **2002**, *67*, 7048–7056.
[162] (a) G. Dai, R. C. Larock, *Org. Lett.* **2001**, *3*, 4035–4038; (b) G. Dai, R. C. Larock, *J. Org. Chem.* **2003**, *68*, 920–928.
[163] (a) Q. Huang, R. C. Larock, *Tetrahedron Lett.* **2002**, *43*, 3557–3560; (b) Q. Huang, R. C. Larock, *J. Org. Chem.* **2003**, *68*, 980–988.
[164] (a) G. Dai, R. C. Larock, *Org. Lett.* **2002**, *4*, 193–196; (b) G. Dai, R. C. Larock, *J. Org. Chem.* **2002**, *67*, 7042–7047.
[165] Y. Murata, L. E. Overman, *Heterocycles* **1996**, *42*, 549–553.
[166] (a) L. E. Overman, A. K. Sarkar, *Tetrahedron Lett.* **1992**, *33*, 4103–4106; (b) H. Arnold, L. E. Overman, M. J. Sharp, M. C. Witschel, *Org. Syn.* **1991**, *70*, 111–119.
[167] L. E. Overman, M. J. Sharp, *J. Am. Chem. Soc.* **1988**, *110*, 612–614.
[168] (a) L. E. Overman, L. A. Robinson, J. Zablocki, *J. Am. Chem. Soc.* **1992**, *114*, 368–369; (b) L. E. Overman, M. J. Sharp, *Tetrahedron Lett.* **1988**, *29*, 901–904.
[169] A. Arcadi, R. Anacardio, G. D'Anniballe, M. Gentile, *Synlett* **1997**, 1315–1317.
[170] S. Kamijo, Y. Yamamoto, *J. Org. Chem.* **2003**, *68*, 4764–4771.
[171] S. Kamijo, Y. Yamamoto, *Angew. Chem., Int. Ed.* **2002**, *41*, 3230–3233; *Angew. Chem.* **2002**, *114*, 3364–3367.
[172] S. Kamijo, Y. Yamamoto, *J. Am. Chem. Soc.* **2002**, *124*, 11940–11945.
[173] L. G. Fedenok, N. A. Zolnikova, *Tetrahedron Lett.* **2003**, *44*, 5453–5455.
[174] N. Asao, K. Sato, Y. Yamamoto, *Tetrahedron Lett.* **2003**, *44*, 5675–5677.
[175] L.-M. Wei, C.-F. Lin, M.-J. Wu, *Tetrahedron Lett.* **2000**, *41*, 1215–1218.
[176] T. Sakamoto, Y. Kondo, N. Miura, K. Hayashi, H. Yamanaka, *Heterocycles* **1986**, *24*, 2311–2314.
[177] P. N. Anderson, J. T. Sharp, *J. Chem. Soc., Perkin Trans. 1* **1980**, 1331–1334.
[178] M. Mizutani, Y. Sanemitsu, Y. Tamaru, Z.-I. Yoshida, *Tetrahedron Lett.* **1985**, *26*, 1237–1240.
[179] For a review, see: J. M. Conia, P. Le Perchec, *Synthesis* **1975**, 1–19.
[180] M. A. Boaventura, J. Drouin, J. M. Conia, *Synthesis,* **1983**, 801–804.
[181] O. Kitagawa, T. Suzuki, T. Inoue, Y. Watanabe, T. Taguchi, *J. Org. Chem.* **1998**, *63*, 9470–9475.
[182] (a) N. Monteiro, G. Balme, J. Gore, *Synlett* **1992**, 227–228;

(b) D. Bouyssi, N. Monteiro, G. Balme, *Tetrahedron Lett.* **1999**, *40*, 1297–1300.

[183] F. E. McDonald, T. C. Olson, *Tetrahedron Lett.* **1997**, *38*, 7691–7692.

[184] (a) P. Cruciani, R. Stammler, C. Aubert, M. Malacria, *J. Org. Chem.* **1996**, *61*, 2699–2708;
(b) P. Cruciani, C. Aubert, M. Malacria, *Tetrahedron Lett.* **1994**, *35*, 6677–6680;
(c) J.-L. Renaud, C. Aubert, M. Malacria, *Tetrahedron* **1999**, *55*, 5113–5128.

[185] N. Monteiro, J. Gore, G. Balme, *Tetrahedron* **1992**, *48*, 10103–10114.

[186] J. J. Kennedy-Smith, S. T. Staben, F. D. Toste, *J. Am. Chem. Soc.* **2004**, *126*, 4526–4527.

[187] (a) X. Marat, N. Monteiro, G. Balme, *Synlett* **1997**, 845–847;
(b) B. Clique, N. Monteiro, G. Balme, *Tetrahedron Lett.* **1999**, *40*, 1301–1304.

[188] G. Fournet, G. Balme, B. Van Hemelryck, J. Gore, *Tetrahedron Lett.* **1990**, *31*, 5147–5150.

[189] D. Bouyssi, G. Balme, J. Gore, *Tetrahedron Lett.* **1991**, *32*, 6541–6544.

[190] C. Nevado, D. J. Cárdenas, A. M. Echavarren, *Chem. Eur. J.* **2003**, *9*, 2627–2635.

[191] (a) K. Maeyama, N. Iwasawa, *J. Am. Chem. Soc.* **1998**, *120*, 1928–1929;
(b) N. Iwasawa, T. Miura, K. Kiyota, H. Kusama, K. Lee, P. H. Lee, *Org. Lett.* **2002**, *4*, 4463–4466.

[192] J. W. Dankwardt, *Tetrahedron Lett.* **2001**, *42*, 5809–5812.

[193] (a) J. Drouin, M.-A. Boaventura, J.-M. Conia, *J. Am. Chem. Soc.* **1985**, *107*, 1726–1729;
(b) J. Drouin, M. A. Boaventura, *Tetrahedron Lett.* **1987**, *28*, 3923–3926;
(c) M.-A. Boaventura, J. Drouin, F. Theobald, N. Rodier, *Bull. Soc. Chim. Fr.* **1987**, 1006–1014;
(d) M.-A. Boaventura, J. Drouin, *Bull. Soc. Chim. Fr.* **1987**, 1015–1026;
(e) C. J. Forsyth, J. Clardy, *J. Am. Chem. Soc.* **1990**, *112*, 3497–3505;
(f) H. Huang, C. J. Forsyth, *Tetrahedron Lett.* **1993**, *34*, 7889–7890;
(g) H. Huang, C. J. Forsyth, *J. Org. Chem.* **1995**, *60*, 2773–2779;
(h) H. Huang, C. J. Forsyth, *J. Org. Chem.* **1995**, *60*, 5746–5747;
(i) A. J. Frontier, S. Raghavan, S. J. Danishefsky, *J. Am. Chem. Soc.* **1997**, *119*, 6686–6687.

[194] K. Imamura, E. Yoshikawa, V. Gevorgyan, Y. Yamamoto, *Tetrahedron Lett.* **1999**, *40*, 4081–4084.

[195] J. Barluenga, G. P. Romanelli, L. J. Alvarez-García, I. Llorente, J. M. González, E. García-Rodríguez, S. García-Granda, *Angew. Chem., Int. Ed.* **1998**, *37*, 3136–3139; *Angew. Chem.* **1998**, *110*, 3332–3334.

[196] H. Inoue, N. Chatani, S. Murai, *J. Org. Chem.* **2002**, *67*, 1414–1417.

[197] N. Chatani, H. Inoue, T. Ikeda, S. Murai, *J. Org. Chem.* **2000**, *65*, 4913–4918.

[198] M. Nishizawa, H. Takao, V. K. Yadav, H. Imagawa, T. Sugihara, *Org. Lett.* **2003**, *5*, 4563–4565.

[199] (a) B. S. Thyagarajan, K. C. Majumdar, D. K. Bates, *J. Heterocycl. Chem.* **1975**, *12*, 59–66;
(b) K. K. Balasubramanian, K. V. Reddy, R. Nagarajan, *Tetrahedron Lett.* **1973**, *14*, 5003–5004;
(c) D. K. Bates, M. C. Jones, *J. Org. Chem.* **1978**, *43*, 3775–3776.

[200] (a) S. J. Pastine, S. W. Youn, D. Sames, *Org. Lett.* **2003**, *5*, 1055–1058;
(b) S. J. Pastine, D. Sames, *Org. Lett.* **2003**, *5*, 4053–4055.

[201] C. Jia, D. Piao, J. Oyamada, W. Lu, T. Kitamura, Y. Fujiwara, *Science* **2000**, *287*, 1992–1995.

[202] A. V. Vasilyev, S. Walspurger, P. Pale, J. Sommer, *Tetrahedron Lett.* **2004**, *45*, 3379–3381.

[203] P. M. Donovan, L. T. Scott, *J. Am. Chem. Soc.* **2004**, *126*, 3108–3112.

[204] K. Sangu, K. Fuchibe, T. Akiyama, *Org. Lett.* **2004**, *6*, 353–355.

[205] A. Fürstner, V. Mamane, *J. Org. Chem.* **2002**, *67*, 6264–6267.

[206] (a) M. B. Goldfinger, T. M. Swager, *J. Am. Chem. Soc.* **1994**, *116*, 7895–7896;
(b) M. B. Goldfinger, K. B. Crawford, T. M. Swager, *J. Am. Chem. Soc.* **1997**, *119*, 4578–4593;
(c) J. D. Tovar, T. M. Swager, *J. Organomet. Chem.* **2002**, *653*, 215–222.

[207] J. Barluenga, J. M. González, P. J. Campos, G. Asensio, *Angew. Chem., Int. Ed.* **1988**, *27*, 1546–1547; *Angew. Chem.* **1988**, *110*, 1604–1605.
[208] T. Yao, M. Campo, R. C. Larock, work in progress.
[209] X. Zhang, R. C. Larock, work in progress.
[210] (a) T. R. Klein, M. Bergemann, N. A. M. Yehia, E. Fanghänel, *J. Org. Chem.* **1998**, *63*, 4626–4631;
(b) T. R. Appel, N. A. M. Yehia, U. Baumeister, H. Hartung, R. Kluge, D. Ströhl, E. Fanghänel, *Eur. J. Org. Chem.* **2003**, 47–53.
[211] T. Kitamura, T. Takachi, H. Kawasato, H. Taniguchi, *J. Chem. Soc., Perkin Trans. 1* **1992**, 1969–1973.
[212] (a) A. S. K. Hashmi, T. M. Frost, J. W. Bats, *J. Am. Chem. Soc.* **2000**, *122*, 11553–11554;
(b) A. S. K. Hashmi, T. M. Frost, J. W. Bats, *Org. Lett.* **2001**, *3*, 3769–3771;
(c) A. S. K. Hashmi, T. M. Frost, J. W. Bats, *Catalysis Today* **2002**, *72*, 19–27.
[213] (a) B. Martín-Matute, D. J. Cárdenas, A. M. Echavarren, *Angew. Chem., Int. Ed.* **2001**, *40*, 4754–4757; *Angew. Chem.* **2001**, *113*, 4890–4893;
(b) B. Martín-Matute, C. Nevado, D. J. Cárdenas, A. M. Echavarren, *J. Am. Chem. Soc.* **2003**, *125*, 5757–5766.
[214] R. Akiyama, S. Kobayashi, *Angew. Chem., Int. Ed.* **2002**, *41*, 2602–2604; *Angew. Chem.* **2002**, *114*, 2714–2716.
[215] For reviews, see: (a) R. Grigg, V. Sridharan, *J. Organomet. Chem.* **1999**, *576*, 65–87; (b) R. Grigg, V. Sridharan, *Pure Appl. Chem.* **1998**, *70*, 1047–1057.
[216] C. H. Oh, S. J. Park, *Tetrahedron Lett.* **2003**, *44*, 3785–3787.
[217] (a) B. Burns, R. Grigg, V. Sridharan, P. Stevenson, S. Sukirthalingam, T. Worakun, *Tetrahedron Lett.* **1989**, *30*, 1135–1138;
(b) R. Grigg, J. M. Sansano, V. Santhakumar, V. Sridharan, R. Thangavelanthum, M. Thornton-Pett, D. Wilson, *Tetrahedron* **1997**, *53*, 11803–11826.
[218] (a) P. Fretwell, R. Grigg, J. M. Sansano, V. Sridharan, S. Sukirthalingam, D. Wilson, J. Redpath, *Tetrahedron* **2000**, *56*, 7525–7539;
(b) S.-H. Min, S.-J. Pang, C.-G. Cho, *Tetrahedron Lett.* **2003**, *44*, 4439–4442.
[219] S. M. A. Rahman, M. Sonoda, K. Itahashi, Y. Tobe, *Org. Lett.* **2003**, *5*, 3411–3414.
[220] N. Saito, H. Saito, M. Anzai, A. Yoshida, T. Fujishima, K. Takenouchi, D. Miura, S. Ishizuka, H. Takayama, A. Kittaka, *Org. Lett.* **2003**, *5*, 4859–4862.
[221] T. Sugihara, C. Copéret, Z. Owczarczyk, L. S. Harring, E. Negishi, *J. Am. Chem. Soc.* **1994**, *116*, 7923–7924.
[222] R. C. Larock, C. K. Reddy, *J. Org. Chem.* **2002**, *67*, 2027–2033.
[223] (a) S. Ma, X. Lu, *J. Org. Chem.* **1991**, *56*, 5120–5125;
(b) S. Ma, G. Zhu, X. Lu, *J. Org. Chem.* **1993**, *58*, 3692–3696;
(c) Q. Zhang, X. Lu, *J. Am. Chem. Soc.* **2000**, *122*, 7604–7605;
(d) C. Muthiah, M. A. Arai, T. Shinohara, T. Arai, S. Takizawa, H. Sasai, *Tetrahedron Lett.* **2003**, *44*, 5201–5204;
(e) S. Ma, X. Lu, *J. Organomet. Chem.* **1993**, *447*, 305–309.
[224] J. Ji, X. Lu, *Synlett* **1993**, 745–747.
[225] J.-E. Bäckvall, Y. I. M. Nilsson, P. G. Andersson, R. G. P. Gatti, J. Wu, *Tetrahedron Lett.* **1994**, *35*, 5713–5716.
[226] N. Asao, T. Shimada, Y. Yamamoto, *J. Am. Chem. Soc.* **1999**, *121*, 3797–3798.
[227] K. Imamura, E. Yoshikawa, V. Gevorgyan, Y. Yamamoto, *J. Am. Chem. Soc.* **1998**, *120*, 5339–5340; erratum **1998**, *120*, 13 284.
[228] H. Huang, C. J. Forsyth, *J. Org. Chem.* **1997**, *62*, 8595–8599.
[229] C. Fernández-Rivas, M. Méndez, A. M. Echavarren, *J. Am. Chem. Soc.* **2000**, *122*, 1221–1222.

3
Addition of Terminal Acetylides to C=O and C=N Electrophiles

Patrick Aschwanden and Erick M. Carreira

3.1
Introduction

For hydrocarbons, acetylenes enjoy rich reaction chemistry. Internal acetylenes not only undergo many of the functionalization reactions of olefins, but are also subject to a plethora of additional transformations [1]. They have been shown to participate readily in amination [2] and carbometalation reactions [3], and can also be isomerized to the terminal isomers under the basic conditions of the zipper reaction [4]. Terminal acetylenes can be subjected to reactions similarly to their internal isomers, albeit with the added benefit that in general their reactions tend to exhibit greater regioselectivity as a consequence of steric and electronic considerations. There is one particular aspect of terminal acetylenes that sets them apart from other hydrocarbons, namely the lability of the terminal proton under a variety of conditions. The ability of terminal acetylenes to be readily deprotonated is believed to be due to a high proportion of s character associated with the carbon atom (Figure 3.1) [5(a), 6]. This feature has been recognized for some time and enables the generation of acetylides capable of participating in C-C bond formation.

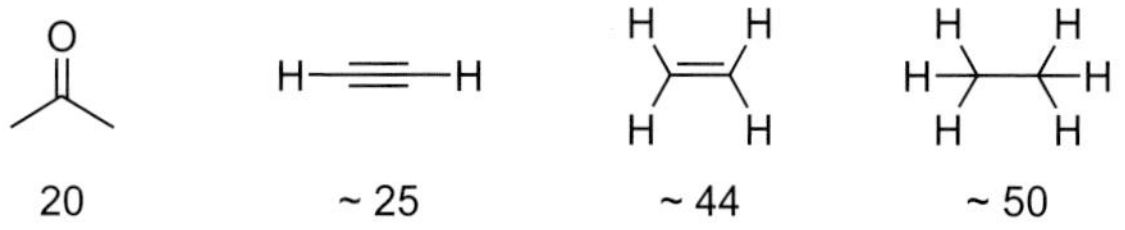

Figure 3.1
pK_a values of acetone and simple hydrocarbons.

The ease with which terminal acetylenes undergo metalation has been critical in a number of key coupling reactions involving terminal acetylenes. Thus, the oxidative dimerization of acetylenes and the Pd-catalyzed coupling with aryl and alkenyl halides are prominently employed in synthetic chemistry. The classic Glaser reaction, first reported in 1869, involves the oxidative coupling of terminal acetylenes in the presence of Cu(I), amine base, and oxygen [7]. This was subsequently to become the cornerstone for the development of the Eglinton, Hay, Cadiot–Chodkie-

Acetylene Chemistry. Chemistry, Biology and Material Science. Edited by F. Diederich, P. J. Stang, R. R. Tykwinski

ISBN 3-527-30781-8

wicz, Castro–Stephens, and Sonogashira coupling reactions, all of which have revolutionized synthetic strategies for C-C bond formation (Figure 3.2) [7]. Additionally, in the presence of simple bases (e.g., KOH), terminal acetylenes will add to aldehydes and ketones. This feature forms the basis of a synthesis of vitamin A (**1**) documented by workers at BASF (Figure 3.3) [8]. The addition of acetylene to methyl heptenone **2** provides access to dehydrolinalool **3** that subsequently leads to β-ionone **4**. The addition of acetylene to **4** furnishes vinyl-β-ionol **5**, a precursor to a key Wittig reagent **6**, which is ultimately coupled to C_5 building block **7** derived from the addition of acetylene to methyl glyoxal dimethylacetal (**8** → **9**).

The traditional methods involving metalation by Ag(I) or Cu(I) furnish alkynylides that appear to lack the necessary reactivity towards a broad range of C=O and C=N electrophiles. Thus, despite the simplicity of the methods described above, alkynylides typically employed in C=O and C=N addition reactions are commonly prepared from an acetylene and bases such as organolithiums (e.g., BuLi) [9], organomagnesiums (e.g., EtMgBr) [10], and amides [11]. Importantly, because the aldehydes and imines that one would wish to use are incompatible with such basic and/or nucleophilic reagents, alkyne deprotonation must necessarily be carried out as a separate step [12].

Glaser Coupling

$$2\ Ph-\!\!\equiv\!\!-H\ +\ 1/2\ O_2 \xrightarrow[\text{NH}_4\text{OH, EtOH}]{\text{cat. CuCl}} Ph-\!\!\equiv\!\!\equiv\!\!-Ph\ +\ H_2O$$

Eglinton Coupling

$$2\ R-\!\!\equiv\!\!-H\ +\ 1/2\ O_2 \xrightarrow[\text{pyridine}]{\text{cat. CuX}} R-\!\!\equiv\!\!\equiv\!\!-R\ +\ H_2O$$

Hay Coupling

$$2\ R-\!\!\equiv\!\!-H\ +\ 1/2\ O_2 \xrightarrow[\text{TMEDA, acetone}]{\text{cat. CuCl}} R-\!\!\equiv\!\!\equiv\!\!-R\ +\ H_2O$$

Cadiot Chodkiewicz

$$R-\!\!\equiv\!\!-H\ +\ X-\!\!\equiv\!\!-R' \xrightarrow[\text{NEt}_3]{\text{cat. CuX}} R-\!\!\equiv\!\!\equiv\!\!-R'$$

Castro Stephens

$$R-\!\!\equiv\!\!-H\ +\ X-Ar \xrightarrow[\text{pyridine}]{\text{cat. CuX}} R-\!\!\equiv\!\!-Ar$$

Sonogashira

$$Ar-X\ +\ H-\!\!\equiv\!\!-R \xrightarrow[\text{cat. CuI, NEt}_3]{\text{cat. PdCl}_2(\text{PPh}_3)_2} Ar-\!\!\equiv\!\!-R$$

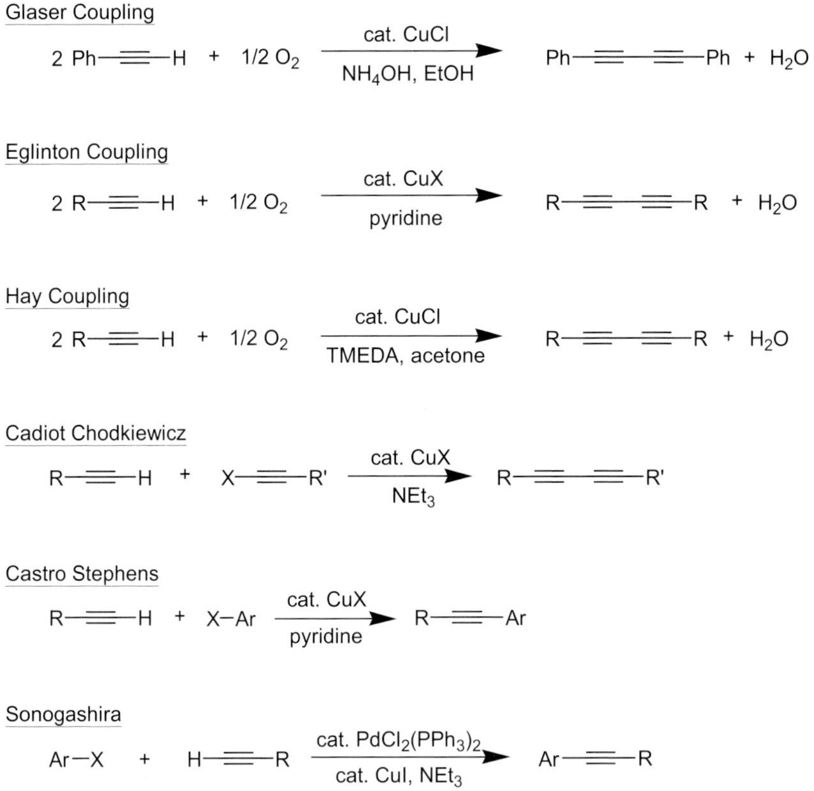

Figure 3.2 The classic Cu(I)-mediated acetylene activation and coupling reactions.

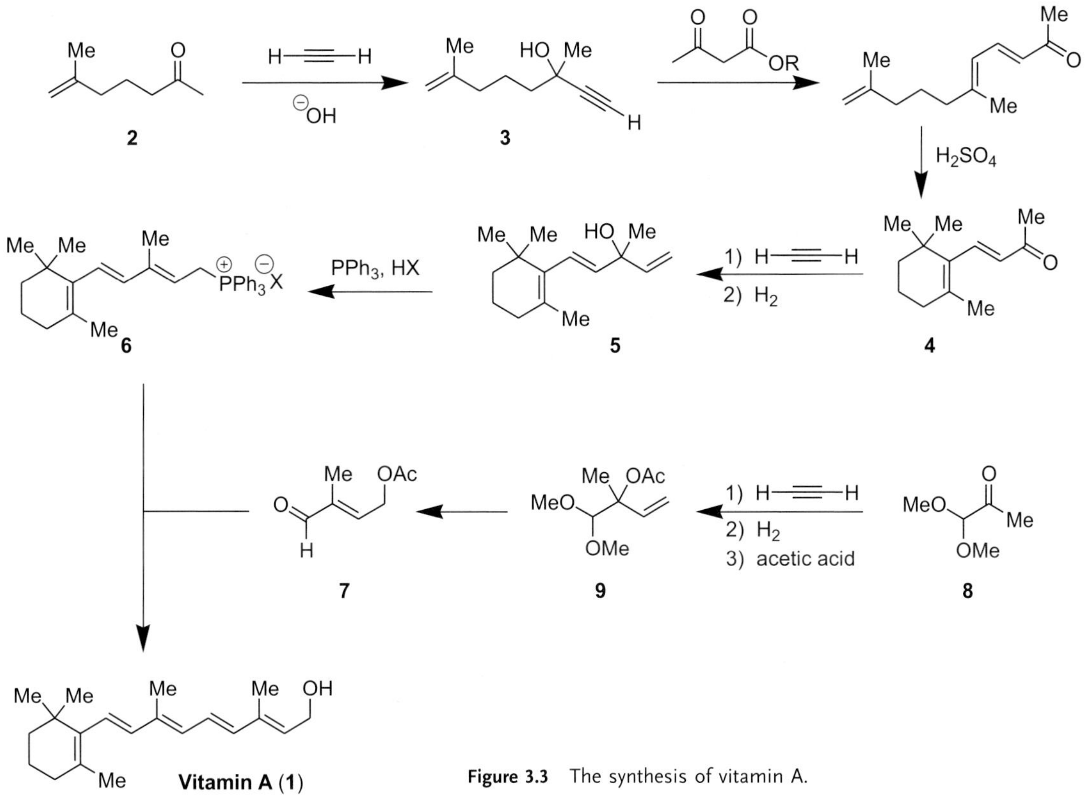

Figure 3.3 The synthesis of vitamin A.

By building upon the known metalation methods and on recent discoveries and developments of novel methods for the activation of terminal acetylenes, a multitude of fresh possibilities for the generation of new processes for nucleophilic additions have been achieved. This chapter focuses on the latest developments in acetylene deprotonation and activation to afford reactive organometallic reagents that have been incorporated into asymmetric C=O and C=N addition reactions.

3.2 Background

As early as 1900, Favorsky and Skosarevsky reported the addition reaction between acetylene and aldehydes or ketones in the presence of KOH [13]. At 5 °C, monoaddition of acetylene to carbonyl acceptors was observed, while at elevated temperatures (25–35 °C) the 1,4-diol adducts were preferentially isolated. Both catalytic and stoichiometric variants of this fundamental process were extensively investigated (Equation 3.1) [14, 15]. In general it was observed that the use of highly polar solvents such as DMSO, *N*-methylpyrrolidone, ammonia, and HMPA was optimal.

The high solubility of acetylene in such media and the existence of complexes involving base and solvents were responsible for the unique aspects of these systems. In 1996, it was shown that, in DMSO, potassium *tert*-butoxide effects deprotonation of acetylenes and the subsequent acetylide participates in ketone addition reactions (Equation 3.2) [16]. Attempts to utilize an enolizable aldehyde in this process, however, resulted in a complex mixture of products. Subsequently, the use of tetraalkylammonium hydroxides in DMSO was shown to permit the addition of terminal acetylenes to aldehydes in yields of up to 96% (Equation 3.3) [17].

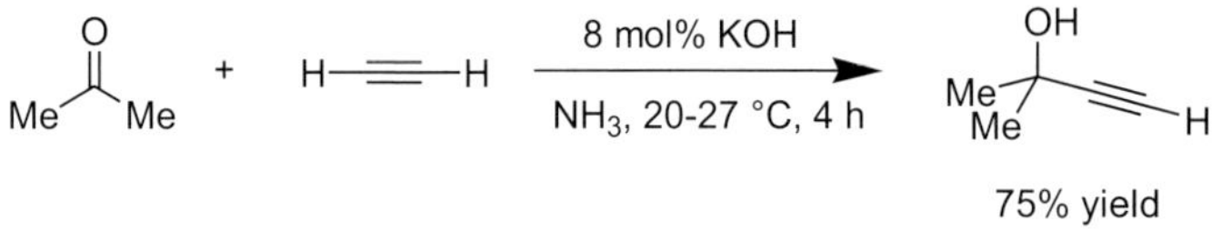

Equation 3-1

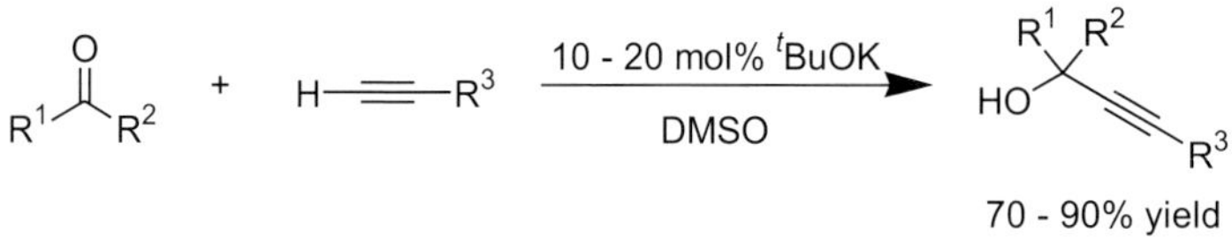

Equation 3-2

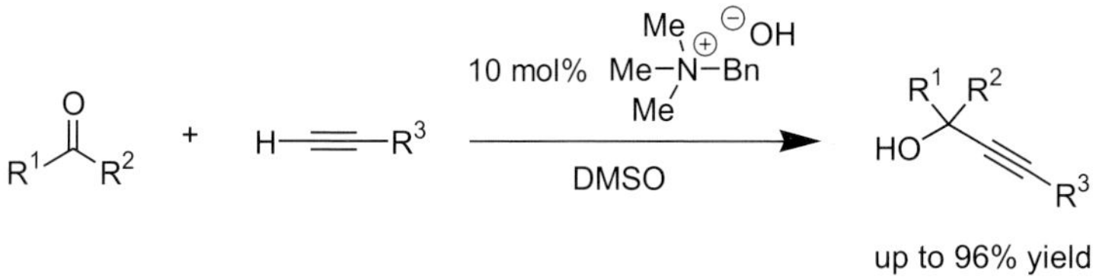

Equation 3-3

In an effort to expand the scope of these straightforward processes, Corey elegantly demonstrated that Cs salts are effective for the conversion of trimethylsilylacetylene into the corresponding cesium acetylides, which can then be employed in carbonyl addition reactions (Equation 3.4) [18]. Subsequent recent work by Knochel has expanded this method to include CsOH in DMSO/THF or THF (Equation 3.5) [19]. These methods for the generation of propargyl alcohols are noteworthy for their simplicity and convenience of execution.

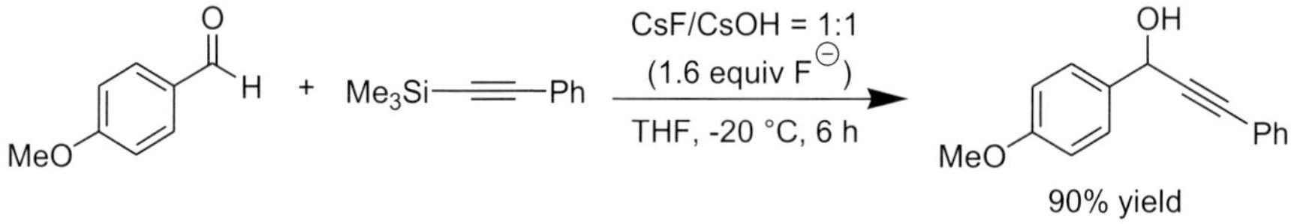

Equation 3-4

$$R^1\overset{O}{\underset{}{\mathord{\text{—}}}}R^2 + H\!=\!\!=\!\!R^3 \xrightarrow[\text{23 °C, 1-5 h}]{\text{CsOH·H}_2\text{O (10-30 mol\%)} \atop \text{THF/DMSO or THF}} R^1\underset{R^2}{\overset{OH}{\mathord{\text{—}}}}\!=\!\!=\!\!R^3$$

66-96% yield

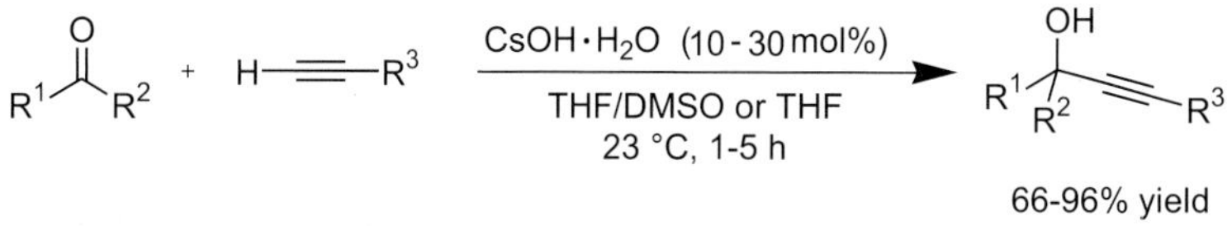

Equation 3-5

Despite the practical aspects of processes involving deprotonation with alkali metal hydroxides and related species, however, the use of such metalation processes in catalytic asymmetric synthesis remains elusive. Although a number of notable processes involving the use of organocesium reagents in asymmetric synthesis have been described [20], in general, understanding and manipulation of organopotassium and -cesium reagents is not as far advanced as that of other organometallic or main group species. There has thus been recent interest in the discovery and study of methods for the metalation of terminal acetylenes under mild conditions to produce species amenable to modification and manipulation by chiral ligands and thus result in the formation of optically active propargyl amines and alcohols.

The prototype for the development of mild methods for acetylene deprotonation and their subsequent use in C-C bond formation has been the chemistry of Cu(I) and terminal acetylenes. Thus, as outlined above, methods that use Cu(I) acetylides enjoy a rich history in synthetic chemistry, because of their reliability in preparative synthesis. However, despite their immense versatility in transition metal-mediated coupling processes, they are considerably less effective with C=O and C=N electrophiles. Reactions of the former are rare [21], while the latter are known, but rather limited in their application (vide infra).

Propargyl alcohols and amines constitute useful and versatile building blocks for chemical synthesis. In particular, the former have served as intermediates for the synthesis of hydroxy acids, steroids, avenaciolide, vitamin E, prostaglandins, pheromones, tetrahydrocerulenin, and other compounds [22]. There has therefore been great interest in the discovery and development of methods for their use as building blocks. The synthesis of optically active propargyl alcohols has so far largely been achieved through the enzymatic resolution of racemic secondary propargyl alcohols or enantioselective reduction of ynones [23, 24, 25].

The capability to prepare propargyl alcohols or amines by direct addition of metalated terminal acetylenes to aldehydes or imines would constitute a complementary approach to the more traditional methods. In the latter methods the propargyl stereocenter is established in a separate step distinct from the C-C bond formation. In contrast, the direct addition of acetylides to C=O or C=N electrophiles should in principle be more efficient, because the ynone starting materials required for the more traditional approach are not in general readily available.

Methods for the enantioselective addition of metal acetylides to aldehydes can be classified into two broad categories: (1) addition reactions involving the stoichiometric use of preformed metalated terminal acetylenes, and (2) those in which metal acetylides are generated in situ and produced in substoichiometric amounts.

3.3 Additions with Stoichiometric Amounts of Metal Acetylides

The pioneering investigations involving asymmetric addition of metalated acetylides are to be found in the late 1980s, with alkynyl boron and zinc reagents. In 1994, Corey and Cimprich documented the effective use of boryl acetylides **10** in enantioselective aldehyde addition reactions (Equation 3.6) [25(b)]. Extensive mechanistic studies of oxazoborolidine-catalyzed borohydride reductions of ketones set the stage and provided a guiding paradigm for the study of acetylide additions [26]. These asymmetric additions of **10** were mediated by the chiral oxazaborolidene **11**, which is readily assembled from a proline-derived amino alcohol and an alkylboronic acid. The reported product enantioselectivities were impressive, with products isolated in preparatively useful yields and in up to 98% *ee*s. A proposed transition state structural model (Figure 3.4) enjoys the important features of dual activation of both the nucleophile (acetylide) and electrophile (aldehyde) that had been proffered in the oxazaborolidine-catalyzed ketone reductions. The precursors to **10** are the corresponding stannylacetylenes. These have been investigated in diastereoselective aldehyde additions in which products are formed with high selectivity (Equation 3.7) [27].

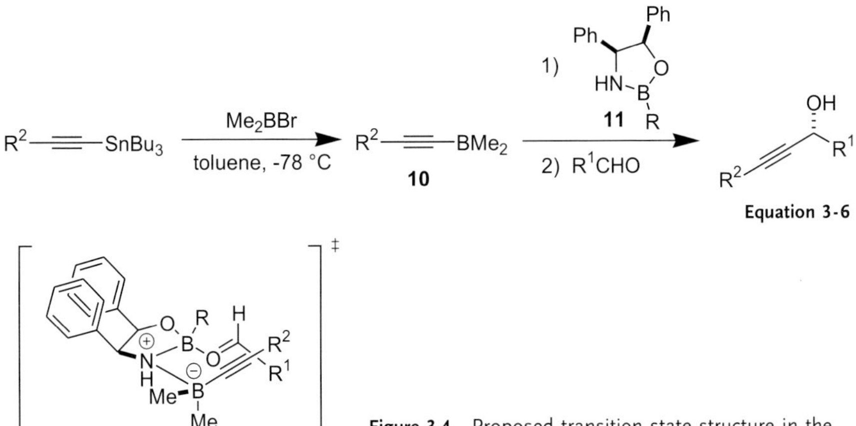

Equation 3-6

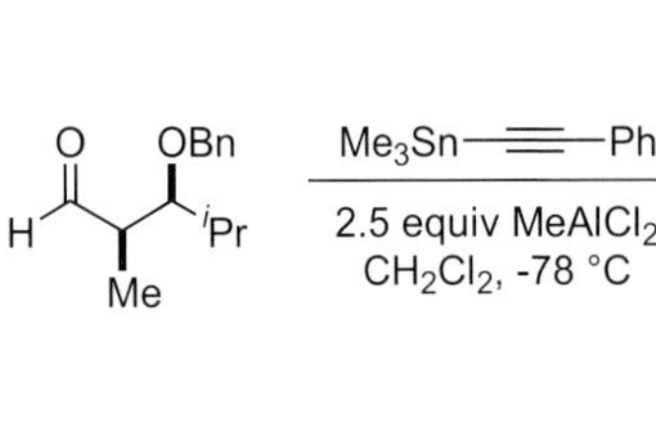

Figure 3.4 Proposed transition state structure in the activation and addition of borylacetylide.

Felkin : anti-Felkin = 4 : 96 (68% yield)

Equation 3-7

Another significant study involves the addition reactions of dialkynylzinc reagents **12** reported by Niwa and Soai (Equation 3.8) [28, 29]. The investigations were based on an earlier report by Mukaiyama and co-workers involving 2-hydroxymethyl-1-[(1-methylpyrolidin-2-yl)methyl]pyrrolidine, which catalyzed the addition of Et_2Zn to benzaldehyde [30]. The metalated acetylene is prepared by allowing the terminal acetylenes to react with Et_2Zn [31]. In the presence of (+)-*N,N*-dibutylnorephedrine (**13**) the dialkynylzinc reagents add to aldehydes to afford propargyl alcohol products with modest enantioselectivities (up to 43% *ee*). In the same report, the complementary addition of dialkylzinc reagents to alkynals proved more effective in furnishing propargyl alcohol adducts in up to 78% *ee*.

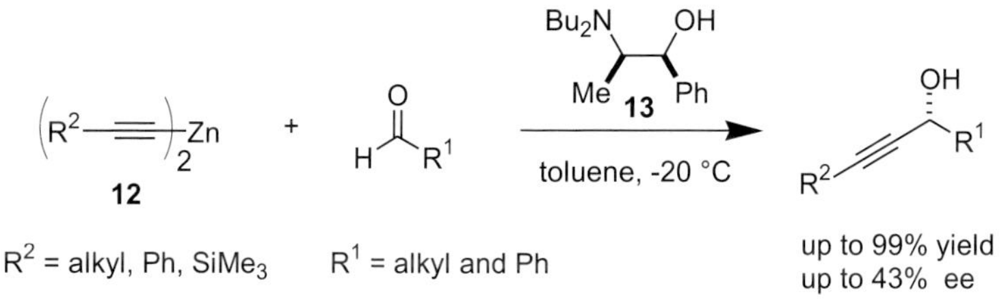

Equation 3-8

Enantioselective additions of PhC≡CZnBr (**14**) were investigated in the context of a synthesis of **15**, an insecticide displaying low toxicity towards fish (Equation 3.9) [32]. The zinc acetylide reagent was generated by transmetalation of lithiated phenyl acetylide (PhC≡CH and BuLi, 0 °C) with $ZnBr_2$. The organometallic species, formulated as PhC≡CZnBr, was treated with the lithium alkoxide salt of (–)-*N*-methylephedrine **16**, and the ensuing complex was shown to undergo addition to aldehyde **17**, giving propargyl alcohol adduct **15** in 88% *ee* and 80% yield. The investigators went on to demonstrate that addition reactions to other aromatic aldehydes furnished adducts with modest to good enantioselectivities.

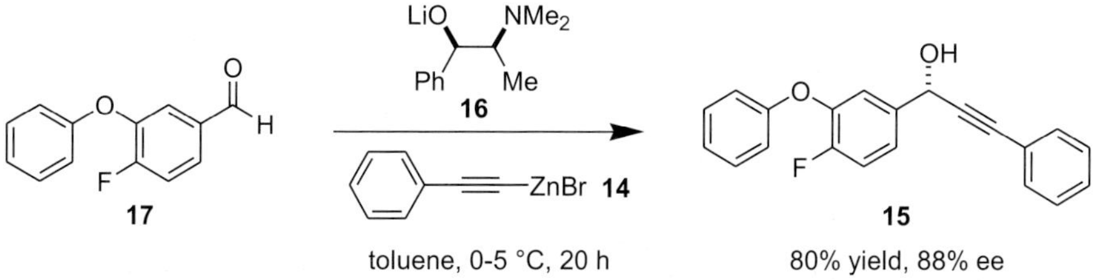

Equation 3-9

Studies have also been carried out with mixed organozinc agents such as ethyl(2-phenylethynyl)zinc (**18**, Equation 3.10) [33], prepared from reactions between terminal alkynes and Et_2Zn. The addition of such species to aldehydes was examined in the presence of 10 mol% pyridinyl alcohol ligand **19**, with optimal selectivities observed at room temperature [34]. The range of aldehydes examined included benzaldehyde and various aliphatic aldehydes (pivaldehyde, cyclohexylcarboxaldehyde, nonaldehyde), and the corresponding adducts were isolated in up to 90% *ee*.

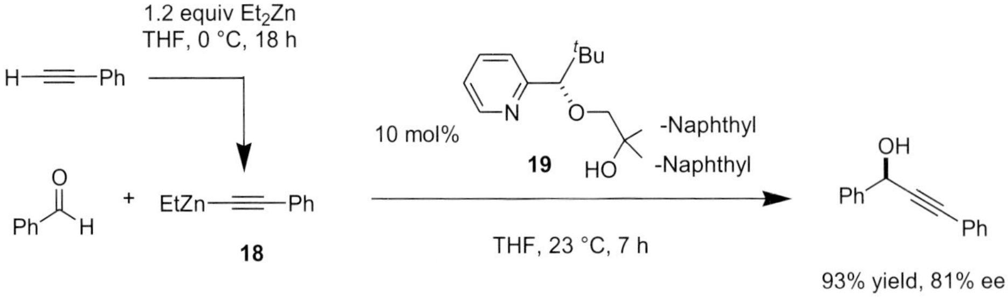

Equation 3-10

Further new developments in the reaction of mixed organozinc reagents R_1ZnR_2 (R_1 = alkyl, R_2 = alkynyl) have recently been documented. The addition reaction of methyl(2-phenylethynyl)zinc (**20**) to aromatic aldehydes has been effected by the use of modified 2-amino-1,2-diphenylethanol-derived ligands (Equation 3.11) [35]. The addition reactions to aromatic aldehydes were conducted over 6–48 h at −20 °C in toluene with 5–20 mol% ligand **21** and 2.2 equiv. Et_2Zn to give adducts with >95% conversion and in up to 93% *ee*. In the course of this study the investigators proposed transition structure **22** (Figure 3.5) to account for the observed absolute configuration of the isolated products.

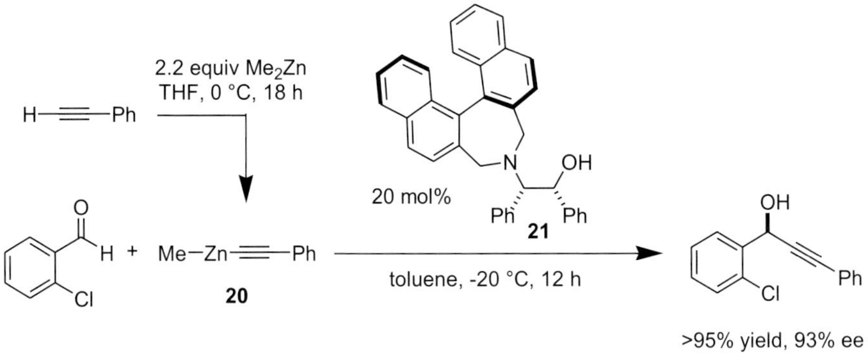

Equation 3-11

The use of BINOL modified by 3,3′-substitution with bulky groups has proven effective in the addition reaction of **18** (generated in situ from phenylacetylene and 2 equiv. Et_2Zn) to aromatic aldehydes such as benzaldehyde, *o*- and *p*-chlorobenzaldehyde, *o*- and *m*-tolualdehyde, *o*- and *p*-methoxybenzaldehyde, and 1-naph-

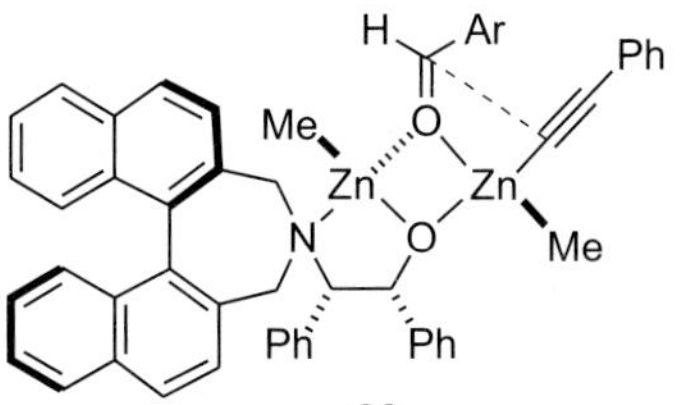

22

Figure 3.5 Proposed transition state structure in the addition of **20** and aromatic aldehydes mediated by ligand **21**.

thaldehyde (Equation 3.12) [36]. The addition reactions are typically conducted with 10 mol% ligand **23** and 2 equiv. Et_2Zn in THF at ambient temperature, affording aryl alkynyl carbinols in 45–75% yield and with up to 94% *ee*.

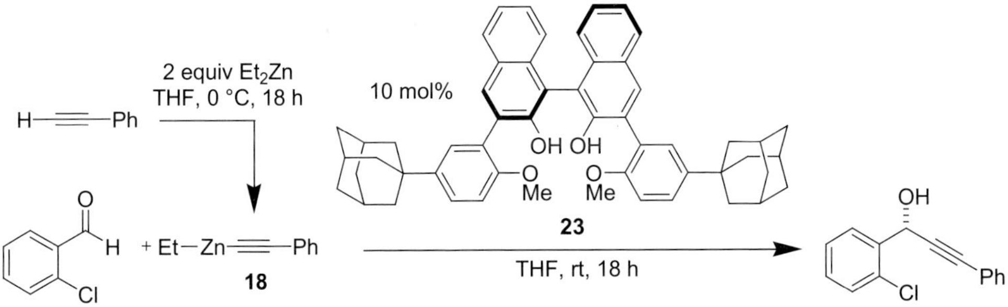

Equation 3-12

There have been a number of studies in which the putative complex formed between BINOL itself and $Ti(O^iPr)_4$ has been utilized in the reactions of **18** or **20** and aldehydes as shown (Equations 3.13–3.16). In the first of these studies [37], Moore and Pu observed optimal selectivities in the additions of aromatic aldehydes and **18**, as well as in an example involving $^iPr_3SiC{\equiv}CZnEt$. The metalation of the alkyne was conducted with 1 equiv. Et_2Zn in refluxing toluene, and the subsequent aldehyde additions were carried out in dichloromethane at ambient temperature. The additions were carried out with 50 mol% $Ti(O^iPr)_4$, and 20 mol% BINOL, giving adducts in 71–81% yields and with 92–98% *ee*s (Equation 3.13). The investigators expanded the scope of the addition of phenylacetylene to include aliphatic carboxaldehydes, including nonanal, octanal, pentanal, isobutyraldehyde, cyclohexylcarboxaldehyde, and dihydrocinnamaldehyde (Equation 3.14) [38]. Extensive studies of the reaction parameters resulted in the optimal procedure, which prescribes the use of 4 equiv. phenylacetylene, 4 equiv. of Et_2Zn, 1 equiv. $Ti(O^iPr)_4$, and 40 mol% BINOL to afford adducts in 58–96% *ee*s and >91% yields [25(a)]. A key aspect of the process is the slow addition of aldehyde to the reaction mixture. This procedure has been employed by Marshall and Bourbeau [39] in the preparation of building blocks of great utility in polyketide synthesis (Scheme 3.1). In related independent work, the benefits of additives in the addition reaction between methyl(phenylethynyl)zinc (**20**; generated in situ from phenylacetylene and 1.2 equiv. Me_2Zn at 0 °C) and aldehydes were demonstrated under conditions involving 15–30 mol % $Ti(O^iPr)_4$ in THF at 0 °C (Equation 3.15) [40]. Particularly importantly, the investigators observed that the addition of 10 mol % phenol can lead to improvements in the enantioselectivity of the products formed. Thus, in a typical example, the addition of phenylacetylene to benzaldehyde in the absence and in the presence of phenol afforded adducts with 90 and 96 % *ees*, respectively. Further developments of the BINOL-catalyzed processes have provided new modes for the preparation of catalytically active systems in situ, with the use of **24** as an additional ligand (Equation 3.16) [40(b)].

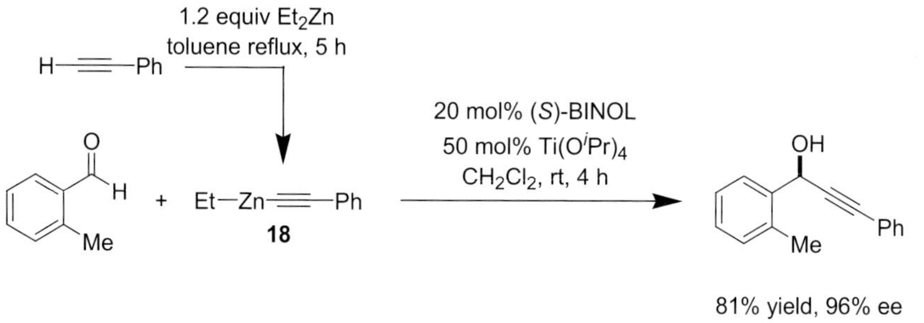

Equation 3-13

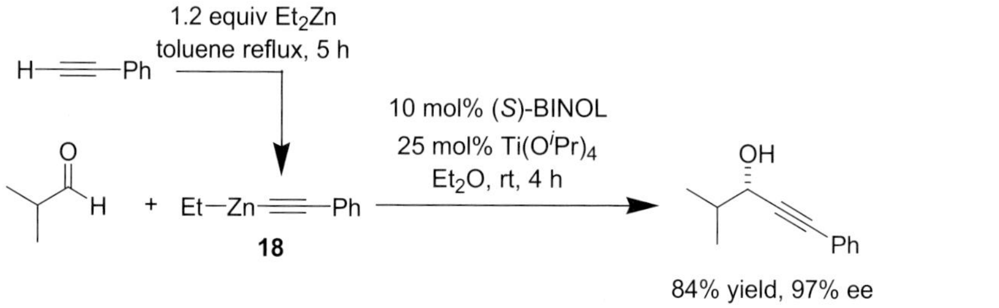

Equation 3-14

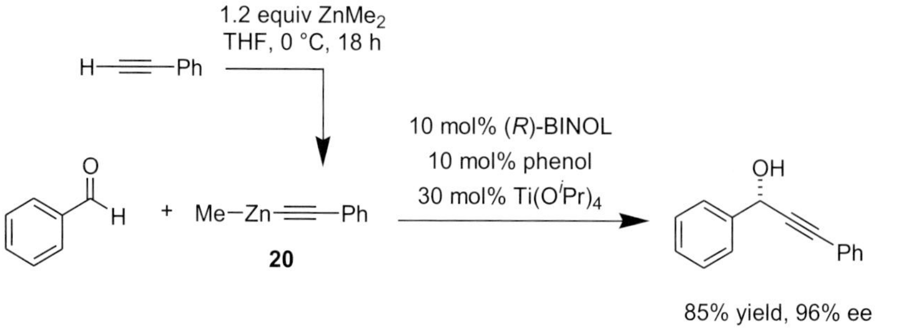

Equation 3-15

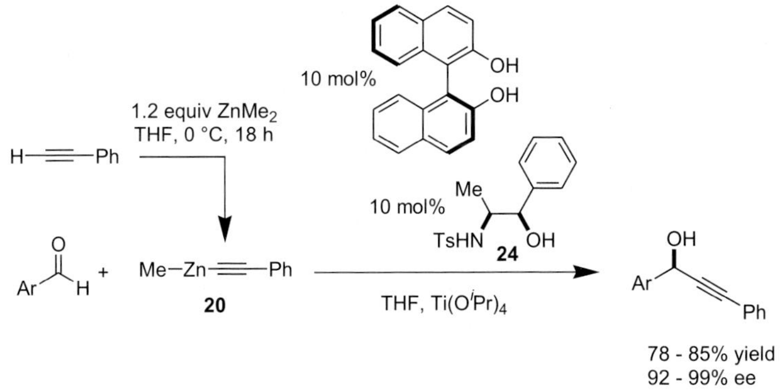

Equation 3-16

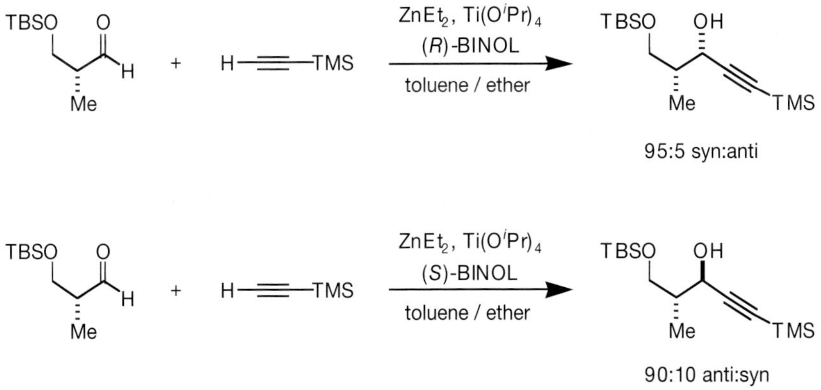

Scheme 3.1 Preparation of propargyl alcohols as building blocks for polyketide synthesis with BINOL, Et_2Zn, and $Ti(O^iPr)_4$.

Additions of metalated terminal acetylenes to ketones have also been investigated. Pioneering studies were carried out at Merck in the context of a practical synthesis of efavirenz (**25**), an HIV reverse transcriptase inhibitor (Scheme 3.2) [41]. This group has reported a key enantioselective addition of lithium-cyclopropylacetylide complex **26** (obtained from **27**) to *p*-methoxybenzyl-protected trifluoromethyl ketoanilide **28** mediated by the lithium alkoxide salt of (–)-*N*-pyrrolidinylnorephedrine **29**, affording adduct **30** in 99% *ee*. Careful mechanistic studies including meticulous spectroscopic work by Collum and co-workers [42] provided detailed understanding and allowed optimization of the reaction. A related process utilizing the zinc acetylide complex **31** (Scheme 3.3) was also developed and shown to give adducts with high selectivity [43].

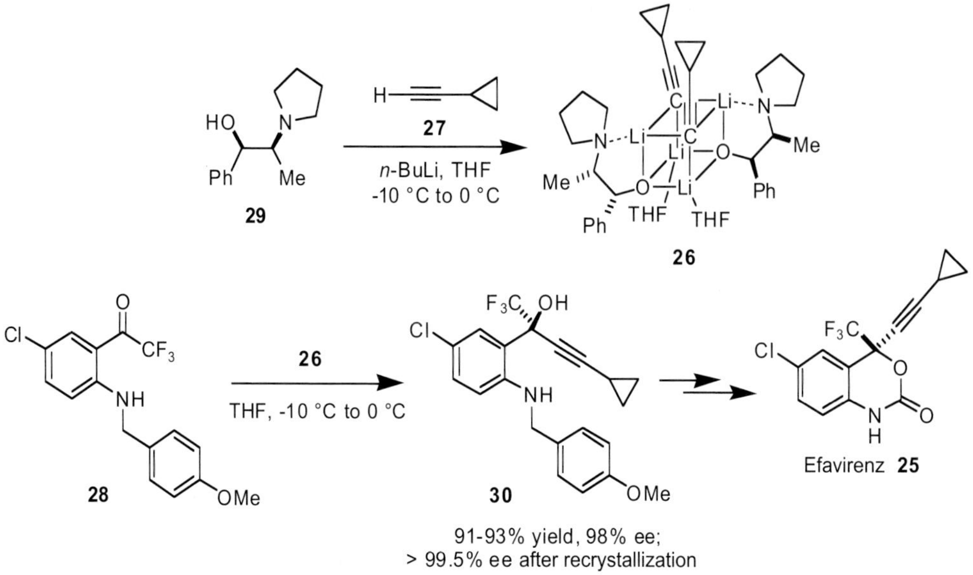

Scheme 3.2 The Merck synthesis of Efavirenz with the aid of a chiral lithium-cyclopropyl acetylide complex.

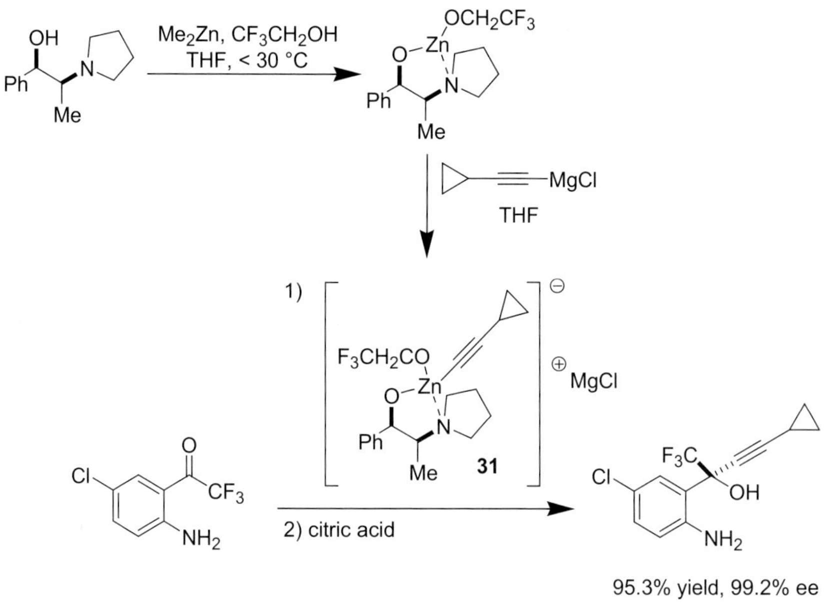

Scheme 3.3 The Merck synthesis of Efavirenz with the aid of an optically active zincate.

Subsequent studies demonstrated that a related reaction process could be employed with several amino alcohol ligands (e. g., **32** and **33**) in the preparation of propargyl alcohols derived from the addition of acetylides to various aromatic aldehydes (Equation 3.17) [44]. The addition of lithium acetylide **34** to benzaldehyde mediated in the presence of 1 equiv. of chiral ligand **35** afforded adduct in 99% *ee* and 84% yield (Equation 3.18) [45].

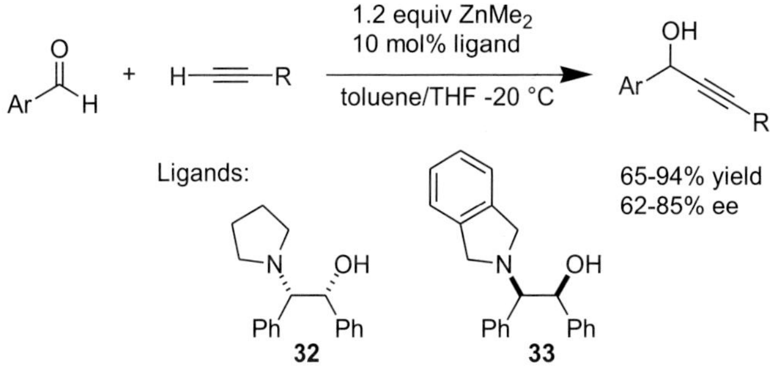

Equation 3-17

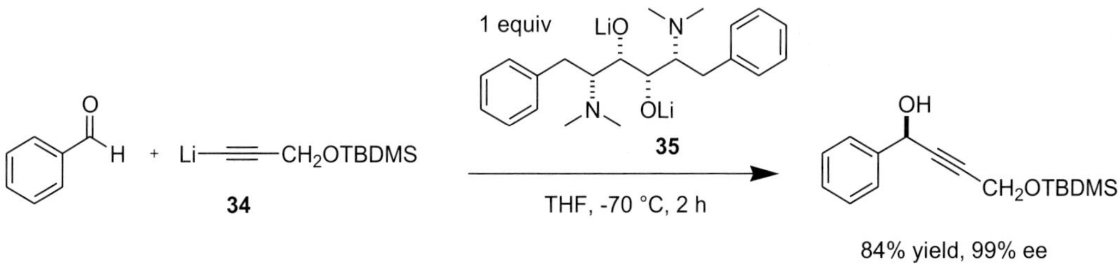

Equation 3-18

There have been additional recent developments in enantioselective addition of mixed methyl alkynylide zinc reagents to ketones [46], Thus, **20** and related species were generated from phenylacetylene, $Me_3SiC{\equiv}CH$, and propargyl chloride by treatment of the terminal acetylenes with 3 equiv. Me_2Zn in toluene. After addition of 20 mol% chiral ligand **36**, the addition reactions are carried out in toluene at ambient temperature over the course of 96 h to afford tertiary alcohols in 40–89% yields and with 32–81% *ees* (Equation 3.19).

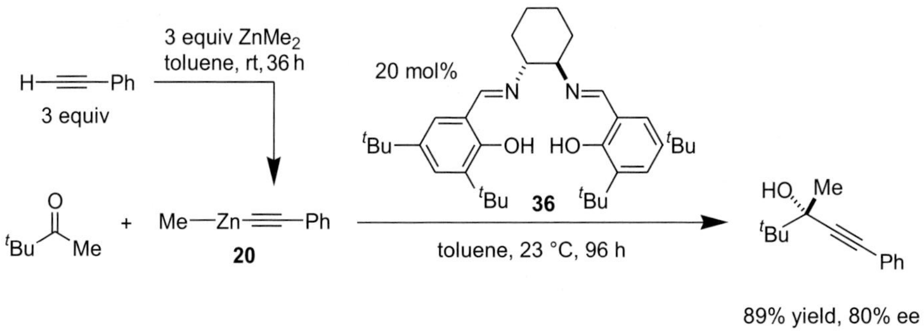

Equation 3-19

3.4
Nucleophilic C=O Additions involving the use of Zn(II) Salts

The majority of processes that have been devised, and continue to be developed, for the preparation and use of acetylide carbanions require the generation of stoichiometric quantities of metalated acetylene. As described above, this has been effected either with strong amide bases or by metalation with alkyllithium (e. g., BuLi) or dialkylzinc, typically in excess. The fact that Cu(I) and Ag(I) acetylides have been generated under exceedingly mild conditions (ambient temperature, amine bases) and that these have been shown to participate in selected additions to iminium electrophiles (vide infra) suggests that similar conditions might be identifiable for other metal salts. Acetylides generated in such a manner could be compatible in catalytic cycles involving C=O and C=N electrophiles and thus present new avenues for further investigation and discovery. In this context, the development of an interesting modern process for acetylide generation is to be found in a report involving the use of amine bases and Sn(II) (Equation 3.20) [47]. In the presence of 1 equiv. $Sn(OTf)_2$ and tributylamine, both aldehydes and ketones were shown to participate in acetylide additions to afford the corresponding propargyl alcohols at ambient temperature in 57–95 % yields.

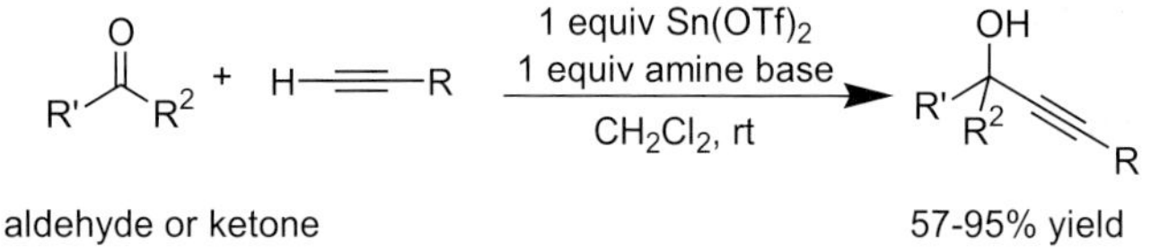

Equation 3-20

In 1999, a novel procedure for the generation of metalated acetylenes under mild conditions was identified, it being shown that terminal acetylenes could be converted rapidly into Zn acetylides in the presence of amines and Zn(II) at room temperature (Equation 3.21) [48]. In subsequent investigations, a mechanistic scheme by which Zn(II) and amine base combine to afford a monoalkynyl zinc species was proposed. This species subsequently participates in C=O and C=N additions. The acetylene can be activated towards deprotonation through either one of two processes (Figure 3.6) [49]. The first involves complexation of the acetylene to the Zn(II) center (**37**), which should increase the ability of the acetylenic proton to form a hydrogen bond with base (**38**) and ultimately to undergo deprotonation. Alternatively, formation of a hydrogen bond between the terminal acetylene and amine (**39**) should lead to a corresponding increase in the propensity to coordinate to Zn(II) and form **40**. The generation of Zn-alkynylides may thus take place through the operation of either pathway.

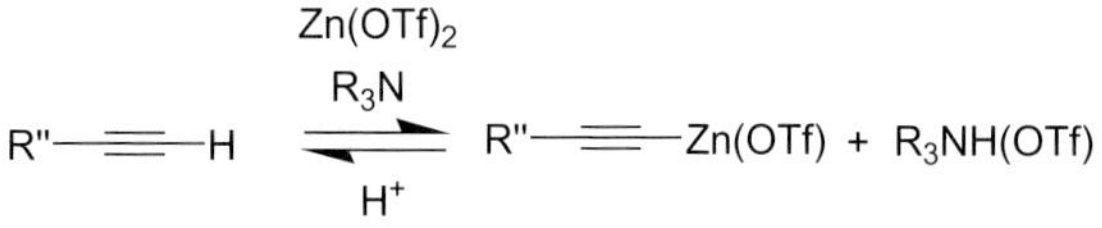

Equation 3-21

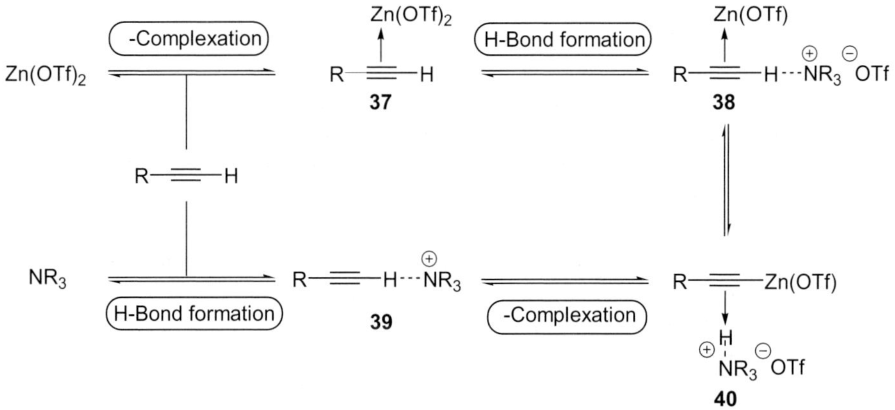

Figure 3.6 Proposed process for the metalation of terminal acetylenes by Zn(II) and amine bases.

Infra-red spectroscopic studies proved critical in providing data consistent with the proposed metalation process (Figure 3.7) [50]. In these experiments, the reversible disappearance of the stretch corresponding to the terminal C-H of phenylacetylene could readily be monitored. When the acetylene and either an amine base (e. g., triethyl amine, Hünig's base) or $Zn(OTf)_2$ were used alone, no evidence of deprotonation was observed. The combination of amine and $Zn(OTf)_2$, however, led to rapid disappearance of the terminal alkyne C-H stretch. Subsequent portionwise addition of triflic acid resulted in the reappearance of the terminal C-H stretch, consistently with a reversible metalation event.

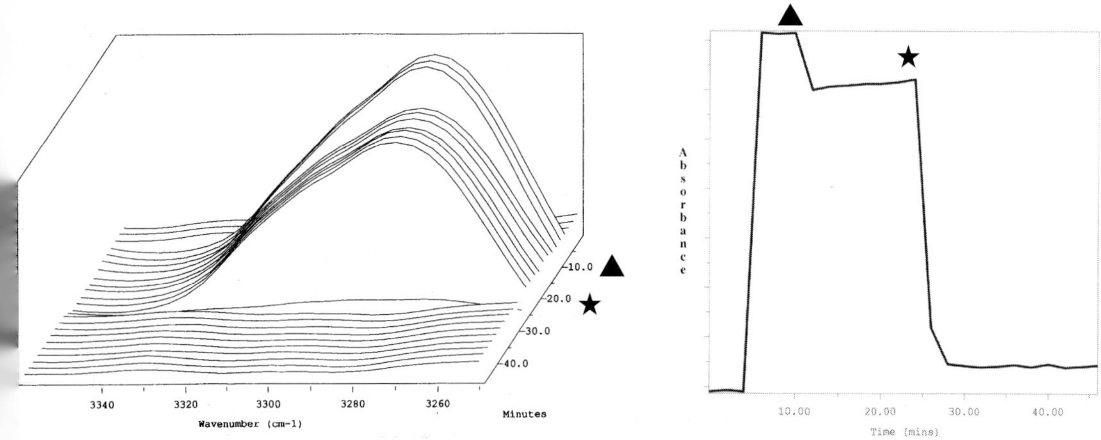

Figure 3.7 ReactIR spectra of the C–H stretch resonance signal of phenylacetylene in CH_3CN. At 10 min (▲) Et_3N was added, and subsequently at 25 min (★) $Zn(OTf)_2$. In the presence of both components the IR band corresponding to the terminal C–H completely disappears within 4 minutes.

The new method for the generation of alkynylzinc reagents was initially applied in additions to nitrones and subsequently aldehydes, to afford propargyl hydroxylamines and alcohols, respectively. As described in the previous section, zinc alkynylides had previously been prepared through the use of stoichiometric metalating agents [51]: (1) most notably by transmetalation of dialkylzinc and terminal acetylenes to give either mixed alkyl alkynyl zinc organometallics or the dialkynylzinc species, or alternatively (2) transmetalation of alkynyl lithium intermediates with ZnX_2 salts. The nature of the species generated in the second of these approaches is far from clear, because any of a number of organometallic reagents (RZnX, $RZnX_2$, $LiZnR_2X$, for example) might result. The observation that acetylenic zinc reagents could be generated directly from terminal acetylenes and simple zinc salts thus not only offers a new organometallic species, but also represents a new reactivity mode not previously observed. Importantly, the method offers new potential for the construction of C-C bonds, as it was subsequently shown that the organozinc reagents displayed reactivity differences with previously investigated metal acetylides.

In the initial applications of the new metalation technique, asymmetric aldehyde addition reactions with stoichiometric quantities of Zn(II) and NEt_3 or Hünig's base were investigated (Equation 3.24–3.27) [52]. Subsequently, it was shown that other conditions can also be employed in the metalation reaction (Equation 3.22). Thus, a study describing the use of $Zn(OTf)_2$ and DBU in ionic liquids – 8-ethyl-1,8-diazabicyclo[5.4.0]-7-undecinium triflate ([EtDBU][OTf]) and 1-butyl-3-methyl-1*H*-imidazolium tetrafluoroborate and hexafluorophosphate ([bmim][BF_4], [bmim][PF_6]) – not only introduced a new base for the metalation, but also implicates the potential of $Zn(BF_4)_2$ and $Zn(PF_6)_2$ salts in the metalation reaction [53]. Additionally, $ZnCl_2$ and $ZnCO_3$ were also shown to be effective in subsequent studies (Equation 3.23) [54].

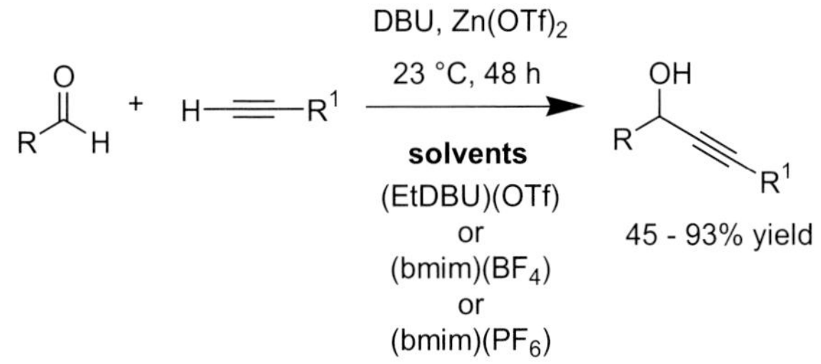

Equation 3-22

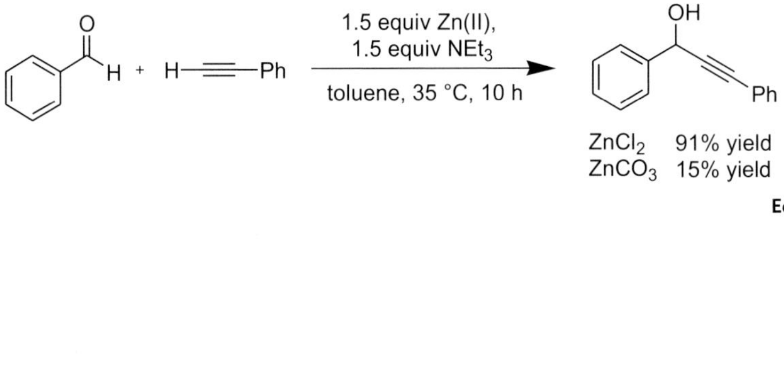

Equation 3-23

The use of stoichiometric amounts of (+)-*N*-methylephedrine (**41**) resulted in the formation of propargyl alcohols at ambient temperatures with up to 99 % *ees* and in preparatively useful yields (Equation 3.24) [52]. The process works with a wide range of aldehyde and acetylene partners, and particularly with Cα-branched aldehydes (Equation 3.25). Also noteworthy are the addition reactions involving functionalized alkynes that not only furnish products that can be employed as versatile building blocks, but also illustrate the mildness of the metalation reaction, which is compatible with functionality (Equation 3.26). In subsequent work, ethyne itself has been shown to participate in the addition reactions, giving products with high enantioselectivity, but at the current level of development the process can be slow (Equation 3.27) [55].

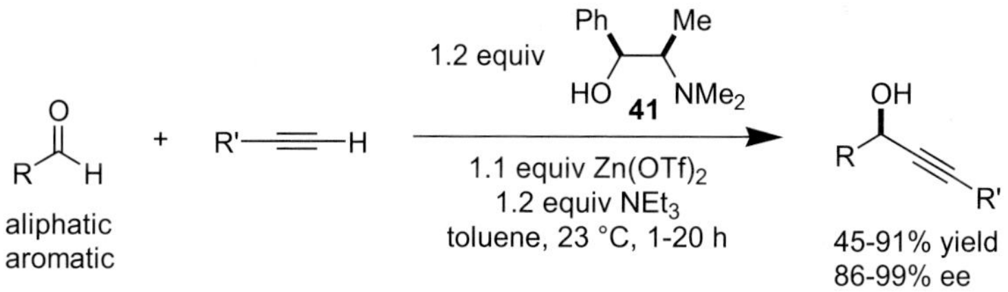

Equation 3-24

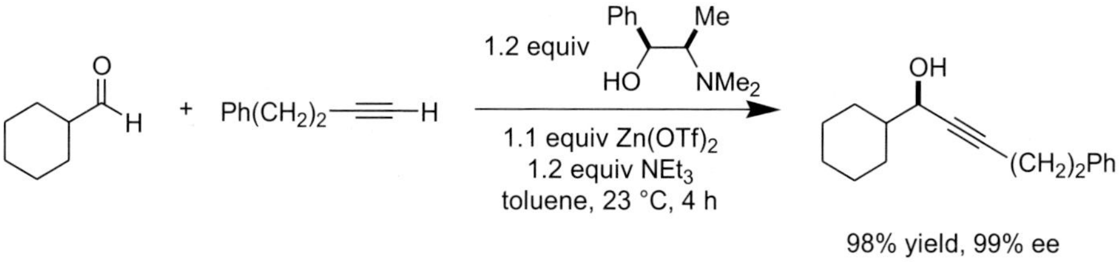

Equation 3-25

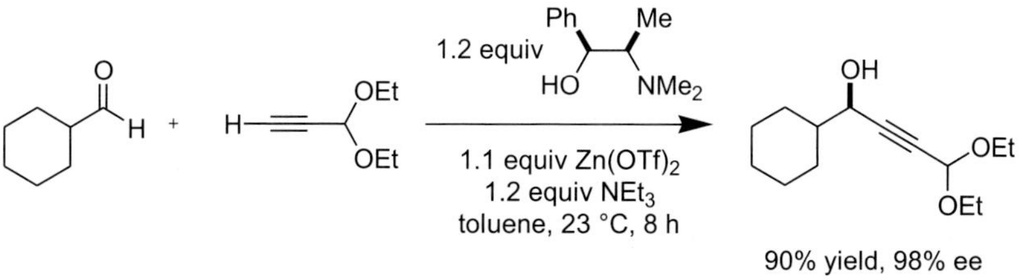

Equation 3-26

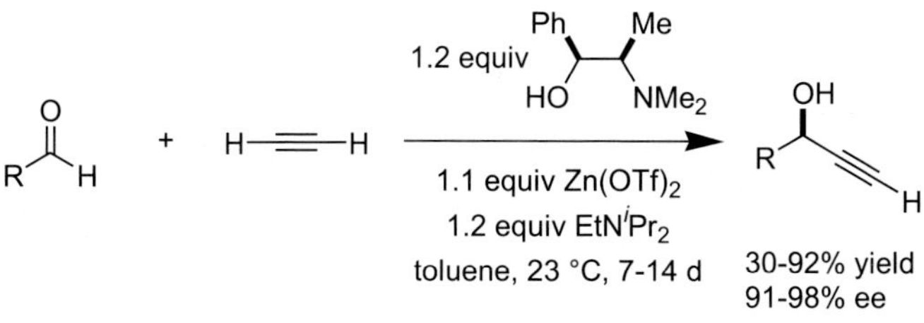

Equation 3-27

As an alternative to ethyne, 2-methyl-3-butyn-2-ol (**42**) can be employed (Equation 3.28); the acetone protecting group on the acetylene can subsequently be conveniently removed (Equation 3.29) [56]. This process permits the synthesis of unsymmetric 1,4-alkynyldiols (Scheme 3.4) [57].

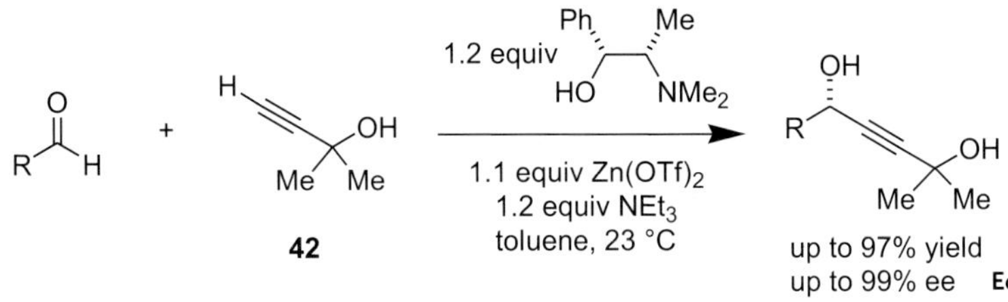

Equation 3-28

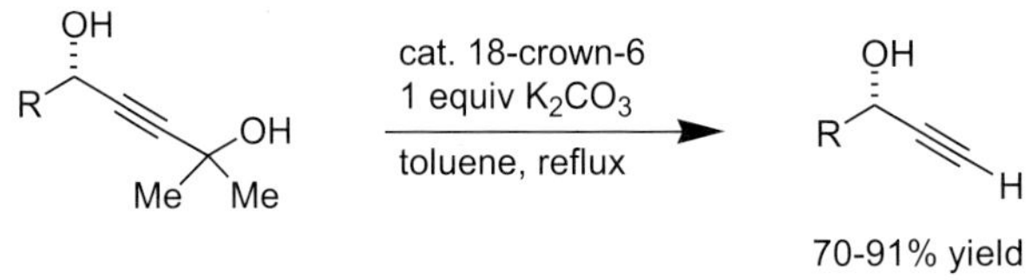

Equation 3-29

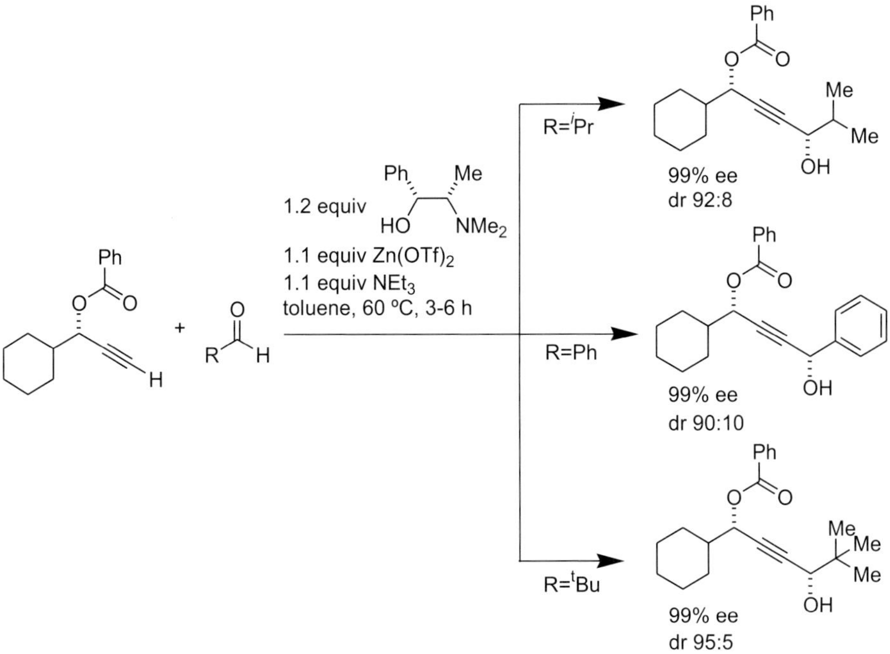

Scheme 3.4 Diastereoselective synthesis of unsymmetrical 1,4-alkynediols.

The reaction has been shown to be tolerant of a wide range of solvents and amenable to variation in the relative stoichiometry of the reagents. Thus, it is easily feasible to optimize yield and enantioselectivity for any substrate by adjustment of any of these variables (Scheme 3.5) [56]. Moreover, the reaction has been employed in the context of complex natural products syntheses, such as those of acetogenins (**43**; via **44**), epothilones A and B (**45** and **46**; via **47**), and leucascandrolide (**48**; via **49**) (Schemes 3.6–3.8) [58]. The syntheses of strongylodiols A (**50**; via **51**) and B (**52**; via **53**) (Schemes 3.9–3.10) [59] expand the scope of the stoichiometric process to include the activation and asymmetric addition of 1,3-diynes such as **54**.

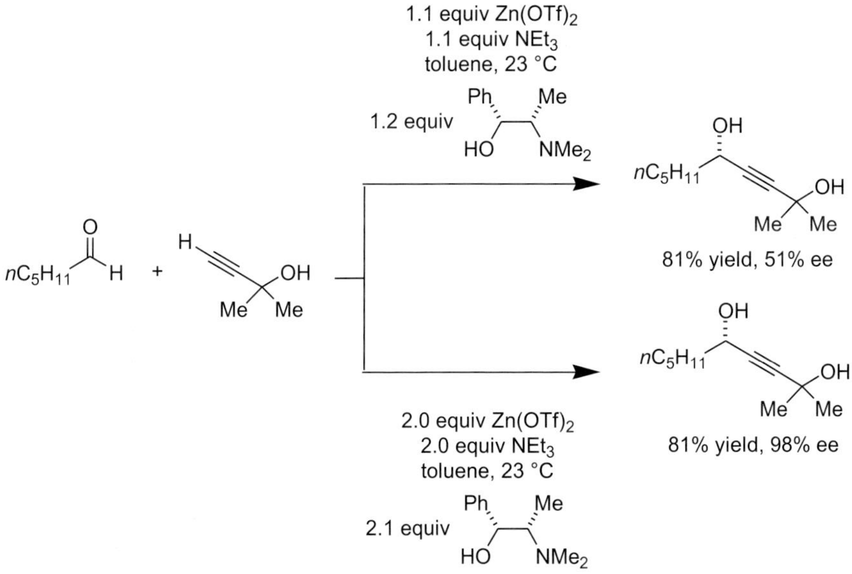

Scheme 3.5 Enantioselective addition of 2-methyl-3-butyn-2-ol to aldehydes.

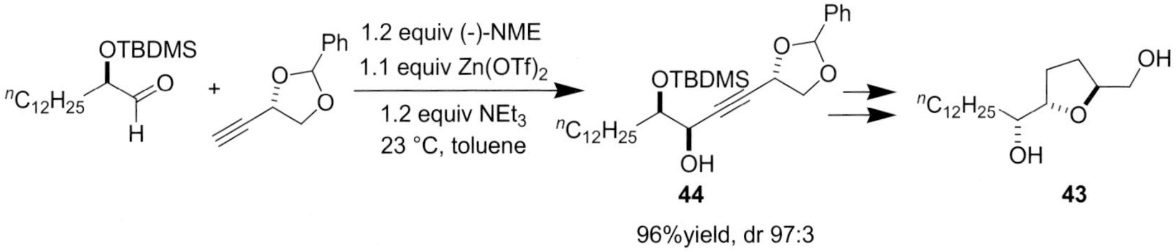

Scheme 3.6 Diastereoselective addition of a zinc acetylide in the synthesis of acetogenin (**43**).

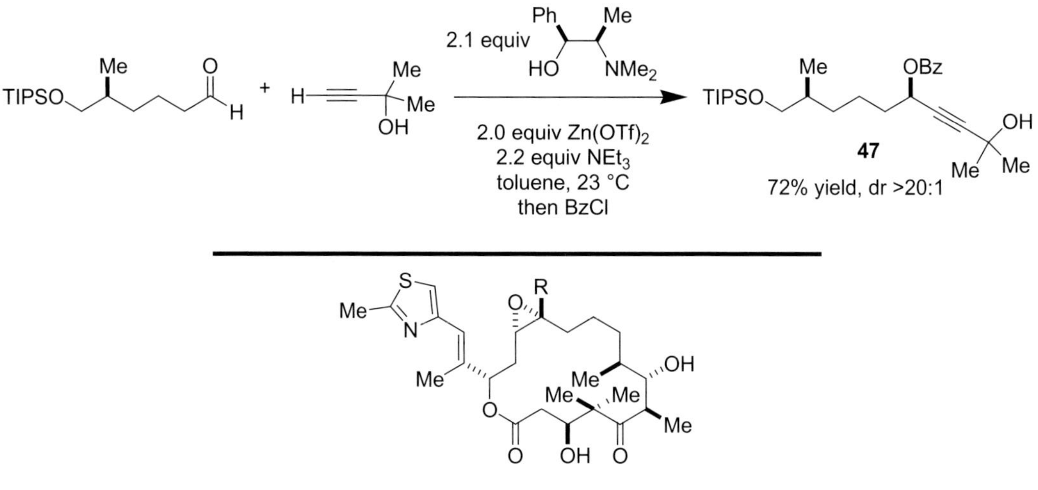

Epothilone A (R = H) (**45**)
Epothilone B (R = Me) (**46**)

Scheme 3.7 Synthesis of propargyl alcohol **47** as an optically active building block in the synthesis of epothilones A and B.

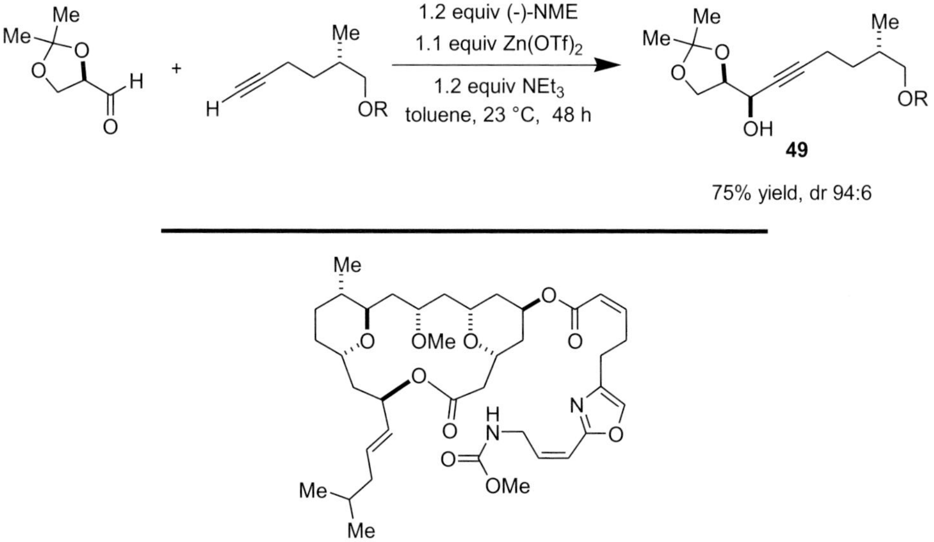

Leucascandrolide A (**48**)

Scheme 3.8 Diastereoselective addition of a zinc acetylide in the synthesis of leucascandrolide (**48**).

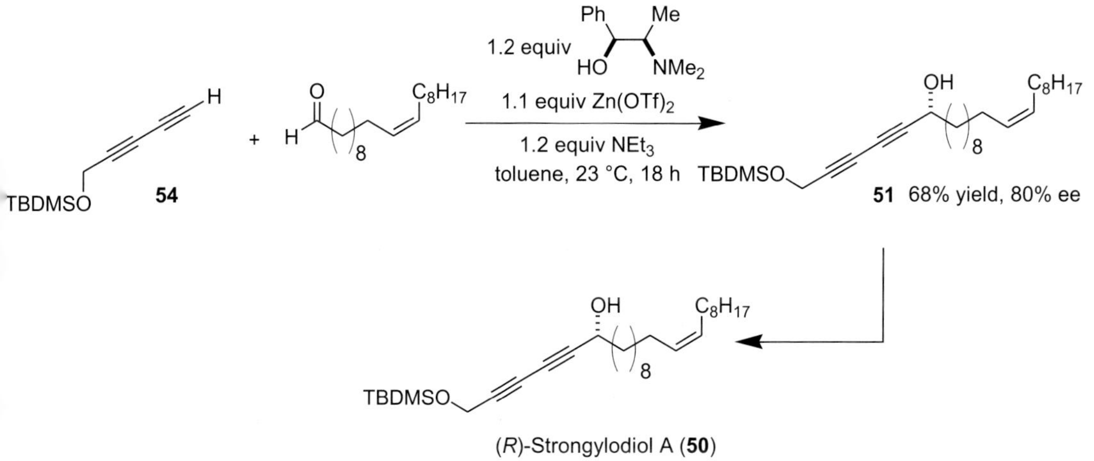

Scheme 3.9 Enantioselective addition of a zinc acetylide as a key step in the synthesis of (*R*)-strongylodiol A (**50**).

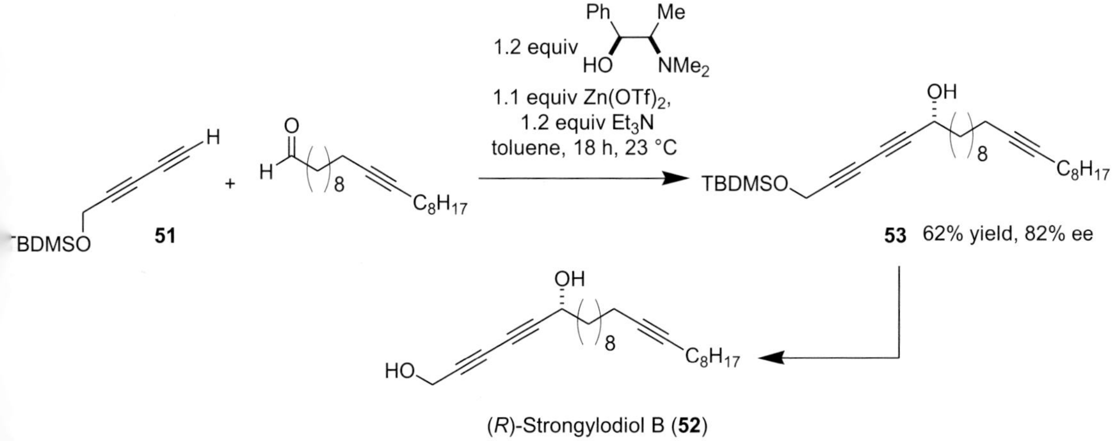

Scheme 3.10 Synthesis of (*R*)-strongylodiol B (**52**) with the use of zinc-mediated asymmetric addition of an alkyne to an aldehyde.

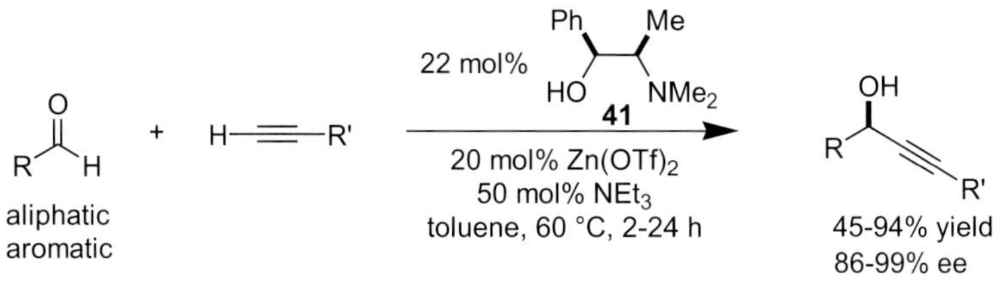

Equation 3-30

Subsequent to the reports of the stoichiometric use of Zn(II), chiral amino alcohol **41**, and amine base, a catalytic version was documented. In the early catalytic work the reaction was reported to proceed with the use of 22 mol% *N*-methylephedrine, 20 mol% $Zn(OTf)_2$, and 50 mol% Et_3N to furnish adducts with high enantioselectivity (Equation 3.30) [60]. The key difference between the earlier stoichiometric version and the catalytic procedure is that the latter is conducted at an elevated temperature (60 °C), which enables catalytic turnover; remarkably, only a small diminution in the enantioselectivity of the products is observed. The catalytic conditions have been utilized in a number of studies for the synthesis of building blocks such as **55** and **56** (Scheme 3.11) [61]. Both the process utilizing stoichiometric Zn(II) and the catalytic version can be conducted with reagent-grade solvent, and neither requires rigorous exclusion of air (Scheme 3.12) [62]. A novel feature of the processes is the fact that the reactions can be carried out under solvent-free conditions, making the process highly atom economic and volumetrically efficient (Equation 3.31). Under these conditions, the reactions can be conducted at considerably lower catalyst loadings (Equation 3.32) [50].

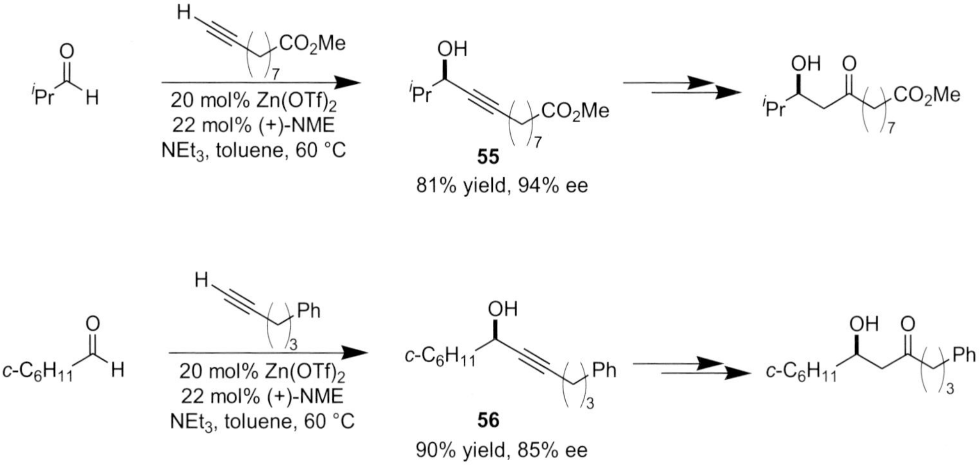

Scheme 3.11 Zinc-catalyzed enantioselective preparation of propargyl alcohols for the synthesis of optically active β-hydroxy ketones.

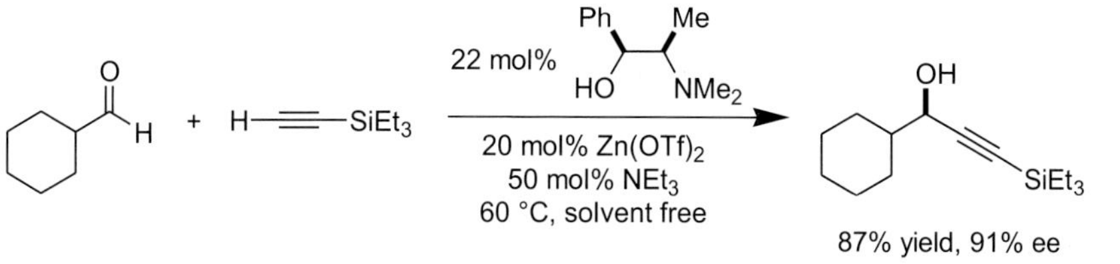

Equation 3-31

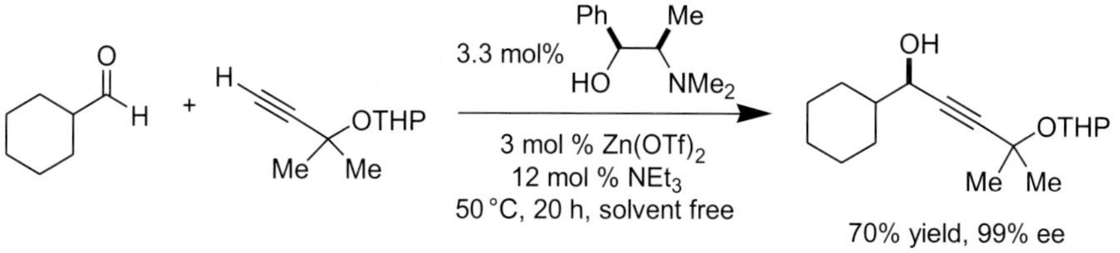

Equation 3-32

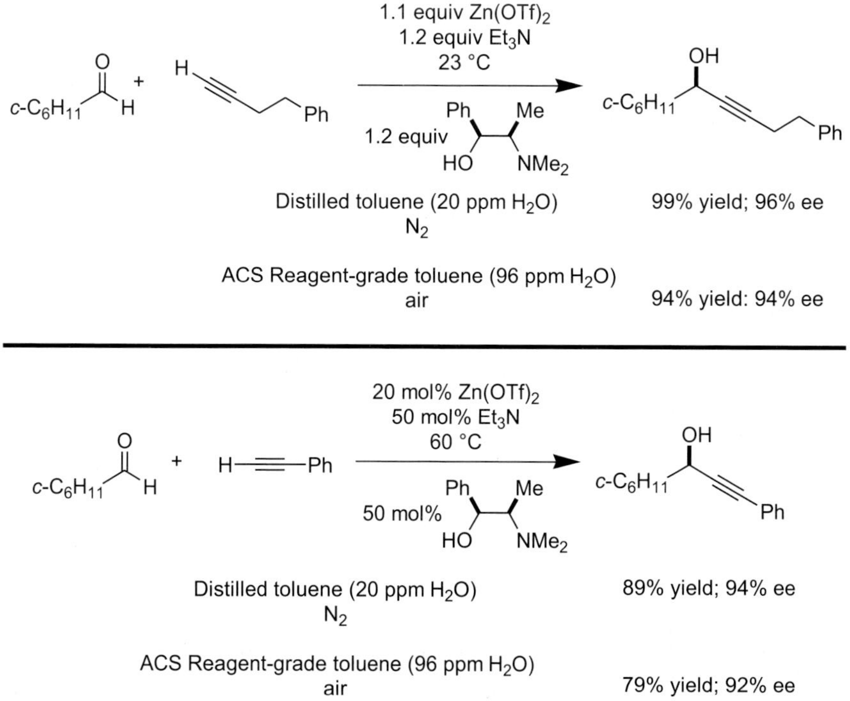

Scheme 3.12 Enantioselective addition of zinc acetylides to aldehydes under operationally convenient conditions

In related work, the use of Zn(II) salts and amino alcohols to effect asymmetric additions of terminal acetylenes to aldehydes has been extended to include other ligands and Zn(II) salts (Equation 3.33) [63]. Thus, Jiang et al. have documented the use of stoichiometric quantities of Zn(II) difluoromethanesulfonate salt and (1*S*,2*S*)-3-(*tert*-butyldimethylsilyloxy)-2-*N*,*N*-dimethylamino-1-(*p*-nitrophenyl)-propan-1-ol (**57**). Both of these components are conveniently available. Difluoromethanesulfonic acid is prepared in a single step from 3,3,4,4-tetrafluoro[1.2]oxathietane, a key monomer in the manufacture of Nafion ion-exchange resin. The amino alcohol is similarly readily accessible, because it has been used in the synthesis of chloramphenicol. This same amino alcohol ligand can also be used with $Zn(OTf)_2$ in enantioselective addition reactions (Equation 3.34) [64].

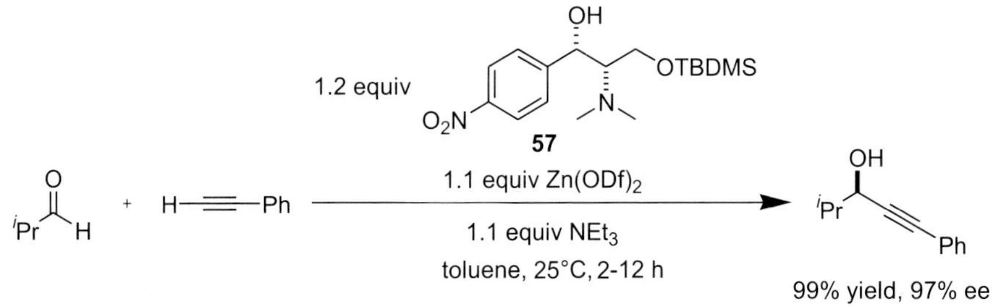

Equation 3-33

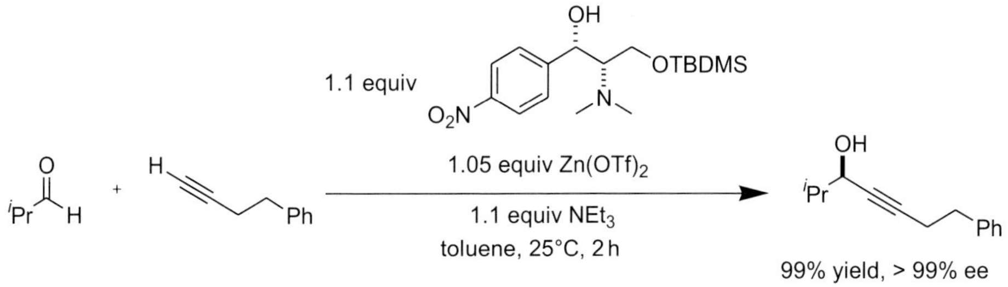

Equation 3-34

A significant recent advancement in the context of these studies is the addition reaction of terminal acetylenes to ketones (Equation 3.35) [65]. Thus, in the presence of 22 mol% **57**, 30% Et_3N, and 20 mol% $Zn(OTf)_2$, the addition of a range of acetylenes to non-enolizable ketones, such as benzoyl formate, heterocyclic glyoxylates and 4,4-dimethyldihydrofuran-2,3-dione, was observed. The reaction was conducted at 70 °C for 2 days, after which tertiary alcohols were isolated in up to 93% yields and 93% *ees*.

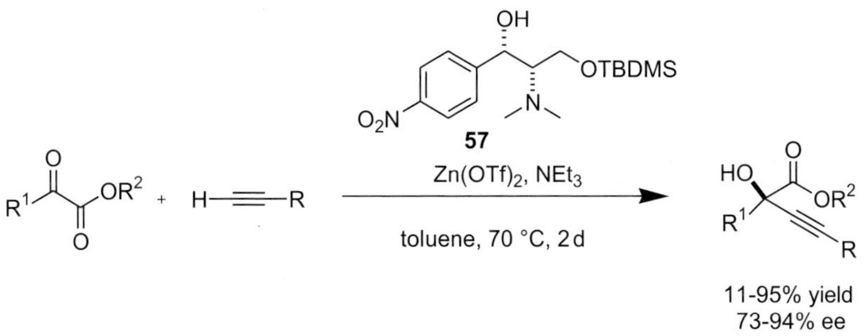

OH
OTBDMS
O_2N
N
57
$Zn(OTf)_2$, NEt_3
toluene, 70 °C, 2 d
O
R^1
OR^2
O
+ H—≡—R
HO
O
OR^2
R^1
R
11-95% yield
73-94% ee

Equation 3-35

3.5 Acetylene Additions to C=N Electrophiles

It has been known for some time that acetylenes and amines form propargyl amines at elevated temperatures. It has been suggested that this reaction proceeds through the addition of a metalated terminal acetylene to an intermediate iminium generated in situ. The enamine is accessed through addition of amine across C≡C in an *anti*-Markovnikov manner [66]. In 1963, investigators at Eastman Kodak made a substantial contribution to the advancement of this reaction process with the observation that CuCl catalyzes the addition of terminal acetylenes to the iminium derivatives generated in situ from the corresponding preformed enamines (Equation 3.36) [67]. Thus, at 100 °C in the presence of catalytic quantities of CuCl for 2–15 h, propargyl amines are formed from terminal alkynes and enamines in 64–89 % yields.

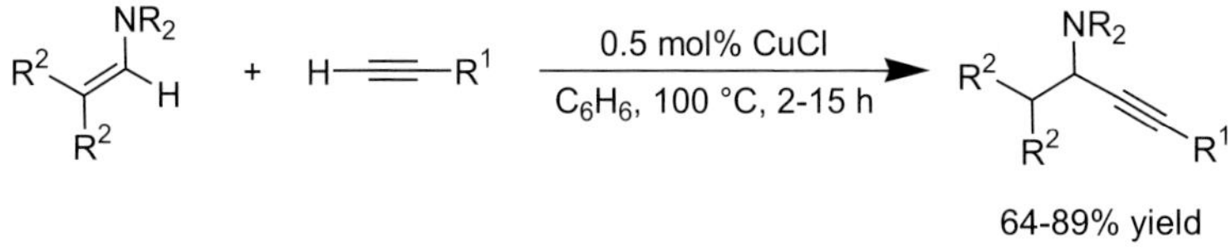

NR_2
R^2
H
R^2
+ H—≡—R^1
0.5 mol% CuCl
C_6H_6, 100 °C, 2-15 h
NR_2
R^2
R^2
R^1
64-89% yield

Equation 3-36

There has recently been a resurgence in the development of methods in which terminal acetylenes are activated in situ under mild deprotonation conditions and are subsequently engaged in C=N addition reactions. This interest was rekindled with the discovery that terminal acetylenes undergo activation by Zn(II) and subsequently participate in nucleophilic additions to nitrones (Equation 3.37). Thus, a mixture consisting of a terminal acetylene, 10 mol % $Zn(OTf)_2$, and 25 mol % Hünig's base afforded propargyl *N*-hydroxylamines in up to 99 % yields [48].

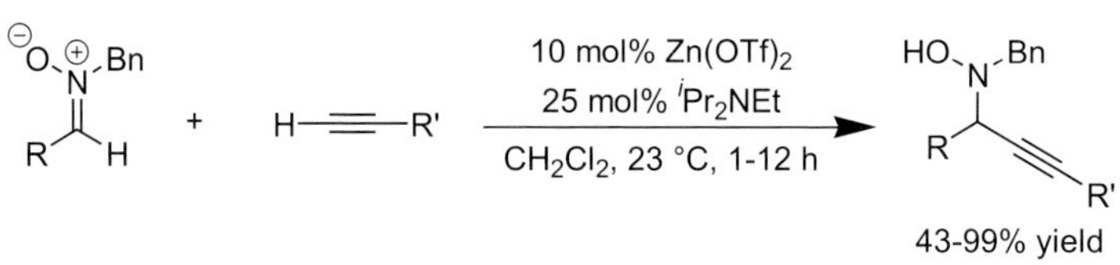

$$\text{R-CH=N(O)-Bn} + \text{H}\equiv\text{-R'} \xrightarrow[\text{CH}_2\text{Cl}_2, 23\,°\text{C}, 1\text{-}12\,\text{h}]{\substack{10\,\text{mol\%}\,\text{Zn(OTf)}_2 \\ 25\,\text{mol\%}\,^i\text{Pr}_2\text{NEt}}} \text{R-CH(N(OH)Bn)-C}\equiv\text{C-R'}$$

43-99% yield   **Equation 3-37**

Optically active N-hydroxylamines can be accessed through the use of auxiliary-controlled diastereoselective addition reactions of terminal acetylenes and chiral nitrones (Scheme 3.13) [68]. Thus, nitrones **58** participate in highly diastereoselective additions with terminal acetylides mediated by 0.5 equiv. Zn(OTf)$_2$, amine base, and 2-(N,N-dimethylamino)ethanol. The addition of this simple amino alcohol was observed to produce significant acceleration in the overall rate of addition. The carbohydrate-derived auxiliary **59** was developed by Vasella and is readily prepared in two steps (85 % overall yield) from mannose [69]. The chiral nitrones are easily obtained. After the nucleophilic additions, the auxiliary can be concomitantly cleaved and regenerated simply by heating the propargyl N-hydroxylamine adduct **60** in an aqueous solution of N-hydroxylamine hydrochloride (Equation 3.38). The optically active propargyl N-hydroxylamines **61** are isolated in high overall yields.

R: aliphatic, aromatic

up to 99% yield
dr 92:8 to 98:2

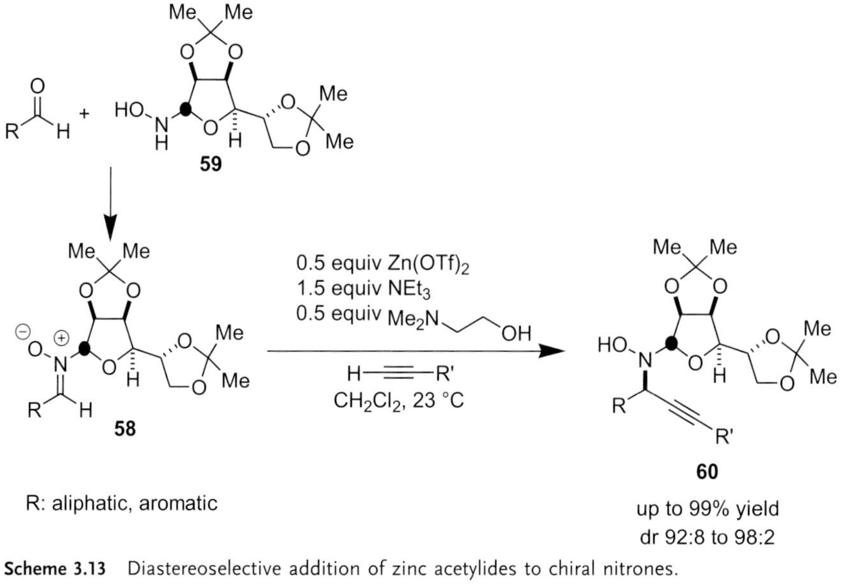

Scheme 3.13 Diastereoselective addition of zinc acetylides to chiral nitrones.

54-99 %yield   **Equation 3-38**

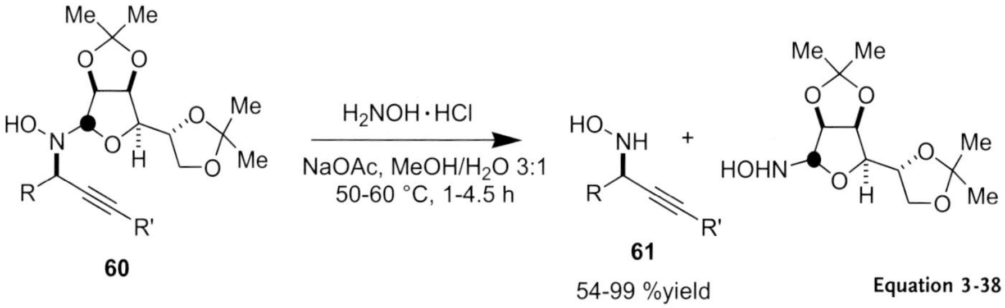

As part of an ongoing program aimed at asymmetric nitrone additions (Equation 3.39) [70], Vallée et al. documented the diastereoselective addition reaction of lithiated terminal acetylenes to nitrones. Thereafter, a procedure using substoichiometric (20 mol%) amounts of Et_2Zn (Equation 3.40) was developed [71]. The addition of a wide range of acetylenes to an equally broad range of nitrones under these mild reaction conditions has been documented, with the product propargyl *N*-hydroxylamines isolated in up to 96% yields. The investigators have speculated that the process involves the formation of a EtZn-alkynylide, which undergoes addition to a nitrone activated by coordination to the Lewis acidic alkynylzinc species. Under these same conditions, additions to nitrones that are *N*-substituted with the previously employed mannose-derived auxiliary **59**, as well as with other chiral controlling groups **62–64**, furnished adduct propargyl *N*-hydroxylamines over 1–72 h with 55:45 to >95:5 diastereoselectivities and in useful yields (Equation 3.41) [72].

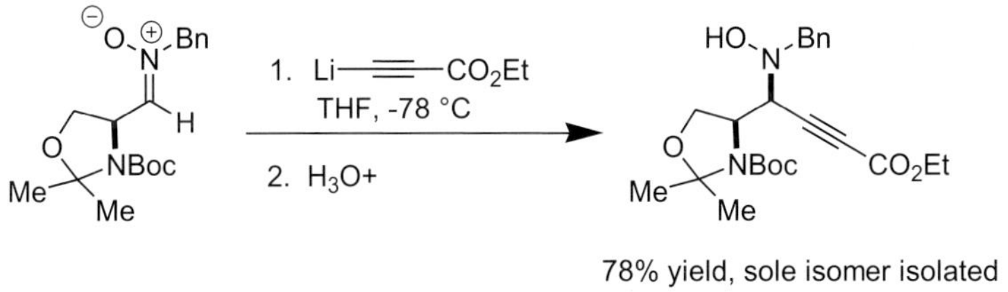

Equation 3-39

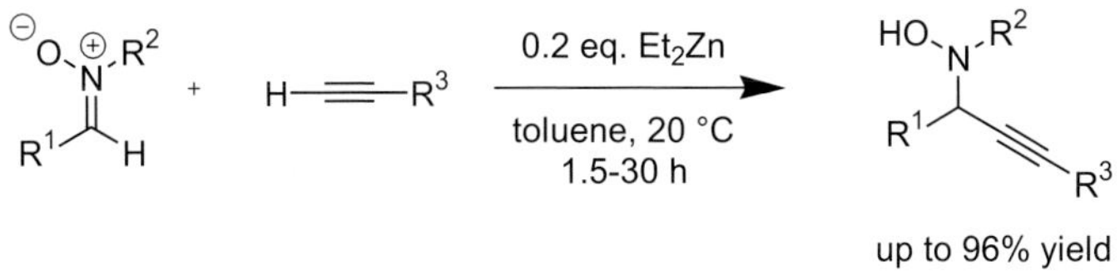

Equation 3-40

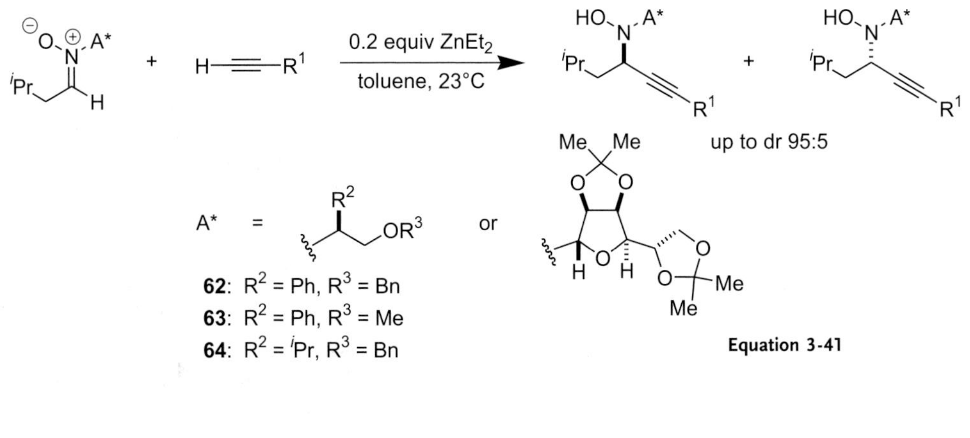

Equation 3-41

The studies on nitrone additions were followed up by exciting new developments in imine addition reactions. One of the earliest reports in this area involves the enantioselective addition of phenylacetylene to N-aryl aldimines derived from aromatic aldehydes (Equation 3.42) [73]. The addition reactions were catalyzed by chiral bisoxazoline ligands, the best being **65**. In the experiment, the 2-phenylethynyl organocopper reagent was generated in the presence of the putative Cu(I) complex in aqueous medium and subsequently added to the imines. The reactions were carried out with 10 mol% **65** at ambient temperature over 4 days to afford adducts with up to 96% ees [74]. The same investigators have also reported a high-yielding (>99%), three-component coupling reaction of aldehydes, secondary amines, and terminal acetylenes utilizing AuCl in water (Equation 3.43) [75]. The reactions probably take place with transient formation of the iminium species in situ.

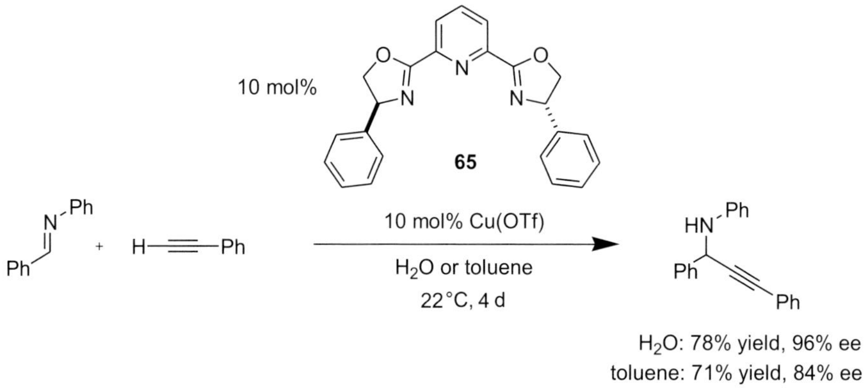

$$\text{Ph-N=CH-Ph} + \text{H-}\equiv\text{-Ph} \xrightarrow[\substack{H_2O \text{ or toluene} \\ 22\,°C,\ 4\ d}]{\substack{10\ \text{mol}\%\ \mathbf{65} \\ 10\ \text{mol}\%\ \text{Cu(OTf)}}} \text{Ph-CH(NHPh)-C}\equiv\text{C-Ph}$$

H$_2$O: 78% yield, 96% ee
toluene: 71% yield, 84% ee

Equation 3-42

$$\text{RCHO} + \text{H}\equiv\text{R}^1 + \text{R}^2{}_2\text{NH} \xrightarrow[H_2O,\ 70\,°C]{1\ \text{mol}\%\ \text{AuCl}} \text{R-CH(NR}^2{}_2\text{)-C}\equiv\text{C-R}^1$$

75-99% yield

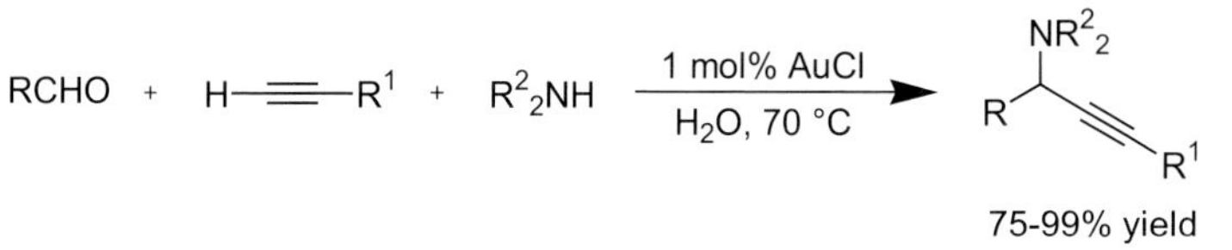

Equation 3-43

There have been recent reports on the use of newer activation methods for the generation of intermediates that undergo additions to imines. Jiang's group has documented the use of ZnCl$_2$ and triethylamine to activate terminal acetylenes and subsequently to carry out additions to N-alkyl arylaldimines in the presence of TMSCl as an additive (Equation 3.44) [76]. An innovative approach to the synthesis of propargylamines involves the generation of N-acyl aldimines **66** in situ from α-phenylsulfonyl N-acyl amines **67** (Equation 3.45) [77]. The reported additions were conducted in water at 40–50 °C with employment of phenylacetylene and 2–3 equiv. CuBr with sonication, and furnished propargylamine adducts in 15–72% yields. The same method has been employed in addition to cyclic N-acyliminium ions. Additionally, there has recently been a report on the use of Ag(I) in water for three-component coupling of terminal acetylenes, aldehydes, and secondary amines [78].

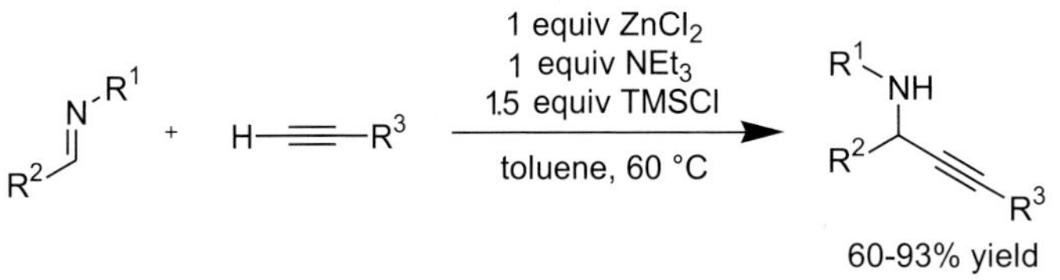

Equation 3-44

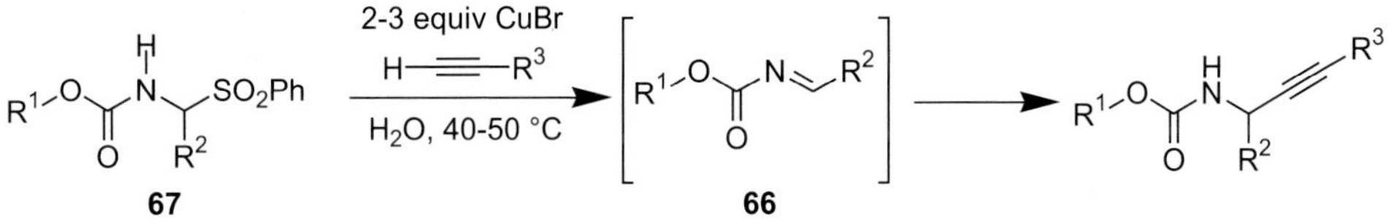

Equation 3-45

Recent investigations have re-examined the reaction first reported by the Eastman Kodak group in which enamines served as precursors to iminium intermediates that subsequently underwent addition by Cu-acetylides. In work by Knochel et al., the coupling reactions are carried out with enamines and terminal acetylenes in toluene at 60–80 °C in the presence of 5 mol% CuBr and 5.5 mol% of the chiral ligand QUINAP (**68**) (Equation 3.46) [79], the tertiary propargylamine products being isolated after 3–24 h in 50–99% yields and with 54–90% *ees*. A variant of the reaction involving three-component coupling of a secondary amine, aldehyde, and terminal acetylene at ambient temperature over 3–6 days under otherwise identical conditions has been documented, furnishing optically active propargylamines in 43–99% yields and with 32–96% *ees* (Equation 3.47) [80].

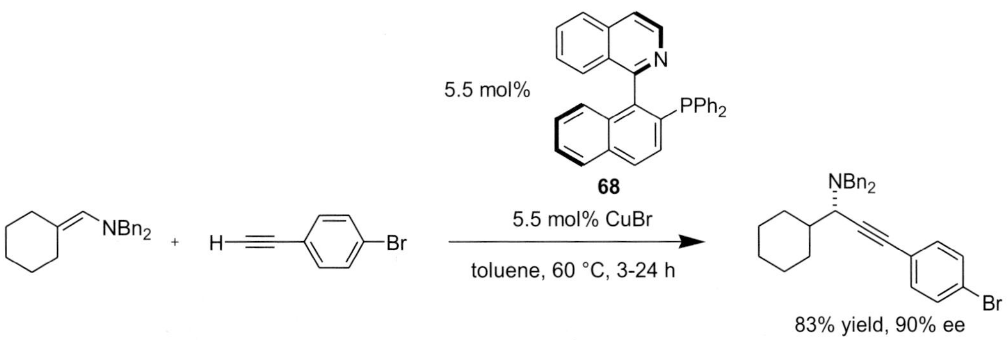

Equation 3-46

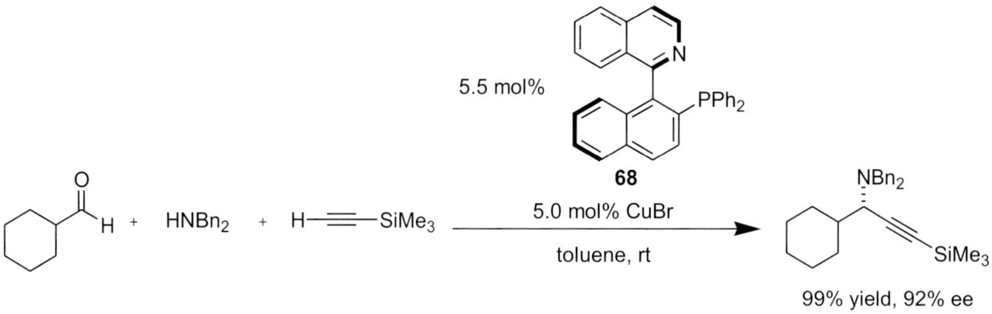

Equation 3-47

In the search for newer metal catalysts that effect acetylene activation and additions to C=N electrophiles, simple Ir(I) complexes have been identified as highly promising [81]. Thus, the addition of trimethylsilylacetylene to a range of structurally different aldimines can be effected at ambient temperature through the use of 4–5 mol% of $[IrCl(COD)]_2$, a commercially available and conveniently handled metal complex (Equation 3.48). Although only silylacetylenes can be employed in the process, the terminal acetylenic products obtained upon desilylative workup serve as versatile building blocks. In contrast, the process displays broad scope with regard to the imine component, so aldimines obtained both from aromatic and aliphatic aldehydes can be employed. Particularly noteworthy is the fact that the protecting group on the imine nitrogen can be varied extensively, including N-aryl (p-anisyl) and N-alkyl (benzyl, benzhydryl, allyl, butyl) moieties. Moreover, the additions can be carried out neat, under solvent-free conditions (Scheme 3.14). As such, the reaction process displays a high degree of atom economy and volumetric efficiency. Ongoing investigations in which the process can be executed at low (0.5 mol% Ir(I)) catalyst loadings have been documented [82].

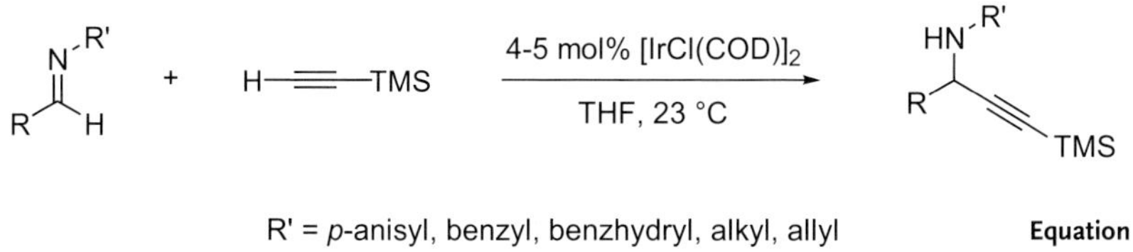

Equation 3-48

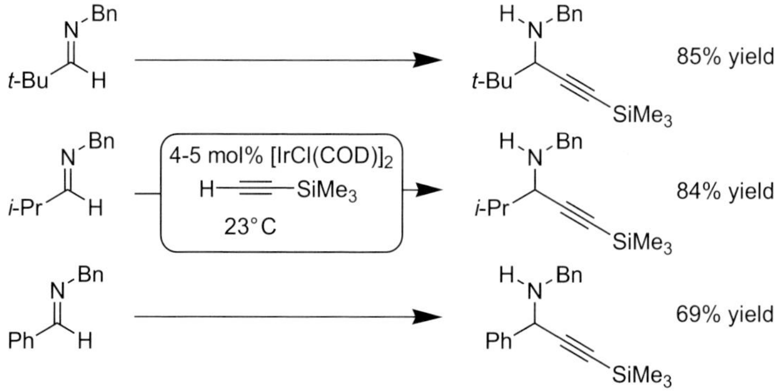

Scheme 3.14 Iridium-catalyzed addition of TMS-acetylene to imines.

3.6 Conclusion

The utility and versatility of terminal acetylenes in C=O and C=N addition reactions has been appreciated for over a century. Concurrently with the birth of the welding industry (1901–03) and the oxyacetylene torch, key experiments were being carried out on the use of acetylene and its higher homologues in synthetically important transformations. The increasing demands on synthetic chemistry vis à vis the environment and economics constantly raises the standards that synthetically useful reactions must meet. These socioeconomic pressures are pervasive throughout science and are not likely to diminish. The focus on acetylene in catalytic and stoichiometric asymmetric synthesis is therefore to be expected and welcome. The inherent ability of terminal acetylenes to undergo metalation under mild, controlled conditions compatible with a host of electrophilic reaction partners makes this functional group an alluring target for researchers. Moreover, the enormous range of reaction chemistry available to the adducts, such as the propargyl amines and alcohols discussed in this chapter, make investigations into its chemistry a wholly worthwhile venture. The evolution of the chemistry of metal acetylides has been impressive and incorporates not only the discovery of new ligands and new reinvestigations of previously documented but under-appreciated species, but also innovation and the discovery of metalation conditions involving new coordination complexes and associated ligands. Both approaches open up new avenues for synthetic chemistry. Additional advances can be anticipated and will be exciting to witness.

3.7 Experimental Procedures

This section presents a selection of general procedures for the alkynylation of aldehydes, ketones, nitrones, and imines.

3.7.1 General Procedure for the Enantioselective Alkynylation of Aldehydes by the use of Stoichiometric Amounts of Zn(OTf)2 (Equation 3.24) [52]

A 10 mL Schlenk flask was charged with $Zn(OTf)_2$ (0.200 g, 0.550 mmol, 1.1 equiv.) and (+)-*N*-methylephedrine (0.108 g, 0.600 mmol, 1.2 equiv.) and was purged with nitrogen for 15 min. Toluene (1.5 mL) and triethylamine (61.0 mg, 0.600 mmol, 1.2 equiv.) were added. The resulting mixture was stirred at 23 °C for 2 h, after which the alkyne (0.600 mmol, 1.2 equiv.) was added in one portion by syringe. After 15 min the aldehyde (0.500 mmol, 1.0 equiv.) was also added in one portion by syringe. The reaction mixture was stirred at 23 °C until full conversion was observed by TLC. The reaction mixture was quenched by addition of NH_4Cl

(sat.) (3 mL) and then poured into a separating funnel containing diethyl ether (10 mL). The layers were separated and the aqueous layer was extracted with diethyl ether (3 × 10 mL). The combined organic layers were washed with NaCl (sat.) (10 mL), dried over anhydrous Na_2SO_4, filtered, and concentrated in vacuo. Purification of the crude product by flash chromatography on silica gel afforded the pure secondary alcohol.

3.7.2
General Procedure for the $Zn(OTf)_2$-Catalyzed Enantioselective Alkynylation of Aldehydes (Equation 3.30) [60]

A 10 mL Schlenk flask was charged with $Zn(OTf)_2$ (36 mg, 0.10 mmol, 20 mol%) and was then heated to 125 °C under vacuum (< 0.5 mbar) for 2 h. After the system had been allowed to cool to 23 °C, (+)-*N*-methylephedrine (20 mg, 0.11 mmol, 22 mol%) was added under nitrogen. The flask was evacuated (< 0.5 mbar) for 30 min, and was then purged with nitrogen prior to addition of toluene (0.50 mL) and triethylamine (25 mg, 0.25 mmol, 50 mol%) by syringe. The resulting mixture was stirred for 2 h at 23 °C, after which the alkyne (0.60 mmol, 1.2 equiv.) was added. After the mixture had been stirred for 15 min at 23 °C, the aldehyde (0.50 mmol, 1.0 equiv.) was added in one portion, and the mixture was then stirred at 60 °C until full conversion was observed by TLC. The reaction mixture was quenched with NH_4Cl (sat.) (3 mL) and then poured into a separating funnel containing diethyl ether (10 mL). The layers were separated, and the aqueous layer was extracted with diethyl ether (3 × 10 mL). The combined organic layers were washed with NaCl (sat.) (10 mL), dried over anhydrous Na_2SO_4, filtered, and concentrated in vacuo. Purification of the crude product by flash chromatography on silica gel afforded the pure secondary alcohol.

3.7.3
General Procedure for the Enantioselective Alkynylation of Ketones Catalyzed by Zn(salen) Complexes (Equation 3.19) [46]

A flask was charged with Me_2Zn (2 M in toluene, 1.5 mL, 3.0 mmol, 3.0 equiv.), phenylacetylene (0.31 g, 3.0 mmol, 3.0 equiv.), and toluene (1.0 mL), and the resulting solution was stirred at 23 °C for 1 h. (*R,R*)-Salen (0.20 mmol, 20 mol%) was added, and the resulting yellow solution was stirred at 23 °C for 1 h before addition of the ketone (1.0 mmol, 1.0 equiv.). The resulting mixture was stirred at 23 °C for 48–96 h, and was then quenched by the addition of water (5 mL). The reaction mixture was diluted with diethyl ether, stirred for 5 min, and then filtered through celite. The layers were separated, and the aqueous phase was extracted with diethyl ether (3 × 3 mL). The combined organic layers were dried over Na_2SO_4, filtered, and concentrated in vacuo. Purification of the crude product by flash chromatography on deactivated silica gel (SiO_2 was deactivated by addition of Et_3N (1% vol) to the eluent during the preparation of the column) afforded the pure tertiary alcohol.

3.7.4
General Procedure for the $Zn(OTf)_2$-Catalyzed Diastereoselective Alkynylation of N-Glycosyl Nitrones (Scheme 3.13) [68]

A 10 mL Schlenk flask was charged with $Zn(OTf)_2$ (91 mg, 0.25 mmol, 50 mol%) and heated with a heat gun for 3 min. The flask was allowed to cool to 23 °C and was then purged with dry nitrogen. Dichloromethane (1.0 mL), triethylamine (76 mg, 0.75 mmol, 1.5 equiv.), and 2-*N,N*-dimethylaminoethanol (22 mg, 0.25 mmol, 50 mol%) were added. The resulting mixture was stirred at 23 °C for 1.5 h, after which the alkyne (0.75 mmol, 1.5 equiv.) was added. After 30 min, the nitrone (0.50 mmol, 1.0 equiv.) in dichloromethane (0.50 mL) was added in one portion by syringe. After full conversion was observed by TLC, the system was quenched with NH_4Cl (sat.) (3.0 mL) and then poured into a separating funnel containing dichloromethane (20 mL). The layers were separated, and the aqueous layer was extracted with dichloromethane (3 × 20 mL). The combined organic layers were washed with NaCl (sat.) (10 mL), dried over anhydrous $MgSO_4$, and concentrated in vacuo. Purification of the crude product by flash chromatography on silica gel afforded the pure propargyl *N*-hydroxylamine.

3.7.5
General Procedure for the Et_2Zn-Catalyzed Diastereoselective Alkynylation of Chiral Nitrones (Equation 3.41) [72]

A solution of diethylzinc in hexane (1.0 M, 0.10 mL, 0.10 mmol) was added to a solution of nitrone (0.50 mmol, 1.0 equiv.) and alkyne (0.65 mmol, 1.3 equiv.) in dry toluene (2 mL). The reaction mixture was stirred at 23 °C until full conversion was observed by TLC, quenched by addition of $NaHCO_3$ (sat.) (4 mL), and extracted with diethyl ether (2 × 10 mL). The combined organic layers were washed with NaCl (sat.) (3 × 15 mL), dried over anhydrous $MgSO_4$, filtered, and concentrated in vacuo. Purification of the crude product by flash chromatography on silica gel afforded the pure propargyl *N*-hydroxylamine.

3.7.6
General Procedure for the CuBr-Catalyzed Enantioselective Preparation of Propargylamines (Equation 3.47) [80]

CuBr (3.6 mg, 25 µmol, 5.0 mol%) and (*R*)-Quinap (12 mg, 27 µmol, 5.5 mol%) were suspended in dry toluene (2 mL) in a dry and argon-flushed 10 mL flask, fitted with a magnetic stirrer and a septum, and the system was stirred for 30 min. MS (4Å, 0.30 g) and *n*-decane (30 mg) were added, followed by the alkyne (0.50 mmol, 1.0 equiv.), the aldehyde (0.50 mmol, 1.0 equiv.), and the secondary amine (0.50 mmol, 1.0 equiv.). The reaction mixture was stirred at 23 °C until GC analysis showed full conversion. The suspension was filtered, and the solid was washed with diethyl ether. The filtrate was concentrated in vacuo and purified by column chromatography on silica gel to afford the pure propargylamine.

3.7.7
General Procedure for the $[IrCl(COD)]_2$-Catalyzed Alkynylation of Imines (Equation 3.48) [81]

A 10 mL Schlenk flask was charged under argon with $[IrCl(COD)]_2$ (16.8 mg, 25.0 μmol, 5 mol%), imine (0.500 mmol, 1.0 equiv.), trimethylsilylacetylene (73.7 mg, 0.750 mmol, 1.5 equiv.), and THF (2.5 mL). The reaction mixture was stirred at 23 °C for 24 h and was then quenched by addition of water (3 mL). The layers were separated, and the aqueous layer was extracted with dichloromethane (3 × 10 mL). The combined organic layers were washed with NaCl (sat.) (10 mL), dried over Na_2SO_4, filtered, and concentrated in vacuo. Purification of the crude product by flash chromatography on silica gel afforded the pure propargylamine.

References

[1] *Modern Acetylene Chemistry* (Eds.: Stang, P. J.; Diederich, F.), VCH, Weinheim, **1995**.

[2] For selected examples, see: (a) Muller, T. E.; Beller, M. *Chem. Rev.* **1998**, *98*, 675–703. (b) Utimoto, K.; Wakabayashi, Y.; Horiie, T.; Inoue, M.; Shishiyama, Y.; Obayashi, M.; Nozaki, H. *Tetrahedron* **1983**, *39*, 967–973. (c) Utimoto, K. *Pure Appl. Chem.* **1983**, 55, 1845–1872. (d) Fukuda, Y.; Utimoto, K.; Nozaki, H. *Heterocycles* **1987**, *25*, 297. (e) Li, Y.; Fu, P. F.; Marks, T. J. *Organometallics* **1994**, *13*, 439–440. (f) McGrane, P. L.; Jensen M.; Livinghouse, T. *J. Am. Chem. Soc.* **1992**, *114*, 5459–5460. (g) McGrane, P. L.; Livinghouse, T. *J. Am. Chem. Soc.* **1993**, *115*, 11 485. (h) Li, Y.; Marks, T. J. *J. Am. Chem. Soc.* **1996**, *118*, 9295–9306.

[3] For selected examples, see: (a) Knochel, P. in *Comprehensive Organic Synthesis* (Ed.: Trost, B. M.), Pergamon Press, Oxford, **1991**, Vol. 4, pp. 865–912. (b) Negishi, E.; Anastasia, L. *Chem. Rev.* **2003**, *103*, 1979–2017. (c) Van Horn, D. E.; Negishi E. *J. Am. Chem. Soc.* **1978**, *100*, 2252–2254. (d) Yoshida T.; Negishi, E. *J. Am. Chem. Soc.* **1981**, *103*, 4985–4987. (e) Negishi E.; Van Horn D. E.; Yoshida, T. *J. Am. Chem. Soc.* **1985**, *107*, 6639–6647. (f) Tessier, P. E.; Penwell, A. J.; Souza, F. E. S.; Fallis, A. G. *Org. Lett.* **2003**, *5*, 2989–2992.

[4] For selected examples, see: (a) Brown, C. A.; Yamashita, A. *J. Am. Chem. Soc.* **1975**, 97, 891–892. (b) Brown, C. A.; Yamashita, A. *J. Chem. Soc. Chem. Comm.* **1976**, 959–960. (c) Macaulay, S. R. *J. Org. Chem.* **1980**, 45, 734–735. (d) Balova, I. A.; Morozkina, S. N.; Knight, D. W.; Vasilevsky, S. F. *Tetrahedron Lett.* **2003**, *44*, 107–109.

[5] (a) Simandi, L. I. in *The Chemistry of Functional Groups, Supplement C, Pt 1* (Eds.: Patai, S.; Rappoport, Z.), Wiley, New York, **1983**, pp. 529–534.
(b) G. Eglinton, A. R. Galbraith, *Chem. & Ind. (London)* **1956**, 737.
(c) Hay, A. S. *J. Org. Chem.*, **1962**, *27*, 3320–3321 (1962).
(d) Chodkiewicz, W. *Ann. Chim. (Paris)* **1957**, *2*, 819.
(e) Cadiot, P.; Chodkiewicz, W. in *Chemistry of Acetylenes* (Eds.: Viehe, H. G.) Decker, New York, **1969**, pp. 597–647.
(f) Sevin, A.; Chodkiewicz, W.; Cadiot, P. *Bull. Soc. Chim. Fr.* **1974**, 913–917.

(g) Stephens, R. D.; Castro, C. E *J. Org. Chem.* **1963**, *28*, 3313–3315.
(h) Sonogashira, K.; Tohda, Y.; Hagihara, N. *Tetrahedron Lett.* **1975**, 4467–4470.

[6] *March's Advanced Organic Chemistry*, 5th Ed. (Eds.: Smith, M. B.; March, J.), Wiley, New York, **2001**, pp. 329–331.

[7] (a) Glaser, C. *Ber. Dtsch. chem. Ges.*, **1869**, *2*, 422–424.
(b) Glaser, C. *Ann. chem. Pharm.*, **1870**, *154*, 137–171.

[8] Reif, W.; Grassner, H. *Chem. Ing. Tech.* **1973**, *45*, 646–652.

[9] Wakefield, B. J. *Organolithium Methods*, Academic Press, London, **1988**, Ch. 3, p. 32.

[10] Wakefield, B. J. *Organomagnesium Methods in Organic Synthesis*, Academic Press, London, **1995**, Ch. 3, pp. 46–48.

[11] Brandsma, L. *Preparative Acetylene Chemistry*, 2nd Ed, Elsevier, Amsterdam, **1988**.

[12] For key references to addition reactions employing stannyl, silylated, and boryl reagents
(a) Yamamoto, Y.; Nishii, S.; Maruyama, K. *J. Chem. Soc., Chem. Commun.* **1986**, 102–103.
(b) Corey, E. J.; Cimprich, K. A. *J. Am. Chem. Soc.* **1994**, *116*, 3151–3152.
(c) Ahn, J. H.; Joung, M. J.; Yoon, N. M. *J. Org. Chem.* **1995**, *60*, 6173–6175.
(d) Evans, D. A.; Halstead, D. P.; Allison, B. D. *Tetrahedron Lett.* **1999**, *40*, 4461–4462.
For general references, see:
(a) Santelli, M.; Pons, J.-M. in *Lewis Acids and Selectivity in Organic Synthesis*, CRC, Boca Raton, **1996**, Ch. 3 and 5.
(b) Marshall, J. A. *Chem. Rev.* **1996**, *96*, 31–48. (c) Yamamoto, Y.; Asao, N. *Chem. Rev.* **1993**, *93*, 2207–2293.

[13] (a) Favorsky, A. E.; Skosarevsky, M. *Russ. J. Phys. Chem. Soc.* **1900**, *32*, 652.
(b) Favorsky, A. E.; Skosarevsky, M. *Bull. Soc. Chim. Fr.* **1901**, *26*, 284.

[14] (a) Zeltner, J.; Genas, M. British Patent 544 221 (April 2, 1942); US Patent 2345170 (March 28, 1944); *Chem. Abstr.* **1944**, *38*, 4273.
(b) Weizman, C., British Patents 573 527 (November 26, 1945); 580921 (September 25, 1946); 580922 (September 25, 1946); *Chem. Abstr.* **1947**, *41*, 2066, 2429.
(c) Smith, E. F. US Patent 2385546 (September 25, 1945); 2385548 (September 25, 1945). Herman, D. F. US Patent 2455058 (December 30, 1948). Weizmann, C. US Patent 2474175 (June 21, 1949). Brotman, A. US Patent 2536028 (January 2, 1951).
(d) Nazarov, I. N. *Izv. Akad. Nauk SSSR, Otd. Khim. Nauk* **1956**, 960. *Chem. Abstr.* **1955**, *49*, 927.

[15] Tedeschi, R. J.; Casey, A. W.; Clark, G. S.; Huckel, R. W.; Kindley, L. M.; Russell, J. P. *J. Org. Chem.* **1963**, *28*, 1740–1743. (b) Tedeschi, R. J. *J. Org. Chem.* **1965**, *30*, 3045–3049. (c) Schachat, N.; Bagnell, J. J. Jr. *J. Org. Chem.* **1962**, *27*, 1498–1504.

[16] Babler, J. H.; Liptak, V. P.; Phan, N. *J. Org. Chem.* **1996**, *61*, 416–417.

[17] Ishikawa, T.; Mizuta, T.; Hagiwara, K.; Aikawa, T.; Kudo, T.; Saito, S. *J. Org. Chem.* **2003**, *68*, 3702–3705.

[18] Busch-Petersen, J.; Bo, Y. X.; Corey, E. J. *Tetrahedron Lett.* **1999**, *40*, 2065–2068.

[19] Tzalis, D.; Knochel, P. *Angew. Chem. Int. Ed.* **1999**, *38*, 1463–1465.

[20] Corey, E. J.; Xu, F.; Noe, M. C. *J. Am. Chem. Soc.* **1997**; *119*, 12 414–12415.

[21] Gouge, M. *Ann. Chim. (Paris)* **1951**, 648–664.

[22] For selected examples of the use of optically active propargyl alcohols in synthesis, see:
(a) Partridge, J. J.; Chadha, N. K.; Uskokovic, M. R. *J. Am. Chem. Soc.* **1973**, *95*, 7171–7172.
(b) Fried, J.; Sih, J. C. *Tetrahedron Lett.* **1973**, 3899–3902.
(c) Chan, K. K.; Cohen, N. C:; DeNoble, J. P.; Specian, A. C. Jr.; Saucy, G. *J. Org. Chem.* **1976**, *41*, 3497–3505.
(d) Pirkle, W. H.; Boeder, C. W. *J. Org. Chem.* **1978**, *43*, 2091–2093.
(e) Mori, K.; Akao, H. *Tetrahedron Lett.* **1978**, 4127–4130.

(f) Sato, K.; Nakayama, T.; Mori, K. *Agric. Biol. chem.* **1979**, *43*, 1571–1575.
(g) Vigneron, J. P.; Blauchard, J. M. *Tetrahedron Lett.* **1980**, *21*, 1739–1742.
(h) Mukaiyama, T.; Suzuki, K. *Chem. Lett.* **1980**, 255–256.
(i) Midland, M. M.; Tramontano, A. *Tetrahedron Lett.* **1980**, *21*, 3549–3552.
(j) Johnson, W. S.; Frei, B.; Gopalan, A. S. *J. Org. Chem.* **1981**, *46*, 1512–1513.
(k) Midland, M. M.; Nguyen, N. H. *J. Org. Chem.* **1981**, *46*, 4107–4108.
(l) Midland, M. M.; Lee, P. E. *J. Org. Chem.* **1981**, *46*, 3933–3934.
(m) Nicolaou, K. C.; Webber, S. E. *J. Am. Chem. Soc.* **1984**, *106*, 5734–5736.
(n) Corey, E. J.; Niimura, K.; Konishi, Y.; Hashimoto, S.; Hamada, Y. *Tetrahedron Lett.* **1986**, *27*, 2199–2202.
(o) Kluge, A. F.; Kertesz, D. J.; O-Yang, C.; Wu, H. Y. *J. Org. Chem.* **1987**, *52*, 2860–2868.
(p) Wender, P. A.; Ihle, N. C.; Correia, C. R. D. *J. Am. Chem. Soc.* **1988**, *110*, 5904–5906.
(q) Trost, B. M.; Hipskind, P. A.; Chung, J. Y. L.; Chan, C. *Angew. Chem. Int. Ed. Engl.* **1989**, *28*, 1502–1504.
(r) Marshall, J. A.; Wang, X. J. *J. Org. Chem.* **1992**, *57*, 1242–1252.
(s) Roush, W. R.; Sciotti, R. J. *J. Am. Chem. Soc.* **1994**, *116*, 6457–6458.
(t) Myers, A. G.; Zheng, B. *J. Am. Chem. Soc.* **1996**, *118*, 4492–4493.

[23] Stoichiometric reductions:
(a) Midland, M. M.; Tramontano, A.; Zderic, S. A. *J. Am. Chem. Soc.* **1977**, *99*, 5211–5213.
(b) Yamaguchi, S.; Mosher, H. S.; Pohland, A. *J. Am. Chem. Soc.* **1972**, *94*, 9254–9255.
(c) Nishizawa, M.; Yamada, M.; Noyori, R. *Tetrahedron Lett.* **1981**, *22*, 247–250.
(d) Brown, H. C.; Ramachandran, P. V. *Acc. Chem. Res.* **1992**, *25*, 16–24.

[24] Catalytic methods:
(a) Helal, C. J.; Magriotis, P. A.; Corey, E. J. *J. Am. Chem. Soc.* **1997**, *118*, 10 938–10939.
(b) Matsumura, K.; Hashiguchi, S.; Ikariya, T.; Noyori, R. *J. Am. Chem. Soc.* **1997**, *119*, 8738–8739.

[25] (a) For pioneering investigations on Ti(IV)-catalyzed dialkylzinc additions to aldehydes, see: Schmidt, B.; Seebach, D. *Angew. Chem. Int. Ed. Engl.* **1991**, *30*, 1321–1323.
(b) Corey, E. J.; Cimprich, K. A. *J. Am. Chem. Soc.* **1994**, *116*, 3151–3152.
(c) Kobayashi, S.; Furuya, M.; Ohtsubo, A.; Mukaiyama, T. *Tetrahedron: Asymmetry* **1991**, *2*, 635–638.
(d) Carreira, E. M.; Singer, R. A.; Lee, W. *J. Am. Chem. Soc.* **1994**, *116*, 8837–8838.
(e) Singer, R. A.; Shepard, M. S.; Carreira E. M. *Tetrahedron* **1998**, *54*, 7025–7032.
For the addition of lithium and magnesium acetylides to trifluoromethyl aryl ketones, see:
(a) Tan, L.; Chen, C.-Y.; Tillyer, R. D.; Grabowski, E. J. J.; Reider, P. *Angew. Chem. Int. Ed.* **1999**, *38*, 711–713.
(b) For a recent report on the addition of aromatic aldehydes with alkynylzinc reagents generated in situ from terminal acetylenes and dimethylzinc, see; Li, Z.; Upadhyay, V.; DeCamp, A. E.; DiMichele, L.; Reider, P. J. *Synthesis* **1999**, 1453–1458.
(c) For a recent study of chelation controlled stannylacetylene additions to aldehydes see: Evans, D. A.; Halstead, D. P.; Allison, B. D. *Tetrahedron Lett.* **1999**, *40*, 4461–4462.

[26] (a) Corey, E. J.; Bakshi, R. K.; Shibata, S. *J. Am. Chem. Soc.* **1987**, *109*, 5551–5553.
(b) Mathre, D. J.; Thompson, A. S.; Douglas, A. W.; Hoogsteen, K.; Carroll, J. D.; Corley, E. G.; Grabowski, E. J. J. *J. Org. Chem.* **1993**, *58*, 2880–2888.
(c) Corey, E. J.; GuzmanPerez, A.; Lazerwith, S. E. *J. Am. Chem. Soc.* **1997**, *119*, 11 769–11776.
(d) Singh, V. K. *Synthesis* **1992**, 605–617. (e) Deloux, L.; Srebnik, M. *Chem. Rev.* **1993**, *93*, 763–784.
(f) Corey, E. J.; Helal, C. J. *Angew Chem. Int. Ed.* **1998**, *37*, 1986–2012.

[27] Evans, D. A.; Halstead, D. P.; Allison, B. D. *Tetrahedron Lett.* **1999**, *40*, 4461–4462.

[28] Niwa, S.; Soai, K. *J. Chem. Soc. Perkin Trans. 1* **1990**, 937–943.

[29] Noyori, R. *Asymmetric Catalysis in Organic Synthesis*, Wiley, New York, **1994**, Ch. 5.

[30] Sato, T.; Soai, K.; Suzuki, K.; Mukaiyama, T. *Chem. Lett.* **1978**, 601–604.

[31] (a) Okhlobystin, O. Y.; Zakharkin, L. I. *J. Organomet. Chem.* **1965**, *3*, 257–258. (b) Xu, F.; Reamer, R. A.; Tillyer, R.; Cummins, J. M.; Grabowski, E. J. J.; Reider, P. J.; Collum, D. B.; Huffman, J. C. *J. Am. Chem. Soc.* **2000**, *122*, 11 212–11218.

[32] Tombo, G. M. R.; Didier, E.; Loubinoux, B. *Synlett* **1990**, 547–548.

[33] Ishizaki, M.; Hoshino, O. *Tertrahedron Asymm.* **1994**, *5*, 1901–1904.

[34] The observation of higher selectivity at ambient temperatures than at 0 °C had been noted previously; see: Chelucci, G.; Conti, S.; Falorni, M.; Giacomelli, G. *Tetrahedron* **1991**, *47*, 8251–8258.

[35] Lu, G.; Li, X.; Zhou, Z.; Chan, W. L.; Chan, A. S. *Tetrahedron Asymm.* **2001**, *12*, 2147–2152.

[36] Xu, M.-H.; Pu, Li *Org. Lett.* **2002**, *4*, 4555–4557.

[37] Moore, D.; Pu, L. *Org. Lett.* **2002**, *4*, 1855–1857.

[38] Gao, G.; Moore, D.; Xie, R.-G.; Pu, L. *Org. Lett.* **2002**, *4*, 4143–4146.

[39] Marshall, J. A.; Bourbeau, M. P. *Org. Lett.* **2003**, *5*, 3197–3199.

[40] (a) Lu, G.; Li, X.; Chen, G.; Chan, W. L.; Chan, A. S. C. *Tetrahedron Asymm.* **2003**, *14*, 449–452. (b) Li, S. S.; Lu, G.; Kwok, W. H.; Chan, A. S. C. *J. Am. Chem. Soc.* **2002**, *124*, 12636–12637.

[41] Pierce, M. E.; Parsons, R. L.; Radesca, L. A.; Lo, Y. S.; Silverman, S.; Moore, J. R.; Islam, Q.; Choudhury, A.; Fortunak, J. M. D.; Nguyen, D.; Luo, C.; Morgan, S. J.; Davis, W. P.; Confalone, P. N.; Chen, C. Y.; Tillyer, R. D.; Frey, L.; Tan, L. S.; Xu, F.; Zhao, D.; Thompson, A. S.; Corley, E. G.; Grabowski, E. J. J.; Reamer, R.; Reider, P. J. *J. Org. Chem.* **1998**, *63*, 8536–8543.

[42] (a) Xu, F.; Reamer, R. A.; Tillyer, R.; Cummins, J. M.; Grabowski, E. J. J.; Reider, P. J.; Collum, D. B.; Huffman, J. C. *J. Am. Chem. Soc.* **2000**, *122*, 11 212–11218. (b) Parsons, R. L.; Fortunak, J. M.; Dorow, R. L.; Harris, G. D.; Kauffman, G. S.; Nugent, W. A.; Winemiller, M. D.; Briggs, T. F.; Xiang, B. S.; Collum, D. B. *J. Am. Chem. Soc.* **2001**, *123*, 9135–9143.

[43] Tan, L.; Chen, C. Y.; Tillyer, R. D.; Grabowski, E. J. J.; Reider, P. J. *Angew. Chem. Int. Ed.* **1999**, *38*, 711–713.

[44] Li, Z.; Upadhyay, V.; DeCamp, A. E.; DiMichele, L.; Reider, P. J. *Synthesis* **1999**, 1453–1458.

[45] Jiang, B.; Feng, Y. *Tetrahedron Lett.* **2002**, *43*, 2975–2977.

[46] Cozzi, P. G. *Angew. Chem. Int. Ed.* **2003**, *42*, 2895–2898.

[47] Yamaguchi, M.; Hayashi, A.; Minami, T. *J. Org. Chem.* **1991**, *56*, 4091–4092.

[48] Frantz, D. E.; Fässler, R.; Carreira, E. M. *J. Am. Chem. Soc.* **1999**, *121*, 11 245–14246.

[49] Frantz, D. E.; Fässler, R.; Tomooka, C. S.; Carreira, E. M. *Acc. Chem. Res.* **2000**, *33*, 373–381.

[50] Fässler, R.; Tomooka, C.; Frantz, D. E; Carreira, E. M. *Proc. Natl. Acad. Sci. U.S.A.* **2004**, *101*, 5843–5845.

[51] Knochel, P.; Jones, P. *Organozinc Reagents: A Practical Approach*, Oxford University Press, New York, **1999**.

[52] Frantz, D. E.; Fässler, R.; Carreira, E. M. *J. Am. Chem. Soc.* **2000**, *122*, 1806–1807.

[53] Kitazume, T.; Kasai, K. *Green Chemistry* **2001**, *3*, 30–32.

[54] Jiang, B.; Si, Y. G. *Tetrahedron Lett.* **2002**, *43*, 8323–8323.

[55] Sasaki, H.; Boyall, D.; Carreira, E. M. *Helv. Chim. Acta* **2001**, *84*, 964–971.

[56] Boyall, D.; Lopez, F.; Sasaki, H.; Frantz, D.; Carreira, E. M. *Org. Lett.* **2000**, *2*, 4233–4236.

[57] (a) Diez, R. S.; Adger, B.; Carreira, E. M. *Tetrahedron* **2002**, *58*, 8341–8344. (b) Amador, M.; Ariza, X.; Garcia, J.; Ortiz, J. *Tetrahedron Lett.* **2002**, *43*, 2691–2694.

[58] (a) Maezaki, N.; Kojima, N.; Asai, M.; Tominaga, H.; Tanaka, T. *Org. Lett.* **2002**, *4*, 2977–2980.
(b) Fettes, A.; Carreira, E. M. *Angew. Chem. Int. Ed.* **2002**, *41*, 4098–4101.
(c) Fettes, A.; Carreira, E. M. *J. Org. Chem.* **2003**, *68*, 9274–9283.
(d) Bode, J. W.; Carreira, E. M. *J. Am. Chem. Soc.* **2001**, *123*, 3611–3612.
(e) Bode, J. W.; Carreira, E. M. *J. Org. Chem.* **2001**, *66*, 6410–6424.

[59] Reber, S.; Knopfel, T. F.; Carreira, E. M. *Tetrahedron* **2003**, *59*, 6813–3817.

[60] Anand, N. K.; Carreira, E. M. *J. Am. Chem. Soc.* **2001**, *123*, 9687–9688.

[61] (a) Trost, B. M.; Ball, Z. T.; Joge, T. *Angew. Chem. Int. Ed.* **2003**, *42*, 3415–3418.
(b) El-Sayed, E.; Anand, N. K.; Carreira, E. M. *Org. Lett.* **2001**, *3*, 3017–3020.

[62] Boyall, D.; Frantz, D. E.; Carreira, E. M. *Org. Lett.* **2002**, *4*, 2605–2606.

[63] Chen, Z. L.; Xiong, W. N.; Jiang, B. *Chem. Commun.* **2002**, 2098–2099.

[64] Jiang, B.; Chen, Z. L.; Xiong, W. N. *Chem. Commun.* **2002**, 1524–1525.

[65] Jiang, B; Chen, Z.; Tang, X. *Org. Lett.* **2002**, *4*, 3451–3453.

[66] (a) Reppe, W. *Liebigs Ann. Chem.* **1955**, *596*, 12–25.
(b) Rose, J. D.; Gale, R. A. *J. Chem. Soc.* **1949**, 792–796. (c) Kruse, C. W.; Kleinschmidt *J. Am. Chem. Soc.* **1961**, *83*, 216–220.

[67] Brannock, K. C.; Burpitt, R. D.; Thweatt, J. G. *J. Org. Chem.* **1963**, *28*, 1462–1464.

[68] Fässler, R.; Frantz, D. E.; Oetiker, J.; Carreira, E. M. *Angew. Chem. Int. Ed.* **2002**, *41*, 3054–3056.

[69] Vasella, A. *Helv. Chim. Acta* **1977**, *60*, 1273–1295.

[70] Denis, J.-N.; Tchertchain, S.; Tomassini, A.; Vallée, Y. *Tetraheron Lett.* **1997**, *38*, 5503–5506.

[71] Pinet, S.; Pandya, S. U.; Chavant, P. Y.; Ayling, A.; Vallée, Y. *Org. Lett.* **2002**, *4*, 1463–1466.

[72] Patel, S. K.; Py, S.; Pandya, S. U.; Chavant, P. Y.; Vallée, Y. *Tetrahedron Asymm.* **2003**, *14*, 525–528.

[73] Wei, C.; Li, C.-H. *J. Am. Chem. Soc.* **2002**, *124*, 5638–5639.

[74] The study states that (+)-propargylamines are isolated in all of the additions, but no indication of the corresponding configuration is given.

[75] Wei, C.; Li, C-J. *J. Am. Chem. Soc.* **2003**, *125*, 9584–9585.

[76] Jiang, B.; Si, Y.-G. *Tetrahedron Lett.* **2003**, *44*, 6767–6768.

[77] Zhang, J.; Wei, C.; Li, C.-H. *Tetrahedron Lett.* **2002**, *43*, 5731–5733.

[78] Wei, C.; Li, Z.; Li, C.-J. *Org. Lett.* **2003**, *5*, 4473–4475..

[79] Koradin, C.; Polborn, K.; Knochel, P. *Angew. Chem. Int. Ed.* **2002**, *41*, 2535–2538.

[80] Gommermann, N.; Koradin, C.; Polborn, K.; Knochel, P. *Angew. Chem. Int. Ed.* **2003**, *42*, 5763–5766.

[81] Fischer, C.; Carreira, E. M. *Org. Lett.* **2001**, *3*, 4319–4321.

[82] Carreira, E. M. Presented at The Knud Lind Larsen Symposium, Danish Academy of Sciences, Copenhagen, DK, January 25 2004.

4
Transition Metal Acetylides

Uwe Rosenthal

Dedicated to Prof. Dr. mult. Günther Wilke on the occasion of his 80th birthday

4.1
Introduction

Acetylides are compounds containing an MC≡CR substructure[1)] (M = complex fragment, R = H, alkyl, aryl, etc.)[2)] and may also be termed *alkynyl complexes* (the shorter name *acetylide*, in principle correct only for the unsubstituted case MC≡CH, has been preferred here). The chemistry of transition metal acetylides developed very rapidly between Nast's first review (based on 200 papers) in 1982 [1a] and 2003, when the last review [1] of this topic (based on 10 000 papers) was published by Long and Williams [1k]. One explanation for this could be that acetylides, metal acetylides in particular, are a fascinating group of compounds for synthetic (organometallic, inorganic, and organic), theoretical, and physical chemistry with great technological potential in materials science (technical chemistry). These overlapping interests of many researchers are not restricted to classical areas of chemistry, but also extend into physics and technology.

This great wealth and the rapid growth of this field prevents a comprehensive review. It is therefore the intention here, after a very short and more general consideration, to highlight the chemistry of transition metal acetylides by presenting examples of recent investigations in the metallocene chemistry of titanium and zirconium [2]. Other special aspects, such as cluster chemistry, applications of metal acetylides in organic syntheses, and π-bonded acetylides, are mostly excluded here.

1) Brackets [ ] have been used only for isolated complexes, not for complex fragments, substructures and structural formulas. Additionally, complexes that were not isolated or were only assumed have been indicated by " ".

2) The following abbreviations are used: M: complex fragment; X: halogenide; R = H, alkyl, aryl; Me: methyl; Ph: phenyl; *t*Bu: tertiary butyl; BTMSA: bis(trimethylsilyl)acetylene; Y: bridging group; Cp: cyclopentadienyl; Cp′: differently substituted cyclopentadienyl; Cp*: pentamethylcyclopentadienyl; ebthi: ethylene-bis-tetrahydroindenyl, Solv.: solvate molecules.

Acetylene Chemistry. Chemistry, Biology and Material Science. Edited by F. Diederich, P. J. Stang, R. R. Tykwinski

ISBN 3-527-30781-8

For more information regarding these and other details of metal acetylide chemistry, some very nice reviews exist in the references cited in [1].

4.2 General Comments

The simplest carbon molecule C_2, as a small part of the linear polyynes [-(C≡C)$_n$-], is extremely reactive and has only been characterized by spectroscopic methods. Its chemistry has been limited to its generation in a carbon arc and condensation of its vapor at low temperatures [1].

The C_2 unit can, however, be stabilized by end-capping of both ends, with formation of various organic acetylenes: HC≡CH, HC≡CR, and RC≡CR. A further alternative is the use of one or two M moieties to produce the mono-acetylides MC≡CH or MC≡CR, or the bis-acetylides MC≡CM.

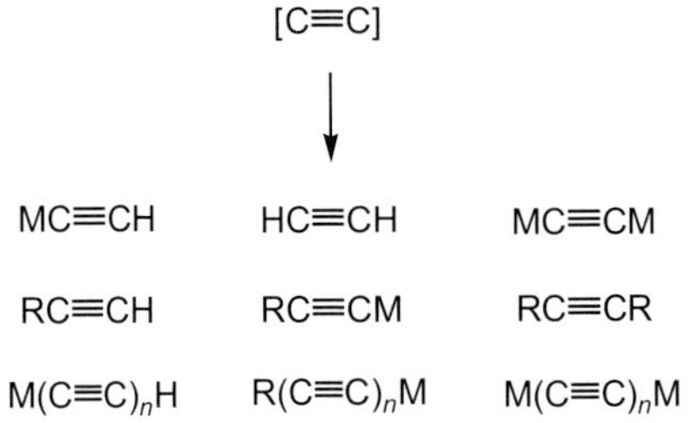

Scheme 4.1 Stabilization of C_2 by end-capping.

For this basic set in the series of metal acetylides, depending on the different numbers n of C_2 units, there are many further combinations, such as M(C≡C)$_n$H, M(C≡C)$_n$R, and M(C≡C)$_n$M. For the mono-acetylides, the number of acetylide ligands can also be larger than one, with compounds M[(C≡C)$_n$H]$_m$ or M[(C≡C)$_n$R]$_m$ being formed. Sometimes the polyyne chain contains bridging groups that, depending on Y and n, may form different linear, non-linear, or branched species such as Y[(C≡C)$_n$M]$_m$. There are many possible combinations and other complicated examples of such compounds, but the smallest subunits MC≡CH, MC≡CR, and MC≡CM are very useful as a basis through which to understand the higher derivatives. An excellent overview of examples of these compounds is provided in two recent contributions by Low and Bruce [1l, 1m].

4.2.1 Structure and Bonding

Transition metal acetylides can be viewed as complexes of the ligand [RC≡C]$^-$, which is isoelectronic with the cyanide ion [C≡N]$^-$; both are regarded as pseudohalides because of the many similarities in the chemical and physical behavior of the corresponding compounds. This is the reason why transition metal acetylides are often viewed as classical coordination compounds rather than as organometallics.

The acetylides are good σ- and π-donor and poor π-acceptor ligands. Acetylides $M{-}C{\equiv}CR$, aryl compounds M-Ar, and alkenyl (vinyl) derivatives $M{-}CH{=}CH_2$ have ligands located, in terms of bonding, between the pure σ-donor (alkyl M-R) and the σ-donor/π-acceptor ligands (CO, RNC, R_3P, etc.). In principle, the acetylides have free π^* orbitals that can interact with filled metal d orbitals, but such interaction seems to be weak. A detailed description of the bonding in metal acetylides based on MO calculations, photoelectron and electronic spectra, X-ray data, and vibrational spectroscopic results has been summarized in a broad review by Manna, John, and Hopkins [1d], in which the metal acetylide interaction is discussed in comparison with other ligands on the basis of carefully collected data for many examples. These results indicate significant overlap of the filled metal d orbitals and the filled π system of the acetylide. The energy difference between the HOMO metal d orbitals and the LUMO π^* acetylide orbitals is too large for π-accepting behavior, which appears only to be possible if strong π-acceptor substituents are conjugated with the acetylide unit [1k]. Nevertheless, there remain some open questions regarding interpretation of structural data for some compounds. The $M{-}C$ bond lengths in M–R, $M{-}CH{=}CH_2$, and M–Ar are nearly identical in some examples, indicating less important π bonding between M and C. In the case of strong back-bonding of acetylides $M(d\pi){\rightarrow}C(p\pi^*)$, larger d(C−C) bond lengths are expected than are in fact found. On the other hand, the polarity of the $M{-}C$ bond also has an influence [1].

4.2.2 Syntheses

There are some "classical" [1], as well as some more "special" [3], routes to transition metal acetylides, which have been used depending on the different characters of the metals, the alkynes, and the products, mostly on the basis of oxidative addition or nucleophilic substitution.

In the first group (Scheme 4.2), the traditional method most widely used is treatment of metal halides with an anionic alkynyl reagent such as alkali metal or magnesium acetylides by *salt elimination*. *Oxidative addition* of 1-alkynes or haloacetylenes to metal complexes is another important method. Alternatively, *dehydrohalogenation* is a direct method starting from metal halides and reactive acetylenes possessing electron-withdrawing substituents. Cu(I) has often been used for this reaction, as well as for *oxidative coupling* and *acetylide metathesis*. In comparison with these terminal alkyne approaches, the *trimethylstannyl method* with corresponding trimethylstannyl alkynes often gives higher reactivity and cleaner products. With transition metal methyl complexes, acetylenes give the corresponding acetylides after *elimination* of methane. The driving force for this reaction is the higher acidity of acetylenes than of methane. The often more strongly coordinating terminal alkynes also form, by elimination of H_2 or N_2, such π-complexes that can give acetylide hydrides or vinylidene complexes by *oxidative addition*. Subsequent abstraction of hydrogen from vinylidenes can also afford acetylides. In the *alkynyliodonium method*, triflates are versatile reagents for the synthesis of acetylides in which "Umpolung" of

Salt Elimination

MX + M'C≡CR $\xrightarrow{- M'X}$ MC≡CR

Oxidative Addition

M + XC≡CR $\longrightarrow$ M(X)C≡CR

Dehydrohalogenation

MX + HC≡CR $\xrightarrow[- HX]{Base}$ MC≡CR

Oxidative Coupling

MC≡CH + HC≡CR $\xrightarrow{- H_2}$ MC≡C—C≡CR

Acetylide Methathesis

MC≡CH + HC≡CR $\xrightarrow{- HC≡CH}$ MC≡CR

Trimethylstannyl Method

MCl + Me$_3$SnC≡CR $\xrightarrow{- Me_3SnCl}$ MC≡CR

Elimination (CH$_4$, H$_2$, N$_2$)

MCH$_3$ + HC≡CR $\xrightarrow{- CH_4}$ MC≡CR

Alkynyliodonium triflate Method

MX + RC≡CI$^{\oplus}$Ph $^{\ominus}$OSO$_2$CF$_3$ $\xrightarrow{- PhI}$ [MC≡CR]$^{\oplus}$[OSO$_2$CF$_3$]$^{\ominus}$ X

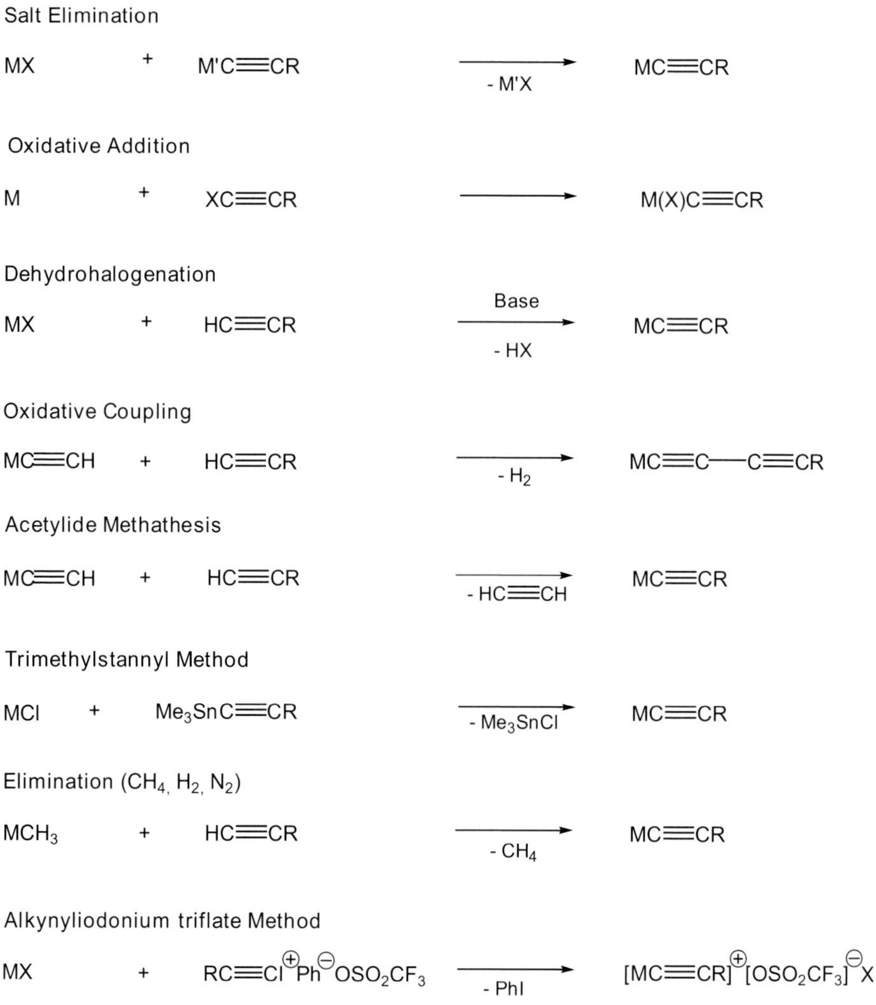

Scheme 4.2 "Classical" methods for the synthesis of acetylides.

the normal alkyne reactivity from RC≡C$^-$ to RC≡C$^+$ has been found, with the transition metal acting as a nucleophile and the iodonium compound as an electrophile.

In addition to these methods, there are also some more special examples for the synthesis of transition metal acetylides (Scheme 4.3).

Cyclic organometallic compounds containing two transition metal–carbon σ-bonds can give acetylides from terminal acetylenes in a *protolysis* reaction without the alkane elimination described above. Dimethylphosphonium bis-methylide complexes, as inner-phosphonium-*ate* complexes, for example, add terminal acetylenes with formation of acetylide-ylide complexes [3a]. Metallacyclopropenes (canonical formula of metal π-alkyne complexes)[3] can also add terminal acetylenes to

---

3) Titanocene and zirconocene complexes of BTMSA can be described by the canonical formulas of π-complexes and metallacyclopropenes; both have been used in this article.

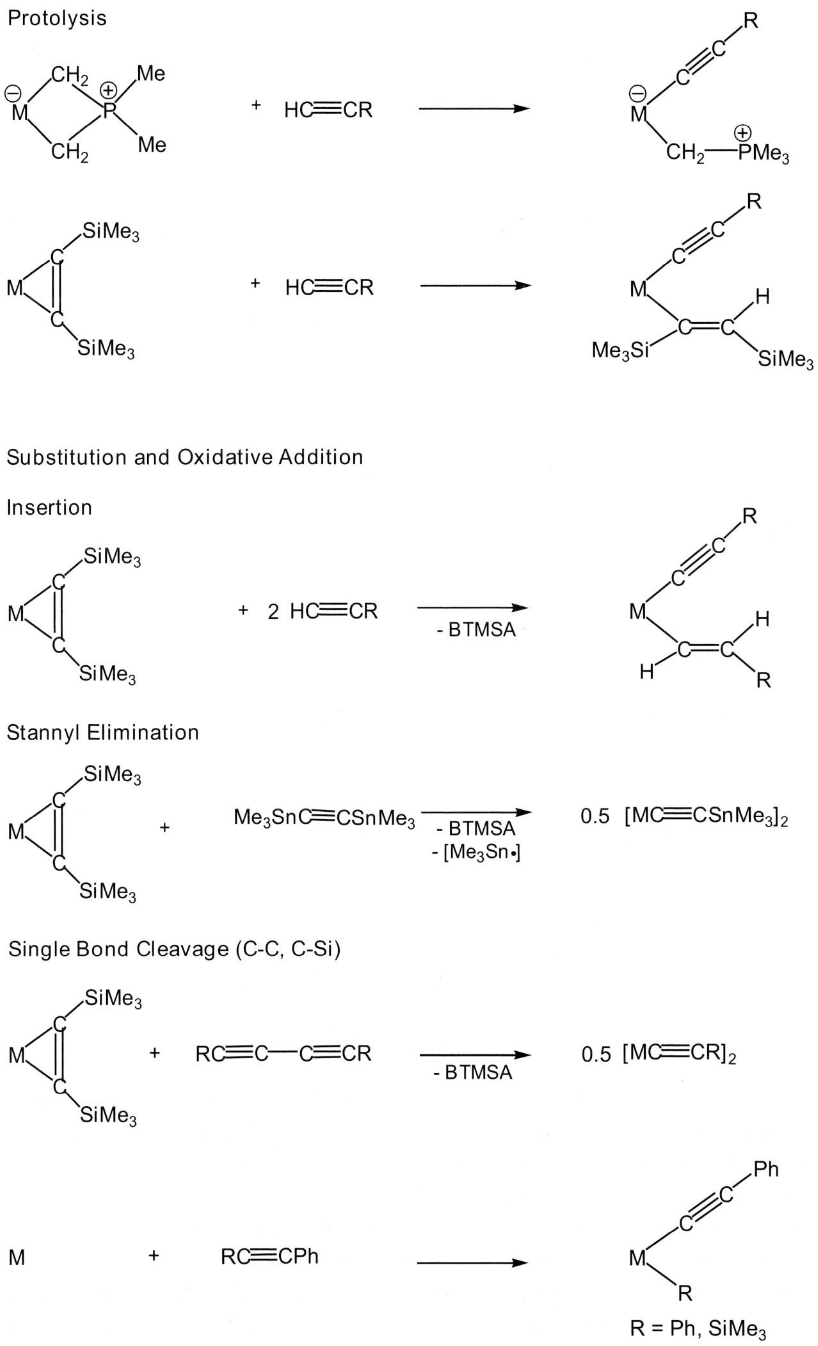

Scheme 4.3 "Special" methods for the synthesis of acetylides.

yield acetylide-alkylidene compounds [3b]. In both cases, the acetylene hydrogen atom remains in the opened former cyclic ligand.

In contrast to this protolysis reaction, metallacyclopropenes also display some *substitution* reactions of the bis(trimethylsilyl)acetylene (BTMSA) followed by *oxidative addition*. Terminal alkynes HC≡CR can yield another type of acetylide-alkenyl complex, via the formed M(H)(C≡CR) and *insertion* of a second terminal acetylene into the metal hydrogen bond [3c].

This reaction differs from that with stannyl-acetylenes, in which the bis(trimethylsilyl)acetylene is also initially substituted, but after oxidative addition of $Me_3SnC{\equiv}CSnMe_3$, via $M(SnMe_3)(C{\equiv}CSnMe_3)$, a *stannyl elimination* subsequently occurs, giving Me_3Sn radicals and σ,π-acetylide-bridged dinuclear complexes $[M(C{\equiv}CSnMe_3)]_2$ [3d].

The same type of complex is obtained if metallacyclopropenes react with 1,3-butadiynes with substitution of the silylalkyne and subsequent oxidative addition with cleavage of the central C−C single bond of the diyne [2]. *C−C single bond cleavage* has also been observed for alkynes such as tolan PhC≡CPh, to acetylide-phenyl complexes [3e]. This reaction is reversible, and was found also for the Si−C bond in alkynylsilanes such as $Me_3SiC{\equiv}CPh$ and $Me_3SiC{\equiv}C{-}C{\equiv}CSiMe_3$ [3f]. This type of reactions is relevant for the dynamics discussed below (see Section 4.4.2).

4.2.3
Reactions

Acetylenes are more acidic than alkanes, alkenes, or arenes, and the corresponding metal σ-acetylide bond M−C≡CR seems to be less reactive than is found for alkyl (M-R), alkenyl ($M{-}CH{=}CH_2$), and aryl (M-Ar) derivatives. Nevertheless, acetylides are frequently explosive, a consequence of instability not of the M−C bond, but of the energy-rich C≡C bond. The stability of transition metal acetylides seems to decrease in the order R = Ar > H > alkyl. In principle, acetylides can give all the typical reactions (*acidolyses, insertions*) of other transition metal carbon bonds (Scheme 4.4). Additionally, they also show the special reactivity of the triple bond (*complexation*). One important feature of the acetylides is their strong tendency to enter into π-complexation to form dinuclear complexes and clusters in which the acetylides have many different bridging functions. Selected examples of compounds of this type are discussed in Section 4.4. Often these di- or polynuclear acetylides are more stable than the monomeric species.

One important point in respect of the topomerization of alkynes, discussed below (Section 4.4.3), is that transition metal acetylides represent very useful starting materials from which to obtain higher carbenes $M{=}C({=}C)_n{=}CR_2$ [1f, 1g, 1l, 1m, 4]. In *reactions with electrophiles*, the acetylides can form vinylidenes M=C=CHR and $M{=}C{=}CR_2$ [4a–c]. Similar products M=C=CHR are also formed if H is present in the coordination sphere of the metal (see also Section 4.4.3) [4d]. In *coupling reactions*, C-ligands instead of H are coupled with acetylides; depending on the reaction conditions either ene-yne complexes (see also Section 4.3.5.1.5) or acetylide-substituted carbene complexes are formed [4e]. After protonation, *elimina-*

Hydrolysis

$MC\equiv CR + H_2O \longrightarrow MOH + HC\equiv CR$

Insertion

$MC\equiv CR + RC\equiv CR \longrightarrow MCR=CR-C\equiv CR$

Complexation

$MC\equiv CR + M' \longrightarrow$ M'—[complex with M, C≡C, R]

Reaction with Electrophiles

$MC\equiv CR \xrightarrow{R^{\oplus}} M=C=CR_2^{\oplus}$

Rearrangement

$(H-)MC\equiv CR \longrightarrow M=C=CHR$

Coupling Reactions

$(RHC=CH-)MC\equiv CR \xrightarrow{H^+}$ M—[C(R)=C, C=C(R,H), H]

$(RHC=C=)MC\equiv CR \longrightarrow$ M—C[=C(R)-H, C≡C-R]

Elimination of Water

$MC\equiv CCR_2OH \xrightarrow{H^+} [M=C=CHCR_2OH]^{\oplus}$

$\downarrow -H_2O$

$[M-C\equiv CCR_2]^{\oplus} \longleftrightarrow [M=C=C=CR_2]^{\oplus}$

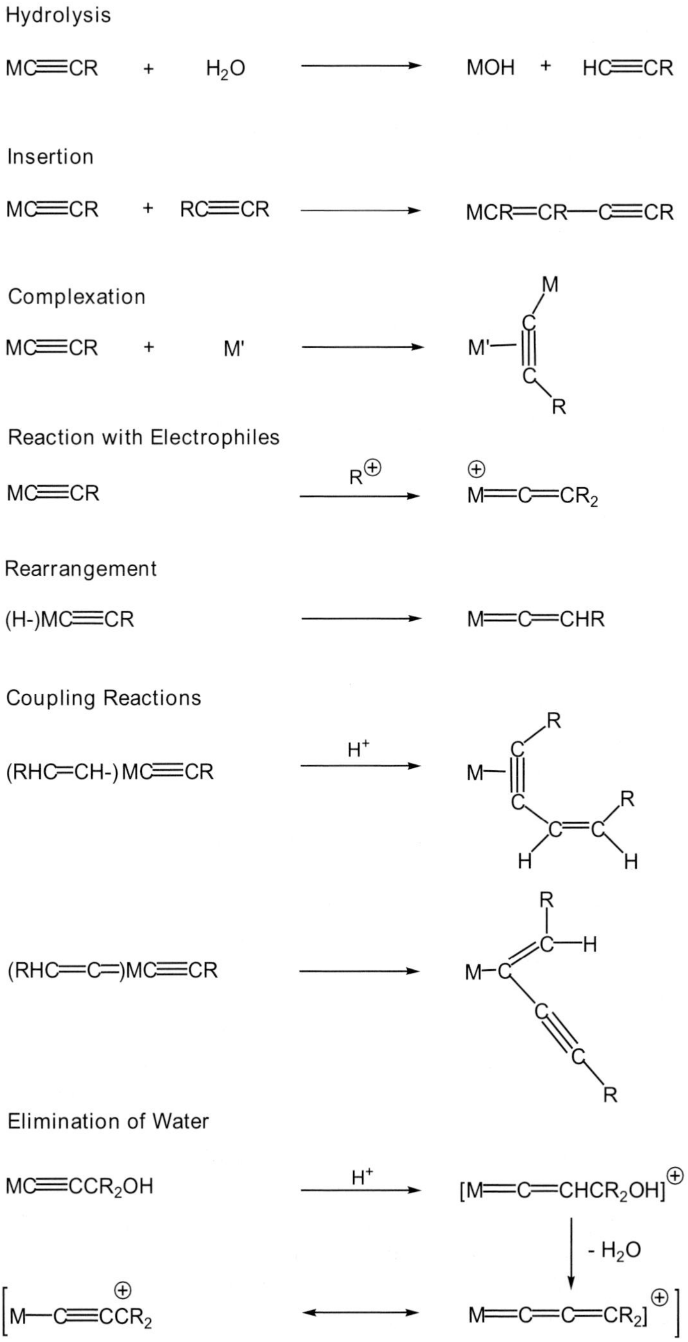

Scheme 4.4 Reactions of acetylides.

tion of water from $MC{\equiv}CCR_2OH$ (via $[M{=}C{=}CH{-}CR_2OH]^+$) is the best method to obtain allenylidenes $[M{=}C{=}C{=}CR_2]^+$, complexes best described by the acetylenic canonical formula $[MC{\equiv}CCR_2]^+$ [4f–h].

4.3
Titanocene- and Zirconocene-Acetylides

There exist so many titanocene- and zirconocene-acetylides that only compounds of the types $[Cp'_2M(\sigma\text{-}C{\equiv}CR)]$ and $[Cp'_2M(\sigma\text{-}C{\equiv}CR)_2]$ with M = Ti, Zr and Cp′= Cp, Cp* are discussed here. No complexes with partially substituted Cp ligands or heterobimetallic combinations of these with other metals are considered here; examples of the latter can be found in the cited reviews [1]. Compounds of this type with M = Ti and Zr in the oxidation states +3 and +4 are important for stoichiometric and catalytic C−C single bond coupling and cleavage reactions [2].

4.3.1
MC≡CR

4.3.1.1 Mono-σ-acetylides $[Cp_2M(\sigma\text{-}C{\equiv}CR)]$

To the best of our knowledge, complexes of the type $[Cp_2M(\sigma\text{-}C{\equiv}CR)]$ do not exist as stable monomeric compounds for M = Ti and Zr, due to their strong tendency to dimerize. In the reaction between $[Cp_2TiCl]_2$ and $NaC{\equiv}CPh$, the product formed was not the expected monomeric $[Cp_2TiC{\equiv}CPh]$ but rather the binuclear $[Cp_2Ti]_2[\mu\text{-}\eta^2(1,3),\eta^2(2,4)PhC_2C_2Ph]$ (**1**),[4] with a bridging butadiyne ligand (see Section 4.3.5.1 and Scheme 4.5) [5a].

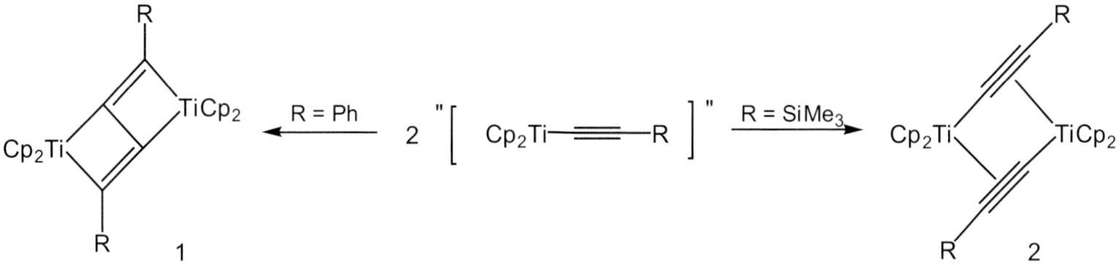

1 2

Scheme 4.5 Stabilization of $[Cp_2TiC{\equiv}CR]$.

Another type of stabilization was found for the assumed, but not isolated, intermediate $[Cp_2TiC{\equiv}CSiMe_3]$, which gives the σ,π-acetylide-bridged Ti(III) complex $[Cp_2Ti(\mu\text{-}\sigma,\pi\text{-}C{\equiv}CSiMe_3)]_2$ (**2**) without any C−C coupling between the metal centers [5b, 5e].

4) In the numbering system used in this article the given number mostly refers not to a single complex, but to a group of isostructural compounds, sometimes with different metals M, ligands Cp′ and substituents R.

Many doubly σ,π-acetylide-bridged Zr(III) complexes $[Cp_2Zr(\mu\text{-}\sigma,\pi\text{-}C{\equiv}CR)]_2$ (**3**) exist, generated either by salt elimination [6a–e] or by reactions of "Cp_2Zr" with different butadiynes RC≡C−C≡CR through C−C single bond cleavage (see Section 4.3.5.1) (Scheme 4.6) [6f–h].

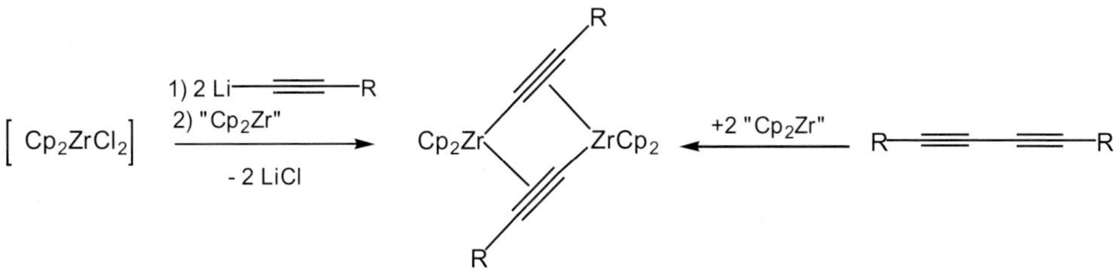

Scheme 4.6 Formation of $[Cp_2ZrC{\equiv}CR]_2$. 3

Heterobimetallic Ti,Zr combinations have been produced in the forms of $[Cp_2Ti(\mu\text{-}\sigma,\pi\text{-}C{\equiv}CR)_2ZrCp_2]$ for R = ^{t}Bu [6i] (see Scheme 4.20, below) and also Me_3Si [6j].

Early-late Ni,Ti- and Ni,Zr-heterobimetallics such as $[Cp_2M(\mu\text{-}\sigma,\pi\text{-}C{\equiv}CR)Ni(PPh_3)(\mu\text{-}\sigma,\pi\text{-}C{\equiv}CSiMe_3)]$ also exist, consisting formally of a metallocene(III)-mono-σ-acetylide $[Cp_2M(\sigma\text{-}C{\equiv}CSiMe_3)]$, stabilized by $[(Ph_3P)Ni(\sigma\text{-}C{\equiv}CSiMe_3)]$ (Section 4.3.5.1) [6k].

Nevertheless, the monomeric metallocene(III)-mono-σ-acetylides $[Cp_2M(\sigma\text{-}C{\equiv}CR)]$ are very important intermediates, in particular in the case of M = Ti, for many stoichiometric and catalytic reactions (see Section 4.3.5).

4.3.1.2 Mono-σ-acetylides $[Cp^*_2M(\sigma\text{-}C{\equiv}CR)]$

Monomeric complexes $[Cp'_2M(\sigma\text{-}C{\equiv}CR)]$ with Cp′ = Cp* are more stable than examples in which Cp′ = Cp, but there are nevertheless only two examples of such monomeric species. The first is $[Cp^*_2TiC{\equiv}CMe]$, obtained without structural characterization by treatment of $[Cp^*_2TiCl]$ with LiC≡CMe [7a]. If $[Cp^*_2TiCl]$ was treated with LiC≡C^{t}Bu, $[Cp^*_2TiC{\equiv}C^tBu]$ (**4**) was formed, but only in *n*-hexane (Scheme 4.7). The compound was characterized by X-ray crystal structure diffraction [7b].

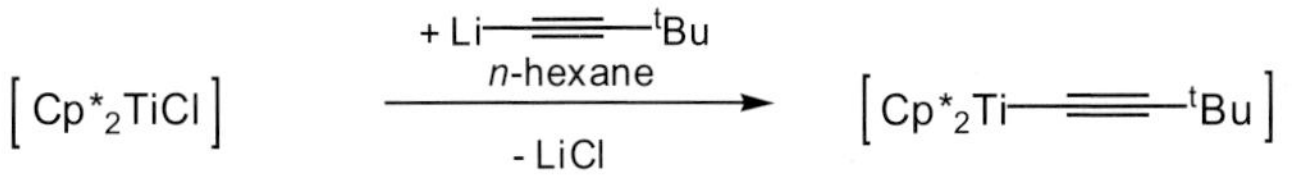

4

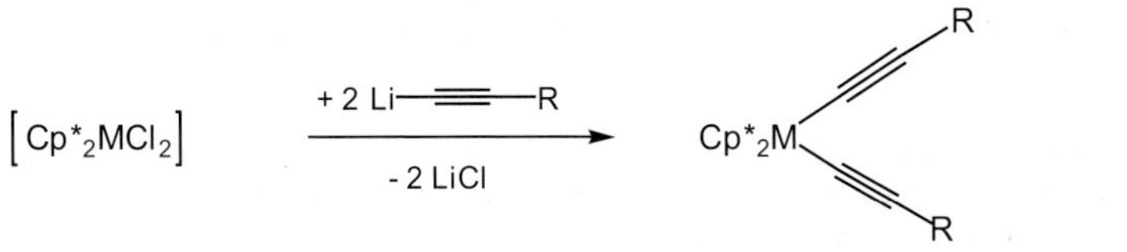

5 Cp' = Cp, Cp*

Scheme 4.7 Synthesis of $[Cp^*_2TiC{\equiv}C^tBu]$ and $[Cp'_2M(C{\equiv}CR)_2]$.

Analogous zirconocene(III)-mono-σ-acetylides [$Cp^*_2Zr(\sigma\text{-}C{\equiv}CR)$] are currently unknown. There are practically no data about monomeric or dimeric complexes of this type.

4.3.2 $M(C{\equiv}CR)_2$

4.3.2.1 Bis-σ-acetylides [$Cp_2M(\sigma\text{-}C{\equiv}CR)_2$]

Such complexes have been well known in the literature for quite a long time, being the most typical examples of group 4 metallocene σ-acetylides and summarized in many reviews [1]. A series of outstanding contributions to this field came from Erker's group [6a–e]. The compounds were typically prepared by lithium salt elimination, starting from the dihalides [Cp_2MX_2] and the corresponding lithium acetylides LiC≡CR [1]. These complexes are mostly very stable and were often used by Lang and others to produce tweezer-like organometallic molecules capable of coordinating many different salts, organometallics, alkali metals, and so on [1c, 1e, 1i, 1j].

One special case is the less stable Ti(IV) complex [$Cp_2Ti(\sigma\text{-}C{\equiv}CSiMe_3)_2$] [5b], which was firstly assumed [5c, 5d], but this complex was in fact the doubly σ,π-acetylide-bridged Ti(III) complex [$Cp_2Ti(\mu\text{-}\sigma,\pi\text{-}C{\equiv}CSiMe_3)$]$_2$ (**2**) mentioned above (Scheme 4.5), formed by reductive elimination of the butadiyne $Me_3SiC{\equiv}C{-}C{\equiv}CSiMe_3$ (see Scheme 4.15 below) [5b, 5e, 5f]. The Zr(IV) complex [$Cp_2Zr(\sigma\text{-}C{\equiv}CSiMe_3)_2$] [5c] also reacts in the same manner to form the doubly σ,π-acetylide-bridged Zr(III) complex [$Cp_2Zr(\mu\text{-}\sigma,\pi\text{-}C{\equiv}CSiMe_3)$]$_2$ [6f–h].

4.3.2.2 Bis-σ-acetylides [$Cp^*_2M(\sigma\text{-}C{\equiv}CR)_2$]

Complexes [$Cp^*_2M(\sigma\text{-}C{\equiv}CR)_2$] (**5**) are relatively rare in relation to the unsubstituted Cp examples discussed above. They have been prepared in the same manner from the dichlorides $Cp^*_2MCl_2$ and the corresponding lithium acetylides LiC≡CR for M = Ti and R = ^{t}Bu [8a], $SiMe_3$ [8b], and Ph [7b], and also for M = Zr and R = ^{t}Bu [7b], $SiMe_3$ [8c], and Ph [8c, 8d]. These complexes are very stable and have also been used as tweezer-like organometallic molecules.

4.3.3 $M(C{\equiv}CR)_3$

The small volume of the acetylide group makes it possible for more than two acetylides to be coordinated to one metal atom [1a]. One interesting question involves the existence of Zr(IV) tris-σ-acetylide *ate*-complexes [$Cp_2Zr(\sigma\text{-}C{\equiv}CPh)$]$_3$[Li(Solv.)] [9a], assumed as products from the reaction between [Cp_2ZrCl_2] and three equivalents of lithium acetylide LiC≡CPh, via the zirconocene(IV) bis-σ-acetylide [$Cp_2Zr(\sigma\text{-}C{\equiv}CR)_2$]. The *ate* complexes give isolated η^2-diyne-zirconate complexes [$Cp_2Zr(\sigma\text{-}C{\equiv}CPh)(\eta^2\text{-}1,2\text{-}PhC_2{-}C{\equiv}CPh)$][Li(Solv.)] [9b] (see Scheme 4.18 below) and, after hydrolysis, *Z*-conjugated enynes [9a].

4.3.4
Products of $[Cp_2M(\eta^2\text{-}RC_2R)]$ and $[Cp^*_2M(\eta^2\text{-}RC_2R)]$ with Acetylenes

The well defined bis(trimethylsilyl)acetylene complexes of the type $[Cp_2M(L)(\eta^2\text{-}Me_3SiC_2SiMe_3)]$ (M = Ti, without L; M = Zr, L = THF or pyridine), as well as the pentamethylcyclopentadienyl complexes $[Cp^*_2M(\eta^2\text{-}Me_3SiC_2SiMe_3)]$ (M = Ti, Zr) and the complexes [*rac*-(ebthi)M(η^2-$Me_3SiC_2SiMe_3$)] (M = Ti, Zr) were found to be excellent starting materials from which to obtain not only the titanocene and zirconocene acetylides described above but also some other compounds possessing the acetylide ligand together with different alkenyl groups, bis-acetylides, diacetylides, and some other very special complexes [2c].

4.3.4.1 RHC=CH-MC≡CR and RHC=CR-MC≡CR

Monosubstituted acetylenes RC≡CH (R = Ph, Me_3Si, $Me(CH_2)_3$, $Me(CH_2)_9$) react with $[Cp^*_2Ti(\eta^2\text{-}Me_3SiC_2SiMe_3)]$ (**6**) to form permethyltitanocene-1-alkenyl-acetylides (**7**) (Schemes 4.3 and 4.8) [3c].

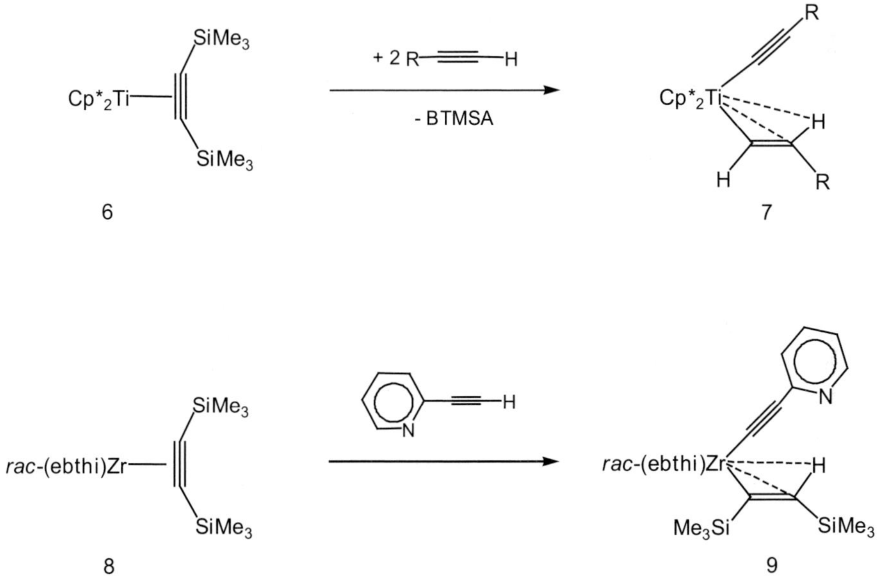

Scheme 4.8 Formation of (RHC=CH)MC≡CR and (RHC=CR)MC≡CR.

The formation steps for such complexes are substitution of $Me_3SiC{\equiv}CSiMe_3$ by RC≡CH to form $[Cp^*_2Ti(\eta^2\text{-}RC_2H)]$, oxidative addition to give "$[Cp^*_2Ti(H)(C{\equiv}CR)]$", and the subsequent insertion of a further RC≡CH to provide [Cp^*_2Ti(σ-(E)−CH=CHR)(σ-C≡CR)] (**7**).

2-Pyridyl−C≡CH, on treatment with [*rac*-(ebthi)Zr(η^2-$Me_3SiC_2SiMe_3$)] (**8**), gave another 1-alkenyl-acetylide, [*rac*-(ebthi)Zr[σ-(E)−C($SiMe_3$)=CH($SiMe_3$)](σ-C≡C-2-

pyridyl)] (**9**), stabilized by an agostic interaction, without substitution of the bis(trimethylsilyl)acetylene (Schemes 4.3 and 4.8) [3b].

4.3.4.2 MC≡CM

Depending on the metals used, compounds possessing the C_2^{2-} group in the μ-η^1:η^2 bridging mode can exist in the electronic structures M–C≡C–M, M=C=C=M, and M≡C–C≡M with central triple, double, or single bonds (see Scheme 4.27 below). The last of these, M≡C–C≡M, can be regarded as the isolobal inorganic counterpart of 1,3-butadiynes RC≡C–C≡CR. Examples with titanocene and zirconocene are very rare, but Binger and co-workers have described complexes such as $[Cp_2Ti(Me_3P)]_2(\mu-C_2)$ with a Ti=C=C=Ti structure [10a] and $[Cp_2Zr(CPh{=}CMePh)]_2(\mu-C{\equiv}C)$ [10b], while Norton et al. have reported the corresponding complex $\{[Cp_2Zr(NH^tBu)]_2(\mu-C{\equiv}C)\}$ [10c].

Treatment of $[Cp_2Zr(THF)(\eta^2\text{-}Me_3SiC_2SiMe_3)]$ (**10**) with unsubstituted acetylene gave (besides coupling to a zirconacyclopentadiene) small amounts of a dinuclear complex with a bridging diacetylide group and agostic 1-alkenyl ligands $\{Cp_2Zr[C(SiMe_3){=}CH(SiMe_3)]\}_2[\mu\text{-}\sigma(1,2)-C{\equiv}C]$ (**11**) (Scheme 4.9) [3c].

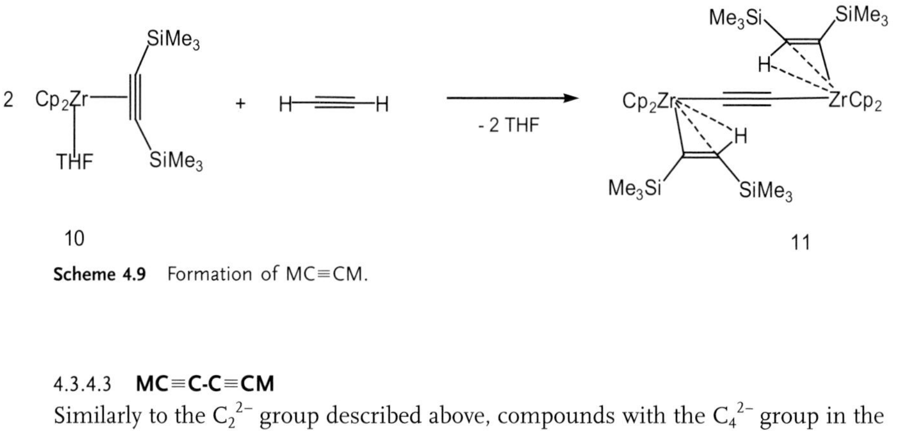

10 11

Scheme 4.9 Formation of MC≡CM.

4.3.4.3 MC≡C-C≡CM

Similarly to the C_2^{2-} group described above, compounds with the C_4^{2-} group in the bridging mode can also exist in the electronic structures M–C≡C–C≡C–M, M=C=C=C=C=M, and M≡C–C≡C–C≡M (i. e., as diynediyls, cumulenes, or dicarbynes). For titanocene and zirconocene diynediyls with the structural subgroup MC≡C–C≡CM have been obtained by starting from RC≡C–C≡C–C≡C–C≡CR with R = Me_3Si for titanocene and zirconocene, but in the case of R = ^{t}Bu only for zirconocene, giving a double C–C single bond cleavage and the formation of complexes of type **12** on treatment with four equivalents of the metallocene (Scheme 4.10) [11].

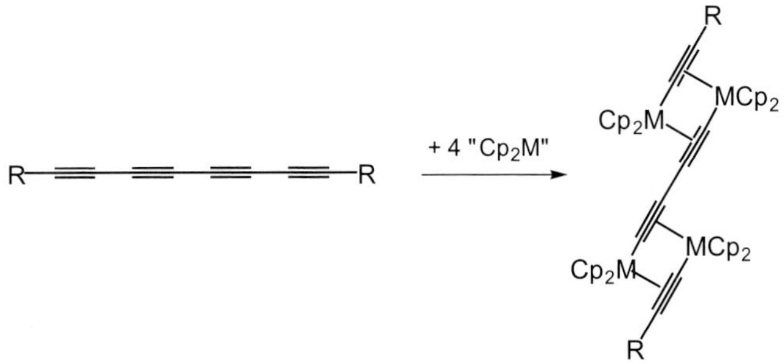

12 R = Me_3Si, M = Ti, Zr
R = ^tBu, M = Zr

Scheme 4.10 Formation of MC≡C–C≡CM.

4.3.4.4 **Bridged Compounds Y(C≡CM)$_3$**

The same type of reaction has been found if the butadiyne moieties are separated by a bridging group Y such as C_6H_3. Tris-butadiynyl-benzenes such as the 1,3,5-(RC≡C–C≡C)$_3$$C_6H_3$ derivative, on treatment with six equivalents of zirconocene, give the triply C–C-cleaved product **13** with the Y(C≡CM)$_3$ structural element (Scheme 4.11) [12].

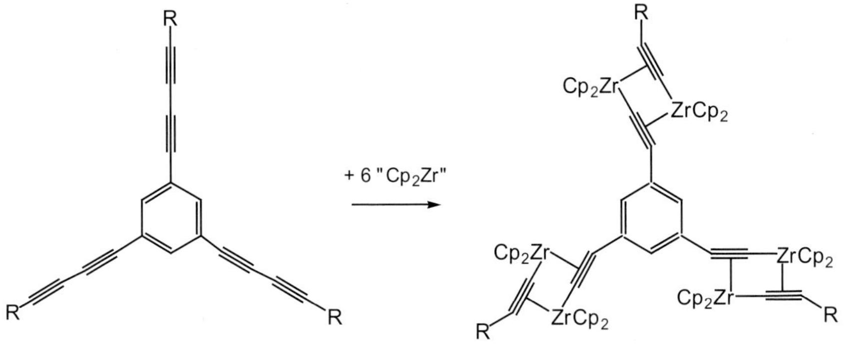

13 R = ^tBu

Scheme 4.11 Formation of [Y(C≡CM)$_3$].

4.3.5
Reactions

Typical of titanocene and zirconocene acetylides are *complexation of the triple bond* to other metals, *insertion reactions* of the acetylide ligand, and *C–C coupling reactions* of two acetylide ligands at one or two metal centers. [Cp^*_2TiC≡C^{t}Bu] (**4**), for example, coordinates to [$Co_2(CO)_8$] to give complex **14** (Scheme 4.12) [13]. In THF, [Cp^*_2TiC≡C^{t}Bu] reacts with LiC≡C^{t}Bu to yield the tweezer-like complex [Cp^*_2Ti(C≡C^{t}Bu)$_2$][Li(THF)] (**15**) [7b].

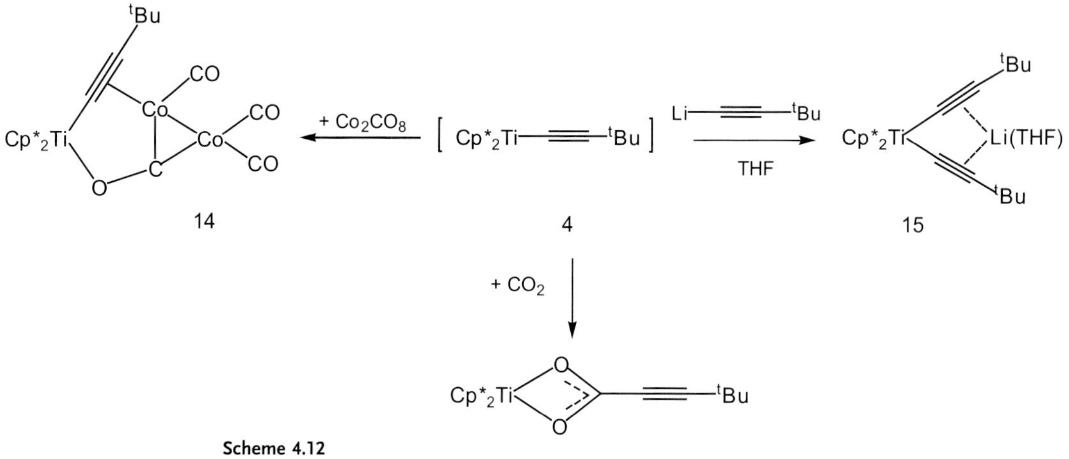

14 4 15 16

Scheme 4.12
Reactions of [Cp*$_2$TiC≡C^{t}Bu].

[Cp*$_2$TiC≡C^{t}Bu] inserts carbon dioxide to form [Cp*$_2$TiO$_2$CC≡C^{t}Bu] (**16**) [7b]. Another insertion proceeds in the reaction of [Cp*$_2$TiCl$_2$] with magnesium and an excess of Me$_3$SiC≡C−C≡CSiMe$_3$. The initial formation of a Ti(III) monoacetylide "[Cp*$_2$TiC≡CSiMe$_3$]" is assumed, with this inserting into the diyne to yield [Cp*$_2$Ti{η^3-Me$_3$SiC$_3$=C(C≡CSiMe$_3$)SiMe$_3$}] (**17**), featuring a hex-3-ene-1,5-diyn-3-yl ligand [14] (Scheme 4.13).

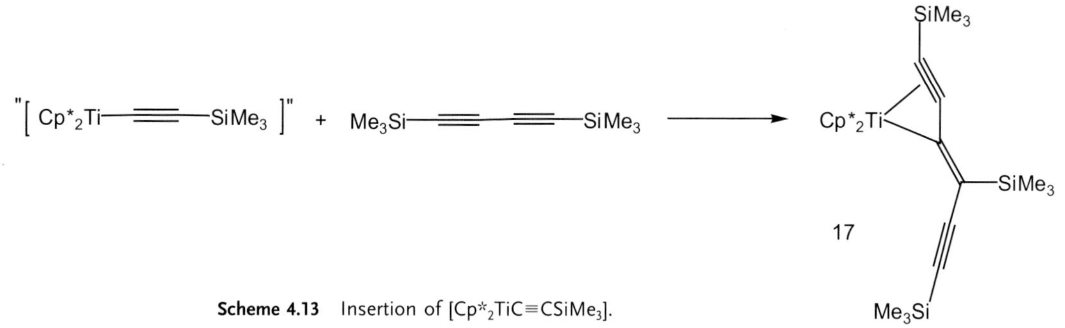

17

Scheme 4.13 Insertion of [Cp*$_2$TiC≡CSiMe$_3$].

Some of these reactions are relevant for catalytic processes. Species such as "[Cp'$_2$TiC≡CR]", for example, are important for photocatalytic C−C single bond metathesis (Section 4.3.5.1), catalytic oligomerization of 1-alkynes, and polymerization of acetylenes (Section 4.3.5.1).

4.3.5.1 **Acetylide C−C Coupling and 1,3-Butadiyne Cleavage**

Compounds with transition metal carbon σ-bonds are often suited for C−C coupling reactions by reductive elimination at one or between more than one metal. One very early example of this is the Glaser reaction, involving the coupling of copper acetylides to give 1,3-butadiynes [15a, 15b]. On the other hand, cleavage of the single bonds of 1,3-butadiynes RC≡C−C≡CR by "Cp_2Ti" or "Cp_2Zr" has also been observed [2, 5e, 6f–h].

Here a brief overview of special aspects of C−C coupling and cleavage reactions in syntheses and conversions of group 4 metallocene mono- and bis-σ-acetylides is presented. Five-membered metallacyclocumulenes $Cp'_2M(\eta^4\text{-}1,2,3,4\text{-}RC_4R)$ (**18**) are the key intermediates both in reactions of *C−C single bond cleavage of RC≡C−C≡CR to produce acetylide groups* and in the opposite reaction of *C−C single bond formation from acetylide groups [RC≡C]⁻ with formation of 1,3-butadiynes* (Scheme 4.14) [2].

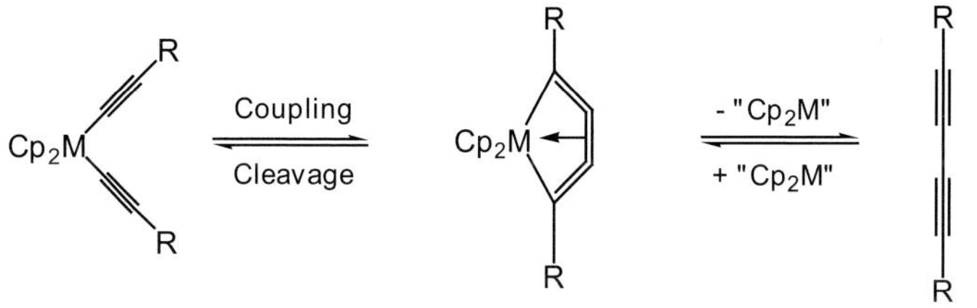

Scheme 4.14 Metallacyclocumulenes in C−C coupling and cleavage reactions.

Whether coupling or cleavage is favored is determined by the natures of the metals M, the substituents R, and the ligands Cp′. In a combination of both reactions, the first C−C single bond metathesis in homogeneous solution was achieved. Photocatalyzed and titanocene-mediated, it proceeds via titanocene monoacetylides that are also interesting species for other stoichiometric and catalytic C−C coupling reactions, such as the oligomerization of 1-alkynes and the polymerization of acetylene.

Stoichiometric coupling of two acetylides

The classical reaction in this field has already been mentioned: Teuben and de Liefde Meijer obtained the binuclear complex $[Cp_2Ti]_2[\mu\text{-}\eta^2(1,3),\eta^2(2,4)PhC_2C_2Ph]$ (**1**), with a bridging butadiyne ligand, from $[Cp_2TiCl]_2$ and NaC≡CPh, as shown above in Scheme 4.5 [5a]. This C−C coupling reaction between two titanium atoms was later investigated by Royo and co-workers [15c] and, together with the reverse cleavage reaction, was subjected to detailed calculations by Jemmis et al. [15d–h] and others [15i]. Other C−C coupling reactions of the acetylides have been found for complexes of type **5**: $[Cp_2Ti(\sigma\text{-}C{\equiv}CSiMe_3)_2]$, for example, eliminates butadiyne $Me_3SiC{\equiv}C{-}C{\equiv}CSiMe_3$, and the intermediate "$[Cp_2Ti(C{\equiv}CSiMe_3)]$" rearranges by dimerization to the more stable σ,π-acetylide-bridged Ti(III) complex $[Cp_2Ti(\mu\text{-}\sigma,\pi\text{-}C{\equiv}CSiMe_3)]_2$ (type **2**) without any further C−C coupling between the metal centers (Scheme 4.15) [5b, 5e].

The corresponding complex $[Cp^*_2Ti(\sigma\text{-}C{\equiv}CSiMe_3)_2]$ is more stable, also giving C−C coupling of the two acetylide groups upon irradiation, but without reduction,

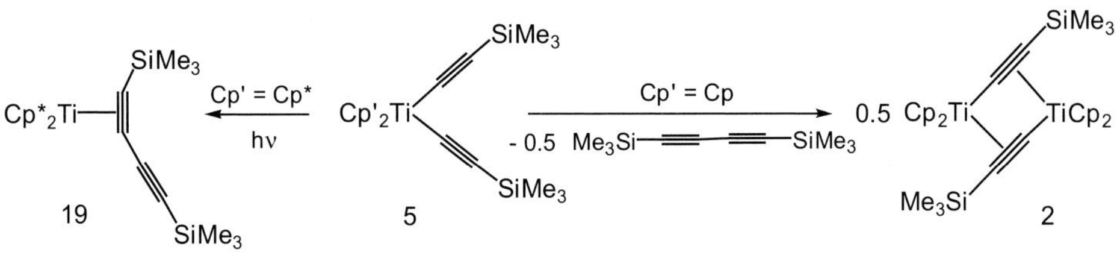

Scheme 4.15 Different stabilities and reactions of $Cp'_2Ti(C{\equiv}CSiMe_3)_2$ (Cp′ = Cp, Cp*).

with the formation of a coordinated 1,3-butadiyne in the titanacyclopropene $[Cp^*_2Ti(\eta^2\text{-}1,2\text{-}Me_3SiC_2{-}C{\equiv}CSiMe_3)]$ (**19**) (Scheme 4.15) [8a–c].

Upon irradiation of $[Cp_2Ti(\sigma\text{-}C{\equiv}C^tBu)_2]$ [16a], C−C coupling of the acetylide groups and the formation of the five-membered titanacyclocumulene $[Cp_2Ti(\eta^4\text{-}1,2,3,4\text{-}^tBuC_4{}^tBu)]$ [6i] was also detected by NMR. In the presence of additional titanocene "Cp_2Ti", $[Cp_2Ti]_2[\mu\text{-}\eta^2(1,3),\eta^2(2,4)\text{-}^tBu{-}C_2C_2\text{-}^tBu]$, with a bridging 1,3-butadiyne, was formed (Scheme 4.16) [16b].

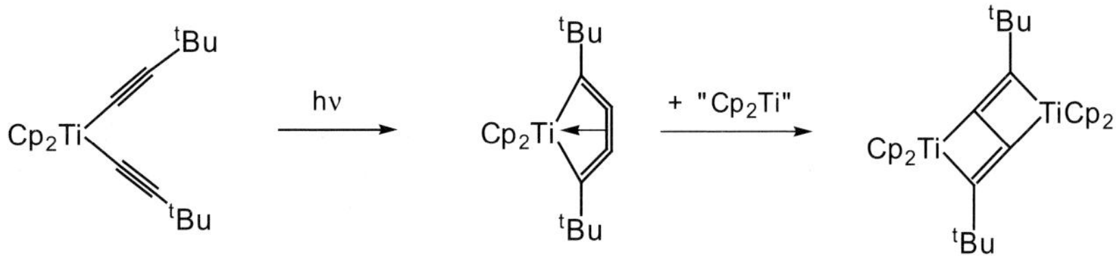

Scheme 4.16 Acetylide coupling of $[Cp_2Ti(C{\equiv}C^tBu)_2]$.

The complexes $[Cp^*_2Zr(\sigma\text{-}C{\equiv}CR)_2]$ (**5**, R = Ph, $SiMe_3$, and Me), when exposed to sunlight, afforded C−C coupling of the acetylide groups to give the 1,3-butadiyne, but no reduction, as found in the case of Ti(III) (as shown for the "Cp_2Ti"-mediated formation of **2**), and no complexation to afford a metallacyclopropene $[Cp^*_2Ti(\eta^2\text{-}1,2\text{-}Me_3SiC_2{-}C{\equiv}CSiMe_3)]$ (as shown for the "Cp^*_2Ti"-mediated formation of **19**) was observed. Instead, the zirconacyclocumulenes $[Cp^*_2Zr(\eta^4\text{-}1,2,3,4\text{-}RC_4R)]$ (**18**, R = Ph, Me, and Me_3Si) were formed in high yields (Scheme 4.17) [8c, 8a].

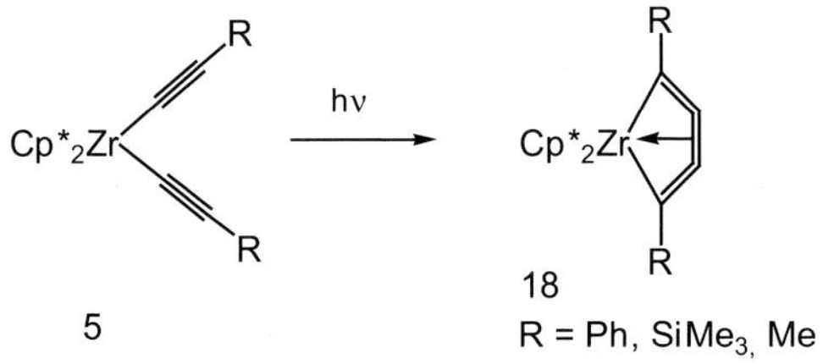

Scheme 4.17 Acetylide coupling of $[Cp^*_2Zr(C{\equiv}CR)_2]$ with R = Ph, $SiMe_3$, and Me.

Coupling of acetylides was also observed after addition of LiC≡CPh to the zirconocene(IV) bis-σ-acetylide $[Cp_2Zr(\sigma\text{-}C{\equiv}CPh)_2]$ (**5**), via the assumed Zr(IV) tris-σ-acetylide *ate*-complex $[Cp_2Zr(\sigma\text{-}C{\equiv}CPh)]_3[Li(Solv.)]$ (**20**) [9a] and the isolated η^2-diyne-zirconate $[Cp_2Zr(\sigma\text{-}C{\equiv}CPh)(\eta^2\text{-}1,2\text{-}PhC_2{-}C{\equiv}CPh][Li(E)]$ (**21**) [9b] (Scheme 4.18), as mentioned above.

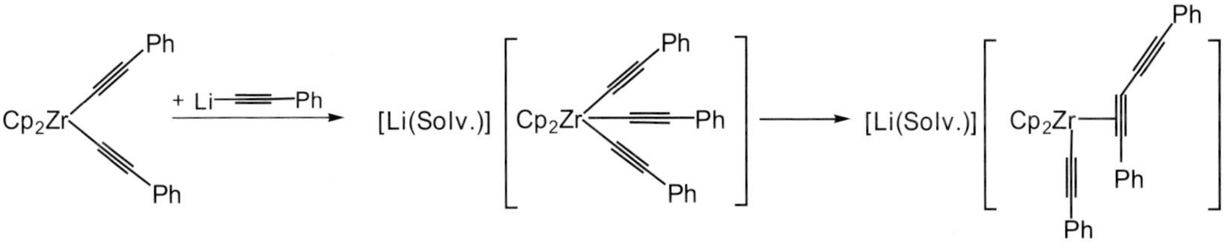

5 20 21

Scheme 4.18 Acetylide coupling of $[[Cp_2Zr(C{\equiv}CPh)_3]Li(Solv.)]$.

Stoichiometric Cleavage of 1,3-Butadiynes

When two equivalents of titanocene or zirconocene react with certain diynes, the products generated are the doubly σ,π-acetylide-bridged metal(III) complexes (**2** and **3**) mentioned above (Scheme 4.19) [2]. One way to obtain these compounds starting from 1,3-butadiynes is C−C single bond activation and cleavage, first investigated for the reaction between "Cp_2Ti" and $Me_3SiC{\equiv}C{-}C{\equiv}CSiMe_3$ [5e]. Here, an unexpected C−C single bond cleavage and the formation of $[Cp_2Ti(\mu\text{-}\sigma,\pi\text{-}C{\equiv}C\text{-}SiMe_3)]_2$ (**2**) was found. In corresponding reactions between "Cp_2Zr" and 1,3-butadiynes RC≡C−C≡CR [6h], other doubly σ,π-acetylide-bridged Zr(III) complexes $[Cp_2Zr(\mu\text{-}\sigma,\pi\text{-}C{\equiv}CR)]_2$ (**3**) (Scheme 4.19) were generated [6h].

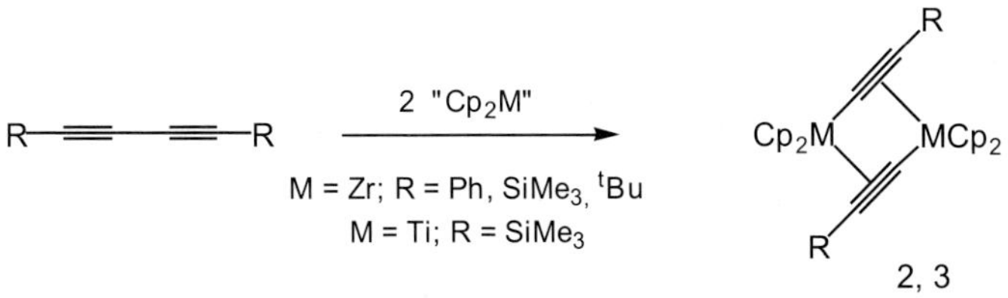

2, 3

Scheme 4.19 Cleavage of RC≡C−C≡CR and formation of $[Cp_2MC{\equiv}CR]_2$.

This cleavage reaction is favored for M = Zr in relation to Ti, an observation supported by theoretical calculations [15d]. Additionally, there is a powerful influence of the substituents R in the butadiynes RC≡C−C≡CR [15e]: the Me_3Si group, for example, seems to weaken the inner C−C single bond by its β effect (withdrawing electron density at the C-β atoms). This explains why the bis(trimethylsilyl) butadiyne is the only diyne substrate that has so far been cleaved in the case of

M = Ti. Nevertheless, this cleavage reaction has also been possible with other substituents attached to the diyne, through combination with a second metal (Zr or Ni) in addition to Ti.

The heterobinuclear complex $[Cp_2Ti(\mu\text{-}\sigma,\pi\text{-}C{\equiv}C^tBu)_2ZrCp_2]$ (**22**) (Scheme 4.20), for example, was isolated both as a product of the reaction between $^tBu{-}C_4{-}^tBu$ and "Cp_2Ti" (via the titanacyclocumulene $[Cp_2Ti(\eta^4\text{-}1,2,3,4\text{-}^tBu{-}C_4\text{-}^tBu)]$) on addition of "$Cp_2Zr$", and also as a product of the reaction between the zirconacyclocumulene $[Cp_2Zr(\eta^4\text{-}1,2,3,4\text{-}^tBu{-}C_4\text{-}^tBu)]$ and "Cp_2Ti" [6i].

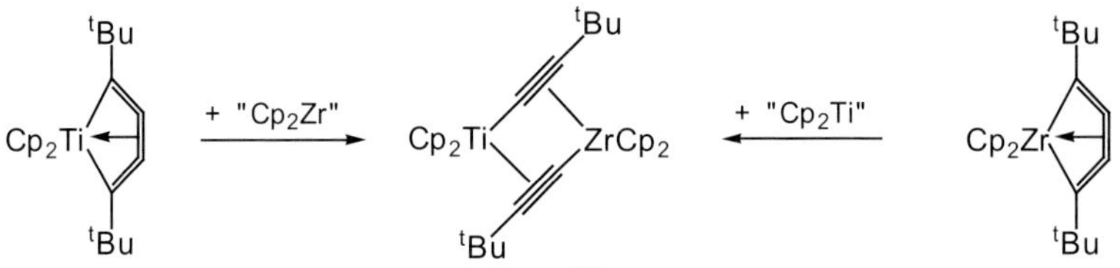

22

Scheme 4.20 Reactions of metallacyclocumulenes to form Ti,Zr-heterobimetallics.

Different C–C single bond cleavage products, depending on R, were obtained in the reactions of 1,3-butadiynes $RC{\equiv}C{-}C{\equiv}CR$ with "Cp_2M" and "$Ni(PPh_3)_2$" [6k, 16a]. Only for R = Ph did this reaction – through metallacyclocumulenes $[Cp_2M(\eta^4\text{-}1,2,3,4\text{-}RC_4R)]$ (**18**) and the "external" complexation by "$Ni(PPh_3)_2$" at the 2,3-double bond – give the heterobimetallic complexes $\{Cp_2M[\mu\text{-}\eta^2\text{-}(1\text{-}4)\text{-}PhC_4Ph]Ni(PPh_3)_2\}$ (**23**) [16a] with an uncleaved bridging diyne, a type of a complex that can be regarded as intermediate for the products **24** and **25**, obtained with other combinations of R. With two Me_3Si groups, the heterodinuclear σ,π-acetylide-bridged complexes $[Cp_2M(\sigma,\pi\text{-}C{\equiv}CSiMe_3)Ni(\sigma,\pi\text{-}C{\equiv}CSiMe_3)(PPh_3)]$ (**24**) was formed (Scheme 4.21), and with one Ph and one Me_3Si group, the tweezer-like complex $[Cp_2Ti(\sigma,\pi\text{-}C{\equiv}CSiMe_3)(\sigma,\pi\text{-}C{\equiv}CPh)Ni(PPh_3)]$ (**25**) was produced [6k].

Catalytic coupling and cleavage reactions

It has been reported for complexes **5** such as $[Cp_2Zr(\sigma\text{-}C{\equiv}CMe)_2]$ that catalytic amounts of strong Lewis acids such as $B(C_6F_5)_3$ catalyze the C–C coupling reaction of the acetylide groups with formation of the 1,3-diyne complexes $[Cp_2Zr(\eta^4\text{-}1,2,3,4\text{-}MeC_4Me)]$ [6a]. It is worth mentioning that the reverse catalytic effect of $B(C_6F_5)_3$ was found in the case of the permethylzirconocene systems, where the C–C bond of the complexed diyne in $[Cp^*_2Zr(\eta^4\text{-}1,2,3,4\text{-}Me_3SiC_4SiMe_3)]$ (**18**) was cleaved with formation of the bis-alkynyls **5**, such as $[Cp^*_2Zr(\sigma\text{-}C{\equiv}CSiMe_3)_2]$ [2d, 6l]. Apparently, an equilibrium between the zirconacycle and the bis-alkynyl is shifted by $B(C_6F_5)_3$ to the cycle for Cp/Me and to the bis-alkynyl complex for Cp*/ Me_3Si. In summary, $B(C_6F_5)_3$ can catalyze either C–C coupling [6a] or cleavage [2d, 6l] in such systems (Scheme 4.22).

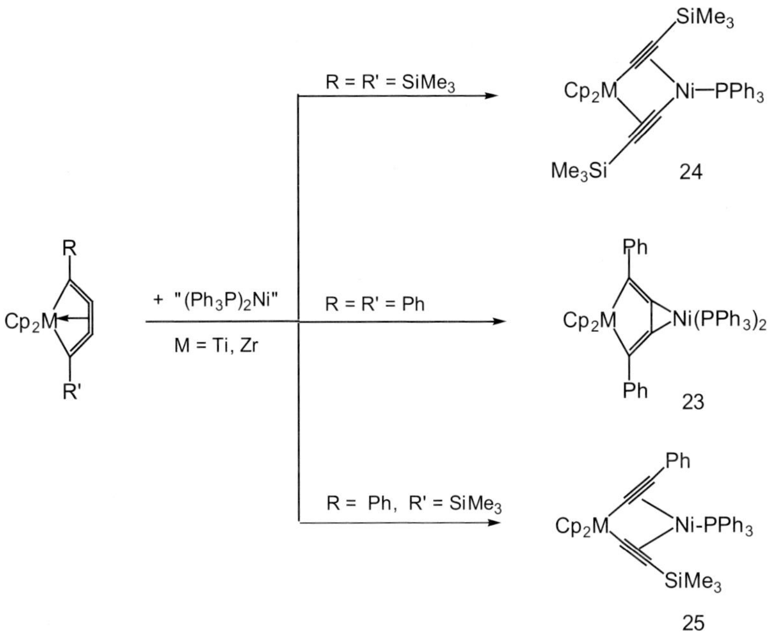

Scheme 4.21 Reactions of metallacyclocumulenes to form Ni,Ti- and Ni,Zr-heterobimetallics.

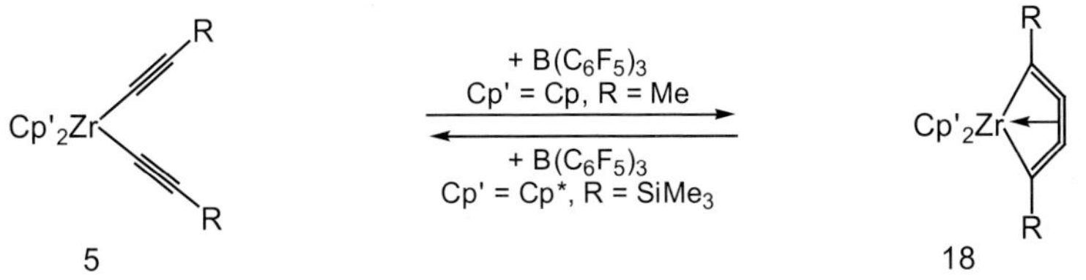

Scheme 4.22 Catalytic C—C coupling of acetylides and cleavage with $B(C_6F_5)_3$.

Photocatalytic C—C single bond metathesis

The interaction of butadiynes with titanocenes and zirconocenes can give cleavage of the diynes and coupling of the cleaved fragments by the metallocene cores [2]. Cleavage of symmetrically substituted butadiynes in combination with a subsequent alternating recombination of the acetylide groups gives rise to a C—C single bond metathesis (Scheme 4.23) [2, 17a].

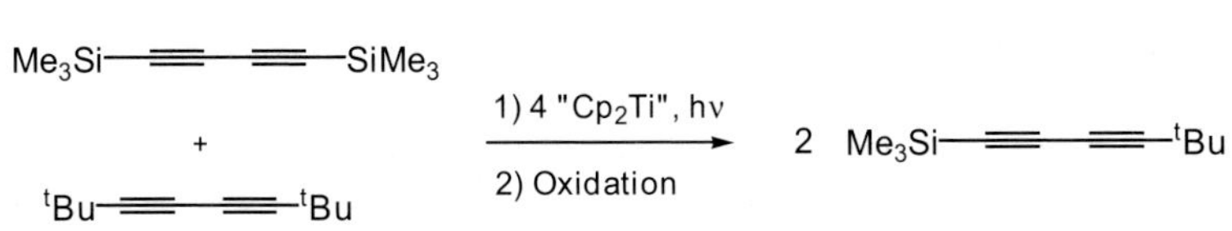

Scheme 4.23 C—C single bond metathesis of 1,3-butadiynes.

If one-to-one mixtures of the butadiynes ${}^tBuC{\equiv}C{-}C{\equiv}C^tBu$ and $Me_3SiC{\equiv}C{-}C{\equiv}CSiMe_3$ are treated with an excess of the "Cp_2Ti" reagent under irradiation conditions, the unsymmetrically substituted diyne ${}^tBuC{\equiv}C{-}C{\equiv}CSiMe_3$ is obtained after oxidative workup in addition to the symmetrically substituted starting diynes. This was the first titanocene-mediated, photocatalyzed C–C single bond metathesis in homogeneous solution. This metathesis cannot be conducted catalytically with regard to the titanium complex, because of coupling reactions that predominate if an excess of diyne is used [2]. The course of the reaction (Scheme 4.24) can be understood in terms of the titanocene forming a binuclear complex with an intact C_4 backbone of type **1** with ${}^tBuC{\equiv}C{-}C{\equiv}C^tBu$ and a σ,π-alkynyl bridged cleavage product of type **2** with $Me_3SiC{\equiv}C{-}C{\equiv}CSiMe_3$.

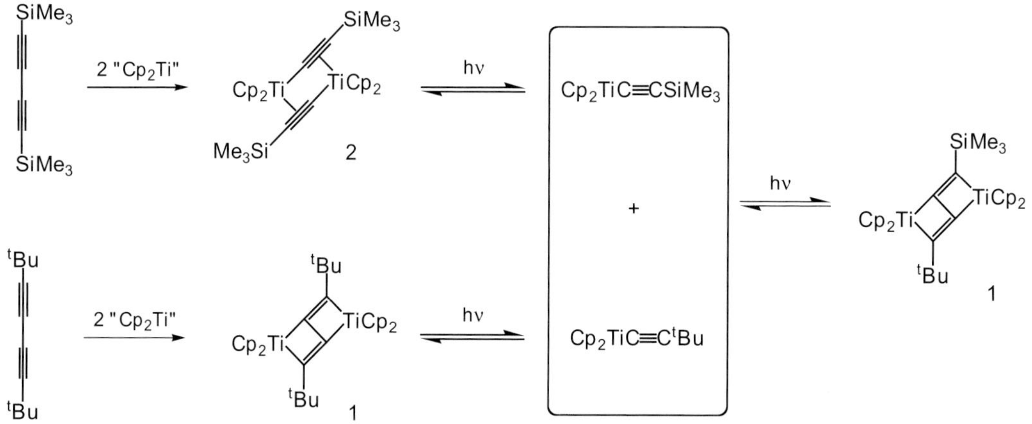

Scheme 4.24 Reaction path for C–C single bond metathesis of 1,3-butadiynes.

In the presence of light, both complexes dissociate into the unstable monomeric Ti(III) complexes "$[Cp_2Ti(\sigma\text{-}C{\equiv}C^tBu)]$" and "$[Cp_2Ti(\sigma\text{-}C{\equiv}CSiMe_3)]$", which then dimerize either to the respective starting complexes or to a differently substituted binuclear complex. The reverse reaction is also accomplishable, as has been verified by dynamic NMR investigations showing that dinuclear C–C-cleaved complexes such as $[Cp_2Ti]_2[(\mu{-}C{\equiv}CSiMe_3)(\mu{-}C{\equiv}C^tBu)]$ are very likely to be present in solutions of the uncleaved complexes $[Cp_2Ti]_2[\mu\text{-}\eta^2(1,3),\eta^2(2,4){}^tBuC_2C_2SiMe_3]$ [17b].

On the basis of all the above results, it is possible to deduce a general reaction scheme capable of explaining both the cleavage of and the coupling to 1,3-butadiynes in a uniform order of events (Scheme 4.25) [2, 15f].

According to this scheme, both the cleavage and the coupling progress *via* metallacyclocumulenes, the *intra*molecular coordination of the inner double bond being replaced, in the subsequent step, by an *inter*molecular one. These intermediates then rearrange to products in which either an intact or a cleaved C_4 linkage is present. The individual energy levels both of the complexes in question and of the intermediates have been determined by calculations, and the results support the

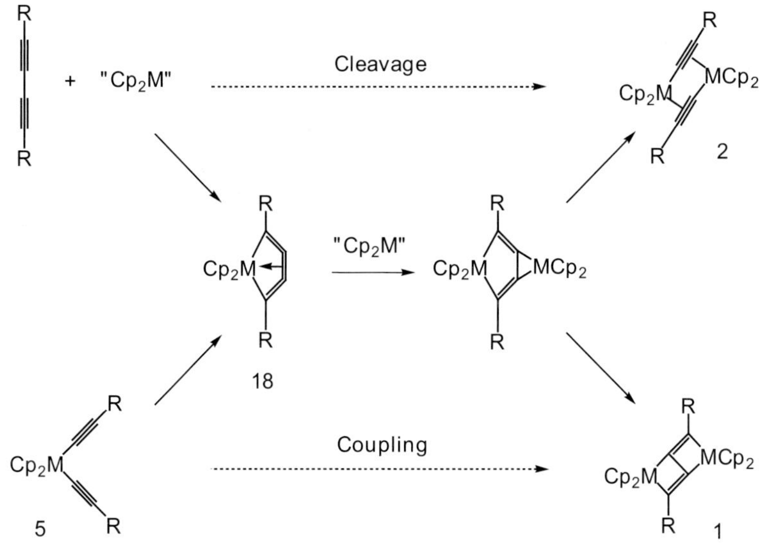

Scheme 4.25 C−C coupling of acetylides and cleavage of 1,3-butadiynes via cyclocumulenes.

experimentally observed relative thermodynamic stabilities. These theoretical results thus emphatically support the documented interpretation of reaction course [15f].

Catalytic oligo- and polymerization of acetylenes

Mach and co-workers reported that the complexes such as $[Cp^*_2Ti(\sigma\text{-}(E)-CH{=}CHR)(\sigma\text{-}C{\equiv}CR)]$ above show photochemical coupling of the alkenyl and the acetylide groups to produce complexes of but-1-ene-3-ynes $[Cp^*_2Ti(3,4\text{-}\eta^2\text{-}RC{\equiv}C-CH{=}CHR)]$ (Scheme 4.4) [18a]. These complexes catalyze the rapid *dimerization of 1-alkynes* to 2,4-disubstituted but-1-ene-3-ynes (head-to-tail-dimers). This coupling is very similar to that described above for acetylide groups in $[Cp^*_2Zr(\sigma\text{-}C{\equiv}CR)_2]$ (see Scheme 4.17 above).

Species such as $Cp^*_2TiC{\equiv}CR$ could also be important in catalytic oligomerization reactions of 1-alkynes [18b–d]. The insertion of a diyne into the acetylide bond with formation of $Cp^*_2Ti[\eta^3\text{-}Me_3SiC_3{=}C(C{\equiv}CSiMe_3)SiMe_3]$ has been mentioned (Scheme 4.13). The Cp* ligands prevent dimerization of the monoalkynyl complexes, which is why Cp^*_2Ti catalysts are better catalysts than the corresponding Cp systems in *oligomerization reactions of 1-alkynes.*

From these considerations regarding the reactivity of $[Cp^*_2Ti(\sigma\text{-}(E)-CH{=}CHR)(\sigma\text{-}C{\equiv}CR)]$, a new mechanism (Scheme 4.26) for the *polymerization of acetylene* [19] was proposed [2c].

The steps consist of π-complexation of acetylene to "$[Cp_2Ti(\eta^2\text{-}HC_2H)]$" (**26**), oxidative addition to give "$[Cp_2Ti(H)(C{\equiv}CH)]$" (**27**), insertion of $HC{\equiv}CH$ to give "$[Cp_2Ti(-CH{=}CH_2)(-C{\equiv}CH)]$" (**28**), and coupling to afford "$[Cp_2Ti(3,4\text{-}\eta^2\text{-}HC{\equiv}C-CH{=}CH_2)]$" (**29**). The substitution of $HC{\equiv}C-CH{=}CH_2$ by $HC{\equiv}CH$

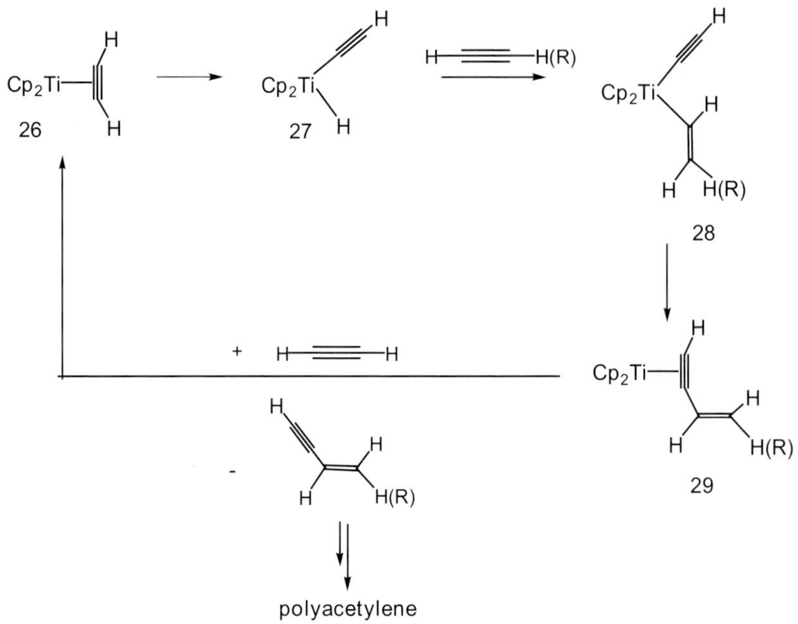

Scheme 4.26 Proposed mechanism for polymerization of acetylene.

opens a new catalytic cycle, again via "[$Cp_2Ti(\eta^2\text{-}HC_2H)$]" and "[$Cp_2Ti(H)(C{\equiv}CH)$]", insertion of $HC{\equiv}C{-}CH{=}CH_2$ to yield "[$Cp_2Ti({-}CH{=}CH{-}CH{=}CH_2)({-}C{\equiv}CH)$]", and coupling to provide "[$Cp_2Ti(3,4\text{-}\eta^2\text{-}HC_2{-}CH{=}CH{-}CH{=}CH_2)$]". Repetitive sequences of substitution, oxidative addition, and coupling ultimately afford *trans*-polyacetylene [19].

4.4 Complexation of MC≡CM

4.4.1 Examples

Depending on the nature of the metal M, three types of complexes – (**A**), (**B**), and (**C**) – have been found for MC≡CM (see Section 4.3.4.3) (Scheme 4.27) [2]. Complexation of MC≡CM with other metals with formation of cluster compounds has frequently been described, but complexes with only one M with formation of M_3C_2 are relatively rare [1l, 1m], Four of the most important interactions (Scheme 4.27) are: the σ,π-bridging mode without metal-metal interaction (**D**), the corresponding type with metal-metal bonds (**E**), and, in (**F**) and (**G**), one metal coordinated at one and the two other metals at the second carbon atom.

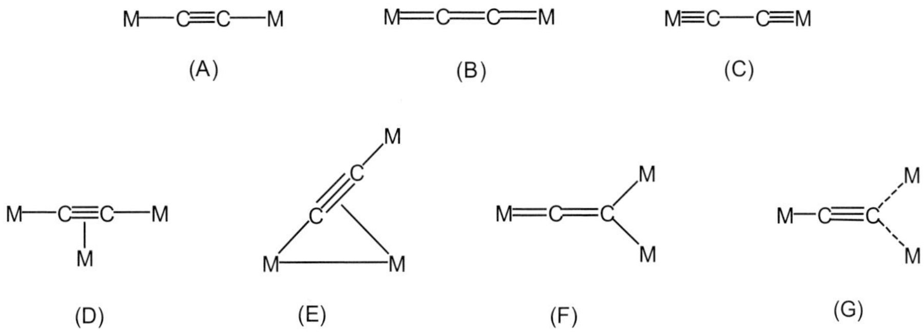

Scheme 4.27 Complexation of MC≡CM.

Many examples of such complexes were published in pioneering contributions in rhenium chemistry carried out by Beck and co-workers [20a–d]. Three representative examples for titanium are listed here (Scheme 4.28). Mach et al. described the π-complexes [Cp'$_2$Ti(η^2-Me$_3$SnC$_2$SnMe$_3$)] (**30**), from which the complexes [Cp$_2$Ti(μ-σ,π-C≡CSnMe$_3$)]$_2$ (**31**) were obtained [3d], and an interesting type of M$_4$C$_8$ complexes with two TiCu$_2$C$_2$ subunits (**32**) was published by Lang, van Koten, and co-workers [20e].

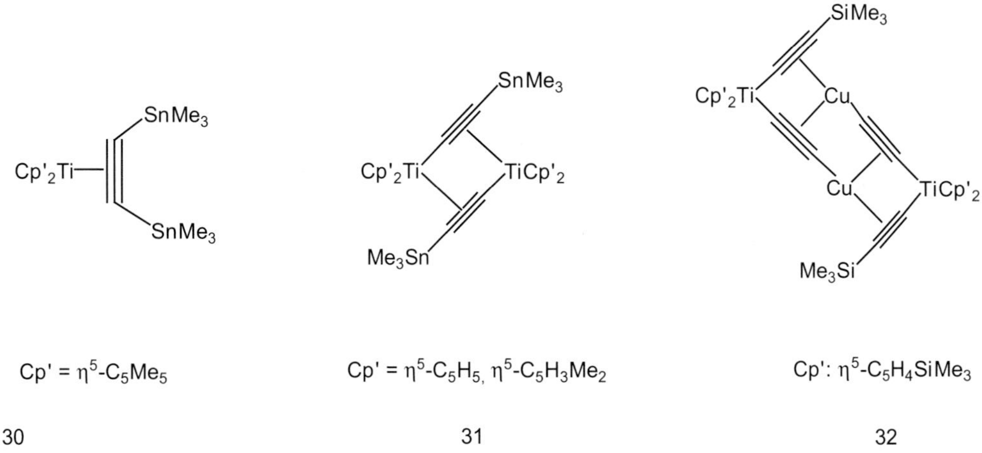

Scheme 4.28 Examples of MC≡CM complexes.

4.4.2 Molecular Dynamics of Acetylides

The most interesting features in the chemistry of the trimetallic complexes are dynamic processes of the C$_2$ ligand. For [{Ru(CO)$_2$(η−C$_5$H$_4$R)}$_3$(μ−C≡C)][BF$_4$] (R = H (**33**) or Me (**34**)), with the Ru$_3$C$_2$ subunit, such a dynamic process of the C$_2$ ligand was recently described as "bearing-like" for the bisacetylide between three metals (Scheme 4.29) [21a].

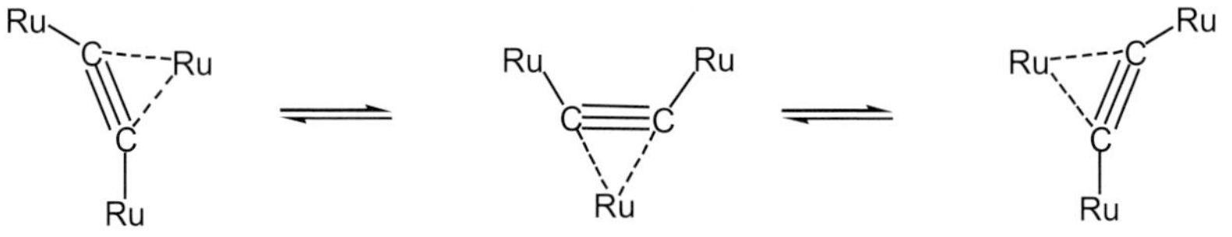

Ru: (η^5-C_5H_5)Ru(CO)$_2$ 33, (η^5-C_5H_4Me)Ru(CO)$_2$ 34

Scheme 4.29 Molecular dynamics of a Ru_3C_2 complex.

A similar fluxional motion of acetylide fragments was also found by Akita et al. for the binuclear complex [{Cp(CO)$_2$Fe}$_2$(μ−C≡CH)]$^+$ (**35**) (Scheme 4.30) [21b–d]. Observations made by solution NMR measurements were interpreted as caused by a 1,2-shift of the hydrogen atom on the carbide bridge via an intermediate according to the example in Scheme 4.29.

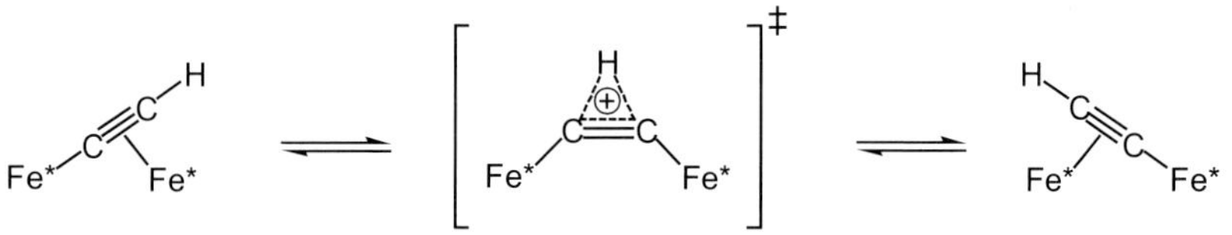

Fe*: Cp*Fe(CO)$_2$, 35

Scheme 4.30 Molecular dynamics of a Fe_2C_2H complex.

More recently, an oscillating C≡C^{2-} unit inside a copper rectangle was published for the complex [Cu$_4$(μ-dppm)(μ^4-η^1,μ^2−C≡C]$^{2+}$ on the basis of NMR studies and DFT calculations [21e].

Prior to all these studies for σ,π-acetylide-bridged complexes of the type **3** M(μ−C≡CR)$_2$M (Scheme 4.31), non-rigid behavior was published by Erker et al. for M_2C_2R, which is isolobal to M_2C_2M [6e].

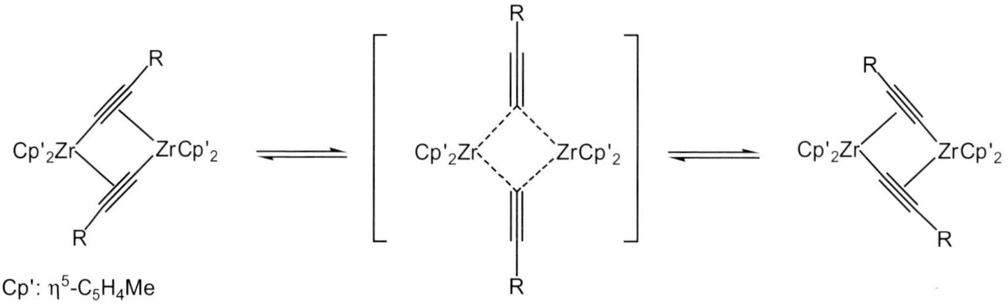

Cp': η^5-C_5H_4Me

σ,π-coordination σ,σ-coordination σ,π-coordination

Scheme 4.31 Molecular dynamics of a $Zr_2(C_2R)_2$ complex.

Some experimental data and theoretical calculations [15d–h] concerning intermediates in this interconversion are important with respect to the C–C coupling reactions of acetylide groups discussed above [2].

4.4.3
Acetylides in the Topomerization of Alkynes

The C_2-ligand is an extremely mobile group in compounds such as M_3C_2 and M_2 (μ-σ,π-C_2R). [2, 6e, 21]. In terms of the isolobal concept, this could also have some more general consequences for the chemistry of alkynes C_2R_2 and their complexes with metals M(π-C_2R_2).

One interesting point in this area is *topomerization of disubstituted alkynes*, in which carbon atom transposition has been observed. This thermal interconversion of alkynes from **36a** to **36b** via vinylidenes **37** at higher temperatures is known as the Roger Brown rearrangement (Scheme 4.32) [22].

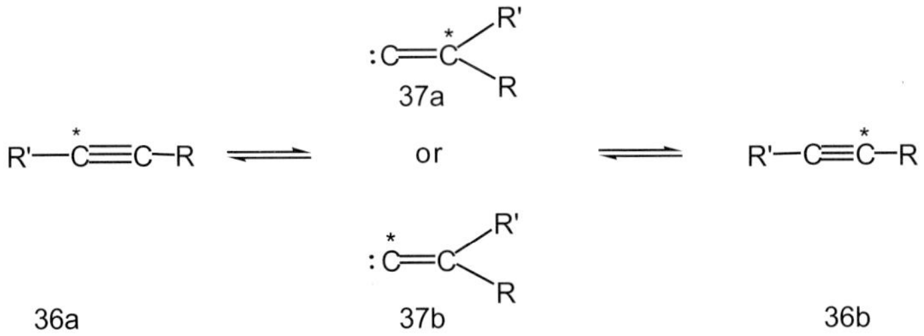

Scheme 4.32 Topomerization of R′C*≡CR and R′C≡C*R via vinylidene species.

At lower temperatures, complexes of vinylidenes M=C=CR_2 (**39**) were produced from metal complexes of disubstituted alkynes M(π-RC_2R) (**38**), as published in some very important papers by Werner et al. (Scheme 4.33) [23a]. Were this process reversible for disubstituted alkynes, a catalytic influence of metals on the above topomerization would be conceivable [23b, 23c].

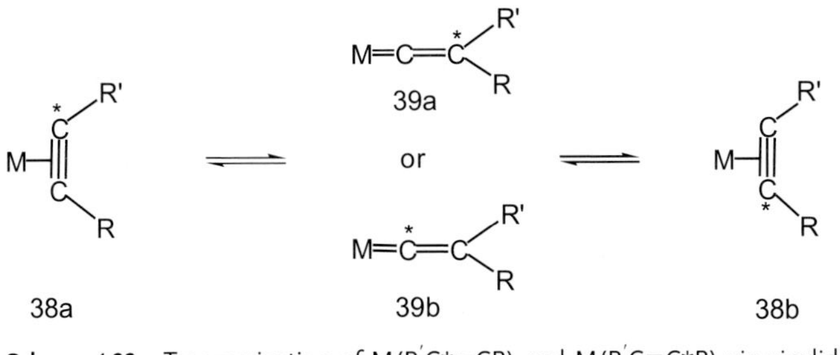

Scheme 4.33 Topomerization of M(R′C*≡CR) and M(R′C≡C*R) via vinylidene complexes.

For acetylene and 1-alkynes, the formation of vinylidene complexes was calculated and explained in terms of a "concerted" metal-slip by 1,2-H-migration [23d–f] rather than a stepwise oxidative addition via M(H)(acetylide) intermediates [23g].

For disubstituted alkynes, C−C or C-Si single bond cleavage of RC≡CPh (R = Ph, Me_3Si) with formation of platinum(II) compounds Pt(σ-R)(σ-C_2Ph) (**40**) in a reversible process was described (Schemes 4.3 and 4.34) [3e, 3f].

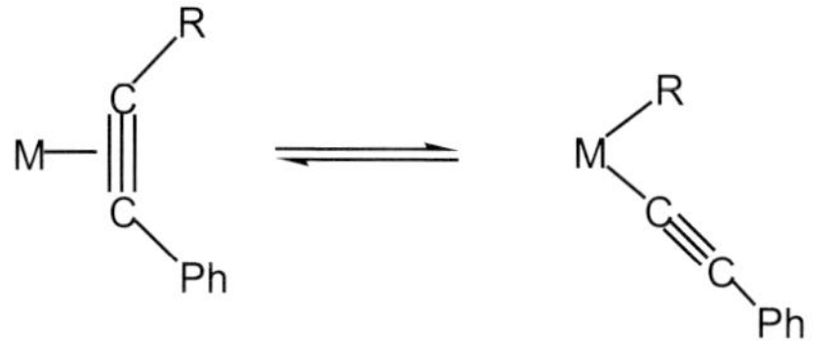

M: (^{i}$Pr_2PCH_2CH_2NMe_2$)Pt, R: Ph, $SiMe_3$, 40

Scheme 4.34 Reversible C−C and C−Si bond cleavage with formation of acetylides.

It is unknown whether the formation of vinylidene complexes such as Rh=C=C($SiMe_3$)(Ph) [23h] with a 1,2-shift of the Me_3Si groups (Scheme 4.33) proceeds via similar species. For Pt(σ-Ph)(σ-C_2Ph) it is not clear whether this complex **40** is formed through *ortho*-metalation via a four-membered metallacycle **41** (Scheme 4.35) [24a]. For the back-reaction, the phenylacetylide can be considered to have a chelate three-electron donor tautomer **42** reacting with a Lewis base L at the electrophilic α-carbon atom. [24b]

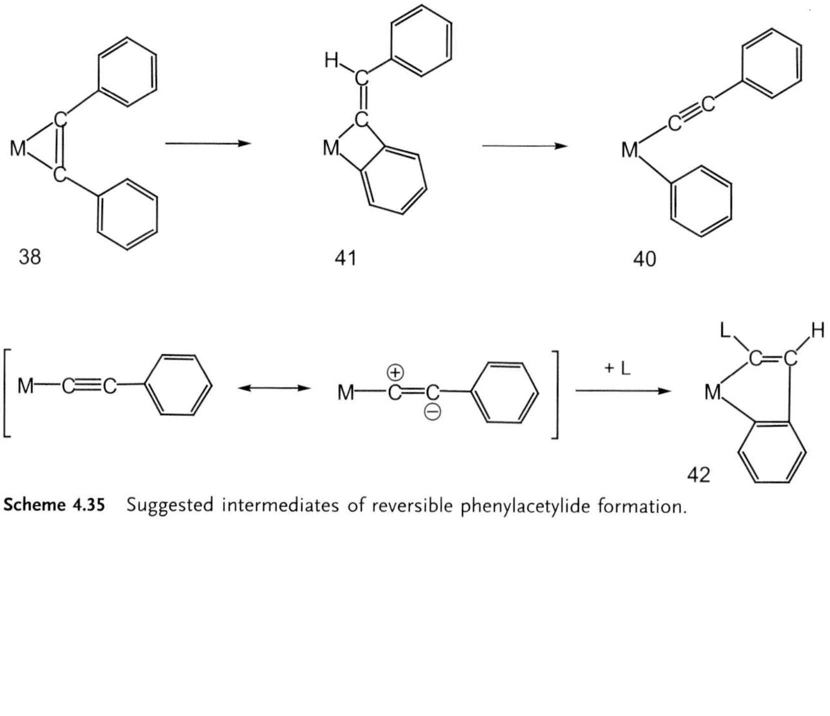

Scheme 4.35 Suggested intermediates of reversible phenylacetylide formation.

As a result of all these considerations, it is reasonable to ask whether species of types **F**, **G**, and **H** (Scheme 4.27) are involved in the carbon atom transposition (Scheme 4.29) of the M_3C_2 compounds mentioned above, giving a formal "rotation" of the bis-acetylide between three metal centers (Scheme 4.36) [25a].

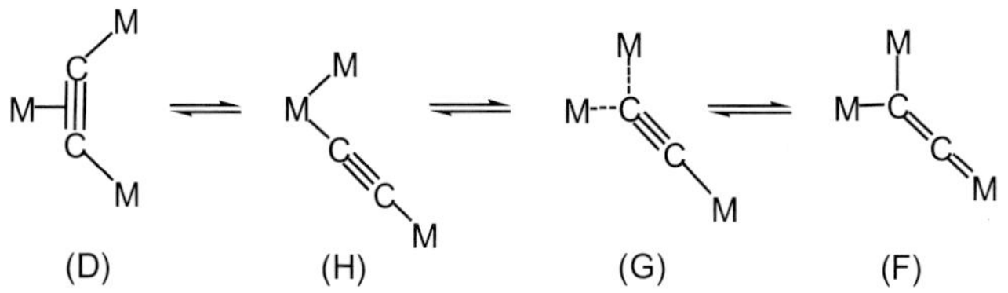

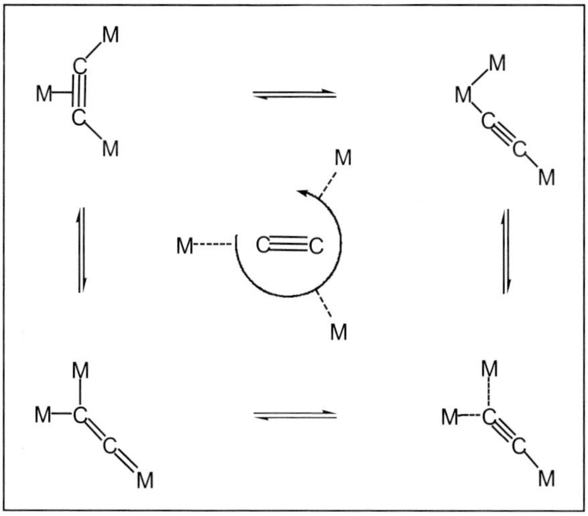

Scheme 4.36 Acetylide complexes in C_2-rotation and alkyne topomerization.

4.5 Summary and Outlook

As examples of the extremely rich chemistry of transition metal acetylides, the selected material on titanocene and zirconocene acetylides presented here should give an impression of the manifold reactivity of species with subgroups such as $MC{\equiv}CR$, $M(C{\equiv}CR)_2$, and $MC{\equiv}CM$. This may also serve as a simple model through which to understand the higher derivatives $M(C{\equiv}C)_nR$, $M[(C{\equiv}C)_nR]_2$, $M(C{\equiv}CR)_n$, $M(C{\equiv}C)_nM$, and so on, which, as carbon-rich organometallics, are of particular interest for synthetic, theoretical, and physical chemistry with great technological potential in materials science.

On the other hand, general fundamental organometallic research with transition metal acetylides in comparison to the corresponding alkyl, alkenyl, and aryl compounds offers some fascinating aspects for C–C coupling and cleavage reactions at one or more metal centers. Here, knowledge of examples such as ($R_2C{=}CR$-)$MC{\equiv}CR$ and $M(C{\equiv}CR)_2$ can lead to better understanding of the corresponding reactions of alkenyl $M(CR{=}CR_2)_2$, aryl $M(Ar)_2$, and alkyl MR_2 compounds [25]. Novel types of transition metal complexes (five-membered cyclocumulenes) and both stoichiometric (C–C single bond cleavage) and catalytic reactions (C–C single bond metathesis) have been demonstrated in this field, showing that the combination of organometallic complexes with suitable substrates has very often led, and will lead in future, to novel chemistry. Transition metal acetylides provide a convincing example for this.

4.6 Typical Experimental Procedures

4.6.1 Synthesis of a Monomeric Ti(III) Monoacetylide [$Cp^*_2TiC{\equiv}C^tBu$] [7b]

3,3-Dimethylbutyne (0.21 mL, 1.7 mmol) in *n*-hexane (10 mL) was cooled to –78 °C and treated with *n*-butyllithium (2.5 M in *n*-hexane, 0.68 mL, 1.7 mmol). After 15 min, the solution was allowed to warm to room temperature and the resulting white suspension was added to a deep blue solution of [Cp^*_2TiCl] (0.606 g, 1.7 mmol) in *n*-hexane (10 mL). The color of the mixture rapidly changed from blue to brown. After filtration and concentration of the solution to 5 mL, crystallization at –78 °C gave greenish-brown crystals, which were separated from the mother liquor, washed with cold *n*-hexane, and dried in vacuo to give 0.430 g (62.9 %), m. p.: 129–131 °C. Anal. Calcd for $C_{26}H_{39}Ti$ (399.48): C, 78.18; H, 9.84. Found: C, 78.24; H, 9.75. IR (nujol) ν (cm^{-1}): 2721 w, 2071 s. MS: (70 eV) EI m/z (u) = 399 $[M]^+$, 317 $[Cp^*_2Ti\text{-}H]^+$.

4.6.2 Synthesis of a Ti(III) Bisacetylide Tweezer [$Cp_2Ti(C{\equiv}C^tBu)_2$][Li(THF)] [7b]

A solution of 3,3-dimethylbutyne (1.30 mL, 10.6 mmol) in THF (15 mL) was cooled to –78 °C and treated with *n*-butyllithium (2.5 M in *n*-hexane, 4.24 mL, 10.6 mmol). After 15 min, the solution was allowed to warm to room temperature, [Cp^*_2TiCl] (1.89 g, 5.3 mmol) was added, the solution was stirred for 24 h, the solvent was removed in vacuo, and the residue was suspended in *n*-hexane (10 mL). After filtration and crystallization at –78 °C, green crystals (2.001 g, 68.0 %) were obtained, and were filtrated and dried in vacuo, m. p.: 178–179 °C. Anal. Calcd for $C_{36}H_{56}LiOTi$ (559.66): C, 77.26; H, 10.09. Found: C, 77.24; H, 9.85. IR (nujol) ν (cm^{-1}): 2030 s; 1237 vs; 1199 m; 1036 s; 883 m. MS: (70eV) EI m/z (u) = 399 $[M\text{–}LiTHF{-}C_2{}^tBu]^+$, 318 $[Cp^*_2Ti]^+$.

4.6.3
Synthesis of a Dinuclear Ti(III) Monoacetylide $[Cp_2TiC_2SiMe_3)]_2$ by C–C Cleavage of a 1,3-Butadiyne [5e]

$[Cp_2Ti(\eta^2\text{-}Me_3SiC_2SiMe_3)]$ (3.47 g, 10 mmol) in pentane (10 mL) was mixed with 1,4-bis(trimethylsilyl)butadiyne (0.970 mg, 5 mmol) in pentane (10 mL), and the mixture was stirred at 40 °C for 1 h. The color of the solution changed from yellow to red. The solvents were then distilled in vacuo, and the residue was suspended in pentane (40 mL). After filtration and crystallization at room temperature, the crystals were washed with pentane and dried in vacuo to afford 2.6 g (95 %), m. p.: 250 °C (dec. under argon). ^1H NMR (THF-d_8) δ/ppm: 0.33 (s, Si*Me*$_3$), 5.20 (s, C_5H_5). ^{13}C NMR (THF-d_8) δ/ppm: 104.6 (Cp), 120.6 (C_5Me_5), 142.8 (C-β), 237.5 (C-α).

4.6.4
Synthesis of a Zr(IV) Bisacetylide $[Cp^*_2Zr(C{\equiv}CSiMe_3)_2]$ [8c]

$Me_3SiC{\equiv}CH$ (432 mg, 4.40 mmol) in toluene (5 mL) was cooled to –78 °C, and one equivalent of *n*-butyllithium (2.5 M in *n*-hexane) was added. After the solution had warmed to room temperature, $[Cp^*_2ZrCl_2]$ (950 mg, 2.20 mmol) was added, the solution was stirred for 24 h, the solvents were then distilled in vacuo, and the residue was suspended in *n*-hexane (10 mL). After filtration and crystallization at –78 °C, the mother liquor was decanted and the crystals were dried in vacuo to afford 450 mg (37 %), m. p.: 162–166 °C (dec. under argon). Anal. Calcd for $C_{30}H_{48}Si_2Zr$ (556.1): C, 64.80; H, 8.70. Found: C, 64.82; H, 8.87. ^1H NMR (C_6D_6) δ/ppm: 0.24 (s, 18H, Si*Me*$_3$), 2.01 (s, 30H, C_5Me_5). ^{13}C NMR (C_6D_6) δ/ppm: 0.6 (Si*Me*$_3$), 12.6 (C_5Me_5), 120.6 (C_5Me_5), 122.3 (C-β), 175.0 (C-α). MS (70 eV) *m/z*: 554 (M^+), 360 ($Cp^*_2Zr^+$). IR (Nujol, cm^{-1}): 2027 m, 1424 s, 1244 vs, 1025 m, 855 vs, 837 vs, 757 m, 683 vs, 604 s.

4.6.5
Synthesis of a Zirconacyclocumulene $[Cp^*_2Zr(\eta^4\text{-}1.2.3.4\text{-}Me_3SiC_4SiMe_3)]$

4.6.5.1 By Photochemical C–C Coupling of Acetylides to a Complexed 1,3-Butadiyne [8c]

$[Cp^*_2Zr(C{\equiv}CSiMe_3)_2]$ (135 mg, 0.24 mmol) was dissolved in toluene (5 mL) under argon. After the solution had been kept standing for four days exposed to sunlight, the color of the solution had turned from light yellow to red. The solvent was removed in vacuo, and the red residue was redissolved in *n*-hexane (1 mL). On standing at –78 °C, orange-red crystals appeared, and were separated, washed with cold *n*-hexane (–75 °C), and dried in vacuo to give 101 mg (79 %), m. p.: 195 °C (dec. under argon). Anal. Calcd for $C_{30}H_{48}Si_2Zr$ (556.1): C, 64.80; H, 8.70. Found: C, 64.57; H, 8.64. ^1H NMR (THF-d_8) δ/ppm: 0.45 (s, 18H, Si*Me*$_3$), 1.61 (s, 30H, C_5Me_5). ^{13}C NMR (THF-d_8) δ/ppm: 3.2 (Si*Me*$_3$), 12.1 (C_5Me_5), 113.4 (C_5Me_5), 144.5 (C-β), 188.0 (C-α). MS (70 eV) *m/z*: 556 (M^+), 360 ($Cp^*_2Zr^+$).

4.6.5.2 By Reduction of [$Cp^*_2ZrCl_2$] in the Presence of a 1,3-Butadiyne [8c]

1,4-Bis(trimethylsilyl)butadiyne (292 mg, 1.50 mmol) was dissolved in THF (10 mL), and the solution was added to [$Cp^*_2ZrCl_2$] (650 mg, 1.50 mmol) and magnesium turnings (36 mg, 1.50 mmol). The mixture was kept at 55–60 °C and stirred for 48 h. The color of the solution changed from light yellow to orange-red. The solution was then filtered, and the solvent was removed in vacuo. The residue was extracted three times with *n*-hexane (5 mL portions). After filtration, concentration of the clear solution to about 5 mL, and crystallization at –78 °C, an quantity of 669 mg (80 %) was produced, and this was filtered and dried in vacuo. Orange-red crystals, data as described under Section 4.6.5.1.

Acknowledgments

This work was supported by the Max-Planck-Gesellschaft, the Deutsche Forschungsgemeinschaft, the Fonds der Chemischen Industrie, and the Bundesland Mecklenburg-Vorpommern. Funding and facilities provided by the Leibniz-Institut für Organische Katalyse at the University of Rostock are gratefully acknowledged. The work reported in this contribution would not have been possible without the excellent efforts of various former Ph. D. students, especially *Andreas Ohff, Siegmar Pulst, Claudia Lefeber, Normen Peulecke, Dominique Thomas, Frank G. Kirchbauer, Thorsten Zippel*, and *Paul-Michael Pellny*, postdoctoral scientists such as *Peer Kosse* and *Stefan Mansel*, and technical staff, in particular *Petra Bartels* and *Regina Jesse*. I am very grateful to my permanent group members *Perdita Arndt, Wolfgang Baumann, Barbara Heller*, and *Anke Spannenberg*, as well as our guest *Vladimir V. Burlakov* for many fruitful discussions and to many other colleagues whose names appear in the list of references.

References

[1] *Examples:* (a) R. Nast, *Coord. Chem. Rev.* **1982**, *47*, 89–124;
(b) W. Beck, B. Niemer, M. Wieser, *Angew. Chem.* **1993**, *105*, 969–996; *Angew. Chem. Int. Ed.* **1993**, *32*, 923–949;
(c) S. Lotz, P. H. Van Rooyen, R. Meyer, *Adv. Organomet. Chem.* **1995**, *37*, 219–320;
(d) J. Manna, K. D. John, M. D. Hopkins, *Adv. Organomet. Chem.* **1995**, *38*, 79–154;
(e) H. Lang, M. Weinmann, *Synlett* **1996**, 1–10;
(f) M. I. Bruce, *Coord. Chem. Rev.* **1997**, *166*, 91–119;
(g) F. Paul, C. Lapinte, *Coord. Chem. Rev.* **1998**, *178-180*, 431–509;
(h) R. Choukroun, P. Cassoux, *Acc. Chem. Res.* **1999**, *32*, 494–502;
(i) H. Lang, G. Rheinwald, *J. Prakt. Chem.* **1999**, *341*, 1–19;
(j) H. Lang, D. S. A. George, G. Rheinwald, *Coord. Chem. Rev.* **2000**, *206-207*, 101–197;
(k) N. J. Long, C. K. Williams, *Angew. Chem.* **2003**, *115*, 2690–2722; *Angew. Chem. Int. Ed.* **2003**, *42*, 2586–2617;

(l) P. J. Low, M. I. Bruce, *Adv. Organomet. Chem.* **2002**, *48*, 71–288;
(m) P. J. Low, M. I. Bruce, *Adv. Organomet. Chem.* **2003**, *50*, 179–444.

[2] *Reviews:* (a) A. Ohff, S. Pulst, C. Lefeber, N. Peulecke, P. Arndt, V. V. Burlakov, U. Rosenthal, *Synlett* **1996**, 111–118;
(b) U. Rosenthal, P.-M. Pellny, F. G. Kirchbauer, V. V. Burlakov, *Acc. Chem. Res.* **2000**, *33*, 119–129;
(c) U. Rosenthal, V. V. Burlakov in *Titanium and Zirconium in Organic Synthesis*, (Ed.: I. Marek) Wiley-VCH, **2002**, 355–389;
(d) U. Rosenthal, V. V. Burlakov, P. Arndt, W. Baumann, A. Spannenberg, *Organometallics* **2003**, *22*, 884–900.

[3] (a) E. Kurras, U. Rosenthal, *J. Organomet. Chem.* **1978**, *160*, 35–40;
(b) D. Thomas, N. Peulecke, V. V. Burlakov, B. Heller, W. Baumann, A. Spannenberg, R. Kempe, U. Rosenthal, R. Beckhaus, *Z. Anorg. Allg. Chem.* **1998**, *624*, 919–924;
(c) R. Beckhaus, M. Wagner, V. V. Burlakov, W. Baumann, N. Peulecke, A. Spannenberg, R. Kempe, U. Rosenthal, *Z. Anorg. Allg. Chem.* **1998**, *624*, 129–134;
(d) V. Varga, K. Mach, J. Hiller, U. Thewalt, P. Sedmera, M. Polasek, *Organometallics* **1995**, *14*, 1410–1416;
(e) C. Müller, C. N. Iverson, R. J. Lachicotte, W. D. Jones, *J. Am. Chem. Soc.* **2001**, *123*, 9718–9719;
(f) C. Müller, R. J. Lachicotte, W. D. Jones, *Organometallics* **2002**, *21*, 1190–1196.

[4] *Examples*: (a) A. Davison, J. P. Selegue, *J. Am. Chem. Soc.* **1978**, *100*, 7763–7765;
(b) M. I. Bruce, R. C. Wallis, *J. Organomet. Chem.* **1978**, *161*, C1–C4;
(c) H. Berke, *Z. Naturforsch. B*, **1980**, *35*, 86–90;
(d) A. Höhn, H. Werner, *J. Organomet. Chem.* **1990**, *382*, 255–272;
(e) J. Gotzig, H. Otto, H. Werner, *J. Organomet. Chem.* **1985**, *287*, 247–254;
(f) J. P. Selegue, *Organometallics* **1982**, *1*, 217–218;
(g) N. Pirio, D. Touchard, L. Toupet, P. H. Dixneuf, *J. Chem. Soc. Chem. Commun.* **1991**, 980–982;
(h) H. Berke, *Chem. Ber.-Recl.* **1980**, *113*, 1370–1376.

[5] (a) J. H. Teuben, H. J. de Liefde Meijer, *J. Organomet. Chem.* **1969**, *17*, 87–93;
(b) G. L. Wood, C. B. Knobler, M. F. Hawthorne, *Inorg. Chem.* **1989**, *28*, 382–384;
(c) A. Sebald, P. Fritz, B. Wrackmeyer, *Spectrochim. Acta, Part A* **1985**, *41A*, 1405–1407;
(d) H. Lang, D. Seyferth, *Z. Naturforsch. B* **1990**, *45B*, 212–220;
(e) U. Rosenthal, H. Görls, *J. Organomet. Chem.* **1992**, *439*, C36–C41;
(f) H. Lang, E. Meichel, T. Stein, S. Back, E. Hovestreydt, *J. Organomet. Chem.* **2001**, *633*, 71–78.

[6] (a) B. Temme, G. Erker, R. Fröhlich, M. Grehl, *Angew. Chem.* **1994**, *106*, 1570–1572; *Angew. Chem. Int. Ed.* **1994**, *33*, 1480–1482;
(b) W. Ahlers, B. Temme, G. Erker, R. Fröhlich, T. Fox, *J. Organomet. Chem.* **1997**, *527*, 191–201;
(c) W. Ahlers, G. Erker, R. Fröhlich, U. Peuchert, *J. Organomet. Chem.* **1999**, *578*, 115–124;
(d) W. Ahlers, B. Temme, G. Erker, R. Fröhlich, F. Zippel, *Organometallics* **1997**, *16*, 1440–1444;
(e) G. Erker, W. Frömberg, R. Benn, R. Mynott, K. Angermund, C. Krüger, *Organometallics* **1989**, *8*, 911–920;
(f) D. P. Hsu, W. M. Davis, S. L. Buchwald, *J. Am. Chem. Soc.* **1993**, *115*, 10394–10395;
(g) N. Metzler, H. Nöth, *J. Organomet. Chem.* **1993**, *454*, C5–C7;
(h) U. Rosenthal, A. Ohff, W. Baumann, R. Kempe, A. Tillack, V. V. Burlakov, *Organometallics* **1994**, *13*, 2903–2906;
(i) V. V. Burlakov, A. Ohff, C. Lefeber, A. Tillack, W. Baumann, R. Kempe, U. Rosenthal, *Chem. Ber.* **1995**, *128*, 967–971;
(j) V. V. Burlakov, N. Peulecke, W. Baumann, A. Spannenberg, R. Kempe, U. Rosenthal, *Coll. ed. Czech. Chem. Commun.* **1997**, *62*, 331–336;

(k) U. Rosenthal, S. Pulst, P. Arndt, A. Ohff, A. Tillack, W. Baumann, R. Kempe, V. V. Burlakov, *Organometallics* **1995**, *14*, 2961–2968;
(l) V. V. Burlakov, P. Arndt, A. Spannenberg, W. Baumann, U. Rosenthal, *Organometallics* **2004**, *23*, 5188–5192.

[7] (a) G. A. Luinstra, L. C. Ten Cate, H. J. Heeres, J. W. Pattiasina, A. Meetsma, J. H. Teuben, *Organometallics* **1991**, *10*, 3227–3237;
(b) F. G. Kirchbauer, P.-M. Pellny, H. Sun, V. V. Burlakov, P. Arndt, W. Baumann, A. Spannenberg, U. Rosenthal, *Organometallics* **2001**, *20*, 5289–5296.

[8] (a) P.-M. Pellny, F. G. Kirchbauer, V. V. Burlakov, W. Baumann, A. Spannenberg, U. Rosenthal, *Chem. Eur. J.* **2000**, *6*, 81–90;
(b) F. G. Kirchbauer, Ph. D. Thesis, University of Rostock, **1999**;
(c) P.-M. Pellny, F. G. Kirchbauer, V. V. Burlakov, W. Baumann, A. Spannenberg, U. Rosenthal, *J. Am. Chem. Soc.* **1999**, *121*, 8313–8323;
(d) Z. Hou, T. L. Breen, D. W. Stephan, *Organometallics* **1993**, *12*, 3158–3167;
(e) S. Pulst, P. Arndt, W. Baumann, A. Tillack, R. Kempe, U. Rosenthal, *J. Chem. Soc. Chem. Commun.* **1995**, 1753–1754.

[9] (a) K. Tagaki, C. J. Rousset, E. Negishi, *J. Am. Chem. Soc.* **1991**, *113*, 1440–1442;
(b) R. Choukroun, J. Zhao, C. Lorber, P. Cassoux, B. Donnadieu, *Chem. Commun.* **2000**, 1511–1512.

[10] (a) P. Binger, P. Müller, P. Philipps, B. Gabor, R. Mynott, A. T. Herrmann, F. Langhauser, C. Krüger, *Chem. Ber.* **1992**, *125*, 2209–2212;
(b) P. Binger, P. Müller, A. T. Herrmann, P. Philipps, B. Gabor, F. Langhauser, C. Krüger, *Chem. Ber.* **1991**, *124*, 2165–2170;
(c) C. J. Harlan, J. A. Tunge, B. M. Bridgewater, J. R. Norton, *Organometallics* **2000**, *19*, 2365–2372.

[11] P.-M. Pellny, N. Peulecke, V. V. Burlakov, A. Tillack, W. Baumann, A. Spannenberg, R. Kempe, U. Rosenthal, *Angew. Chem.* **1997**, *109*, 2728–2730; *Angew. Chem. Int. Ed.* **1997**, *36*, 2615–2617.

[12] P.-M. Pellny, V. V. Burlakov, W. Baumann, A. Spannenberg, R. Kempe, U. Rosenthal, *Organometallics* **1999**, *18*, 2906–2909.

[13] A. Spannenberg, K. Dallmann, V. V. Burlakov, U. Rosenthal, *Z. Kristallogr.* **2002**, *217*, 241–243.

[14] P.-M. Pellny, F. G. Kirchbauer, V. V. Burlakov, A. Spannenberg, K. Mach, U. Rosenthal, *Chem. Commun.* **1999**, 2505–2506.

[15] (a) C. Glaser, *Ber. Dtsch. Chem. Ges.* **1869**, *2*, 422–424;
(b) C. Glaser, *Ann. Chem. Pharm.* **1870**, *154*, 137–171;
(c) T. Cuenca, R. Gomez, P. Gomez-Sal, G. M. Rodriguez, P. Royo, *Organometallics* **1992**, *11*, 1229–1234;
(d) P. N. V. Pavan Kumar, E. D. Jemmis, *J. Am. Chem. Soc.* **1988**, *110*, 125–131;
(e) E. D. Jemmis, K. T. Giju, *Angew. Chem.* **1997**, *109*, 633–635; *Angew. Chem. Int. Ed.* **1997**, *36*, 606–608;
(f) E. D. Jemmis, K. T. Giju, *J. Am. Chem. Soc.* **1998**, *120*, 6952–6964;
(g) E. D. Jemmis, A. K. Phukan, U. Rosenthal, *J. Organomet. Chem.* **2001**, *635*, 204–211;
(h) E. D. Jemmis, A. K. Phukan, K. T. Giju, *Organometallics* **2002**, *21*, 2254–2261;
(i) G. Aullon, S. Alvarez, *Organometallics* **2002**, *21*, 2627–2634.

[16] (a) S. Pulst, P. Arndt, B. Heller, W. Baumann, R. Kempe, U. Rosenthal, *Angew. Chem.* **1996**, *108*, 1175–1178; *Angew. Chem. Int. Ed.* **1996**, *35*, 1112–1115;
(b) U. Rosenthal, A. Ohff, A. Tillack, W. Baumann, H. Görls, *J. Organomet. Chem.* **1994**, *468*, C4–C8.

[17] (a) S. Pulst, F. G. Kirchbauer, B. Heller, W. Baumann, U. Rosenthal, *Angew. Chem.* **1998**, *110*, 2029–2031; *Angew. Chem. Int. Ed.* **1998**, *37*, 1925–1927;
(b) W. Baumann, P.-M. Pellny, U. Rosenthal, *Magn. Reson. Chem.* **2000**, *38*, 515–519.

[18] (a) P. Štěpnička, R. Gyepes, I. Císařová, M. Horáček, J. Kubišta, K. Mach, *Organometallics* **1999**, *18*, 4869–4880; (b) M. Horáček, I. Císařová, J. Čejka, J. Karban, L. Petrusová, K. Mach, *J. Organomet. Chem.* **1999**, *577*, 103–112; (c) M. Akita, H. Yasuda, A. Nakamura, *Bull. Chem. Soc. Jpn.* **1984**, *57*, 480–487; (d) V. Varga, L. Petrusová, J. Čejka, V. Hanuš, K. Mach, *J. Organomet. Chem.* **1996**, *509*, 235–240.

[19] A. Ohff, V. V. Burlakov, U. Rosenthal, *J. Mol. Catal. A*, **1996**, *108*, 119–123.

[20] (a) J. Heidrich, M. Steimann, M. Appel, W. Beck, J. R. Phillips, W. C. Trogler, *Organometallics* **1990**, *9*, 1296–1300; (b) T. Weidmann, V. Weinrich, B. Wagner, C. Robl, W. Beck, *Chem. Ber.* **1991**, *124*, 1363–1368; (c) S. Mihan, T. Weidmann, V. Weinrich, D. Fenske, W. Beck, *J. Organomet. Chem.* **1997**, *541*, 423–439; (d) S. Mihan, K. Sünkel, W. Beck, *Chem. Eur. J.* **1999**, *5*, 745–753; (e) M. D. Janssen, M. Herres, L. Zsolnai, D. M. Grove, A. L. Spek, H. Lang, G. van Koten, *Organometallics* **1995**, *14*, 1098–1100.

[21] (a) C. S. Griffith, G. A. Koutsantonis, B. W. Skelton, A. H. White, *J. Chem. Soc. Chem. Commun.* **2002**, 2174–2175; (b) M. Akita, Y. Moro-oka, *Bull. Chem. Soc. Jpn.* **1995**, *68*, 420–432; (c) M. Akita, M. Terada, S. Oyama, Y. Moro-oka, *Organometallics* **1990**, *9*, 816–825; (d) M. Akita, S. Sugimoto, H. Hirikawa, S. I. Kato, M. Terada, M. Tanaka, Y. Moro-oka, *Organometallics* **2001**, *20*, 1555–1568; (e) W.-Y. Lo, C.-H. Lam, W. K.-M. Fung, H.-Z. Sun, V. W.-W. Yam, D. Balcells, F. Maseras, O. Eisenstein, *Chem. Commun.* **2003**, 1260–1261.

[22] (a) R. F. C. Brown, *Eur. J. Org. Chem.* **1999**, 3211–3222; (b) J. Mabry, R. P. Johnson, *J. Am. Chem. Soc.* **2002**, *124*, 6497–6501.

[23] (a) H. Werner, *Nachr. Chem. Tech. Lab.* **1992**, *40*, 435–444; (b) V. Cadierno, M. P. Gamasa, J. Gimeno, E. Pérez-Carreño, S. García-Granda, *Organometallics* **1999**, *18*, 2821–2832; (c) R. S. Bly, Z. Zhong, C. Kane, R. K. Bly, *Organometallics* **1994**, *13*, 899–905; (d) J. Silvestre, R. Hoffmann, *Helv. Chim. Acta* **1985**, *68*, 1461–1506; (e) R. Stegmann, G. Frenking, *Organometallics* **1998**, *17*, 2089–2095; (f) F. De Angelis, A. Sgamellotti, N. Re, *Organometallics* **2002**, *21*, 2715–2723; (g) N. E. Kolobova, A. B. Antonova, P. V. Perovsky, O. M. Khitrova, M. Yu. Antipin, Yu. T. Struchkov, *J. Organomet. Chem.* **1977**, *137*, 69–78; (h) D. Schneider, H. Werner, *Angew. Chem.* **1991**, *103*, 710–712; *Angew. Chem. Int. Ed.* **1991**, *30*, 700–702.

[24] (a) F. Torres, E. Sola, A. Elduque, A. P. Martínez, F. J. Lahoz, L. A. Oro, *Chem. Eur. J.* **2000**, *6*, 2120–2128; (b) M. A. Esteruelas, F. J. Fernandez, A. M. Lopez, E. Onate, *Organometallics* **2003**, *22*, 1787–1789.

[25] (a) U. Rosenthal, *Angew. Chem.* **2003**, *115*, 1838–1842; *Angew. Chem. Int. Ed.* **2003**, *42*, 1794–1798; (b) U. Rosenthal, P. Arndt, W. Baumann, V. V. Burlakov, A. Spannenberg, *J. Organomet. Chem.* **2003**, *670*, 84–96; (c) E. D. Jemmis, A. K. Phukan, H. Jiao, U. Rosenthal, *Organometallics* **2003**, *22*, 4958–4965.

5
Acetylenosaccharides

Bruno Bernet and Andrea Vasella

5.1
Introduction

Acetylenosaccharides are combinations of two quite different, but both versatile, structural elements: an alkynyl and a carbohydrate moiety. The alkynyl moiety is hydrophobic, achiral, rigid, and linear, and it is easily and selectively transformed into a variety of functional groups. The carbohydrate portion is highly oxidized, hydrophilic, chiral, occurs in a number of acyclic and cyclic isomers, and is far less easily selectively modified. The combination of these divergent elements in a single molecule makes acetylenosaccharides attractive. Nature does not produce alkynylated carbohydrates, except for polyacetylenes that combine cumulated alkynyl groups and a short alditol portion (tetritol or shorter chain).

Although the first acetylenosaccharides were prepared from non-carbohydrate precursors as early as 1925, by Lespieau [1], and in 1949, by Raphael [2] (see [3] for an early review on the preparation of carbohydrates from acetylenic precursors), the first transformation of a carbohydrate into an acetylenosaccharide was published by Zelinski and Meyer in 1958 [4]. Since 1970, growing interest in acetylenosaccharides has led to about 700 publications. This development is reflected in the nomenclature of carbohydrates, as the infix "-yn-" was added only late, in the 1996 recommendations. Prior to this date, a glycynose had to be named as a tetradehydro-glycose, whereas olefinic saccharides were already called glycenoses.

The goal of this article is to provide a comprehensive overview of the chemistry of acetylenosaccharides. The review is divided into four main sections: isolation from natural sources, preparation of mono- and dialkynylated saccharides, synthetic transformations of alkynylated saccharides, and biological and medicinal applications.

In this review, acetylenosaccharides are defined as glycopyranoses or -furanoses possessing a C-alk-1-ynyl or a C-alk-2-ynyl substituent or compounds formally derived from them and containing a 3,4,5-trihydroxypent-1-ynyl or a 4,5,6-trihydroxyhex-1-ynyl moiety (see **1** and **2** in Scheme 5.1; R = any substituent). Thus, 2,3-epoxyalk-4-ynols **3** are included in the discussion, but not 2,3-dihydroxyalk-4-ynoates **4**,

Acetylene Chemistry. Chemistry, Biology and Material Science. Edited by F. Diederich, P. J. Stang, R. R. Tykwinski

ISBN 3-527-30781-8

which are readily accessible by aldol reactions of propargylic aldehydes. Also excluded are 2-ethynylated glycerines **5**, C-alkynylated tartaric acids **6**, and alkynylated compounds in which the alkynyl group is attached at a heteroatom of the glycosyl moiety (such as propargyl ethers and glycosides).

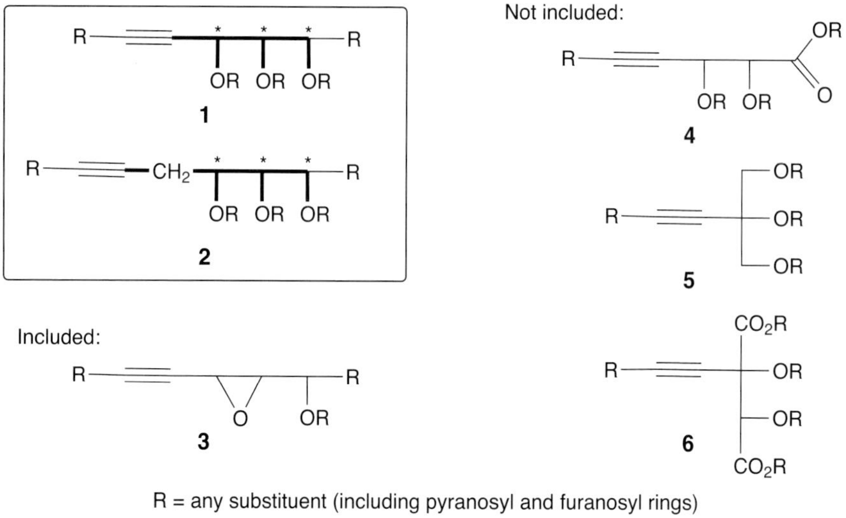

Scheme 5.1 Acetylenosaccharides defined as compounds of structures **1** and **2**.

5.2 Isolation of Acetylenosaccharides from Natural Sources

Some alk-4-yne-1,2,3-triols and alk-5-yne-1,2,3,4-tetrols have been isolated from *Basidiomycete*, *Hyphomycete*, and *Actinomycete* fungi. They contain from two to four cumulated triple bonds and are representatives of antifungal polyacetylene antibiotics. The first described polyacetylene, biformin or biformyne (**7**), an epoxytriynol, was isolated from *Polyporus biformis* in 1947 [5]; its relative configuration was assigned by Anchel and Cohen in 1954 [6] and by Jones and co-workers in 1963 [7] (Scheme 5.2). Further polyacetylenes from *Basidiomycete* fungi were described later on, in the forms of the epoxytriynol **8** [8], the C_2-symmetric diepoxydiynediol **9** ((+)-repandiol) [9], the epoxydiynol **10** (nitidon) [10], the diastereoisomeric diynetriols **11** [11] and **12** [12], and the diynetetrol **13** [13]. The unusual (chlorovinyl group) enediynetriol **14** was extracted from a *Hyphomycete* fungus [14]. Finally, a 4,5-disubstituted triacetylenic dioxolone was isolated from *Actinomycete* fungi [15]. The absolute configuration was determined by chemical correlation of **11–13**, a Sharpless asymmetric epoxidation [16] affording (+)-repandiol **9** [10], and a Sharpless kinetic resolution [17] providing **15** [18]. Only the relative configurations were assigned to **7**, **8**, and **10** (all *trans*), and to **14** (*erythro*).

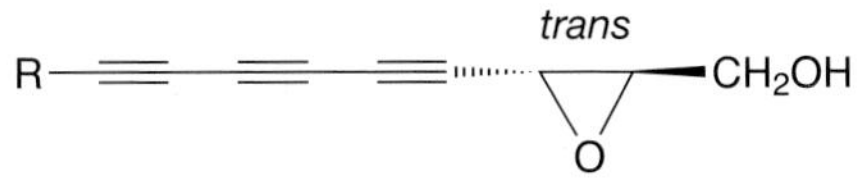

7 R = H (*Polyporus biformis* and *Coprinus quadrifidus*)
8 R = Me (*Trametes pubescens*)

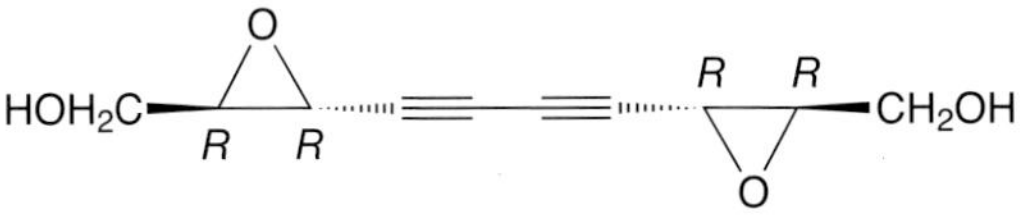

9 (*Hydnum repandum*)

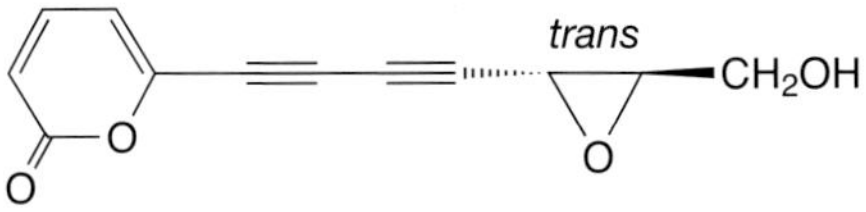

10 (*Junghuhnia nitida*)

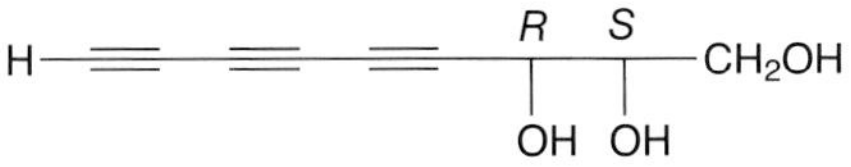

11 (*Coprinus quadrifidus*)

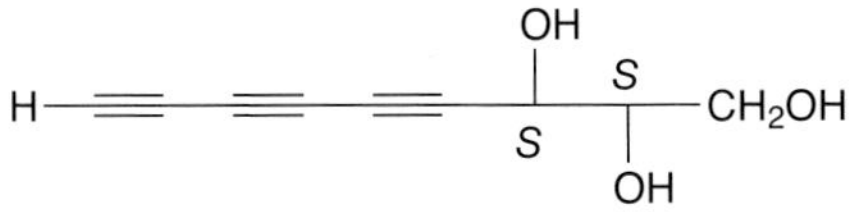

12 (*Aleurodiscus roseus*)

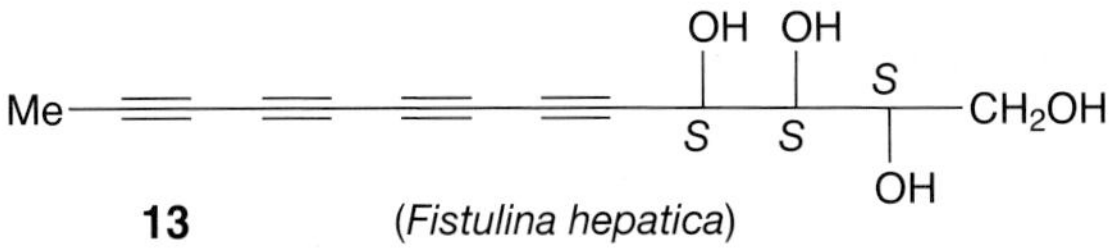

13 (*Fistulina hepatica*)

14 (*Pneumatospora obcoronata*)

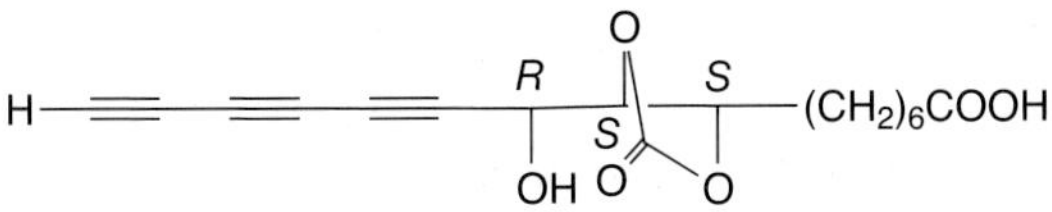

15 (*Microbispora* sp.)

Scheme 5.2 Polyacetylenes isolated from *Basilidiomycete* (**7–13**), *Hyphomycete* (**14**), and *Actinomycete* (**15**).

Osirisyne F [19] and haliclonyne [20] are two highly oxygenated C_{47} polyacetylenes isolated from the marine sponge *Haliclona*. They each contain a 4,5,6-trihydroxyhex-2-ynoic acid fragment **16** (Scheme 5.3). The configuration of the glycerol moiety was not assigned.

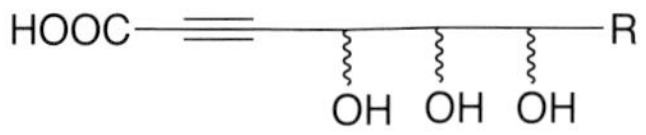

R = polyoxygenated, unsaturated C_{41} linear chain

16 (*Haliclona* sp.)

Scheme 5.3 Polyacetylenes isolated from sponges.

A C_{10} polyacetylene, the antifeedant **17**, was isolated from a *Compositae* plant [21–23] (Scheme 5.4). Unlike the fungal polyacetylene epoxyalcohols, **17** is a *cis*-substituted epoxide. Plant polyacetylenes are usually C_{13} to C_{17} 1,3-disubstituted glycerols or 1,4-disubstituted tetritols, such as the 1,2,3-trihydroxyhepta-4,6-diyne

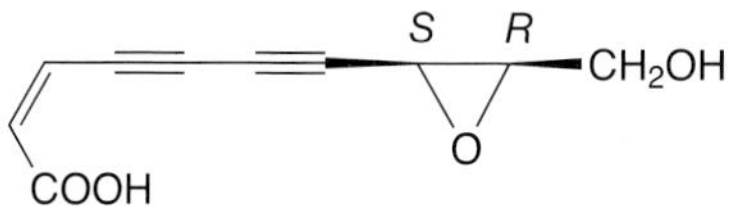

17 (*Chrysothamnus nauseosus*)

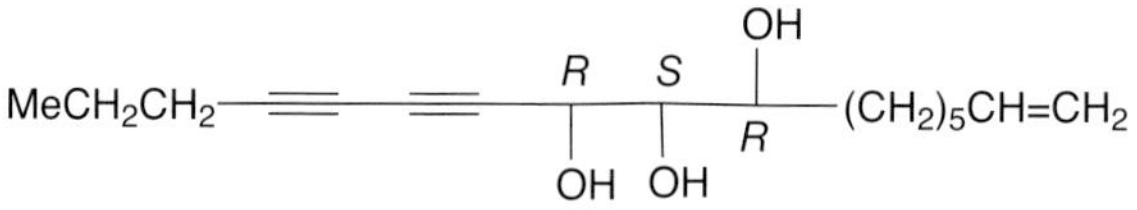

18 (*Cirsium* spp., *Ptilostemon* spp., or *Vladimira denticulata*)

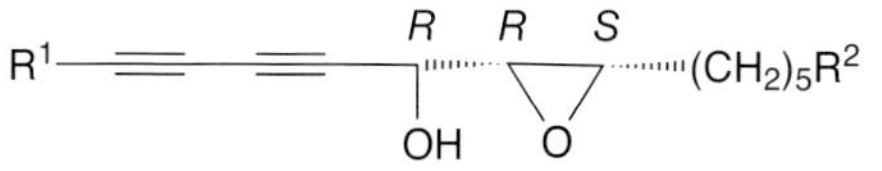

19a R^1 = (*Z*)-CH=CHMe, R^2 = $CH{=}CH_2$ (*Anthemis rudolfiana*)
19b R^1 = CH_2CH_2Me, R^2 = $CH{=}CH_2$ (*Cirsium japonicum*)
19c R^1 = $CHOHCH{=}CH_2$, R^2 = CH_2Me (*Panax quinquefolium*)

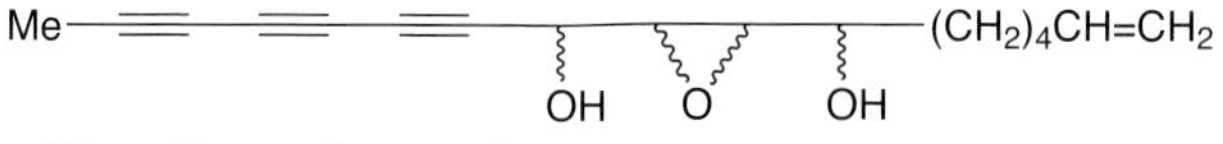

20 (*Cacosmia rugosa*)

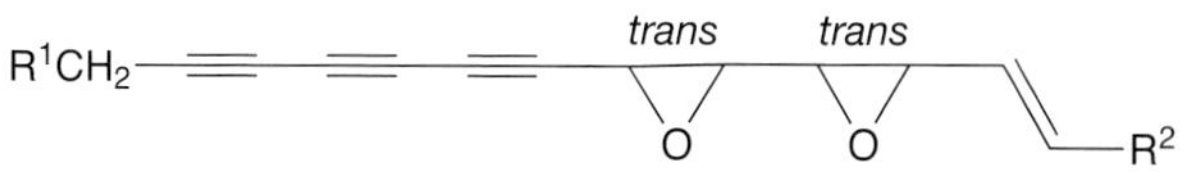

21a R^1 = R^2 = H (*Artemisia annua*)
21b R^1 = OH, R^2 = CH_2OAc (*Adenia gummifera*)

Scheme 5.4 Polyacetylenes **17–21** isolated from plants.

18 [24–27], the 1,2-epoxy-3-hydroxyhept-4,6-diynes **19a** [28], **19b** [25, 29, 30], and **19c** [31], the 2,3-epoxy-1,4-dihydroxydeca-5,7,9-triynes **20** [32], and the 1,2:3,4-diepoxydeca-5,7,9-triynes **21a** (annuadiepoxide) [33] and **21b** (gummiferol) [34]. The (8*R*,9*S*,10*R*) configuration of **18** is based on chemical correlation [35] and is at variance with the (8*S*,9*R*,10*S*) configuration postulated earlier [36] on the basis of Horeau's method [37]. The epoxides **19a–c** are *cis*-substituted. The absolute configuration of **19a** [28] was assigned by chemical correlation and that of **19b** by an asymmetric synthesis [38]. Only the *trans* configurations of the two oxirane rings of **21a** and **21b** were assigned.

The enediynes neocarzinostatin (**22**) [39, 40] and N1999A2 (**23**) [41, 42] were isolated from *Streptomyces* sp. (Scheme 5.5). They each contain a 1,2-dialkynylated 1,2-anhydrotetritol moiety. Surprisingly, opposite absolute configurations were deduced for **22** and for **23** (D-*ribo* *vs.* L-*ribo*). Strained unsaturated 13-oxatricyclo[7.3.1^{4,5}]trideca-8,12-diene-2,6-diynes are a popular synthetic target (see, for example, [43–47]).

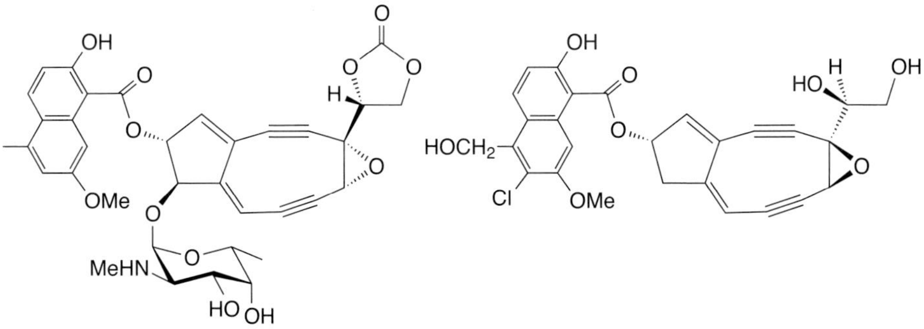

22 (Neocarzinostatin from *Streptomyces carzinostaticus*)

23 (N1999-A2 from *Streptomyces* sp. AJ9493)

Scheme 5.5 Enediynes **22** and **23** isolated from *Streptomyces* species.

5.3 Preparation of Monoalkynylated Acetylenosaccharides

Degradation, double elimination, and C,C-bond formation to terminal C-atoms of monosaccharides afford linear acetylenosaccharides, whereas C,C-bond formation to internal C-atoms yields branched-chain acetylenosaccharides.

5.3.1 Preparation of Linear Acetylenosaccharides

5.3.1.1 By Degradation

Treatment of the *N*-nitroso GlcNAc derivative **24** with KOH in ^{i}PrOH/Et_2O followed by acetylation gave the pent-4-ynitol **25** [48] (Scheme 5.6). The transformation proceeds via an intermediate diazo derivative, as illustrated by the transformation of the diazoester **26** into the acetylene **27**. We presume that **26** reacts by fragmentation, to generate a vinyldiazo compound that is transformed into the alkyne via an alkenylidene carbene. This degradation is of little synthetic interest, since pent-4-ynitols are easily accessible by addition of acetylene to glyceraldehydes.

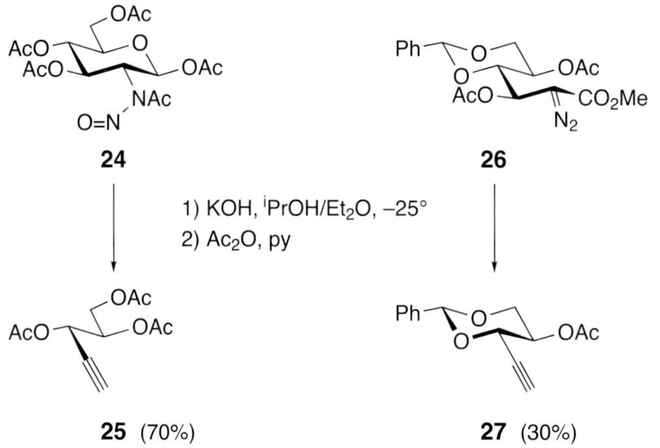

Scheme 5.6 Degradation of the nitrosoacetamide **24** and the diazo ester **26**.

5.3.1.2 By Elimination

Double elimination of fully protected 1-chloroalditols and 6-chloro-D-glucofuranosides gives terminal acetylenes [49]. Thus, treatment of the D-xylitol derivative **28** and the D-glucofuranose **30** with $LiNH_2$ in liquid NH_3 gave the hydroxyacetylenes **29** and **31**, respectively (Scheme 5.7). Application of this reaction to 1-chlorohex-2-ynitols, -hept-2-ynitols, and -oct-2-ynitols gave the corresponding aldo-1,3-diynitols

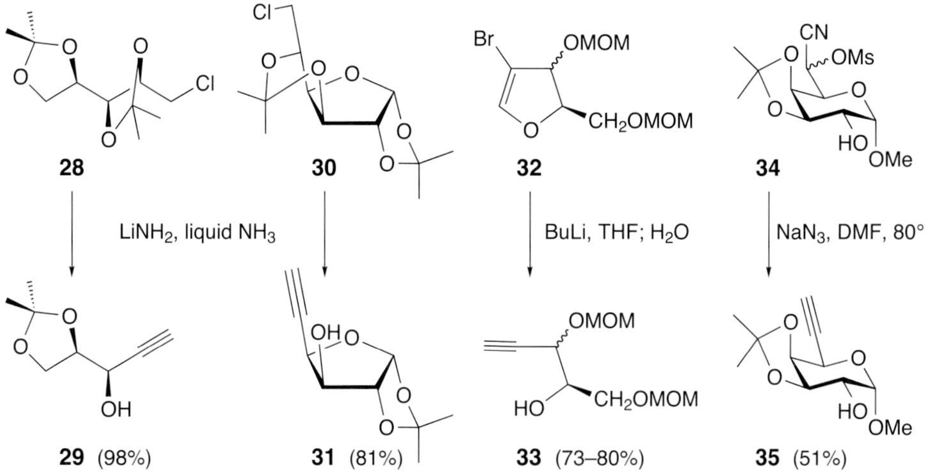

Scheme 5.7 Double elimination of **28** and **30**, reductive elimination of **32**, and transformation of **34** into **35**.

[50]. Double elimination of 2,3-epoxychlorides (readily accessible by Sharpless asymmetric epoxidation) provided chiral propargylic alcohols [51]. To the best of our knowledge, no double elimination of 6-halopyranosides and 5-halofuranosides to provide acyclic aldehydo-aldynoses is yet known,

Reductive elimination (BuLi in THF) of a protected 1,5-anhydro-2-bromo-2-deoxy-D-*arabino*-hexitol (a 2-bromo-D-glucal) [52] and of 1,4-anhydro-2-bromo-2-deoxy-D-pentitols [53] resulted in ring-opening and a terminal acetylene. Higher yields were obtained in the reductive elimination of the 1,4-anhydropent-1-enitol **32** (73–80%; Scheme 5.7) than in that of the analogous, six-membered 2-bromo-glucal (37%).

An unusual acetylene formation (**35**) was observed upon treatment of the cyano-hydrin mesylate **34** with NaN_3 in DMF [54] (Scheme 5.7). The reaction probably proceeds through the intermediate formation of a cyano-azide and a 1,2,3,4-tetrazine. We are not aware of other examples of this reaction.

5.3.1.3 By Wittig-Type Reactions

In 1972, Corey and Fuchs published a powerful transformation of aldehydes into acetylenes via dibromoethenes [55]; treatment of the dibromoethene with BuLi in THF then provides the acetylene. In the second step, the intermediate acetylide anion may be intercepted by electrophiles, such as H_2O, Br_2, I_2, carbonyl compounds including acid derivatives, and epoxides. Since aldehydo derivatives of carbohydrates are easily accessible by selective protection or by oxidation of the primary OH group of (protected) glycosides, the Corey–Fuchs method was widely applied in the carbohydrate field (more than 60 papers), first by Tronchet and co-workers (see, for example, [56]). The method was used for the preparation of selectively labeled 6-[^{2}H]-D-glucoses (reduction of **36**, epoxidation, and opening of the oxirane ring [57]), for the preparation of C-disaccharides (deoxygenation and hydrogenation of **37** [58–60]; Scheme 5.8), and for the preparation of 4′-ethynylated nucleosides (e. g., **38** [61–63]; for analogues of nucleic acids derived from **38**, see [64, 65]). Replacement of BuLi by the less nucleophilic $NaN(SiMe_3)_2$ in the elimination

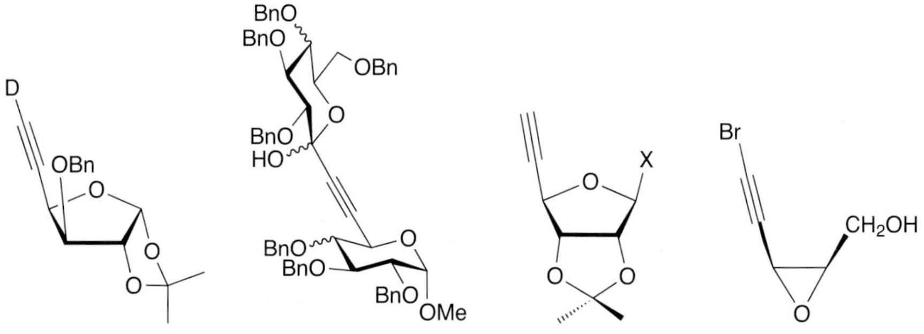

36 **37** **38** X = OR or nucleobase **39**

Scheme 5.8 Examples of linear acetylenosaccharides obtained by the Corey–Fuchs method.

step allowed the epoxyalkyne **39** to be prepared from the corresponding aldehyde [22]. Toma and Lièvre and their co-workers found conditions for the transformation of unprotected or partially protected aldopyranoses and -furanoses into terminal acetylenosaccharides; excess BuLi in THF at −70° was used for the elimination step [66, 67]. A Corey–Fuchs-type reaction was also described for lactones; lactones were transformed into dichloromethylene derivatives and further by treatment of the halide with Li sand in THF into hydroxyacetylenes (82–88 %) [68].

The direct transformation of aldehydes into terminal acetylenes is feasible with dimethyl (diazomethyl)phosphonate and KO^{t}Bu in THF [69]. The first preparation of an acetylenosaccharide in low yields (17–22 %) by the use of dimethyl (diazomethyl)phosphonate and BuLi as base was described by Tronchet and co-workers as early as 1974 [70]. Later on, both fully protected aldehydo-sugars [71, 72] and partially protected hemiacetals [73] were transformed (48–87 %) into terminal acetylenosaccharides by this method.

5.3.1.4 By Substitution

Primary OH groups of carbohydrates have been replaced by ethynyl moieties by substitution of a sulfonate, and by ring-opening of an epoxide or a cyclic sulfate. Whereas the 2-deoxytosylate **40** was easily transformed into the acetylene **41** [74] (Scheme 5.9), tosylates, bromides, and iodides possessing β-oxygen substituents

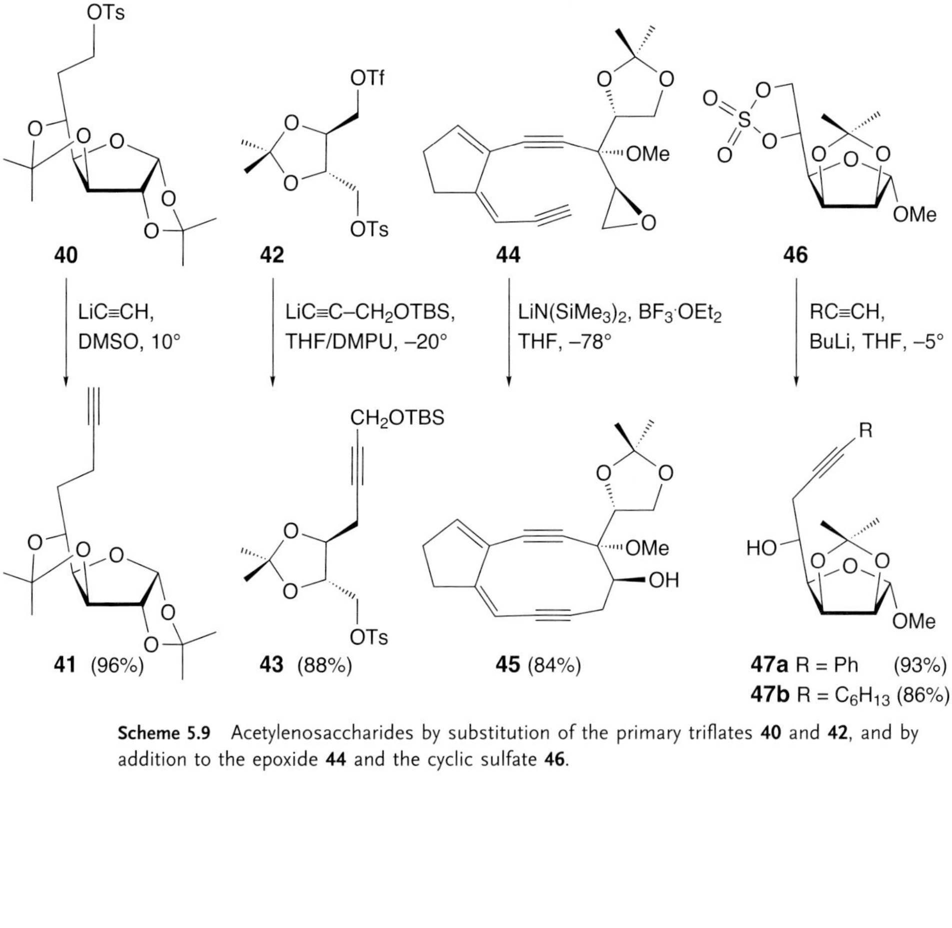

Scheme 5.9 Acetylenosaccharides by substitution of the primary triflates **40** and **42**, and by addition to the epoxide **44** and the cyclic sulfate **46**.

are much less reactive [75]. Indeed, only the more reactive triflyloxy group of the disulfonate **42** was selectively replaced [75] (for other substitutions of primary TfO groups of saccharides, see [76, 77]). A 6-iodo-α-D-glucopyranoside was ethynylated by radicals generated by irradiation of (trifluoromethyl)sulfonylated acetylenes with UV light [78].

Terminal epoxides and cyclic sulfates of carbohydrates are attacked by nucleophiles at the less hindered methylene group. Such epoxides have been opened with acetylide anions in polar solvents such as liquid NH_3 or HMPA [79]. More convenient is the reaction in THF at low temperature and in the presence of a Lewis acid [80, 81]. Thus, the moderately strained ten-membered ring of the homo-neocarzinostatin analogue **45** was closed by an intramolecular nucleophilic substitution of the terminal epoxide **44** [82] (Scheme 5.9; see neocarzinostatin in Scheme 5.5). The readily accessible *manno*-configured cyclic sulfate **46** reacted exclusively at the primary carbon to yield the linear acetylenes **47a** and **47b** [83].

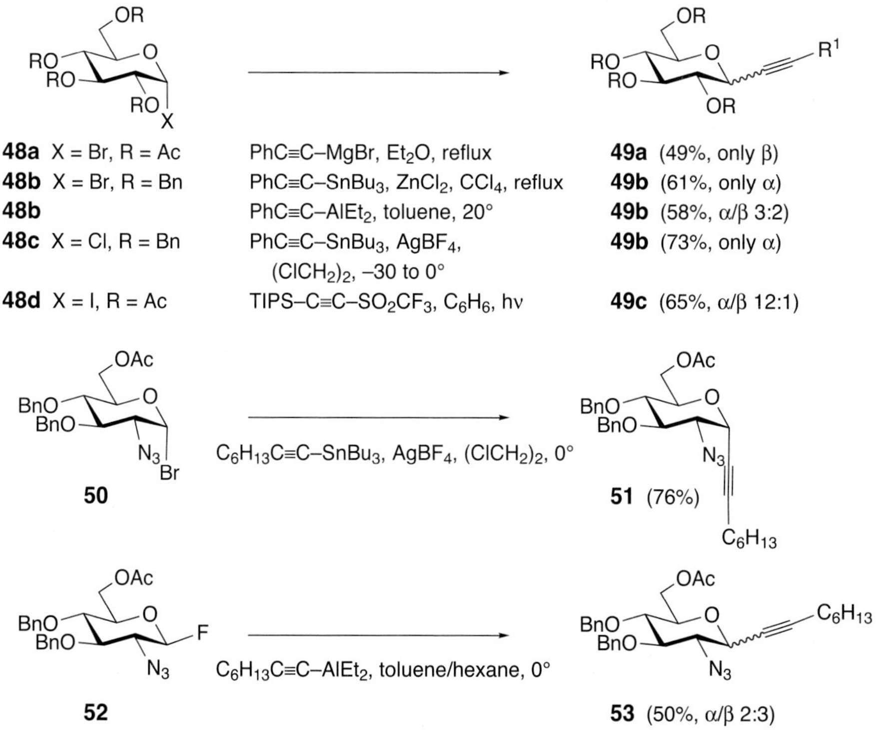

Scheme 5.10 Acetylenosaccharides by substitution of glucopyranosyl and 2-azido-2-deoxyglucopyranosyl halides.

Nucleophilic attack by alkynes on pyranosyl or furanosyl halides may a priori give mixtures of anomeric glycosylacetylenes. The diastereoselectivity depends upon the leaving group X, the nature of the protecting groups R (especially

RO−C(2)), and the reaction conditions (Schemes 5.10 and 5.11). In a (predominantly) S_N2 reaction, the acetylated 1,2-*cis*-configured α-D-glucopyranosyl bromide **48a** reacted with PhC≡C−MgBr to provide the β-D-glucopyranosylacetylene **49a** exclusively [4], while the analogous more reactive benzylated halides **48b** and **48c** reacted with PhC≡C−$SnBu_3$ in the presence of a Lewis acid, or with PhC≡C−$AlEt_2$ by a S_N1 mechanism, to yield, either exclusively or predominantly, the α-D-glucopyranosylacetylene **49b** [84–86] (Scheme 5.10). The retention of configuration and the axial attack of the acetylene are evidence of the formation of an intermediate oxycarbenium cation. Replacement of the C(2)-benzyloxy group by an azido group has no influence upon the diastereoselectivity; **50** was transformed into the α-D-glucopyranosylacetylene **51** in 76 % yield [87]. The β-D-glucopyranosyl fluoride **52**, however, reacted with $C_6H_{13}C$≡C−$AlEt_2$ to give a mixture of anomers **53** predominantly formed with retention of configuration [87], hinting at neighboring group participation by the azido group [88]. The α-D-anomer is favored in the radical ethynylation of the acetylated iodide **48d** [78].

α-D-Hexopyranosylacetylenes were exclusively obtained from TMSOTf-catalyzed reactions between α-D-hexopyranosyl acetates and Me_3Si−C≡C−$SnBu_3$ [89–92]; the $SnBu_3$ group is essential, as Me_3Si−C≡C−$SiMe_3$ did not react under a variety of conditions.

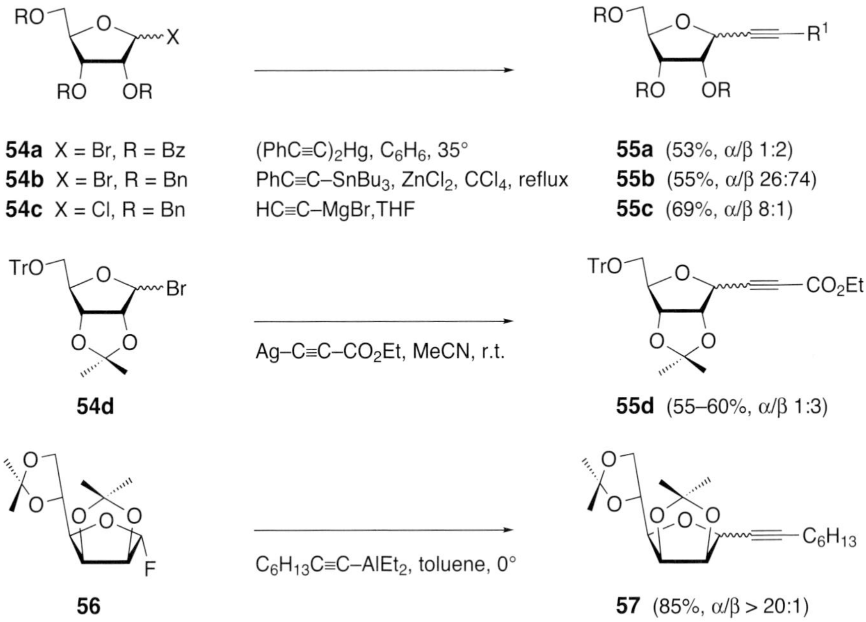

Scheme 5.11 Acetylenosaccharides by substitution of furanosyl halides.

A series of ribofuranosylacetylenes were prepared as intermediates in a synthesis of C-nucleosides featuring a 1,3-dipolar cycloaddition to these alkynes. The anomeric mixtures of the furanosyl halides **54a** [93], **54b** [84], and **54c** [94, 95], prepared in situ, were transformed into mixtures of anomeric ribofuranosylacetylenes **55** (Scheme 5.11). In S_N1 processes, the isopropylidenated ribofuranosyl bromide **54d** [96] and the mannofuranosyl fluoride **56** [97] were mostly attacked from their less hindered *exo* sides to give **55d** and **57**, respectively.

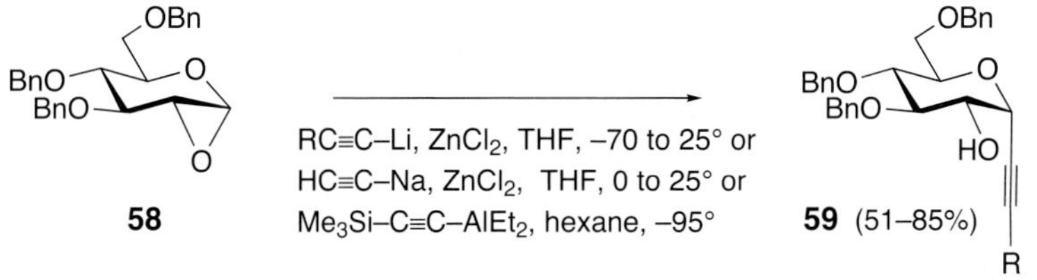

Scheme 5.12 α-D-Glucopyranosylacetylenes from the 1,2-anhydro-α-D-glucopyranose **58**.

In the presence of a Lewis acid, the 1,2-anhydro-α-D-glucopyranose **58** was selectively transformed into the α-D-glucopyranosylacetylenes **59** with retention of configuration [98–100] (Scheme 5.12).

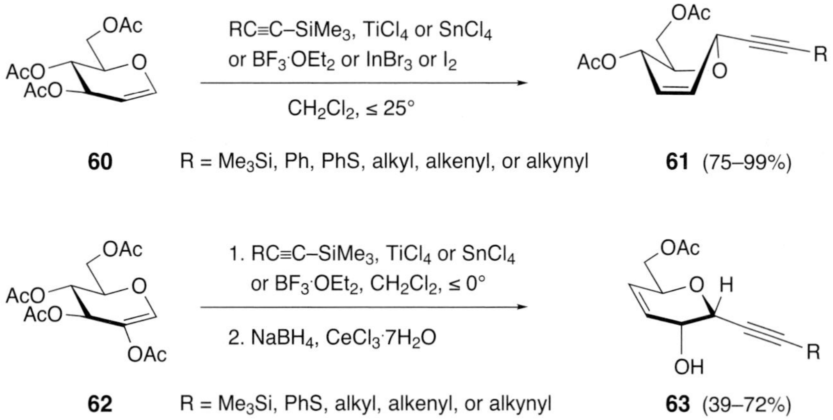

Scheme 5.13 3,7-Anhydro-D-octenynitols from glucals.

The Ferrier rearrangement of glucals in the presence of silylated acetylenes afforded 3,7-anhydro-D-*arabino*-oct-4-en-1-ynitols, as illustrated for the transformation of **60** into **61** in Scheme 5.13 [101–103] (for reviews see [89, 90]). The S_N1 reaction is initiated by loss of AcO–C(3), resulting in an intermediate allyloxycarbenium ion that is intercepted by a pseudoaxially attacking silylated acetylene. The α-D-hex-2-enopyranosylacetylene **61** was epimerized to the β-D-anomer by an acid-catalyzed isomerization of its dicobalt hexacarbonyl complex [104]. Under the same conditions as used for **60**, the 2-acetoxyglucal **62** was transformed into moderately stable 3,7-anhydro-D-*threo*-oct-5-en-1-yn-5-uloses and, thence, by reduction to the 3,7-anhydro-D-*arabino*-oct-5-en-1-ynitols **63** [102].

5.3.1.5 By Addition to Aldehydes and Hemiacetals

Unlike the Corey–Fuchs reaction, the addition of ethynyl or propargyl anions to carbohydrate-derived aldehydes and hemiacetals results, as a rule, in mixtures of epimers. Despite this disadvantage, there are over one hundred papers concerned with additions to aldehydes and over fifty with additions to hemiacetals. The pioneering work in additions to fully protected aldehydo-saccharides was by Horton and co-workers [105], a typical example being the addition of HC≡C−MgBr to the isopropylidenated glyceraldehyde **64** to provide a 44:56 mixture of the propargylic alcohols **65a** and **65b** [106] (Scheme 5.14). The yield of the desired isomer may be increased by judicious choice of the reaction conditions, as illustrated by

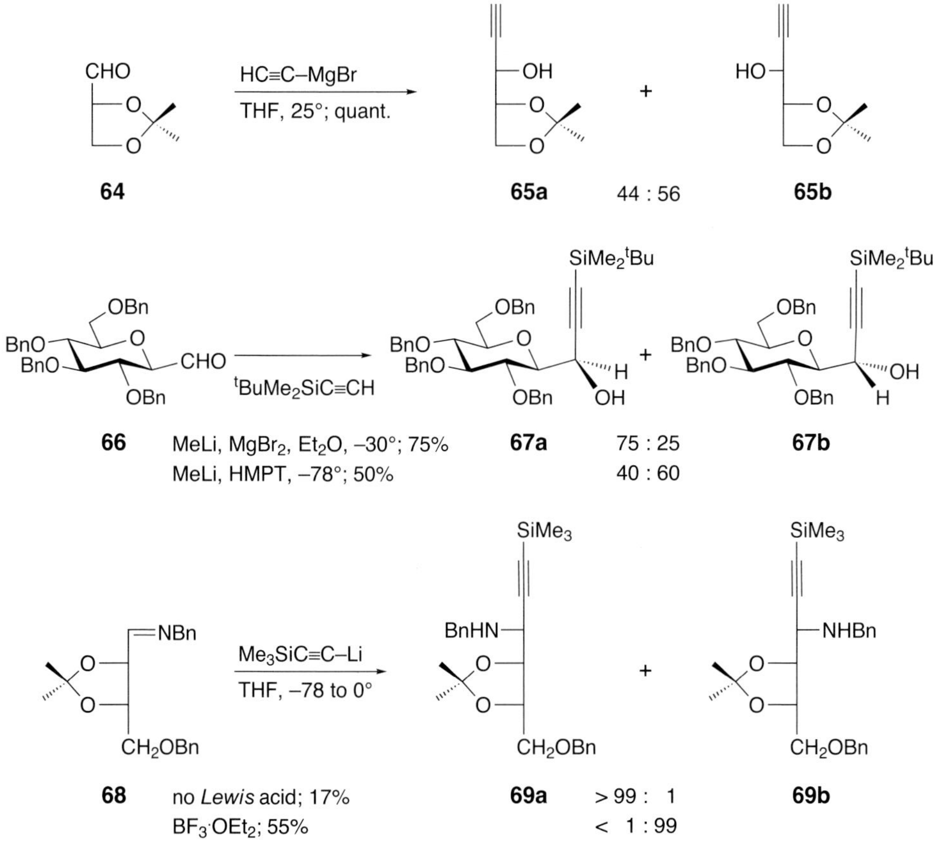

Scheme 5.14 Acetylenosaccharides by addition to protected aldehydo- and iminosaccharides.

the transformation of **66** into **67a** and **67b** in ratios ranging from 75:25 to 40:60 [107], by the transformation of the unwanted isomer into the desired isomer through a Mitsunobu reaction (see, for example [108–111]), or by oxidation of the epimeric alcohols to the ynone followed by reduction with Selectrides [112]. The propargylic amines **69a** and **69b** were obtained with excellent diastereoselectivity, albeit in low to moderate yields, by the addition of $Me_3Si-C{\equiv}C-Li$ to the aldimine **68** [113].

In the context of the preparation of oligonucleotide analogues in which the phosphate is replaced by an ethynediyl bridge between C(5′) and either C(6) of an adjacent uridyl or C(8) of an adjacent adenyl unit, we added $Me_3Si-C{\equiv}C-MgBr$ to the aldehydes **70a** [114] and **70b** [115] (Scheme 5.15). The yield and the ratio of the D-*allo*- and the L-*talo*-configured products depend on the nucleobase (U: 61–63 %, **71a**/**71b** 2:1; N^6-Bz-A: 41–42 %, **71c**/**71d** 1:1), but not on the silyl substituent. Independently and in another context, the addition to **70b** was also described by Matsuda et al. [116].

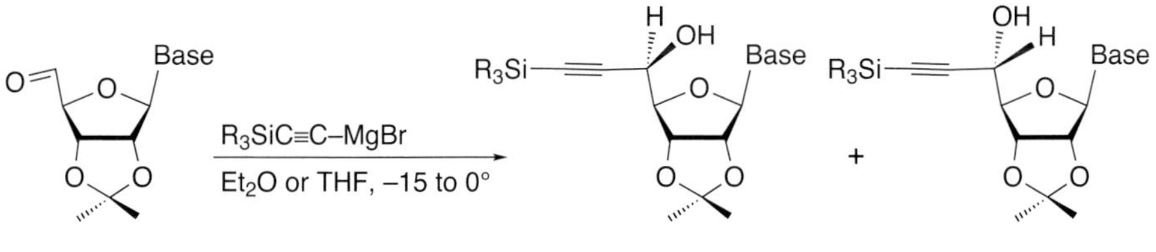

70a Base = U, R = Me or Et; 61–63% **71a** 2 : 1 **71b**
70b Base = N^6-Bz-A, R = Me or Et; 41–42% **71c** 1 : 1 **71d**

Scheme 5.15 Acetylenosaccharides by addition to β-D-*ribo*-pentodialdofuranosyl nucleosides **70a** and **70b**.

The addition of acetylide anions to hemiacetals produces epimeric mixtures of acyclic dihydroxyacetylenes. Although the first experiment was described by Chilton and co-workers [117], the reaction was investigated thoroughly by Buchanan's group [94, 118, 119]. In a typical example, the mannofuranose **72** gave the epimeric diols **73a** and **73b** in 65 and 5 % yields, respectively (Scheme 5.16). Upon treatment with TsCl in pyridine, the more reactive propargylic hydroxy group of **73a** was selectively tosylated. Intramolecular substitution then provided (with inversion of configuration) the β-D-mannofuranosylacetylene **74**. Although additions to pyranoses have also been described (see, for example, [60, 120, 121]), the ring-closing reaction has only been used for the preparation of glycofuranosylacetylenes. The glucopyranosyl hydroxylamine **75** was transformed via the tautomeric hydroxynitrone into the propargylic hydroxylamines **76a** and **76b** [122], and the 2-hydroxypyrrolidine **77** was selectively converted into the hydroxy tosylamide **78** [123].

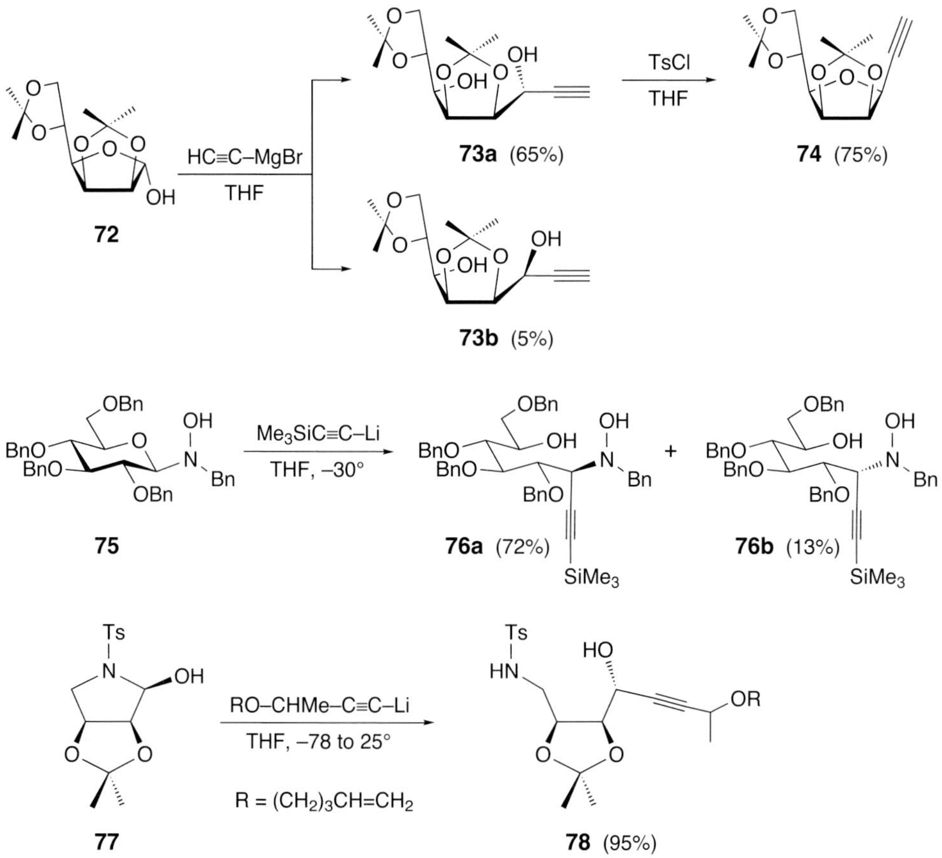

Scheme 5.16 Acetylenosaccharides by addition to hemiacetals and to N,O-acetals.

5.3.1.6 By Addition to Lactones and Lactams

Addition of acetylides to lactones gives hemiacetals, the first additions having been performed with γ-lactones [124]. Interest in this reaction increased when Kraus and co-workers published a mild and selective deoxygenation of such hemiacetals by treatment with Et_3SiH and $BF_3 \cdot OEt_2$ in CH_2Cl_2 at –78° [125]. Thus, Sinaÿ and co-workers transformed the gluconolactone **79** selectively into the β-D-glucopyranosylacetylene **80** [126] (Scheme 5.17). Glucono- and galactono-1,5-lactones yield β-D-pyranosylacetylenes exclusively and mannonolactones predominantly (α/β ca. 2:5) [92, 127, 128]. This route to β-D-pyranosylacetylenes thus complements that based on reactions between acetylenes and glycopyranosyl acetates, which exclusively affords α-D-pyranosylacetylenes (see Section 5.3.1.4). The reduction of a furanose-derived hemiacetal with Et_3SiH and $BF_3 \cdot OEt_2$ predominantly resulted in a 1,2-*trans* glycofuranosylacetylene (*trans*/*cis* 4:1 [129]). Addition of RC≡C−Li to a pentono-1,4-lactam followed by $NaBH_4$ reduction provided a ca. 3:1 mixture of the corresponding 6-aminated 3-hydroxy-hept-1-ynitols [130].

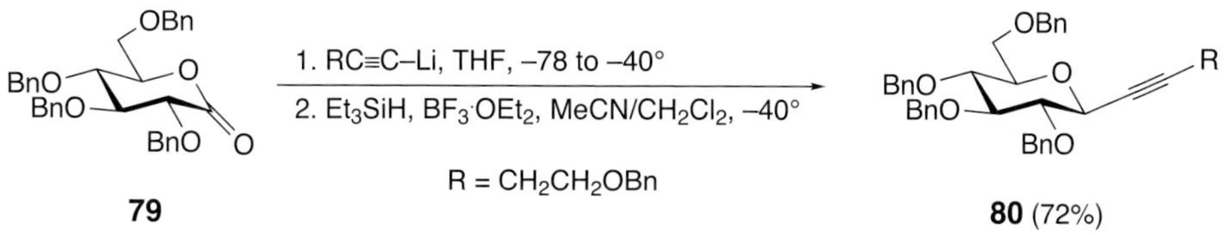

Scheme 5.17 Transformation of the lactone **79** into the β-D-glucopyranosylacetylene **80** by addition and Et_3SiH reduction.

We prepared the β-D-glucopyranosylacetylene and -buta-1,3-diyne moieties of **81** [131] and **82** [132] by addition of silylated acetylene and buta-1,3-diyne to a glucono-1,5-lactone and subsequent reduction with Et_3SiH (Scheme 5.18). A Bergman–Masamune–Sondheimer rearrangement of the di(glucosylacetylene) **81** gave 3,4-di-(β-D-glucopyranosyl)naphthalenes; the intermediate diradical regioselectively abstracted two H-atoms from the $PhCH_2O-C(2)$ groups. The 1,8-dialkynylated anthraquinone **82** (n = 6), bearing two parallel cellooligoside chains, is a model for cellulose I. The rigid C≡C and C≡C–C≡C linkers fix the chains in a parallel orientation and also mimic the phase shift between the chains. Indeed, the CP-MAS solid-state ^{13}C NMR spectrum of **82** (n = 6) resembles that of native cellulose (mixture of celluloses I_α and I_β) [133].

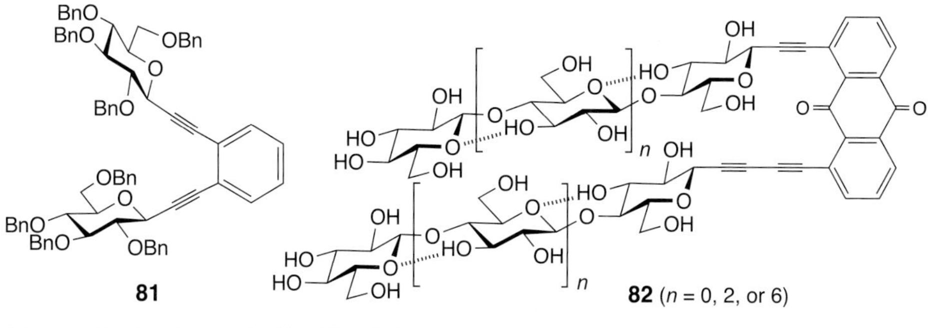

Scheme 5.18 Acetylenosaccharides **81** and **82**.

5.3.1.7 From Non-Carbohydrate Precursors

Raphael transformed the pentenynol **83a** into a mixture of DL- and *meso*-pent-4-yne-1,2,3-triols by oxidation with performic acid [2] (Scheme 5.19). Independently of one another, we [134] and Oehlschlager and Czyzewska [135] were the first to demonstrate that the triple bonds in the pent-2-en-4-yn-1-ols **83a** and **83b** are not attacked under the conditions of the Sharpless epoxidation [16]. Later on, several 2,3-epoxy-pent-4-yn-1-ols were prepared by Sharpless epoxidation (see, for example, [136, 137]) and 1,2-epoxy-pent-4-yn-3-ols by Sharpless kinetic resolution [17] (see, for example, [138, 139]). Optical active pent-4-yn-1,2,3-triols were also ob-

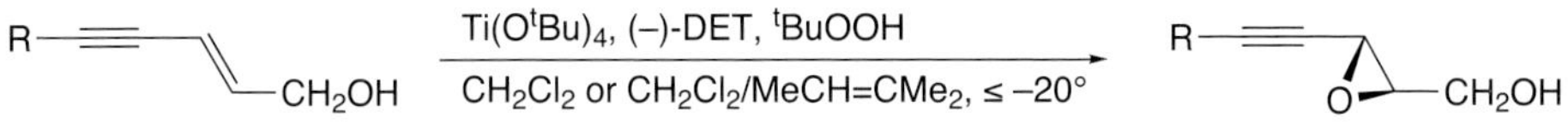

83a R = H
83b R = $C_{13}H_{27}$

84a R = H (43%)
84b R = $C_{13}H_{27}$ (72%)

Scheme 5.19 Sharpless epoxidation of the enynols **83a** and **83b**.

tained by a Sharpless asymmetric dihydroxylation [140] of pent-2-en-4-ynoates and subsequent reduction of the ester moiety [141].

Aldol reactions between propargylic aldehydes and 2-oxygenated ketene silyl *O,O*- or *O,S*-acetals with subsequent reduction of the ester or transformation of the thioester into an aldehyde, followed by a second aldol reaction, provided pent-4-yne-1,2,3-triols [142, 143].

5.3.2 Preparation of Branched-Chain Acetylenosaccharides

Introduction of an alkynyl moiety at an internal C-atom of a saccharide is, in general, more difficult than at a terminal C-atom. Alkynyl groups have been introduced at internal C-atoms in saccharides by substitution of iodides, opening of epoxides, addition to carbonyl groups, and through Wittig-type reactions of branched-chain aldehydo-saccharides.

5.3.2.1 By Substitution

Substitution of a saccharide-derived secondary triflate or halide by the poorly nucleophilic acetylide anions is hampered by the adjacent *O*-substituents, so it is not surprising that radical reactions have been used for the introduction of alkynyl groups. Upon irradiation with UV light (300 nm), the 3-iodo-allofuranose **85** reacted with TIPS−C≡C−SO_2CF_3 to afford the *gluco*-configured acetylene **86** [78] (Scheme 5.20). In radical atom-transfer reactions, the iodides **87** and **89** were transformed into 5-[(trimethylsilyl)iodomethylene]-1-sil-2-oxolanes and thence, by treatment with TBAF, into the alcohols **88** and **90**, respectively [144]. This intramolecular alkynyl transfer always results in 1,2-*cis* hydroxyacetylenes and complements the alkynylating opening of epoxides.

Opening of the D-*lyxo* epoxides **91** with LiC≡CH · $(H_2NCH_2)_2$ gave rise to the D-*arabino* hydroxyacetylenes **92** in 66–74% yields [145–147] (Scheme 5.21). Regioselective addition of acetylides to 2,3- and 3,4-epoxy-ribopyranosides [148–151] and to an epoxy-*epi*-inositol [152] have also been described. The homopropargyl alcohol **94** was obtained in moderate yields from the spiroepoxide **93** [153]. Glucopyranosides ethynylated at C(2) and C(4) were best prepared by substitution of 1,6:2,3-dianhydro-β-D-mannopyranoses and 1,6:3,4-dianhydro-β-D-galactopyranoses, respectively, as illustrated by the transformation of **95** into **96** [154] (see also Section 5.4.2 for a route to diethynylated monosaccharides).

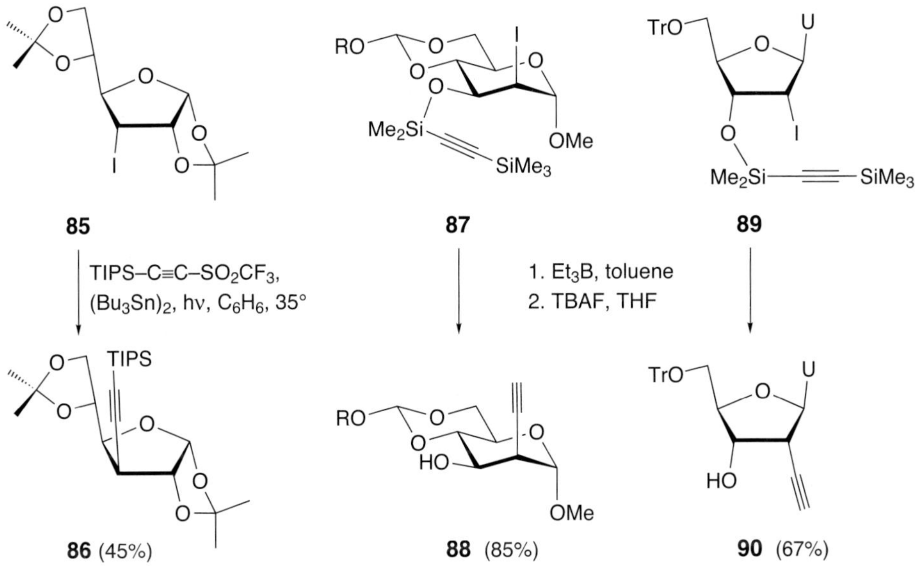

Scheme 5.20 Homolytic substitution of the iodides **85**, **87**, and **89**, affording branched-chain acetylenosaccharides.

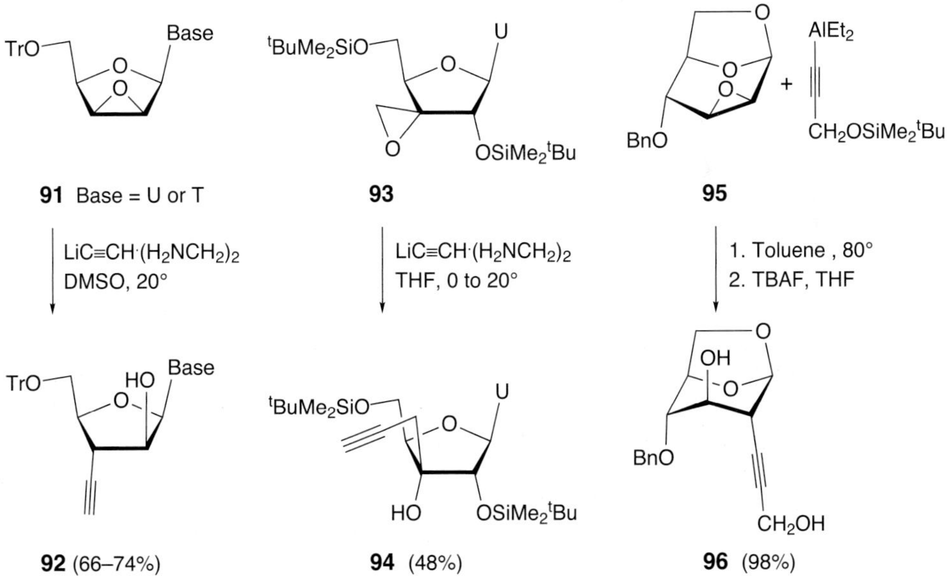

Scheme 5.21 Branched-chain acetylenosaccharides from epoxides.

5.3.2.2 By Addition to C=O Groups

About 50 publications dealing with the addition of acetylenes to aldosulopyranosides and -furanosides have been published since the first reports by Overend and co-workers in 1965 and 1970 [155]. Mixtures of epimeric propargylic alcohols are expected, but the (mostly steric) influence of adjacent substituents may give diastereoselective addition. Thus, *exo*-attack is strongly preferred in the addition to 1,2-*O*-isopropylidene-ald-3-ulofuranoses [156–161] (such as **97**; Scheme 5.22) and to 1,2:4,5-di-*O*-isopropylidene-β-D-*erythro*-hex-2,3-diulopyranose [162, 163]. Selective deuteriation and modification of **98** followed by cleavage or degradation of the glucosyl moiety has allowed optically active glycine, 3-alkylmalic acids, and monodeuteriated glycerols to be prepared [164]. Potassium ascorbate reacted with propargyl bromide to provide the 2-C-propargyl derivative [165].

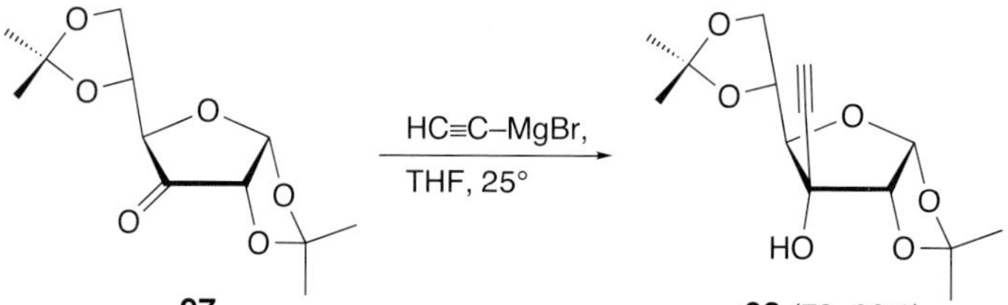

97 **98** (79–86%)

Scheme 5.22 Diastereoselective *exo*-addition of HC≡C−MgBr to the hexulose **97**.

Ethynylated nucleosides have gained ample interest as antiviral, antitumor, and anticancer agents (see Section 5.6), and much effort has been devoted to their selective preparation. Addition of $Me_3SiC{\equiv}C{-}Li$ to the fully protected pent-3-ulofuranosyl nucleosides **99a** gave a 3:7 to 4:6 mixture of the D-*ribo*- and D-*xylo*-configured propargylic alcohols **100a** and **100b** [166] (Scheme 5.23). High yields of the D-*xylo* alcohols **100b** and **101b** were obtained through the use of $Me_3SiC{\equiv}C{-}CeCl_2$ [166] and HC≡C−MgBr [167] as nucleophiles. As a result of neighboring group participation, however, the D-*ribo* diols **102a** were nearly exclusively obtained from the reaction between $Me_3SiC{\equiv}C{-}CeCl_2$ and the pent-3-ulofuranosyl nucleosides **99b** unprotected at HO−C(5′) [166]. This reaction is the key step of a laboratory-scale synthesis of the anticancer drug 3′-C-ethynylcytidine [168] (see [169] for its 4′-thio analogue). $Me_3SiC{\equiv}C{-}CeCl_2$ attacks fully protected 2-deoxy-D-*glycero*-pent-3-ulofuranosyl nucleosides and 1,3,5-tri-*O*-benzoyl-α-D-*erythro*-pent-2-ulofuranose exclusively from the less hindered side [170, 171].

Deoxygenation of non-anomeric propargylic alcohols with Et_3SiH and $BF_3 \cdot OEt_2$ (see Section 5.3.1.6) was sluggish, yielding less than 10% of the desired product [172]. Matsuda and co-workers found a stereoselective method for the deoxygenation of such non-anomeric propargylic alcohols [173]. The alcohol **104** was transformed into the mixed (methoxy)oxalyl ester **105** (Scheme 5.24). Treatment of this ester with Bu_3SnH and AIBN afforded the D-arabinofuranosyluracil **106** with inversion of configuration at the propargylic center. This reduction with

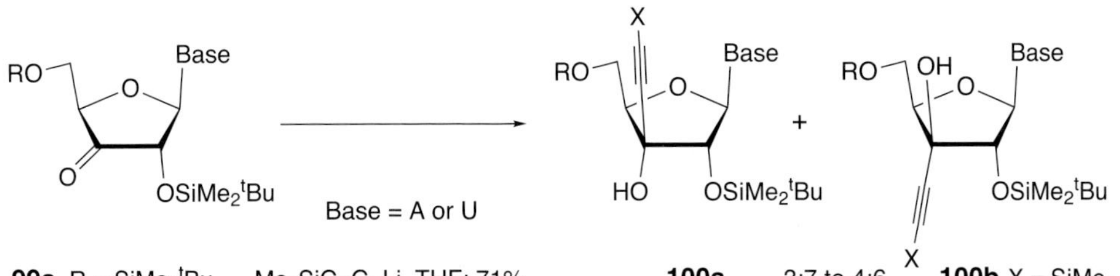

99a R = $SiMe_2{}^tBu$	$Me_3SiC{\equiv}C{-}Li$, THF; 71%	**100a**	3:7 to 4:6	**100b** X = $SiMe_3$
99a	$Me_3SiC{\equiv}C{-}CeCl_2$, THF, −78°; 81%	**100a**	≤ 5 : 95	**100b**
99a	$HC{\equiv}C{-}MgBr$, THF, 25°; 78%	**101a**	7 : 93	**101b** X = H
99b R = H	$Me_3SiC{\equiv}C{-}CeCl_2$, THF, −78°; 75%	**102a**	> 98 : 2	**102b** X = $SiMe_3$

Scheme 5.23 Dependence of the diastereoselectivity upon the nature of the nucleophile and the protection at HO−C(5′) in **99**.

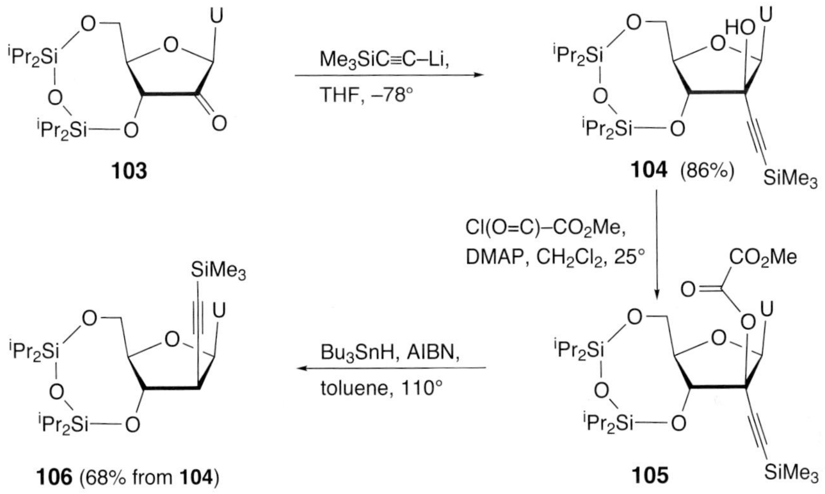

Scheme 5.24 Inverting deoxygenation of the propargylic alcohol **104**.

Bu_3SnH has been applied to other systems, and as far as we know, has always proceeded with inversion of configuration [174, 175].

5.3.2.3 By Sequential Formylation/Wittig-Type Alkynylation

Alkynyl moieties have been introduced at C(4) of furanosides by application of the Corey–Fuchs reaction to branched-chain 4-formyl-furanosides [176–178] as illustrated by the transformation of **107** into the ethynylated uridine **110** via **108** and **109** (Scheme 5.25). The Corey–Fuchs reaction has also been used for the preparation of 3′-deoxy-3′-ethynylthymidine [179], but to the best of our knowledge no example of a similar alkynylation at C(5) of pyranosides has been described.

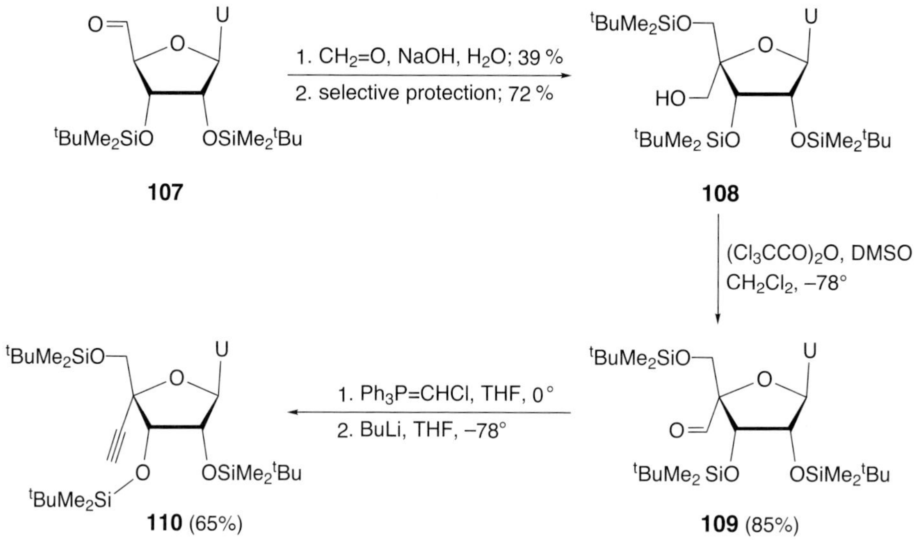

Scheme 5.25 Introduction of an ethynyl group at C(4′) in the nucleoside **107**.

The 4-deoxy-4-ethynyl-α-D-glycopyranosides **113** were prepared by addition of $Me_3SiC{\equiv}C{-}Li$ to the hex-4-ulopyranoside **111**, resulting in a 2:3 mixture of the gluco- and galactopyranosides **112a** and **112b** [172] (Scheme 5.26). Deoxygenation of this epimeric mixture gave the desired products **113a** and **113b**, though in very poor yields (< 10 % of a 1:1 mixture). Better results were obtained by application of the Corey–Fuchs reaction [172]. The required aldehyde **115** was prepared from **111** in four steps via **114** (olefination, hydroboration, oxidation, and Swern oxidation). Treatment of **115** with Et_3N shifted the position of the *gluco/galacto* equilibrium of **115** (1:2) in favor of the *gluco* isomer, and only the *gluco*-configured alkyne **113a** was observed as a product of the subsequent Corey–Fuchs reaction. Thus, **113a** was obtained from **111** in six steps and in 29 % overall yield. We have described a shorter route to the analogue **118** from the galactoside **116** [180]: **118** was obtained by substitution with CN^- to give **117**, reduction to the aldehyde, and Corey–Fuchs reaction in four steps and in 65 % overall yield. This route is still lengthy, however, if allowance for the preparation of **116** from galactose is made, though a more convenient route from levoglucosan has been described (see Section 5.4.2.1).

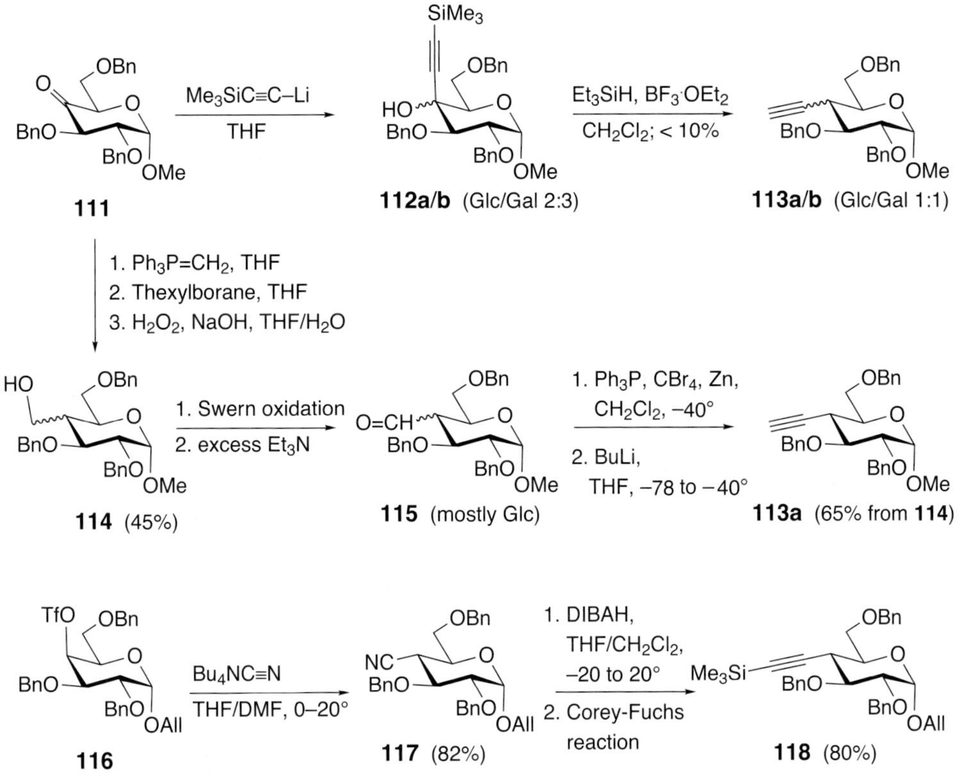

Scheme 5.26 Preparation of 4-deoxy-4-ethynyl-α-D-glucopyranosides.

5.4 Preparation of Dialkynylated Acetylenosaccharides

Dialkynylated monosaccharides can possess either a linear or a branched carbon chain. The former are obtained by formal introduction of an alkynyl moiety at both ends of the monosaccharide chain and the latter by formal addition of at least one alkynyl group to a non-terminal C-atom.

5.4.1 Linear Dialkynylated Acetylenosaccharides

Application of the alkynylating ring-opening of epoxides has been restricted to C_2-symmetric diepoxides: namely the 1,2:5,6-dianhydro-D-mannitol **119** [181–184] (Scheme 5.27) and 1,2:4,5-dianhydro-3-*O*-benzyl-D-arabinitol [185]. This avoids the problem of regioselectivity. The mannitol derivative **120** is an easily accessible, enantiomerically pure starting material for the synthesis of leukotrienes; cleavage of the C(3)–C(4) bond of the corresponding 2,5-di-*O*-protected diol with $NaIO_4$

yielded two equivalents of an ethynylated aldehyde. Treatment of **119** with the hexa-1,5-diyne **121** derived from diethyl L-tartrate (reduction and Corey–Fuchs reaction) gave the C_2-symmetric cyclododeca-1,5-diyne **122**, albeit in only 30% yield [184].

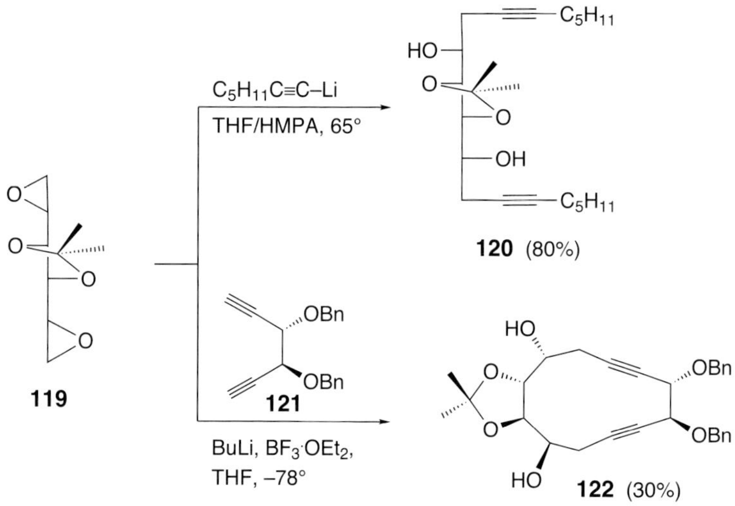

Scheme 5.27 Diethynylated saccharides **120** and **122** from the C_2-symmetric diepoxide **119**.

5.4.2
Branched Dialkynylated Acetylenosaccharides

Gossauer and co-workers described the preparation of the 3′,5′-diethynylated nucleoside analogues **124a** and **124b** (Scheme 5.28) in 1995 [186]. The known 3′-ethynyluridine **92** (Scheme 5.21) was transformed into the aldehydes **123a** and **123b** by standard reactions, and thence, by a Corey–Fuchs reaction, to the 3′,5′-diethynylated nucleosides **124a** (40% overall yield) and **124b** (12% overall yield). Although the oligomerization of **124b** has been announced, no publication describing such a reaction has yet appeared.

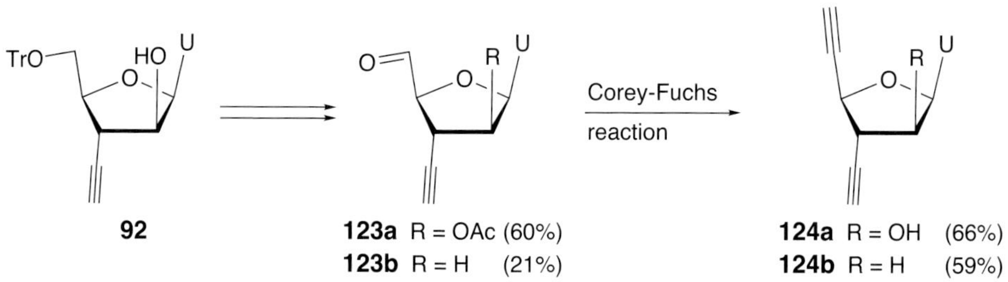

Scheme 5.28 Preparation of the diethynylated nucleoside analogues **124a** and **124b**.

Together with our 4-*O*-alkynyl-β-D- and -α-D-glycopyranosylacetylenes (see Sections 5.4.2.1 and 5.4.2.2; for a review see [187]) and an announcement of the synthesis of 1,4-ethynylated oligomeric glucopyranosyl derivatives [188] published only in a Ph. D. thesis [189], **124a** and **124b** are the only branched-chain dialkynylated acetylenosaccharides described so far.

5.4.2.1 4-O-Alkynyl-β-D-glucopyranosylacetylenes

Intraresidue, intrachain-interresidue, interchain, and even intersheet hydrogen bonds are found in celluloses (see [132, 190] and quoted references). To prepare an oligomeric cellulose model devoid of intrachain-interresidue H-bonds, we replaced O−C(4) by a longer and rigid linker to prevent the formation of the (strong) interresidue H-bond between HO−C(3) and O−C(5) of the adjacent glucosyl unit. The buta-1,3-diynediyl linker has the advantage over the ethynediyl linker of allowing the preparation of oligomers by heterocoupling of 4-ethynyl-β-D-glucopyranosylacetylenes; a binomial synthesis indeed proved to be particularly effective.

Preparation of the diethynylated monomer from levoglucosan appeared convenient [154]. In a first approach, addition of $Me_3SiC{\equiv}C{-}Li$ to the easily accessible 1,6:3,4-dianhydro-β-D-galactopyranose **125** (three steps from levoglucosan) gave the 4-ethynylated 1,6-anhydro-β-D-glucopyranose **126** (69 %), which was transformed into the gluconolactone **129** (70 % overall yield) in five steps (via **127** and **128**) [180] (Scheme 5.29). Under standard conditions ($Me_3SiC{\equiv}C{-}Li$, $TiCl_4$, THF, −78° and Et_3SiH, $BF_3 \cdot OEt_2$, $MeCN/CH_2Cl_2$), **129** yielded 86 % of the desired monomer **130**.

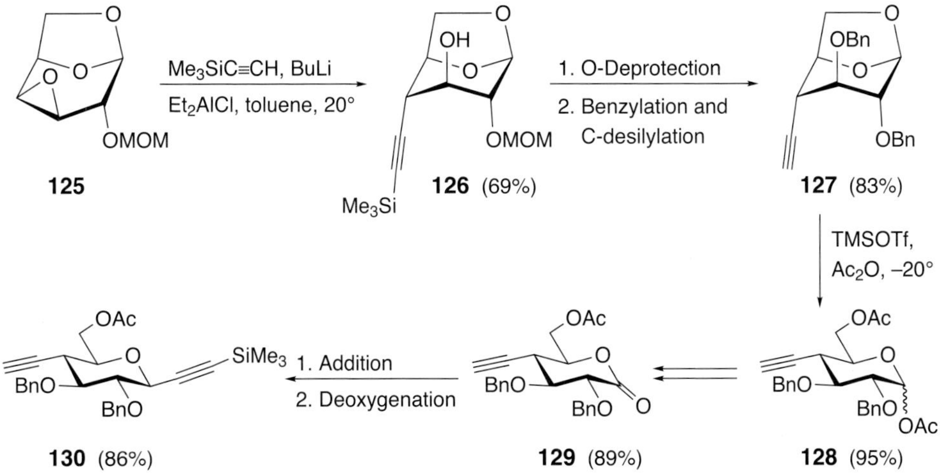

Scheme 5.29 Preparation of the 4-ethynyl-β-D-glucopyranosylacetylene **130**.

The route to **130** was shortened by combining the inverting alkynylation at C(4) with a retaining alkynylation at the anomeric center. Treatment of **131** with an eightfold excess of $Me_3SiC{\equiv}C{-}Li$ and Et_2AlCl in toluene/THF at 90° gave an 81% yield of the disilylated diacetylene **132a** [180] (Scheme 5.30). The orthogonally protected analogue **132b** was similarly prepared in two steps and 53% overall yield from **131**, via **133** [191]. The exclusive formation of the β-D-glucopyranosylacetylene was explained by the postulated binding of alkynylated Lewis acid to HO−C(3) and O−C(6) to afford the intermediate **A**. Ring-opening to the conformationally biased oxycarbenium ion **B** and intramolecular acetylide transfer selectively affords **132a** and **132b**.

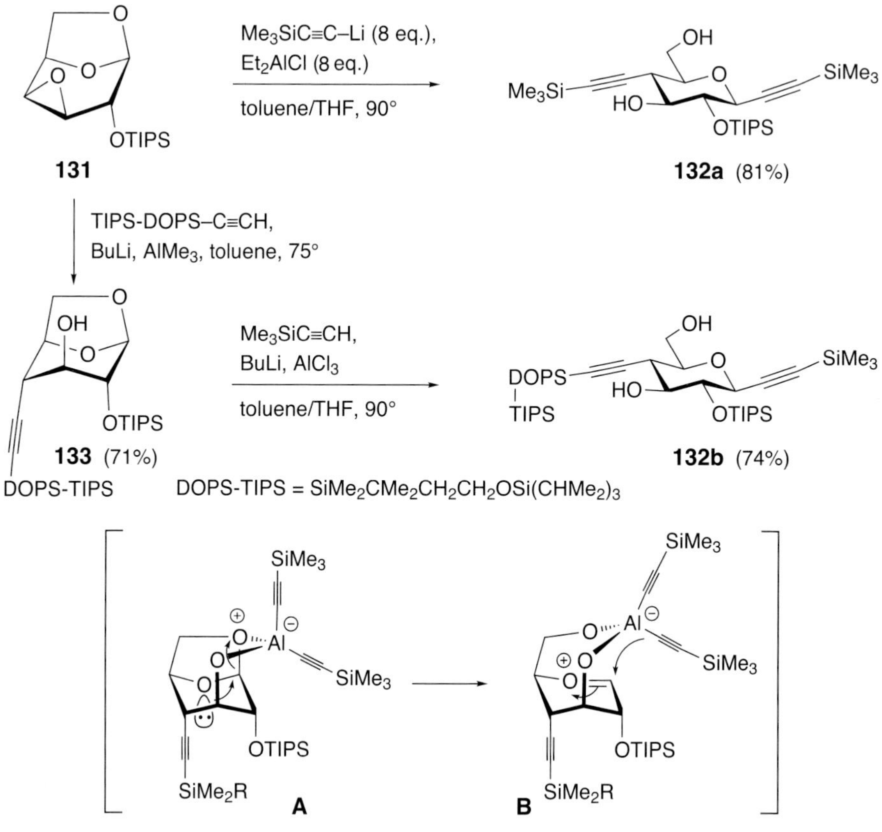

Scheme 5.30 Improved preparation of 4-ethynyl-β-D-glucopyranosylacetylenes.

Conditions for orthogonal, regioselective monodesilylation of **132a**, based on the different electrophilic and nucleophilic properties of the anomeric (propargylic) and the C(4) (homopropargylic) ethynyl groups, have been reported (Scheme 5.31). Treatment with BuLi in THF cleaved the anomeric (trimethylsilyl)ethynyl group (yielding 70% of **134a** besides 20% of the completely C-desilylated product), whereas use of $AgNO_2$ and KCN in $MeOH/H_2O$ cleaved the non-anomeric (trimethylsilyl)ethynyl group (95% of **135**) [192]. The orthogonally protected diacetylene **132b** (substituted with the newly developed DOPS group) was selectively monodesilylated to provide both **134b** (0.25 N NaOMe in MeOH; 94%) and, in two steps, **135** (0.01 N HCl in EtOH and cat. BuLi in THF at −78°; 94%) [191]. Treatment of equimolar amounts of **134a** and **135** with CuI in pyridine gave only the homocoupled dimers, while Cadiot–Chodkiewicz coupling between an *O*-protected iodoacetylene derived from **134a** and **135** and subsequent deprotection gave the heterocoupled dimer **136a** (64% of the hetero-dimer together with 20% of the homo-dimer derived from **135**) [192].

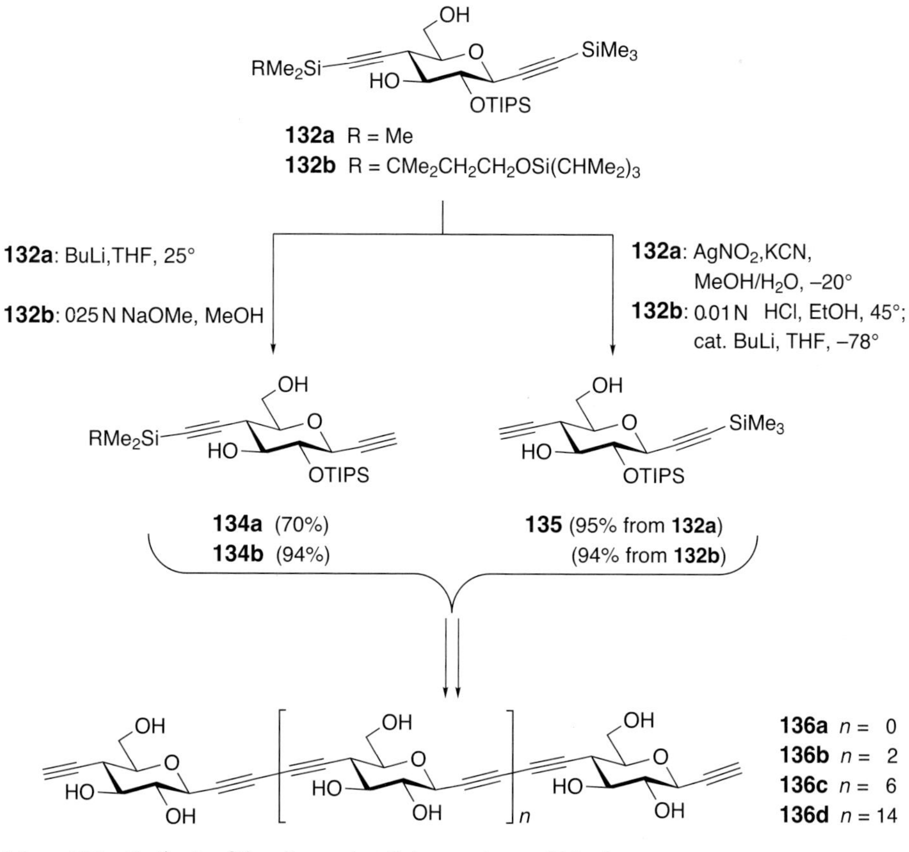

Scheme 5.31 Synthesis of the oligomeric cellulose analogues **136a–d**.

The elaboration of orthogonal protecting groups for the diacetylene moieties (compare **132b**) and improved reaction conditions allowed the preparation of the tetramer **136b**, the octamer **136c**, and the hexadecamer **136d** in a binomial synthesis [191, 193]. The desired heterooligomer was always accompanied by minor amounts of the homo-dimer derived from the ethynyl component. The reactivity of the oligomers decreased with increasing chain length, resulting in a decrease in the overall yield of a binomial cycle from 63 % for the tetramer to 59 % for the octamer and to 49 % for the hexadecamer. The hetero-dotriacontamer could not be prepared because of the poor reactivity of the hexadecameric coupling partners, but the isomeric homo-dimer of the ethynylated hexadecamer was obtained in 35 % yield. The hexadecamer **136d** has a rod-like structure with a length of 150 Å. High-resolution (transmission) electron microscopy (HREM) of **136d** showed domains of molecules in a parallel arrangement, but the resolution did not allow to assign a parallel or antiparallel orientation of the chains, or a phase shift.

The repeating unit of celluloses is the cellobiosyl unit. In a second model series, we therefore prepared the oligomers **140b–d**, in which the cellobiosyl units are separated by buta-1,3-diynediyl moieties [194] (Scheme 5.32). The monomeric dialkyne **139**, characterized by ethynyl groups in adjacent glucosyl units, was conveniently prepared by glycosidation of two suitably monoethynylated glucosides. The glycosyl donor **137** was obtained by addition of acetylene to 1,6:3,4-dianhydro-β-D-galactopyranose, followed by transformation of the product into the β-D-thioglucopyranoside, while the glycosyl acceptor **138** was prepared from an appropriately protected glucono-1,5-lactone or by an ethynylating ring-opening of a levoglucosan derivative. A binomial synthesis provided the dimer **140b**, the tetramer **140c**, and

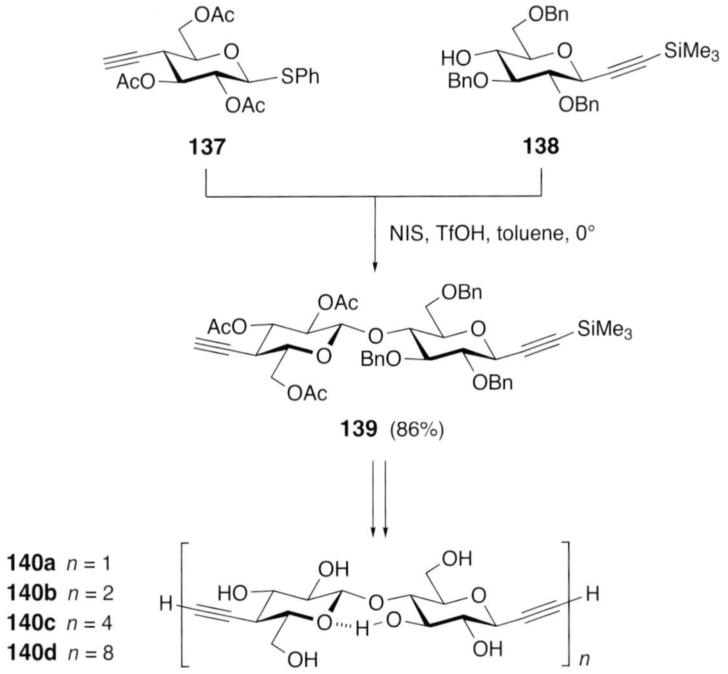

Scheme 5.32 Preparation of the diethynylated cellobiose **139** for the synthesis of the oligomeric cellulose analogues **140b–d**.

the octamer **140d**. Force-field calculations suggest that **140d** is not a good mimic for celluloses, as it tends to adopt a linear, corrugated, or a helical conformation with six to seven monomers per turn.

5.4.2.2 4-O-Alkynyl-α-D-glucopyranosylacetylenes

4-Ethynyl-α-D-glucopyranosylacetylenes were used for the preparation of cyclodextrin analogues free of inter-residue hydrogen bonds [195]. A monomeric unit was prepared from the 4-ethynylated levoglucosan **141** which was transformed into an 1:2 α/β mixture of the glucopyranosyl chlorides **142** [196] (Scheme 5.33). Under standard conditions (see Section 5.3.1.4), **142** gave the desired diacetylene **143** in moderate yields (31 %) together with 29 % of the α-D-glucopyranosylbenzene derived from **142** by formation of a C–C bond between C(1) and an *ortho*-C-atom of BnO–C(2) (compare with the intramolecular Friedel–Crafts alkylation of 2-*O*-benzylated glycosides described by Martin et al. [197]). An inverting alkynylating ring-opening of levoglucosans gave better results. Upon Lewis-acid promoted ring-opening of **145**, obtained from **141** via **144**, the C(2) *O*-silyl substituent was transferred to the anomeric center with inversion of configuration. *O*-Desilylation yielded 87 % of the triol **146**.

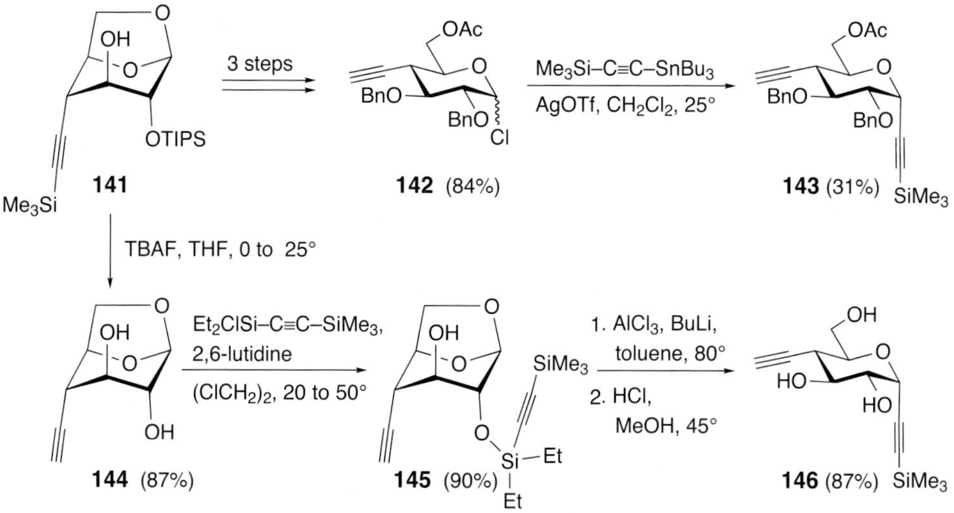

Scheme 5.33 Preparation of 4-ethynyl-α-D-glucopyranosylacetylenes.

O-Protection and orthogonal transformations of the acetylene groups of **146** allowed sequential Cadiot–Chodkiewicz cross-couplings and resulted in the construction of linear oligomers. Intramolecular Cadiot–Chodkiewicz coupling of the oligomers provided the C_3-symmetric trimer **147**, the C_1-symmetric trimer **148**, and several tetra-, penta-, and hexamers, such as the D_2-symmetric **149**

[195] (Scheme 5.34). These cyclic acetylenosaccharides may, like cyclodextrins, act as hosts for small molecules, as illustrated by the formation of a complex between **149** and D- or L-adenosine.

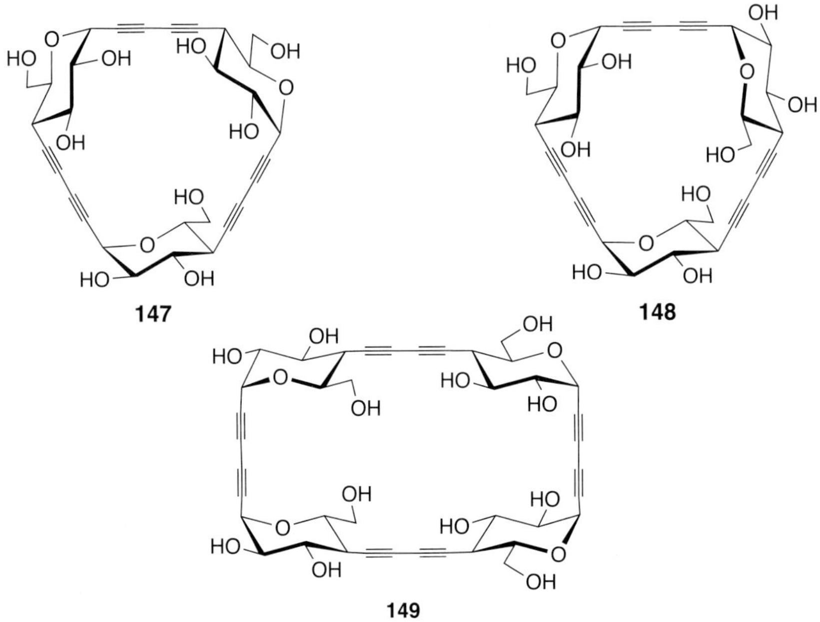

Scheme 5.34 Buta-1,3-diynylated analogues of cyclodextrins.

Replacement of single glycosidic O atoms in cyclodextrins by a buta-1,3-diynediyl group affords cyclodextrin analogues in which the cooperative HO−C(2) and HO−C(3) flip-flop H-bond network is interrupted, as evidenced by downfield shifts of the two HO groups adjacent to the buta-1,3-diynediyl group in DMSO-d_6 [198, 199]. The γ-cyclodextrin analogue **150** (Scheme 5.35) was prepared by an improved selective acetolysis of one glycosidic bond of α-cyclodextrin (compare [200]), two glycosidation reactions of the intermediate hexaoside (first acting as glycosyl acceptor to a 4-ethynylglucoside and then as glycosyl donor to an α-D-glucopyranosylacetylene), and a ring-closure of the resulting diacetylene with $Cu(OAc)_2$ in MeCN/pyridine at 80° (50 % yield in the ring-closing step; Scheme 5.35). The glycosidation gave access to isomers of **150** possessing one or two cellobiosyl units. The β-cyclodextrin analogue **151** was obtained by selective monoacetolysis of α-cyclodextrin, followed by introduction of an *O*-propargyl moiety, glucosidation with a 4-ethynyl-thioglucoside, and ring-closure. Hexa-2,4-diyne-1,6-dioxy analogues were also prepared by introduction of two propargyl moieties followed by intramolecular coupling.

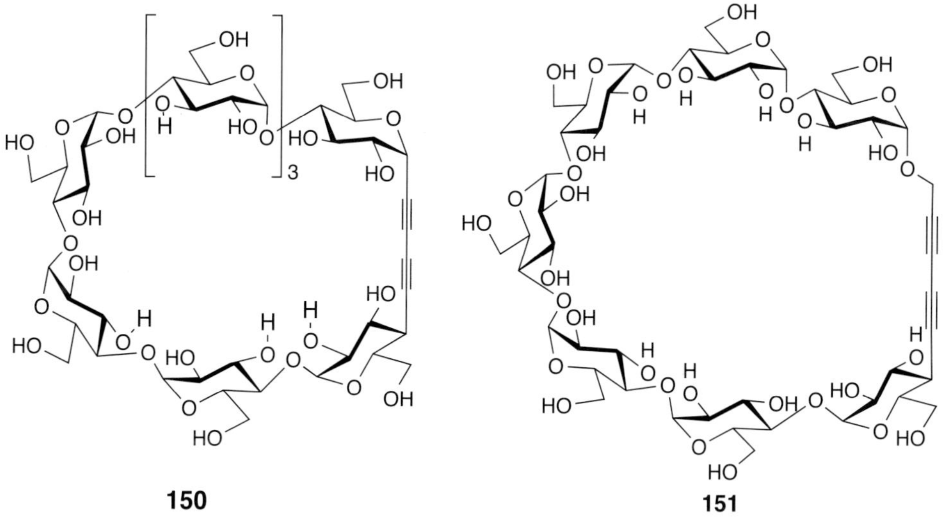

Scheme 5.35 Monobuta-1,3-diynylated analogues of γ- and β-cyclodextrin.

5.4.2.3 4-O-Alkynyl-α/β-D-hexopyranosylacetylenes derived from Mannose and from GlcNAc

1,4-Diethynylated GlcNAc derivatives could not be prepared from levoglucosan, as 1,6-anhydro-2-azido-4-deoxy-4-ethynyl-β-D-glucopyranose and the corresponding amine did not react to afford the expected pyranosylacetylenes under a variety of conditions [201]. These GlcNAc derivatives were therefore prepared from 4-ethynyl-D-mannopyranosylacetylenes. Treatment of the 1,6:3,4-dianhydro-β-D-talopyranose **152** with 3 equivalents of $Me_3Si{-}C{\equiv}C{-}Li/AlMe_3$ at 65° gave the *manno*-configured acetylene **153** which was *O*-silylated to provide **154** (Scheme 5.36). With 10 equivalents of the same reagent and at 80°, **154** was transformed into the α-D-mannopyranosylacetylene **155**. *C*-Desilylation to **156**, followed by treatment with $Cu(OAc)_2$ in pyridine, selectively gave the C_2-symmetric dimer **157** (77 %), which surprisingly cyclized to the strained cyclodimer **158** at 100° (71 %; structure established by X-ray analysis); no trace of the expected tetramer was found.

The 1,4-diethynylated GlcNAc derivative **161** was obtained from the disilyl ether **156** by complete desilylation to afford **159**, selective silylation of HO−C(6) and the equatorial HO−C(3), triflation of HO−C(2), substitution by azide, and reduction of the resulting azide **160** followed by acetylation (Scheme 5.37). Upon treatment with $Cu(OAc)_2$ in pyridine, **161** reacted to give the corresponding 1,4-dipyranosylbuta-1,3-diyne selectively (83 %), the C(4)-ethynyl group of **161** presumably being protected by the bulky TIPSO−C(3) substituents. This hypothesis is in keeping with the observation that oxidation of the alcohol **162** provided the C_1-symmetric trimer **163**, albeit in only 33 % yield.

We expected that $Me_3Si{-}C{\equiv}C{-}Li/AlMe_3$ should react with HO−C(3) of the mannoside **153** to afford β-D-mannopyranosylacetylenes with retention of confi-

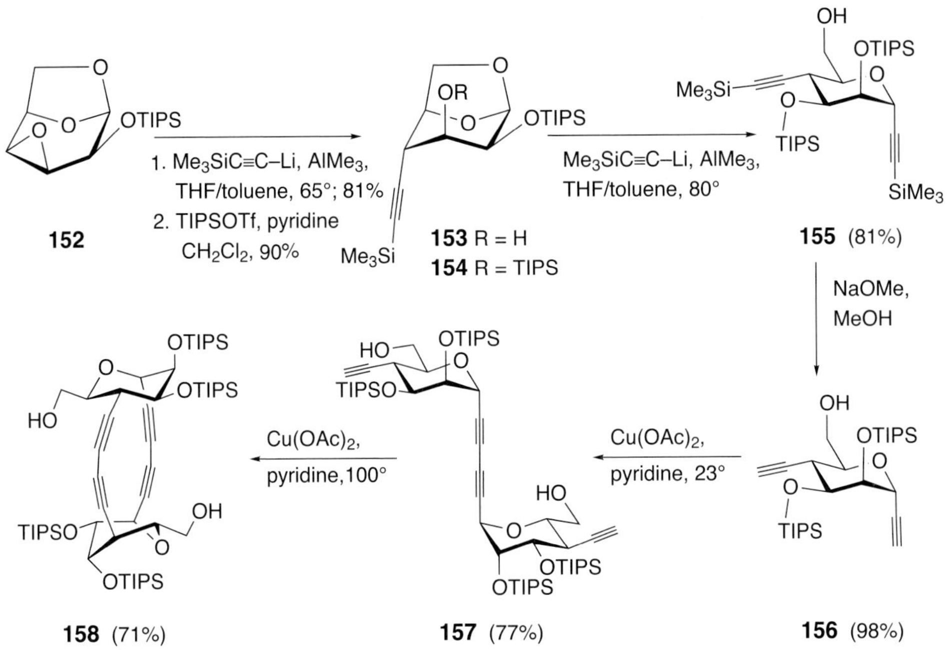

Scheme 5.36 Preparation of the 4-ethynyl-α-D-mannopyranosylacetylene **156** and its transformation into the dimer **158**.

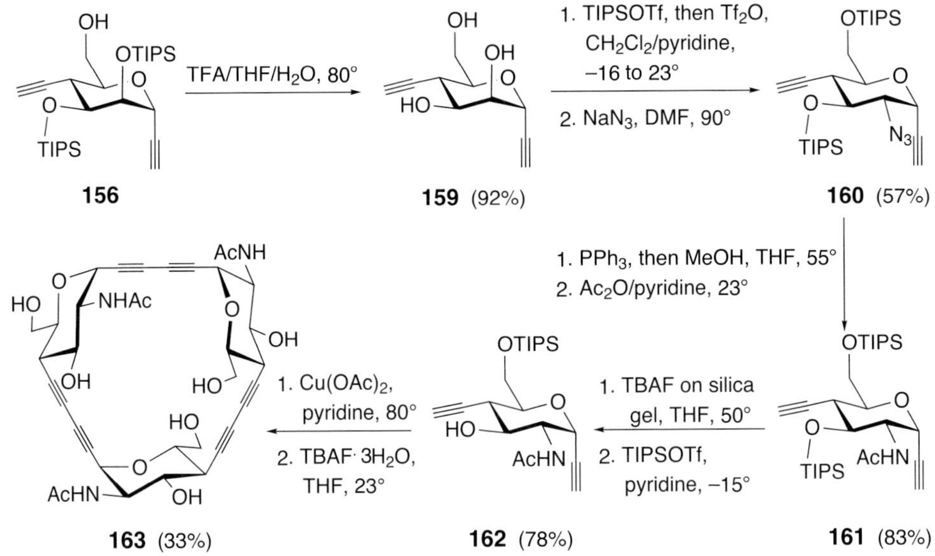

Scheme 5.37 Preparation of the GlcNAc-derived 4-ethynyl-α-D-pyranosylacetylene **161** and its transformation into the trimer **163**.

guration at the anomeric center (Scheme 5.38). The desired diacetylene **165** was indeed obtained (65 %), together with the enyne **164** (17 %). Under the reaction conditions, TIPSO−C(2) was partially cleaved and HO−C(2) allowed or promoted the attack of $Me_3Si{-}C{\equiv}C{-}Li$ on the anomeric ethynyl group. The analogous reaction sequence as described for the α-D-anomer **156** transformed **166** into the diethynylated GlcNAc derivative **167**.

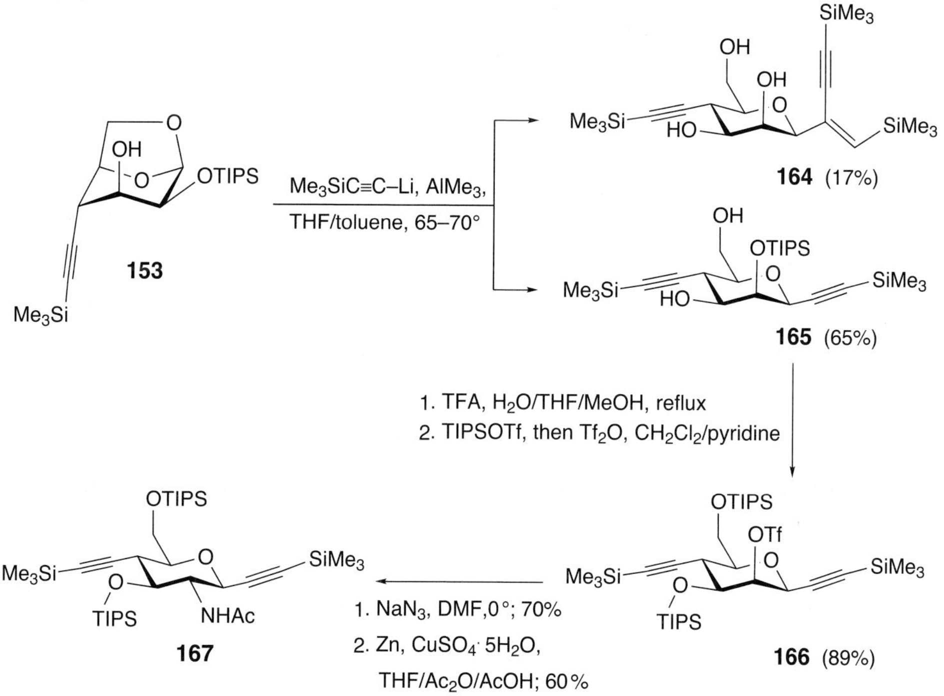

Scheme 5.38 Preparation of 4-ethynyl-β-D-glucopyranosylacetylenes of mannose and GlcNAc.

5.5 Transformations of Acetylenosaccharides

Much of the value of alkynes other than that deriving from their structural properties results from their ready transformation, by selective reduction to *cis*- and *trans*-alkenes, for example, or by addition of HX (X = halogen, OR, SiR_3, SnR_3, SR, and others), cycloaddition, radical cyclization, addition to electrophiles, and cross-coupling (see [202] for a review). This section focuses on ring-forming and coupling reactions, while simple transformations of acetylenosaccharides are not discussed.

5.5.1
Ring-Forming Reactions

5.5.1.1 Cycloadditions and Cyclocondensations

Intermolecular cycloadditions and cyclocondensations of acetylenosaccharides have been widely used, especially for the preparation of C-nucleosides. Thus, 1,3-dipolar cycloadditions to azides, diazoalkanes, azomethine imines, and nitrile oxides have provided triazoles, pyrazoles, pyrazolines, and isoxazoles, respectively [93, 96, 203–210] (for a recent monograph on 1,3-dipolar cycloadditions, see [211]). Cyclocondensations of propiolaldehydes and 2-ynones with hydrazines or hydroxylamine have given pyrazoles and isoxazoles [212–215]. The propiolate **168** reacted with guanidine to give the *C*-ribofuranosylpyrimidine **169** [96] (Scheme 5.39). In the context of the elucidation of the mechanism of β-glucosidases, the triazoles **171a**, **171b**, and **173** were prepared through intramolecular 1,3-dipolar cycloaddition of the corresponding azidoacetylenes (primary products from **170** and **172**, respectively) [216, 217].

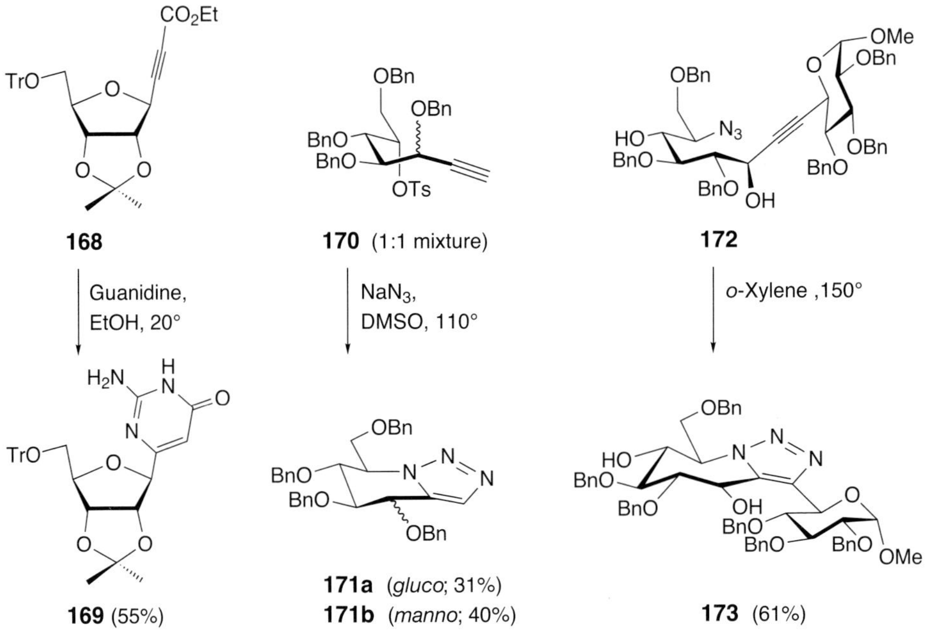

Scheme 5.39 Preparation of **169** by cyclocondenstaion and of **171** and **173** by intramolecular 1,3-dipolar cycloaddition.

An inverse electron demand Diels–Alder cycloaddition between 3,6-bis(trifluoromethyl)-1,2,4,5-tetrazine and the β-D-ribofuranosylacetylene **174**, with subsequent cycloreversion and loss of N_2, gave a 77% yield of the pyridazine **175** [218] (Scheme 5.40). The cyclohexadiene **177** was obtained from **176** through an intramolecular Diels–Alder addition [219], whereas the dihydro-oxepin **179** was prepared in low yields from **178** by means of a [2+2] cycloaddition followed by ring-expansion [220]. Rh-catalyzed cyclotrimerization of the propargyl 1-C-ethynyl-α/β-D-glucopyranosides **180** with acetylene gave an 89% yield of the spiroacetals **181** [221].

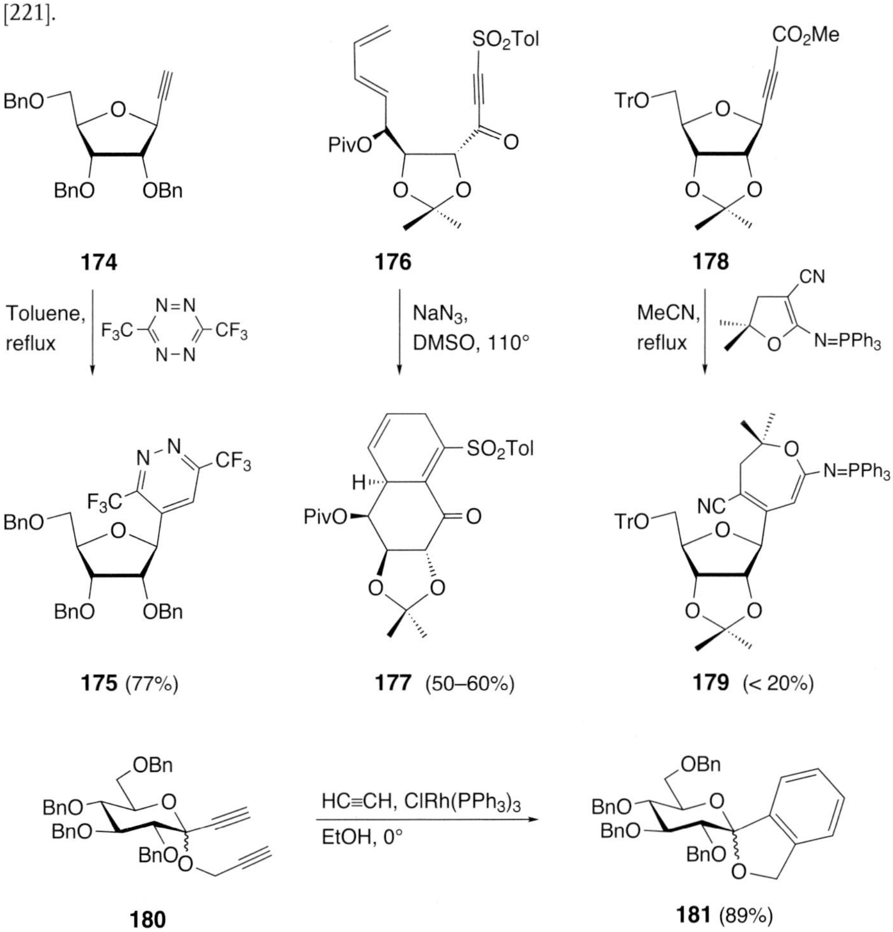

Scheme 5.40 Preparation of **175** and **177** by Diels–Alder cycloaddition, of **179** by [2+2] cycloaddition followed by ring-expansion, and of **181** by cyclotrimerization.

5.5.1.2 Radical Cyclizations

Radical cyclizations of acetylenosaccharides have been well investigated. Radicals generated by the treatment of thionocarbonates [222, 223] and iodides [215, 224] with Bu_3SnH added intramolecularly to an ethynyl group to afford alkenes, as illustrated in Scheme 5.41 by the transformations of **182** and **184** to yield **183a/b** and **185**, respectively. Alk-5-ynyl, alk-6-ynyl, and alk-7-ynyl radicals usually undergo 5-*exo-dig*, 6-*exo-dig*, and 7-*exo-dig* cyclizations, respectively. Disubstituted alkynes usually give *E*/*Z* mixtures of the exocyclic alkenes. Secondary radicals gave mixtures of epimers (e. g., **183a**/**183b** 4:6). The use of *O*-benzyl protecting groups should be avoided, as they may interact with the resulting alkenyl radicals [225]. The 6-*endo-dig* cyclization products **187** were selectively obtained from **186**, since a 5-*exo-dig* cyclization would give rise to strained *trans*-2,4-dioxabicyclo[3.3.0] octanes [226].

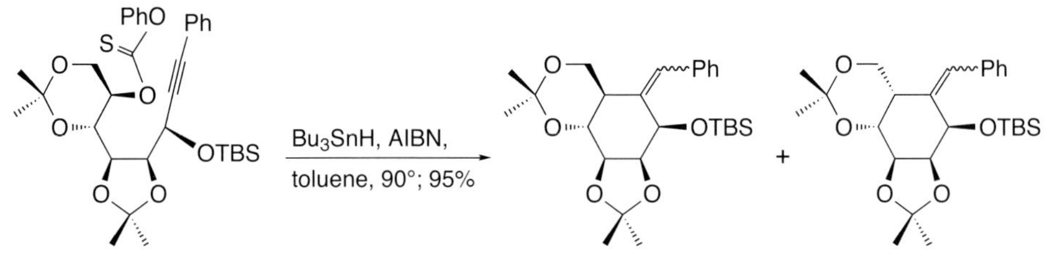

182 **183a** 4 : 6 **183b**

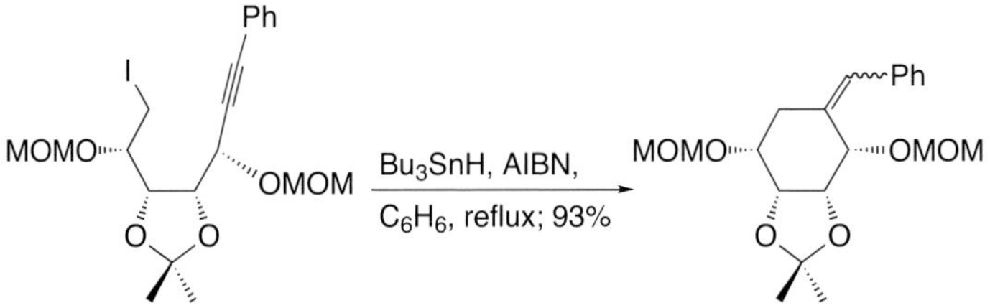

184 **185** (1:1 mixture)

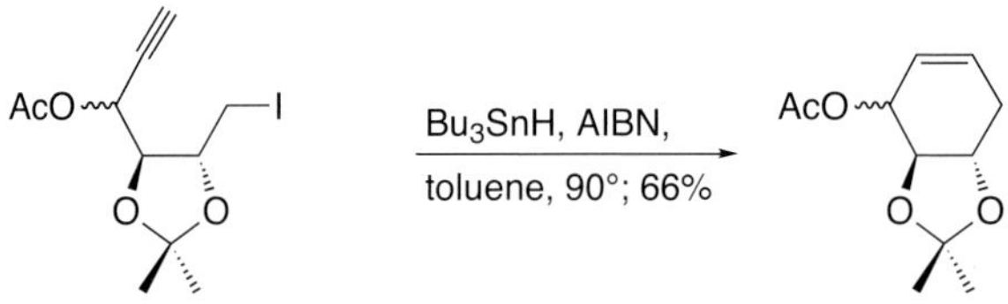

186 (D-*xylo*/L-*lyxo* ≈ 2:1) **187** (D-*xylo*/L-*lyxo* 1:1)

Scheme 5.41 Radical cyclization of acetylenic thionocarbonates and iodides.

As illustrated by the transformation of the hept-6-ynose **188** (Scheme 5.42), Bu_3SnH may react (in the absence of an iodo or phenoxythiocarbonyloxy substituent) with a terminal alkynyl group to form a 1-stannylated alk-1-en-2-yl radical. Diastereoselective 5- and 6-*exo-trig* additions of such radicals to a variety of C=X double bonds (X = O [227], NOBn [228], $NNMe_2$ [229], CH_2 [230], and $CHCO_2Et$ [99]) have been described. Thus, **188** reacted selectively to give a 5:1 *E*/*Z* mixture of the 2-(tributylstannylmethylene)cyclohexanol **189**, which was readily destannylated under acidic conditions to afford the *exo*-methylene cyclitol **190**. Similarly, a 6-arylated hex-5-ynose oxime reacted with thiophenol under photolytic conditions to provide a 3-benzyloxyamino-1-(phenylthio)cyclohex-1-ene [231].

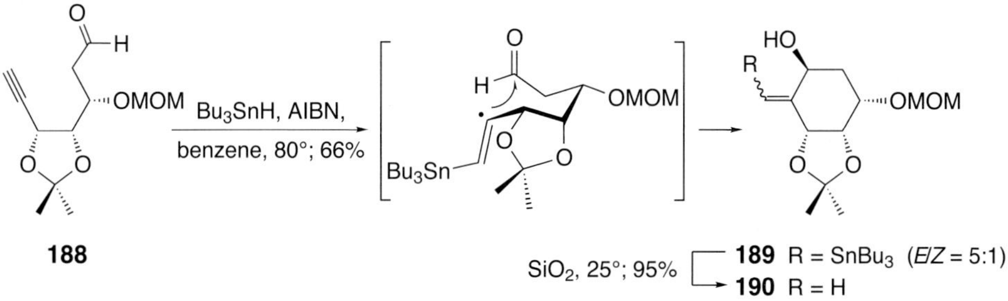

Scheme 5.42 Diastereoselective radical cyclization of the hept-6-ynose **188**.

The pent-4-ynylamine **192** was readily obtained from the bromide **191** [232] (Scheme 5.43). Its cyclization to the (+)-isofogamine precursor **193** proceeded via an intermediate α-(trimethylsilyl)methylamine radical cation.

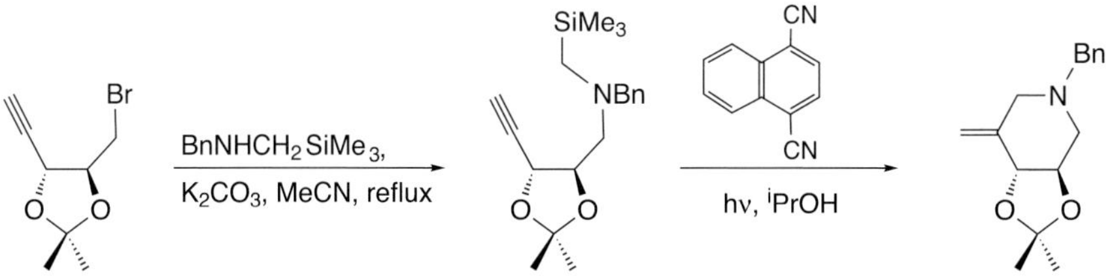

191 **192** (65%) **193** (60%)

Scheme 5.43 Radical cyclization of the pent-4-ynylamine **192**.

5.5.1.3 Transition-Metal Promoted Ring-Closing Reactions

The enyne **194** was transformed into the cyclopentanes **195a** and **195b** in a zirconocene-mediated cyclization [233], while a cobalt-mediated cyclization of the propargyl allyl ether **196** gave the dihydrofuran **197** [234] (Scheme 5.44). The octynulose **198** cyclized to the cyclopentanol **199** via an allenyl metal species [235] (see [236] for the isomerization of acetylenosaccharides into allenes). The SmI_2/Pd(0)-promoted reaction of the glucoside **200** provided the 2-ethynylcyclopentanols **201a** and **201b**; related hept-6-ynopyranosyl acetates reacted similarly, and a hex-5-ynopyranosyl acetate gave a 2-ethynylcyclobutanol, albeit in lower yield (23%) [235]. Acid was sufficient for the transformation of the 7-C-(trimethylsilyl)-oct-6-ynose **202** into the 2-ethenylidenecyclohexanol **203** in high yield (91%) [237]. 2-Metalated oxolanes (metal = Cr or W) were obtained from homopropargylic alcohols [238]. Treatment of β-D-glucopyranosylacetylenes with chromium carbene complexes gave the glucosylated α-naphthol **204** and the hydroquinone **205** [239] (see Section 5.3.1.6 for the preparation of naphthalenes by Bergman–Masamune–Sondheimer rearrangements).

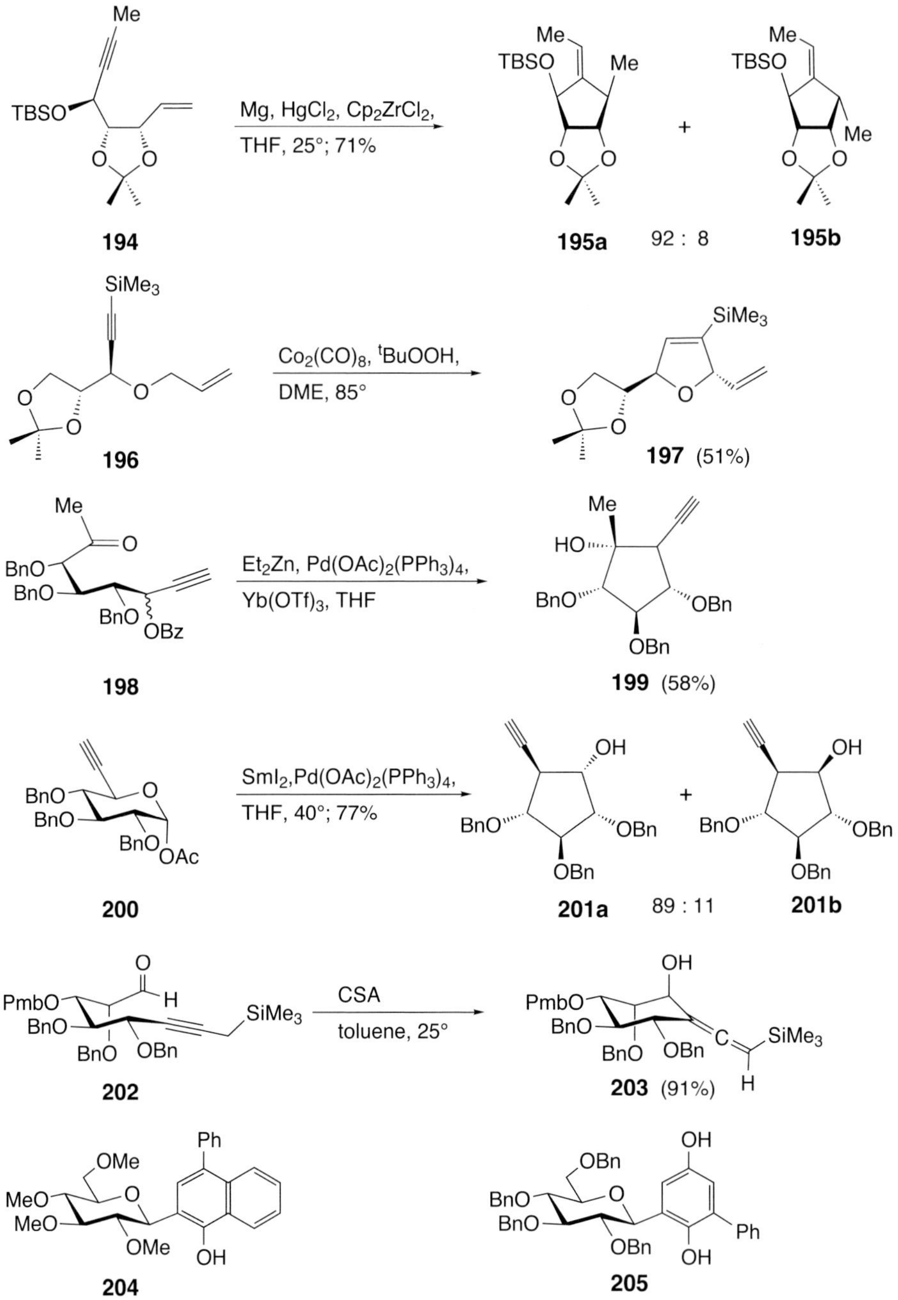

Scheme 5.44 Cycloalkanes, dihydrofurans, and phenols by transition metal- and acid-promoted cyclizations.

5.5.1.4 Enyne Metathesis

Like the ring-closing metathesis of dialkenes (see [240] for a review of RCM of carbohydrates), ring-closing enyne metathesis to afford 1,3-butadienes [241] has been applied to carbohydrates for the preparation of polyhydroxylated 1-vinylcyclohex-1-enes and 1-vinylcyclopent-1-enes, as illustrated by the transformation of **206** into **207** [242, 243] (Scheme 5.45). The propargyl allyl ether **208** was similarly transformed into the spiroacetal **209** [244].

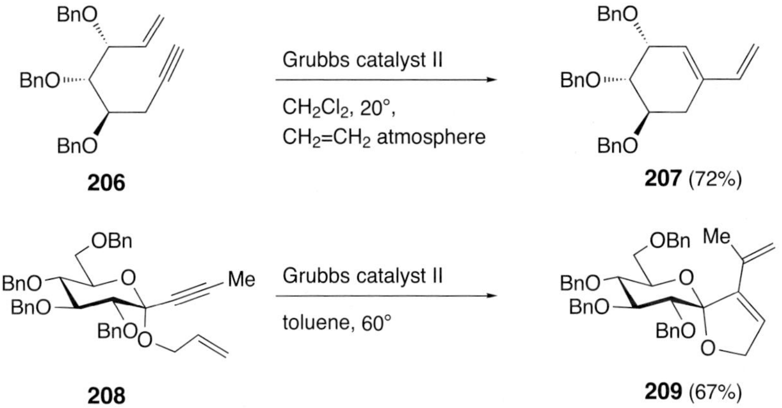

Scheme 5.45 Enyne metathesis of acetylenosaccharides.

5.5.1.5 Pauson–Khand Reaction

The Pauson–Khand reaction [245], a $Co_2(CO)_8$-promoted transformation of enynes into cyclopentenones, has been applied to acetylenosaccharides [244, 246–248]. The cyclopentenones **211** and **212** were thus obtained from the enynes **210** and **208**, respectively (Scheme 5.46). Yields depend heavily upon substrate and conditions.

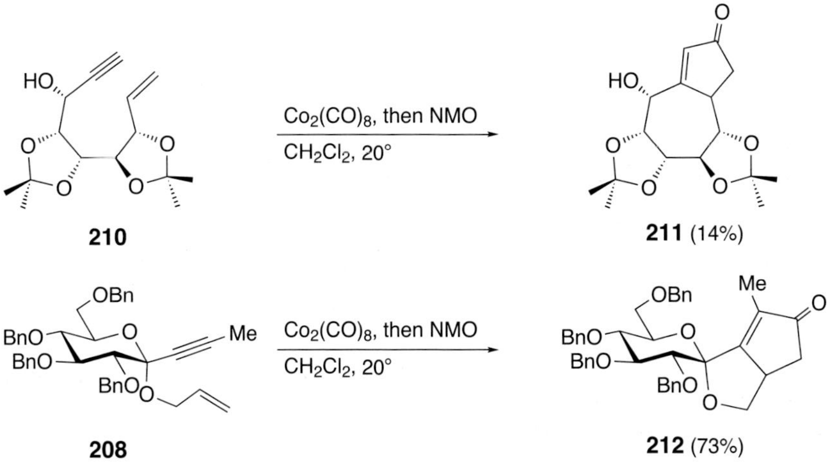

Scheme 5.46 Pauson–Khand reactions of acetylenosaccharides.

5.5.1.6 **Alkynol Cycloisomerization**

Hydroxylated oligoacetylenes undergo base-catalyzed 5-*exo-dig* cyclizations to afford 2-ethylidene-oxolanes [7, 11–13, 249], this reaction having been used for determination of the absolute configurations of oligoacetylenes. Treatment of **11** with aqueous NaOH gave the diyne-enol ether **213** (Scheme 5.47; configuration of the double bond not determined, but probably *Z* as in **214**, **216**, and **217**). The enol ether **213** was then oxidized to L-erythronolactone. The 5-*exo-dig* alkynol cycloisomerization is favored by cumulated triple bonds, by the presence of a 1-bromo, 1-chloro, or 1-(het)aryl substituent or of an additional ring in the starting material (as in **39**), and by proximity of the reacting groups [22, 199, 250, 251]. Activation of an alkynyl group resulting from coordination also promotes its reaction with nucleophiles; pent-4-yne-1,3-diol was thus transformed into 2-methylene-oxolan-3-ol by Lewis-acid catalysis (Ag_2CO_3 in refluxing benzene; 99% yield [248]).

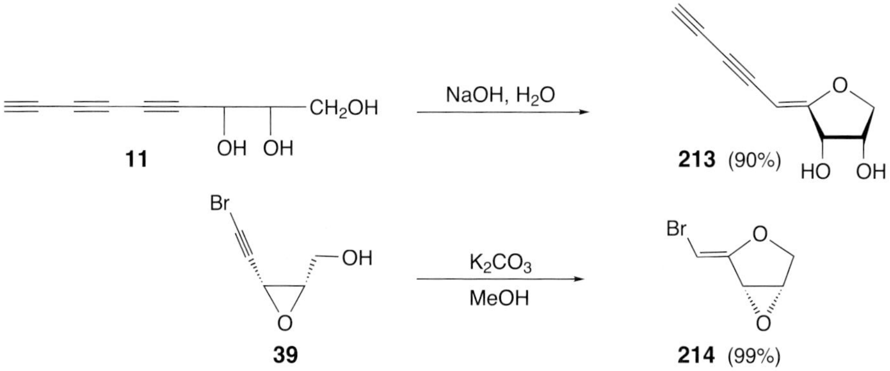

Scheme 5.47 5-*exo-dig* alkynol cycloisomerization of 1-alkynylated or 1-brominated alk-1-yn-5-ols.

The 4,4′-unprotected di(α-D-glucopyranosyl)buta-1,3-diyne **215** reacted with K_2CO_3 in $MeOCH_2CH_2OH$ to give the monoenol ether **216** selectively within 5.5 h (70%). Under the same conditions, **216** was transformed within 24 h, though in a distinctly lower yield (37%), into the (*Z*,*Z*)-configured dienol ether **217** [199, 251] (Scheme 5.48). In contradistinction to this diastereoselectivity, the *E*-configured spiroacetals (*E*)-**219** were mostly formed by base-treatment of the propargyl aldehyde **218** [252]. For a Pd-catalyzed cyclization to a closely related spiroacetal, see [253].

Alk-3-ynols and alk-4-ynols undergo formal 5-*endo-dig* and 6-*endo-dig* cyclizations, respectively, via transition metal (Mo, Cr, W, Ru) vinylidene intermediates (see [81, 254, 255] and [256] for a recent review). Thus, pent-4-yne-1,2-diol (**220**) gave a 9:1 mixture of the dihydrofuran **221** and the dihydropyran **222** (Scheme 5.49). The regioselectivity of such ring-closures is influenced by the configuration of the alkynols and more strongly by the *O*-protecting groups. Sequential cyclization and addition of alkynols to the resulting glycals allowed the preparation of oligosaccharides [254, 255].

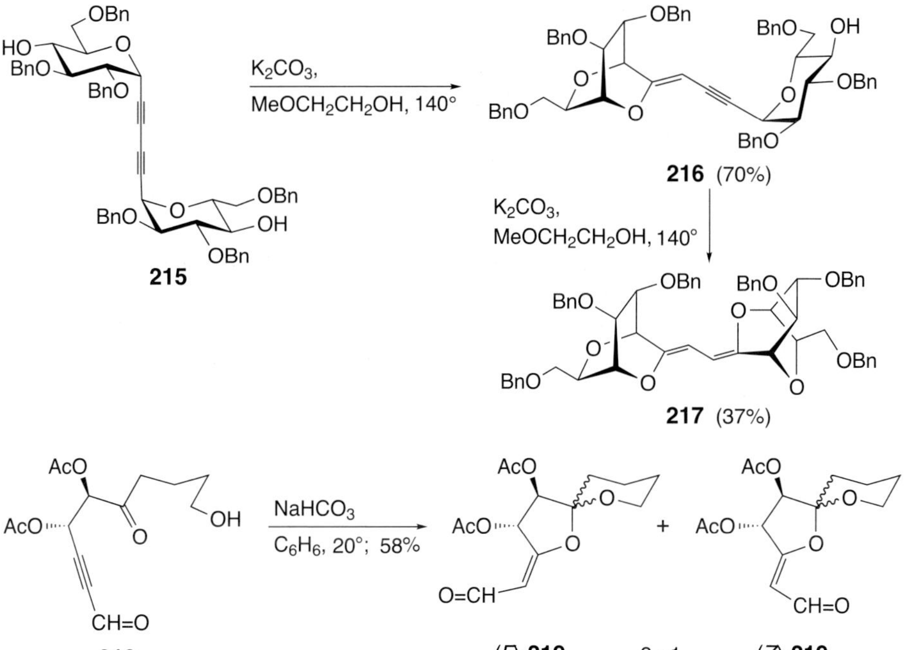

215

218

Scheme 5.48 6-*exo-dig* and 5-*exo-dig* alkynol cycloisomerization of **215**, **216**, and **218**.

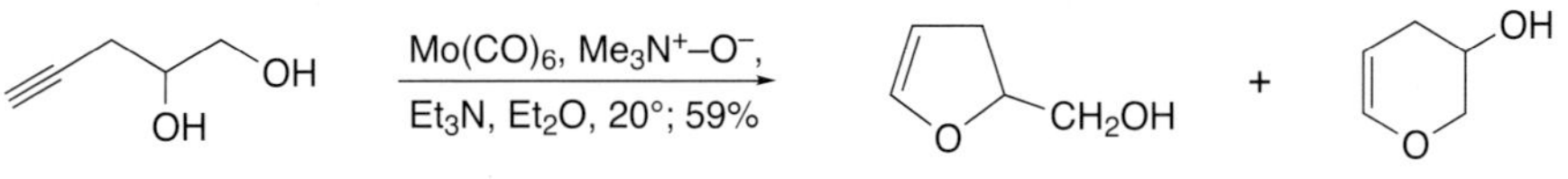

220 221 222

Scheme 5.49 4,5-Dihydrofurans and 5,6-dihydro-2*H*-pyrans from alk-1-yne-4,5-diols via transition metal vinylidene species.

5.5.1.7 Miscellaneous Cyclizations

The ring-closure of **223** on treatment with I_2 and TsOI(OH)Ph to give the 2-(diiodomethylene)oxolane **224** is a 5-*exo-dig* cyclization [159] (Scheme 5.50). The cleavage of the methyl ether of **223** was especially noteworthy. The ring-closure of the α-D-mannopyranosylacetylene **225** to afford the indole **226** [91] and the transformation of 1-(1,2-epoxycycloalkyl)propargyl alcohols into 2-(ω-oxoalkyl)furans by treatment with yellow HgO and H_2SO_4 in acetone [257] are also 5-*endo-dig* cyclizations.

The ready formation of carbocations of hexacarbonyldicobalt complexes (Nicholas reaction) allowed Isobe and co-workers to transform glycopyranosylacetylenes by ring-opening and ring-closing into medium-sized (n = 7–10) cyclic ethers or to form additional medium-sized rings [90, 258]. Thus, the pyranosylacetylene hexacarbonyl cobalt complex **227** was transformed in several steps (via **228** and **229**)

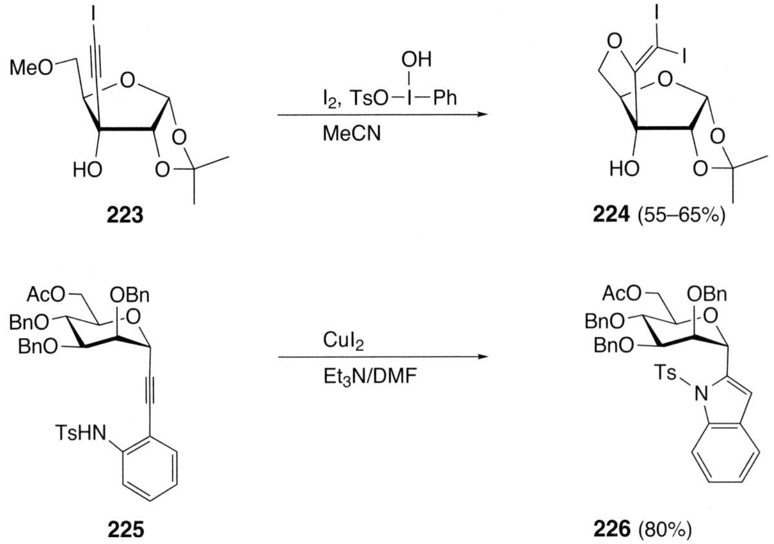

Scheme 5.50 5-*exo-dig* and 5-*endo-dig* ring-closure of **223** and **225**, respectively.

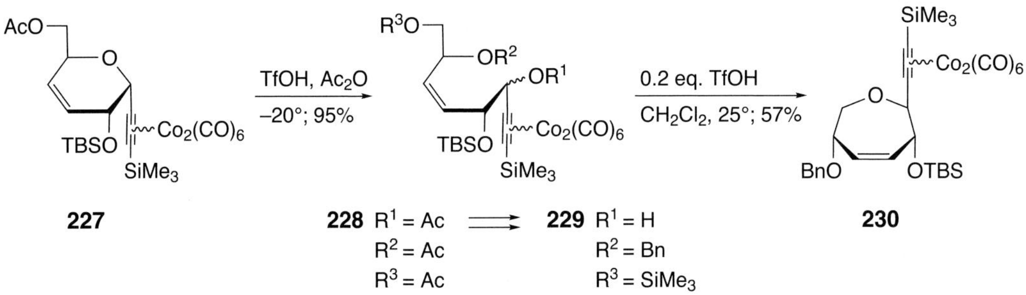

Scheme 5.51 Transformation of the pyranosylacetylene cobalt complex **227** into the septanosylacetylene cobalt complex **230**.

into the septanosylacetylene complex **230** (Scheme 5.51). The Nicholas reaction has also been used for the transformation of alk-5-yne-1,4-diols into furanosylacetylenes [259].

5.5.2
Coupling Reactions

Since Glaser's first dimerizations in 1869 [260], acetylenic coupling has become a powerful synthetic tool for the construction of C−C bonds (see [261, 262] for recent reviews of acetylenic couplings).

5.5.2.1 **Homocoupling of Acetylenosaccharides**

Glaser–Hay (CuCl, O_2, TMEDA, acetone) and Eglinton ($Cu(OAc)_2$, MeOH/pyridine) couplings of terminal acetylenosaccharides give C_2-symmetric 1.4-diglycosylated buta-1,3-diynes. Such couplings have provided access to hexadecitols [117, 263] and to branched-chain nucleoside dimers [264], while a cyclophane has been prepared by intramolecular homocoupling of two α-D-ribofuranosylacetylene moieties [265].

5.5.2.2 **Heterocoupling of Acetylenosaccharides**

Cadiot–Chodkiewicz heterocoupling of a terminal acetylene with a bromo- or iodoacetylene (typically with CuCl, $EtNH_2$, $NH_2OH \cdot HCl$ in EtOH) and Sonagashira-type reactions (typically with CuCl and $(Ph_3P)_2PdCl_2$ in Et_2NH or pyrrolidine) were used for the preparation of the tridecadialdoside **233** from **231** and **232** [266], the buta-1,3-diynyl-bridged nucleoside dimer **236** from **234** and **235** [264], and of epoxyoligoynes [22, 23] (Scheme 5.52). Our homo- and heterocouplings of diethynylated saccharides are discussed in Section 5.4.2; the best conditions for the Cadiot–Chodkiewicz cross-coupling were $Pd(dba)_2$, CuI, $P(fur)_3$, and Et_3N in DMSO [191]. The $Pd(OAc)_2$-catalyzed coupling of $PhC{\equiv}C{-}I^+{-}PhBF_4^-$ with a terminal acetylenosaccharide proceeded in high yields [267].

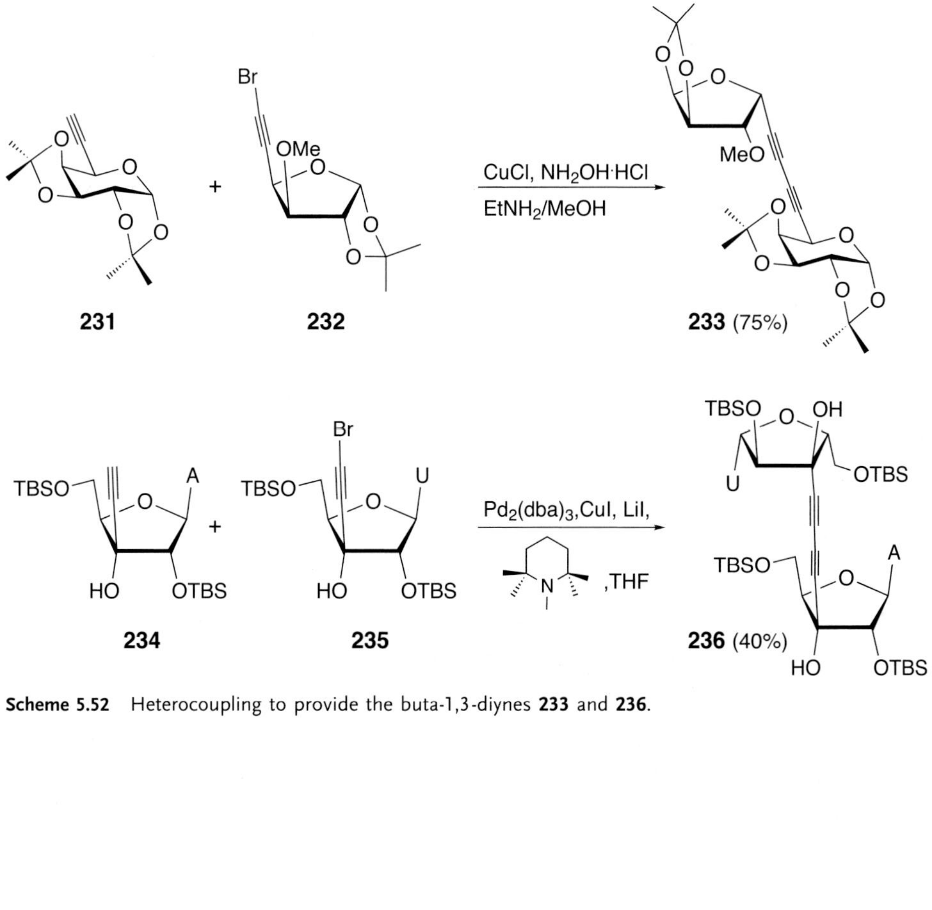

Scheme 5.52 Heterocoupling to provide the buta-1,3-diynes **233** and **236**.

5.5.2.3 Coupling of Acetylenosaccharides with Aromatic and Olefinic Moieties

Sonogashira coupling of acetylenosaccharides with iodo- and (triflyloxy)benzenes gave mono- to tetraalkynylated benzenes (see **81** in Scheme 5.18, and **237** and **238** in Scheme 5.53) [121, 131, 268, 269]. Enynes were obtained by Sonogashira coupling of iodoalkenes with acetylenosaccharides [270, 271].

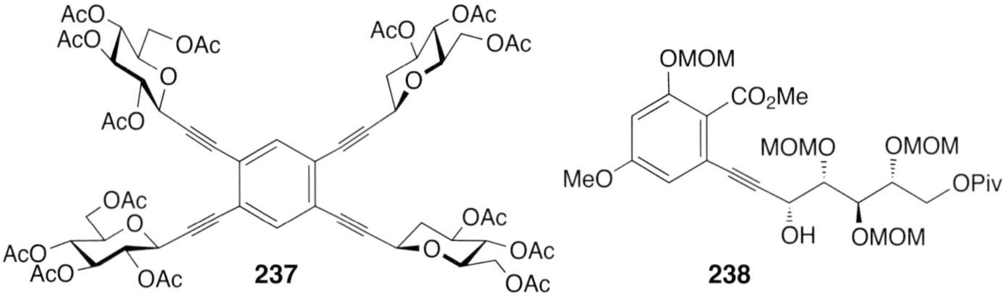

Scheme 5.53 Phenylated acetylenosaccharides obtained by Sonogashira coupling.

Sonogashira coupling of 6,6′-dibromo-2,2′-bipyridine with a 4-ethynyl-α-D-glucopyranosyl(trimethylsilylacetylene) gave the corresponding 6,6′-diethynylated bipyridine [272]. Intramolecular Eglinton coupling and *O*-deprotection provided the C_2-symmetric hybrid **239**, which complexes a range of metal ions (best Cu^{2+} and Zn^{2+}) [272] (Scheme 5.54). The *peri*-dialkynylated anthraquinones **82** (Scheme 5.18) were prepared by sequential Sonogashira couplings with (triflyloxy) anthraquinones [132, 133].

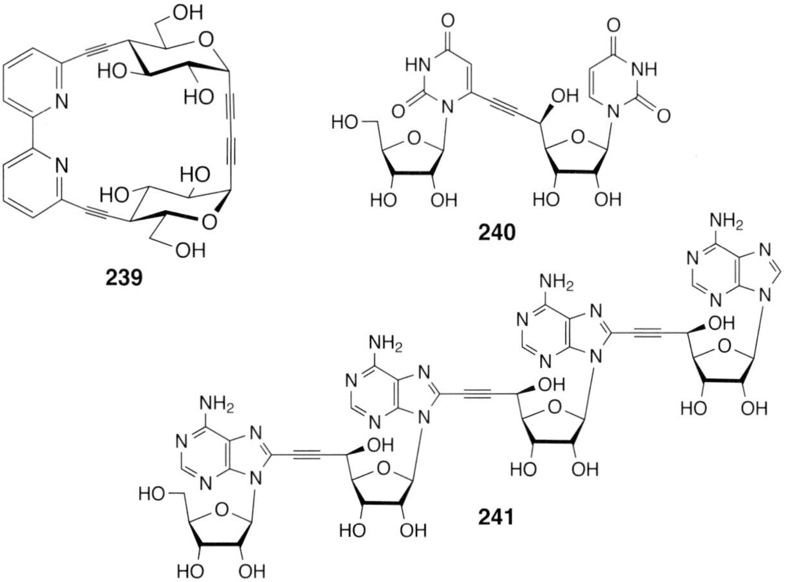

Scheme 5.54 Preparation of **239**, **240**, and **241** by Sonogashira coupling from the corresponding iodinated bipyridine and nucleosides.

The key step in the syntheses of the oligonucleotide analogues **240** and **241** (Scheme 5.54) from the precursors **71a** and **71c** (Scheme 5.15) is an intermolecular Sonogashira coupling ($Pd_2(dba)_3$, CuI, $P(furyl)_3$ in Et_3N/toluene) between an iodinated nucleobase and a D-*allo*-hept-5-ynofuranose [114, 115] (Scheme 5.54).

5.5.2.4 Coupling of Acetylenosaccharides with Amino Acid Derivatives

Sonogashira coupling of glycopyranosylacetylenes with *para*- or *meta*-iodinated phenylalanines gave *C*-linked glycopeptides [127]. Key steps of other approaches to *C*-linked glycopeptides are the addition of glycopyranosylacetylenes to aldehydes derived from amino acids [92] and the addition of acetylenes derived from amino acids to glyconolactones [128, 273].

5.6 Biological and Medicinal Uses of Acetylenosaccharides

A short enzymatic synthesis gave a 57 % yield of 6,7-dideoxy-L-*lyxo*-hept-6-ynulofuranose [274]. Several acetylenosaccharides inhibit enzymes: there is a polyacetylene that inhibits a cytosolic acetoacetyl-CoA thiolase [275] and a 6-ethynylated GPD-mannose analogue that inhibits GPD-mannose dehydrogenase [276], while 9-(5′,6′-dideoxy-β-D-*ribo*-hex-5′-ynofuranosyl)adenine and its 6′-halo derivatives inhibit *S*-adenosyl-L-homocysteine hydrolase [63, 277–279].

A systematic investigation of the antimicrobial activity of polyacetylenes has been published in Korean [280].

4′-Branched-chain- and 5′-ethynylated nucleoside analogues show antiviral activity [63, 278, 279, 281, 282], some against HIV viruses [176, 178, 283].

3′-Ethynylated nucleosides, especially the anticancer drug 3′-*C*-ethynylcytidine, possess antitumor and cytostatic activities [284–286]. Neocarcinostatin (**22** in Scheme 5.5) is a potent antitumor antibiotic [39].

As glycosylacetylenes are not degraded by glycosidases, *ortho*-carboranes derived from an α-L-*galacto*-hept-6-ynopyranose [287], some glycopyranosylacetylenes [288], and 3′-ethynylyted and propargylated thymidines [179] have been prepared for the treatment of cancer by boron neutron capture therapy (see [289] for an *ortho*-carborane derived from an ethynylated amino acid bound to a glucuronamide).

5.7 Experimental Protocols

5,9-Anhydro-6,7,8,10-tetra-O-benzyl-1,2,3,4-tetradeoxy-1-(trimethylsilyl)-D-glycero-D-gulo-*deca-1,3-diynitol* (**80a**) [132]. A solution of bis(trimethylsilyl)butadiyne (2.65 g, 13.6 mmol) in THF (35 ml) was cooled to 0° under Ar, treated dropwise with MeLi/LiBr in hexane (1.4 M, 10.22 ml) over a period of 0.5 h, and stirred for 4.5 h at 20°. The deep brown solution was transferred by cannula into a cooled (–78°) solution

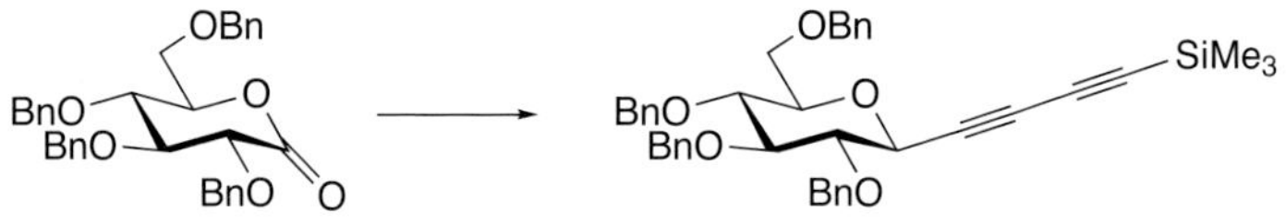

79 **80a**

of **79** (7.0 g, 13 mmol) in THF (35 ml) over a period of 10 min. The resulting pale brown solution was stirred for 1 h, treated with saturated aqueous NH_4Cl solution (20 ml), stirred for 0.5 h, and extracted with Et_2O. The organic layer was washed with H_2O, dried ($MgSO_4$), and evaporated to afford a brownish syrup of crude hemiacetals (α/β 3:2, 8.55 g).

A solution of $BF_3 \cdot Et_2O$ (17.8 ml, 142 mmol) and Et_3SiH (11.4 ml, 72 mmol) in CH_2Cl_2/MeCN (1:1, 30 ml) was cooled to 0° and added by cannula to a stirred and chilled (0°) solution of the crude hemiacetals (8.5 g) in CH_2Cl_2/MeCN (1:1, 40 ml) over a period of 0.5 h. The mixture was stirred at 0° for 2.5 h, treated with saturated aqueous $NaHCO_3$ solution (28.5 ml), stirred for 15 min, and extracted with AcOEt. The organic layer was washed with H_2O, dried ($MgSO_4$), and evaporated. FC (hexane/AcOEt 16:1) gave **80a** (6.28 g, 76%) as a colorless to pale yellow oil.

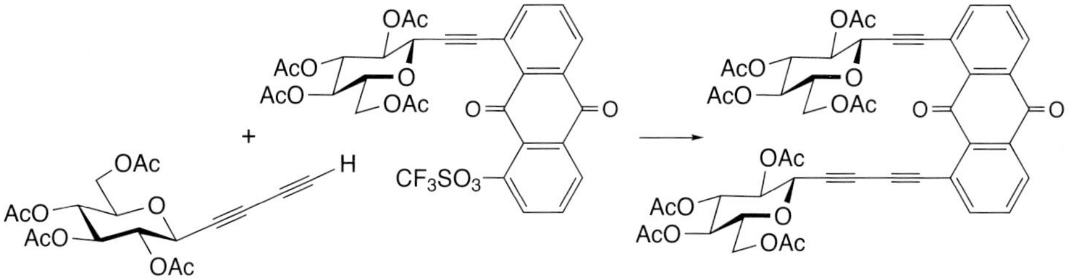

80b **82a** **82b**

*6,7,8,10-Tetra-O-acetyl-5,9-anhydro-1,2,3,4-tetradeoxy-1-C-[8-{4,5,6,8-tetra-O-acetyl-3,7-anhydro-1,2-dideoxy-*D-glycero-D-gulo-*oct-1-ynitol-1-yl)-9,10-anthraquinon-1-yl]-*D-glycero-D-gulo-*deca-1,3-diynitol* (**82b**) [132]. A stirred suspension of **82a** (1.0 g, 1.47 mmol), $Pd(PPh_3)_2Cl_2$ (49.4 mg, 0.07 mmol), and CuI (40.2 mg, 0.2 mmol) in degassed Et_3N/DMF 1:5 (10 ml) was treated dropwise under Ar with a solution of **80b** (1.07 g, 2.8 mmol) in Et_3N/DMF 1:5 (10 ml) over a period of 9 h at 21–23°. After stirring for an additional 12 h, the mixture was diluted with AcOEt, washed with brine, dried ($MgSO_4$), and evaporated. FC (AcOEt/hexane 1:1) gave light yellow **82b** (1.0 g, 76%), which was recrystallized in CH_2Cl_2/MeOH to afford colorless crystals (m. p. 205.1–206.3° (dec)).

HC≡C−DOPS-TIPS [191a]. A solution of $Me_3Si{-}C{\equiv}C{-}SiMe_2CMe_2CH_2CH_2OH$ (7.29 g, 30.0 mmol) and 2,6-lutidine (8.7 ml, 75.0 mmol) in dry CH_2Cl_2 (100 ml) was cooled to 0°, treated dropwise with TIPSOTf (10.5 ml, 39.1 mmol), stirred for 2 h, and treated with H_2O (40 ml). Extraction with CH_2Cl_2, washing

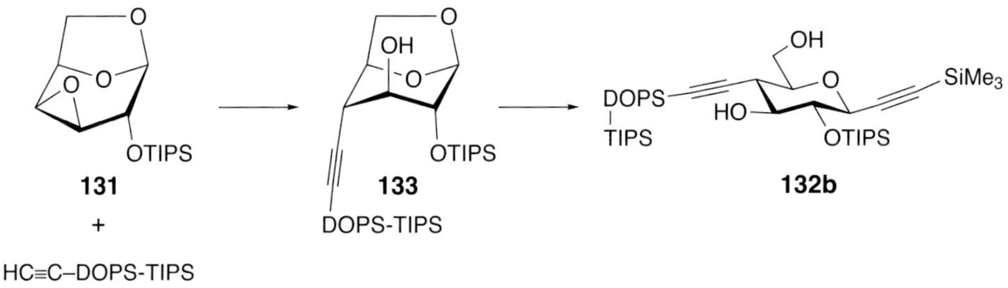

of the organic layer with H_2O and brine, drying ($MgSO_4$), evaporation at 300 Torr, and FC (hexane) gave $Me_3Si{-}C{\equiv}C{-}$DOPS-TIPS (10.68 g, 96 %) as a transparent oil.

A solution of $Me_3Si{-}C{\equiv}C{-}$DOPS-TIPS (14.58 g, 36.6 mmol) in dry, freshly distilled MeOH (240 ml) was treated with freshly prepared NaOMe in MeOH (0.25 M, 30 ml), and the mixture was stirred for 3 h at 22°. After addition of Amberlite IR-120 (H^+ form), the mixture was filtered, and the filtrate was evaporated at 300 Torr. Distillation (0.4 mbar, 54°) gave $HC{\equiv}C{-}$DOPS-TIPS (11.76 g, 98 %) as a transparent oil.

1,6-Anhydro-4-deoxy-4-C-{2-{dimethyl[1,1-dimethyl-3-(triisopropylsilyloxy)propyl]silyl}ethynyl}-2-O-(triisopropylsilyl)-β-D-glucopyranose (**133**) [191a]. A solution of $HC{\equiv}C{-}$DOPS-TIPS (16.9 g, 51.7 mmol) in dry toluene (30 ml) was cooled to –15°, treated dropwise with BuLi in hexane (2.3 M, 21 ml, 51.7 mmol), warmed to 20°, stirred for 30 min, cooled to –15°, treated dropwise with Me_3Al in toluene (2 M, 26 ml, 51.7 mmol), warmed to 20°, and stirred for 60 min (→ white precipitate). The suspension was heated to 75° and treated with a solution of **131** (10.4 g, 34.5 mmol) in toluene (30 ml) by double-ended needle. The mixture was stirred for 2 h at 75°, cooled to 0°, and slowly treated with a saturated NH_4Cl solution (5 ml). After filtration over Celite, the filtrate was diluted with AcOEt, washed with H_2O, dried ($MgSO_4$), and evaporated. FC (hexane/AcOEt 15:1) gave **133** (15.3 g, 71 %) as a transparent syrup.

*3,7-Anhydro-1,2,6-trideoxy-6-C-{2-{dimethyl[1,1-dimethyl-3-(triisopropylsilyloxy)propyl]silyl}ethynyl}-4-O-(triisopropylsilyl)-1-C-(trimethylsilyl)-*D-glycero-D-gulo-*oct-1-ynitol* (**132b**) [191a]. A solution of $Me_3Si{-}C{\equiv}CH$ (7.5 ml, 55.2 mmol) in dry toluene (50 ml) was cooled to –15°, treated dropwise with BuLi in hexane (2.3 M, 23.5 ml, 55.2 mmol), warmed to 21°, stirred for 30 min, diluted with THF (2 ml), and added to a cooled (–15°), mechanically stirred suspension of $AlCl_3$ (7.27 g, 55.2 mmol) in dry toluene (40 ml) by double-ended needle. Upon stirring at 21° for 45 min, a white precipitate was formed. The suspension was heated to 90° (bath temperature) and treated dropwise with a solution of **133** (11.39 g, 18.2 mmol) in dry toluene (80 ml), which resulted in dissolution of the precipitate. The solution was stirred at 90° for 18 h, cooled to 0°, and treated with a saturated NH_4Cl solution (10 ml). After extraction with AcOEt, the organic layer was washed with H_2O, dried ($MgSO_4$), and evaporated. FC (AcOEt/hexane 1:15) gave **132b** (9.77 g, 74 %; m. p. 82°) as a white solid.

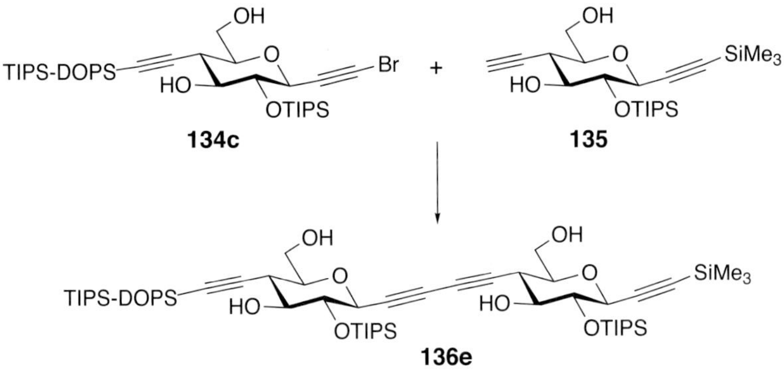

*5,9-Anhydro-1,2,3,4,8-pentadeoxy-8-C-{2-{dimethyl[1,1-dimethyl-3-(triisopropylsilyl-oxy)propyl]silyl}ethynyl}-6-O-(triisopropylsilyl)-*D-glycero-D-gulo-*deca-1,3-diynitol-1-yl-(1→6-C)-3,7-anhydro-1,2,6-trideoxy-4-O-(triisopropylsilyl)-1-C-(trimethylsilyl)-*D-glycero-D-gulo-*oct-1-ynitol* (**136e**) [191a]. A solution of **134c** (3.45 g, 4.71 mmol), **135** (2.00 g, 4.71 mmol), $Pd_2(dba)_3$ (129.3 mg, 0.14 mmol), CuI (26.9 mg, 0.14 mmol), and $P(fur)_3$ (54.7 mg, 0.236 mmol) in DMSO (70 ml) was degassed in a flame-dried *Schlenk* flask for 15 min, treated with dry Et_3N (2.0 ml, 14.13 mmol), and stirred in the dark for 10 h. The mixture was poured onto ice/H_2O and neutralized with HCl (1 N). After extraction with Et_2O, the organic layer was washed with H_2O, dried ($MgSO_4$), and evaporated. FC (AcOEt/hexane 1:10→3:17) gave the C_2-symmetric dimer derived from **134c** (116.7 mg, 2%), **136e** (4.00 g, 79%; m. p. 85–86°), and the C_2-symmetric dimer derived from **135** (31.1 mg, < 1%) as white solids.

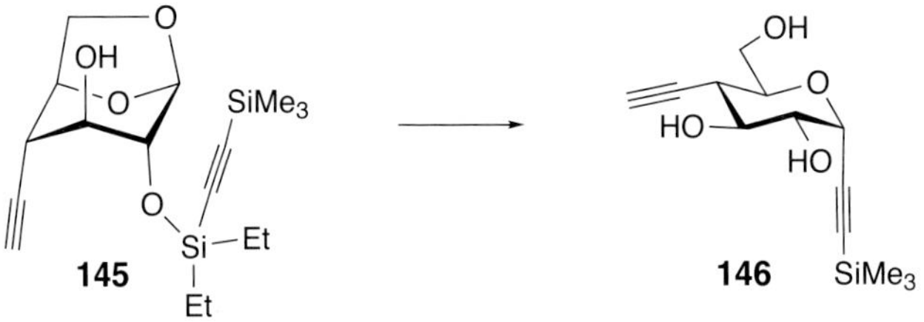

*2,6-Anhydro-3,7,8-trideoxy-3-C-ethynyl-8-C-(trimethylsilyl)-*D-glycero-L-gulo-*oct-7-ynitol* (**146**) [196]. Under N_2, a suspension of $AlCl_3$ (4.39 g, 32.9 mmol) in toluene (40 ml) was cooled to 0°, treated with BuLi in hexane (1.9 M, 17.3 ml, 32.9 mmol), warmed to 23°, stirred for 30 min, heated to 80°, treated with a solution of **145** (3.20 g, 9.1 mmol) in toluene (25 ml), stirred vigorously for 30 min, cooled to 0°, diluted with CH_2Cl_2, and filtered (washing with CH_2Cl_2). The combined filtrate and washings were washed with saturated $NaHCO_3$ solution and H_2O, dried ($MgSO_4$), and evaporated. The yellow residual oil was treated with HCl in MeOH (0.1 M, 10 ml), heated to 45°, stirred for 2 h, and evaporated. FC (toluene/AcOEt 6:4→ 1:1) gave **146** (2.19 g, 90%; m. p. 89–90°) as a colorless solid.

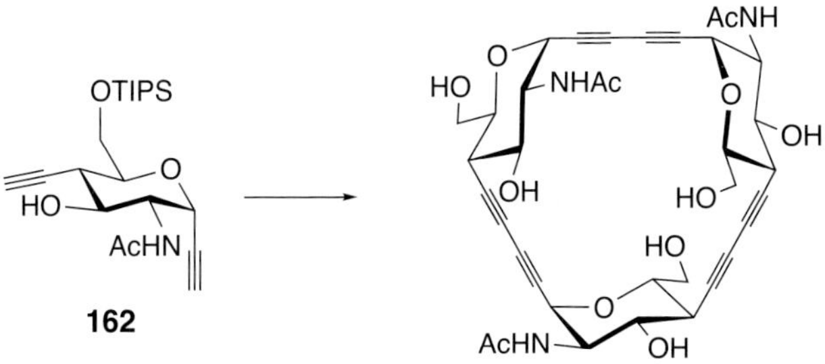

162 163

*3,3'-(Buta-1,3-diyne-1,4-diyl){2,6:11,15-dianhydro-14-C-[5-acetamido-2,6-anhydro-3,5,7,8,9,10-hexadeoxy-*D-glycero-L-gulo-*deca-7,9-diynitol-10-yl]-3,5,7,8,9,10,12,14-octadeoxy-*D*-erythro-*L*-ido-*L-gulo-*hexadeca-7,9-diynitol* (**163**) [201b]. A solution of **162** (395 mg, 1.0 mmol) in pyridine (1.0 l) was heated to 80°, treated with $Cu(OAc)_2$ (1.00 g, 5 mmol), stirred for 2 h, cooled to 20°, diluted with AcOEt (1.5 l) and Et_2O (1.5 l), and washed with H_2O (1.0 l). The aqueous phase was extracted with Et_2O (3 × 100 ml). The combined organic layers were washed with brine (3 × 50 ml), dried ($MgSO_4$), and evaporated at 40° and 12 mbar. Filtration over silica gel (70 g, AcOEt/MeOH 20:1) gave a brown foam (220 mg), which was dissolved in THF (30 ml), treated with TBAF · $3H_2O$ (390 mg, 1.2 mmol), stirred for 4 h, treated with MeOH (5.0 ml), and stirred for 1 h. Evaporation at 23° and 12 mbar and FC (70 g, AcOEt/H_2O/MeOH 10:3:2) of the red oil (660 mg) gave slightly impure (^1H NMR) **163** (112 mg, ca. 48 %). Crystallization from MeOH (3.0 ml) gave pure, colorless **163** (78 mg, 33 %; m. p. > 250° (dec.)).

Acknowledgements

We thank the Swiss National Science Foundation and F. Hoffmann-La Roche AG, Basel, for ongoing generous support.

References

[1] a) R. Lespieau, *Compt. Rend.* **1925**, *181*, 557–558.
b) R. Lespieau, *Bull. Soc. Chim.* **1928**, *43*, 657–662.

[2] R. A. Raphael, *J. Chem. Soc.* **1949**, S44–S48.

[3] J. K. N. Jones, W. A. Szarek, in '*The Total Synthesis of Natural Products*', Ed. J. ApSimon, Wiley-Interscience, New York, 1973, Vol. 1, p. 1–80.

[4] R. Zelinski, R. E. Meyer, *J. Org. Chem.* **1958**, *23*, 810–813.

[5] W. J. Robbins, F. Kavanagh, A. Hervey, *Proc. Natl. Acad. Sci. U. S. A.* **1947**, *33*, 176–182.

[6] a) M. Anchel, M. P. Cohen, *J. Biol. Chem.* **1954**, *208*, 319–326.
b) M. Anchel, *Trans. New York Acad. Sci.* **1954**, *16*, 337–342.

[7] E. R. H. Jones, J. S. Stephenson, W. B. Turner, M. C. Whiting, *J. Chem. Soc.* **1963**, 2048–2055.

[8] E. Dagne, S. Asmellash, *J. Nat. Prod.* **1994**, *57*, 390–392.

[9] A. Takahashi, T. Endo, S. Nozoe, *Chem. Pharm. Bull.* **1992**, *40*, 3181–3184.

[10] A. Gehrt, G. Erkel, T. Anke, O. Sterner, *Z. Naturforsch.* **1998**, *53*, 89–92.

[11] E. R. H. Jones, J. S. Stephenson, *J. Chem. Soc.* **1959**, 2197–2203.

[12] R. C. Cambie, A. Hirschberg, E. R. H. Jones, G. Lowe, *J. Chem. Soc.* **1963**, 4120–4130.

[13] E. R. H. Jones, G. Lowe, P. V. R. Shannon, *J. Chem. Soc. C* **1966**, 139–144.

[14] D. H. G. Crout, P. J. Fisher, V. S. B. Gaudet, H. Stoeckli-Evans, A. E. Anson, *Phytochemistry* **1986**, *25*, 1224–1226.

[15] a) M. Patel, M. Conover, A. Horan, D. Loebenberg, J. Marquez, R. Mierzwa, M. S. Puar, R. Yarborough, J. A. Waitz, *J. Antibiot.* **1988**, *41*, 794–797.
b) J. J. Wright, M. S. Puar, B. Pramanik, A. Fishman, *J. Chem. Soc., Chem. Commun.* **1988**, 413–414.

[16] T. Katsuki, K. B. Sharpless, *J. Am. Chem. Soc.* **1980**, *102*, 5974–5976.

[17] V. S. Martin, S. S. Woodard, T. Katsuki, Y. Yamada, M. Ikeda, K. B. Sharpless, *J. Am. Chem. Soc.* **1981**, *103*, 6237–6240.

[18] M. D. Lewis, R. Menes, *Tetrahedron Lett.* **1987**, *28*, 5129–5132.

[19] J. Shin, Y. Seo, K. W. Cho, J.-R. Rho, V. J. Paul, *Tetrahedron* **1998**, *54*, 8711–8720.

[20] L. Chill, A. Miroz, Y. Kashman, *J. Nat. Prod.* **2000**, *63*, 523–526.

[21] A. F. Rose, B. A. Butt, T. Jermy, *Phytochemistry* **1980**, *19*, 563–566.

[22] D. Grandjean, P. Pale, J. Chuche, *Tetrahedron Lett.* **1992**, *33*, 5355–5358.

[23] D. Grandjean, P. Pale, J. Chuche, *Tetrahedron* **1993**, *49*, 5225–5236.

[24] F. Bohlmann, J. Ziesche, *Phytochemistry* **1980**, *19*, 692–696.

[25] Y. Takaishi, T. Okuyama, A. Masuda, K. Nakano, K. Murakami, T. Tomimatsu, *Phytochemistry* **1990**, *29*, 3849–3852.

[26] P. Janackovic, V. Tesevic, P. D. Marin, S. M. Milosavljevic, B. Petkovic, M. Sokovic, *Biochem. Syst. Ecology* **2002**, *30*, 69–71.

[27] J. Zhou, B. Ning, Y. Gao, Q. Fang, *Zhongguo Yaoxue Zazhi* **2002**, *37*, 574–577.

[28] F. Bohlmann, K. M. Kleine, *Chem. Ber.* **1966**, *99*, 2096–2103.

[29] K. Yano, *Phytochemistry* **1980**, *19*, 1864–1866.

[30] K. Kawazu, Y. Nishii, S. Nakajima, *Agric. Biol. Chem.* **1980**, *44*, 903–906.

[31] a) Y. Fujimoto, M. Satoh, N. Takeuchi, M. Kirisawa, *Chem. Pharm. Bull.* **1991**, *39*, 521–523.
b) Y. Fujimoto, H. Wang, M. Kirisawa, M. Satoh, N. Takeuchi, *Phytochemistry* **1992**, *31*, 3499–3501.

[32] F. Bohlmann, K. H. Knoll, H. Robinson, R. M. King, *Phytochemistry* **1980**, *19*, 599–602.

[33] D. Manns, R. Hartmann, *J. Nat. Prod.* **1992**, *55*, 29–32.

[34] F. Fullas, D. M. Brown, M. C. Wani, M. E. Wall, T. E. Chagwedera, N. R. Farnesworth, J. M. Pezzuto, A. D. Kinghorn, *J. Nat. Prod.* **1995**, *58*, 1625–1628.

[35] N.-I. Baek, J. D. Park, Y. H. Lee, S. Y. Jeong, S. I. Kim, *Yakhak Hoechi* **1995**, *39*, 268–275.

[36] Y. Takaishi, T. Okuyama, K. Nakano, K. Murakami, T. Tomimatsu, *Phytochemistry* **1991**, *30*, 2321–2324.

[37] A. Horeau, in '*Stereochemistry: Fundamentals and Methods*', Ed. H. B. Kagan, Georg Thieme, Stuttgart, 1977, Vol. 3, p. 51–94.

[38] T. Sugiyama, K. Yamashita, *Agric. Biol. Chem.* **1980**, *44*, 1983–1984.

[39] N. Ishida, K. Miyazaki, K. Kumagai, M. Rikimaru, *J. Antibiot.* **1965**, *18*, 68–76.

[40] K. H. Kim, B. M. Kwon, A. G. Myers, D. C. Rees, *Science* **1993**, *262*, 1042–1046.

[41] T. Ando, M. Ishii, T. Kajiura, T. Kameyama, K. Miwa, Y. Sugiura, *Tetrahedron Lett.* **1998**, *39*, 6495–6498.

[42] a) S. Kobayashi, R. S. Reddy, Y. Sugiura, D. Sasaki, N. Miyagawa, M. Hirama, *J. Am. Chem. Soc.* **2001**, *123*, 2887–2888.
b) S. Kobayashi, S. Ashizawa, Y. Takahashi, Y. Sugiura, M. Nagaoka, M. J. Lear, M. Hirama, *J. Am. Chem. Soc.* **2001**, *123*, 11294–11295.

[43] P. Magnus, M. Davies, *J. Chem. Soc., Chem. Commun.* **1991**, 1522–1524.

[44] K. Nakatani, K. Arai, S. Terashima, *Tetrahedron* **1993**, *49*, 1901–1912.

[45] K. Toshima, K. Ohta, K. Yanagawa, T. Kano, M. Nakata, M. Kinoshita, S. Matsumura, *J. Am. Chem. Soc.* **1995**, *117*, 10825–10831.

[46] a) A. G. Myers, M. Hammond, Y. Wu, J.-N. Xiang, P. M. Harrington, E. Y. Kuo, *J. Am. Chem. Soc.* **1996**, *118*, 10006–10007.
b) A. G. Myers, R. Glatthar, M. Hammond, P. M. Harrington, E. Y. Kuo, J. Liang, S. E. Schaus, Y. Wu, J.-N. Xiang, *J. Am. Chem. Soc.* **2002**, *124*, 5380–5401.

[47] K. Toyama, S. Iguchi, H. Sakazaki, T. Oishi, M. Hirama, *Bull. Chem. Soc. Jpn.* **2001**, *74*, 997–1008.

[48] a) D. Horton, W. Loh, *Carbohydr. Res.* **1974**, *36*, 121–130.
b) D. Horton, W. Loh, *Carbohydr. Res.* **1974**, *38*, 189–203.
c) Y. Gelas-Mialhe, D. Horton, *J. Org. Chem.* **1978**, *43*, 2307–2310.

[49] a) J. S. Yadav, M. C. Chander, B. V. Joshi, *Tetrahedron Lett.* **1988**, *29*, 2737–2740.
b) J. S. Yadav, M. C. Chander, C. S. Rao, *Tetrahedron Lett.* **1989**, *30*, 5455–5458.
c) J. S. Yadav, M. C. Chander, *Tetrahedron Lett.* **1990**, *31*, 4349–4350.
d) J. S. Yadav, M. C. Chander, K. K. Reddy, *Tetrahedron Lett.* **1992**, *33*, 135–138.

[50] J. S. Yadav, A. Maiti, *Tetrahedron Lett.* **2001**, *42*, 3909–3912.

[51] J. S. Yadav, P. K. Deshpande, G. V. M. Sharma, *Tetrahedron* **1990**, *46*, 7033–7046.

[52] C. D. Hurd, H. Jenkins, *Carbohydr. Res.* **1966**, *2*, 240–250.

[53] M. Obayashi, M. Schlosser, *Chem. Lett.* **1985**, 1715–1718.

[54] S. Czernecki, J.-M. Valéry, *J. Carbohydr. Chem.* **1986**, *5*, 235–240.

[55] E. J. Corey, P. L. Fuchs, *Tetrahedron Lett.* **1972**, *13*, 3769–3772.

[56] a) J. M. J. Tronchet, C. Cottet, B. Gentile, E. Mihaly, J. B. Zumwald, *Helv. Chim. Acta* **1973**, *56*, 1802–1806.
b) J. M. J. Tronchet, A. P. Bonenfant,

F. Perret, A. Gonzalez, J. B. Zumwald, E. M. Martinez, B. Baehler, *Helv. Chim. Acta* **1980**, *63*, 1181–1189.
c) J. M. J. Tronchet, A. P. Bonenfant, *Carbohydr. Res.* **1981**, *93*, 205–217.

[57] a) K. Kakinuma, *Tetrahedron Lett.* **1977**, *18*, 4413–4416.
b) K. Kakinuma, *Tetrahedron* **1984**, *40*, 2089–2094.

[58] a) D. Rouzaud, P. Sinaÿ, *J. Chem. Soc., Chem. Commun.* **1983**, 1353–1354.
b) Y.-C. Xin, Y.-M. Zhang, J.-M. Mallet, C. P. J. Glaudemans, P. Sinaÿ, *Eur. J. Org. Chem.* **1999**, 471–476.

[59] M. A. Leeuwenburgh, S. Picasso, H. S. Overkleeft, G. A. van der Marel, P. Vogel, J. H. van Boom, *Eur. J. Org. Chem.* **1999**, 1185–1189.

[60] H. Streicher, A. Geyer, R. R. Schmidt, *Chem. Eur. J.* **1996**, *2*, 502–510.

[61] R. A. Sharma, M. Bobek, *J. Org. Chem.* **1978**, *43*, 367–369.

[62] S. F. Wnuk, C.-S. Yuan, R. T. Borchardt, J. Balzarini, E. De Clercq, M. J. Robins, *J. Med. Chem.* **1994**, *37*, 3579–3587.

[63] M. J. Robins, S. F. Wnuk, X. Yang, C.-S. Yuan, R. T. Borchardt, J. Balzarini, E. De Clercq, *J. Med. Chem.* **1998**, *41*, 3857–3864.

[64] a) J. Lebreton, A. De Mesmaeker, A. Waldner, *Synlett* **1994**, 54–56.
b) S. Wendeborn, C. Jouanno, R. M. Wolf, A. De Mesmaeker, *Tetrahedron Lett.* **1996**, *37*, 5511–5514.

[65] N. Solladié, M. Gross, *Tetrahedron Lett.* **1999**, *40*, 3359–3362.

[66] M. Mella, L. Panza, F. Ronchetti, L. Toma, *Tetrahedron* **1988**, *44*, 1673–1678.

[67] F. Dolhem, C. Lièvre, G. Demailly, *Tetrahedron* **2003**, *59*, 155–164.

[68] J. S. Yadav, V. Prahlad, M. C. Chander, *J. Chem. Soc., Chem. Commun.* **1993**, 137–138.

[69] J. C. Gilbert, U. Weerasooriya, *J. Org. Chem.* **1983**, *48*, 448–453.

[70] J. M. J. Tronchet, A. Gonzalez, J.-B. Zumwald, F. Perret, *Helv. Chim. Acta* **1974**, *57*, 1505–1510.

[71] a) J. D. White, M. A. Holoboski, N. J. Green, *Tetrahedron Lett.* **1997**, *38*, 7333–7336.
b) J. D. White, P. R. Blakemore, N. J. Green, E. B. Hauser, M. A. Holoboski, L. E. Keown, C. S. N. Kolz, B. W. Phillips, *J. Org. Chem.* **2002**, *67*, 7750–7760.

[72] a) K. A. Scheidt, A. Tasaka, T. D. Bannister, M. D. Wendt, W. R. Roush, *Angew. Chem., Int. Ed.* **1999**, *38*, 1652–1655.
b) K. A. Scheidt, T. D. Bannister, A. Tasaka, M. D. Wendt, B. M. Savall, G. J. Fegley, W. R. Roush, *J. Am. Chem. Soc.* **2002**, *124*, 6981–6990.

[73] J.-C. Thiéry, C. Fréchou, G. Demailly, *Tetrahedron Lett.* **2000**, *41*, 6337–6339.

[74] K. S. Åkerfeldt, P. A. Bartlett, *J. Org. Chem.* **1991**, *56*, 7133–7144.

[75] H. Kotsuki, I. Kadota, M. Ochi, *Tetrahedron Lett.* **1990**, *31*, 4609–4612.

[76] Q. Shen, D. G. Sloss, D. B. Berkowitz, *Synth. Commun.* **1994**, *24*, 1519–1530.

[77] L. M. Mikkelsen, S. L. Krintel, J. Jiménez-Barbero, T. Skrydstrup, *J. Org. Chem.* **2002**, *67*, 6297–6308.

[78] J. Xiang, P. L. Fuchs, *Tetrahedron Lett.* **1998**, *39*, 8597–8600.

[79] a) J. S. Yadav, P. R. Krishna, M. K. Gurjar, *Tetrahedron* **1989**, *45*, 6263–6270.
b) J. S. Yadav, D. Rajagopal, *Tetrahedron Lett.* **1990**, *31*, 5077–5080.

[80] A. Herunsalee, M. Isobe, S. Pikul, T. Goto, *Synlett* **1991**, 199–201.

[81] F. E. McDonald, C. B. Connolly, M. M. Gleason, T. B. Towne, K. D. Treiber, *J. Org. Chem.* **1993**, *58*, 6952–6953.

[82] a) K. Nakatani, K. Arai, S. Terashima, *J. Chem. Soc., Chem. Commun.* **1992**, 289–291.
b) K. Nakatani, K. Arai, N. Hirayama, F. Matsuda, S. Terashima, *Tetrahedron* **1992**, *48*, 633–650.

[83] T. Gourlain, A. Wadouachi, D. Beaupère, *Synthesis* **1999**, 290–294.

[84] D. Zhai, W. Zhai, R. M. Williams, *J. Am. Chem. Soc.* **1988**, *110*, 2501–2505.

[85] G. A. Tolstikov, N. A. Prokhorova, A. Yu. Spivak, L. M. Khalilov, V. R. Sultanmuratova, *J. Org. Chem. USSR* **1991**, *27*, 1858–1863.

[86] J. Désiré, A. Veyrières, *Carbohydr. Res.* **1995**, *268*, 177–186.

[87] a) L. Jobron, C. Leteux, A. Veyrières, J.-M. Beau, *J. Carbohydr. Chem.* **1994**, *13*, 507–512.
b) C. Leteux, A. Veyrières, *J. Chem. Soc., Perkin Trans. 1* **1994**, 2647–2655.

[88] a) A. Streitwieser, S. Pulver, *J. Am. Chem. Soc.* **1964**, *86*, 1587–1588.
b) S. Hanessian, *Carbohydr. Res.* **1965**, *1*, 178–180.

[89] a) M. Isobe, *J. Synth. Org. Chem. Jpn.* **1994**, *52*, 968–979.
b) M. Isobe, R. Saeeng, R. Nishizawa, M. Konobe, T. Nishikawa, *Chem. Lett.* **1999**, 467–468.

[90] M. Isobe, R. Nishizawa, S. Hosokawa, T. Nishikawa, *Chem. Commun.* **1998**, 2665–2676.

[91] T. Nishikawa, M. Ishikawa, M. Isobe, *Synlett* **1999**, 123–125.

[92] a) A. Dondoni, G. Mariotti, A. Marra, *Tetrahedron Lett.* **2000**, *41*, 3483–3487.
b) A. Dondoni, G. Mariotti, A. Marra, *J. Org. Chem.* **2002**, *67*, 4475–4486.

[93] a) K. Arakawa, T. Miyasaka, N. Hamamichi, *Chem. Lett.* **1974**, 1305–1308.
b) K. Arakawa, T. Miyasaka, N. Hamamichi, *Nucleic Acids Res.* **1976**, *2*, 1–4.
c) K. Arakawa, T. Miyasaka, N. Hamamichi, *Chem. Lett.* **1976**, 1119–1122.
d) N. Hamamichi, T. Miyasaka, K. Arakawa, *Chem. Pharm. Bull.* **1978**, *26*, 898–907.

[94] J. G. Buchanan, A. R. Edgar, M. J. Power, *J. Chem. Soc., Chem. Commun.* **1972**, 346–347.

[95] a) J. G. Buchanan, A. R. Edgar, M. J. Power, *J. Chem. Soc., Perkin Trans. 1* **1974**, 1943–1949.
b) J. G. Buchanan, A. R. Edgar, M. J. Power, G. C. Williams, *J. Chem. Soc., Chem. Commun.* **1975**, 501–502.
c) J. G. Buchanan, M. E. Chacon-Fuertes, A. Stobie, R. H. Wightman, *J. Chem. Soc., Perkin Trans. 1* **1980**, 2561–2566.

[96] a) S. Y.-K. Tam, F. G. De Las Heras, R. S. Klein, J. J. Fox, *Tetrahedron Lett.* **1975**, *16*, 3271–3274.
b) F. G. De Las Heras, S. Y.-K. Tam, R. S. Klein, J. J. Fox, *J. Org. Chem.* **1976**, *41*, 84–90.

[97] G. H. Posner, S. R. Haines, *Tetrahedron Lett.* **1985**, *26*, 1823–1826.

[98] a) M. A. Leeuwenburgh, H. S. Overkleeft, G. A. van der Marel, J. H. van Boom, *Synlett* **1997**, 1263–1264.
b) M. A. Leeuwenburgh, C. M. Timmers, G. A. van der Marel, J. H. van Boom, J.-M. Mallet, P. G. Sinaÿ, *Tetrahedron Lett.* **1997**, *38*, 6251–6254.

[99] M. A. Leeuwenburgh, R. E. J. N. Litjens, J. D. C. Codée, H. S. Overkleeft, G. A. van der Marel, J. H. van Boom, *Org. Lett.* **2000**, *2*, 1275–1277.

[100] a) J. D. Rainier, J. M. Cox, *Org. Lett.* **2000**, *2*, 2707–2709.
b) S. P. Allwein, J. M. Cox, B. E. Howard, H. W. B. Johnson, J. D. Rainier, *Tetrahedron* **2002**, *58*, 1997–2009.
c) M. A. Leeuwenburgh, G. A. van der Marel, H. S. Overkleeft, J. H. van Boom, *J. Carbohydr. Chem.* **2003**, *22*, 549–564.

[101] K. C. Nicolaou, C. K. Hwang, M. E. Duggan, *J. Chem. Soc., Chem. Commun.* **1986**, 925–926.

[102] a) Y. Ichikawa, M. Isobe, M. Konobe, T. Goto, *Carbohydr. Res.* **1987**, *171*, 193–199.
b) T. Tsukiyama, M. Isobe, *Tetrahedron Lett.* **1992**, *33*, 7911–7914.
c) T. Tsukiyama, S. C. Peters, M. Isobe, *Synlett* **1993**, 413–414.

[103] a) J. S. Yadav, B. V. S. Reddy, A. K. Raju, C. V. Rao, *Tetrahedron Lett.* **2002**, *43*, 5437–5440.
b) J. S. Yadav, B. V. S. Reddy, C. V. Rao, M. S. Reddy, *Synthesis* **2003**, 247–250.

[104] S. Tanaka, T. Tsukiyama, M. Isobe, *Tetrahedron Lett.* **1993**, *34*, 5757–5760.

[105] a) D. Horton, J. B. Hughes, J. M. J. Tronchet, *J. Chem. Soc., Chem. Commun.* **1965**, 481–483.
b) D. Horton, J. M. J. Tronchet, *Carbohydr. Res.* **1966**, *2*, 315–327.
c) J. L. Godman, D. Horton, J. M. J. Tronchet, *Carbohydr. Res.* **1967**, *4*, 392–400.
d) R. Hems, D. Horton, M. Nakadate, *Carbohydr. Res.* **1972**, *25*, 205–216.

[106] D. Horton, J. B. Hughes, J. K. Thomson, *J. Org. Chem.* **1968**, *33*, 728–734.

[107] V. Michelet, K. Adiey, S. Tanier, G. Dujardin, J.-P. Genêt, *Eur. J. Org. Chem.* **2003**, 2947–2958.
[108] S. H. Kang, W. J. Kim, *Tetrahedron Lett.* **1989**, *30*, 5915–5918.
[109] S. F. Martin, M. C. Hillier, *Tetrahedron Lett.* **1998**, *39*, 2929–2932.
[110] F. Compostella, L. Franchini, G. B. Giovenzana, L. Panza, D. Prosperi, F. Ronchetti, *Tetrahedron: Asymmetry* **2002**, *13*, 867–872.
[111] J. Yu, J.-Y. Lai, J. Ye, N. Balu, L. M. Reddy, W. Duan, E. R. Fogel, J. H. Capdevila, J. R. Falck, *Tetrahedron Lett.* **2002**, *43*, 3939–3941.
[112] T. Umino, N. Minakawa, A. Matsuda, *Tetrahedron Lett.* **2000**, *41*, 6419–6423.
[113] M. Shimizu, M. Kawamoto, Y. Niwa, *Chem. Commun.* **1999**, 1151–1152.
[114] S. Eppacher, N. Solladié, B. Bernet, A. Vasella, *Helv. Chim. Acta* **2000**, *83*, 1311–1330.
[115] a) H. Gunji, A. Vasella, *Helv. Chim. Acta* **2000**, *83*, 1331–1345.
b) H. Gunji, A. Vasella, *Helv. Chim. Acta* **2000**, *83*, 2975–2992.
c) H. Gunji, A. Vasella, *Helv. Chim. Acta* **2000**, *83*, 3229–3245.
d) P. K. Bhardwaj, A. Vasella, *Helv. Chim. Acta* **2002**, *85*, 699–711.
[116] A. Matsuda, H. Kosaki, Y. Saitoh, Y. Yoshimura, N. Minakawa, H. Nakata, *J. Med. Chem.* **1998**, *41*, 2676–2678.
[117] a) W. S. Chilton, W. C. Lontz, R. B. Roy, C. Yoda, *J. Org. Chem.* **1971**, *36*, 3222–3225.
b) R. B. Roy, W. S. Chilton, *J. Org. Chem.* **1971**, *36*, 3242–3243.
[118] a) J. G. Buchanan, A. D. Dunn, A. R. Edgar, *Carbohydr. Res.* **1974**, *36*, C5–C7.
b) J. G. Buchanan, A. D. Dunn, A. R. Edgar, *J. Chem. Soc., Perkin Trans. 1* **1975**, 1191–1200.
c) J. G. Buchanan, A. D. Dunn, A. R. Edgar, *J. Chem. Soc., Perkin Trans. 1* **1976**, 68–75.
[119] J. G. Buchanan, A. R. Edgar, R. J. Hutchison, A. Stobie, R. H. Wightman, *J. Chem. Soc., Perkin Trans. 1* **1980**, 2567–2571.
[120] J. G. Buchanan, M. L. Quijano, R. H. Wightman, *J. Chem. Soc., Perkin Trans. 1* **1992**, 1573–1576.
[121] K. Tatsuta, S. Takano, T. Sato, S. Nakano, *Chem. Lett.* **2001**, 172–173.
[122] A. Dondoni, D. Perrone, *Tetrahedron* **2003**, *59*, 4261–4273.
[123] F. Chouteau, K. Addi, M. Bénéchie, T. Prangé, F. Khuong-Huu, *Tetrahedron* **2001**, *57*, 6229–6238.
[124] a) H. Ogura, H. Takahashi, T. Itoh, *J. Org. Chem.* **1972**, *37*, 72–75.
b) H. Ogura, H. Takahashi, *Synth. Commun.* **1973**, *3*, 135–143.
[125] a) G. A. Kraus, K. A. Frazier, B. D. Roth, M. J. Taschner, K. Neuenschwander, *J. Org. Chem.* **1981**, *46*, 2417–2419.
b) G. A. Kraus, S. Liras, T. O. Man, M. T. Molina, *J. Org. Chem.* **1989**, *54*, 3137–3139.
[126] J.-M. Lancelin, P. H. A. Zollo, P. Sinaÿ, *Tetrahedron Lett.* **1983**, *24*, 4833–4836.
[127] T. Lowary, M. Meldal, A. Helmboldt, A. Vasella, K. Bock, *J. Org. Chem.* **1998**, *63*, 9657–9668.
[128] A. Dondoni, G. Mariotti, A. Marra, A. Massi, *Synthesis* **2001**, 2129–2137.
[129] E. Calzada, C. A. Clarke, C. Roussin-Bouchard, R. H. Wightman, *J. Chem. Soc., Perkin Trans. 1* **1995**, 517–518.
[130] a) H. Yoda, T. Oguchi, K. Takabe, *Tetrahedron: Asymmetry* **1996**, *7*, 2113–2116.
b) H. Yoda, T. Oguchi, K. Takabe, *Tetrahedron Lett.* **1997**, *38*, 3283–3284.
[131] a) J. Xu, A. Egger, B. Bernet, A. Vasella, *Helv. Chim. Acta* **1996**, *79*, 2004–2022.
b) B. Bernet, R. Bürli, J. Xu, A. Vasella, *Helv. Chim. Acta* **2002**, *85*, 1800–1811.
[132] K. V. S. N. Murty, A. Vasella, *Helv. Chim. Acta* **2001**, *84*, 939–963.
[133] T. Xie, K. V. S. N. Murty, B. Bernet, B. Meier, A. Vasella, in preparation.
[134] a) B. Bernet, A. Vasella, *Tetrahedron Lett.* **1983**, *24*, 5491–5494.
b) R. Julina, T. Herzig, B. Bernet, A. Vasella, *Helv. Chim. Acta* **1986**, *69*, 368–373.
[135] A. C. Oehlschlager, E. Czyzewska, *Tetrahedron Lett.* **1983**, *24*, 5587–5590.
[136] a) L. L. Vasiljeva, T. A. Manukina, P. M. Demin, M. A. Lapitskaja, K. K. Pivnitsky, *Tetrahedron* **1993**, *49*, 4099–4106.

b) P. M. Demin, D. M. Kochev, T. A. Manukina, A. C. R. Pace, K. K. Pivnitsky, *Bioorg. Khim.* **1998**, *24*, 778–786.
c) M. A. Lapitskaya, L. L. Vasiljeva, D. M. Kochev, K. K. Pivnitsky, *Russ. Chem. Bull.* **2000**, *49*, 549–556.

[137] a) F.-Y. Dupradeau, S. Allaire, J. Prandi, J.-M. Beau, *Tetrahedron Lett.* **1993**, *34*, 4513–4516.
b) F.-Y. Dupradeau, J. Prandi, J.-M. Beau, *Tetrahedron* **1995**, *51*, 3205–3220.

[138] R. E. Dolle, K. C. Nicolaou, *J. Am. Chem. Soc.* **1985**, *107*, 1691–1694.

[139] a) A. V. R. Rao, A. P. Khrimian, P. R. Krishna, P. Yadagiri, J. S. Yadav, *Synth. Commun.* **1988**, *18*, 2325–2330.
b) A. V. R. Rao, P. R. Krishna, J. S. Yadav, *Tetrahedron Lett.* **1989**, *30*, 1669–1670.
c) J. S. Yadav, P. Radhakrishna, *Tetrahedron* **1990**, *46*, 5825–5832.

[140] K. B. Sharpless, W. Amberg, Y. L. Bennani, G. A. Crispino, J. Hartung, K.-S. Jeong, H.-L. Kwong, K. Morikawa, Z.-M. Wang, D. Xu, X.-L. Zhang, *J. Org. Chem.* **1992**, *57*, 2768–2771.

[141] a) L. He, H.-S. Byun, R. Bittman, *J. Org. Chem.* **2000**, *65*, 7627–7633.
b) J. Chun, H.-S. Byun, R. Bittman, *Tetrahedron Lett.* **2002**, *43*, 8043–8045.

[142] a) S. Kobayashi, T. Hayashi, T. Kawasuji, *Tetrahedron Lett.* **1994**, *35*, 9573–9576.
b) S. Kobayashi, T. Hayashi, S. Iwamoto, T. Furuta, M. Matsumura, *Synlett* **1996**, 672–674.
c) S. Kobayashi, M. Horibe, *Chem. Eur. J.* **1997**, *3*, 1472–1481.
d) S. Kobayashi, M. Matsumura, T. Furuta, T. Hayashi, S. Iwamoto, *Synlett* **1997**, 301–303.
e) S. Kobayashi, T. Furuta, *Tetrahedron* **1998**, *54*, 10275–10294.
f) S. Kobayashi, T. Furuta, T. Hayashi, M. Nishijima, K. Hanada, *J. Am. Chem. Soc.* **1998**, *120*, 908–919.

[143] a) C. Mukai, O. Kataoka, M. Hanaoka, *Tetrahedron Lett.* **1994**, *35*, 6899–6902.
b) C. Mukai, O. Kataoka, M. Hanaoka, *J. Org. Chem.* **1995**, *60*, 5910–5918.

[144] M. Sukeda, S. Ichikawa, A. Matsuda, S. Shuto, *Angew. Chem., Int. Ed.* **2002**, *41*, 4748–4750.

[145] A. Mete, J. B. Hobbs, D. I. C. Scopes, R. F. Newton, *Tetrahedron Lett.* **1985**, *26*, 97–100.

[146] M. Ashwell, A. S. Jones, R. T. Walker, *Nucleic Acids Res.* **1987**, *15*, 2157–2166.

[147] P. Herdewijn, J. Balzarini, M. Baba, R. Pauwels, A. Van Aerschot, G. Janssen, E. De Clercq, *J. Med. Chem.* **1988**, *31*, 2040–2048.

[148] A. A. J. Feast, W. G. Overend, N. R. Williams, *J. Chem. Soc.* **1965**, 7378–7388.

[149] a) T. Inghardt, T. Freid, *Synthesis* **1990**, 285–291.
b) T. Inghardt, T. Frejd, *Tetrahedron* **1991**, *47*, 6483–6492.

[150] Y. Al-Abed, T. H. Al-Tel, M. S. Shekhani, W. Voelter, *Nat. Prod. Lett.* **1994**, *4*, 273–277.

[151] M. V. Rao, M. Nagarajan, *J. Org. Chem.* **1988**, *53*, 1184–1191.

[152] S. V. Ley, L. L. Yeung, *Synlett* **1992**, 291–292.

[153] S. P. Auguste, D. W. Young, *J. Chem. Soc., Perkin Trans. 1* **1995**, 395–404.

[154] a) L. Magdzinski, B. Cweiber, B. Fraser-Reid, *Tetrahedron Lett.* **1983**, *24*, 5823–5826.
b) L. Magdzinski, B. Fraser-Reid, *Can. J. Chem.* **1988**, *66*, 2819–2825.

[155] a) J. S. Burton, W. G. Overend, N. R. Williams, *J. Chem. Soc.* **1965**, 3433–3445.
b) W. G. Overend, A. C. White, N. R. Williams, *Carbohydr. Res.* **1970**, *15*, 185–195.

[156] J. Lehmann, H. Schäfer, *Chem. Ber.* **1972**, *105*, 969–974.

[157] D. C. Baker, D. K. Brown, D. Horton, R. G. Nickol, *Carbohydr. Res.* **1974**, *32*, 299–319.

[158] A. Gonzalez, M. Orzaez, E. Martinez, V. Custardoy, R. Mestres, *Anal. Quim.* **1974**, *70*, 1073–1076.

[159] a) E. Djuardi, E. McNelis, *Tetrahedron Lett.* **1999**, *40*, 7193–7196.
b) M. Blandino, E. McNelis, *Org. Lett.* **2002**, *4*, 3387–3390.

[160] G. V. M. Sharma, J. J. Reddy, M. H. V. R. Rao, N. Gallois, *Tetrahedron: Asymmetry* **2002**, *13*, 1599–1607.
[161] M. Nomura, T. Sato, M. Washinosu, M. Tanaka, T. Asao, S. Shuto, A. Matsuda, *Tetrahedron* **2002**, *58*, 1279–1288.
[162] P. C. M. Herve du Penhoat, A. S. Perlin, *Carbohydr. Res.* **1979**, *71*, 135–148.
[163] G. Dorey, P. Léon, S. Sciberras, S. Léonce, N. Guilbaud, A. Pieré, G. Atassi, D. C. Billington, *Bioorg. Med. Chem. Lett.* **1996**, *6*, 3045–3050.
[164] a) K. Kakinuma, N. Imamura, Y. Saba, *Tetrahedron Lett.* **1982**, *23*, 1697–1700.
b) K. Kakinuma, H.-Y. Li, *Tetrahedron Lett.* **1989**, *30*, 4157–4160.
c) K. Kakinuma, Y. Iihama, I. Takagi, K. Ozawa, N. Yamauchi, N. Imamura, Y. Esumi, M. Uramoto, *Tetrahedron* **1992**, *48*, 3763–3774.
d) T. Eguchi, T. Koudate, K. Kakinuma, *Tetrahedron* **1993**, *49*, 4527–4540.
e) K. Kakinuma, H. Terasawa, H.-Y. Li, K. Miyazaki, T. Oshima, *Biosci. Biotechnol. Biochem.* **1993**, *57*, 1916–1923.
[165] A. J. Poss, R. K. Belter, *Synth. Commun.* **1988**, *18*, 417–423.
[166] a) P. M. J. Jung, A. Burger, J.-F. Biellmann, *Tetrahedron Lett.* **1995**, *36*, 1031–1034.
b) P. M. J. Jung, A. Burger, J.-F. Biellmann, *J. Org. Chem.* **1997**, *62*, 8309–8314.
[167] S. Huss, F. G. De Las Heras, M. J. Camarasa, *Tetrahedron* **1991**, *47*, 1727–1736.
[168] P. S. Ludwig, R. A. Schwendener, H. Schott, *Synthesis* **2002**, 2387–2392.
[169] N. Minakawa, D. Kaga, Y. Kato, K. Endo, M. Tanaka, T. Sasaki, A. Matsuda, *J. Chem. Soc., Perkin Trans. 1* **2002**, 2182–2189.
[170] S. L. Bender, K. K. Moffett, *J. Org. Chem.* **1992**, *57*, 1646–1647.
[171] a) R. E. Harry-O'kuru, J. M. Smith, M. S. Wolfe, *J. Org. Chem.* **1997**, *62*, 1754–1759.
b) R. E. Harry-O'kuru, E. A. Kryjak, M. S. Wolfe, *Nucleos. Nucleot.* **1997**, *16*, 1457–1460.
[172] S. M. Daly, R. W. Armstrong, *Tetrahedron Lett.* **1989**, *30*, 5713–5716.
[173] a) Y. Yoshimura, T. Iino, A. Matsuda, *Tetrahedron Lett.* **1991**, *32*, 6003–6006.
b) T. Iino, Y. Yoshimura, A. Matsuda, *Tetrahedron* **1994**, *50*, 10397–10406.
[174] A. Kakefuda, Y. Yoshimura, T. Sasaki, A. Matsuda, *Tetrahedron* **1993**, *49*, 8513–8528.
[175] a) R. Buff, H. Stoeckli-Evans, J. Hunziker, *Acta Crystallogr., Sect. C: Cryst. Struct. Commun.* **1998**, *54*, 1860–1862.
b) R. Buff, J. Hunziker, *Bioorg. Med. Chem. Lett.* **1998**, *8*, 521–524.
c) R. Buff, J. Hunziker, *Synlett* **1999**, 905–908.
d) R. Buff, J. Hunziker, *Nucleos. Nucleot.* **1999**, *18*, 1387–1388.
[176] a) I. Sugimoto, S. Shuto, S. Mori, S. Shigeta, A. Matsuda, *Bioorg. Med. Chem. Lett.* **1999**, *9*, 385–388.
b) M. Nomura, S. Shuto, M. Tanaka, T. Sasaki, S. Mori, S. Shigeta, A. Matsuda, *J. Med. Chem.* **1999**, *42*, 2901–2908.
[177] a) S. Kohgo, H. Horie, H. Ohrui, *Biosci. Biotechnol. Biochem.* **1999**, *63*, 1146–1149.
b) S. Kohgo, E. Kodama, S. Shigeta, M. Saneyoshi, H. Machida, H. Ohrui, *Nucl. Acids Symp. Ser.* **1999**, *42*, 127–128.
c) R. Yamaguchi, T. Imanishi, S. Kohgo, H. Horie, H. Ohrui, *Biosci. Biotechnol. Biochem.* **1999**, *63*, 736–742.
d) H. Ohrui, S. Kohgo, K. Kitano, S. Sakata, E. Kodama, K. Yoshimura, M. Matsuoka, S. Shigeta, H. Mitsuya, *J. Med. Chem.* **2000**, *43*, 4516–4525.
e) S. Kohgo, H. Mitsuya, H. Ohrui, *Biosci. Biotechnol. Biochem.* **2001**, *65*, 1879–1882.
[178] S. Kohgo, K. Yamada, K. Kitano, S. Sakata, H. Hayakawa, D. Nameki, E. Kodama, M. Matsuoka, H. Mitsuya, H. Ohrui, *Nucleosides Nucleotides Nucleic Acids* **2003**, *22*, 887–889.
[179] J. Yan, C. Naeslund, A. S. Al-Madhoun, J. Wang, W. Ji, G. Y. Cosquer, J. Johnsamuel, S. Sjoberg, S. Eriksson, W. Tjarks, *Bioorg. Med. Chem. Lett.* **2002**, *12*, 2209–2212.

[180] J. Alzeer, C. Cai, A. Vasella, *Helv. Chim. Acta* **1995**, *78*, 242–264.

[181] a) Y. Le Merrer, A. Duréault, C. Gravier, D. Languin, J. C. Depezay, *Tetrahedron Lett.* **1985**, *26*, 319–322.
b) Y. Le Merrer, C. Gravier, D. Languin-Micas, J. C. Depezay, *Tetrahedron Lett.* **1986**, *27*, 4161–4164.
c) Y. Le Merrer, C. Gravier-Pelletier, D. Micas-Languin, F. Mestre, A. Duréault, J. C. Depezay, *J. Org. Chem.* **1989**, *54*, 2409–2416.
d) M. Sanière, Y. Le Merrer, B. Barbe, T. Koscielniak, J. Dumas, D. Micas-Languin, J. C. Depezay, *Angew. Chem.* **1989**, *101*, 645–647.
e) M. Sanière, Y. Le Merrer, B. Barbe, T. Koscielniak, J. C. Depezay, *Tetrahedron* **1989**, *45*, 7317–7328.

[182] G. B. Dreyer, J. C. Boehm, B. Chenera, R. L. DesJarlais, A. M. Hassell, T. D. Meek, T. A. Tomaszek, Jr., M. Lewis, *Biochemistry* **1993**, *32*, 937–947.

[183] T. K. M. Shing, K. H. Gibson, J. R. Wiley, C. I. F. Watt, *Tetrahedron Lett.* **1994**, *35*, 1067–1070.

[184] C. Mukai, E. Kasamatsu, T. Ohyama, M. Hanaoka, *J. Chem. Soc., Perkin Trans. 1* **2000**, 737–744.

[185] a) S. L. Schreiber, T. Sammakia, D. E. Uehling, *J. Org. Chem.* **1989**, *54*, 15–16.
b) M. Nakatsuka, J. A. Ragan, T. Sammakia, D. B. Smith, D. E. Uehling, S. L. Schreiber, *J. Am. Chem. Soc.* **1990**, *112*, 5583–5601.

[186] M. A. Amin, H. Stoeckli-Evans, A. Gossauer, *Helv. Chim. Acta* **1995**, *78*, 1879–1886.

[187] A. Vasella, *Pure Appl. Chem.* **1998**, *70*, 425–430.

[188] D. P. Sutherlin, R. W. Armstrong, *Tetrahedron Lett.* **1993**, *34*, 4897–4900.

[189] S. M. Daly, Ph. D. thesis, University of California, Los Angeles, CA, USA, 1992.

[190] B. Bernet, J. Xu, A. Vasella, *Helv. Chim. Acta* **2000**, *83*, 2072–2114.

[191] a) T. V. Bohner, R. Beaudegnies, A. Vasella, *Helv. Chim. Acta* **1999**, *82*, 143–160.
b) T. V. Bohner, O.-S. Becker, A. Vasella, *Helv. Chim. Acta* **1999**, *82*, 198–228.

[192] J. Alzeer, A. Vasella, *Helv. Chim. Acta* **1995**, *78*, 177–193.

[193] a) C. Cai, A. Vasella, *Helv. Chim. Acta* **1995**, *78*, 732–757.
b) J. Alzeer, A. Vasella, *Helv. Chim. Acta* **1995**, *78*, 1219–1237.
c) C. Cai, A. Vasella, *Helv. Chim. Acta* **1996**, *79*, 255–268.

[194] a) A. Ernst, A. Vasella, *Helv. Chim. Acta* **1996**, *79*, 1279–1294.
b) A. Ernst, L. Gobbi, A. Vasella, *Tetrahedron Lett.* **1996**, *37*, 7959–7962.
c) A. Ernst, W. B. Schweizer, A. Vasella, *Helv. Chim. Acta* **1998**, *81*, 2157–2189.

[195] a) R. Bürli, A. Vasella, *Helv. Chim. Acta* **1997**, *80*, 1027–1052.
b) R. Bürli, A. Vasella, *Angew. Chem., Int. Ed.* **1997**, *36*, 1852–1853.
c) R. Bürli, A. Vasella, *Helv. Chim. Acta* **1997**, *80*, 2215–2237.

[196] R. Bürli, A. Vasella, *Helv. Chim. Acta* **1996**, *79*, 1159–1168.

[197] a) O. R. Martin, *Tetrahedron Lett.* **1985**, *26*, 2055–2058.
b) O. R. Martin, *Carbohydr. Res.* **1987**, *171*, 211–222.
c) O. R. Martin, S. P. Rao, K. G. Kurz, H. A. El-Shenawy, *J. Am. Chem. Soc.* **1988**, *110*, 8698–8700.
d) O. R. Martin, C. A. V. Hendricks, P. P. Deshpande, A. B. Cutler, S. A. Kane, S. P. Rao, *Carbohydr. Res.* **1990**, *196*, 41–58.
e) O. R. Martin, S. P. Rao, C. A. V. Hendricks, R. E. Mahnken, *Carbohydr. Res.* **1990**, *202*, 49–66.

[198] a) B. Hoffmann, D. Zanini, I. Ripoche, R. Bürli, A. Vasella, *Helv. Chim. Acta* **2001**, *84*, 1862–1888.
b) B. Hoffmann, B. Bernet, A. Vasella, *Helv. Chim. Acta* **2002**, *85*, 265–287.

[199] B. Hoffmann, Thesis 14832, ETH Zürich, 2002.

[200] N. Sakairi, L.-X. Wang, H. Kuzuhara, *J. Chem. Soc., Perkin Trans. 1* **1995**, 437–443.

[201] a) J. Stichler-Bonaparte, A. Vasella, *Helv. Chim. Acta* **2001**, *84*, 2355–2367.
b) J. Stichler-Bonaparte, B. Bernet, A. Vasella, *Helv. Chim. Acta* **2002**, *85*, 2235–2257.

[202] G. V. Boyd, in '*The Chemistry of Functional Groups, Supplement C2: The Chemistry of Triple-Bonded Functional Groups*', Ed. S. Patai, John Wiley & Sons, Chichester, 1994, p. 287–374.

[203] a) H. El Khadem, D. Horton, M. H. Meshreki, *Carbohydr. Res.* **1971**, *16*, 409–418.
b) D. Horton, A. Liav, *Carbohydr. Res.* **1976**, *47*, 81–90.
c) D. Horton, J.-H. Tsai, *Carbohydr. Res.* **1978**, *67*, 357–370.
d) D. Horton, J.-H. Tsai, *Carbohydr. Res.* **1979**, *75*, 141–150.

[204] a) M. T. Garcia-Lopez, G. Garcia-Munoz, R. Madronero, *J. Heterocycl. Chem.* **1971**, *8*, 525.
b) G. Alonso, M. T. Garcia-Lopez, G. Garcia-Munoz, R. Madronero, *Anal. Quim.* **1976**, *72*, 987–990.

[205] J. G. Buchanan, A. R. Edgar, M. J. Power, G. C. Williams, *Carbohydr. Res.* **1977**, *55*, 225–238.

[206] J. M. J. Tronchet, A. P. Bonenfant, K. D. Pallie, F. Habashi, *Helv. Chim. Acta* **1979**, *62*, 1622–1625.

[207] a) S. Nagai, T. Ueda, N. Oda, J. Sakakibara, *Heterocycles* **1983**, *20*, 995–1000.
b) D. L. Swartz, H. S. El Khadem, *Carbohydr. Res.* **1983**, *112*, C1–C3.

[208] B. Stanovnik, B. Jelen, M. Zlicar, *Farmaco* **1993**, *48*, 231–242.

[209] Z. Maqbool, M. Hasan, K. T. Pott, A. Malik, T. A. Nizami, W. Voelter, *Z. Naturforsch. B, Chem. Sci.* **1997**, *52*, 1383–1392.

[210] N. Al-Masoudi, N. A. Hassan, Y. A. Al-Soud, P. Schmidt, A. E.-D. M. Gaafar, M. Weng, S. Marino, A. Schoch, A. Amer, J. C. Jochims, *J. Chem. Soc., Perkin Trans. 1* **1998**, 947–954.

[211] A. Padwa, W. H. Pearson, '*Synthetic Applications of 1,3-Dipolar Cycloaddition Chemistry Toward Heterocycles and Natural Products*', John Wiley & Sons, Hoboken, 2002.

[212] J. G. Buchanan, A. D. Dunn, A. R. Edgar, R. J. Hutchison, M. J. Power, G. C. Williams, *J. Chem. Soc., Perkin Trans. 1* **1977**, 1786–1791.

[213] a) A. Gonzalez, A. Llamas, M. A. Pascual, R. Mestres, *Carbohydr. Res.* **1977**, *59*, 248–254.
b) A. Llamas, A. Gonzalez, E. Martinez, *Carbohydr. Res.* **1978**, *67*, 515–521.

[214] a) M. W. Logue, S. Sarangan, *Proc. North Dakota Acad. Sci.* **1980**, *34*, 6.
b) M. W. Logue, S. Sarangan, *Nucleos. Nucleot.* **1982**, *1*, 89–98.

[215] A. D. Rycroft, G. Singh, R. H. Wightman, *J. Chem. Soc., Perkin Trans. 1* **1995**, 2667–2668.

[216] a) T. D. Heightman, M. Locatelli, A. Vasella, *Helv. Chim. Acta* **1996**, *79*, 2190–2200.
b) N. Panday, M. Meyyappan, A. Vasella, *Helv. Chim. Acta* **2000**, *83*, 513–538.

[217] K. Tezuka, P. Compain, O. R. Martin, *Synlett* **2000**, 1837–1839.

[218] M. Richter, G. Seitz, *Arch. Pharm.* **1994**, *327*, 365–370.

[219] K. Kim, U. S. M. Maharoof, J. Raushel, G. A. Sulikowski, *Org. Lett.* **2003**, *5*, 2777–2780.

[220] H. Wamhoff, H. Warnecke, P. Sohár, A. Csámpai, *Synlett* **1998**, 1193–1194.

[221] F. E. McDonald, H. Y. H. Zhu, C. R. Holmquist, *J. Am. Chem. Soc.* **1995**, *117*, 6605–6606.

[222] J. J. Gaudino, C. S. Wilcox, *J. Am. Chem. Soc.* **1990**, *112*, 4374–4380.

[223] a) A. M. Gómez, G. O. Danelón, E. Moreno, S. Valverde, J. C. López, *Chem. Commun.* **1999**, 175–176.
b) A. M. Gómez, E. Moreno, S. Valverde, J. C. López, *Tetrahedron Lett.* **2002**, *43*, 5559–5562.
c) A. M. Gómez, E. Moreno, S. Valverde, J. C. López, *Tetrahedron Lett.* **2002**, *43*, 7863–7866.
d) A. M. Gómez, E. Moreno, S. Valverde, J. C. López, *Synlett* **2002**, 891–894.
e) A. M. Gómez, E. Moreno, G. O. Danelón, S. Valverde, J. C. López, *Tetrahedron: Asymmetry* **2003**, *14*, 2961–2974.

[224] a) J. Marco-Contelles, E. de Opazo, *Tetrahedron Lett.* **2000**, *41*, 5341–5345.
b) J. Marco-Contelles, E. de Opazo, *J. Org. Chem.* **2002**, *67*, 3705–3717.

[225] a) I. Rochigneux, M.-L. Fontanel, J.-C. Malanda, A. Doutheau, *Tetrahedron Lett.* **1991**, *32*, 2017–2020.
b) J.-C. Malanda, A. Doutheau, *J. Carbohydr. Chem.* **1993**, *12*, 999–1016.

[226] J. Marco-Contelles, M. Bernabé, D. Ayala, B. Sánchez, *J. Org. Chem.* **1994**, *59*, 1234–1235.

[227] J. S. Yadav, A. Maiti, A. R. Sankar, A. C. Kunwar, *J. Org. Chem.* **2001**, *66*, 8370–8378.

[228] a) J. Marco-Contelles, C. Destabel, J. L. Chiara, M. Bernabé, *Tetrahedron: Asymmetry* **1995**, *6*, 1547–1550.
b) J. Marco-Contelles, C. Destabel, P. Gallego, J. L. Chiara, M. Bernabé, *J. Org. Chem.* **1996**, *61*, 1354–1362.

[229] J. Marco-Contelles, M. Rodríguez, *Tetrahedron Lett.* **1998**, *39*, 6749–6750.

[230] A. M. Gómez, G. O. Danelón, S. Valverde, J. C. López, *J. Org. Chem.* **1998**, *63*, 9626–9627.

[231] a) G. E. Keck, T. T. Wager, *J. Org. Chem.* **1996**, *61*, 8366–8367.
b) G. E. Keck, T. T. Wager, J. F. D. Rodriquez, *J. Am. Chem. Soc.* **1999**, *121*, 5176–5190.

[232] a) G. Pandey, M. Kapur, *Tetrahedron Lett.* **2000**, *41*, 8821–8824.
b) G. Pandey, M. Kapur, *Synthesis* **2001**, 1263–1267.

[233] T. V. Rajan Babu, W. A. Nugent, D. F. Taber, P. J. Fagan, *J. Am. Chem. Soc.* **1988**, *110*, 7128–7135.

[234] A. Ajamian, J. L. Gleason, *Org. Lett.* **2001**, *3*, 4161–4164.

[235] a) J. M. Aurrecoechea, B. López, M. Arrate, *J. Org. Chem.* **2000**, *65*, 6493–6501.
b) J. M. Aurrecoechea, J. H. Gil, B. López, *Tetrahedron* **2003**, *59*, 7111–7121.

[236] a) S.-K. Kang, S.-G. Kim, D.-G. Cho, *Tetrahedron: Asymmetry* **1992**, *3*, 1509–1510.
b) A. G. Myers, B. Zheng, *J. Am. Chem. Soc.* **1996**, *118*, 4492–4493.

[237] a) D. L. J. Clive, X. He, M. H. D. Postema, M. J. Mashimbye, *Tetrahedron Lett.* **1998**, *39*, 4231–4234.
b) D. L. J. Clive, X. He, M. H. D. Postema, M. J. Mashimbye, *J. Org. Chem.* **1999**, *64*, 4397–4410.

[238] a) K. H. Dötz, O. Neuss, M. Nieger, *Synlett* **1996**, 995–996.
b) R. Ehlenz, O. Neuss, M. Teckenbrock, K. H. Dötz, *Tetrahedron* **1997**, *53*, 5143–5158.
c) B. Weyershausen, M. Nieger, K. H. Dötz, *Organometallics* **1998**, *17*, 1602–1607.
d) B. Weyershausen, M. Nieger, K. H. Dötz, *J. Org. Chem.* **1999**, *64*, 4206–4210.

[239] a) D. Paetsch, K. H. Dötz, *Tetrahedron Lett.* **1999**, *40*, 487–488.
b) E. Janes, M. Nieger, P. Saarenketo, K. H. Dötz, *Eur. J. Org. Chem.* **2003**, 2276–2285.

[240] M. Jørgensen, P. Hadwiger, R. Madsen, A. E. Stütz, T. M. Wrodnigg, *Curr. Org. Chem.* **2000**, *4*, 565–588.

[241] A. Kinoshita, M. Mori, *Synlett* **1994**, 1020–1022.

[242] C. S. Poulsen, R. Madsen, *J. Org. Chem.* **2002**, *67*, 4441–4449.

[243] F. Dolhem, C. Lièvre, G. Demailly, *Eur. J. Org. Chem.* **2003**, 2336–2342.

[244] M. A. Leeuwenburgh, C. C. M. Appeldoorn, P. A. V. van Hooft, H. S. Overkleeft, G. A. van der Marel, J. H. van Boom, *Eur. J. Org. Chem.* **2000**, 873–877.

[245] P. L. Pauson, *Tetrahedron* **1985**, *41*, 5855–5860.

[246] M. Isobe, S. Takai, *J. Organomet. Chem.* **1999**, *589*, 122–125.

[247] J. Marco-Contelles, E. de Opazo, *J. Carbohydr. Chem.* **2002**, *21*, 201–218.

[248] A. Pal, A. Bhattacharjya, *J. Org. Chem.* **2001**, *66*, 9071–9074.

[249] F. Bohlmann, P. Herbst, H. Gleinig, *Chem. Ber.* **1961**, *94*, 948–957.

[250] a) D. Grandjean, P. Pale, J. Chuche, *Tetrahedron Lett.* **1992**, *33*, 4905–4908.
b) V. Dalla, P. Pale, *New J. Chem.* **1999**, *23*, 803–805.
c) P. Pale, J. Chuche, *Eur. J. Org. Chem.* **2000**, 1019–1025.

[251] B. Hoffmann, Z. W. Miao, M. Xu, B. Bernet, A. Vasella, in preparation.

[252] H. Toshima, H. Aramaki, A. Ichihara, *Tetrahedron Lett.* **1999**, *40*, 3587–3590.

[253] N. Miyakoshi, C. Mukai, *Org. Lett.* **2003**, *5*, 2335–2338.

[254] a) F. E. McDonald, C. C. Schultz, *J. Am. Chem. Soc.* **1994**, *116*, 9363–9364.
b) F. E. McDonald, M. M. Gleason, *J. Am. Chem. Soc.* **1996**, *118*, 6648–6659.
c) F. E. McDonald, K. S. Reddy, Y. Diaz, *J. Am. Chem. Soc.* **2000**, *122*, 4304–4309.
d) F. E. McDonald, K. S. Reddy, *Angew. Chem., Int. Ed.* **2001**, *40*, 3653–3655.
f) F. E. McDonald, K. S. Reddy, *J. Organomet. Chem.* **2001**, *617–618*, 444–452.
g) F. E. McDonald, M. Wu, *Org. Lett.* **2002**, *4*, 3979–3981.

[255] B. M. Trost, Y. H. Rhee, *J. Am. Chem. Soc.* **2002**, *124*, 2528–2533.

[256] P. Wipf, T. H. Graham, *J. Org. Chem.* **2003**, *68*, 8798–8807.

[257] a) C. M. Marson, S. Harper, R. Wrigglesworth, *J. Chem. Soc., Chem. Commun.* **1994**, 1879–1880.
b) C. M. Marson, S. Harper, *J. Org. Chem.* **1998**, *63*, 9223–9231.

[258] a) S. Tanaka, N. Tatsuta, O. Yamashita, M. Isobe, *Tetrahedron* **1994**, *50*, 12883–12894.
b) C. Yenjai, M. Isobe, *Tetrahedron* **1998**, *54*, 2509–2520.
c) K. Kira, M. Isobe, *Tetrahedron Lett.* **2000**, *41*, 5951–5955.
d) K. Kira, A. Hamajima, M. Isobe, *Tetrahedron* **2002**, *58*, 1875–1888.
e) T. Baba, G. Huang, M. Isobe, *Tetrahedron* **2003**, *59*, 6851–6872.

[259] D. D. Díaz, M. A. Ramírez, J. P. Ceñal, J. R. Saad, C. E. Tonn, V. S. Martín, *Chirality* **2003**, *15*, 148–155.

[260] a) C. Glaser, *Ber. Dtsch. Chem. Ges.* **1869**, *2*, 422–424.
b) C. Glaser, *Ann. Chem. Pharm.* **1870**, *154*, 137–171.

[261] K. Sonogashira, in '*Metal-Catalyzed Cross-Coupling Reactions*', Eds. F. Diederich and P. J. Stang, Wiley-VCS, Weinheim, 1998, p. 203–229.

[262] P. Siemsen, R. C. Livingston, F. Diederich, *Angew. Chem., Int. Ed.* **2000**, *39*, 2632–2657.

[263] D. Horton, J.-H. Tsai, *Carbohydr. Res.* **1979**, *75*, 151–174.

[264] F. Jung, A. Burger, J.-F. Biellmann, *Org. Lett.* **2003**, *5*, 383–385.

[265] R. R. Bukownik, C. S. Wilcox, *J. Org. Chem.* **1988**, *53*, 463–471.

[266] a) J. M. J. Tronchet, A. Bonenfant, *Helv. Chim. Acta* **1977**, *60*, 892–895.
b) J. M. J. Tronchet, F. Habashi, O. R. Martin, A. P. Bonenfant, B. Baehler, J. B. Zumwald, *Helv. Chim. Acta* **1979**, *62*, 894–898.
c) J. M. J. Tronchet, A. P. Bonenfant, *Helv. Chim. Acta* **1981**, *64*, 1893–1901.

[267] S.-K. Kang, H.-W. Lee, S.-B. Jang, P.-S. Ho, *Chem. Commun.* **1996**, 835–836.

[268] M. Suzuki, M. Kambe, H. Tokuyama, T. Fukuyama, *Angew. Chem., Int. Ed.* **2002**, *41*, 4686–4688.

[269] A. Dondoni, A. Marra, M. G. Zampolli, *Synlett* **2002**, 1850–1854.

[270] T.-Z. Liu, M. Isobe, *Tetrahedron* **2000**, *56*, 5391–5404.

[271] S. Takahashi, T. Nakata, *J. Org. Chem.* **2002**, *67*, 5739–5752.

[272] R. Bürli, A. Vasella, *Helv. Chim. Acta* **1999**, *82*, 485–493.

[273] a) J. L. Koviach, M. D. Chappell, R. L. Halcomb, *J. Org. Chem.* **2001**, *66*, 2318–2326.
b) J. W. Lane, R. L. Halcomb, *Tetrahedron* **2001**, *57*, 6531–6538.
c) J. W. Lane, R. L. Halcomb, *J. Org. Chem.* **2003**, *68*, 1348–1357.

[274] W.-D. Fessner, C. Gosse, G. Jaeschke, O. Eyrisch, *Eur. J. Org. Chem.* **2000**, 125–132.

[275] M. D. Greenspan, J. B. Yudkovitz, J. S. Chen, D. P. Hanf, M. N. Chang, P. Y. Chiang, J. C. Chabala, A. W. Alberts, *Biochem. Biophys. Res. Commun.* **1989**, *163*, 548–553.

[276] N. Elloumi, B. Moreau, L. Aguiar, N. Jaziri, M. Sauvage, C. Hulen, M. L. Capmau, *Eur. J. Med. Chem.* **1992**, *27*, 149–154.

[277] R. J. Parry, A. Muscate, L. J. Askonas, *Biochemistry* **1991**, *30*, 9988–9997.

[278] N. J. Prakash, G. F. Davis, E. T. Jarvi, M. L. Edwards, J. R. McCarthy, T. L. Bowlin, *Life Sci.* **1992**, *50*, 1425–1435.

[279] X. Yang, D. Yin, S. F. Wnuk, M. J. Robins, R. T. Borchardt, *Biochemistry* **2000**, *39*, 15234–15241.

[280] H.-J. Park, N.-D. Sung, *Han'guk Nonghwa Hakhoechi* **1998**, *41*, 258–263.

[281] L. M. Nutter, S. P. Grill, G. E. Dutschman, R. A. Sharma, M. Bobek, Y. C. Cheng, *Antimicrob. Agents Chemotherapy* **1987**, *31*, 368–374.

[282] A. Matsuda, H. Kosaki, Y. Yoshimura, S. Shuto, N. Ashida, K. Konno, S. Shigeta, *Bioorg. Med. Chem. Lett.* **1995**, *5*, 1685–1688.

[283] a) S. Kohgo, H. Ohrui, K. Kitano, E. Kodama, H. Mitsuya, *Tennen Yuki Kagobutsu Toronkai Koen Yoshishu* **2000**, *42nd*, 835–840.
b) E.-I. Kodama, S. Kohgo, K. Kitano, H. Machida, H. Gatanaga, S. Shigeta, M. Matsuoka, H. Ohrui, H. Mitsuya, *Antimicrob. Agents Chemotherapy* **2001**, *45*, 1539–1546.

[284] a) A. Matsuda, H. Hattori, M. Tanaka, T. Sasaki, *Bioorg. Med. Chem. Lett.* **1996**, *6*, 1887–1892.
b) H. Hattori, M. Tanaka, M. Fukushima, T. Sasaki, A. Matsuda, *J. Med. Chem.* **1996**, *39*, 5005–5011.
c) S. Tabata, M. Tanaka, A. Matsuda, M. Fukushima, T. Sasaki, *Onc. Rep.* **1996**, *3*, 1029–1034.
d) S. Tabata, M. Tanaka, Y. Endo, T. Obata, A. Matsuda, T. Sasaki, *Cancer Lett.* **1997**, *116*, 225–231.
e) H. Hattori, E. Nozawa, T. Iino, Y. Yoshimura, S. Shuto, Y. Shimamoto, M. Nomura, M. Fukushima, M. Tanaka, T. Sasaki, A. Matsuda, *J. Med. Chem.* **1998**, *41*, 2892–2902.
f) S. Takatori, S. Tsutsumi, M. Hidaka, H. Kanda, A. Matsuda, M. Fukushima, Y. Wataya, *Nucleos. Nucleot.* **1998**, *17*, 1309–1317.
g) A. Matsuda, M. Fukushima, Y. Wataya, T. Sasaki, *Nucleos. Nucleot.* **1999**, *18*, 811–814.
h) S. Takatori, H. Kanda, K. Takenaka, Y. Wataya, A. Matsuda, M. Fukushima, Y. Shimamoto, M. Tanaka, T. Sasaki, *Cancer Chemother. Pharmacol.* **1999**, *44*, 97–104.
i) A. Azuma, A. Matsuda, T. Sasaki, M. Fukushima, *Nucleos. Nucleot. Nucl. Acids* **2001**, *20*, 609–619.
j) Y. Shimamoto, A. Fujioka, H. Kazuno, Y. Murakami, H. Ohshimo, T. Kato, A. Matsuda, T. Sasaki, M. Fukushima, *Jpn. J. Cancer Res.* **2001**, *92*, 343–351.
k) T. Naito, T. Yokogawa, H.-S. Kim, M. Futagami, Y. Wataya, A. Matsuda, M. Fukushima, Y. Kitade, T. Sasaki, *Nucleic Acids Res. Suppl.* **2002**, *2*, 241–242.

[285] V. Holl, P. Jung, D. Weltin, J. Dauvergne, A. Burger, D. Coelho, P.ufour, A.-M. Aubertin, P. L. Bischoff, J.-F. Biellmann, *Anticancer Res.* **2000**, *20*, 1739–1742.

[286] W. Plunkett, V. Gandhi, *Cancer Chemother. Biol. Resp. Mod.* **2001**, *19*, 21–45.

[287] P. Basak, T. L. Lowary, *Can. J. Chem.* **2002**, *80*, 943–948.

[288] L. F. Tietze, U. Griesbach, I. Schuberth, U. Bothe, A. Marra, A. Dondoni, *Chem. Eur. J.* **2003**, *9*, 1296–1302.

[289] C. Thimon, L. Panza, C. Morin, *Synlett* **2003**, 1399–1402.

6
Semiconducting Poly(arylene ethylene)s

Timothy M. Swager

6.1
Introduction

The alkyne functionality has long been a cornerstone for the formation of electronic polymers. The formation of highly conductive polyacetylene [1] and the topochemical solid-state polymerization of diacetylenes [2] are discoveries of monumental importance in establishing the field. In these polymerizations an acetylene bond is necessarily sacrificed to produce two new carbon-carbon bonds that interconnect the monomers. Despite the obvious utility of alkynes in electronic polymers, poly(aryl ethynylene)s (PArEs) [3] were largely ignored until the mid-1990s, with the principle focus of electronic polymer research being directed toward polyacetylenes, polyarylenes, and materials containing alkene linkages between aromatic moieties such as poly(arylene vinylenes) [4]. This late interest in PArEs is the result of nascent efficient approaches to their synthesis through organometallic chemistry and inaccurate perception of their inferior electronic properties. The prejudice was in part founded on a prevailing interest in the creation of electronic materials capable of high delocalization and charge stabilization, goals better served by poly(aryl vinylene)s and polyarylenes. Indeed, polyheterocycles such as polythiophenes and polypyrroles were found to be best for the formation of highly oxidized structures, and computational analysis correctly predicted that the acetylene groups in poly(phenylene ethynylene) (PPE) should have larger band-gaps and narrower bandwidths than its close relative poly(phenylene vinylene) (PPV) [5].

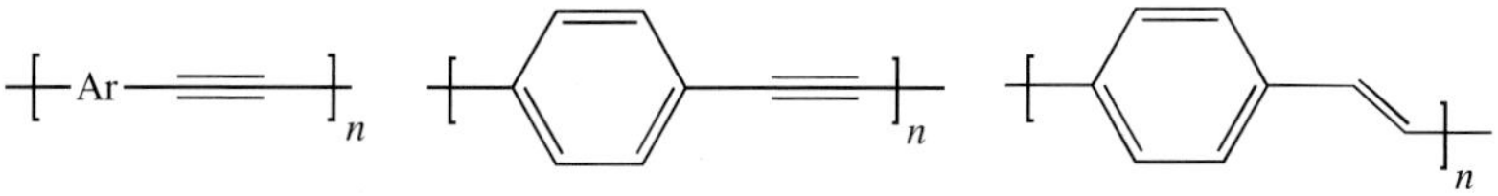

PArE **PPE** **PPV**

Acetylene Chemistry. Chemistry, Biology and Material Science. Edited by F. Diederich, P. J. Stang, R. R. Tykwinski

ISBN 3-527-30781-8

In spite of their delayed emergence, PArEs are now recognized as one of the most important classes of electronic polymers, and the carbon-carbon triple bond is a critical element that sets these materials apart from others. Of particular note is the ability of an alkyne linkage to be more accommodating than an alkene to steric and conformational constraints. However, the virtues of an alkyne are more profound than simple sterics. Steric interactions between the aromatics and the alkenes in poly(arylene vinylene)s, for example, will cause non-planar conformations and produce a dramatic reduction in delocalization. In contrast, direct steric interactions with alkynes can result in a bending distortion, but conjugation will be maintained due to their cylindrical electronic symmetry. PArEs are also of tremendous interest from the standpoint of their shape-persistent rigid structures, useful for the construction of nanostructures [6].

This chapter presents a summary of the electronic properties and some applications of PArEs. Because of the current vast body of work on these materials, I will draw largely upon my personal publications and will further restrict the discussion to materials with well defined extended conjugation. This will exclude branched structures and those with other non-conjugating linkages such as 1,3-phenylene groups in their backbone. It is my intent to convey critical insight into the properties of these materials to the reader rather than to produce a systematic catalog of published works. There are of course many other excellent researchers with intense interest in PArEs. I cannot do justice to all of their work, and interested readers are directed to other reviews [7, 8].

6.2 Synthesis

The preferred and most versatile synthesis of poly(arylene ethynylene)s makes use of the Sonogashira cross-coupling of acetylenes (Equation 6.1). Polymerizations of this type involve bond-forming reactions between difunctional monomers and are classified as step-growth polymerizations. The degree of polymerization (DP) in a step-growth process is governed by the equation DP = 1/(1–extent of reaction). Hence, for a reaction that proceeds to 90 % completion (extent of reaction = 0.9) a DP of 10 is expected and only oligomers are produced. To produce high molecular weight polymers (DP>100) the coupling reaction must proceed to greater than 99 % yield. A further challenge to the synthetic chemist is that this condition must be met with a strict 1:1 stoichiometry of the monomers. Hence, one reagent cannot be used in excess to drive a reaction to completion, as is often the case for published yields in small-molecule coupling literature. Sonogashira cross-coupling reactions are one of a handful of reactions that can satisfy this stringent criterion for a broad range of monomers, and in my laboratory polymers are routinely synthesized with molecular weights exceeding 100,000 Daltons. To achieve the highest molecular weight poly(polyarylene ethynylene)s by this method (the world record is 5,000,000 Daltons [9]), it is necessary to compensate for a minor ubiquitous side reaction, namely the formation of diacetylene linkages. Indeed the combination of amines, Cu$^+$ ions, a terminal alkyne, and an oxidant is the standard recipe for the efficient formation of diacetylene linkages. In my experience there is always at least a fraction of a percent of acetylene homo-couplings even if a researcher goes to great lengths to exclude oxygen and to purify solvents. Therefore, to produce the highest molecular weight polymers, a slight excess of the acetylene monomer should be added to offset the acetylene homo-couplings and to ensure a 1:1 stoichiometry for the Sonogashira hetero-couplings.

$$\equiv\!\!-\!\text{Ar}\!-\!\!\equiv\ +\ \text{X}\!-\!\text{Ar'}\!-\!\text{X}\ \xrightarrow[\text{Cu}^+,\ \text{R}_3\text{N}]{\text{Pd}^0\text{L}_x}\ \left[\!-\!\!\equiv\!\!-\!\text{Ar}\!-\!\!\equiv\!\!-\!\text{Ar'}\!-\!\right]_n$$

X = I, Br

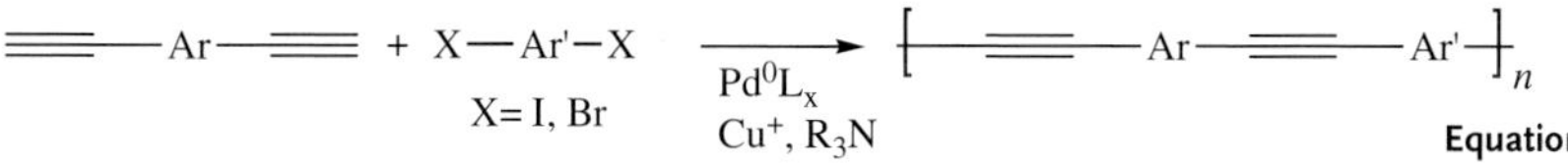

Equation 6-1

A second additional significant synthetic method for poly(arylene ethynylene)s makes use of alkyne metathesis [10]. This approach has, by and large, been developed and practiced by Bunz and involves the reaction shown in Equation 6.2, which liberates 2-butyne as a byproduct [11]. Alkyne metathesis catalysts can tolerate some functional groups and have also attracted recent interest in catalytic transformations for small molecule synthesis [12]. As an alternative to well defined isolated catalysts, which are difficult to prepare and are air-sensitive, Bunz has developed procedures from commercial air-stable chemicals that appear to produce a catalytic molybdenum alkylidyne composition in situ [13]. This method has been successful in preparing a number of polymers. However, the generated catalyst is less functional group tolerant than is found in palladium cross-coupling reactions, with similar limitations to earlier olefin metathesis catalysts formed by in situ procedures [14]. Polymers produced by this method can produce extraordinarily high molecular weights [15]. However, given that the catalyst mixture probably produces multiple organometallic complexes, it should not be assumed that these high molecular weight polymers are structurally simple. Indeed, as mentioned above, other related pre-catalyst mixtures give metal alkylidenes that have long been known to polymerize acetylenes through a chain-growth polymerization to produce substituted polyacetylenes [16]. Furthermore, very recent studies with well defined molybdenum alkylidyne catalysts with phenolic additives have also revealed the formation of polyacetylenes [17]. In the case of ultra-high molecular weight materials it is therefore possible, if not likely, that these materials have comb architectures with polyacetylene backbones. With long extended PPE side groups, and a highly twisted substituted polyacetylene backbone, this type of structural defect would be difficult to detect by standard spectroscopic methods.

$$\equiv\!\!-\!\text{Ar}\!-\!\!\equiv\ \xrightarrow{\text{ML}_x\!\!\equiv\!\!-\!\text{R}}\ \left[\!\!\equiv\!\!-\!\text{Ar}\!-\!\!\equiv\!\right]_n$$

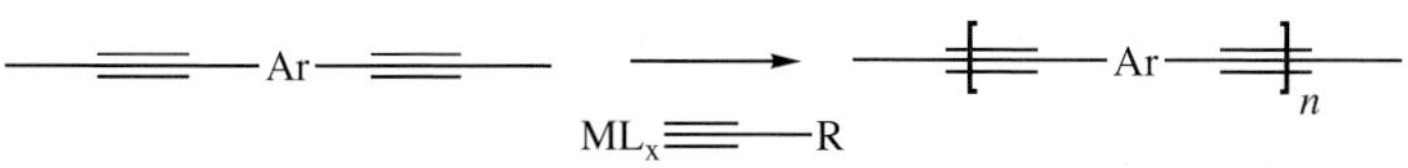

Equation 6-2

6.3
Conducting Properties of PArEs

Poly(phenylene ethynylene)s were originally thought to very be poor conductors, due to early studies on a limited number of structures. This assumption was not unexpected given that conventional wisdom favors more delocalized systems. Like typical organic electronic polymers, PArEs are wide bandgap semiconductors that are highly resistive (insulating) in their neutral (uncharged) state. To produce conductive forms of these materials, mobile charges must be injected, generally by removal or addition of electrons to the polymer's π-electron system. For a good conductor this process (which can often result in resistivity changes as high as 10^{13}) is referred to as doping, in analogy to terminology used in the semiconductor field. However, it should be noted that, in conventional inorganic semiconductors, doping involves replacement of atoms in a solid, as in the example of the substitution of a silicon atom with boron or phosphorous. In organic polymers, though, doping involves a much higher number of carriers than is typically possible in classical inorganic semiconductors. In heavily doped polyacetylene, for instance, the dopant counter-ions can constitute the majority of the material's weight.

In early conductivity investigations using chemical doping, researchers found that PPE oxidation required forcing conditions, since alkynes do not readily stabilize carbocations. Hence, to achieve high doping levels, the parent PPE was initially doped with AsF_5, a very powerful oxidation agent, to give a material (most probably degraded) with a conductivity of 10^{-3} S cm^{-1} [18]. Other studies yielded even lower conductivities, with AsF_5 doping generating a conductivity of 10^{-7} S cm^{-1} and I_2 doping producing values of 10^{-5} S cm^{-1} [19]. High conductivities of 70 S cm^{-1} were reported with treatment with SO_3 vapor, but it is likely that the polymers are severely degraded during this treatment, and no structural characterization of the final products was performed [20]. Poly(thienyl ethynylene) (**1**) was synthesized in an effort to incorporate thiophene residues, which can better stabilize charges, into the structures. However, these studies produced only highly resistive materials with I_2 doping [21]. The poor conductivity of PArEs under these conditions served to "confirm" erroneously that these materials were incapable of high conductivity. This was consistent with their narrower energy band widths, which should produce lower carrier mobilities.

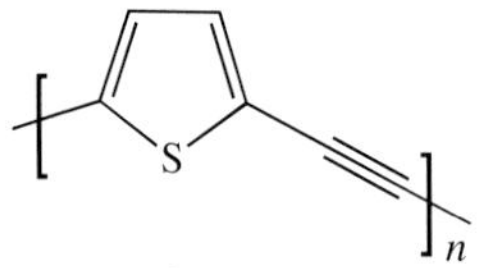

1

Electrochemical methods are generally preferred for doping polymers, since this approach readily reveals whether chemical reversibility accompanies the doping process. As indicated, it is certain that PArEs are extensively degraded by treatment with highly reactive agents such as AsF_5 and SO_3. Doped polymers prepared in this way have little relevance to the true electronic structure. Furthermore, electrochemical studies performed on PPEs bearing donating alkyloxy side chains (**2**; Figure 6.1) demonstrated this tendency to degrade in the presence even of a weakly nucleophilic solvent such as acetonitrile [22]. This same investigation concluded that in order to obtain stable forms of these materials in their highly doped states, electrochemical studies must be performed under exotic conditions with use of the extremely non-nucleophilic solvent SO_2. Given the low boiling point of SO_2, the electrochemical behavior was studied at −70 °C, where it was found to be reversible and the polymers were found to oxidize at 0.5 V vs. a poly(vinylferrocene) reference electrode. The reversibility of this process clearly demonstrated that the polymer does not degrade with oxidation and that proper doping is occurring.

The conductivity of any conducting polymer is not only dependent on its chemical stability but is also highly sensitive to strong electronic interactions/delocalization of both an intrapolymer and interpolymer nature. The order (crystallinity) of the polymer chains therefore greatly influences the magnitude of conductivity in PPEs. The attachment of long side chains ($C_{16}H_{33}$) to **2** produced a highly organized phase in which the polymer chains were organized into lamellar sheets (Figure 6.1) separated by crystalline or disordered alkanes [22]. The lamellar structure of **2** R = $C_{16}H_{33}$ displays a diffraction pattern with seven orders of diffraction

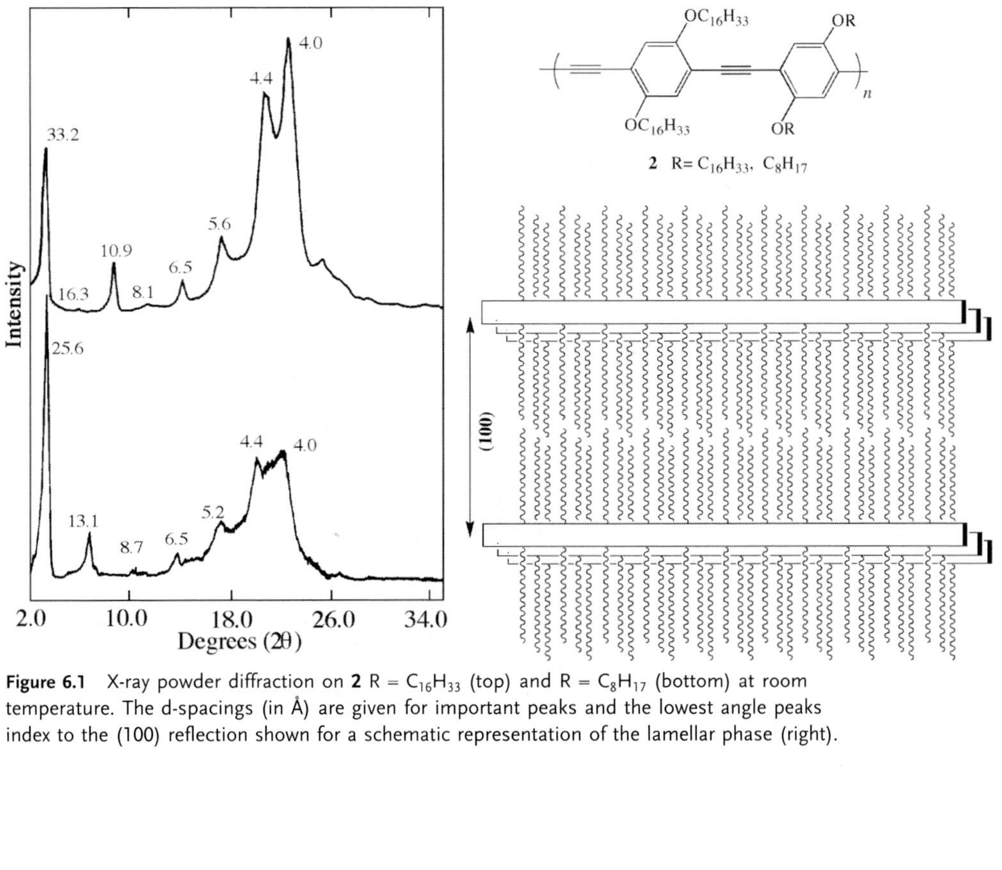

Figure 6.1 X-ray powder diffraction on **2** R = $C_{16}H_{33}$ (top) and R = C_8H_{17} (bottom) at room temperature. The d-spacings (in Å) are given for important peaks and the lowest angle peaks index to the (100) reflection shown for a schematic representation of the lamellar phase (right).

indicating long range registry of the lamellae. In addition, the diffraction pattern also showed a reduction in the intensity of the even reflections, consistent with a highly ordered lamellar structure. These electrochemically doped PPEs **2** revealed that they can be highly conducting with conductivities ranging from 0.2 to 5 S cm^{-1} depending upon R [22]. These values are truly remarkable when one considers that these materials are as much as 80% saturated (insulating) alkane by weight. Hence, if doped under the right conditions, PPEs have conductivities comparable to those of elaborated polythiophenes or polypyrroles, structures well known for their high conductivities.

The high reactivity of doped PPEs does, however, represent a practical limitation on the utility of their conducting properties. The stability can be significantly increased by placement of highly electron-donating groups into the polymer backbone [23]. This is demonstrated by Zotti's studies, in which polymers **3**, **4**, and **5** with bithienyl linkages were stable to typical cyclic voltammetric conditions in the presence of nucleophilic solvent. However, these materials exhibited only low conductivities of 10^{-3} S cm^{-1}, and EPR spectroscopic studies suggested that these low values are the result of a high localization of the charge on the bithienyl linkages.

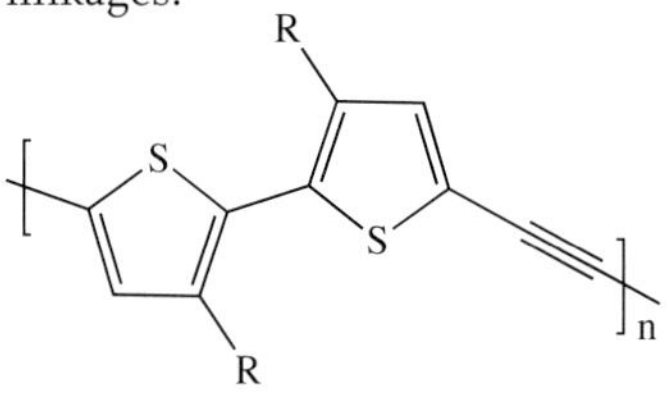

3 R = H, C_8H_{17}, OCH_3

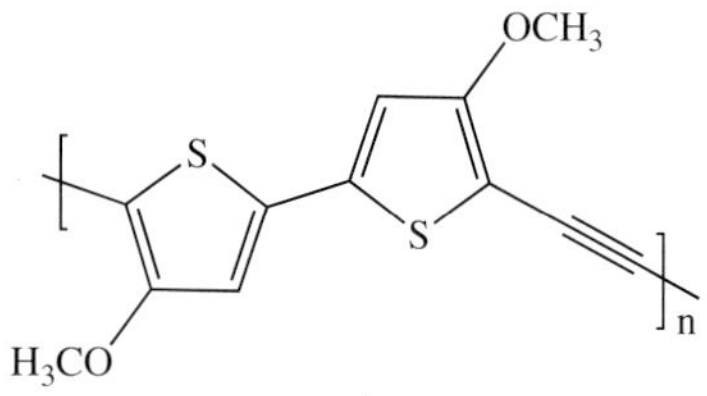

4

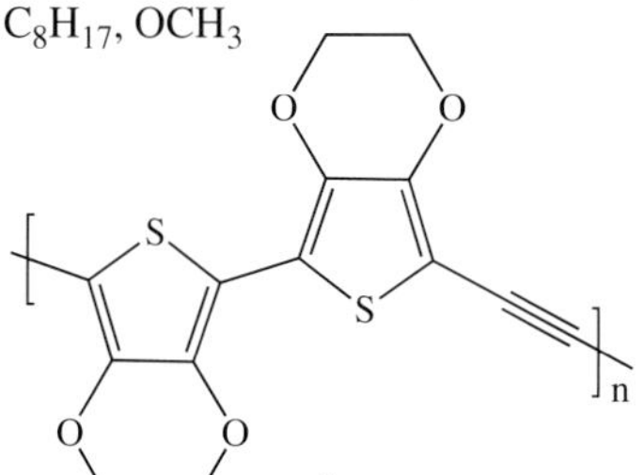

5

6.4
Photophysical Properties and Interpolymer Electronic Interactions

The most striking property of poly(arylene ethynylene)s is the fact that many of them display extremely efficient fluorescence in solution. However, simple PPEs such as **2** (Figure 6.1) exhibit pronounced self-quenching in thin films. This effect can be attributed to a strong tendency to exhibit interpolymer electronic interactions [24] that are much stronger than in their close relatives the PPVs. The reason for the stronger interchain electronic couplings in PPEs can be qualitatively understood by considering that their electronic structures are more localized in nature than those of PPVs. This greater localization (or lower delocalization) is also reflected in their narrower energy bands. More localized electronic states in PPEs allow for a larger orbital coefficient to be used in cofacial π-bonding between neighboring polymer chains. In contrast, a more diffuse orbital in a PPV has an overall reduced overlap and a weaker interaction.

The study of polymers is always complicated by the fact that single crystals without defects are impossible, although the lamellar phases (Figure 6.1) represent a good approximation. However, studies of solids of this type cannot be used to identify the intrinsic properties of isolated chains and the perturbations associated with the π stacking that occur in bulk. The complex conformational possibilities associated with polymers have long presented a challenge to basic understanding of the roles of intrachain conformation and interchain electronic interactions in electronic polymers. To resolve these issues comprehensively we have made extensive use of Langmuir–Blodgett methods [25]. This approach has a number of important features. (1) Unlike in a solid surface, the polymers are assembled on a liquid surface that is dynamic, thereby allowing for free molecular motions. (2) The two-dimensional nature of the film dramatically restricts accessible conformational states for the polymers, which assists in their assembly into well defined structures. (3) The ability to apply an anisotropic compression to the films allows for phase transitions, mechanical annealing, and chain alignment. (4) It is possible to perform in situ UV/Vis absorbance and fluorescence spectroscopy on well defined monolayers.

It has long been understood that strong interchain interactions need to be prevented to reduce self-quenching in thin films of emissive semiconductive polymers. Prevention of these deleterious interactions has often been accomplished by incorporation of bulky side chains and/or by the creation of structures that prevent crystallization. It is self-evident that enforcing large interchain spacing between polymer chains will produce systems that have photophysical properties closely resembling those in solution. However, too large an interchain spacing is undesirable as it will also reduce the transport of charges and excited states. In an effort to determine the minimal interelectronic spacing between PPEs chains, we studied Langmuir films of PPEs **6**, prepared with alternating monomers possessing a surfactant character with a hydrophilic acylated glycol and a hydrophobic $C_{16}H_{33}O$- group (Figure 6.2). This structure creates monolayers with an edge-on orientation of the polymer at the air/water interface [26]. The alkyl groups on the non-surfactant monomer were chosen to present variable steric interactions that result in different interchain separations. The relevant interpolymer spacings shown in Figure 6.2 were determined by two independent methods: by the area per molecule determined on the Langmuir–Blodgett trough and by X-ray diffraction. With increased steric bulk of the alkyl side chains, both the interchain spacing and the fluorescence quantum yield increase. We find that for interchain spacings >4.3 Å the polymer chains behave spectroscopically as though they were in solution with no interchain electronic coupling.

An important issue in electronic polymers that has traditionally presented an intractable research problem lies in understanding which relative perturbations to the electronic structure are best ascribed to intrachain conformational properties and which are truly interchain in origin. There have been many efforts to correlate different processing conditions and spectroscopic features of thin films of semiconducting polymers in order to understand these effects better [27]. However, these approaches all suffer from the fact that intrapolymer and interpolymer interactions are interdependent and often inseparable. In other words, a polymer exhibiting strong interpolymer π-electronic couplings will generally favor a conformation that may not be displayed in the absence of such interchain association. It is intuitive that as a polymer begins to assemble into a structure with strong interchain electronic interactions the backbone will favor a more planar structure to maximize these interactions. To deconvolute these different factors, we designed a series of polymers that have different equilibrium two-dimensional liquid crystalline and crystalline structures at the air/water interface and studied reversible pressure-induced 2D-phase transitions of these materials [28]. Four examples of polymers used in these investigations and their respective structures are shown schematically in Figure 6.3 (polymers **7–10**). We have referred to the different two-dimensional structures as face-on, zipper, and edge-on. The edge-on structure has already been discussed (Figure 6.2) and has strong cofacial interactions between polymer chains and a highly planarized structure. This structure (polymer **10**) is properly represented as a two-dimensional crystal of polymer chains and is corroborated by the close resemblance of its spectroscopic features to those of the lamellar crystalline structures produced in bulk (Figure 6.1). A second highly planarized structure (polymer **7**) that lacks interchain interactions is produced by the face-on sur-

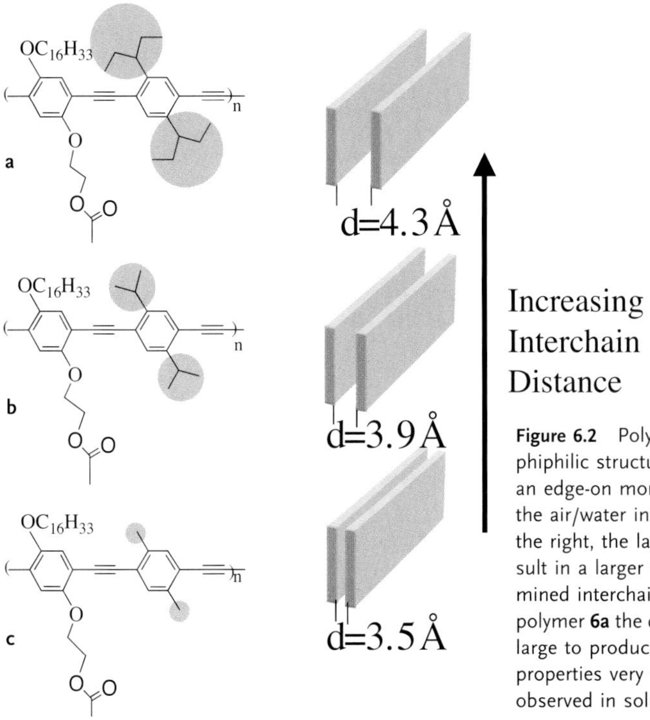

Figure 6.2 Polymers having an amphiphilic structure that organize into an edge-on monolayer structure at the air/water interface. As shown on the right, the larger alkyl groups result in a larger experimentally determined interchain distances. For polymer **6a** the distance is sufficiently large to produce spectroscopic properties very similar to those observed in solution.

face assembly of polymers with alternating sequences of hydrophilic and hydrophobic side-chain substituted monomers. By virtue of its confinement to the two-dimensional air/water interface, this structure does not exhibit cofacial π-interactions between chains, and its properties are consistent with a two-dimensional liquid crystal. In between these two extremes we have developed polymers that exhibit an intermediate structure that we have referred to as the zipper phase, in which alternate monomers have edge-on and face-on organizations. The zipper structure displays an interlocking structure between polymer chains resembling that of a conventional zipper. To produce this highly correlated structure it is of crucial importance to minimize the number of diacetylene linkages in the polymers, because even at low concentrations they introduce defects that dramatically change the character of the film in the alternating structure. As shown in Figure 6.3, polymer **8** was designed to display a face-on structure at low pressure and with increased pressure it transforms into a zipper structure. Alternatively we have also produced polymers (polymer **9**) that display equilibrium zipper structures that can be transformed to edge-on structures at higher pressures.

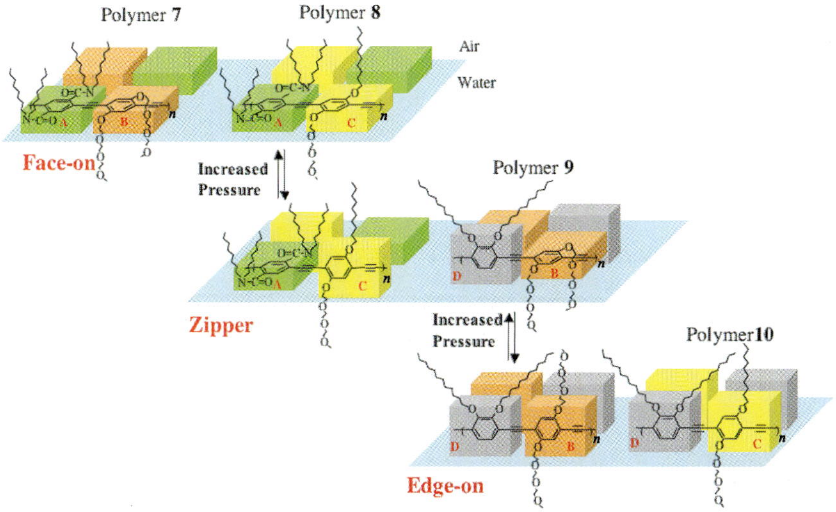

Figure 6.3 Well defined phases of PPEs arranged at the air/water interface. The polymers are made up from monomers **A**, **B**, **C**, and **D**, with different organizational preferences at the interfaces. With applied pressure, monomers **C** and **B** in polymers **8** and **9**, respectively, can be made to rotate into an edge-on organization.

To establish the structures and spectroscopic features of the edge-on, face-on, and zipper phases we analyzed a multitude of spectra and structure property relationships. The volume of data and discussion required to make definitive assignments is beyond the scope of this review and the interested reader is referred to our primary publications [28, 29]. Nevertheless, the consistency of the results is readily revealed by simple inspection of the UV/Vis absorption spectra (Figure 6.4) taken from the different monolayer phases at the air/water interface. The absorption intensity scales with the density of the polymer chromophores at the surface, and hence the face-on structure has the lowest absorption cross-section and the edge-on structure the highest. A very surprising conclusion from this study is that solution spectra most closely resemble the zipper phases. We can thus infer from this that in solution most PPEs have considerable conformational disorder and lack extended conjugation.

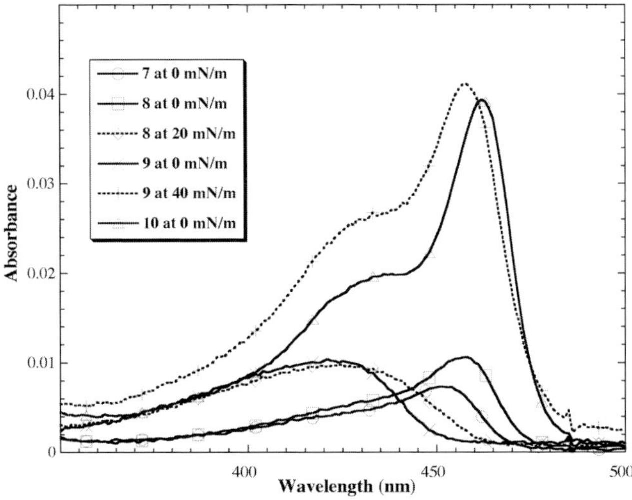

Figure 6.4 Comparison of the UV/Vis absorption spectra of the monolayers shown in Figure 6.3 at different surface pressures (mN m^{-1}). The PPEs have slightly different band-gaps due to their different structures, but the correlations between the spectra of the polymers having the same phase are clear. The face-on structures of polymers **7** and **8** have very similar spectra at pressures of 0 mN m^{-1}, and at 20 mN m^{-1} polymer **8** adopts the same higher band-gap zipper structure as displayed by polymer **9** at 0 mN m^{-1}. At higher pressure polymer **9** transforms to the edge-on structure that is displayed by polymer **10** at 0 mN m^{-1}.

As discussed earlier, to produce highly luminescent PArEs in thin films the universal rule has been to avoid strongly interacting chains. This is generally accomplished by appending the polymers with large, bulky chains. However, to produce sufficient entropies in solution, the side chains generally need to be flexible and electronically inactive. These large side chains have two negative consequences: they dilute the active electronic elements, namely the conjugated chains, and they make for softer, lower modulus materials that often structurally evolve to produce solids with small amounts of interpolymer associations capable of producing self-quenching and broadened emissions. To create non-interacting chains that are immune from these limitations we designed polymer **11** containing pentiptycene functionalities [30]. As a consequence of the 2.2.2 fused ring system the scaffold is completely inflexible, and spin-coated films could not be made to display interchain electronic interactions even with extended heating [31]. A secondary consequence of this scaffold is the generation of high degrees of free volume that allow for galleries to exist within the polymer film. This structure was demonstrated to have size exclusion characteristics in sensory experiments. The free volume also provides a novel mechanism to produce greater stability from oxidative degradation. Consistently with Winstein's classical model for the electronic stabilization of carbocations by neighboring π-bonds [32], site isolation of the polymer's alkynes endows films of polymer **11** with extraordinary stability among PPEs.

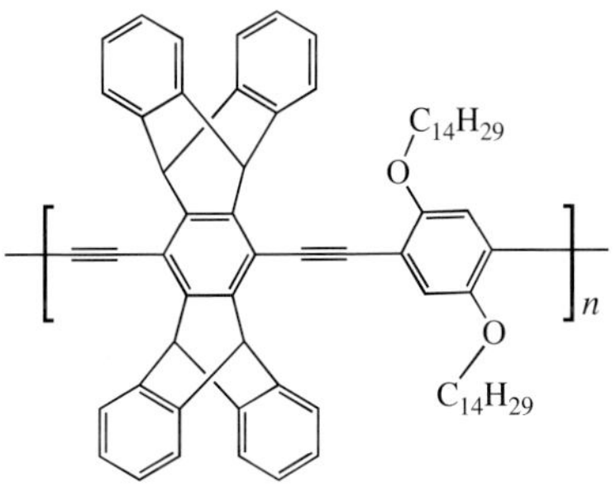

11

Variations in the electronic structures of PArEs can be used to produce materials with different band-gaps (emission colors). A well known example is to include an anthracene group, as in polymer **12**, which produces a more red-shifted emissive material [33]. This bathochromic shift is understood to arise because the center ring of the anthracene is less aromatic than a simple phenyl moiety. Aromatic stabilization tends to localize the electronic structures of materials; hence, because of the anthracene group, polymer **12** is more delocalized than a typical PPE and has a reduced band-gap. A less well known electronic design principle involves formation of systems with longer excited state lifetimes. Extended excited state lifetimes are of importance for applications that make use of the ParEs' ability to transport excited states over long distances. Lifetime engineering can be accomplished by incorporating large polycyclic chromophores into PArEs, which themselves have long excited state lifetimes. Two systems that accomplish this are the polymers **13** [34] and **14** [35]. Triphenylene and dibenzo[*g,p*]chrysene derivatives tend to have much longer lifetimes than PPEs (generally <600 ps) and the incorporation of these groups into the polymers generates materials with extended lifetimes of 700 ps and 2.6 ns, respectively. A key consideration when implementing this design principle is that the polycyclic aromatic moiety must have high aromatic character in order to preserve its electronic character and thereby transfer some of its original photophysical properties to the PArE.

Although random interchain electronic couplings generally produce non-emissive states, there are arrangements by which these interactions can exhibit an emissive state. In these systems the emission efficiency is controlled by the relative orientation of the transition moments of the neighboring polymer chains and there is a general tendency for PArEs and other electronic polymers to exhibit interpoly-

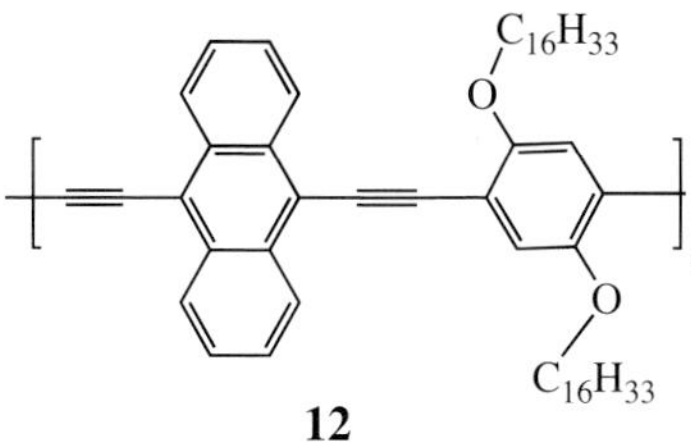

12

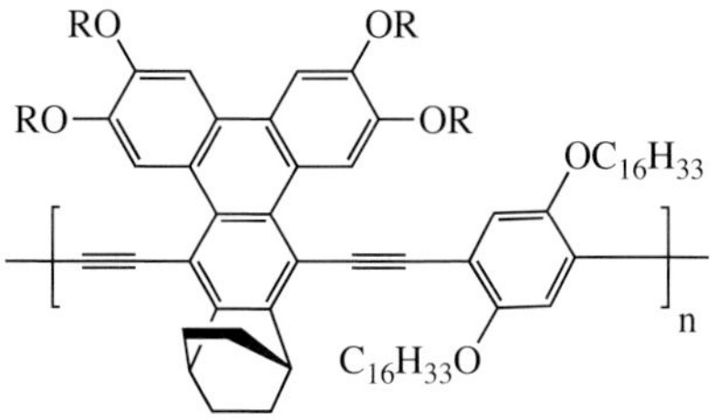

13 R = 2-ethylhexyl

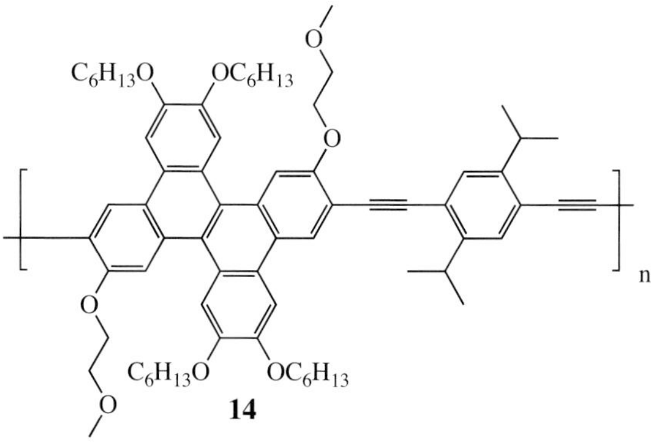

14

mer associations with chain alignment. This organizational propensity is driven by orbital overlap and anisotropic polarizability, with the resultant interpolymer electronic couplings dependably producing low-energy traps with decreased emission efficiencies. This situation has interesting parallels to classical descriptions of exciton coupled chromophores, in which a parallel eclipsed alignment of small molecules results in a cancellation of the aggregate's transition moment [36]. However, these classical models do not provide a completely satisfactory description of the properties of semiconducting polymers with extended electronic systems, and more recent theoretical studies on PPV model systems have articulated the differences associated with semiconducting polymers [37]. Nevertheless, it has been proposed that, similarly to classical exciton theory, an oblique arrangement (neither parallel or perpendicular) of coupled electronically polymeric chains will produce an emissive state. In recognition of this theoretical framework we shortly thereafter proposed that an unexpectedly highly emissive PPE aggregate derived its extraordinary properties from isolated oblique aggregates [38]. We reasoned that polymer **15** was coerced to adopt isolated oblique electronic interactions between the polymer chains by the cyclophane macrocycles that prevent strong interpolymer electronic interactions in parallel interchain geometries (Figure 6.5). Consistent with this explanation, similar aggregates could not be produced when films of **15** were produced by a Langmuir–Blodgett method. Under these deposition conditions the polymers were forced to have parallel chains originating from their assembly at the air/water interface.

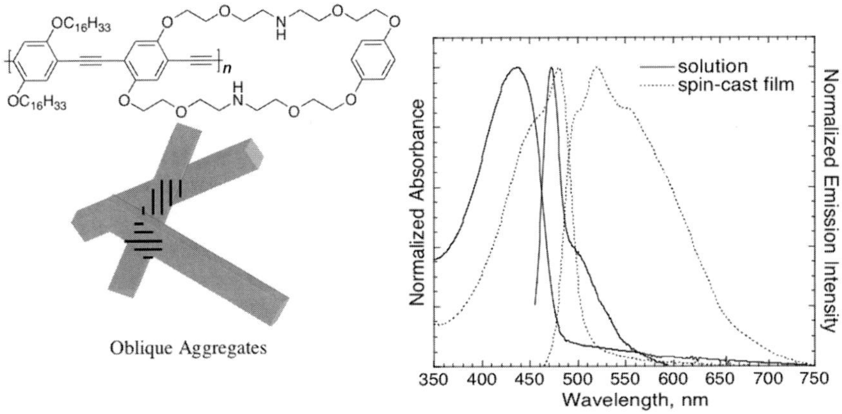

Figure 6.5 Proposed oblique aggregate of polymer **15** giving rise to the red-shifted absorption and emission spectra in spin-cast films. The broad nature of the thin film emission spectrum is typical of an aggregate, although unlike typical aggregates this one has a high quantum yield of 21%.

Our understanding that an oblique arrangement of polymers can give an emissive state with strong interchain interactions was a guiding principle in our design of highly fluorescent PPEs with truly three-dimensional electronic structures. Unfortunately, cyclophane residues as in **15** yield only metastable oblique aggregates. We have therefore introduced chiral side chains onto the polymers in order to generate extended structures with non-canceling transition moments reproducibly, by producing a torque between associated chains [39]. In general, we seek to assemble the polymers into a helical structure with the polymer chains oriented perpendicularly to the helical axis. Simple substitution of chiral side chains in polymer **16** was found to introduce this helical arrangement only as a metastable transient state. The final assembled state, which evolves from the addition of a poor solvent that induces aggregation, was non-emissive and lacked the chiral-optical properties of a helical assembly. These studies revealed that in order to stabilize the oblique structure it is necessary to prevent the formation of more favored assemblies of parallel aligned polymer chains. To accomplish this we made use of the pentiptycene group used earlier in the design of chiral polymer **17**, which is incapable of parallel interchain electronic interactions [40]. This approach was very successful. As shown in Figure 6.6, the rigid structure of the pentiptycene produces a notched polymer structure and a very stable interlocking grid. The assembly of the grid is kinetically slow, as it requires a considerable amount of organization. Also shown schematically in Figure 6.6, spectroscopic studies reveal that the assembly of the grid proceeds through an initial association of less ordered polymer chains and structurally evolves to give the final highly emissive aggregate.

The three-dimensional electronic structure of polymer **17**'s chiral grid produces a highly emissive state and has important implications for the applications of electronic polymers. These design principles enhance the sensory properties [40] and pave

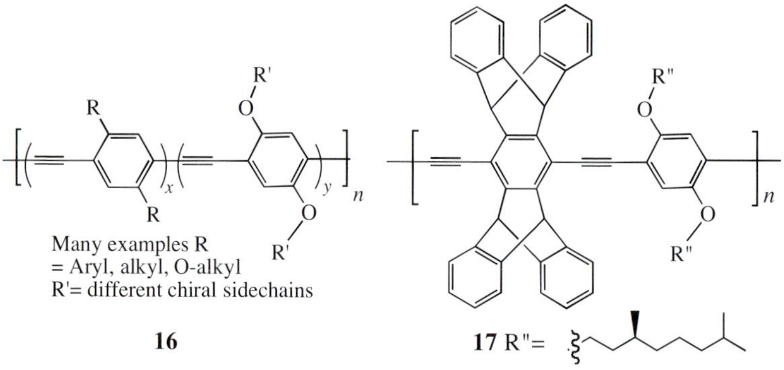

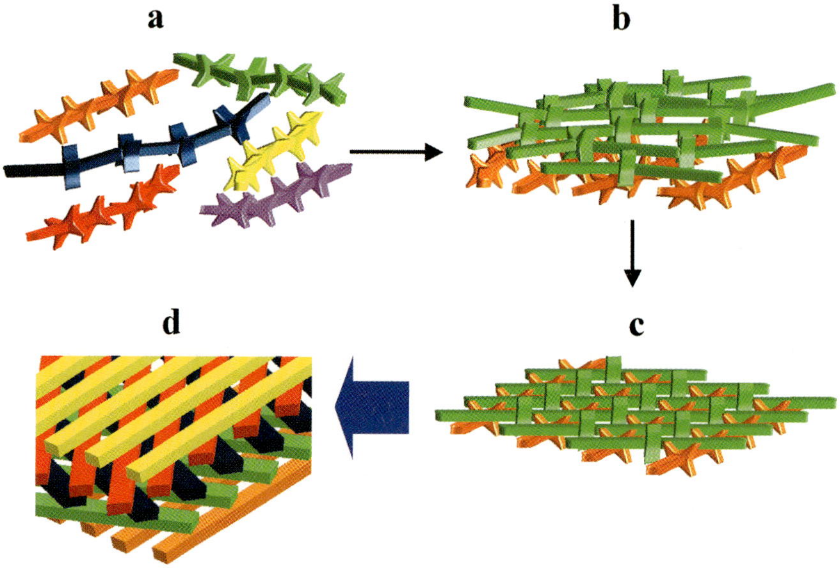

Figure 6.6 The process for the assembly of polymer **17** into the chiral helical structure **d**. The polymers begin in solution as non-interacting polymer chains (**a**) and with the addition of a poor solvent polymer **17** begins to associate. This initially produces a less organized phase **b**, which structurally evolves into a more regular interlocking grid **c** with an extended helical structure **d**.

the way to new photovoltaic materials, which require long-lived mobile excited states and three-dimensional electronic structures to transport charge (electrons and holes) efficiently.

6.5
Sensor Applications

Much of the present interest in PArEs is related to their use in the amplification of sensory responses [41]. The general ability of an electronic material to produce gain (amplification) is derived from its transport properties and for electronic polymers this has been demonstrated for the transport of both charge and energy [42]. We initially embarked on our sensory studies of PPEs in order to establish that individual molecular wires could be made to amplify a transduction event. The notion of using an optical sensory signal, rather than an electrical signal, was particularly attractive to us because of the difficulty in making resistivity measurements on individual wires. By using fluorescence spectroscopy it is possible to study dilute solutions of isolated molecular wires.

The initial fundamental issue to be addressed in our molecular wire proposal was: "will the hard wiring of individual sensory molecules together with molecular circuitry produce an enhanced sensory response?" To answer this question we designed monomeric (**18**) and polymeric (**19**) fluorescent sensory materials with a cyclophane macrocycle [43] known to bind methyl viologen [44], a well known electron-accepting fluorescent quenching agent (Figure 6.7). The amplification in this system is a result of the high mobility of the photon-induced excited state generated on the polymer. As in all organic molecules there is a geometric relaxation around this excited state. This distortion is localized and has a physical dimension. As a result these species are best described by a quasiparticle formalization, and the excited states are called excitons. When an exciton encounters a receptor occupied by methyl viologen, a highly efficient electron transfer reaction occurs and the polymer is returned to its ground state without emission of a photon. This process gives an effective binding constant that is the product of the number of binding sites visited by the exciton and the binding constant of an individual receptor. By design, the receptors on the polymer and the monomeric compound exhibited the same binding constant to methyl viologen, and hence Stern–Volmer quenching studies represented the net amplification provided by the exciton migration [43]. In these studies we were careful to factor out the binding constant from the response to achieve a true amplification factor. Systems lacking well defined receptors can provide for additional complications, in which the form of the polymer influences the effective binding constant. This has led to inconsistencies in the definition of the amplification factors. In more recent studies in which researchers also studied quenching by viologens, for example, they included the binding constants in their reported amplification [45], thereby reporting amplification factors that may be more than 10,000 times the amplification that can be ascribed to exciton transport.

By using discrete methyl viologen receptors wired in series, we were able to determine the diffusion length of an exciton on an isolated PPE chain in solution unambiguously [43]. To accomplish this we investigated a series of polymers of varying molecular weights. When the length of the polymers exceeded the diffusion length of the exciton, we found that further increases in molecular weight did not extend the exciton diffusion length. By this method we found that the exciton

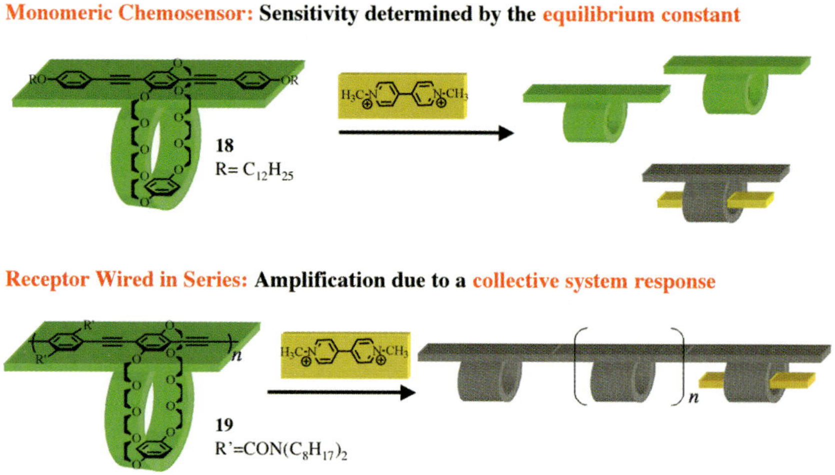

Figure 6.7 Monomeric (top) and polymeric (bottom) fluorescent receptors for methyl viologen. In solution the monomeric system gives a static 1:1 quenching with methyl viologen binding, but for polymers with $n \leq 67$ a single binding event quenches the entire polymer.

could visit 67 receptor units or 134 phenylene ethynylene units. Although this represents a significant amplification, we have shown that this effective one-dimensional (intrapolymer) diffusion of excitons represents a small fraction of the true ability of PArEs to amplify sensory responses. To understand this limitation one must consider that the excitons in our initial experiments were constrained to diffuse on a single polymer in a one-dimensional random walk. The statistics of such a diffusion are such that for an exciton to diffuse over 134 sites the exciton must make $(134)^2$ random stepwise movements. In this situation the majority of the exciton's motions are unproductive, as it may visit some receptor sites many times.

A straightforward and simple approach to improved sensitivity is to put the sensory polymers into bulk films in which the excitons are free to diffuse in three dimensions. The enhancement with increased dimensionality has been studied in detail by systematic studies of thin films built in consecutive layers by Langmuir–Blodgett deposition methods [46]. As a result of the improved energy migration in three dimensions, luminescent thin films exhibit considerably greater amplifying characteristics. A particularly noteworthy application of the extreme signal enhancement that is possible is the high sensitivity of polymer **11** to trinitrotoluene (TNT) [30, 31]. As shown in Figure 6.8, TNT is a very good electron-accepting fluorescence quencher, and polymer **11** and related materials provide a basis for new explosives detection systems. An advantage of the higher sensitivity PArE sensor systems over conventional ion mobility systems deployed at airports is that PArE detectors have sufficient sensitivity to detect TNT vapors. Conventional ion mobility spectrometers currently deployed in airports require much larger quantities of explosives, and hence explosive particles must be present. The most demanding explosives detection application is perhaps the detection of buried landmines. Given

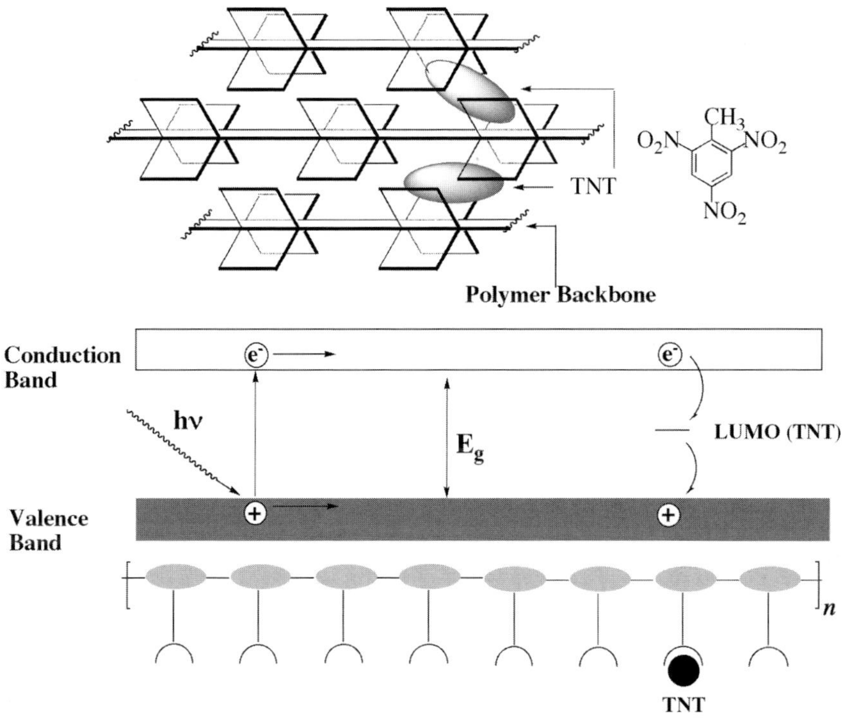

Figure 6.8 Thin films of polymer **11** shown schematically have a porous structure that binds TNT and produces an amplified sensory response. The electron-hole pairs (excitons) migrate throughout the structure and, when they encounter TNT, electron transfer quenching occurs. The fact that the exciton can visit many ($>10^4$) repeating units increases the probability of encountering TNT and produces an ultra-sensitive sensor.

that the equilibrium vapor pressure of TNT is only 7 ppb, the explosive vapor that exists in the head-space over a landmine is indeed a challenging signal to detect. Nevertheless, sensors using PArEs have been able to detect landmines in the field [47].

6.6 Superstructures

The properties of electronic polymers universally depend on their environment and their conformational properties. Although most of the PArEs investigated are linear rigid-rod polymers, in reality they have random coil structures in solution with large persistence lengths [48]. The entropy afforded by a random coil structure assists in making the polymers soluble and generally their solubility will be greatly reduced if the polymers are made sufficiently rigid to force a completely linear structure. Nevertheless, the formation of highly organized chain-extended structures is not only key to fundamental investigations of PArEs, but is also important for the production of materials with optimized properties.

The confinement of PArE chains to a two-dimensional air/water interface dramatically reduces a material's conformational space and favors a more extended structure. The face-on structure, which avoids strong enthalpic interactions between neighboring polymers, produces a two-dimensional liquid crystal phase that is readily aligned by anisotropic compression and flow. Studies at the air/water interface and in highly aligned deposited films have been indispensable for establishing fundamental spectroscopic and transport investigations [46]. The ability to add one layer at a time has further utility for the formation of nanostructures. In this regard we have been interested in producing films for directed energy transport, a process with broad applicability in many photonic technologies. Langmuir–Blodgett processes offer an efficient approach to these systems by allowing sequential addition of polymer layers of varying band-gaps. To promote dipolar energy transfer between layers the emission of a donating polymer should have high spectral overlap with the absorption of the acceptor. Three different PArEs that satisfy this criteria are polymers **20**, **21**, and **22**. Stratified thin films of these materials prepared by Langmuir–Blodgett deposition displayed extremely efficient vectorial energy transfer as shown schematically in Figure 6.9 [49].

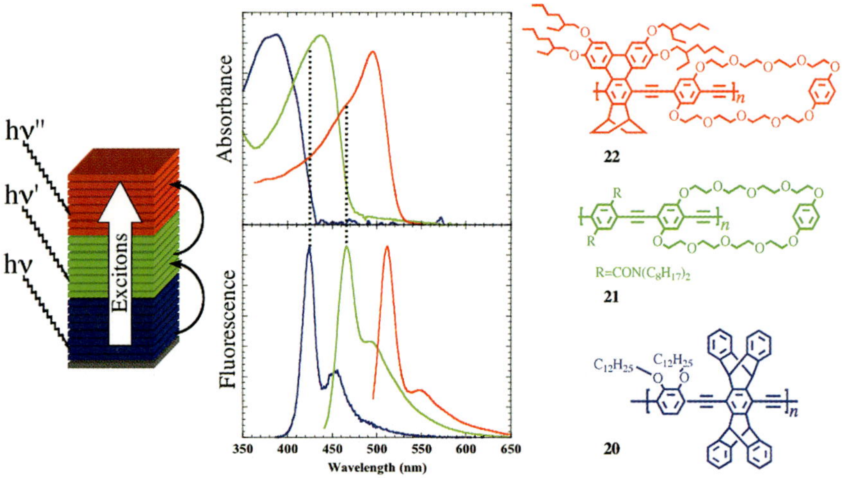

Figure 6.9 Polymers **20**, **21**, and **22** are assembled into a multilayer structure for vectorial energy transport. The color-coded emission and absorption spectra illustrate the overlap between the emission of the donor polymers with the absorption of the acceptors. As shown on the left, a stratified assembly of these materials in multilayer films gives a system capable of harvesting a broad spectrum of light and directionally transporting the excitons to the lowest band-gap polymer **22**.

Arrays of aligned nanofibrils are also of interest for study of electronic transport in polymers in restricted arrangements and for investigation of their nanomechanics. Nanofibrils can be readily prepared with bulk alignment by a surface reconstruction that we have observed in aligned PPE monolayers deposited by the Langmuir–Blodgett method [50]. These nanostructures form spontaneously when amphiphilic PPEs are deposited upon a non-polar interface. As schematically illustrated in Figure 6.10, the thus deposited polymers have high surface energies (polar functionality exposed to the air) and weak anchoring to the surface (hydrophobic interactions). To lower their energy, the polymers roll into cylinders with their polar groups aligned inward for improved dipolar stabilization. Critical elements of this assembly process are the rigid rod structures and the high degrees of chain alignment of the PPEs. Without these factors the films would separate into irregular globular domains, and indeed such behavior is achieved if the films are heated to a temperature at which the alignment is lost due to polymer dynamics.

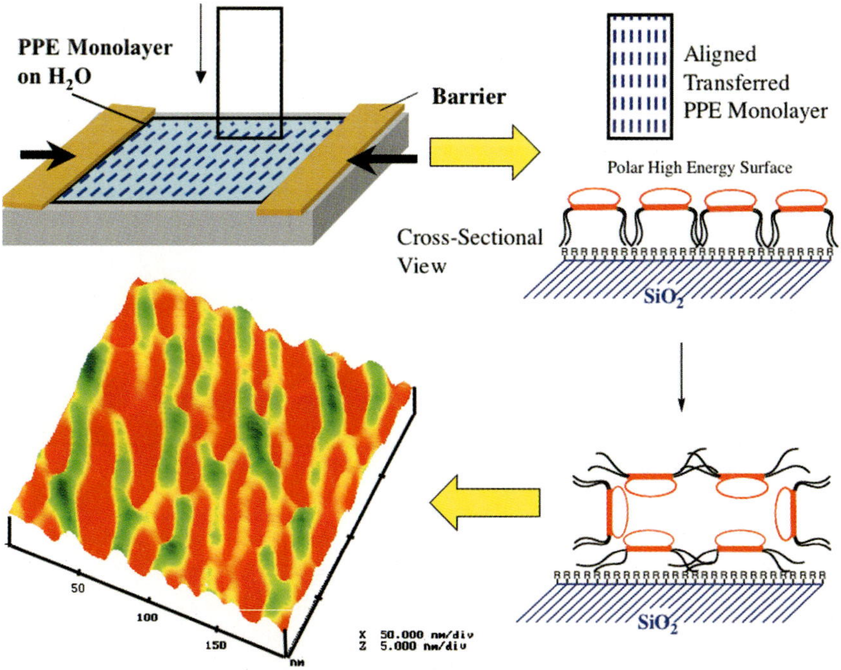

Figure 6.10 Assembly process for formation of nano-filaments from PPEs such as polymer **19**. The polymers are first assembled into an aligned two-dimensional liquid crystal film at the air/water interface of a Langmuir–Blodgett trough and are then transferred to a hydrophobic substrate (functionalized silica). The resultant high-energy polar surface then reconstructs and the polymers roll up into nano-filaments that are imaged with atomic force microscopy (bottom left) and are observed to preserve the bulk directional alignment produced in the deposition.

The power of the Langmuir–Blodgett method is that it allows great precision in the study and organization of PArEs. However, there is the need to process these materials from solution. We have been interested in developing methods capable of producing aligned polymers in solution that can perhaps be used to produce very precise thin films in mass. To this end we have been actively developing other methods to align polymers in solution or in bulk films. One method makes use of nematic liquid crystalline solvents that can be readily aligned with fields or interfaces over macroscopic distances. The anisotropic one-dimensional order in a nematic liquid crystal solvent favors chain-extended conformations of linear rigid-rod PArEs rather than the random coils that dominate in isotropic solvents. To accomplish this alignment it is necessary to produce PArEs with high solubility in the liquid crystalline solvents. This property is not as easily achieved as might be imagined at first glance. A cursory screening of a number of PArE and PPV polymers (which are readily soluble in typical non-polar or polar organic solvents) revealed that few of these materials are soluble in thermotropic liquid crystals. The instability of most polymer liquid crystalline solvents can be understood when one considers that the polymers are being forced into chain-extended conformations that have reduced entropy. These conformations are also poised to aggregate or crystallize.

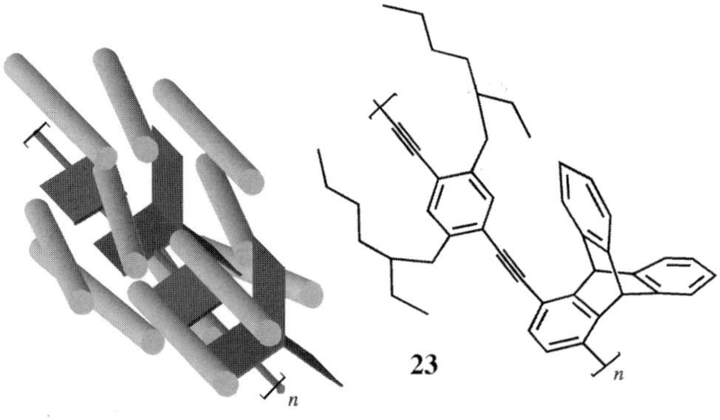

Figure 6.11 Polymer **23** and a schematic representation of its interaction with a nematic liquid crystalline solution. The liquid crystalline molecules force an alignment of the polymer to produce a structure that best fills space around the triptycene groups and the polymer backbone.

Therefore, to create stable solutions of conjugated polymers we needed to add functionality to the polymer to enhance interactions with the liquid crystal solvent and also to prevent strong interchain interactions. Both criteria are met by polymer **23** (Figure 6.11), which contains pendant triptycene groups [51]. The three-dimensional shape of the triptycene units prevents strong interchain π-interactions, similarly to the pentiptycene-containing PPEs (polymer **11**). This triptycene structure also has the novel effect of producing a high degree of interaction and directional coupling with the liquid crystalline solvent [52]. This latter effect is a result of the concave clefts formed by the aromatic surfaces of the triptycene, which, when dissolved in a liquid crystalline solvent, induce a preferred organization of the liquid crystalline molecules about the polymer (Figure 6.11). As is readily apparent from simple inspection of the polarized absorption spectra from an aligned liquid crystalline solution, the PPE chains are highly aligned and in a chain-extended conformation. An additional feature apparent from comparisons of the polymer's absorption spectrum in a variety of media is that the conjugation length is considerably enhanced in the liquid crystalline solution relative to that observed in an isotropic solvent (methylene chloride in this case). In solution, semiconducting polymers generally have a broad emission that reflects a composite absorption from a number of different conjugation lengths, and the polymers can best be considered as a polychromophoric mixture. This distribution of chromophores generally sums to give the observed spectrum with a Gaussian character, as is the case in the methylene chloride solution of the PPE (Figure 6.12). In contrast, PPE dissolved in the liquid crystal is distinctly non-Gaussian and has a very distinct nearly vertical absorption profile at its low energy band edge. This unusual spectral shape indicates that the polymer is in an extended and highly planarized conformation and is consistent with what would be calculated from its one-dimensional band structure with an infinite conjugation length. For PPEs, as in the case for all one-dimensional systems, the highest density of states (most orbitals per unit energy) occurs at the band edges [53]. Given the allowed nature of the transitions between the valence band (HOMO band) and the conduction band (LUMO band), we expect that the highest absorption coefficient should be the lowest energy transition. This is exactly what we observe, and the liquid crystal solutions of PPEs therefore not only yield chain extension and macroscopic alignment, but also produce an optimized electronic structure. This ability to organize electronic polymers with nearly perfect electronic structures opens exciting new possibilities for studying electronic transport in these systems and is the subject of ongoing investigations.

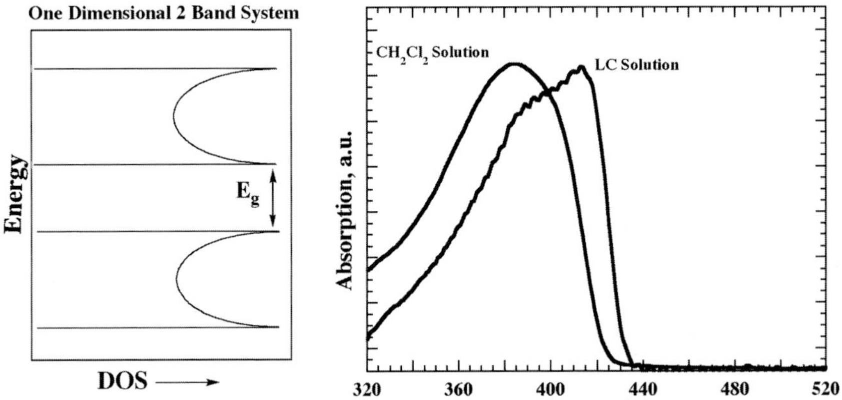

Figure 6.12 Shown on the left is a schematic generic representation of the density of states (DOS) for a one-dimensional system in which there is a peak at the band edge. On the right is the absorption spectrum of polymer **23** in a nematic liquid crystalline phase (1-(*trans*-4-hexylcyclohexyl)-4-isothiocyanatobenzene T_m = 12.4 °C, T_{NI} = 42.4 °C) and in methylene chloride solution. Notice that the steep rise of the absorption is what would be expected on the basis of a simple DOS argument.

Polymer hosts can also be used to create highly aligned PPEs. This was first demonstrated by mixing PPEs in high molecular weight polyethylene, which can be aligned by mechanical stretching [54]. By this method, materials with very high optical anisotropy were obtained, and it was suggested that these materials could find applications in the brightening of liquid crystal displays. A further extension of this approach was to mix small molecule dyes that exhibit a random orientation into the polyethylene. These could then absorb polarizations other than those aligned with the transition moment of the polymer. In this way the dyes transfer energy to the polymer films and thereby harvest light of all polarizations but emit light of one dominant polarization.

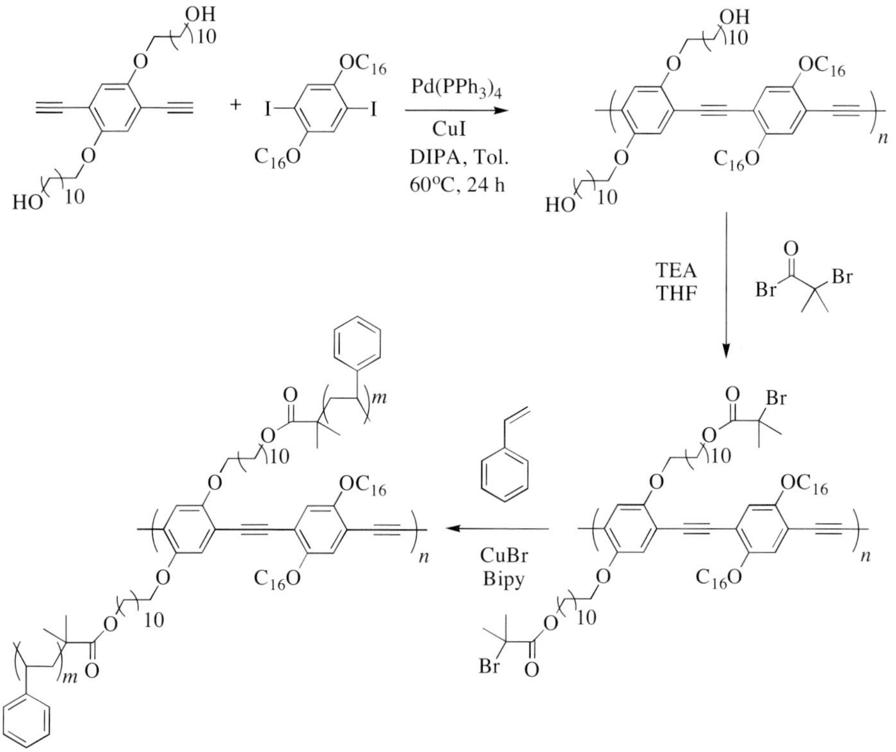

Scheme 6.1 Synthesis of polystyrene grafted PPE by ATRP polymerization.

We have developed an alternative method for the alignment of PPEs, based upon their confinement to one phase of a nanostructured monodisperse block polymer. In doing so we have made use of the bulk alignment of cylinder phases of block copolymers by using the roll-casting method developed by Thomas [55]. To test this scheme we sought to confine PPE guests selectively in the minor (cylinder) component of a cylinder phase of a low polydispersity polystyrene-polyisoprene-polystyrene block copolymer (M_n = 101,000, PDI = 1.05) that could be readily roll-cast into highly ordered films. To produce the desired phase partitioning of the PPE into the minor polystyrene component we initially functionalized polymers as shown in Scheme 6.1 with initiating groups for an atom transfer radical polymerization (ATRP) [56]. The ATRP method then allowed us to initiate polystyrene chains from the polymer backbone in a controlled fashion that should generate grafted chains of moderate to low polydispersity [57]. The overall alignment of the PPEs is confirmed by polarized UV/Vis spectroscopy of the bulk films. Interestingly, the emission indicated a higher polarization than the absorption, which suggests that the same exciton migration that produces enhanced sensory responses is operative in these materials. As illustrated in Figure 6.13, this behavior can be accounted for by considering that the most extended highly conjugated polymer chains are best aligned with the polystyrene cylinders. However, at the interfaces of the cylinders the polymers have more of a coiled higher band-gap structure. In this way excitons created on the coiled portion of the polymers migrate to the highly aligned lower band-gap regions. In so doing, the polymer harvests light from both parallel and perpendicular polarizations and can emit more dominantly with polarization parallel to the cylinders.

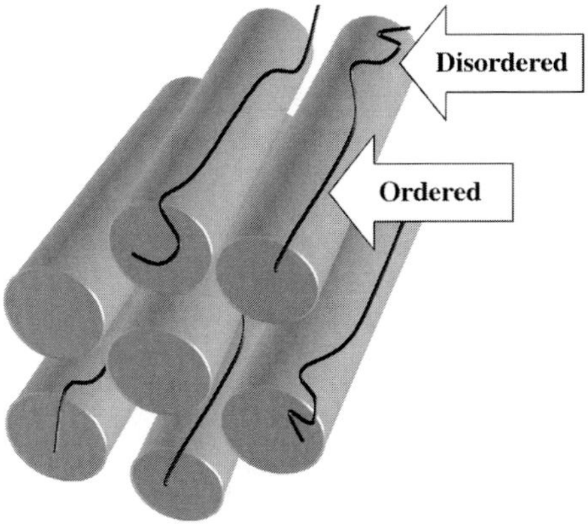

Figure 6.13 Schematic representation of polystyrene cylinders of an aligned block copolymer film with dissolved PPE chains. There are ordered and disordered regions of the PPE, with the ordered regions having a lower band-gap and being better aligned with the cylinders. Absorption of photons in the higher disordered regions creates excitons that migrate to the lower band-gap regions and hence the emission polarization is higher than the absorption.

6.7 Summary

Poly(aryleneethynylenes) have emerged as one of the most important classes of semiconducting polymers. Among their important properties are their rigid shape-persistent nature, their outstanding photophysics, their structural diversity, and sensory properties. It is clear that this class of materials will be an important element in many new emerging technologies.

6.8 General Procedures for Synthesis of PPEs

The synthesis of PPEs is relatively straightforward but the conditions required can vary considerably, depending on the solubility of the monomers and polymers and the reactivity of the organic halide. As mentioned earlier, to create polymers of the highest molecular weight a slight excess of the acetylene monomer is required. The quantity of excess acetylene monomer required to compensate for diacetylene formation will vary (1–3 %) depending upon the conditions and monomers.

The following is a procedure for the formation of polymer **11**, the long reaction time used reflecting the low solubility of the pentiptycene monomer. The monomers were used in a strict 1:1 ratio to provide a molecular weight of 144,000 (PDI = 2.6). Under an atmosphere of argon, diisopropylamine/toluene (2:3, 2.5 mL) solvent was added to a 25 mL Schlenk flask fitted with a Teflon^{TM} valve capable of holding high vacuum, containing the pentiptycene compound appended centrally with two terminal alkynes (40 mg, 0.084 mmol), 1,4-bis(tetradecanyloxyl)-2,5-diiodobenzene (63 mg, 0.084 mmol), CuI (10 mg, 0.053 mmol), and $Pd(PPh_3)_4$ (10 mg, 0.0086 mmol). This mixture was sealed with the Teflon^{TM} valve, heated at 65 °C for three days, and then subjected to a $CHCl_3/H_2O$ workup. The combined organic phases were washed with NH_4OH solution, NH_4Cl solution, and water, and then dried ($MgSO_4$). The solvent was removed in vacuo, and the residue was dissolved in CH_2Cl_2, precipitated in methanol, and washed with excess methanol. The solid was then dried and subjected twice more to the same precipitation purification procedure. The resultant polymer **11** was a yellow solid with a strong green emission (76 mg, 75 %).

Acknowledgements

I have been privileged to collaborate with many talented students and postdoctoral workers who have shared my love affair with PArEs. Our initial work was funded by the National Science Foundation and the Office of Naval Research. The results discussed in this review were also funded by the Department of Energy, the Army Research Office, the National Institutes of Health, the National Aeronautics and Space Administration, and the Defense Advanced Research Projects Agency.

References

[1] C. K. Chiang, C. R. Fincher Jr., Y. W. Park, A. J. Heeger, H. Shirakawa, E. J. Louis, S. C. Gau, A. G. MacDiarmid, *Phys. Rev. Lett.* **1977**, *39*, 1098–1101.

[2] (a) "Polydiacetylenes: Synthesis, Structure, and Electronic Properties" *NATO Advanced Science Institutes Series, Series E, Applied Sciences*, No. 102, Eds, D. Bloor, R. R. Chance, Martinus Nijhoff Publishers, Dordrecht, **1985**; (b) G. Wegner, *Pure Appl. Chem.* **1977**, *49*, 443–454.

[3] I use PArE, rather than PAE as has been often done in the literature, because the latter acronym is generally associated with poly(aryl ether)s, a large and well known class of materials.

[4] *Handbook of Conducting Polymers*, 2nd edition, Revised and Expanded, Eds. T. A. Skotheim, R. L. Elsenbaumer, J. R. Reynolds, Marcel Dekker, New York, **1997**.

[5] J.-L. Brédas, R. R. Chance, R. H. Baughman, R. Silbey, *J. Chem. Phys.* **1982**, *76*, 3673–3678.

[6] J. K. Young, J. S. Moore, *Modern Acetylene Chemistry*, Eds. P. J. Stang and F. Diederich, VCH, Weinheim, **1995**, pp. 415–442.

[7] R. Giesa, *Rev. Macromol. Chem. Phys.* **1996**, *C36*, 631–670.

[8] U. H. F. Bunz, *Chem. Rev.* **2000**, *100*, 1605–1644.

[9] These materials have been studied extensively to confirm their linear natures and their persistence lengths. P. M. Cotts, T. M. Swager, Q. Zhou, *Macromolecules* **1996**, *29*, 7323–7328.

[10] R. R. Schrock, *Polyhedron* **1995**, *14*, 3177–3195.

[11] K. Weiss, A. Michel, E.-M. Auth, U. H. F. Bunz, T. Mangel, K. Müllen, *Angew. Chem. Int. Ed. Engl.* **1997**, *36*, 506–509; *Angew. Chem.* **1997**, *109*, 522–525.

[12] A. Fürstner, C. Mathes, C. W. Lehmann, *J. Am. Chem. Soc.* **1999**, *121*, 9453–9454.

[13] G. Brizius, N. G. Pschirer, W. Steffen, K. Stitzer, H. C. zur Loye, U. H. F. Bunz, *J. Am. Chem. Soc.* **2000**, *122*, 12435–12440.

[14] For many examples see: K. J. Ivin, *Olefin Metathesis*, Academic Publishers, New York, **1983**.

[15] L. Kloppenburg, D. Jones, U. H. F. Bunz, *Macromolecules* **1999**, *32*, 4194–4203.

[16] T. Masuda, T. Higashimura, *Adv. Polym. Sci.* **1987**, *81*, 121–165.

[17] W. Zhang, S. Kraft, J. S. Moore, *J. Am. Chem. Soc.* **2004**, *126*, 329–335.

[18] M. V. Lakshmikantham, J. Vartikar, K. Y. Jen, M. P. Cava, W. S. Huang, A. G. MacDiarmid, *Poly. Prepr.* **1983**, *24*, 75.

[19] K. Sanechika, T. Yamamoto, A. Yamamoto, *Bull. Chem. Soc. Jpn.* **1984**, *57*, 752–755.

[20] M. Tateishi, H. Nishihara, K. Aramaki, *Chem. Lett.* **1987**, 1727–1728.

[21] G. V. Tormos, P. N. Nugara, M. V. Lakshmikantham, M. P. Cava, *Synth. Met.* **1993**, *53*, 271–281.

[22] D. Ofer, T. M. Swager, M. S. Wrighton, *Chem. Mater.* **1995**, *7*, 418–425.

[23] G. Zotti, G. Schiavon, S. Zecchin, A. Berlin, *Synth. Met.* **1998**, *97*, 245–254.

[24] These attractive interchain interactions produce strong electronic coupling in both the ground and excited states. They have often erroneously been referred to as excimers in the literature. The term "excimer" refers to an attractive association that exists in the excited state with no interactions in the ground state.

[25] G. Wegner, *Thin Solid Films* **1992**, *216*, 105–116.

[26] D. T. McQuade, J. Kim, T. M. Swager, *J. Am. Chem. Soc.* **2000**, *122*, 5885–5886.

[27] For a recent review see: B. J. Schwartz, *Annu. Rev. Phys. Chem.* **2003**, *54*, 141–172.

[28] J. Kim, T. M. Swager, *Nature* **2001**, *411*, 1030–1034.

[29] J. Kim, I. A. Levitsky, D. T. McQuade, T. M. Swager, *J. Am. Chem. Soc.* **2002**, *124*, 7710–7718.

[30] J.-S. Yang, T. M. Swager, *J. Am. Chem. Soc.* **1998**, *120*, 5321–5322.

[31] J.-S. Yang, T. M. Swager, *J. Am. Chem. Soc.* **1998**, *120*, 11864–11873.

[32] S. Winstein, R. Adams, *J. Am. Chem. Soc.* **1948**, *70*, 838–840.

[33] T. M. Swager, C. J. Gil, M. S. Wrighton, *J. Phys. Chem.* **1995**, *99*, 4886–4893.

[34] A. Rose, C. G. Lugmair, T. M. Swager, *J. Am. Chem. Soc.* **2001**, *123*, 11298–11299.

[35] S. Yamaguchi, T. M. Swager, *J. Am. Chem. Soc.* **2001**, *123*, 12087–12088.

[36] M. Kasha, *Spectroscopy of the Excited State*, Ed., B. Di Bartolo, Plenum Press, New York, **1976**, pp. 337–363.

[37] J.-L. Brédas, J. Cornil, D. Beljonne, D. A. dos Santos, Z. Shuai, *Acc. Chem. Res.* **1999**, *32*, 267–276.

[38] R. Deans, J. Kim, M. Machacek, T. M. Swager, *J. Am. Chem. Soc.* **2000**, *122*, 8565–8566.

[39] In our original cyclophane system the oblique interchain associations and minor species are probably limited to two polymer chains. To create a three-

dimensional electronic structure we need extended oblique interactions. If these interactions produce centrosymmetric electronic states, there could still be cancellation of the material's overall transition dipole moment. However, if the system is made to be chiral then such a cancellation is not possible.

[40] S. Zahn, T. M. Swager, *Angew. Chem. Int. Ed. Engl.* **2002**, *41*, 4225–4230; *Angew. Chem.* **2002**, *114*, 4399–4404.

[41] D. T. McQuade, A. E. Pullen, T. M. Swager, *Chem. Rev.* **2000**, *100*, 2537–2574.

[42] T. M. Swager, *Acc. Chem. Res.* **1998**, *31*, 201–207.

[43] (a) Q. Zhou, T. M. Swager, *J. Am. Chem. Soc.* **1995**, *117*, 7017–7018; (b) Q. Zhou, T. M. Swager, *J. Am. Chem. Soc.* **1995**, *117*, 12593–12602.

[44] P. L. Anelli, P. R. Ashton, R. Ballardini, V. Balzani, M. Delgado, M. T. Gandolfi, T. T. Goodnow, A. E. Kaifer, D. Philp, M. Pietraszkiewicz, L. Prodi, M. V. Reddington, A. M. Z. Slawin, N. Spencer, J. F. Stoddart, C. Vicent, D. J. Williams, *J. Am. Chem. Soc.* **1992**, *114*, 193–218.

[45] L. Chen, D. W. McBranch, H.-L. Wang, R. Helgeson, F. Wudl, D. G. Whitten, *Proc. Natl. Acad. Sci. USA* **1999**, *96*, 12287–12292.

[46] I. A. Levitsky, J. Kim, T. M. Swager, *J. Am. Chem. Soc.* **1999**, *121*, 1466–1472.

[47] C. J. Cumming, C. Aker, M. Fisher, M. Fox, M. J. la Grone, D. Reust, M. G. Rockley, T. M. Swager, E. Towers, V. Williams, *IEEE Transactions on Geoscience and Remote Sensing* **2001**, *39*, 1119–1128.

[48] The persistence length for a worm-like chain model was determined to be 150 Å by detailed light scattering studies (Ref. 9).

[49] J. Kim, D. T. McQuade, A. Rose, Z. Zhu, T. M. Swager, *J. Am. Chem. Soc.* **2001**, *123*, 11488–11489.

[50] J. Kim, S. K. McHugh, T. M. Swager, *Macromolecules* **1999**, *32*, 1500 –1507.

[51] Z. Zhu, T. M. Swager, *J. Am. Chem. Soc.* **2002**, *124*, 9670–9671.

[52] We had previously demonstrated for small molecules that triptycene functionality could be used to create guest dyes that exhibit higher degrees of interaction than their non-triptycene analogues with their liquid crystalline hosts: T. M. Long, T. M. Swager, *J. Am. Chem. Soc.* **2002**, *124*, 3826–3827.

[53] R. Hoffmann, *Solids and Surfaces: A Chemist's View of Bonding in Extended Structures*, VCH Publishers, New York, **1988**.

[54] (a) C. Weder, C. Sarwa, A. Mantoli, C. Bastiaansen, P. Smith, *Science* **1998**, *279*, 835–837; (b) A. Mantoli, C. Bastiaansen, P. Smith, C. Weder, *Nature* **1998**, *392*, 261–264.

[55] R. J. Albalak, E. L. Thomas, *J. Polym. Sci., Polym. Phys. Ed.* **1994**, *32*, 341–350.

[56] C. A. Breen, T. Deng, T. Breiner, E. L. Thomas, T. M. Swager, *J. Am. Chem. Soc.* **2003**, *125*, 9942–9943.

[57] H. G. Börner, K. Beers, K. Matyjaszewski, S. S. Sheiko, M. Moller, *Macromolecules* **2001**, *34*, 4375–4383.

7
Polyynes via Alkylidene Carbenes and Carbenoids

Sara Eisler and Rik R. Tykwinski

7.1
Introduction

Interest in polyalkynyl-containing structures has exploded over the last two decades, and sp-hybridized carbon building blocks have been incorporated into a great number of carbon-rich molecules [1]. In addition to their fascinating structures, many of these molecules have unique electronic, mechanical, and structural properties that make them attractive as advanced materials [2]. Cyclic carbon allotropes consisting entirely of sp-hybridized carbon have also been objects of intense interest for many years [3–4], while carbyne, the hypothetical linear form of sp-hybridized carbon, has continued to be a significant synthetic challenge [5–6].

The desire to incorporate alkynyl moieties into carbon-rich scaffolding requires synthetic methods that can accomplish this goal in a facile manner. The three methods most commonly used to form polyynic structures are the Cadiot–Chodkiewicz reaction [7], the Sonogashira cross-coupling reaction [8], and oxidative homocoupling by the Glaser, Hay, or Eglinton/Galbraith techniques [9]. The common link between these methods is the need for a terminal acetylene as a starting material; such molecules can be unstable and difficult to manipulate.

In general, the formation of extended polyyne systems can be problematic, due to the usual correlation between a dramatic decrease in stability and an increase in the number of conjugated acetylene units. Several strategies to circumvent the need for terminal alkynes as precursors to larger systems have been developed. Approaches that avoid a separate deprotection step, for example, include the desilylation of a protected alkyne in situ [10], and the direct cross-coupling of a trialkylsilyl-protected acetylene [11]. Alternatively, the polyyne framework can be formed in the final synthetic step either through elimination or through extrusion of a suitable functional group [12]. While these routes have afforded a number of interesting derivatives, their generality has in many cases yet to be established.

The work described in this chapter outlines an emerging synthetic route to polyynes that circumvents the need for unstable terminal alkynes as precursors through the use of alkylidene carbene and carbenoid species. In varying degrees

Acetylene Chemistry. Chemistry, Biology and Material Science. Edited by F. Diederich, P. J. Stang, R. R. Tykwinski

ISBN 3-527-30781-8

of detail it also surveys some of the unique properties displayed by these newly obtained materials, including solid-state structure and stability, and the linear and nonlinear optical properties of conjugated polyynes.

7.2
Alkylidene Carbene and Carbenoid Species

Alkylidene carbenes/carbenoids (**1** and **2**) are reactive intermediates capable of undergoing several distinct reactions (Scheme 7.1) [13]. Carbenoid **1** and alkylidene carbene **2** are distinguished by the association of the carbon in **1** with a metal and a halogen (or another leaving group), while the free carbene **2** has no such ties. The two extremes can vary greatly in reactivity. Within this range of reactivity, solvent, β-substituent, and temperature can all affect the reaction outcome. For carbenoid intermediates, both the metal and the leaving group can also greatly influence the reaction [13e].

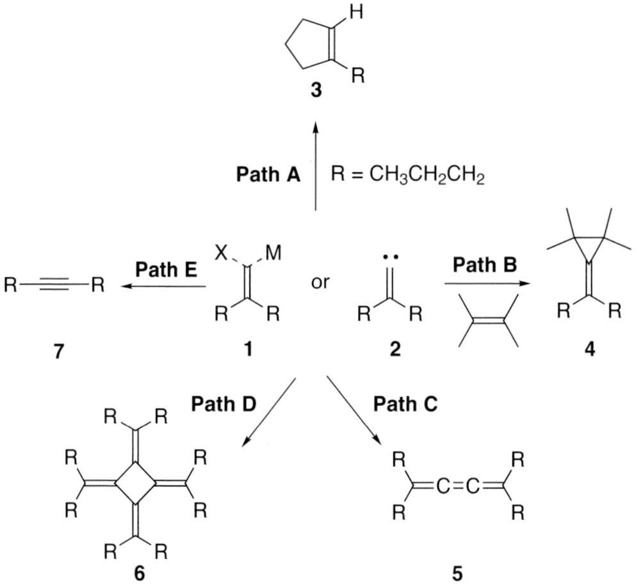

Scheme 7.1 Reactions of alkylidene carbenes/carbenoids.

Alkylidene carbenes **2** are perhaps most easily recognized as a convenient route to cyclopentenyl rings **3** through 1,5-insertion into an unactivated C−H bond (**Path A**). They are also progenitors to cyclopropylidene rings **4** by carbene addition into an olefin (**Path B**) [13a, 13b]. On the other hand, alternative (predominantly carbenoid) pathways can be quite effective for the formation of carbon-rich materials through oligomerization reactions, such as, for example, 1,2,3-butatrienes **5** (**Path C**) [14] and radialenes **6** (**Path D**) [15–16]. It is, however, **Path E** – the rearrangement of a carbenoid **1** (or carbene **2**) to an alkyne **7** – that represents a potential route to extended conjugated polyynes.

7.3
Alkyne Formation from Carbenes and Carbenoids

Three German chemists, Fritsch, Buttenberg, and Wiechell, demonstrated the ability of a carbenoid intermediate **1** to collapse to an alkyne through 1,2-migration of a pendent aryl moiety in 1894. The synthesis of an acetylene from a carbene or carbenoid precursor has thus commonly become known as the Fritsch–Buttenberg–Wiechell (FBW) rearrangement [17]. Since its discovery, a broad range of substituted tolans have been formed by application of the FBW rearrangement, as well as many acetylene derivatives with alkyl and vinyl substituents [13b]. While many of these reactions have been reviewed [13], it is worth surveying several examples to demonstrate the breadth of this reaction for alkyne formation.

7.3.1
Synthesis of Acetylenes: the Fritsch–Buttenberg–Wiechell Rearrangement

Fritsch, Buttenberg, and Wiechell's initial studies involved the treatment of a monohaloolefin such as **8** (Scheme 7.2) with a base at rather high temperatures to give the carbenoid intermediate **9**, with subsequent migration of one of the aryl groups to produce the tolan product **10**. Modern variants of the FBW rearrangement often involve treatment of a dihaloalkene such as **11** with an alkyllithium base (typically BuLi or PhLi) to form the carbenoid intermediate **9**, although a multitude of modifications to this general method have been reported [13].

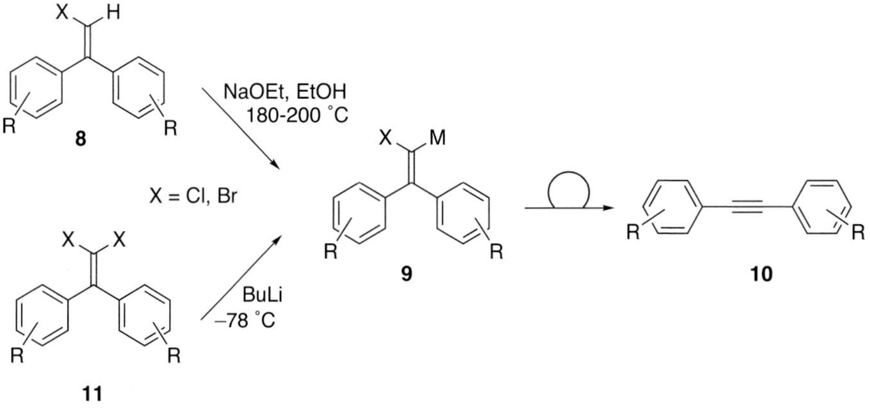

Scheme 7.2 The Fritsch–Buttenberg–Wiechell rearrangement.

As well as aryl groups as demonstrated in Scheme 7.2, heteroaryl moieties such as thiophenes also demonstrate a high migratory potential in the FBW reaction [18]. A clever one-pot route for symmetrical or unsymmetrical alkyne formation, including the formation of dithienylacetylene **12**, has been reported by Savignac and co-workers (Scheme 7.3). Addition of phosphonate **13** and ketone **14** to a solution of LiHMDS gave dichloroalkene **15**, and the subsequent addition of two equiv. of BuLi effected the high-yielding FBW rearrangement of **15** to **12**.

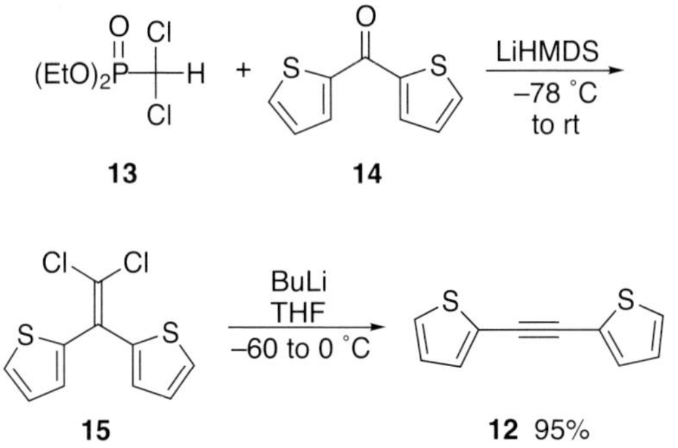

Scheme 7.3 Savignac's one-pot synthesis of acetylenes.

While there have been few systematic studies [19], empirically it has been established that aryl groups have a stronger tendency than alkyl moieties towards migration in FBW rearrangements. Nonetheless, methods that induce alkyl groups to undergo migration have been established. Normant and co-workers, for example, have recently demonstrated that substituted zinc carbenoids **16a** smoothly rearrange to form dialkyl-substituted alkynes such as **17** (Equation 7.1) [20]. Strategic ^{13}C labeling of the precursors **16** also afforded the identity of the migrating group in this study. Interestingly, the successful rearrangement of **16a** contrasts that of the analogous lithium carbenoid **16b**, which fails to form the acetylene **17** in greater than 10% yield [20b].

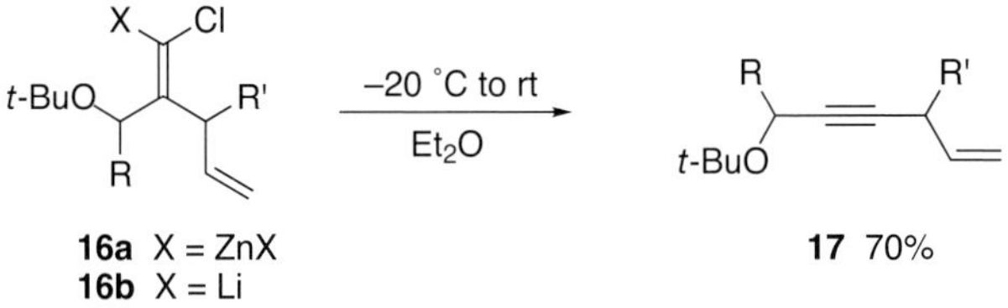

Equation 7-1

It is perhaps not surprising that the cyclopropyl group has a moderate migratory aptitude, intermediate between that of an alkyl and an aryl moiety [21]. This has been explored by Köbrich, who has shown that treatment of chloroolefin **18** with BuLi at low temperature effects rearrangement to **19** in decent yield (Equation 7.2).

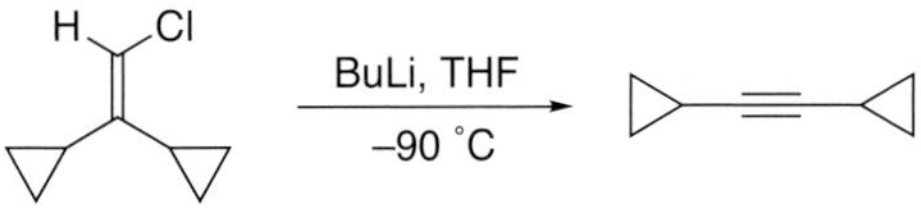

18 **19** 77% **Equation 7-2**

Likewise, the vinyl moiety has a reasonable migratory aptitude and is a willing participant in the FBW rearrangement under the appropriate conditions. For example, treatment of **20** with BuLi generates the carbenoid intermediate, followed by rearrangement to the enyne derivative **21** (Equation 7.3) [22]. It is interesting to note that the analogous rearrangement with the *Z*-isomer of the monochloroolefin **20** yields only 14% of the desired enyne **21**.

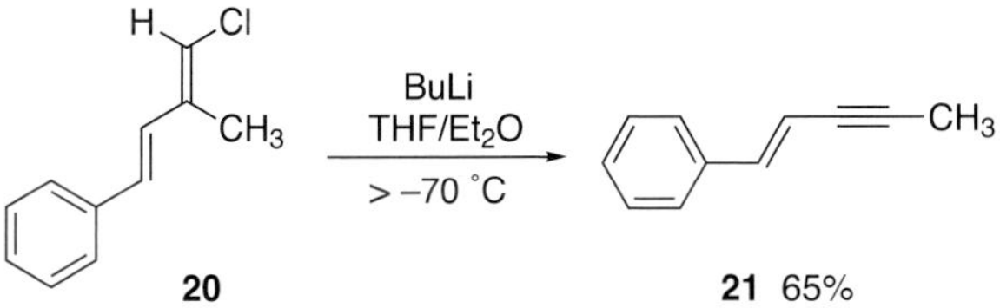

20 **21** 65% **Equation 7-3**

The intramolecular migration of a pendent group in a FBW rearrangement has long been a route for the formation of cyclic alkynes difficult or impossible to produce by more traditional means. The synthesis of incredibly strained cycloalkynes can be achieved by this approach, and the formation of cyclopentyne **22** represents a good example (Equation 7.4). Treatment of (bromomethylene)cyclobutane **23** with *t*-BuOK [23] or of (dibromomethylene)cyclobutane **24** with PhLi [24] provides a carbenoid intermediate that rapidly rearranges to give cyclopentyne. While unstable to isolation, the transient intermediate **22** has been trapped by a number of different reagents, affording products consistent with its formation. Larger cycloalkynes are also readily provided by this and analogous methods [25].

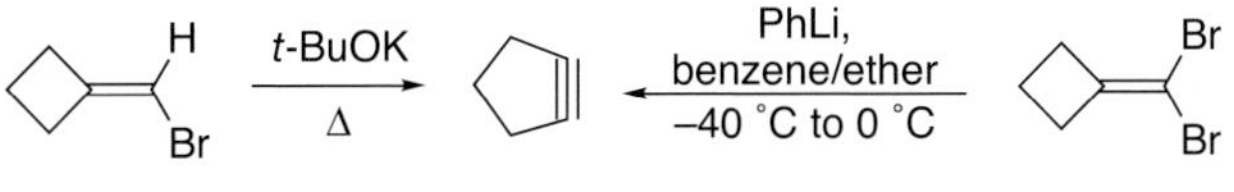

23 **22** **24** **Equation 7-4**

Whereas the formation of cyclopentyne **22** involves the migration of an alkyl group, migration of an aryl moiety can be even more effective in the formation of strained cyclic products. Treatment of monochloroolefin **25** with PhLi at ambient temperature, for example, resulted in the formation of the cycle **26** (Equation 7.5), presumably through the FBW rearrangement of a carbenoid intermediate. While the strained cyclic alkyne **26** could not be isolated, its existence was supported by the formation of products consistent with its reaction with PhLi in situ [26].

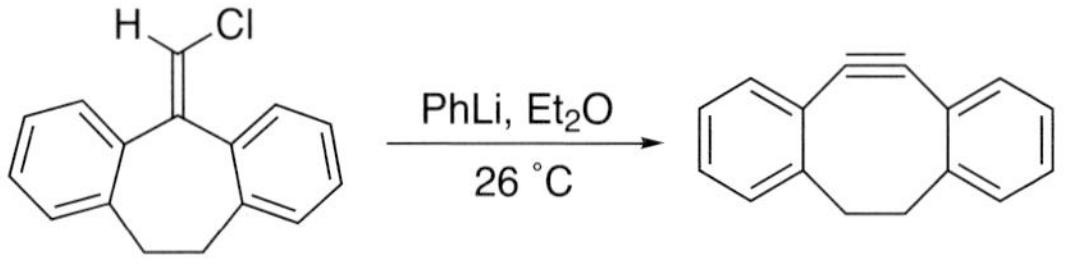

25 26 Equation 7-5

A clever variation of the FBW process was used by Nakagawa and co-workers to afford cyclic tolans with increasingly strained frameworks (Scheme 7.4). Two different series were assembled, based on the connectivity of the ethereal bridging unit. In the first case, rearrangement of the dichloroolefins **27** with BuLi at low temperature gave the *p,p*-bridged cycles **28**, molecules in which the phenyl rings are held in a coplanar conformation [27]. In a similar manner, dibromoolefins **29a** and monobromoolefins **29b** (X = H), were rearranged with BuLi in moderate yields to provide the *o,p*-bridged cycles **30**. In this situation, the two phenyl rings of the increasingly strained molecules are now twisted away from coplanarity [28].

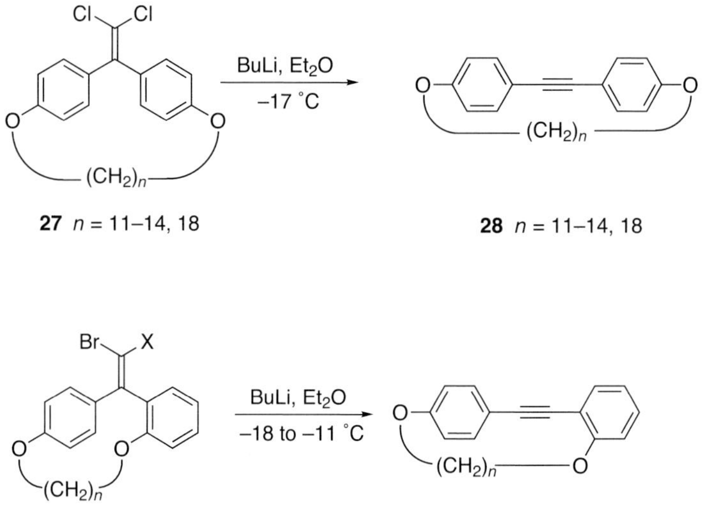

27 $n = 11–14, 18$ **28** $n = 11–14, 18$

29a X = Br, $n = 10, 11, 12$ **30** $n = 7–12$
29b X = H, $n = 7, 8, 9$

Scheme 7.4 Nakagawa's cyclic tolans by the FBW rearrangement.

A double FBW rearrangement of the tetrachloro-precursor **31** has been successfully used by Oda and co-workers to assemble the planar, strained dehydroannulene **32** in a surprisingly good yield of 30% (Equation 7.6) [29]. Attempts to obtain **32** by other methods resulted only in trace formation of the desired macrocycle, the major product being the larger, trimeric analogue. X-ray crystallographic analysis of

32 showed that the alkyne moieties introduced into the conjugated framework through the carbenoid rearrangement are substantially distorted by ring strain, with the average C−C≡C bond angle reduced to 166°.

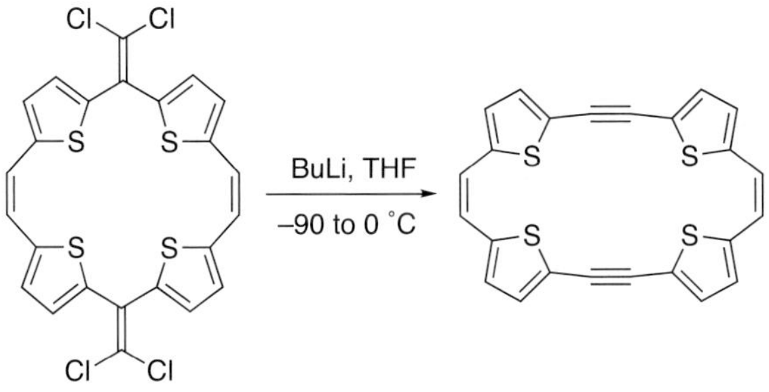

31 **32** 30% **Equation 7-6**

Alkyl, vinyl, and aryl functionalities all have established potential as migrating groups in Fritsch–Buttenberg–Wiechell rearrangements. Alkynes, on the other hand, had not been shown to undergo migration in either a free carbene or carbenoid intermediate until very recently. It is this process that presents a potentially useful route for the formation of extended, conjugated polyynic systems.

7.3.2 Synthesis of 1,3–Butadiynes

Mixed FBW rearrangements in which one of the potential migrating groups is an alkyne and the other a vinyl, aryl, or heteroaryl group should offer a convenient route for the formation of butadiynes. The key to diyne formation is access to the dibromoolefinic precursors, and two routes have commonly been used (Scheme 7.5). **Path A** uses the addition of a lithium acetylide to an aldehyde **33** to generate the alcohol **34** [30], followed by oxidation to ketone **35**, typically with PCC, MnO_2, or $BaMnO_4$. Both reactions generally proceed cleanly and often without the need for column chromatography. Alternatively, ketones **35** can be accessed directly by Friedel–Crafts acylation by use of an acid chloride **36**, **Path B** [31]. Commercially available aryl or vinyl carboxylic acids are easily transformed into acid chlorides **36** by treatment with thionyl chloride, and can then be taken on into the reaction with trimethylsilyl acetylenes in the presence of $AlCl_3$ in CH_2Cl_2. The ketones **35** are formed in good to excellent yields, and can typically be purified by passing the crude product through a plug of silica gel.

The well known dibromoolefination reaction, introduced by Ramirez and made famous by Corey and Fuchs [32], then provides dibromide precursors **37**. This reaction can usually be carried out at room temperature with less hindered ketones; elevated temperatures are, however, sometimes required for more sterically hindered substrates [32c, 33].

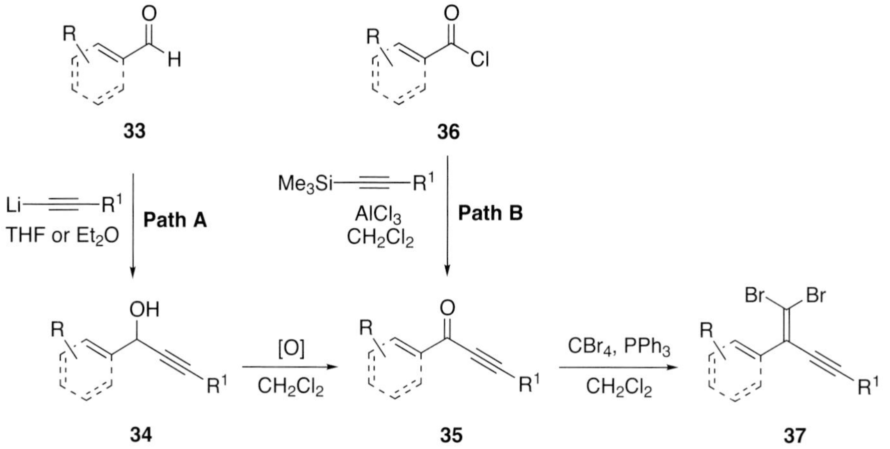

Scheme 7.5 Formation of mixed dibromoolefins.

Conversion of a dibromoolefin **37** into the desired diyne **38** has been accomplished by treatment with BuLi at −78 °C in rigorously dried hexanes (Table 7.1) [34], diyne formation typically being complete by the time the reaction mixture has warmed to ca. −40 °C. After quenching of the reaction at low temperature, the product can often be isolated pure simply by passing the crude reaction mixture through a short plug of silica gel. This method affords a range of functionalized diynes, formed in good to excellent yields. A sampling of the products can be found in Table 7.1, and illustrates that electron-rich and electron-poor aryl groups, heteroaryl, vinyl, and azobenzene groups are all tolerated by the reaction conditions. Several different alkynes have also been appended, with both silyl (entries a–k) and alkyl (entries l–n) substitution.

The greatest determinant of success of the FBW rearrangement in these reactions is the presence of water. Experimental evidence suggests that lithium/halogen exchange is faster than in situ quenching of the BuLi by adventitious water. As a result, any water present in the reaction medium results in protonation of the carbenoid intermediate, which severely complicates purification, due to very similar retention times on chromatographic supports.

Table 7.1 Summary of synthetic yields for diynes **38**.

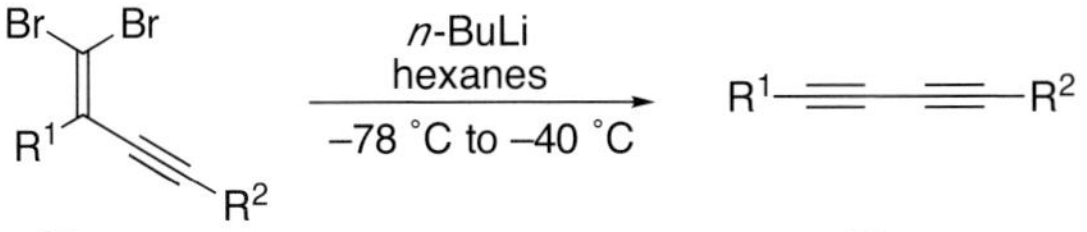

37a–n **38a–n**

Cmpd.	*R¹*	*R²*	*38[%]*	*Route*[a]	*[Ref.]*
a	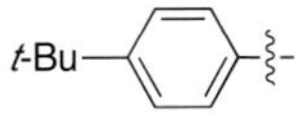	Me_3Si	88	Path B	[34]
b	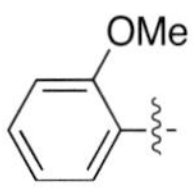	Me_3Si	43	Path A	[34]
c	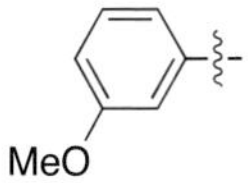	Me_3Si	75	Path A	[34]
d	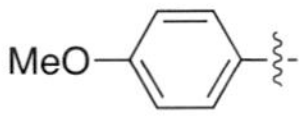	Me_3Si	82	Path A	[34]
e	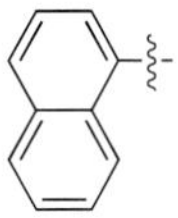	Me_3Si	93	Path A	[34]
f	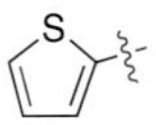	Me_3Si	91	Path A	[34]
g	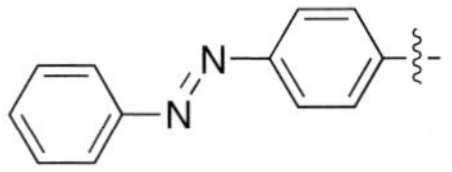	Me_3Si	61	Path B	[34]
h	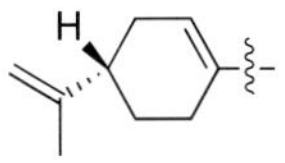	Me_3Si	46	Path A	[34]
i	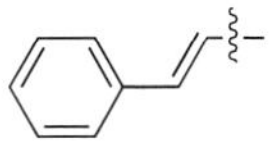	Me_3Si	95	Path A	[34]
j	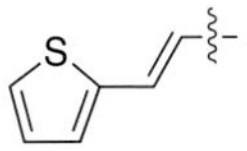	Me_3Si	55	Path B	[34]
k	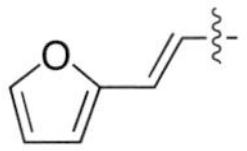	Me_3Si	70	Path B	[35]
l	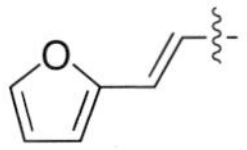	Me	59	Path B	[34]
m	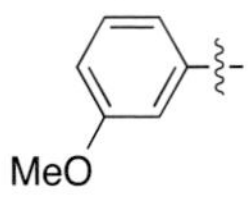	*n*-Bu	92	Path A	[34]
n	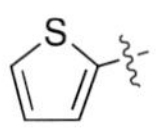	*n*-Bu	86	Path A	[34]

[a]Dibromoolefin formation as in Scheme 7.5.

7.3.3 Synthesis of 1,3,5–Hexatriynes

For the "mixed" FBW systems described in the previous section, the question that quickly arises relates to the identity of the migrating group, that is to say: is the alkynyl group a willing participant in the rearrangement, or is it consistently the aryl or vinyl moiety that undergoes 1,2-migration in the carbenoid intermediate? While preliminary evidence using ^{13}C labeling in several of the above examples shows that 1,2-migration of an alkynyl group is quite common [36], the facile rearrangements of enediynes described in the next section clearly show the high migratory potential of the alkynyl group.

7.3.3.1 Triynes from Enediynes

1,1-Dibromoolefins also serve as precursors for the FBW rearrangement to triynes, and three general routes based on adaptation of known procedures are readily available (Scheme 7.6). **Path A** uses condensation between a lithium acetylide and an α,β-ethynyl aldehyde to afford the unsymmetrical alcohol **39** ($R^1 \neq R^2$) [37]. **Path B** forms symmetrical alcohols **39** ($R^1 = R^2 = R$) through the addition of 2 equiv. of a lithium acetylide to ethyl formate [38]. In both cases, the alcohols can generally be isolated pure after workup, and oxidation to ketone **40** is easily effected by use of PCC. Alternatively, Friedel–Crafts acylation is a convenient method for direct formation of ketones **40** from acetylenic carboxylic acid chlorides and trimethylsilyl-protected acetylenes (**Path C**) [31]. The dibromoolefins **41** are then formed by Corey and Fuchs's method [32, 37a], and yields for this step are generally quite good.

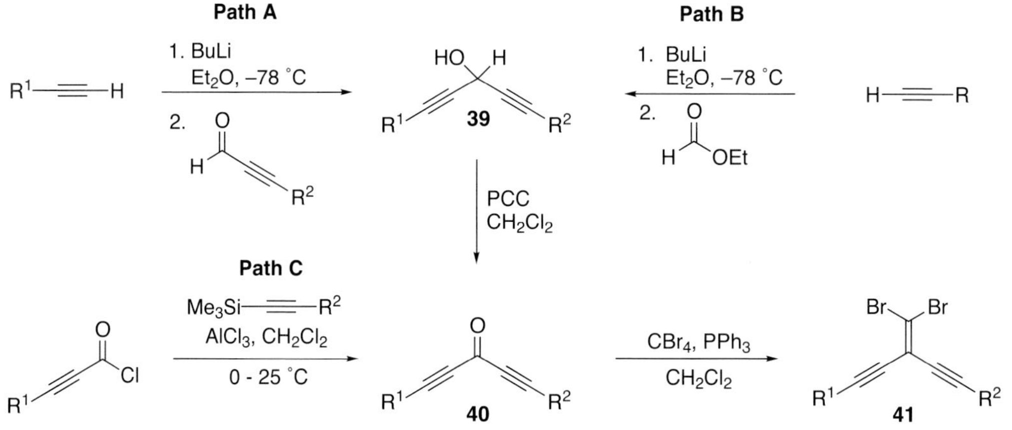

Scheme 7.6 Methods of forming dibromoolefins **41**.

Initial attempts to transform a 2,2-dialkynyl-1,1-dibromoolefin into a triyne involved treating **41a** with BuLi in THF or Et_2O at –78 °C, common FBW rearrangement conditions (Scheme 7.7) [39]. Lithium/halogen exchange between BuLi and dibromoolefin **41a** resulted in the formation of carbenoid **42**, but these reaction conditions did not afford appreciable amounts of the triyne **43a**. Mass spectral analysis of the product mixtures did suggest the formation of trace amounts of the desired product, but the major isolated product was the monobromoolefinic species **44**, resulting from protonation of the lithiated intermediate **42** upon workup.

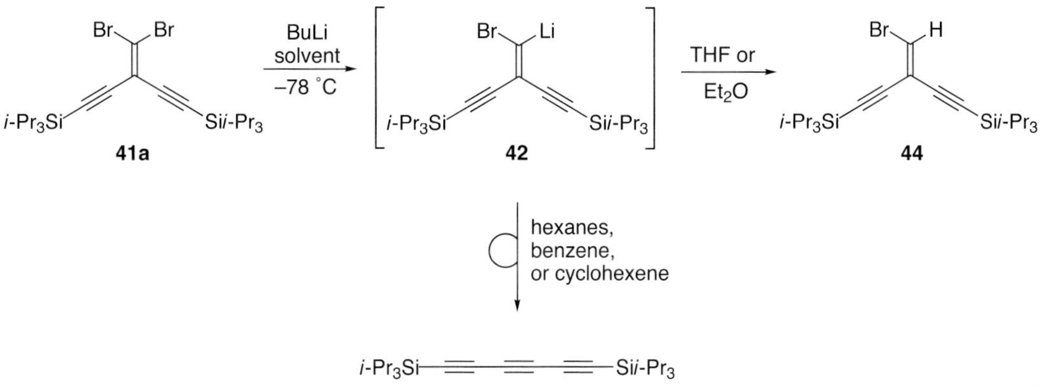

Scheme 7.7 Rearrangement of enediyne **41a**. **43a** 65–80%

It has been established that the reactivity of a carbenoid species can be very sensitive to reaction conditions, and coordinating solvents such as Et_2O and THF can stabilize a carbenoid intermediate [13e, 40]. Assuming that the solvating ability of Et_2O and THF was a stabilizing factor that prevented or retarded collapse of the vinyl lithium intermediate **42**, the use of less polar and noncoordinating solvents was explored. Thus, addition of BuLi at low temperature to a solution of **41a** in a range of solvents such as hexanes, benzene, and cyclohexene gratifyingly resulted in the formation of a single product: triyne **43a** [41].

From these initial studies, standard conditions allowing formation of a wide variety of triynes have been established: 1.2 equiv. of BuLi is added at –78 °C to the dibromoolefin in hexanes, and the solution is allowed to warm to ca. –10 °C and then quenched. If strictly anhydrous conditions are maintained, the desired triyne product is typically the sole product as observed by TLC analysis. Table 7.2 provides a summary of yields for a selection of triynes, the dibromoolefins **41** having been prepared by the synthetic routes outlined in Scheme 7.6.

A variety of substitution patterns are quite easily constructed by this basic reaction sequence. Symmetrically substituted dibromoolefins with heteroaryl (**41m**), alkyl (**41g**, **41h**), and silyl (**41a**, **41b**) groups cleanly afford the corresponding triynes **43** in good yields. On the other hand, the trialkylsilyl protecting groups of unsymmetrical polyynes (**43c–f**, **43i–l**) allow for additional elaboration through protiodesilylation to give a terminal alkyne. The formation of the larger carbon-rich skeletons such as **43j** and **43k** in a few steps is also noteworthy.

Table 7.2 Summary of synthetic yields for polyynes **43**.

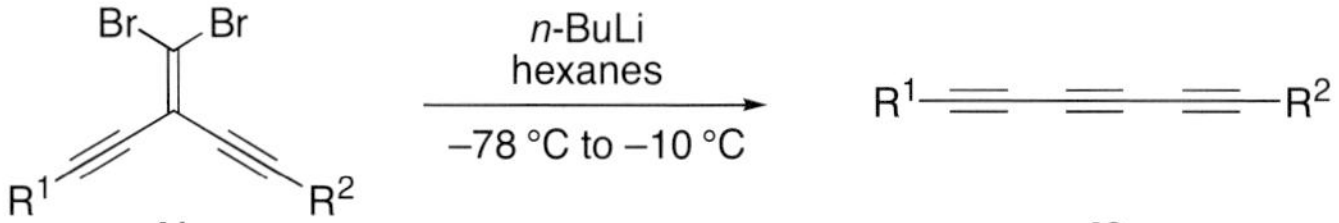

41 **43**

Cmpd.	*R*[1]	*R*[2]	*43 [%]*	*Route*[a]	*[Ref.]*
a	*i*-Pr$_3$Si	*i*-Pr$_3$Si	61	Path A	[41]
b	Me$_3$Si	Me$_3$Si	50	Path B	[41]
c	Me$_3$Si	*i*-Pr$_3$Si	61	Path A	[41]
d	Me$_3$Si	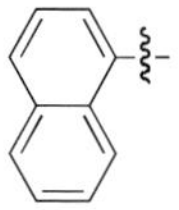	70	Path A	[41]
e	*i*-Pr$_3$Si	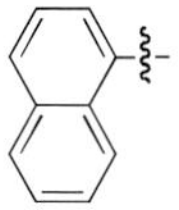	62	Path A	[41]
f	*i*-Pr$_3$Si	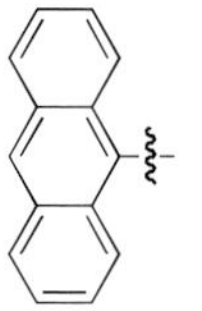	41	Path A	[42]
g	*n*-Bu	*n*-Bu	80	Path B	[41]
h	*n*-octyl	*n*-octyl	66	Path B	[41]
i		Me$_3$Si	84	Path C	[41]
j	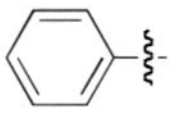	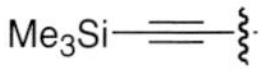	64	Path C	[41]
k	Me$_3$Si	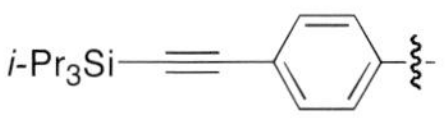	61	Path A	[41]
l	Me$_3$Si	*n*-Bu	82	Path A	[41]
m	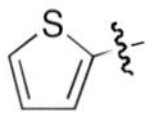	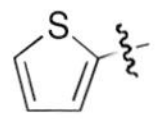	64	Path B	[41]

[a]Dibromoolefin formation as in Scheme 7.6.

7.3.3.2 **Triynes from Acid Chlorides**

A complementary route to unsymmetrical triynes utilizes aryl or alkenyl acid chlorides in Friedel–Crafts acylation reactions (Scheme 7.8 and Table 7.3) [31]. This can be a particularly attractive route to triynes because of the diverse range of carboxylic acids available either commercially or by facile synthetic procedures. Thus, treatment of acid chlorides **36** with bis(trimethylsilyl)butadiyne in the presence of $AlCl_3$ gave diynones **45**. In cases in which the pendant functionality was an aryl group (**45a–c**), the resulting ketones were surprisingly unstable and were converted directly into dibromoolefins (**46a–c**). Ketones with alkenyl substitution (**45d–g**)

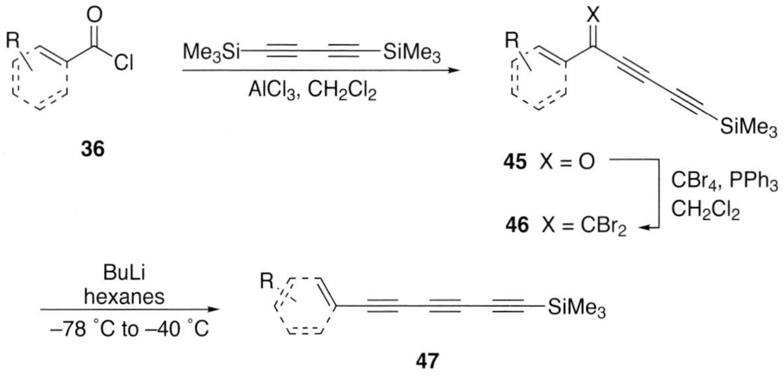

Scheme 7.8 Triyne formation from acid chlorides.

showed greater stability and could be isolated pure and fully characterized prior to elaboration to **46d–g**. Rearrangement of **46a–g** in hexanes gave conjugated triynes **47a–g** under standard conditions [34–35].

Table 7.3 Summary of yields for compounds **47**.

Cmpd.	*R*	*47 [%]*	*Ref.*
a	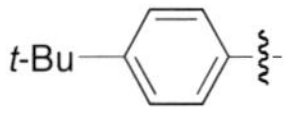	98	[34]
b	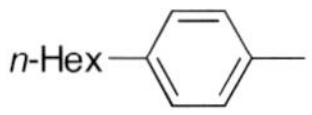	41	[34]
c	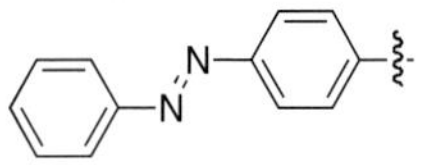	46	[34]
d	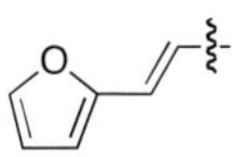	82	[35]
e	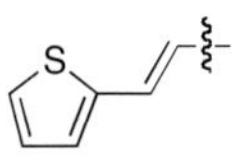	74	[34]
f	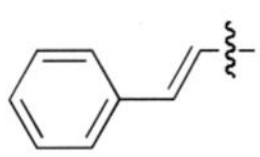	82	[34]
g	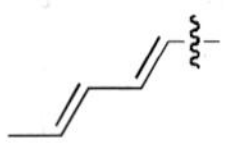	76	[34]

7.3.3.3 Carbene or Carbenoid?

Mechanistically, a question naturally arises from FBW rearrangements as an alkyne formation method: does 1,2-migration occur through the intermediacy of a free alkylidene carbene or a carbenoid species? A number of preliminary experiments that involve alkyne migration have been conducted in order to determine the nature of the intermediate under the standard conditions employed for this reaction (–78 °C, hexanes). A free alkylidene carbene is a highly reactive species, and evidence of its existence as an intermediate is generally obtained through the isolation of products common to this species but less likely to be formed from a carbenoid intermediate [13b]. For example, cyclopropylidene formation through the addition of an alkylidene carbene to an olefin is well known, and the rearrangement of **41a** was thus conducted in cyclohexene as solvent (Equation 7.7). This procedure gave only triyne **43a**, with no addition product **48** indicated by ^{1}H NMR or mass spectral analysis [41].

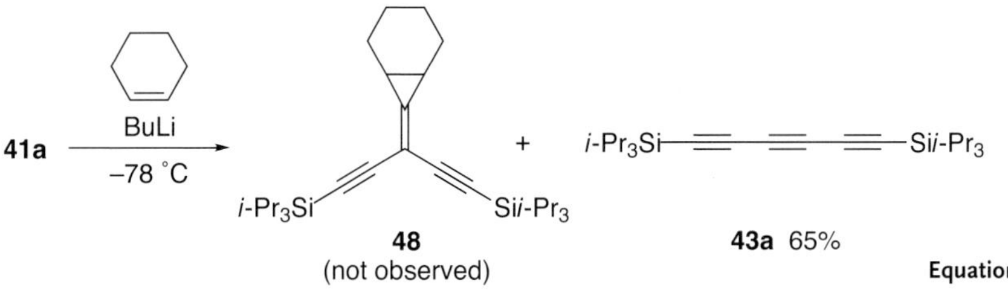

Equation 7-7

Unsaturated carbenes are known to undergo intermolecular insertion into Si–H bonds [13b]. When the rearrangement of **41a** was conducted in the presence of excess triethylsilane, however, no evidence of the insertion product **49** was observed and, again, only the triyne **43a** was isolated from the reaction (Equation 7.8) [43].

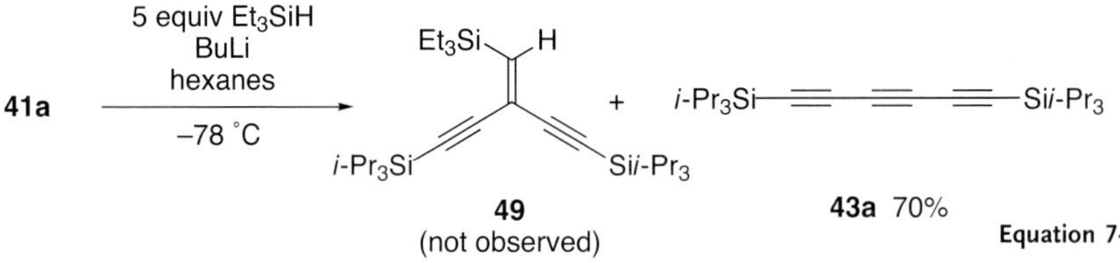

Equation 7-8

Intramolecular 1,5-C–H bond insertion is a typical reaction pathway for carbenes, but is much less likely to occur in a carbenoid intermediate. Appropriate substrates to test this possibility, dibromides **50**, were easily assembled from the appropriate carboxylic acids, and were subjected to the standard FBW rearrangement conditions (Equation 7.9). ^{1}H NMR spectroscopic analysis gave no evidence for the formation of cyclopentenyl derivatives **51**, and diynes **52** were isolated in good yields as the only products [43].

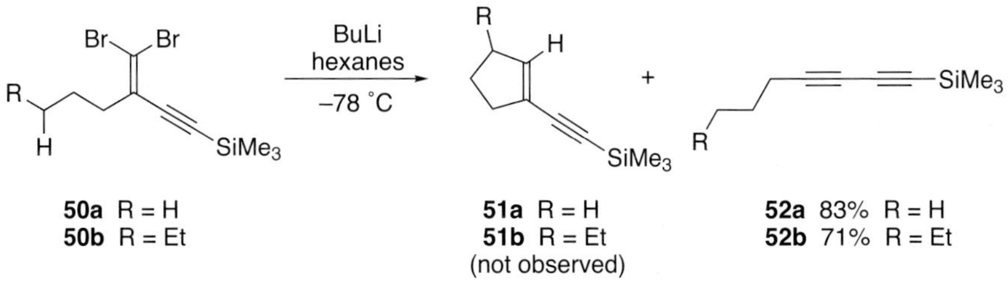

Equation 7-9

While by no means exhaustive, the results outlined in Equations 7.7–7.9 suggest that the intermediates formed from lithium/halogen exchange at the dibromoolefin moiety in these FBW reactions demonstrate chemical behavior more consistent with a carbenoid species than a free alkylidene carbene. These results are also in agreement with previous mechanistic studies on similar FBW rearrangements [13].

7.3.4
Tri- and Pentaynes from Free Alkylidene Carbenes

Tobe and co-workers have recently designed an elegant new method to form polyynes that exploits the 1,2-migration of an alkynyl moiety in a free alkylidene carbene intermediate (Scheme 7.9) [4, 44]. The requisite precursors, bicyclo[4.3.1]decatrienes **53**, are readily synthesized in several steps from dihydroindane. Spectroscopic characterization is consistent with the open form of **53** rather than its valence isomer **54**. Photolysis of **53** in hexanes at 0 °C results in the [2+1] cheletropic extrusion of indane from **54** to give the free alkylidene carbene **55**. Migration of one of the ethynyl moieties then gives the desired triynes (**42b** and **56a**). This process has also been successful for the formation of the longer polyyne, pentayne **56b**, in 59 % yield, through migration of a butadiynyl group.

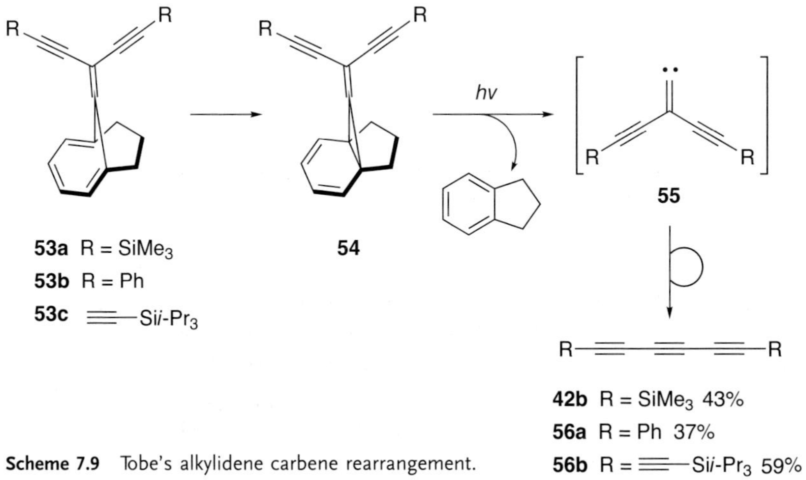

Scheme 7.9 Tobe's alkylidene carbene rearrangement.

7.4
Toward applications

Alkynes, conjugated oligo- and polyynes in particular, are a common substructure in molecules of interest for applications spanning from medicine to materials. The use of the FBW rearrangement for construction of conjugated polyyne segments has allowed the production of a number of useful new molecules for these applications, and several examples are illustrative of this process.

7.4.1
Natural Products Synthesis

Naturally occurring polyynes make up a diverse group of biologically active compounds. Polyacetylenes have been isolated from fungi, bacteria, and plants, as well as marine sponges, and many possess a range of medicinal properties including antibacterial, anticancer, and pesticidal characteristics, to name just a few [45]. The Cadiot–Chodkiewicz reaction [7] has traditionally been the most likely method of choice for assembling the di-, tri-, or tetrayne cores of naturally occurring polyynes [46]. The modified FBW rearrangement, however, presents a convenient and complementary method for this task.

First isolated from species of *Chrysanthemum* by Bohlmann [47–48], the enetriyne **57** (Scheme 7.10) was subsequently found to be highly phototoxic towards mosquito larvae [49]. Bohlmann's initial synthesis of **57** relied on the derivatization of a highly reactive triynal precursor [48]. A reported alternative synthesis of compound **57** involves only three steps from acid chloride **58**, which is generated quantitatively from the commercially available carboxylic acid [35]. A typical sequence of Friedel–Crafts acylation (**59**) and dibromoolefination provides the necessary precursor **60**, and the final FBW rearrangement with **60** proceeds nicely to give enetriyne **57** in 84 % yield. Overall, this represents a 21 % yield of **57** from **58**.

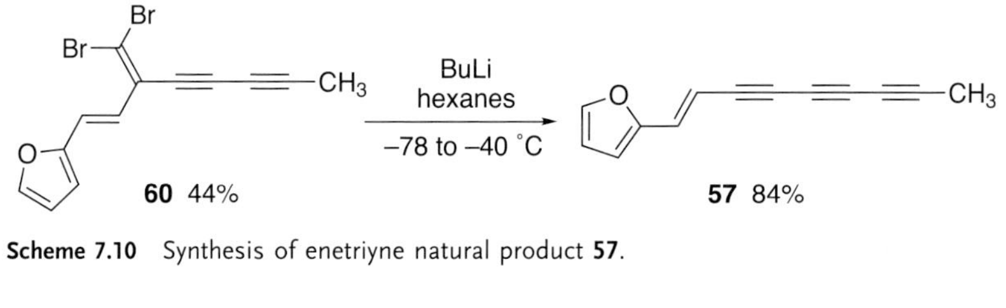

Scheme 7.10 Synthesis of enetriyne natural product **57**.

Diene-diyne **61**, atractylodin, has been isolated from the rhizomes of plants of the genus *Atractylodes*, plants that have been used in homeopathic medicine to treat a range of symptoms including fatigue, fever, and lack of appetite [50]. The initial synthesis of atractylodin reported by Yosioka et al. was used to determine the (*E,E*) stereochemistry about the olefins. These authors utilized an oxidative homocoupling method, which ultimately afforded three products requiring a difficult separation procedure [51]. Atractylodin has now been readily synthesized in a few steps by utilizing a FBW rearrangement to form the diene-diyne skeleton, as outlined in Scheme 7.11. The initial step was a one-pot reaction consisting of the elimination of dibromoolefin **62** to provide acetylide **63** [52] and condensation with aldehyde **64** to give alcohol **65** in 54% yield. With **65** to hand, a standard sequence of oxidation (**66**), dibromoolefination (**67**), and FBW rearrangement ultimately provided diyne **61** in 72% yield, a 13% yield overall from commercially available **64** [35].

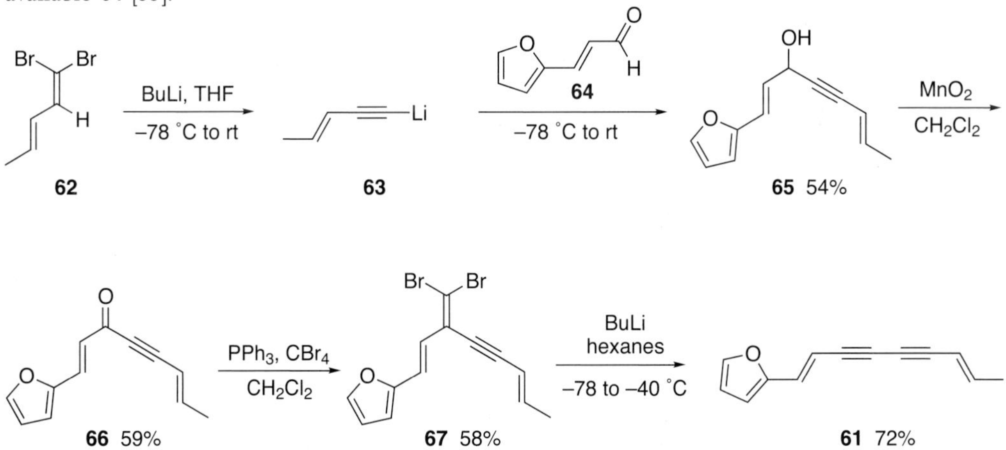

62 **63** **65** 54% **66** 59% **67** 58% **61** 72%

Scheme 7.11 Synthesis of atractylodin.

1-Phenylhepta-1,3,5-triyne **68** (PHT) has been isolated from a range of plant species and has been extensively studied for biological activity. For example, **68** is a major component in the leaves and flowers of *Bidens pilosa* L., which have been used in Mexican traditional medicine as treatment for stomach disorders, hemorrhoids, and diabetes [53]. The triyne core of PHT (**68**) was first constructed by Cadiot–Chodkiewicz coupling [54]. As outlined in Scheme 7.12, **68** has also recently been assembled through a FBW rearrangement [35]. A concise reaction sequence consisting of Friedel–Crafts acylation of phenylpropioloyl chloride **69** with trimethylsilylpropyne to provide ketone **70** (56%) [55], dibromoolefination to **71** (67%), and rearrangement with BuLi ultimately gave the triyne **68** in 18% overall yield from readily available **69**.

The triyne (–)-ichthyothereol (**72**) and its acetate **73** (Scheme 7.13) have been isolated from the plants *Dahlia coccinea* and *Ichthyother terminals* [56], the latter compound known to natives of the lower Amazon Basin as a fish poison [57]. While the

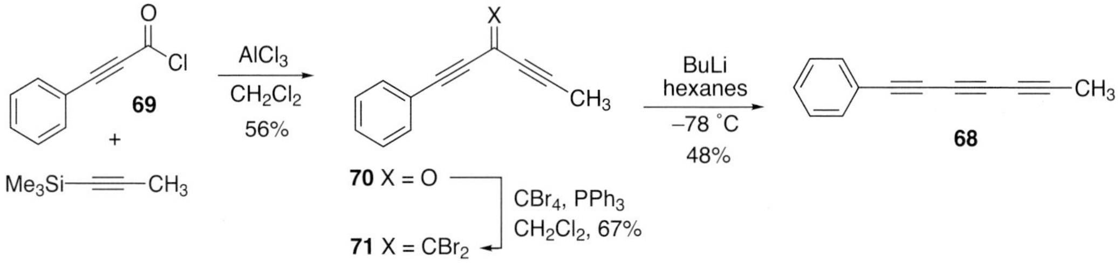

Scheme 7.12 Synthesis of phenylheptatriyne (PHT) **68**.

absolute configurations of **72** and **73** were confirmed by Jones and co-workers in 1970 [58], it was not until 2001 that the total stereoselective synthesis of (–)-ichthyothereol and its acetate was achieved by Mukai et al. [59]. This efficient synthesis relies on the FBW rearrangement of **74** to assemble the triyne core **75**. After conversion of the alkynylsilane **75** into stannane **76**, a Stille coupling reaction between **76** and vinyl iodide **77** completes the conjugated framework **78**. Finally, removal of the TBDMS protecting group quantitatively yields **72**, and acetylation gives **73**.

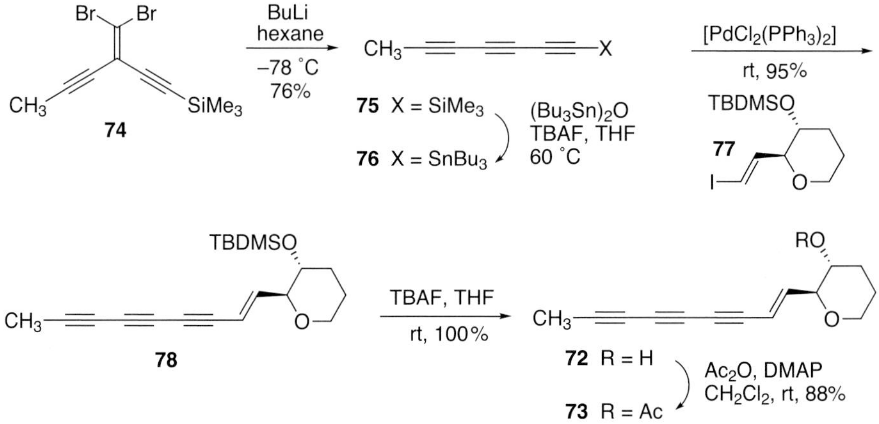

Scheme 7.13 Mukai's synthesis of (–)-ichthyothereol **72** and its acetate **73**.

7.4.2
Extended Arylenethynylene Derivatives

Exploitation of the FBW rearrangement for the formation of extended carbon-rich networks was a primary motivation behind the development of this methodology, and the rearrangement of two or more dibromoolefin moieties within the same molecule was viewed as a potentially useful strategy for the formation of aryl-polyyne building blocks. The general sequence of reactions used for the synthesis of functionalized diynes and triynes outlined above could, in theory, be efficiently applied to larger systems, with multiple reactions being carried out at each step.

For example, commercially available terephthalaldehyde (**79**) was carried through the three-step sequence consisting of condensation with a lithium acetylide to give **80**, oxidation to **81**, and dibromoolefination to give an excellent yield of tetrabromide **82** (Scheme 7.14). A dual FBW rearrangement with **82** then gave tetrayne **83** in 87 % yield, representing an impressive 95 % yield for each rearrangement event [34]. This reaction could be scaled up to provide 5 g of **83** without any compromise in yield.

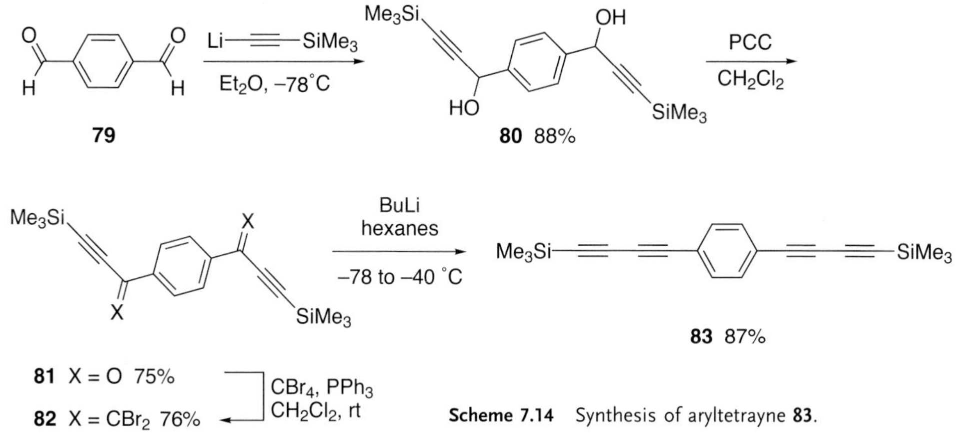

Scheme 7.14 Synthesis of aryltetrayne **83**.

Functionalized aryltetraynes have similarly been synthesized in a few steps (Scheme 7.15), starting with the appropriate *p*-diethynylbenzene precursors (not shown) [34]. The transformation of thienyl-capped tetrabromide **84** proceeded in comparatively low yield, due to the limited solubilities both of the precursor **84** and of the tetrayne product **85**. The enhanced solubility originating from the octyloxy and *tert*-butyl groups appended to tetrabromide **86**, on the other hand, greatly facilitated the rearrangement to **87**, which was obtained in almost quantitative yield. Frustratingly, while the dibromoolefination to **84** could be accomplished in reasonable yield in spite of the limited solubility (35 %), formation of **86** was less efficient (~20 % yield) [60].

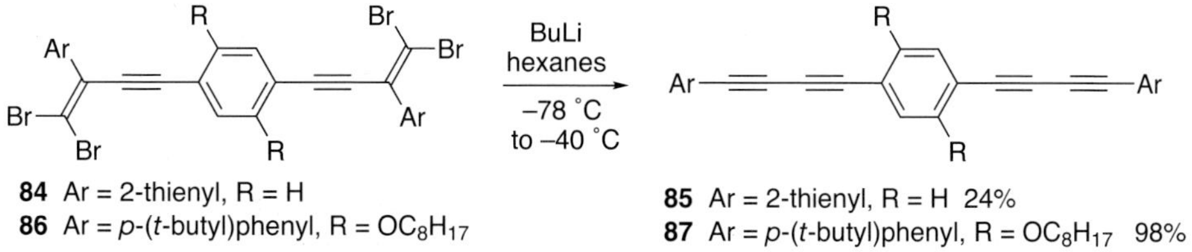

Scheme 7.15 Synthesis of functionalized aryltetraynes **85** and **87**.

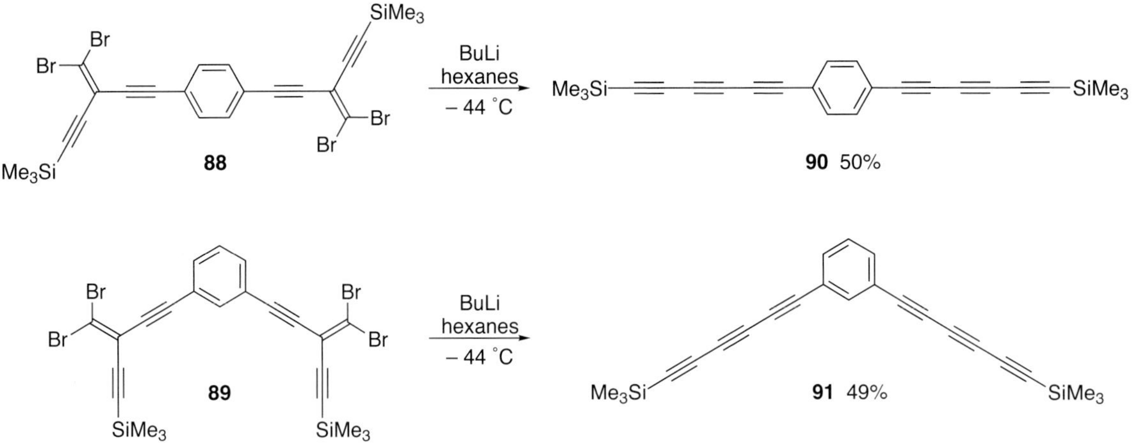

Scheme 7.16 Synthesis of arylhexaynes **90** and **91**.

Aryl hexaynes such as the 1,4- and 1,3-bis(1,3,5-hexatriynyl)benzene derivatives shown in Scheme 7.16 have been formed in a manner similar to that used for the aryltetraynes (**83**, **85**, **87**), with the tetrabromides **88** and **89** derived from either 1,4- or 1,3-diethynyl benzene, respectively. As **88** and **89** proved to be quite insoluble in hexanes at –78 °C, the reaction mixture was in each case warmed to about –44 °C, at which point the solution became homogeneous. BuLi was then added, the reaction mixtures were allowed to warm to ~ –5 °C, and hexaynes **90** [41] and **91** [61] were isolated in 50 % and 49 % yields, respectively.

Hexayne **90** is a nicely crystalline and stable solid when precipitated from hexanes. Surprisingly, however, crystals of **90** grown from a concentrated hexanes/CH_2Cl_2 solution through diffusion of MeOH at 4 °C turned brown in a matter of hours at ambient temperature. A structural analysis, with respect to topochemical polymerization, of the solid-state packing parameters [62] of crystals of **90** grown under the latter conditions has provided insight into this solid-state instability of this highly unsaturated molecule [41].

Along the crystallographic *b* axis, a pseudo-stacking situation is observed for neighboring polyynes (Figure 7.1). Polymerization of the triyne moieties in a 1,6-manner to give a polytriacetylene product is not possible, as $R_{1,6} > 4$ Å (Figure 7.2) [63–64]. The individual molecules, however, are aligned in a manner suitable for a 1,4-addition process, with close contacts for $R_{1,4} = 4.0$ Å and $R_{3,6} = 3.9$ Å, both within the desired range of ≤ 4 Å. Furthermore, the stacking angle ($\Theta = 45$) and the stacking distance ($d = 5.5$ Å) are also nearly optimal for a 1,4-addition. It is therefore likely that this solid-state alignment of the triyne arms for **90** results in polydiacetylene formation at ambient temperatures, albeit in a non-regioselective manner because of competing/concurrent 1,4- and 3,6-polymerization processes.

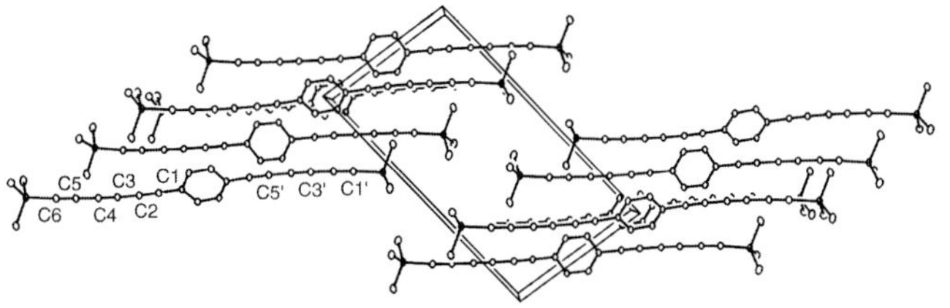

Figure 7.1 Illustration of crystal packing for compound **90** as viewed along the crystallographic *b* axis.

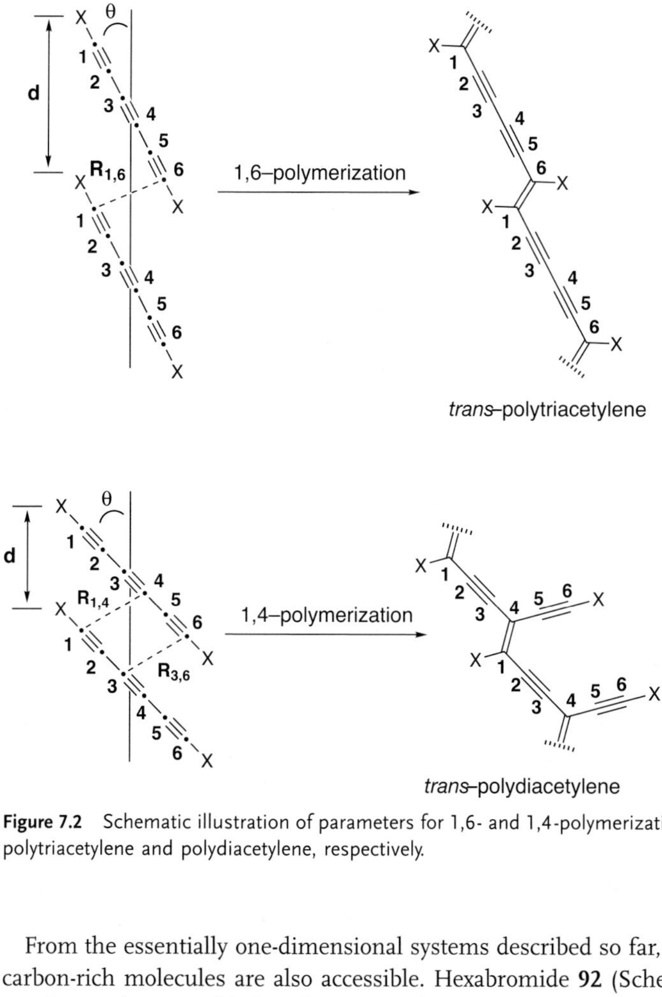

Figure 7.2 Schematic illustration of parameters for 1,6- and 1,4-polymerization of triynes to give polytriacetylene and polydiacetylene, respectively.

From the essentially one-dimensional systems described so far, two-dimensional carbon-rich molecules are also accessible. Hexabromide **92** (Scheme 7.17), for example, can be assembled in three steps from 1,3,5-triethynyl benzene and is surprisingly soluble (unlike **88** and **90**) in hexanes at low temperatures [41]. Treatment with BuLi at –78 °C thus effected three simultaneous/consecutive rearrangements to produce nonayne **93**, which could be isolated pure simply by passing the crude reaction mixture through a short column of silica gel. While the overall isolated yield for **93** was only 35 %, this still amounts to about a 70 % yield per rearrangement event.

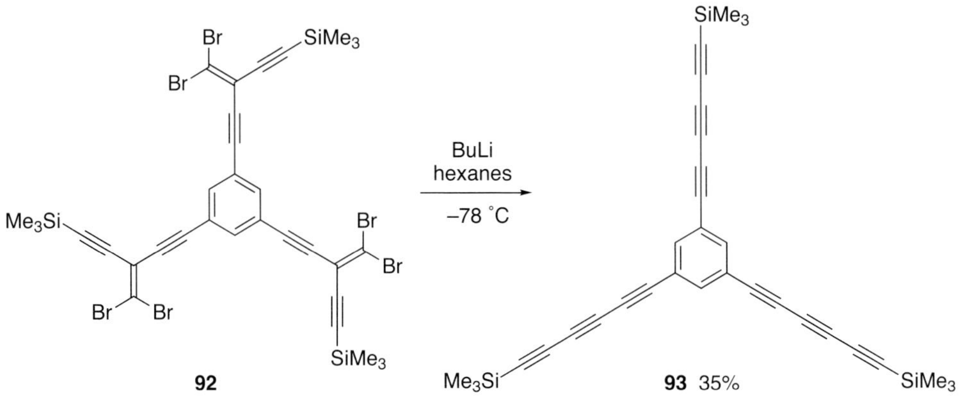

Scheme 7.17 Threefold FBW rearrangement leading to nonayne **93**.

Diol **94** (a by-product in the synthetic sequence leading to **92**, Scheme 7.17) presented an opportunity to explore the efficiency of the FBW reaction method in the presence of a terminal alkyne moiety (Scheme 7.18) [41]. Oxidation of **94** to **95** and conversion into dibromoolefin **96** proceeded in reasonable yields. The acidity of the terminal acetylene presented a potential problem in the final step, but use of excess BuLi to effect the α-elimination cleanly gave heptayne **97**. Unstable as a neat solid, compound **97** was expeditiously carried on without isolation into an oxidative coupling reaction and purification, and the extended arylalkynyl system **98** was formed in 22 % yield over the final two steps. While it is a large (2 × 2.4 nm Si to Si) and highly unsaturated molecule, tetradecayne **98** is a surprisingly stable solid and could be fully characterized spectroscopically. The ^{13}C NMR spectrum is particularly interesting, as all eight unique sp–hybridized carbons are well resolved.

Single crystals of **98** suitable for X-ray crystallographic analysis were obtained, and an ORTEP drawing is shown in Figure 7.3a [41]. The two aryl groups are nearly coplanar, and the four hexatriynyl arms bend gently above and below this plane. The packing diagram reveals that neighboring molecules are paired face-to-face (Figure 7.3b). Each pair of neighboring molecules is nearly coplanar and related by a center of inversion. The aryl groups of neighboring molecules are separated by a distance of 3.4 Å, and offset in a manner expected for face-to-face π-stacking. This alternating packing motif has the effect of causing neighboring hexatriynyl arms to be offset in space, accommodating the trimethylsilyl groups. Crystalline **98** is surprisingly stable under ambient conditions, which can be explained by two factors. Firstly, the large offset distance between neighboring triyne groups negates any possibility for solid-state reaction between these moieties. Secondly,

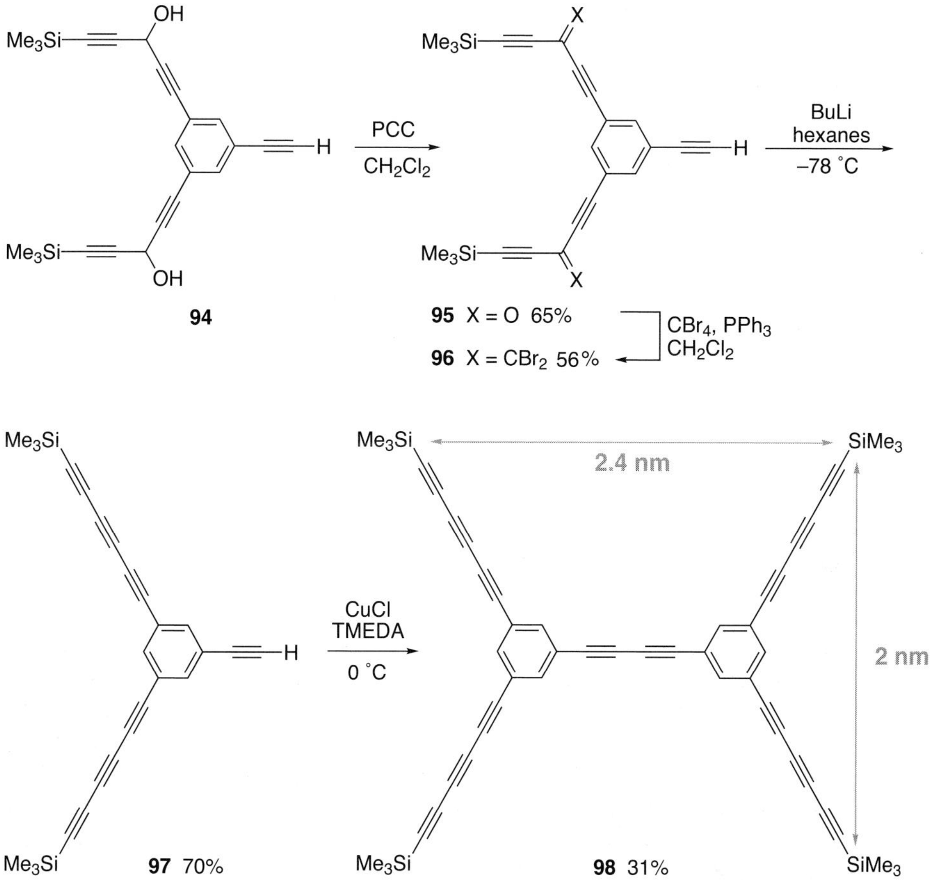

Scheme 7.18 Synthesis of carbon-rich network **98**.

the closest intermolecular interactions are between the central butadiynyl segments of neighboring molecules. The distance between carbon atoms ($R_{1,4}$ = 4.2 Å) and the stacking angle (Θ = 70°) are, however, just outside the values necessary ($R_{1,4} < 4$ Å and Θ = 90°) for 1,4–topochemical polymerization to afford a *cis*–polydiacetylene structure.

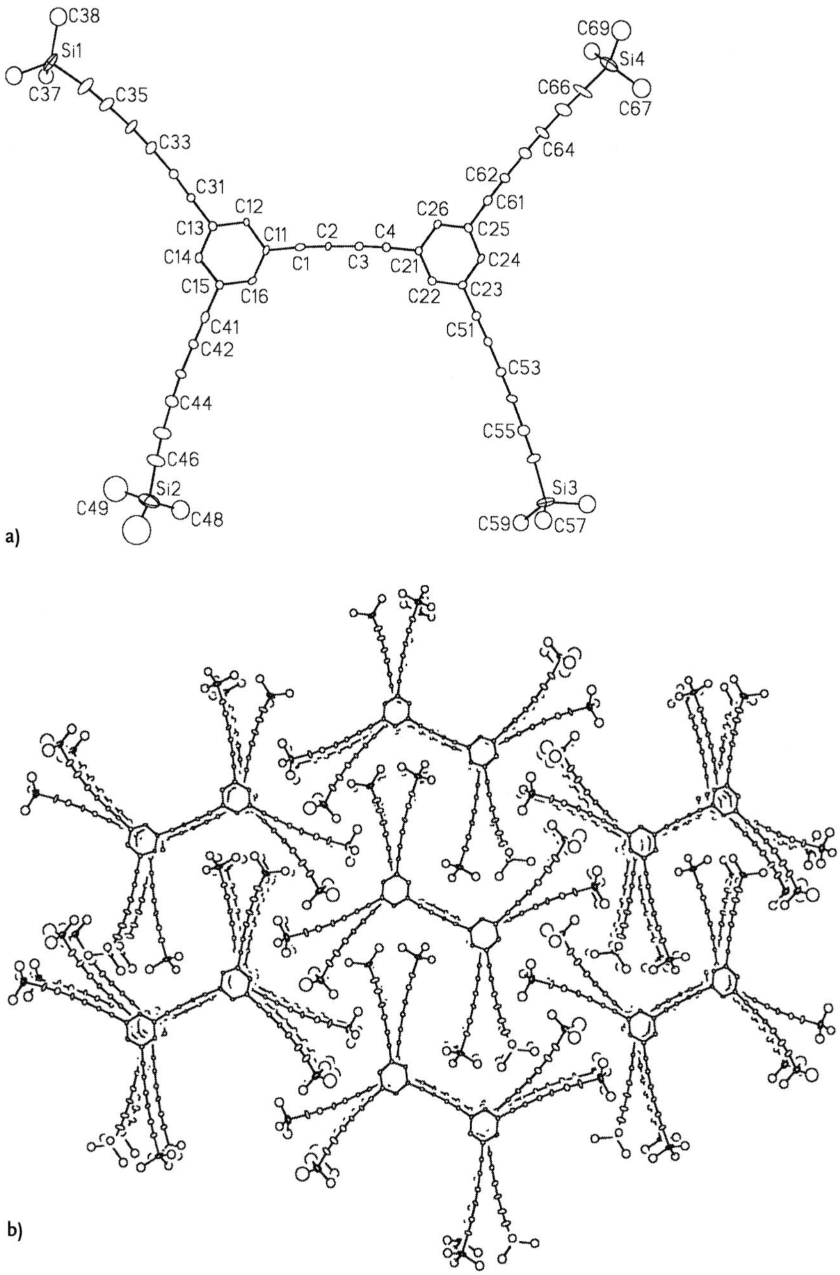

Figure 7.3 a) ORTEP drawing (20% probability level) for compound **98** and b) crystal packing diagram for **98** viewed approximately along the crystallographic *a* axis.

7.4.3
Cyclo[*n*]carbons

For nearly two decades, the synthesis of monocyclic all-carbon molecules – the cyclo[*n*]carbons – has captured the imagination of numerous research groups, most notably in the work of Diederich [3] and Tobe [4]. In addition to the esthetic beauty of these molecules, the cyclo[*n*]carbons also serve a more practical purpose in that they are potential precursors to fullerenes and two-dimensional carbon networks [2, 4]. A number of different routes toward the cyclo[*n*]carbons have been explored, leading both to success and to frustration. One of the most challenging aspects in the synthesis of these molecules is the introduction of the final acetylene units from suitably functionalized, macrocyclic precursors. In this regard, decarbonylation [65], retro-[4+2] [66] and retro-[2+2] cycloaddition [67], and the decomposition of aminotriazoles [68] have all shown promise.

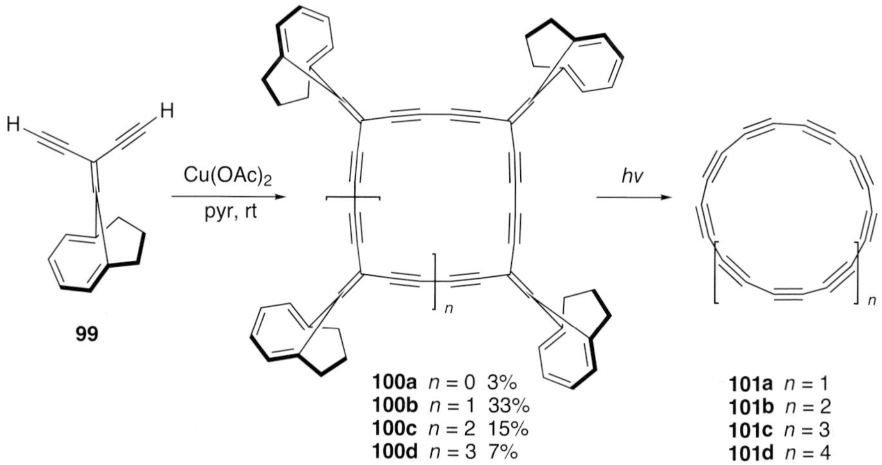

Scheme 7.19 Tobe's vinylidene rearrangement to cyclo[*n*]carbons **101a–d**.

It has recently been shown that the alkylidene-to-acetylene rearrangement, as outlined in Section 7.3.4, is an excellent way to access an acetylene unit [44]. This method relies on the cheletropic extrusion of indane to form the alkylidene carbene, which then rearranges to introduce the final alkyne segments. Tobe has very recently shown that this is a expedient route for the synthesis of the cyclo[*n*]carbons (Scheme 7.19) [4]. Copper-mediated homocoupling of enediyne **99** gave the necessary precursors, the expanded radialenes **100a–d**, as a separable mixture in a combined yield of 58 %. The individual radialenes were then subjected to laser desorption time of flight (LD-TOF) mass spectral analysis, conditions that effected the gas-phase extrusion of the indane fragments. Subsequent rearrangement of the intermediate alkylidene carbenes gave the cyclo[*n*]carbons C_{18}, C_{24}, C_{30}, and C_{36} (**101a–d**). In the negative mode, the LD-TOF spectra clearly showed signals of the corresponding cyclo[n]carbon anions, as well as signals consistent with the stepwise loss of indane fragments. Notably, this route was the first successful formation of C_{36} (**101d**) from an organic precursor.

7.5 Linear Conjugated Polyynes

The formation of carbyne, the hypothetical linear form of carbon consisting entirely of sp-hybridized carbon (Figure 7.4), has been an intriguing goal for many years [5, 69]. To date, however, carbyne has presented a significant synthetic challenge, and the allure of this elusive allotrope continues to stimulate interest in end-capped sp-hybridized carbon chains. Polyynes are the oligomeric cousins of carbyne, and it would be expected that the properties of carbyne might be extrapolated from trends observed in the spectroscopic data of polyynes. sp-Hybridized carbon oligomers are also interesting in their own right thanks to their unique electronic, optical, and physical properties [70].

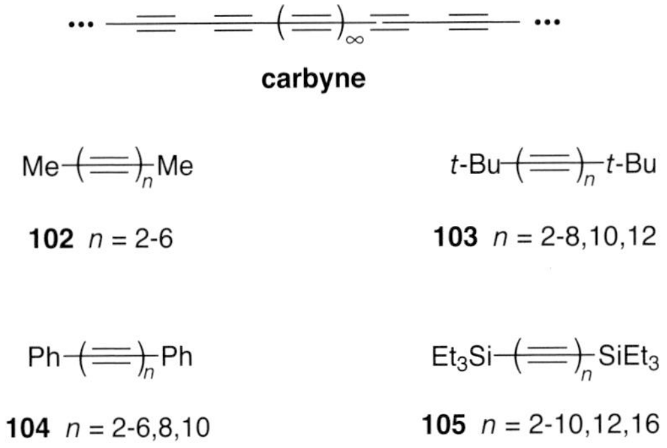

Figure 7.4 Carbyne and several known series of polyynes.

A firm foundation for the synthesis and study of polyynes was laid in the 1950s, 60s, and 70s (Figure 7.4), with the production of several impressive series of molecules terminated with methyl (**102**) [71], *t*-Bu (**103**) [72], phenyl (**104**) [72c, 73], and triethylsilyl (**105**) [72c, 74] groups [75]. In comparison with these "classic" works, more recent syntheses of polyynes have focused on three main goals: (1) experimentation with end-groups to improve stability and to manipulate the polyynes' properties, (2) development of improved synthetic methodology, and (3) use of modern methods of analysis to explore the properties of polyyne chains more fully.

While several research groups have worked with metal end-capping groups on polyynes [76–77], the longest polyynes in this class known to date are those synthesized by Gladysz and co-workers (Figure 7.5). Rhenium end-capped polyynes (**106**), for example, are highly crystalline and quite stable, allowing for extensive characterization of their structural, electronic, and optical properties [6a]. Platinum-terminated polyynes such as **107** are also stable, and a comprehensive analysis of

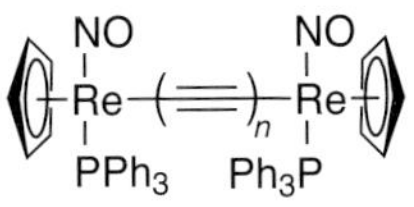

106 n = 3–6,8,10

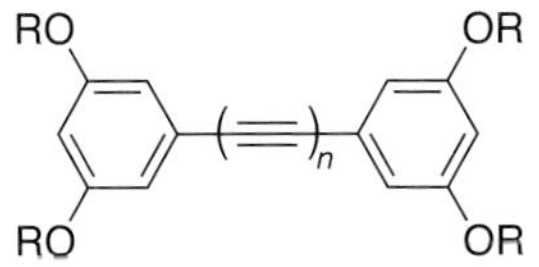

109 n = 2,4,6,8,10

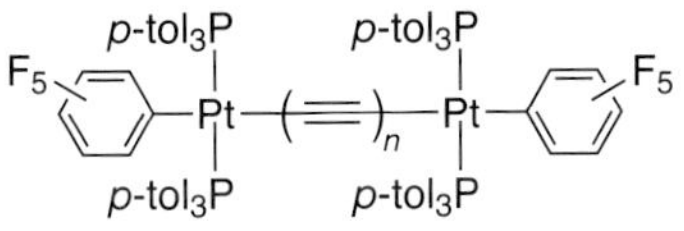

107 n = 2–5,8,10,12

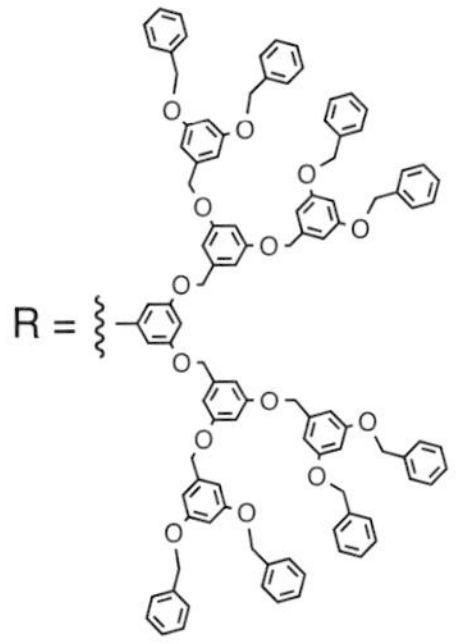

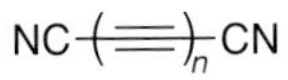

108 n = 3–8

Figure 7.5 Polyynes series recently reported by Gladysz (**106–107**) and Hirsch (**108–109**).

their physical characteristics (up to n = 8) has been reported, including the solid-state structures of several extended derivatives (vide infra) [78–79].

While not nearly as stable as their metal end-capped relatives **106–107**, dicyano-polyynes **108** (Figure 7.5) have also recently been synthesized and studied [80]. Substantially more resilient are the polyyne chains **109** reported by Hirsch's group, in which bulky dendrimeric end-groups endow the conjugated alkynyl structures with stability and solubility up to the decayne C_{20} [6b].

In spite of many efforts over the past half-century, there remain surprisingly few ways to assemble extended, conjugated polyynes, despite the obvious interest of their formation and study. In the majority of cases, oxidative coupling reactions have been used, but these methods have unfortunately been plagued by low yields, unstable intermediates, and for the longer carbon chains, the formation of polyynic by-products that complicate purification [6b]. One of the greatest problems associated with the study of polyyne molecules has been the inability to isolate reasonable quantities of the longest compounds, especially derivatives of C_{16}, C_{20}, and beyond. Current interest in polyynes thus continues to be motivated both by the synthetic challenge and the desire to explore their fundamental properties.

7.5.1
Synthesis of Triisopropylsilyl End-Capped Polyynes

The series of triisopropylsilyl (TIPS) end-capped polyynes shown in Figure 7.6 was targeted by Tykwinski and co-workers to probe the physical characteristics of polyynes as a function of length. The bulky TIPS groups provide both solubility and stability in the polyyne products, as well as in the necessary precursors. Dimer and tetramer **110** and **111**, respectively, were synthesized by an oxidative homo-

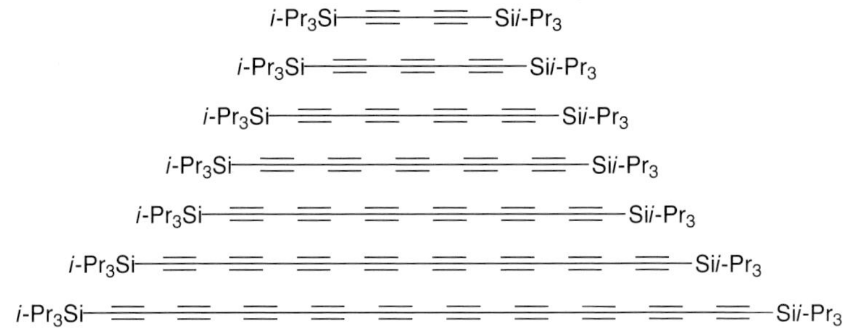

Figure 7.6 Tykwinski and Eisler's triisopropylsilyl-end-capped polyynes.

coupling procedure as described [81–82], while the synthesis of tryine **43a** by a FBW rearrangement is discussed in Section 7.3.3.1.

The synthesis of pentayne **56b** (Equation 7.10) has previously been reported by Tobe (Section 7.3.4) and also by Diederich [12c]. In the latter approach, the central acetylene unit is introduced in the final synthetic step by flash vacuum pyrolysis (FVP) of the diyne precursor **115**.

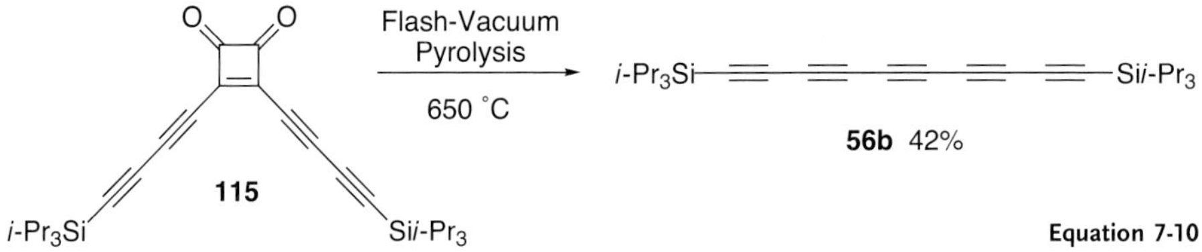

Equation 7-10

To avoid the necessity to assemble a FVP apparatus, the FBW approach was used as an alternative for the synthesis of pentayne **56b**. In an approach consisting of three steps from the readily available enediyne **41c** [37a] (Scheme 7.20), the key step involved lithiation of **116** with BuLi, followed by a condensation reaction with the TIPS propargyl aldehyde to give alcohol **117**. Oxidation (**118**) and dibromoolefination (**119**) proceeded uneventfully, and a two-fold FBW rearrangement by the standard procedure gave pentayne **56b** [70].

The synthesis of hexayne **112** was straightforward, thanks to the symmetry about the central single bond, a bond easily introduced by an oxidative homocoupling procedure (Scheme 7.21). Thus, deprotected enediyne **116** was subjected to the Hay procedure to provide tetrabromide **120** as the only isolable product in 55% yield. It is likely that the modest yield of this step is the result of a competitive cross-coupling reaction between the terminal acetylene and the vinyl bromide moiety in the presence of Cu(I/II) (Castro-Stephens coupling) giving rise to polymeric material [83]. The subsequent rearrangement of **120** provided the hexayne **112** as a pale yellow solid in 70% yield [41].

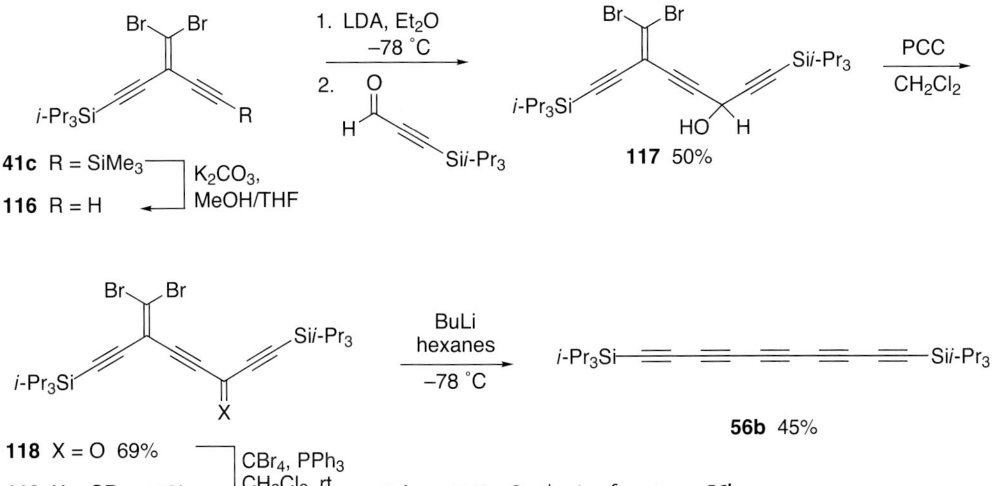

Scheme 7.20 Synthesis of pentayne **56b**.

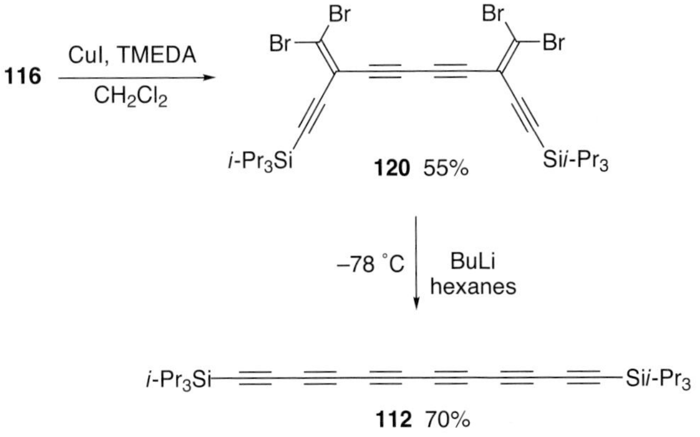

Scheme 7.21 Synthesis of hexayne **112**.

Friedel–Crafts acylation has become a powerful method for the direct formation of conjugated ketones, and treatment of acid chloride **121** with bis(trimethylsilyl) butadiyne represented a key step towards the formation of octayne **113** (Scheme 7.22). The unstable ketone **122** was formed in excellent yield and was directly carried on to the dibromoolefin **123**. Protiodesilylation of **123** and Hay coupling gave tetrabromide **124**, a surprisingly stable yellow solid. Under the general rearrangement conditions, octayne **113** was produced as a stable, light orange/yellow solid. While the yield of this two-fold reaction was lower than in the case of the penta- and the hexayne, an overall yield of 10% was obtained, still representing a 33% yield for each rearrangement event [70].

As described, polyynes of the series up to this point can be assembled fairly easily, and even the octayne **113** can be constructed in reasonable quantities in a few days. The longer the polyyne, however, the more elaborate the synthesis.

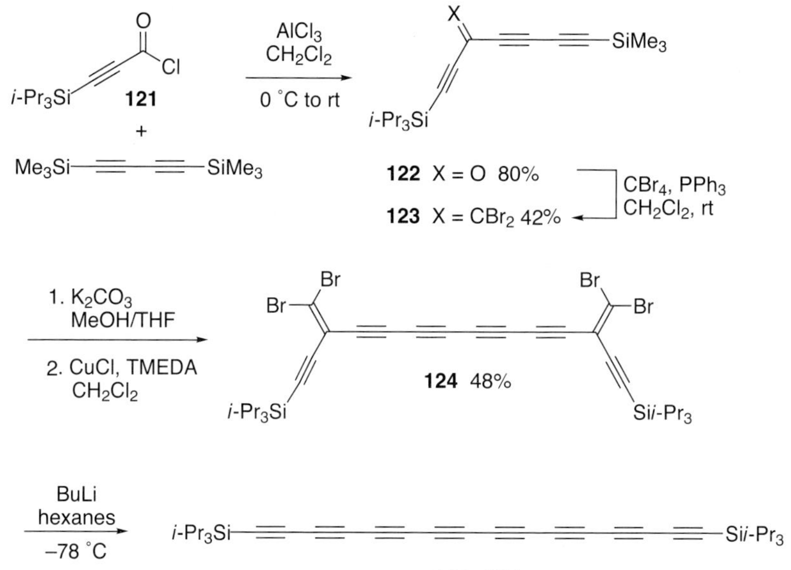

Scheme 7.22 Synthesis of octayne **113**.

This reality is reflected in the formation of the final compound in the series: decayne **114** (Scheme 7.23). The initial stages of this synthesis were analogous to those used in the formation of pentayne **56b** (Scheme 7.20), culminating in the tetrabromide **125**. To complete the 20-carbon framework for **114**, tetrabromide **125** was selectively deprotected and homocoupled under Hay conditions to give octabromide **126** in 25 % yield [70].

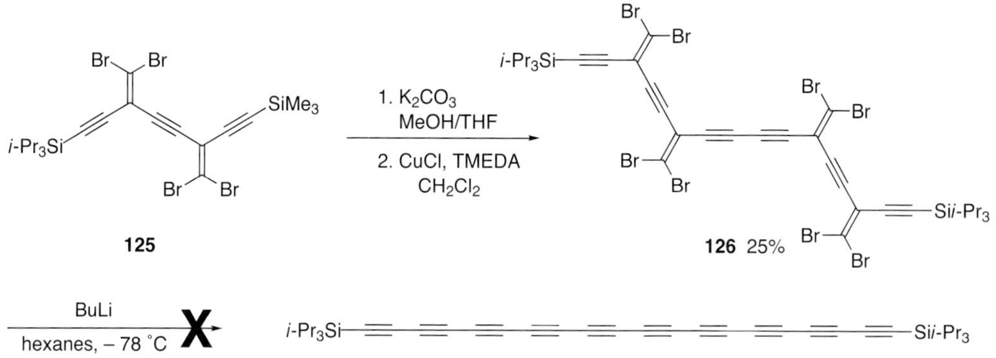

Scheme 7.23 Attempted synthesis of decayne **114**.

Although numerous attempts to accomplish the quadruple rearrangement of **126** → **114** have been made, all have been frustratingly ineffective, producing only intractable black solids. An alternative approach was therefore explored, as outlined in Scheme 7.24. The 10-carbon segment **125** was first rearranged to provide the differentially protected pentayne **127**, which, somewhat surprisingly, was found to be much less stable than the TIPS-protected analogue **56b**. Pentayne **127** was therefore carried on directly into the desilylation step, followed immediately by oxidative homocoupling to afford the decayne **114**. Unexpectedly, the nonayne **128** was also formed as ~10% of the product mixture when the homocoupling reaction was performed at ambient temperature [84]. It was subsequently determined that the formation of the nonayne could be almost completely suppressed by lowering the temperature of the homocoupling reaction to –10 °C. The bright orange decayne **114** shows limited stability under certain conditions, such as during chromatography (silica gel and alumina) as well as under ambient light and temperatures for extended periods of time. It does, however, demonstrate a fair degree of kinetic stability when kept in the freezer and protected from light. Overall, 51 mg of decayne **114** was isolated pure from this reaction, representing a yield of 30% over the last two steps.

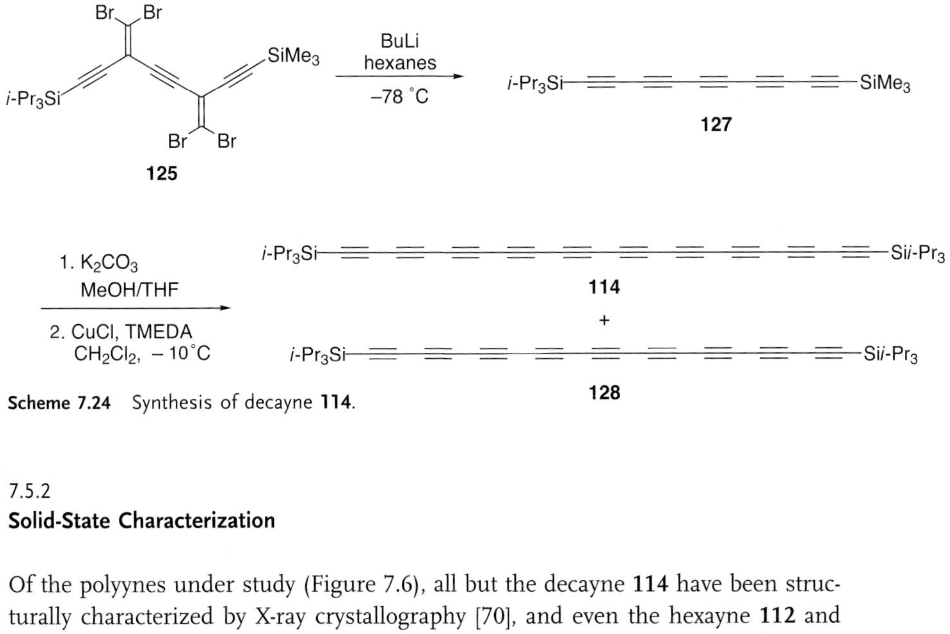

Scheme 7.24 Synthesis of decayne **114**.

7.5.2
Solid-State Characterization

Of the polyynes under study (Figure 7.6), all but the decayne **114** have been structurally characterized by X-ray crystallography [70], and even the hexayne **112** and the octayne **113** exhibit surprising stability as crystalline solids, easily affording crystals suitable for analysis. While the general solid-state structural characteristics of polyynes has been elegantly and comprehensively reviewed by Szafert and Gladysz only last year [85], a brief discussion of the solid-state characteristics of compounds **112** and **113** here is valuable, as they represent the only structurally characterized hexa- and octayne without metal-containing end-groups (Figure 7.7).

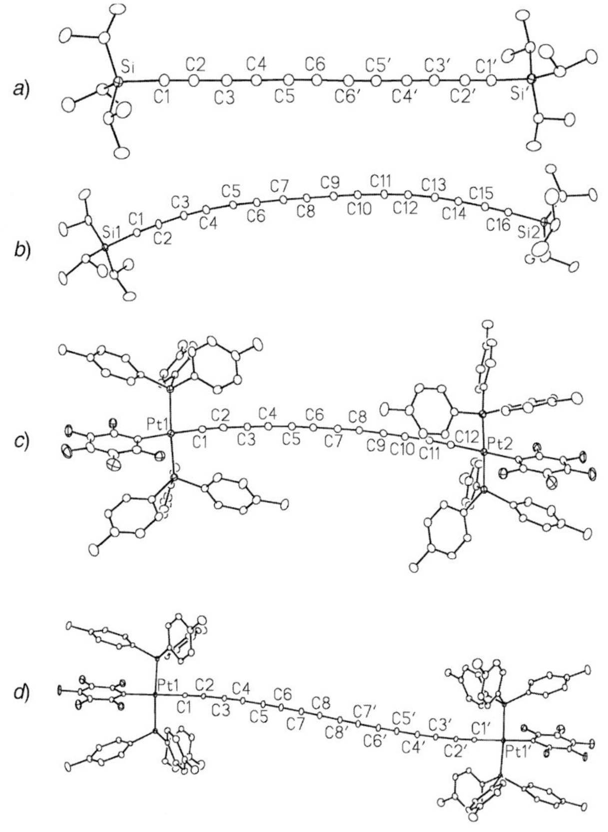

Figure 7.7
ORTEP diagrams of:
a) hexayne **112**,
b) octayne **113**,
c) hexayne **129** from Gladysz et al. [78], and
d) octayne **130** from Gladysz et al. [78]
(all shown at 20% probability level).

Of the two structures, the octayne **113** has the more unusual shape (Figure 7.7b). Each individual acetylene unit bears only a slight deviation from linearity, with an average C≡C−X (X = C or Si) bond angle of 176.9°. Overall, however, the cumulative effect of these deviations results in a rather dramatic curvature to give an unsymmetrical bow shape. The shape of **113** most closely resembles that of hexayne **129** reported by Gladysz and co-workers [78], which shows a symmetric bow conformation with a slightly lower average C≡C−X bend of 174.6° (Figure 7.7c). Interestingly, octayne **130** (Figure 7.7d), the only other structurally characterized octayne, shows a much more linear conformation, much like that of TIPS-end-capped hexayne **112** (Figure 7.7a).

Within this series of molecules, from the diyne **110** through to the decayne **114**, a reduction in the degree of bond-length alternation would be predicted as one proceeds to longer and longer derivatives [85]. Similarly to the study of series **107** (including **129** and **130**), however, no trend in the experimental bond lengths was discernible for the TIPS-end-capped derivatives including **112** and **113**.

The crystal packing diagram for **113** highlights the bending of the molecules as viewed down the *a* axis (Figure 7.8). In the solid state, the curved molecules clearly

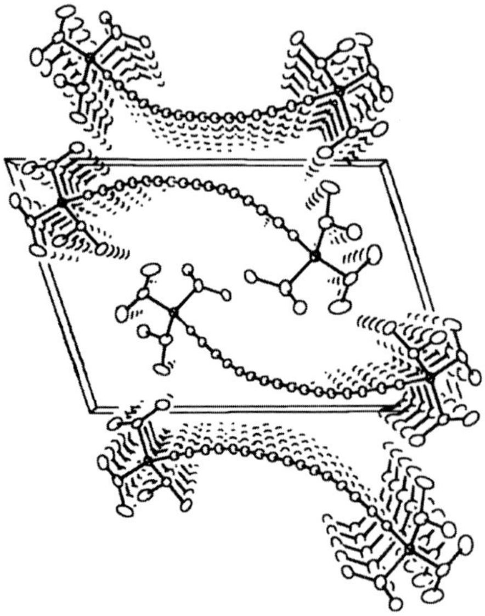

Figure 7.8 Crystal packing diagram of octayne **113** viewed down the crystallographic *a* axis.

pack in an alternating fashion to accommodate the bulky TIPS groups, and there is also alignment of the gently curving structures down the *a* axis. As a result of this alignment of the polyyne cores, there are a number of close contacts in the 3.5–4 Å range between neighboring molecules, potentially suitable for topochemical polymerization [63–64]. Furthermore, the packing orientation is well within the range expected to afford 1,6-polymerization (see Figure 7.2) [64]. Nonetheless, octayne **113** remains free from polymerization at room temperature, presumably due to a significant kinetic barrier to this solid-state reaction. Differential scanning calorimetry (DSC) analysis of a crystalline sample, however, clearly shows a narrow exotherm initiated at ca. 130 °C, consistent with thermally induced polymerization [70].

7.5.3
Linear Optical Properties

UV/Vis spectra of the TIPS polyynes are presented in Figure 7.9a. In contrast with the situation in most other conjugated organic molecules, the high-energy regions of the UV spectra (220–270 nm) of the longest polyynes, octayne **113** and decayne **114**, are nearly transparent. This rare characteristic could allow for interesting optical applications of polyynes within this window of transparency. At lower energies, vibrational fine structures are clearly visible, appearing as a series of narrow absorption peaks with steadily increasing intensity toward the visible region. There is a corresponding increase in the molar absorptivity (ϵ) as the chain length increases, and the TIPS-end-capped system shows some of the highest molar absorptivity values measured for polyynes, with octayne **113** at ϵ = 603,000 cm^{-1} M^{-1} and

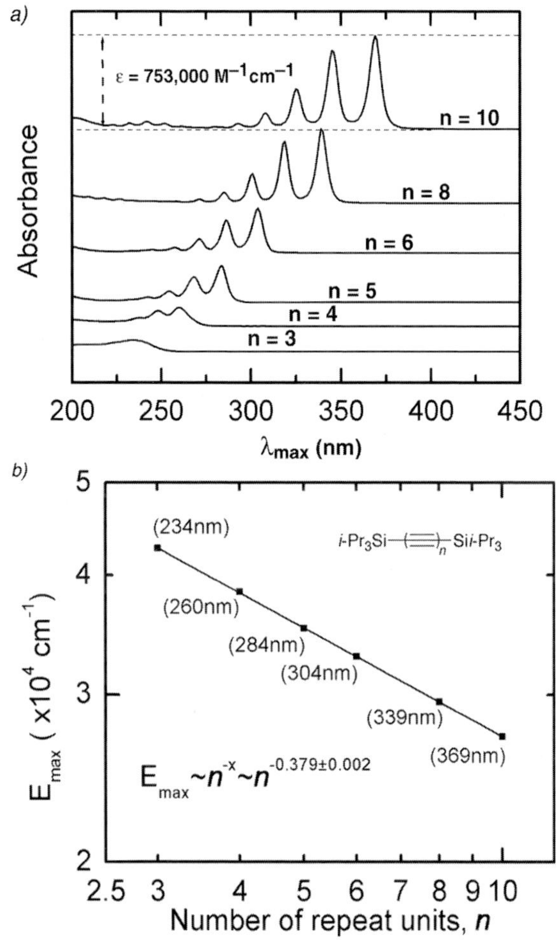

Figure 7.9
a) UV/Vis absorption spectra for polyynes in hexanes as a function of the number of C≡C units (n). Molar absorptivity (ϵ) at λ_{max} is shown for the decayne **114**, and b) power-law plot of E_{max} versus n ($E_{max} = 1/\lambda_{max}$), with corresponding λ_{max} values given in parentheses adjacent to the respective data points. The solid line represents the line-of-best-fit to the data.

decayne **114** at $\epsilon = 753{,}000\ \text{cm}^{-1}\ \text{M}^{-1}$ (Table 7.4) [86]. A comparison of the ϵ value at λ_{max} for **114** to that of other known decaynes highlights the dramatic dependence of the oscillator strength on the nature of the end-group: rhenium acetylide complex **106** ($n = 10$) shows $\epsilon = 190{,}000\ \text{cm}^{-1}\ \text{M}^{-1}$ [6a], dendrimer-terminated decayne **109** ($n = 10$) shows $\epsilon = 605{,}000\ \text{cm}^{-1}\ \text{M}^{-1}$ [6b], while the hydrocarbon-terminated decayne **103** ($n = 10$) shows $\epsilon = 850{,}000\ \text{cm}^{-1}\ \text{M}^{-1}$ [72b].

A red-shift in λ_{max} is clearly visible as the length of the carbon rods is increased, indicating a decrease in the HOMO→LUMO energy gap (Figure 7.9a). At a particular chain length, however, saturation of this effect would be expected, and λ_{max} would converge to a minimum and constant value. This particular chain length would then represent the effective conjugation length for this series of polyyne oligomers, and this minimum value of λ_{max} would also provide an estimate of the HOMO→LUMO gap of carbyne. Considering electron correlation effects, the empirical power-law $E_{max} = 1/\lambda_{max} \sim n^{-x}$ describes the relationship between E_{max}, λ_{max},

and n (where E_{max} is the energy of the HOMO→LUMO gap in cm^{-1}) [87]. This relationship is depicted in Figure 7.9b, and shows a consistent power-law decrease in E_{max} through at least C_{20}. Analysis of this data affords, with high precision, the relationship of $E_{max} \sim n^{-0.379 \pm 0.002}$. While this relationship is different from that found by Gladysz and co-workers, who report a relationship of $E_{max} \sim n^{-1}$ for rhenium end-capped polyynes **106** [6a], it is close to the well established Lewis–Calvin relationship for $E_{max} \sim n^{-0.5}$ observed for many polyenic materials [88]. Since the power-law decrease in E_{max} depicted in Figure 7.9b includes all of the oligomers, saturation of the HOMO–LUMO gap has clearly not yet begun.

It is well known that the relationship between chain length and λ_{max} for a homologous family of molecules can be used to estimate the λ_{max} of the infinite-length polymer of the series [6]. This is based on the assumption that λ_{max} values of the oligomers asymptotically approach the value of the polymer. By this method, it is interesting to extrapolate toward the electronic absorption properties of the infinite-length oligomer, which would be representative of carbyne. A plot of E_{max} against $1/n$ for the TIPS-end-capped polyynes is shown in Figure 7.10. Extrapolation to the y-intercept predicts λ_{sat} = 570 nm, the expected value for $i\text{-Pr}_3\text{Si–}(C{\equiv}C)_\infty\text{–Si}i\text{-Pr}_3$ [70]. This, then, suggests the HOMO–LUMO gap for carbyne, and this estimate is quite similar to the values of 565 nm [6a] and 569 nm [6b] predicted for the rhenium and dendrimer end-capped series **106** and **109**, respectively. While the above study gives an estimate for λ_{max} of carbyne, it does not answer the question "how long is carbyne?" By the following analysis, however, the length (n) of carbyne can be predicted.

The power-law relationship $E_{max} \sim n^{-0.379}$ (from Figure 7.9 above) can be transformed into terms of polyyne length [70], yielding (Equation 7.11):

$$n = (\lambda_{max}/154\ \text{nm})^{2.638} \qquad (7.11)$$

where λ_{max} is in nm. Substitution of the estimated HOMO–LUMO limit of λ_{max} = 570 nm predicted from Figure 7.10 into Equation 7.11 yields a value of $n \sim 32$.

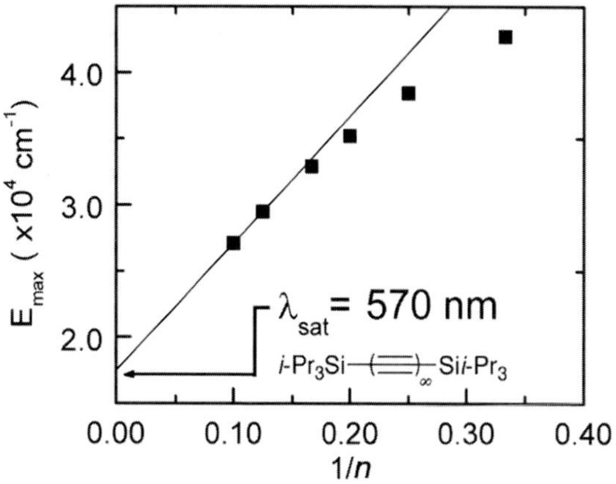

Figure 7.10 A plot of E_{max} (in cm^{-1}) versus $1/n$ for the TIPS-end-capped polyynes, extrapolated to provide λ_{max} for $i\text{-Pr}_3\text{Si–}(C{\equiv}C)_\infty\text{–Si}i\text{-Pr}_3$.

Thus, if it is assumed that saturation occurs rapidly for this polyyne system, 64 carbons represents the effective conjugated length for this series of polyynes (i-Pr_3Si–$(C{\equiv}C)_{32}$–$Si$$i$-$Pr_3$) and provides a reasonable experimental estimate of the *minimum* length of carbyne [70].

7.5.4
Third-Order Nonlinear Optical Properties

Polyynes are perhaps the simplest conjugated organic oligomers, and they offer an opportunity to probe electronic communication free of the configurational limitations often imposed by rotation about single bonds. This characteristic distinguishes polyynes from other typical organic oligomers and polymers, for which bond rotation can result in an interruption in conjugation along the molecular framework [89]. While the third order NLO properties of polyynes have been explored in numerous theoretical studies [90], it was not until recently that these properties were measured experimentally [86].

The molecular second hyperpolarizabilities (γ) of the polyynes shown in Figure 7.6 have been evaluated by use of a differential optical Kerr effect (DOKE) detection setup [86, 91]. The third-order response for these polyynes is dominated by an ultrafast (< 100 fs) electronic hyperpolarizability, and Table 7.4 lists γ values for the individual members of the series. While the second hyperpolarizabilities for shorter polyynes ($n \leq 6$) show only moderate values, γ values of the longest polyynes (e. g., **114**: $\gamma = (6.5 \pm 0.3) \times 10^{-34}$ esu) are substantial for relatively small molecules [92]. By way of comparison, the conjugated polyene molecule β-carotene, which has 11 consecutive single-double bonds, has a γ value of $(7.9 \pm 0.8) \times 10^{-34}$ esu, very similar to that of decayne **114** (as measured in THF with the DOKE setup).

Table 7.4 Linear and nonlinear optical characteristics of polyynes.

Compound	*n*	λ_{max} [nm][a]	ε_{max} [$M^{-1}cm^{-1}$][a]	$\gamma \times 10^{-36}$ [esu][b]
110	2	<210	–	2.75 ± 0.28
43a	3	234	92 700	6.99 ± 0.70
111	4	260	157 000	12.5 ± 2.1
56b	5	284	293 000	35.3 ± 1.2
112	6	304	359 000	64.3 ± 2.9
113	8	339	603 000	238.0 ± 47
114	10	369	753 000	646.0 ± 27

[a] As measured in hexanes. [b] As measured in THF by ultrafast Kerr spectroscopy (DOKE) [86].

Various mathematical approaches have been established to explain the relationship between the molecular hyperpolarizability and length for conjugated oligomers, and they show that γ values generally increase superlinearly as a function

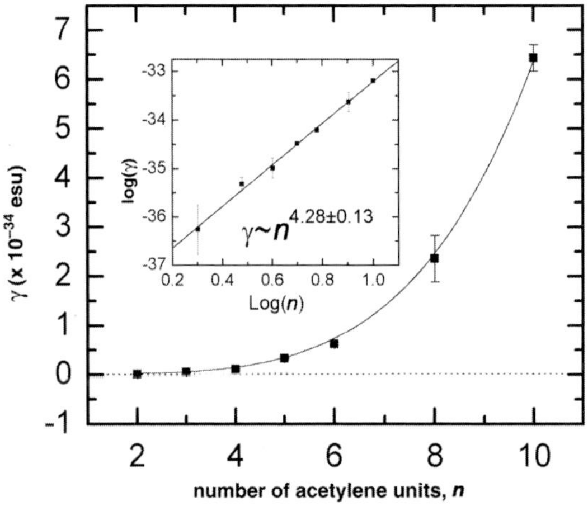

Figure 7.11 Polyyne molecular second hyperpolarizability (γ) as a function of the number of repeat units (n) in the oligomer chain. The solid line is a fit of the form $\gamma = a + bn^c$, where a is an offset due to end-group effects, b is a constant, and c is the power law exponent. The inset shows a log-log plot of the same data with the coefficient a subtracted, yielding a power law exponent (slope of the solid line) of $c = 4.28 \pm 0.13$ for the polyynes.

of length for a series of structurally similar molecules (Figure 7.11) [87, 93]. As with most conjugated oligomers, γ values of the polyynes can be fitted to a power-law relationship of $\gamma \sim n^c$, and this analysis gives a value of $c = 4.28 \pm 0.13$ (Figure 7.11, inset). This trend is continuous with the longest polyyne studied, the C_{20} chain **114**, and shows no indication that saturation of the second hyperpolarizability has begun.

The analysis of γ values for polyynes compares very favorably with that for other conjugated oligomers. The third-order optical nonlinearities of various substituted and unsubstituted polyenes have been investigated, for example, and experimental power-law exponents ranging from $c = 2.3$ to 3.6 have been reported (Figure 7.12) [94–95]. The third-order NLO response of polytriacetylenes [96], perhaps the closest structural relative to polyynes, shows power-law behavior with a reported exponent of $c = 2.5$ that is substantially smaller than that for the polyynes. Conjugated oligomers such as oligo(1,4-phenyleneethynylene)s also show an exponent of $c = 2.5$ [97], whereas values as high as $c = 4.05$ have been reported for polythiophenes [98].

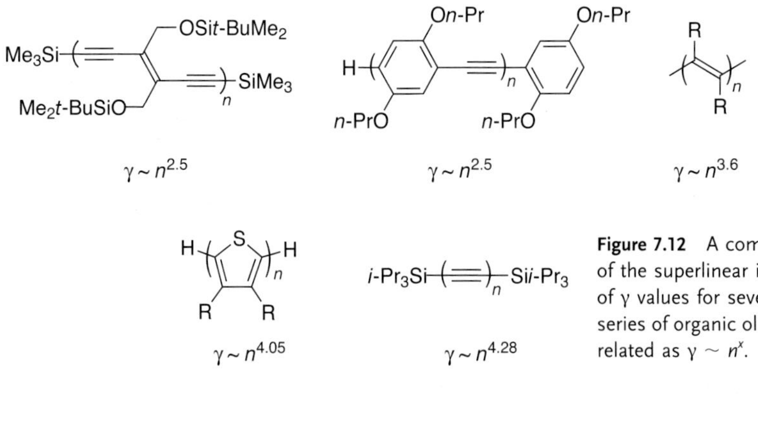

Figure 7.12 A comparison of the superlinear increase of γ values for several series of organic oligomers, related as $\gamma \sim n^x$.

Hegmann and Tykwinski report that, to the best of their knowledge, the exponent $c = 4.28 \pm 0.13$ is the highest exponent observed for a series of nonaromatic, conjugated oligomers [86]. The potential of polyynes as NLO materials would therefore seem to be quite exciting, particularly in light of recent success toward stabilization and protection of the typically reactive polyyne core. This concept has been ingeniously demonstrated by Gladysz, with the reactive polyyne being encased in a sheath of insulating alkyl groups (Figure 7.13) [79]. Future studies on even longer polyynes will no doubt continue to shed light on the unique physical and optical properties of these one-dimensional, carbon-rich systems.

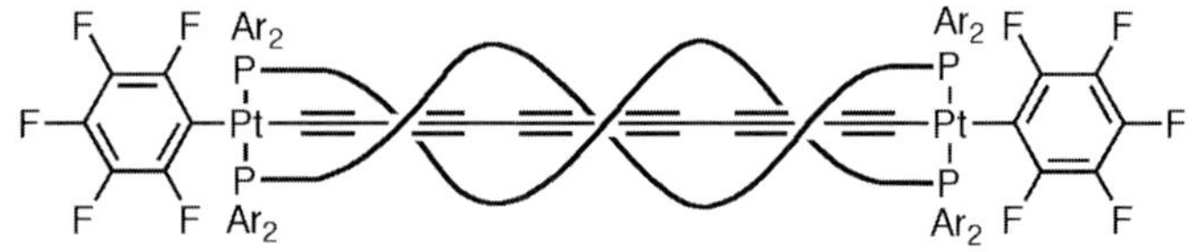

Figure 7.13 Schematic representation of a hexayne insulated by alkyl chains as per Gladysz et al. [79].

7.6 Conclusions

Alkylidene carbenes/carbenoids undergo a number of useful transformations, allowing for the formation of butatrienes and radialenes, as well as cyclopentyl and cyclopropylidene ring systems. It is their rearrangement to form an acetylene fragment, however, that has recently emerged as a particularly useful route to conjugated acetylenic molecules. From its discovery in 1894, the Fritsch–Buttenberg–Wiechell rearrangement has been employed as an efficient synthetic route to a plethora of alkynes appended with alkyl, vinyl and aromatic groups. It has also afforded an intriguing array of highly strained cycloalkynes. It has recently been demonstrated that the usefulness of the FBW rearrangement can be extended well beyond the formation of simple tolans. These efforts have shown that alkylidene carbenes/carbenoids allow for the facile formation of symmetrical and unsymmetrical di-, tri-, tetra-, and pentayne-containing structures. Nanoscale carbon-rich assemblies and new carbon allotropes (the cyclo[*n*]carbons) are afforded by FBW rearrangements, and extended polyynes (currently up to $R–(C{\equiv}C)_{10}–R$) can also be constructed in a few steps. These polyynes can now be produced in macroscopic quantities, a factor that greatly facilitates evaluation of their unique optical and electronic properties. In view of the tolerance of the FBW rearrangement to a range of substrates, coupled with the ability to accomplish multiple rearrangements in a single step, this method offers a facile synthetic route to polyyne molecules with applications spanning from natural products to advanced materials.

7.7 Experimental procedures

7.7.1 General procedure for Friedel–Crafts Acylation

$SOCl_2$ (28 mmol) was added to the carboxylic acid (7.0 mmol) in a dry flask protected from moisture with a drying tube containing $CaCl_2$, and the mixture was stirred overnight at rt. The excess $SOCl_2$ was removed in vacuo to provide the acid chloride. Distilled CH_2Cl_2 (50 mL) was added to the acid chloride, and the temperature of the solution was lowered to 0 °C. Bis(trimethylsilyl)acetylene (7.0 mmol) or bis(trimethylsilyl)butadiyne (7.0 mmol) and $AlCl_3$ (8.0 mmol) were added, and the reaction mixture was allowed to warm to rt. over 3 h. The reaction was carefully quenched by pouring of the reaction mixture into a beaker containing HCl (10 %, 50 mL) in ice (50 mL). Et_2O (75 mL) was added, the organic layer was separated, washed with satd. aq. $NaHCO_3$ (2 × 20 mL) and NaCl (2 × 20 mL), and dried ($MgSO_4$), and the solvent was removed in vacuo. Column chromatography (silica gel) provided the pure ketones [31].

7.7.2 General Procedure for Dibromoolefination.

CBr_4 (4.0 mmol) and PPh_3 (8.0 mmol) were added to distilled CH_2Cl_2 (100 mL), and the system was stirred for 5 min at rt. until the mixture turned bright orange. The ketone (3.0 mmol) in CH_2Cl_2 (5 mL) was added to the CBr_4/PPh_3 mixture over a period of 1 min. The reaction mixture turned a darker red/orange color upon addition of the ketone and was stirred until TLC analysis indicated consumption of the ketone. Solvent was reduced to ca. 5 mL, hexanes were added (100 mL), the inhomogeneous mixture was filtered through silica gel, and the solvent was removed. Column chromatography (silica gel), if necessary, provided the pure product [32, 34, 37a].

7.7.3 General FBW Rearrangement Procedure

A solution of the dibromoolefin (0.40 mmol) in distilled hexanes (12 mL) was cooled to –78 °C. BuLi (1.2 equiv. per dibromoolefin moiety) was slowly added over a period of ca. 2 min. The reaction mixture turned a pale yellow/orange color. While TLC analysis typically indicated that the reaction was complete soon after addition of base, warming of the reaction solution in the TLC capillary could influence this analysis (the R_f of the polyyne product is nearly always just greater than that of the dibromoolefinic precursor). Thus, the reaction mixture was warmed to between –40 °C and –10 °C, depending on the example, over a period of 1 h and then quenched with satd. aq. NH_4Cl (10 mL). Et_2O (10 mL) was added, the organic layer was separated, washed with satd. aq. NH_4Cl (2 × 20 mL), dried ($MgSO_4$), and the solvent was removed in vacuo. The crude reaction was passed through a plug of silica to remove baseline material. Column chromatography (silica gel), if necessary, gave the desired products. As outlined in the text, rigorously dry conditions are essential to the success of these reactions.

7.7.4
General Oxidative Coupling Procedure.

A mixture of the trimethylsilyl-protected alkyne (0.15 mmol) and K_2CO_3 (0.030 mmol) in wet THF/MeOH (30 mL, 1:1 v/v) was stirred for 2 h. Et_2O (30 mL) and satd. aq. NH_4Cl (30 mL) were added, the organic phases were separated, washed with satd. aq. NH_4Cl (2 × 20 mL), and dried ($MgSO_4$), and the solvent was reduced to ca. 2 mL. The terminal acetylene was added to a solution of the Hay catalyst [99] (CuCl (0.30 mmol) and TMEDA (0.60 mmol) in CH_2Cl_2 (60 mL), stirred until homogeneous) that had previously been oxygenated by passing O_2 through for 30 min. This mixture was stirred under air at rt. until TLC analysis no longer showed the starting material (ca. 3 h). Et_2O (30 mL) and satd. aq. NH_4Cl (30 mL) were added, the organic phase was separated, washed with satd. aq. NH_4Cl (2 × 20 mL), dried ($MgSO_4$), and the solvent was removed in vacuo. Column chromatography (silica gel) and/or recrystallization gave the desired product.

Acknowledgements

I (RRT) am indebted to the talented group of undergraduate and graduate students who have made this work possible, as well as to the fruitful collaboration with Professor Frank Hegmann's NLO group of at the University of Alberta. Financial support from the National Science and Engineering Research Council (NSERC), the University of Alberta (Dissertation Award to SE), and Petro Canada (Young Innovator Award to RRT) is gratefully acknowledged.

References

[1] *Modern Acetylene Chemistry* (Eds.: P. J. Stang, F. Diederich), VCH, Weinheim, 1995. *Top. Curr. Chem.* (Ed. A. De Meijere), Springer, Berlin, 1999, Vol. 201.

[2] (a) F. Diederich, Y. Rubin, *Angew. Chem., Int. Ed. Engl.* **1992**, *31*, 1101–1123; *Angew. Chem.* **1992**, *104*, 1123–1146; (b) U. H. F. Bunz, Y. Rubin, Y. Tobe, *Chem. Soc. Rev.* **1999**, 107–119.

[3] F. Diederich, Y. Rubin, O. L. Chapman, N. S. Goroff, *Helv. Chim. Acta* **1994**, *77*, 1441–1457.

[4] Y. Tobe, R. Umeda, N. Iwasa, M. Sonoda, *Chem. Eur. J.* **2003**, *9*, 5549–5559.

[5] (a) P. P. K. Smith, P. R. Buseck, *Science (Washington, D. C.)* **1982**, *216*, 984–986;
(b) F. Cataldo, *Polym. Int.* **1997**, *44*, 191–200.

[6] (a) R. Dembinski, T. Bartik, B. Bartik, M. Jaeger, J. A. Gladysz, *J. Am. Chem. Soc.* **2000**, *122*, 810–822;
(b) T. Gibtner, F. Hampel, J. P. Gisselbrecht, A. Hirsch, *Chem. Eur. J.* **2002**, *8*, 408–432.

[7] P. Cadiot, W. Chodkiewicz, in *Chemistry of Acetylenes* (Ed.: H. G. Viehe), Marcel Dekker, New York, 1969, Chapter 9.

[8] (a) K. Sonogashira, *J. Organomet. Chem.* **2002**, *653*, 46–49;
(b) K. Sonogashira, in *Handbook of Organopalladium Chemistry for Organic Synthesis* (Ed.: E. Negishi), Wiley, New York, 2002, pp 493–529.

[9] P. Siemsen, R. C. Livingston, F. Diederich, *Angew. Chem. Int. Ed.* **2000**, *39*, 2633–2657; *Angew. Chem.* **2000**, *112*, 2740–2767.

[10] (a) M. A. Heuft, S. K. Collins, G. P. A. Yap, A. G. Fallis, *Org. Lett.* **2001**, *3*, 2883–2886;
(b) M. L. Bell, R. C. Chiechi, C. A. Johnson, D. B. Kimball, A. J. Matzger, W. B. Wan, T. J. R. Weakley, M. M. Haley, *Tetrahedron* **2001**, *57*, 3507–3520.

[11] (a) T. Hiyama, in *Metal-Catalyzed Cross-Coupling Reactions* (Eds.: F. Diederich, P. J. Stang), Wiley-VCH, Weinheim, 1997, Chapter 11;
(b) Y. Nishihara, K. Ikegashira, K. Hirabayashi, J. Ando, A. Mori, T. Hiyama, *J. Org. Chem.* **2000**, *65*, 1780–1787.

[12] (a) A. Orita, N. Yoshioka, P. Struwe, A. Braier, A. Beckmann, J. Otera, *Chem. Eur. J.* **1999**, *5*, 1355–1363;
(b) Y. Tobe, T. Fujii, K. Naemura, *J. Org. Chem.* **1994**, *59*, 1236–1237;
(c) Y. Rubin, S. S. Lin, C. B. Knobler, J. Anthony, A. M. Boldi, F. Diederich, *J. Am. Chem. Soc.* **1991**, *113*, 6943–6949.

[13] (a) W. Kirmse, *Angew. Chem., Int. Ed. Engl.* **1997**, *36*, 1164–1170; *Angew. Chem.* **1997**, *109*, 1212–1218;
(b) P. J. Stang, *Chem. Rev.* **1978**, *78*, 383–405;
(c) H. D. Hartzler, in *Carbenes* (Eds.: R. A. Moss, M. Jones, Jr.), Wiley and Sons, New York, 1975, Vol. 2, pp. 43–100;
(d) G. Köbrich, *Angew. Chem., Int. Ed. Engl.* **1972**, *11*, 473–485; *Angew. Chem.* **1972**, *84*, 557–596;
(e) G. Köbrich, P. Buck, in *Chemistry of Acetylenes* (Ed.: H. G. Viehe), Marcel Dekker, New York, 1969, Chapter 2;
(f) M. Braun, *Angew. Chem. Int. Ed.* **1998**, *37*, 430–451; *Angew. Chem.* **1998**, *110*, 444–465.

[14] *Inter alia*, (a) J.-D. van Loon, P. Seiler, F. Diederich, *Angew. Chem., Int. Ed. Engl.* **1993**, *32*, 1187–1189; *Angew. Chem.* **1993**, *105*, 1235–1238;
(b) P. A. Morken, P. C. Bachand, D. C. Swenson, D. J. Burton, *J. Am. Chem. Soc.* **1993**, *115*, 5430–5439.

[15] H. Hopf, G. Maas, *Angew. Chem., Int. Ed. Engl.* **1992**, *31*, 931–954; *Angew. Chem.* **1992**, *104*, 953–977.

[16] (a) M. Iyoda, N. Nakamura, M. Todaka, S. Ohtsu, K. Hara, Y. Kuwatani, M. Yoshida, H. Matsuyama, M. Sugita, H. Tachibana, H. Inoue, *Tetrahedron Lett.* **2000**, *41*, 7059–7064;
(b) M. Iyoda, H. Otani, M. Oda, Y. Kai, Y. Baba, N. Kasai, *J. Am. Chem. Soc.* **1986**, *108*, 5371–5372.

[17] (a) H. Wiechell, *Liebigs Ann. Chem.* **1894**, *279*, 337–344;
(b) W. P. Buttenberg, *Liebigs Ann. Chem.* **1894**, *279*, 324–337;
(c) P. Fritsch, *Liebigs Ann. Chem.* **1894**, *279*, 319–323.

[18] V. Mouriès, R. Waschbüsch, J. Carran, P. Savignac, *Synthesis* **1998**, 271–274.

[19] *Inter alia*: (a) F. Bertha, J. Fetter, K. Lempert, M. Kajtar–Peredy, G. Czira, E. Koltai, *Tetrahedron* **2001**, *57*, 8889–8895;
(b) W. G. von der Schulenburg, H. Hopf, R. Walsh, *Angew. Chem. Int. Ed.* **1999**, *38*, 1128–1130; *Angew. Chem.* **1999**, *111*, 1200–1203.

[20] (a) H. Rezaei, S. Yamanoi, F. Chemla, J. F. Normant, *Org. Lett.* **2000**, *2*, 419–421;

(b) I. Creton, H. Rezaeï, I. Marek, J. F. Normant, *Tetrahedron Lett.* **1999**, *40*, 1899–1902.
[21] G. Köbrich, D. Merkel, K.-W. Thiem, *Chem. Ber.* **1972**, *105*, 1683–1693.
[22] H. Fienemann, G. Köbrich, *Chem. Ber.* **1974**, *107*, 2797–2803.
[23] For an excellent analysis of the subtleties of this mechanism, see: Z. Du, M. J. Haglund, L. A. Pratt, K. L. Erickson, *J. Org. Chem.* **1998**, *63*, 8880–8887.
[24] L. Fitjer, S. Modaressi, *Tetrahedron Lett.* **1983**, *24*, 5495–5498.
[25] K. L. Erickson, J. Wolinsky, *J. Am. Chem. Soc.* **1965**, *87*, 1142–1143.
[26] D. Y. Curtin, W. H. Richardson, *J. Am. Chem. Soc.* **1959**, *81*, 4719–4728.
[27] T. Ando, M. Nakagawa, *Bull. Chem. Soc. Jpn.* **1971**, *44*, 172–177.
[28] M. Kataoka, T. Ando, M. Nakagawa, *Bull. Chem. Soc. Jpn.* **1971**, *44*, 177–184.
[29] T. Kawase, H. R. Darabi, R. Uchimiya, M. Oda, *Chem. Lett.* **1995**, 499–500.
[30] R. F. C. Brown, W. R. Jackson, T. D. McCarthy, *Tetrahedron* **1994**, *50*, 5469–5488, and references therein.
[31] D. R. M. Walton, F. Waugh, *J. Organomet. Chem.* **1972**, *37*, 45–56.
[32] (a) F. Ramirez, N. B. Desai, N. McKelvie, *J. Am. Chem. Soc.* **1962**, *84*, 1745–1747.
(b) E. J. Corey, P. L. Fuchs, *Tetrahedron Lett.* **1972**, *13*, 3769–3772.
(c) G. H. Posner, G. L. Loomis, H. S. Sawaya, *Tetrahedron Lett.* **1975**, *16*, 1373–1376.
[33] A recent report offers an alternative route to dibromoolefins that may circumvent this limitation, see: H. Rezaei, J. F. Normant, *Synthesis* **2000**, 109–112.
[34] A. L. K. Shi Shun, E. T. Chernick, S. Eisler, R. R. Tykwinski, *J. Org. Chem.* **2003**, *68*, 1339–1347.
[35] A. L. K. Shi Shun, R. R. Tykwinski, *J. Org. Chem.* **2003**, *68*, 6810–6813.
[36] P. Bichler, A. L. K. Shi Shun, E. T. Chernick, S. Eisler, R. R. Tykwinski, unpublished results.
[37] (a) J. Anthony, A. M. Boldi, Y. Rubin, M. Hobi, V. Gramlich, C. B. Knobler, P. Seiler, F. Diederich, *Helv. Chim. Acta* **1995**, *78*, 13–45;
(b) B. A. Kulkarni, S. Chattopadhyay, A. Chattopadhyay, V. R. Mamdapur, *J. Org. Chem.* **1993**, *58*, 5964–5966;
(c) D. H. Wadsworth, S. M. Geer, M. R. Detty, *J. Org. Chem.* **1987**, *52*, 3662–3668.
[38] H. Hauptmann, M. Mader, *Synthesis* **1978**, 307–309.
[39] S. Eisler, R. R. Tykwinski, *J. Am. Chem. Soc.* **2000**, *122*, 10736–10737.
[40] G. Köbrich, H. Heinemann, W. Zündorf, *Tetrahedron* **1967**, *23*, 565–584.
[41] S. Eisler, N. Chahal, R. McDonald, R. R. Tykwinski, *Chem. Eur. J.* **2003**, *9*, 2542–2550.
[42] M. A. Heuft, S. K. Collins, G. P. A. Yap, A. G. Fallis, *Org. Lett.* **2001**, *3*, 2883–2886.
[43] S. Eisler, R. R. Tykwinski, unpublished results.
[44] Y. Tobe, N. Iwasa, R. Umeda, M. Sonoda, *Tetrahedron Lett.* **2001**, *42*, 5485–5488.
[45] *Chemistry and Biology of Naturally-Occurring Acetylenes and Related Compounds (NOARC)* (Eds.: J. Lam, H. Breteler, T. Arnason, L. Hansen), *Bioactive Molecules*, Elsevier, New York, 1988, Vol. 7.
[46] F. Bohlmann, H. Burkhardt, C. Zdero, *Naturally Occurring Acetylenes*, Academic Press, New York, 1973.
[47] F. Bohlmann, L. Fanghänel, K.-M. Kleine, H.-D. Kramer, H. Mönch, J. Schuber, *Chem. Ber.* **1965**, *98*, 2596–2604.
[48] F. Bohlmann, W. von Kap-Herr, L. Fanghänel, C. Arndt, *Chem. Ber.* **1965**, *98*, 1411–1415.
[49] J. T. Arnason, B. J. R. Philogène, C. Berg, A. MacEachern, J. Kaminski, L. C. Leitch, P. Morand, J. Lam, *Phytochemistry* **1986**, *25*, 1609–1611.
[50] M. Resch, J. Heilmann, A. Steigel, R. Bauer, *Planta Med.* **2001**, *67*, 437–442.
[51] I. Yosioka, H. Hikino, Y. Sasaki, *Chem. Pharm. Bull.* **1960**, *8*, 957–959.
[52] H. J. Bestmann, H. Frey, *Liebigs Ann. Chem.* **1980**, 2061–2071.

[53] L. Alvarez, S. Marquina, M. L. Villareal, E. Alonso, G. Delgado, *Planta Med.* **1996**, *62*, 355–357.

[54] J. Meier, W. Chodkiewicz, P. Cadiot, A. Willemart, *C. R. Hebd. Séances Acad. Sci.* **1957**, *245*, 1634–1636.

[55] G. N. Rule, M. R. Detty, J. E. Kaeding, J. A. Sinicropi, *J. Org. Chem.* **1995**, *60*, 1665–1673.

[56] C. Chin, E. R. H. Jones, V. Thaller, R. T. Aplin, L. J. Durham, S. C. Cascon, W. B. Mors, B. M. Tursch, *J. Chem. Soc., Chem. Commun.* **1965**, 152–154.

[57] S. C. Cascon, W. B. Mors, B. M. Tursch, R. T. Aplin, L. J. Durham, *J. Am. Chem. Soc.* **1965**, *87*, 5237–5241.

[58] C. Chin, M. C. Cutler, E. R. H. Jones, J. Lee, S. Safe, V. Thaller, *J. Chem. Soc. C* **1970**, 314–322.

[59] C. Mukai, N. Miyakoshi, M. Hanaoka, *J. Org. Chem.* **2001**, *66*, 5875–5880.

[60] This is an example in which the dibromoolefination could not be optimized to give a reasonable yield, presumably due to steric interactions.

[61] T. R. Rankin, R. R. Tykwinski, unpublished results.

[62] V. Enkelmann, *Adv. Polym. Sci.* **1984**, *63*, 91–136.

[63] V. Enkelmann *Chem. Mater.* **1994**, *6*, 1337–1340.

[64] J. Xiao, M. Yang, J. W. Lauher, F. W. Fowler, *Angew. Chem. Int. Ed.* **2000**, *39*, 2132–2135; *Angew. Chem.* **2000**, *112*, 2216–2219.

[65] (a) Y. Rubin, C. B. Knobler, F. Diederich, *J. Am. Chem. Soc.* **1990**, *112*, 1607–1617;
(b) Y. Rubin, M. Kahr, C. B. Knobler, F. Diederich, C. L. Wilkins, *J. Am. Chem. Soc.* **1991**, *113*, 495–500.

[66] F. Diederich, Y. Rubin, C. B. Knobler, R. L. Whetten, K. E. Schriver, K. N. Houk, Y. Li, *Science (Washington, D. C.)* **1989**, *245*, 1088–1090.

[67] Y. Tobe, T. Fujii, H. Matsumoto, K. Tsumuraya, D. Noguchi, N. Nakagawa, M. Sonoda, K. Naemura, Y. Achiba, T. Wakabayashi, *J. Am. Chem. Soc.* **2000**, *122*, 1762–1775.

[68] G. A. Adamson, C. W. Rees, *J. Chem. Soc., Perkin Trans. 1* **1996**, 1535–1543.

[69] R. J. Lagow, J. J. Kampa, H.-C. Wei, S. L. Battle, J. W. Gegne, D. A. Laude, C. J. Harper, R. Bau, R. C. Stevens, J. F. Haw, E. Munson, *Science (Washington, D. C.)* **1995**, *267*, 362–367.

[70] S. Eisler, A. D. Slepkov, E. Elliott, T. Luu, R. McDonald, F. A. Hegmann, R. R. Tykwinski, *J. Am. Chem. Soc.*, in press.

[71] C. L. Cook, E. R. H. Jones, M. C. Whiting, *J. Chem. Soc.* **1952**, 2883–2891.

[72] (a) F. Bohlmann, *Chem. Ber.* **1953**, *86*, 657–667;
(b) E. R. H. Jones, H. H. Lee, M. C. Whiting, *J. Chem. Soc.* **1960**, 3483–3489;
(c) T. R. Johnson, D. R. M. Walton, *Tetrahedron* **1972**, *28*, 5221–5236.

[73] J. B. Armitage, N. Entwistle, E. R. H. Jones, M. C. Whiting, *J. Chem. Soc.* **1954**, 147–154.

[74] R. Eastmond, T. R. Johnson, D. R. M. Walton, *Tetrahedron* **1972**, *28*, 4601–4616.

[75] For an excellent summary of older work in this area see: H. Hopf, *Classics in Hydrocarbon Synthesis*, Wiley-VCH, Weinheim, 2000, Chapter 8.

[76] N. J. Long, C. K. Williams, *Angew. Chem. Int. Ed.* **2003**, *42*, 2586–2617; *Angew. Chem.* **2003**, *115*, 2690–2722.

[77] For recent examples, see: (a) G.-L. Xu, G. Zou, Y.-H. Ni, M. C. DeRosa, R. J. Crutchley, T. Ren, *J. Am. Chem. Soc.* **2003**, *125*, 10057–10065;
(b) A. B. Antonova, M. I. Bruce, B. G. Ellis, M. Gaudio, P. A. Humphrey, M. Jevric, G. Melino, B. K. Nicholson, G. J. Perkins, B. W. Skelton, B. Stapleton, A. H. White, N. N. Zaitseva, *Chem. Commun.* **2004**, 960–961.

[78] W. Mohr, J. Stahl, F. Hampel, J. A. Gladysz, *Chem. Eur. J.* **2003**, *9*, 3324–3340.

[79] J. Stahl, J. C. Bohling, E. B. Bauer, T. B. Peters, W. Mohr, J. M. Martin-Alvarez, F. Hampel, J. A. Gladysz, *Angew. Chem. Int. Ed.* **2002**, *41*, 1872–1876; *Angew. Chem.* **2002**, *114*, 1951–1957.

[80] G. Schermann, T. Grösser, F. Hampel, A. Hirsch, *Chem. Eur. J.* **1997**, *3*, 1105–1112.

[81] J. C. Bottaro, R. J. Schmitt, C. D. Bedford, R. Gilardi, C. George, *J. Org. Chem.* **1990**, *55*, 1916–1919.
[82] J. Hlavaty, J. Kavan, J. Kubista, *Carbon* **2002**, *40*, 345–349.
[83] R. D. Stephens, C. E. Castro, *J. Org. Chem.* **1963**, *28*, 3313–3315.
[84] During the course of our study, Hirsch and co-workers have reported the analogous formation of a nonayne product during the homocoupling of a pentayne precursor, see reference 6b.
[85] S. Szafert, J. A. Gladysz, *Chem. Rev.* **2003**, *103*, 4175–4205.
[86] A. D. Slepkov, F. A. Hegmann, S. Eisler, E. Elliott, R. R. Tykwinski, *J. Chem. Phys.* **2004**, *120*, 6807–6810.
[87] C. Bubeck, in *Electronic Materials – the Oligomer Approach* (Eds.: K. Müllen, G. Wegner), Wiley-VCH, Weinheim, 1998, Chapter 8.
[88] (a) G. N. Lewis, M. Calvin, *Chem. Rev.* **1939**, *25*, 273–326;
(b) A. Mathy, K. Ueberhofen, R. Schenk, H. Gregorius, R. Garay, K. Müllen, C. Bubeck, *Phys. Rev. B* **1996**, *53*, 4367–4376.
[89] (a) I. Ledoux, I. D. W. Samuel, J. Zyss, S. N. Yaliraki, F. J. Schattenmann, R. R. Schrock, R. J. Silbey, *Chem. Phys.* **1999**, *245*, 1–16;
(b) G. Rossi, R. R. Chance, R. Silbey, *J. Chem. Phys.* **1989**, *90*, 7594–7601.
[90] M. Schulz, S. Tretiak, V. Chernyak, S. Mukamel, *J. Am. Chem. Soc.* **2000**, *122*, 452–459, and references therein.
[91] A. D. Slepkov, F. A. Hegmann, Y. Zhao, R. R. Tykwinski, K. Kamada, *J. Chem. Phys.* **2002**, *116*, 3834–3840.
[92] For a general description of third order NLO processes in organic materials, see: U. Gubler, C. Bosshard, *Advances in Polymer Science* (Ed.: K.-S. Lee), Springer, Berlin, **2002**, Vol. 158, pp. 123–191.
[93] J.-L. Brédas, C. Adant, P. Tackx, A. Persoons, *Chem. Rev.* **1994**, *94*, 243–278.
[94] (a) I. D. W. Samuel, I. Ledoux, C. Dhenaut, J. Zyss, H. H. Fox, R. R. Schrock, R. J. Silbey, *Science (Washington, D. C.)* **1994**, *265*, 1070–1072;
(b) G. S. W. Craig, R. E. Cohen, R. R. Schrock, R. J. Silbey, G. Puccetti, I. Ledoux, J. Zyss, *J. Am. Chem. Soc.* **1993**, *115*, 860–867.
[95] H. S. Nalwa, in *Nonlinear Optics of Organic Molecules and Polymers* (Eds.: H. S. Nalwa, S. Miyata), CRC Press, Boca Raton, FL, 1997, Chapters 9 and 11.
[96] R. E. Martin, U. Gubler, C. Boudon, V. Gramlich, C. Bosshard, J.–P. Gisselbrecht, P. Günter, M. Gross, F. Diederich, *Chem. Eur. J.* **1997**, *3*, 1505–1512.
[97] H. Meier, D. Ickenroth, U. Stalmach, K. Koynov, A. Bahtiar, C. Bubeck, *Eur. J. Org. Chem.* **2001**, 4431–4443.
[98] M.-T. Zhao, B. P. Singh, P. N. Prasad, *J. Chem. Phys.* **1988**, *89*, 5535–5541.
[99] A. S. Hay, *J. Org. Chem.* **1962**, *27*, 3320–3321.

8
Macrocycles Based on Phenylacetylene Scaffolding

Carissa S. Jones, Matthew J. O'Connor, and Michael M. Haley

8.1
Introduction

The last decade of the 20th century witnessed a tremendous resurgence in the chemistry of the carbon-carbon triple bond [1]. The advent of novel synthetic methodology tailored towards the construction of the alkyne moiety, combined with the use of organotransition metal complexes for carbon-carbon bond formation [2], has revolutionized the assembly of acetylene-containing systems. In particular, recent advances in Pd-mediated cross-coupling reactions have allowed for the production of phenylacetylene derivatives from an alkyne sp-carbon atom and an sp^2 center either of an arene or of an alkene [3], while the construction of a butadiyne moiety through homo- or heterocoupling of terminal acetylene units has become routine practice in the laboratory [4]. Consequently, the assembly of novel acetylene-containing compounds, including alkyne-rich macrocycles, previously a laborious process, can now be accomplished with relative ease.

Phenylacetylene macrocycle (PAM), phenyldiacetylene macrocycle (PDM), phenyltriacetylene macrocycle (PTrM), phenyltetraacetylene macrocycle (PTeM), and larger phenyloligoacetylene macrocycles are not only of theoretical importance but have also been shown to exhibit myriad interesting physical and chemical properties, which bodes well for their use in materials applications [5]. The range of fascinating properties displayed by this family of macrocycles includes nonlinear optical and discotic liquid crystalline behavior, fluorescence, and the ability to form host-guest complexes, while some even explode to furnish ordered carbon nanostructures. The relative ease with which phenylacetylene macrocycles can now be constructed has allowed the bench chemist to prepare a wide variety of novel derivatives that can be readily functionalized and thus allow their physical properties and chemical reactivity to be tailored.

Although a tremendous number of related aryleneethynylene macrocycles have been prepared in recent years, this report focuses on systems containing benzene and acetylene moieties only. The reader is referred to several excellent reviews regarding acetylenic macrocycles containing other arenes, metal-coordinated arenes,

Acetylene Chemistry. Chemistry, Biology and Material Science. Edited by F. Diederich, P. J. Stang, R. R. Tykwinski

ISBN 3-527-30781-8

and heteroarenes [6]. Our review surveys the main synthetic strategies employed to prepare phenyl- and acetylene-containing macrocycles, provides a synopsis of the compounds prepared to date, and also offers an overview of the properties and reactivity of these novel molecules.

8.2
Synthetic Strategies

Two main synthetic strategies exist for the preparation of phenylacetylene macrocycles and may be classified depending upon whether the ring is constructed either in an *intermolecular* or in an *intramolecular* sense. The classical approach utilizes an *intermolecular* reaction in which two or more monomers are coupled together to form the macrocycle; the ring-closure thus involves formation of at least *two* new bonds. Whereas the starting materials are prepared with relative ease, numerous cyclooligomeric by-products are obtained in addition to the target molecule, which significantly reduces the yield of the desired molecule. Conversely, the *intramolecular* approach involves building the macrocyclic precursor in a stepwise fashion. Subsequent ring-closure affords the target molecule with concomitant creation of only *one* new bond per ring formed. Although numerous steps are often required to prepare the macrocyclic synthon, the ring-closure typically proceeds in good yield and the formation of by-products is minimized, and this has been the method of choice for phenylacetylene macrocycle construction in recent years [5]. The following subsections contain selected examples of each of these methods of macrocycle assembly.

8.2.1
Intermolecular Approach

The initially developed (and still utilized) intermolecular approach is the cyclooligomerization technique in which several (n) monomers (X) are coupled together to form the macrocycle (Z; i. e., $n\text{X} = \text{Z}$). In the second approach, two different components, X and Y, are coupled together to form the ring Z such that $\text{X} + \text{Y} = \text{Z}$. While the former technique can afford either monoyne- or diyne-linked products, this latter route has only been successfully used to prepare monoyne structures.

8.2.1.1 $n\text{X} = \text{Z}$

The first example of a phenyldiacetylene macrocycle (PDM) reported in the literature was described by Eglinton and co-workers in the late 1950s [7]. The authors believed they had isolated the trimeric [18]annulene **1** after oxidatively coupling the terminal acetylene moieties of *o*-diethynylbenzene (**2**) through treatment with $Cu(OAc)_2$ in pyridine under high-dilution conditions (Scheme 8.1). However, they subsequently reported that the reaction product was the strained dimeric

[12]annulene **3**, and no evidence to support the formation of higher macrocyclic analogues such as **1** was obtained [8]. Although cyclodimer **3** proved somewhat difficult to manipulate, the authors obtained a low-resolution X-ray structure of the compound that showed the "bowed" diacetylenic linkages, which in turn impart significant strain upon the molecule [9]. Not surprisingly, the energy-rich hydrocarbon decomposes explosively upon grinding or upon heating above 80 °C [8b]. Although yellow crystals of **3** blacken at ambient temperature within a few days, presumably due to auto-polymerization, it can be kept in a dilute solution of benzene or pyridine for extended periods of time, especially when refrigerated and stored under an inert atmosphere.

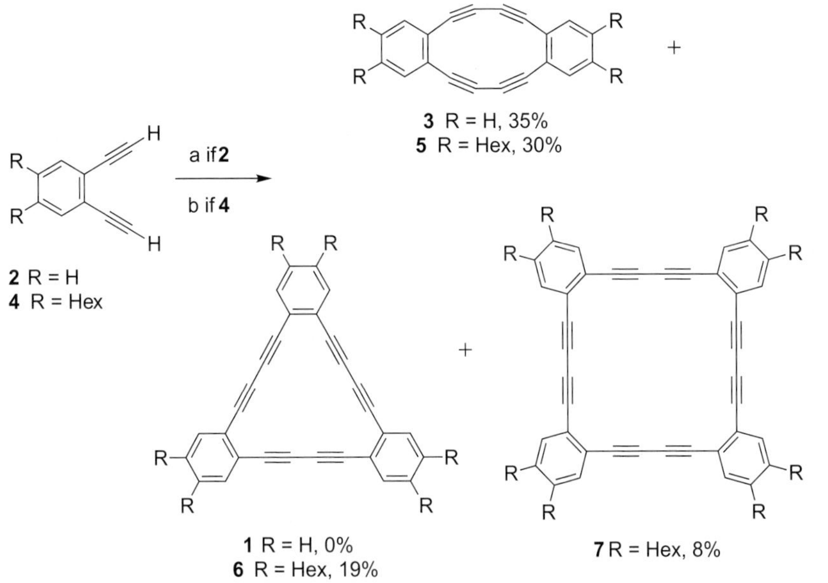

Scheme 8.1 Intermolecular synthesis of PDMs. Reagents and conditions: (a) $Cu(OAc)_2$, py, MeOH; (b) CuCl, TMEDA, ODCB, O_2.

In 1994 the Swager group oxidatively homocoupled *o*-diethynylbenzene derivative **4**, which possesses two hexyl groups, to form dimer **5** together with isolable amounts of trimer **6** and tetramer **7** (Scheme 8.1) [10]. The successful formation of the last two compounds can be attributed to the more strongly solubilizing nature of the hexyl groups and thus represents the first time that larger PDMs had been characterized from an intermolecular cyclooligomerization strategy where $n > 2$ for $nX = Z$.

In 1966, Eglinton et al. [11] used the Castro–Stephens reaction [12] to prepare phenylacetylene macrocycle (PAM) **8** in 26 % yield (Scheme 8.2). The cyclization of synthon **9**, in which a Cu-acetylide group is *ortho* to a halogen, afforded only monoyne-containing macrocyclic compounds. Repetition of this work some 20 years later by Youngs and co-workers afforded the same PAM, but in a somewhat higher 48 % yield along with the higher macrocycles **10** and **11** [13]. Huynh and Linstrumelle have also prepared PAM **8** (36 % yield) through a Pd-mediated cross-coupling reaction of **12** [14]. In addition to **8**, the latter reaction also yields small quantities of the higher homologues **10** and **11** [15].

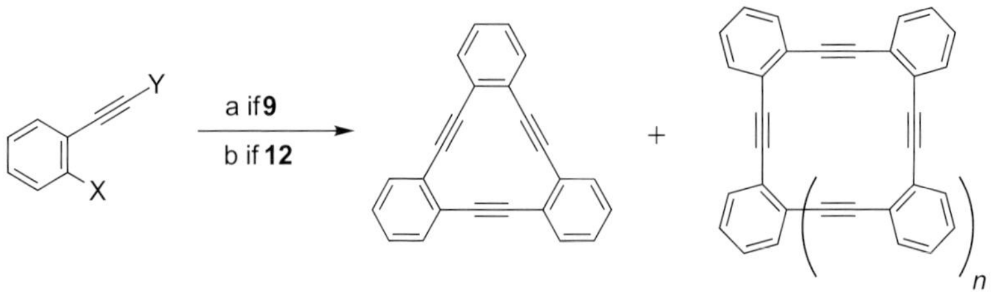

9 X = I; Y = Cu
12 X = Br; Y = $C(CH_3)_2OH$

8 26-48%

10 *n* = 1
11 *n* = 3

Scheme 8.2 Synthesis of PAMs **8**, **10**, and **11**. Reagents and conditions: (a) py, reflux; (b) $[Pd(PPh_3)_4]$, CuI, KOH, BTEACl, PhH, 85 °C.

A recent report by Vollhardt and co-workers describes the synthesis of trimeric macrocycle **8** in moderate yield. Alkyne metathesis of *o*-dipropynylated arene **13** was effected with the tungsten reagent $[(tBuO)_3W{\equiv}CtBu]$ to give the [12]annulene in 54 % isolated yield (Scheme 8.3) [16].

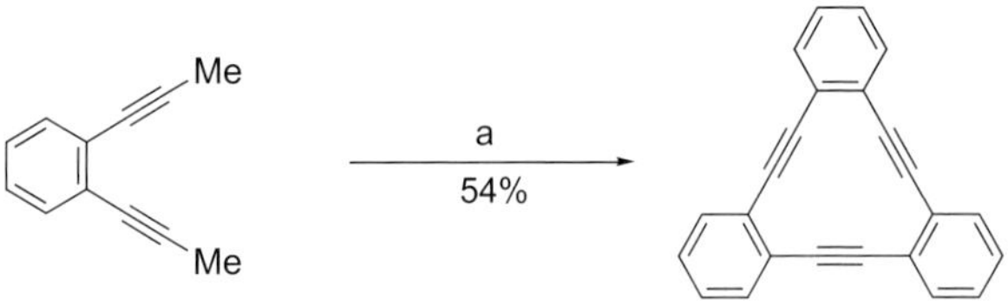

13 **8**

Scheme 8.3 Vollhardt synthesis of PAM **8**. Reagents and conditions: (a) $[(tBuO)_3W{\equiv}CtBu]$, toluene, 80 °C.

8.2.1.2 **X + Y = Z**

The first example of an X + Y = Z intermolecular synthesis of a phenylacetylene macrocycle was reported by Staab and Graf in 1966 (Scheme 8.4) [17]. X and Y were the bis(ylide) derived from **14** and *o*-phthalaldehyde, which, after a double Wittig reaction, formed intermediate **15**. Bromination of this compound and subsequent didehydrobromination with strong base gave **8** in an overall yield of 9 %.

Iyoda and co-workers prepared PAM **8** in a single step by way of the intermolecular X + Y = Z approach, cross-coupling 1,2-diodobenzene and excess acetylene gas under Pd catalysis conditions to give the target molecule in a modest 39 % yield [18].

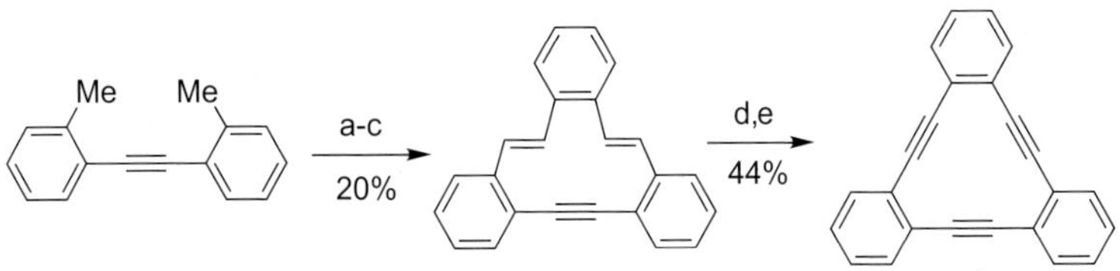

14 15 8

Scheme 8.4 Staab synthesis of PAM **8**. Reagents and conditions: (a) Br_2, CCl_4, NBS, BPO; (b) PPh_3, PhH; (c) i] PhLi, THF, ii] phthalaldehyde, THF; (d) Br_2, CCl_4; (e) *t*-BuOK, THF.

The synthesis of larger oligo(phenylacetylene)s such as **16** (Scheme 8.5) was accomplished by Youngs et al. through the iterative Sonogashira cross-coupling of diiodide **17** and polyyne **18** [19]. In addition to isolating the 40-membered macrocycle **16** in 25 % yield, the authors obtained evidence to support the formation of trace amounts of the higher derivatives $C_{160}H_{80}$, $C_{240}H_{120}$, $C_{320}H_{160}$, and $C_{400}H_{200}$.

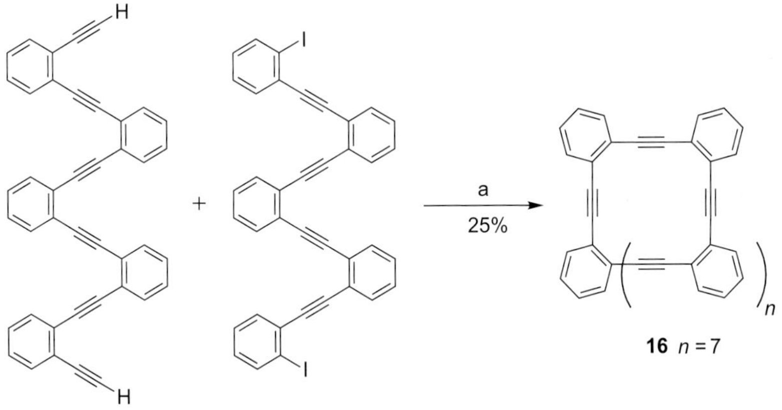

18 17

Scheme 8.5 Synthesis of PAM **16**. Reagents and conditions: (a) $[Pd(PPh_3)_4]$, CuI, NEt_3, toluene.

8.2.2 Intramolecular Approach

The main advantage of the intramolecular synthesis of a phenylacetylene macrocycle is that it provides a rational route towards a *single* product. Consequently, the formation of oligomeric by-products is minimized, which greatly facilitates the purification of the target molecule. Although the number of steps required to prepare the macrocyclic precursor is typically high, this method allows unsymmetrical systems to be assembled. The majority of recently prepared phenylacetylene macrocycles have been synthesized by this approach.

The intramolecular synthesis can be divided into two main classes: *linear* or *convergent*. In the linear approach the molecule is assembled until the ring is ready for final intramolecular ring-closure. In the convergent approach, two or more components are cross-coupled together to form the macrocyclic synthon, which is subjected to intramolecular ring-closure to give the target molecule.

8.2.2.1 **Linear**

The linear approach requires the greater number of steps needed for the synthesis of intermediates, most of which are obtained by cross-coupling reactions. Haley and co-workers prepared PAM **8** by this strategy in a 10-step sequence (Scheme 8.6) [20]. 2-Iodoaniline was diazotized and quenched with Et_2NH to give the triazene, which was cross-coupled with trimethylsilylacetylene (TMSA) to give the corresponding acetylene derivative. Half of this material was desilylated to give the free alkyne, while the remaining material was converted into the aryl iodide by heating in MeI. Coupling of the last compound with the free alkyne gave the advanced intermediate **19**, which was converted in two steps into the triyne **20**. Finally, conversion of the triazene into the iodoarene, deprotection of the alkyne, and intramolecular ring-closure afforded the target molecule in a modest 35 % overall yield for the 10 steps. Not surprisingly, due to the significant effort required to prepare the cyclization precursor, this method is seldom used to prepare phenylacetylene macrocyclic derivatives.

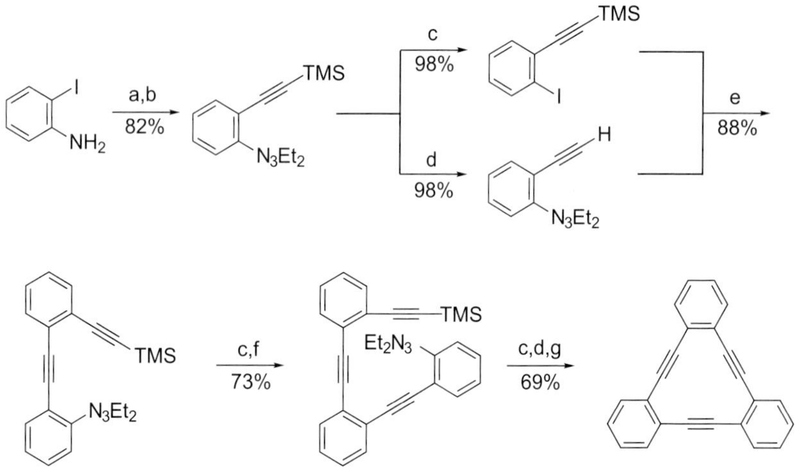

19 **20** **8**

Scheme 8.6 Stepwise synthesis of PAM **8**. Reagents and conditions: (a) i] $NaNO_2$, HCl, MeCN, H_2O, ii] Et_2NH, K_2CO_3, H_2O; (b) TMSA, $[PdCl_2(PPh_3)_2]$, CuI, Et_3N; (c) MeI, 120 °C; (d) K_2CO_3, THF, MeOH; (e) $[PdCl_2(PPh_3)_2]$, CuI, Et_3N; (f) *N,N*-diethyl-2-ethynylphenyltriazene, $[PdCl_2(PPh_3)_2]$, CuI, Et_3N; (g) $[Pd(dba)_2]$, PPh_3, CuI, Et_3N.

8.2.2.2 Convergent

The first example of a convergent synthesis of a phenylacetylene macrocycle was reported by Eglinton in 1964 as corroboration of the structure of the strained PDM **3** [21]. Alkyne dimerization of the starting material with $Cu(OAc)_2$ in pyridine precluded the formation of cyclic products (Scheme 8.7). Didehydrobromination of the bromo intermediate furnished the terminal bisalkyne **21**, which was intramolecularly homocoupled to afford **3**. Youngs later prepared the acetylene synthon **21** from arene **22** and repeated the cyclization reaction (Scheme 8.7) in the presence of CuCl instead of $Cu(OAc)_2$ and with aeration of the solution. This modification afforded not only **3**, but also the higher PDM **23**, formed in 65 and 20% yields, respectively [22]. Swager and his group obtained the [12]- and [24]annulenes **5** and **7**, respectively, from synthon **24**, which was prepared from butadiyne **25** [10]. It should be noted that use of the Hay catalyst (CuCl, TMEDA) in this latter reaction afforded a greater yield of the tetrameric PDM **7** (45%) versus the dimeric PDM **5** (13%).

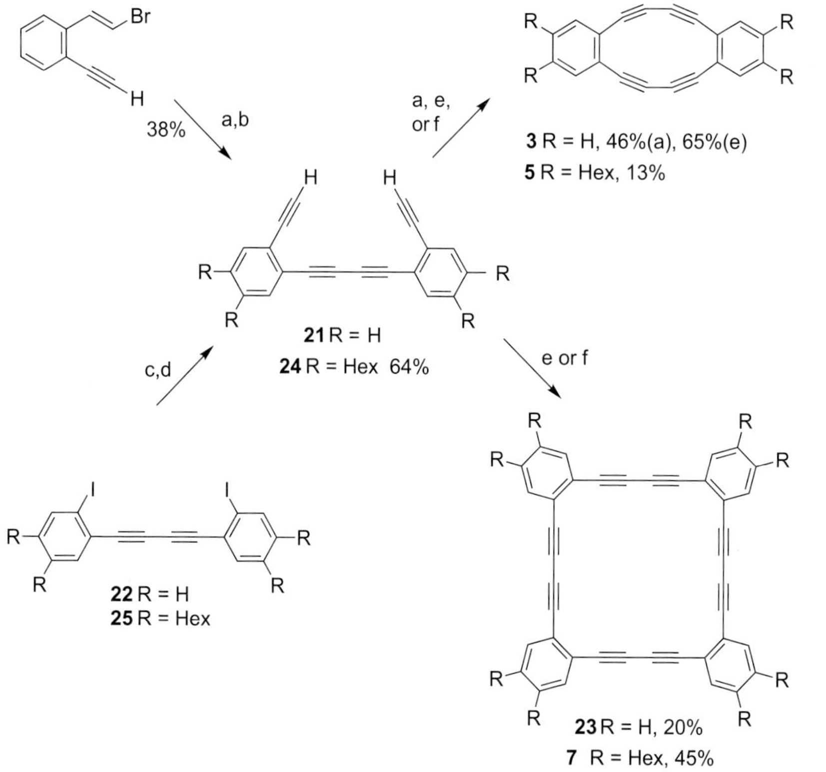

Scheme 8.7 Intra/intermolecular synthesis of PDMs. Reagents and conditions: (a) $Cu(OAc)_2$, py, MeOH; (b) *t*-BuOK, *t*-BuOH; (c) TMSA, $[PdCl_2(PPh_3)_2]$, CuI; (d) KF or KOH, H_2O, THF, MeOH; (e) CuCl, py, O_2; (f) CuCl, TMEDA, ODCB, O_2.

By the convergent intramolecular approach it is possible to prepare a variety of novel macrocycles inaccessible by the other routes. One such example is the C_{2v}-symmetric [14]annulene **26**, which was prepared in an excellent 83 % yield from the cyclization of tetrayne **27** with CuCl in dilute pyridine/MeOH solution (Scheme 8.8) [23].

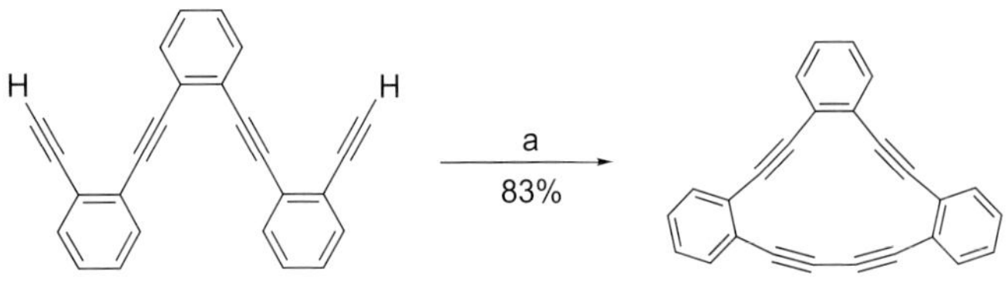

27 **26**

Scheme 8.8 Synthesis of PDM **26**. Reagents and conditions: (a) CuCl, py, MeOH.

The Haley group very recently used a convergent intramolecular approach for the selective synthesis of bis[14]- and bis[15]annulene derivatives [24]. The authors effected macrocycle formation under the traditional Eglinton conditions ($Cu(OAc)_2$, py), but more importantly, ring-closure was also achieved under milder Pd-mediated homocoupling conditions similar to those used in Sonogashira reactions to provide the appropriate PDMs in high yield. Cross-coupling of diyne **28** with tetraiodobenzene, for example, furnished octayne **29**, which upon subsequent removal of the TIPS groups and metal-mediated ring-closure could afford either PDM **30** or **31**, or a mixture of both (Scheme 8.9). Interestingly, under Cu-mediated conditions, the bis[15]annulene **30** was formed as the sole product. Conversely, use of $[PdCl_2(PPh_3)_2]$ provided a mixture of **30** together with the bis[14] derivative **31**, while acetylenic homocoupling with $[PdCl_2(dppe)]$ furnished **31** exclusively.

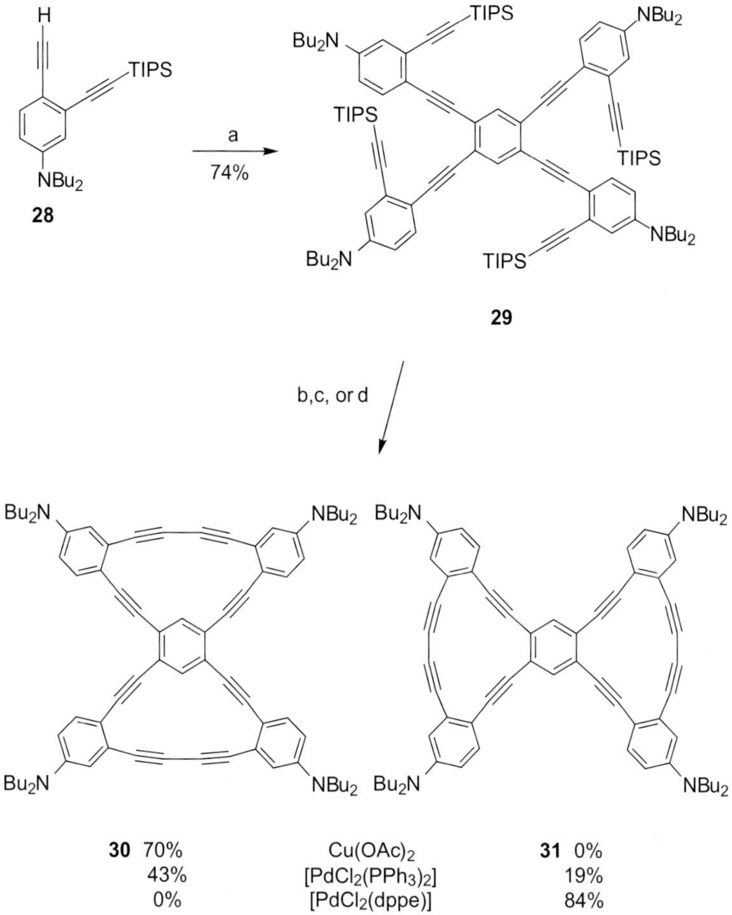

Scheme 8.9 Selective synthesis of PDMs **30** and **31**. Reagents and conditions: (a) 1,2,4,5-tetraiodobenzene, $[Pd(PPh_3)_4]$, CuI, *i*-Pr_2NH, THF, 40 °C; (b) i] TBAF, MeOH, THF, ii] $Cu(OAc)_2$, py, 60 °C; (c) i] TBAF, MeOH, THF, ii] $[PdCl_2(PPh_3)_2]$, CuI, I_2, *i*-Pr_2NH, THF, 50 °C; (d) i] TBAF, MeOH, THF, ii] $[PdCl_2(dppe)]$, CuI, I_2, *i*-Pr_2NH, THF, 50 °C.

8.2.3 Comparison of the Two Pathways

Phenylacetylene macrocycles can be synthesized by either an inter- or an intramolecular approach. An advantage of the intermolecular approach is that the starting materials can typically be prepared in relatively few steps, but the probability of forming numerous cyclooligomeric macrocycles is high. Unfortunately, this reduces the yield of the desired product and can make purification of the target molecule very difficult.

While the main disadvantage of the intramolecular approach is that a greater number of steps is typically required to assemble the more complex precursor molecules, this particular pathway for macrocycle synthesis allows one to prepare the target compounds in a systematic manner, which greatly improves the yield of the final products. Oligomeric by-products are minimized, so purification is greatly simplified. Additionally, this route permits the preparation of novel macrocycles that would otherwise be nearly impossible to construct by intermolecular strategies, such as macrocycles possessing little or no symmetry, products possessing functionality at predetermined positions, and compounds with various numbers of acetylenic linkages.

8.3 Phenylacetylene Macrocycles

8.3.1 *Ortho* PAMs

The simplest *o*-phenylacetylene macrocycle, highly strained diyne **32** (Scheme 8.10), was first reported by Sondheimer and co-workers 30 years ago [25]. The [8]annulene was prepared by bromination of *sym*-dibenzocyclooctatetraene (**33**) with two molar equivalents of bromine, to give **34** or **35**. Treatment of **35** with an excess of 1,5-diazabicyclo[4.3.0]non-5-ene in boiling benzene gave **34** as a mixture of isomers. Subsequent treatment of either with *t*-BuOK furnished the target hydrocarbon as a yellow solid that decomposes at ~110 °C. Although the X-ray structure of **32** shows the acetylenic linkages are distorted from the normal linear arrangement by some 24.3°, the molecule decomposes only slowly over 2 days when stored under ambient conditions. Although PAM **32** has also been synthesized by other methods [26], that shown in Scheme 8.10 still remains the predominant route for its preparation.

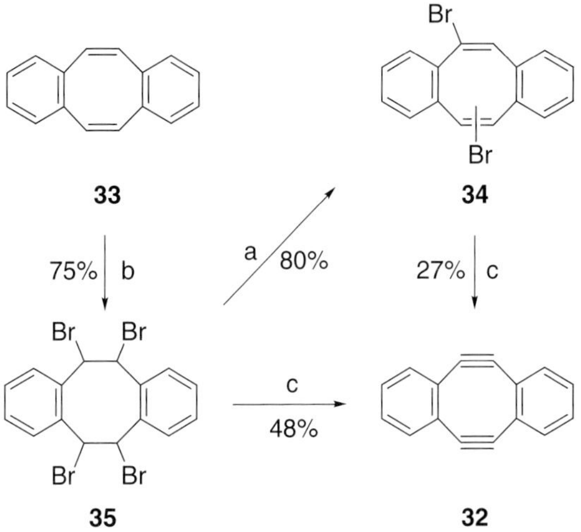

Scheme 8.10 Synthesis of PAM **32**. Reagents and conditions: (a) 1,5-diazabicyclo[4.3.0]non-5-ene, PhH, reflux; (b) 2 equiv. Br_2, CCl_4, *hν*; (c) *t*-BuOK, THF, rt, 30 min.

As mentioned previously, PAM **8**, which contains three phenyl rings *ortho*-linked by the same number of acetylene moieties, was prepared independently by the groups of Eglinton and Staab in 1966 by use of an intermolecular synthetic approach (see Schemes 8.1 and 8.4). Since then, two strategies have been evolved for the assembly of **8**. The first has focused on improvement of the Eglinton route by modification of the reaction conditions, either by changing the method of Cu-acetylide formation [13] or by inclusion of Pd as a catalyst [14]. The second strategy has been to develop completely new synthetic routes [16, 18, 20]. Very recently, Iyoda and his group prepared **8** in moderate yield through an intermolecular cross-coupling reaction between 1,2-diiodobenzene and triyne **36** (Scheme 8.11) [27]. Heating a DMF solution of **36** and 30 mol% of CuI and PPh_3 at reflux in the presence of K_2CO_3 afforded **8** in 51% yield. Interestingly, substitution of K_2CO_3 with either $CsCO_3$ or $CaCO_3$ failed to afford significant amounts of **8**, while the use of Na_2CO_3 gives the target macrocycle in only 20% yield. The authors also prepared **8** in 26% yield from 1,2-diethynylbenzene **2** and bis(2-iodophenyl)acetylene (**37**) by a similar approach (Scheme 8.11).

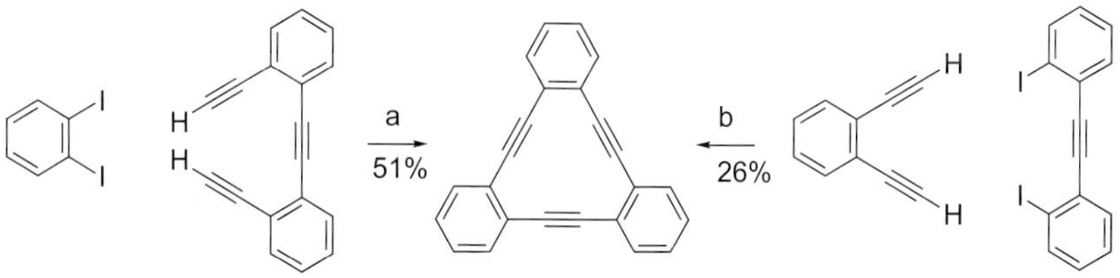

36 **8** **2** **37**

Scheme 8.11 Iyoda synthesis of PAM **8**. Reagents and conditions: (a) CuI (30 mol%), K_2CO_3 (3 equiv.), PPh_3 (30 mol%), DMF, reflux; (b) CuI (60 mol%), K_2CO_3 (3 equiv.), PPh_3 (60 mol%), DMF, reflux.

Vollhardt and co-workers recently reported the synthesis of **8** in 54% yield by alkyne metathesis of dialkynylarene **13** and the tungsten reagent [$(t\text{BuO})_3\text{W}{\equiv}\text{C}t\text{Bu}$] (Scheme 8.12) [16]. Analogous treatment of other appropriately substituted 1,2-dipropynylated arenes (**38**–**40**) afforded the novel hexasubstituted PAMs **41**–**43**, respectively. Unfortunately, it was not possible to effect the above transformation with substituents *ortho* to the acetylene groups in synthon **13**, which accordingly limits the synthetic utility of the procedure.

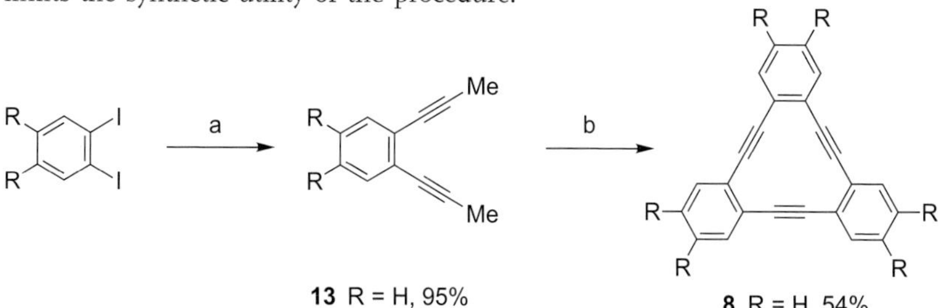

13 R = H, 95%
38 R = Me, 57%
39 R = OMe, 81%
40 R = Br, 71%

8 R = H, 54%
41 R = Me, 27%
42 R = OMe, 28%
43 R = Br, 12%

Scheme 8.12 Synthesis of PAMs **8** and **41**–**43**. Reagents and conditions: (a) $HC{\equiv}CH_3$, $[PdCl_2(PPh_3)_2]$, CuI, NEt_3 (or NEt_3/DMF), Δ; (b) $[(t\text{-BuO})_3W{\equiv}Ct\text{-Bu}]$, toluene, 80 °C.

In addition to the synthesis of **8**, that of hexamethyl[12]annulene **41** and hexafluoro[12]annulene **44** has been accomplished by Iyoda and his group through cyclization of the appropriate 1,2-disubstituted-4,5-diiodobenzene derivative with acetylene gas in the presence of [$PdCl_2(PPh_3)_2$] and CuI (Scheme 8.13) [18]. The yield of product was affected dramatically by the choice of solvent; thus, when 4,5-diiodo-*o*-xylene (**45**) was cyclized in morpholine, PAM **41** was obtained in trace quantities, but with the DMF/NEt_3 solvent system the yield increased to 11%. Analogous treatment of 1,2-difluoro-4,5-diiodobenzene (**46**) gave the hexasubstituted macrocycle **44** in 32% yield in morpholine, but in a somewhat reduced 24% yield in DMF/NEt_3.

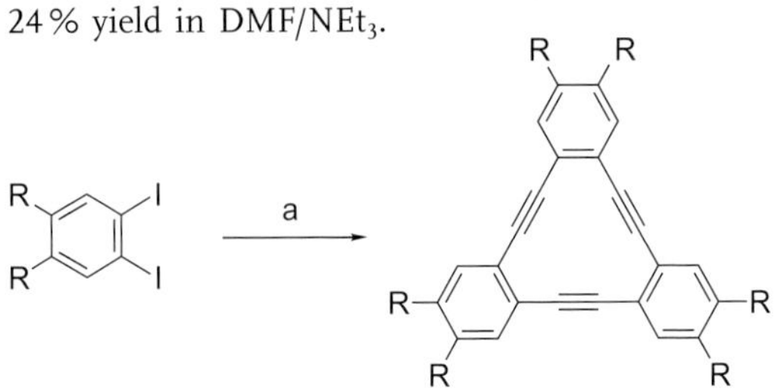

45 R = Me
46 R = F

41 R = Me, trace-11%
44 R = F, 24-32%

Scheme 8.13 Synthesis of PAMs **41** and **44**. Reagents and conditions: (a) HC≡CH, [$PdCl_2(PPh_3)_2$], CuI, morpholine or NEt_3, DMF.

The most commonly used method for *ortho*-PAM synthesis is the intermolecular cyclotrimerization of *ortho*-ethynyliodoarenes [27], which can be effected either with catalytic or with stoichiometric amounts of Cu salts (Scheme 8.14). Treatment of such arenes with CuI, K_2CO_3, and either PPh_3 or P(2-furyl)$_3$ in DMF at 160 °C gave the hexasubstituted PAMs **8**, **41**, **42**, and **47** in low to moderate yields [27a]. It should be noted that the choice of phosphine ligand had a significant impact upon the outcome of the reaction. For example, use of PPh_3 with *ortho*-ethynyliodobenzene gave macrocycle **8** in 55% yield, while use of P(2-furyl)$_3$ afforded the same compound in only 24% yield. Alternatively, cyclotrimerization by a modified Castro-Stephens reaction (generation of the Cu-acetylide in situ) furnished trimers **48** and **49** together with equal amounts of tetramers **50** and **51** in ca. 30% combined yield [27b]. Although designed to behave as discotic mesogens, PAMs **48–51** exhibit no liquid crystalline phases.

Vollhardt and his group have reported the synthesis of hexaethynylated [12]annulenes (**52–55**) [28]. The synthons for cyclization were alkyne derivatives **56–58**, which were in turn prepared in a stepwise fashion from 1,2,3,4-tetrabromobenzene (Scheme 8.15). Regioselective alkylation of the tetrahalogenated substrate at the 1- and 4-positions afforded diacetylene derivatives **59–61**, which were monoalkynylated at the 2-position to afford compounds **62–64**. Halogen/lithium exchange of the latter, quenching of the resultant anion with molecular iodine, and subsequent removal of the trimethylsilyl protecting group with base furnished **56–58**. Conversion of **56–58** into the Cu-acetylide derivatives, followed by heating at reflux in pyridine, afforded PAMs **52–54**. The parent macrocycle **55** was generated by fluoride-induced desilylation of **54**.

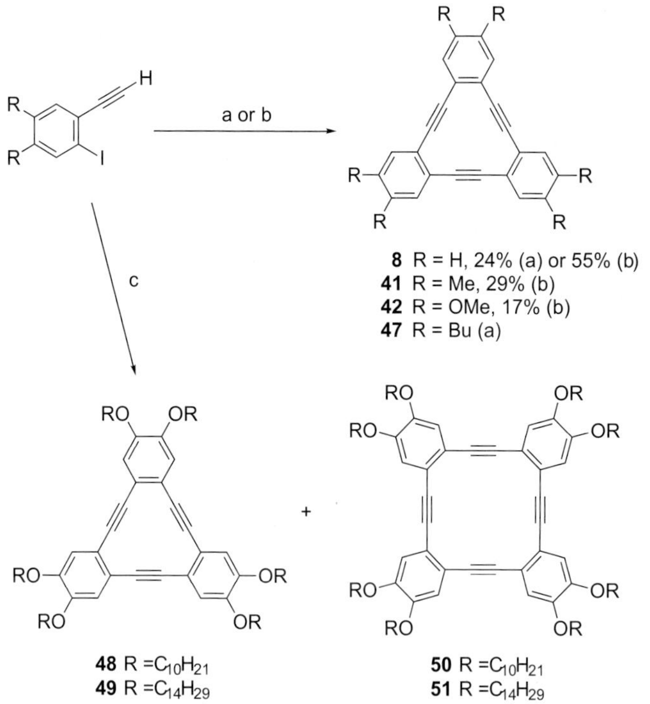

Scheme 8.14 PAM synthesis by cyclooligomerization. Reagents and conditions: (a) CuI, PPh_3, K_2CO_3, DMF, 160 °C; (b) CuI, P(2-furyl)$_3$, K_2CO_3, DMF, 160 °C; (c) i] *t*-BuOK, py, ii] CuCl, reflux.

The crystal structure of [12]annulene **52** shows the molecule to be locally C_2-symmetric [28a]. As a result of severe steric congestion caused by the bulky acetylenic substituents, the macrocyclic core of **52** is highly distorted, with the triple bonds deviating from linearity by an average of 5.7° (cyclic) and 4.5° (exocyclic). In contrast, the crystal structure of the unsubstituted hexaethynyl macrocycle **55** shows the molecule to be essentially planar and D_3-symmetric [28b]. Interestingly, the proximal alkynyl CH bonds of **55** chelate THF with supramolecular organization of the macrocycle around the occluded solvent molecules. This augers well for use of **55** as a supramolecular synthon for crystal engineering.

The hexamethoxy[12]annulene **65** has been prepared by Youngs et al. from substrate **66** [29]. Treatment of alkyne **67** with 'superbase' and subsequently with $Br_2Mg \cdot OEt_2$ and quenching with molecular iodine gave **66** (Scheme 8.16). Formation of the Cu-acetylide from the terminal acetylene moiety of **66** and subsequent Castro–Stephens cyclotrimerization in pyridine at reflux furnished **65** in an excellent 80% yield. The cyclization could also be effected by heating of **66** with $[PdCl_2(PPh_3)_2]$ in pyridine at reflux, although the yield of **65** was very poor (5%). Double demethylation of one of the arene rings of **65** could be effected with ceric ammonium nitrate (CAN) to give quinone **68** in 91% yield. Further oxidation to the corresponding all-quinone derivative was not possible.

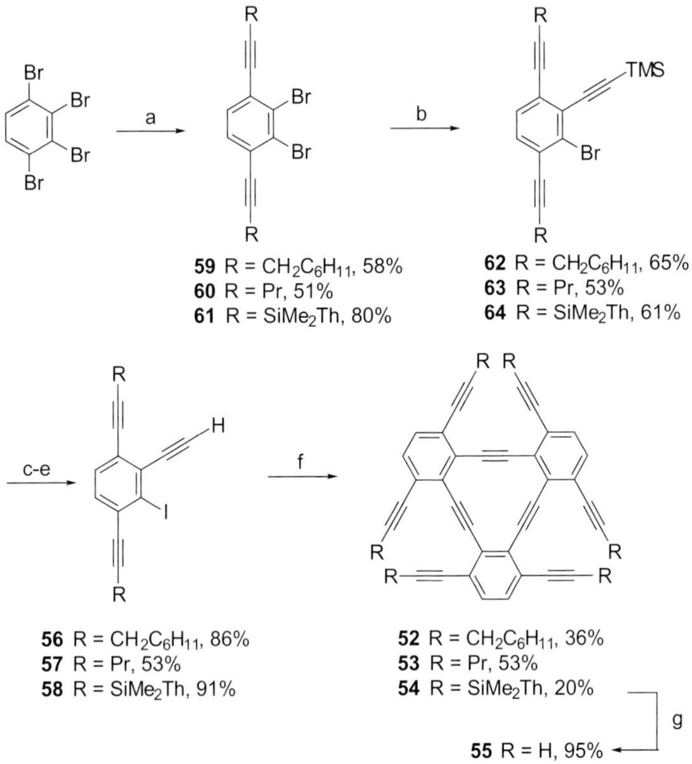

Scheme 8.15 Synthesis of ethynyl PAMs **52–55**. Reagents and conditions: (a) RC≡CH (2 equiv.), [PdCl$_2$(PPh$_3$)$_2$], CuI, NEt$_3$; (b) TMSA (1 equiv.), [PdCl$_2$(PPh$_3$)$_2$], CuI, NEt$_3$; (c) BuLi, Et$_2$O; (d) I$_2$, Et$_2$O; (e) K$_2$CO$_3$, MeOH; (f) CuCl, NH$_4$OH, EtOH, py; (g) TBAF, THF, MeCN.

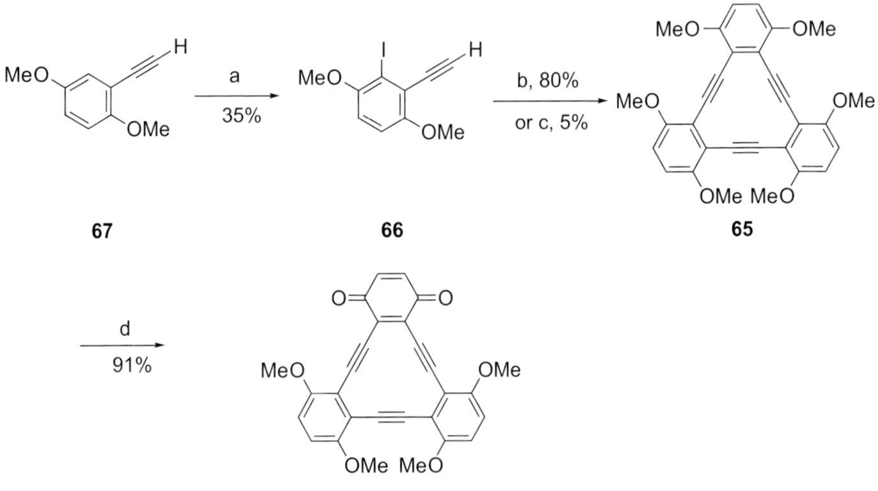

Scheme 8.16 Synthesis of PAM **65**. Reagents and conditions: (a) i] *t*-BuOK, BuLi, THF, ii] MgBr$_2$ · Et$_2$O, iii] I$_2$, THF; (b) i] CuCl, NH$_4$OH, EtOH, ii] py; (c) [PdCl$_2$(PPh$_3$)$_2$], py; (d) CAN, MeCN.

The ready availability of PAM **8** through several syntheses has allowed elucidation of the chemistry displayed by the molecule. In particular, a variety of organometallic species in which the π electron-rich **8** participates as the ligand have been prepared. Treatment of **8** with $Co_2(CO)_8$ furnished the 66-electon cluster **69**, which was characterized by X-ray crystallography [30]. The authors inferred that the binding of cobalt to the macrocycle resembled the postulated transition state of a metal-mediated [2+2+2] cyclotrimerization of alkynes. Treatment of **8** and the hexamethoxy derivative **65** with $[Ni(COD)_2]$ resulted in incorporation of the Ni(0) atom into the cavity of the macrocycle, to give complexes **70** and **71**, respectively, in which the metal is bound by π alkyne/transition metal bonding interactions [31]. Treatment of a saturated solution of **8** with $[Cu_2(C_6H_6)(OTf)_2]$ provided complex **72**, in which the Cu(I) ion is chelated by the three alkyne moieties [32]. Conversely, mixing of stoichiometric amounts of **8** and $[Cu_2(C_6H_6)(OTf)_2]$ generated complex **73**, in which three Cu centers are bound to the face of the macrocycle [32]. Treatment of **8** with half an equivalent of AgOTf furnished the novel macrocyclic sandwich complex **74** [33]. This last complex was found to dissociate in solution, though the sandwich complex formed from **8** and $AgBF_4$ is stable enough to be characterized by NMR spectroscopy [18]. The hexabutyl-substituted macrocycle **47** also formed a sandwich complex with the silver cation [26]. The treatment of **47** with $AgBF_4$ generated a sandwich species, which has been characterized by NMR spectroscopy.

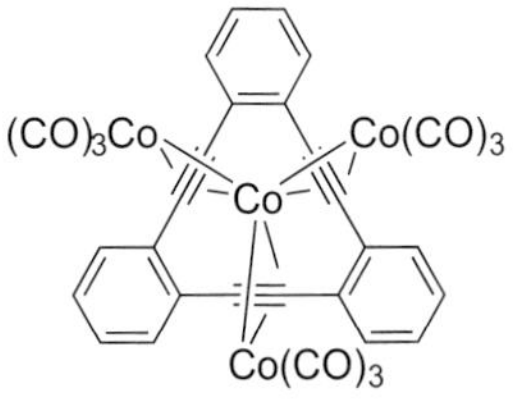

69

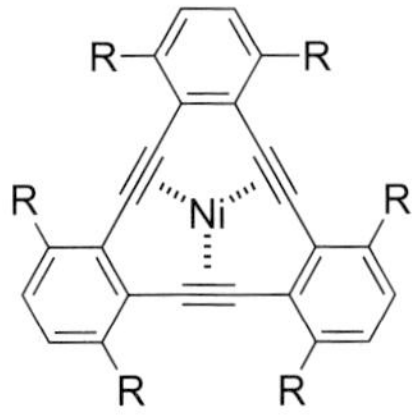

70 R = H
71 R = OMe

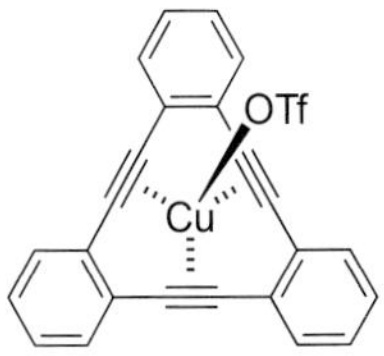

72

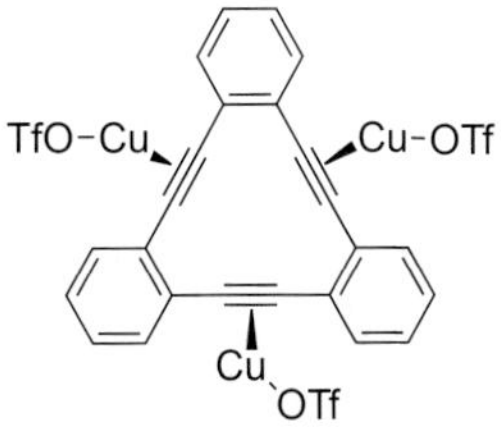

73

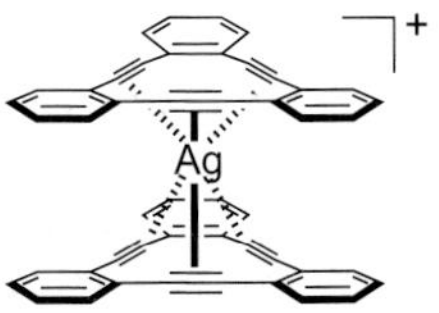

74

Very recently, Iyoda and his group have shown that macrocycle **8** undergoes ruthenium-catalyzed oxidation in the presence of a variety of oxidants to give both dione **75** and the unusual polycyclic hexaketone monohydrate **76** (Scheme 8.17) [34].

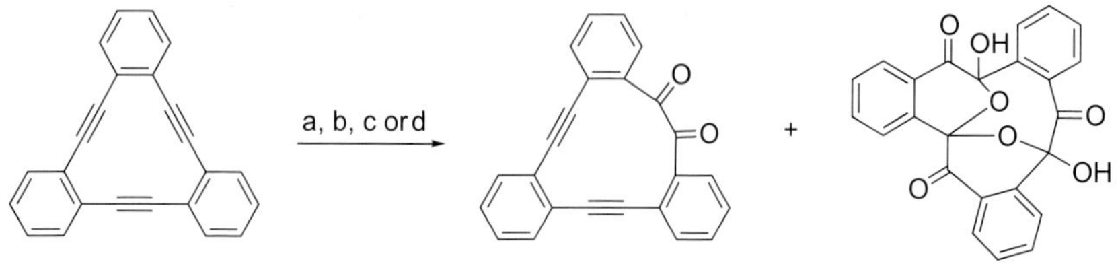

8 **75** **76**

	75	76
	a, 27%	a, 13%
	b, 6%	b, 31%
	c, 21%	c, trace
	d, 26%	d, trace

Scheme 8.17 Oxidation of PAM **8**. Reagents and conditions: (a) $RuO_2 \cdot 2H_2O$, $NaIO_4$, CCl_4, MeCN, H_2O; (b) $RuCl_3 \cdot 3H_2O$, PhIO, Me_2CO, H_2O; (c) $[RuCl_2(PPh_3)_2]$, CH_2Cl_2; (d) $[RuCl_2(MeCN)_2(PPh_3)_2]$, PhIO, CH_2Cl_2.

The behavior of macrocycle **8** with Li metal has been studied [35]. Addition of four molar equivalents of Li to a THF solution of **8** produced the highly air-sensitive dilithiate **77**, the structure of which was confirmed by X-ray crystallography (Scheme 8.18). Youngs et al. proposed that the first two equivalents of Li generate the diradical dianion, which then collapses to form the central six-membered ring. Further reduction by two more equivalents of alkali metal and subsequent protonation by the solvent affords **77**.

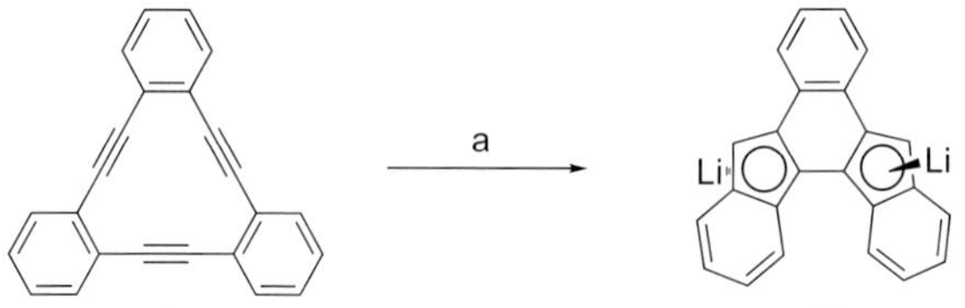

8 **77**

Scheme 8.18 Reduction of PAM **8**. Reagents and conditions: (a) Li, THF.

The syntheses of more complicated *ortho*-PAMs have been achieved recently. The driving force behind the preparation of these multiple [12]annulene derivatives has been predictions of interesting materials properties for oligomeric/polymeric structures based on **8**. In 1987 Baughman et al. calculated the structures of several hypothetical all-carbon networks and predicted that graphyne (**78**), comprised equally of sp and sp^2 carbons, should display promising third-order nonlinear optical properties. The material was also calculated to be a large bandgap semiconductor and, once suitably doped with alkali metals, to exhibit metallic behavior [36]. Recent semiempirical INDO/S and AM1 molecular orbital calculations have shown that molecular chains derived from [12]annulene (e. g., **79**) would also be expected to display nonlinear optical properties [37]. Despite these predictions, graphyne-related research has been hampered by synthetic accessibility, and so most efforts in this area have focused on smaller, soluble, discrete substructures (vide infra).

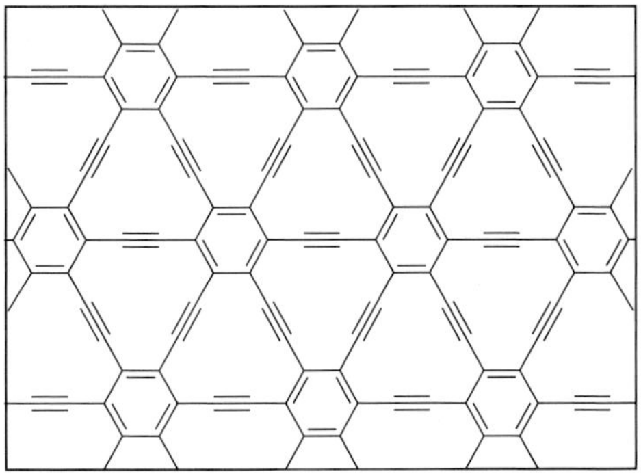

78

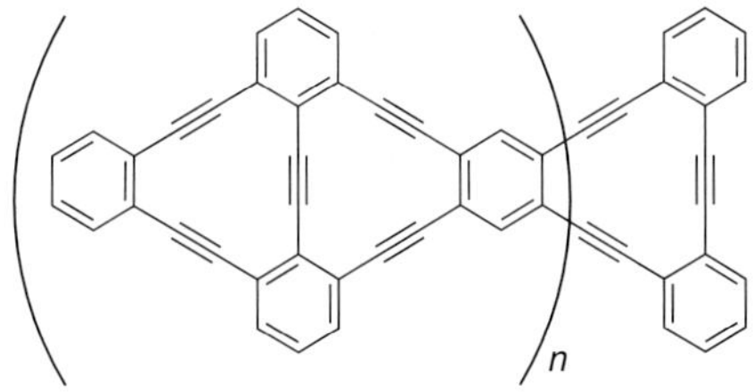

79

n = 1-4

The Haley group has prepared "bow-tie" bis[12]annulene **80** by an intramolecular stepwise strategy (Scheme 8.19) [20]. Using an set of reactions analogous to those by which macrocycle **8** was prepared (see Scheme 8.6), the Haley group synthesized the advanced intermediate **81**. Conversion of the triazene moieties into the corresponding iodo derivatives and subsequent deprotection of the TMS-protected alkynes and cross-coupling gave PAM **80** as a sparingly soluble, yellow solid. Although the authors did not obtain NMR data for **80**, they were able to obtain UV/Vis, IR, and MS data fully in accord with the proposed structural assignment.

More recently, Vollhardt et al. prepared **80** by a sextuple metathesis reaction between two equivalents of **13** and tetrayne **82** (Scheme 8.20) [16]. Not only did the authors obtain macrocycle **80** in 6 % yield, which is impressive since six metathesis reactions must occur, but they were also able to secure proton NMR data for the compound.

Iyoda and co-workers have prepared the octasubstituted bow-tie PAM **83** by Cu-mediated cross-coupling of 1,2,4,5-tetraiodobenzene and the substituted triyne **84** (Scheme 8.21) [27]. Although the yield of formation of bismacrocycle **83** is extremely low (~1 %), the authors were able to characterize the first example of a substi-

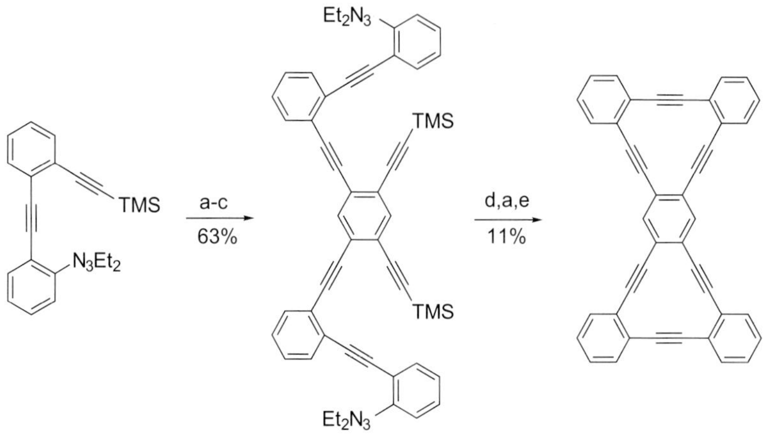

19 81 80

Scheme 8.19 Stepwise synthesis of PAM **80**. Reagents and conditions: (a) K_2CO_3, MeOH, THF; (b) 1,5-dibromo-2,4-diiodobenzene, $[PdCl_2(PPh_3)_2]$, CuI, Et_3N; (c) TMSA, $[PdCl_2(PPh_3)_2]$, CuI, Et_3N; (d) MeI, 120 °C; (e) $[Pd(dba)_2]$, PPh_3, CuI, Et_3N.

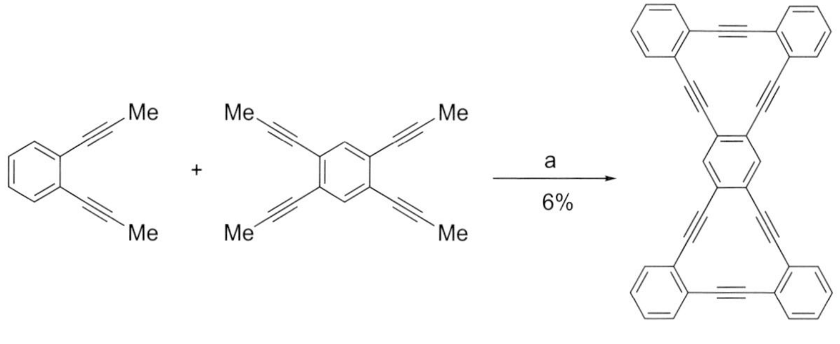

13 82 80

Scheme 8.20 Metathesis synthesis of PAM **80**. Reagents and conditions: (a) $[(t\text{-BuO})_3W{\equiv}Ct\text{-Bu}]$, toluene, 80 °C.

tuted derivative of **80**. Not surprisingly, the eight butyl substituents aided product solubility significantly.

The novel PAM **85**, which contains three [12]annulenes fused to a benzene core, was prepared very recently by Tobe and his group [38]. Sextuple Pd-catalyzed cross-coupling of hexabromobenzene with phenylacetylene **86** afforded **87** in 60 % yield (Scheme 8.22). Treatment of this with $[VCl_3[THF]_3]$/Zn furnished the all-*erythro* polycycle **88**. Chlorination/dehydrochlorination gave the target macrocycle **85** in 10 % yield for the two steps combined.

Haley and co-workers prepared the 'diamond' bisPAM **89**, containing solubilizing *tert*-butyl groups, by a stepwise intramolecular approach (Scheme 8.23) [20]. 4-*tert*-Butylaniline was converted into iodoarene **90** in four steps. Part of this material was carried forward to **91**, which was then desilylated and cross-coupled

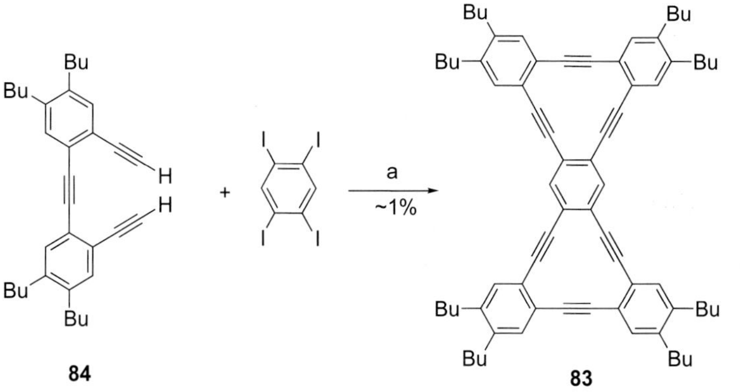

Scheme 8.21 Synthesis of PAM **83**. Reagents and conditions: (a) CuI, PPh_3, K_2CO_3, DMF, 160 °C.

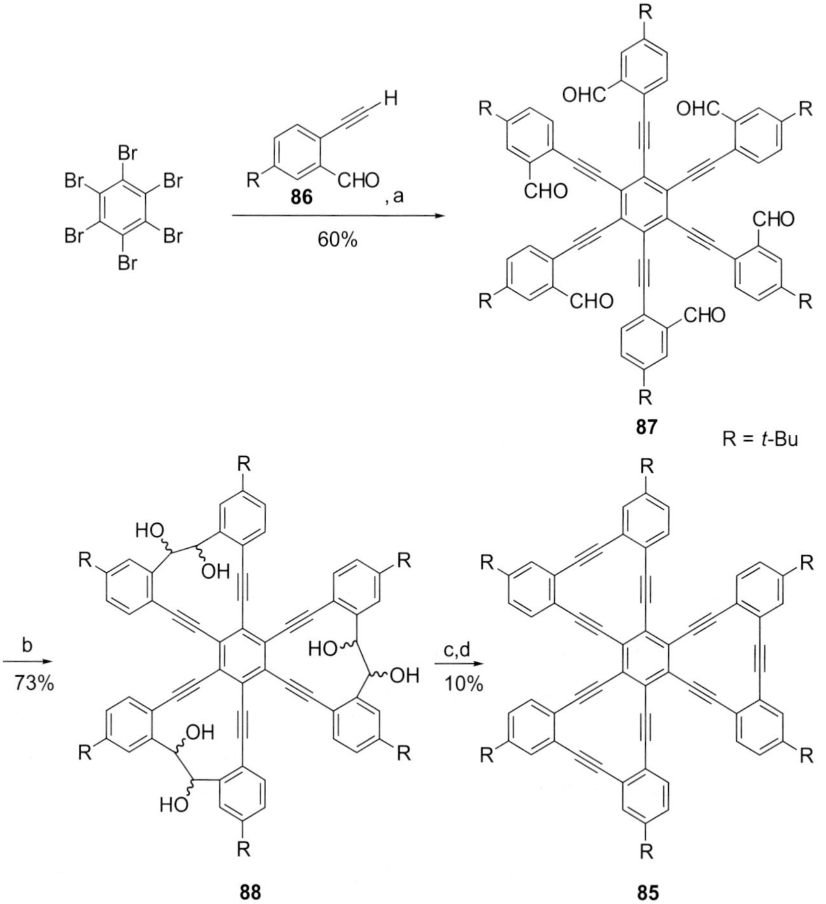

Scheme 8.22 Synthesis of PAM **85**. Reagents and conditions: (a) [Pd(PPh_3)$_4$], CuI, Et_3N, toluene, THF, 80 °C; (b) [VCl_3[THF]$_3$], Zn, DMF, CH_2Cl_2; (c) $SOCl_2$, py, DCE, 50 °C; (d) *t*-BuOK, THF, 50 °C.

with **90** to afford **92**. Iodination, desilylation, and double intramolecular cross-coupling furnished the desired macrocycle **89** as a bright yellow solid in an overall yield of 0.6 %.

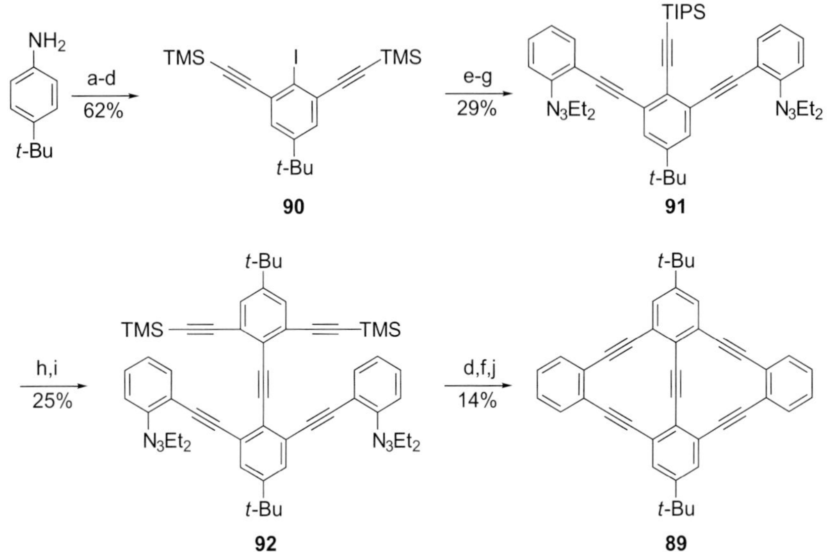

Scheme 8.23 Stepwise synthesis of PAM **89**. Reagents and conditions: (a) $BTEAICl_2$, $CaCO_3$, CH_2Cl_2, MeOH; (b) i] $NaNO_2$, HCl, MeCN, H_2O, ii] Et_2NH, K_2CO_3, H_2O; (c) TMSA, $[PdCl_2(PPh_3)_2]$, CuI, Et_3N; (d) MeI, 120 °C; (e) TIPSA, $[PdCl_2(PPh_3)_2]$, CuI, Et_3N; (f) K_2CO_3, THF, MeOH; (g) *N,N*-diethyl-2-iodophenyltriazene, $[PdCl_2(PPh_3)_2]$, CuI, Et_3N; (h) TBAF, THF, EtOH; (i) **90**, $[PdCl_2(PPh_3)_2]$, CuI, Et_3N; (j) $[Pd(dba)_2]$, PPh_3, CuI, Et_3N.

Tobe and his group very recently prepared the hexasubstituted 'diamond' PAM **93** by an intermolecular approach (Scheme 8.24) [38]. 4-Decylaniline was converted into iodide **94** in two steps in 61 % yield. Part of this material was treated with 2-methyl-3-butyne-2-ol and then with KOH to afford terminal acetylene **95**, which was cross-coupled with **94** to furnish the diphenylacetylene **96**. Heating of this compound at reflux in basic media under Pd catalysis with the diacetylene **97**, itself prepared in one step from 1,2-didecyl-4,5-diiodobenzene and 2-methyl-3-butyne-2-ol, afforded the target macrocycle **93** in 9 % yield.

The only *ortho*-PAMs in which the macrocycle contains more than twelve carbon atoms to have been prepared and characterized formally are the [16]-, [24]-, and [40]annulenes **10**, **11**, and **16**, respectively [13, 15, 19]. An X-ray structure has been obtained on PAM **16** and shows that the molecule contains no symmetry other than a C_1 axis [19b]. Similarly to **8**, PAM **10** reacts with $Co_2(CO)_8$ to give a tetracobalt cluster and undergoes a Li-induced 'zipper' cyclization reaction [39].

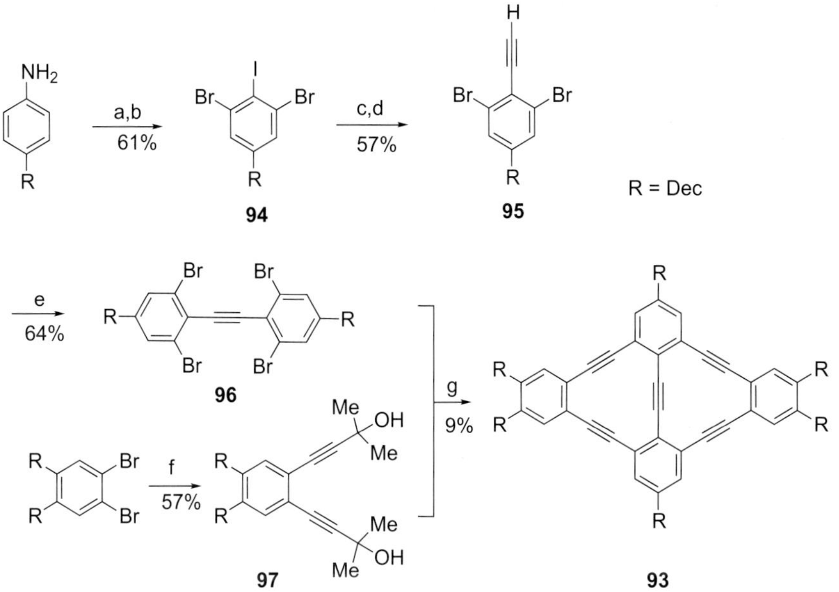

Scheme 8.24 Synthesis of PAM **93**. Reagents and conditions: (a) $BTMABr_3$, K_2CO_3, CH_2Cl_2; (b) isopentyl nitrite, I_2, PhH; (c) 2-methyl-3-butyn-2-ol, $[Pd(PPh_3)_4]$, CuI, Et_3N; (d) KOH, PhH; (e) **94**, $[Pd(PPh_3)_4]$, CuI, Et_3N; (f) 2-methyl-3-butyn-2-ol, $[Pd(PPh_3)_4]$, CuI, piperidine; (g) $[Pd(PPh_3)_4]$, CuI, PPh_3, KOH, PhH, $CH_3N(C_8H_{17})_3Cl$.

8.3.2
Meta-PAMs

The smallest known *m*-phenylacetylene macrocycle is the [15]annulene **98** (Scheme 8.25), which was prepared by Oda and his group in 1997 and has a computed strain energy of 48 kcal mol^{-1} [40]. The synthon for the trimeric macrocycle **98** is triene **99**, which was synthesized by a Ti-mediated intramolecular reductive coupling of the formylstyryl benzene derivative **100**. Treatment of **99** with excess bromine and subsequent dehydrobromination yielded macrocycle **98** as a moderately stable crystalline material that decomposed thermally above 180 °C. An X-ray structure of **98** shows the molecule to be nearly planar and C_2-symmetric [40]. The benzene rings are distorted from a regular hexagon by 3.3° and the torsion angles between the benzene rings and triple bonds by less than 3.5°. The alkyne moieties, which impart significant strain on the molecule, are highly distorted from linearity with an average sp bond angle of 158.6°, a value comparable to those in *sym*-dibenzocyclooctatetraene **32** (159.1°).

PAM **98** was found to undergo facile Diels–Alder reaction with cyclopentadiene at room temperature to afford a mixture of the *syn*/*anti* diadducts **101** in an approximately 1:1 ratio (Scheme 8.26). Surprisingly, no evidence to support the formation of either the mono- or the triadduct was obtained.

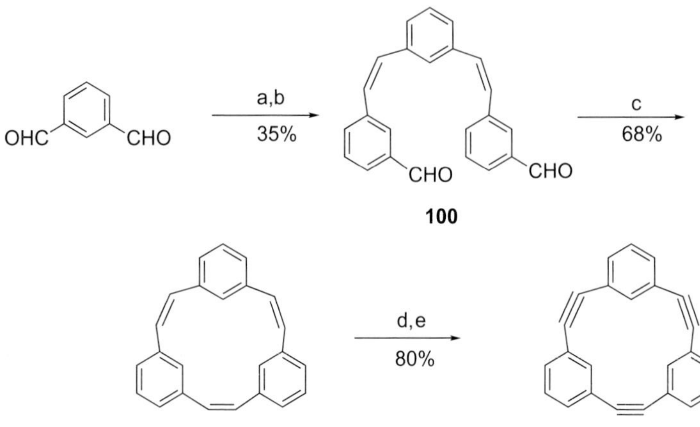

100

99 **98**

Scheme 8.25 Synthesis of PAM **98**. Reagents and conditions: (a) m-$BrC_6H_4CH_2PPh_3^+Br^-$, t-BuOK, DMSO; (b) i] BuLi, THF, ii] DMF; (c) $TiCl_4$, Zn, DME; (d) Br_2, $CHCl_3$; (e) t-BuOK, Et_2O.

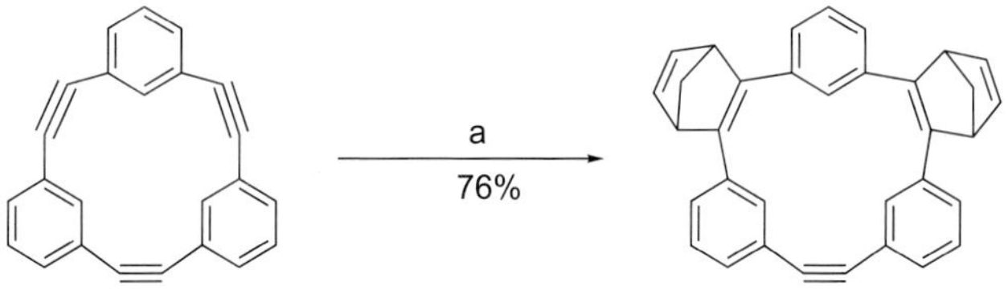

98 **101**

Scheme 8.26 Cycloaddition of PAM **98**. Reagents and conditions: (a) cyclopentadiene, $CHCl_3$.

The synthesis of the m-phenylacetylene macrocycle **102** (Scheme 8.27), containing four alternating acetylene and phenyl moieties, had been reported the previous year by Oda and colleagues [41]. McMurry coupling of dialdehyde **103** with low-valent titanium metal afforded a mixture of (*Z,Z,E,E*) and (*E,Z,E,Z*) isomers of **104**; neither the (*Z,Z,Z,Z*) isomer nor the intramolecular coupling product from **103** were detected. Treatment of **104** with molecular bromine, followed by dehydrobromination of the resulting adduct with t-BuOK, afforded PAM **102** in 55 % yield.

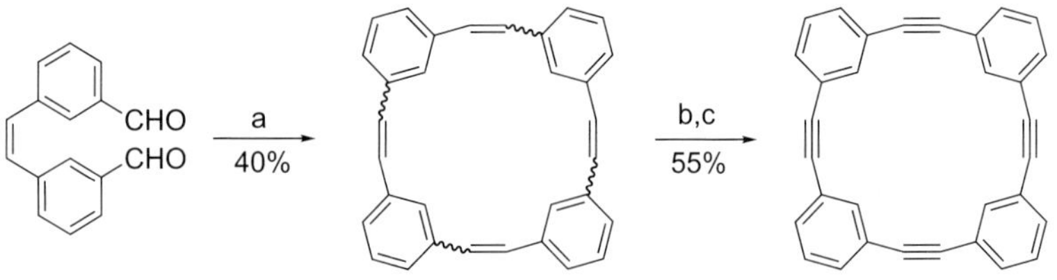

103 **104** **102**

Scheme 8.27 Synthesis of PAM **102**. Reagents and conditions: (a) $TiCl_4$, Zn, DME; (b) Br_2, $CHCl_3$; (c) t-BuOK, THF.

The X-ray structure of **102** reveals the molecule to be non-planar, and the twist angle of the benzene rings from the plane of the macrocycle is less than 2° [41]. As anticipated, the triple bonds are distorted from linearity (167.7–169.9°), which imparts strain upon the molecule. Although the strain energy of the macrocycle has been calculated as ca. 11 kcal mol^{-1}, it is remarkably stable and decomposes above 300 °C on attempted melting. In contrast to triyne **98**, macrocycle **102** is inert towards cycloaddition reactions either with furan or with cyclopentadiene.

PAM **105**, containing four intraannular methoxy substituents, was synthesized analogously to the parent compound **102** (Scheme 8.28) [42]. McMurry coupling of dimethoxystilbene **106** afforded a mixture of cyclic dimers (**107**) from which the (*E,Z,E,Z*)-tetraene separated out as crystals in 19 % yield. Bromination/dehydrobromination of this compound furnished **105** as a stable solid in a modest 62 % yield.

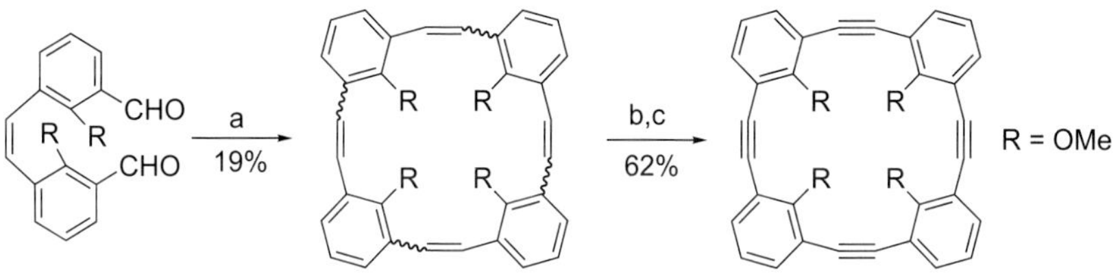

106 **107** **105**

Scheme 8.28 Synthesis of PAM **105**. Reagents and conditions: (a) $TiCl_4$, Zn, DME; (b) Br_2, $CHCl_3$; (c) *t*-BuOK, THF.

In contrast to that of the nearly planar parent macrocycle **102**, the X-ray structure of **105** reveals the molecule to be non-planar with C_2 symmetry, the first and third benzene rings lying in the plane whilst the second and forth rings are twisted out of the plane by 42° [42]. As expected, the triple bonds of **105** are bent, with an average angle of 166.9°, and as a result of steric congestion, the methoxy moieties of rings one and four point upward while the other two point downward. The X-ray structure shows that macrocycle **105** has an organized cavity with a diameter of 1.2 Å, which augers well for the molecule to act as an ionophore. Indeed, when **105** was treated with the alkali metal picrates (PicM), where M = Li^+, Na^+, K^+, Rb^+, and Cs^+, encapsulation took place for all but the last ion. The association constants are large, ranging from 8.23 to 10.59, and the ability of the macrocycle to bind K^+ ions is comparable to that of 18-crown-6.

The only reported example of a *meta*-PAM containing five alternating phenyl and acetylene units is the *n*-butyl ester-substituted **108** [43]. The pentayne was prepared by a convergent stepwise approach from ester **109** (Scheme 8.29). Treatment of this ester either with weak base or with MeI furnished **110** and **111**, respectively, and these were cross-coupled to afford **112**. Repetition of the above transformations generated the linear pentamer **113**, and treatment of this with MeI and subsequent base-induced desilylation of the protected acetylene moiety afforded synthon **114**.

Cyclization was accomplished under pseudo-high-dilution conditions by slow addition of **114** to an active solution of Pd catalyst to give the macrocycle **108** in moderate yield.

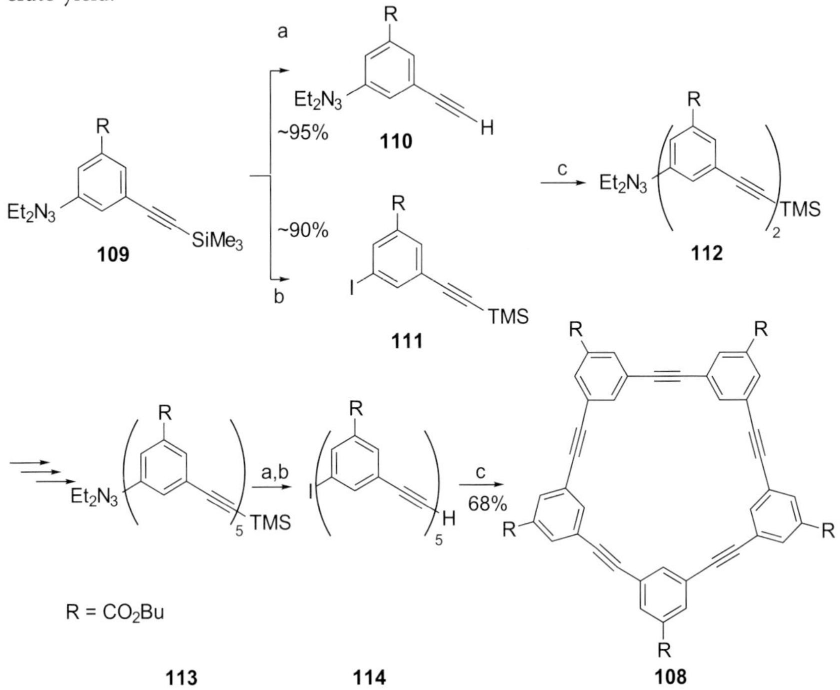

Scheme 8.29 Synthesis of PAM **108**. Reagents and conditions: (a) K_2CO_3, MeOH; (b) MeI, 110 °C; (c) $[Pd(dba)_2]$, PPh_3, CuI, NEt_3.

The parent hexamer **115** was first reported in the literature 30 years ago, by Staab and Neuenhoeffer [44]. The authors prepared the requisite macrocycle though a sextuple Castro-Stephens coupling of Cu-acetylide **116** in 4.6% yield (Scheme 8.30). Synthon **116** was, in turn, generated from treatment of acetylene **117** with CuCl in aqueous ammonia/ethanol.

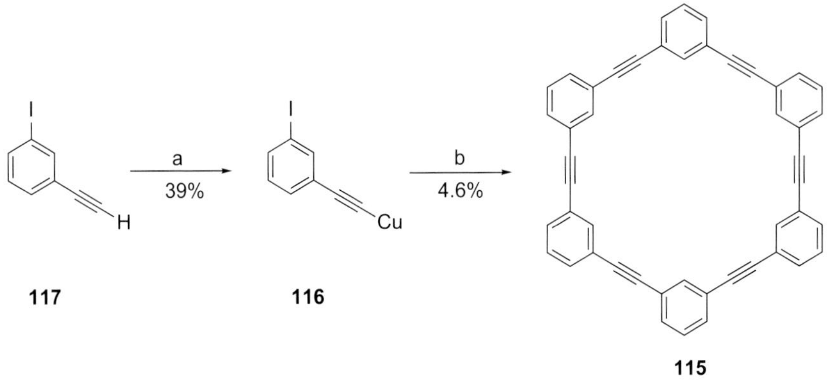

Scheme 8.30 Synthesis of PAM **115**. Reagents and conditions: (a) CuCl, NH_3, H_2O, EtOH; (b) py, Δ.

Bunz and co-workers prepared the hexasubstituted *meta*-phenylacetylene derivatives **118** and **119** by a one-pot cyclooligomerization strategy (Scheme 8.31) [45]. Treatment of the substituted benzene derivatives **120** and **121** with the molybdenum reagent [$Mo(CO)_6$] in 4-chlorophenol at elevated temperature afforded the sextuple alkyne metathesis hexamers **118** (6 %) and **119** (1.2 %), respectively. Treatment of the former macrocycle with [$(MeCN)_2Os_3(CO)_{10}$] resulted in coordination of one acetylene moiety to metal atoms to form a triosmiumdecacarbonyl cluster complex in 11 % yield. As evidenced by an X-ray structure, the attachment of the osmium cluster to the acetylene moiety disrupts the planarity of the ring significantly. An X-ray crystal structure of **118** shows the PAM to be nearly planar and essentially strain-free [45]. The molecule was found to include two equivalents of disordered hexane per molecule in the solid state, one passing through the center of the ring while the other is located between the stacks of rings. The rings of **118** are stacked and align in columns.

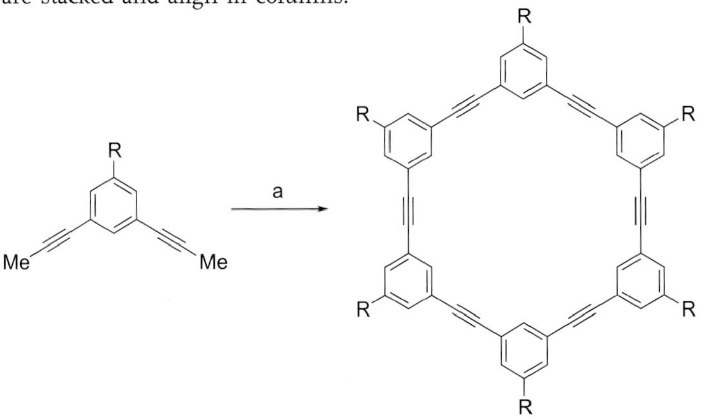

120 R = *t*-Bu
121 R = Hex

118 R = *t*-Bu, 6%
119 R = Hex, 1.2%

Scheme 8.31 Synthesis of PAMs **118** and **119**. Reagents and conditions: (a) $Mo(CO)_6$, 4-chlorophenol, 150 °C.

A vast array of substituted hexameric (and other) PAMs (e. g., **122–129**) have been prepared since the early 1990s by Moore and his group [5c, 43, 46]. By use of an iterative route it was possible to synthesize appropriately functionalized phenylacetylene macrocycles by sequential addition of derivatized monomers and subsequent cyclization of the linear hexamers to afford well defined macrocycles with precise substituent placement on the periphery of the PAM (Scheme 8.32). Thus, the coupling of monomers **130** and **131** and subsequent treatment of the derived adduct with base gave dimer **132**. Treatment of this compound with **133** and removal of the TMS moiety of the resultant species with base furnished tetramer **134**. Repetition of this sequence of events with dimer **135** then afforded hexamer **136**. Conversion of the triazene moiety of this last compound into the corresponding aryl iodide, protiodesilylation, and Pd-mediated ring-closure afforded the geometrically well defined PAMs **122–129**.

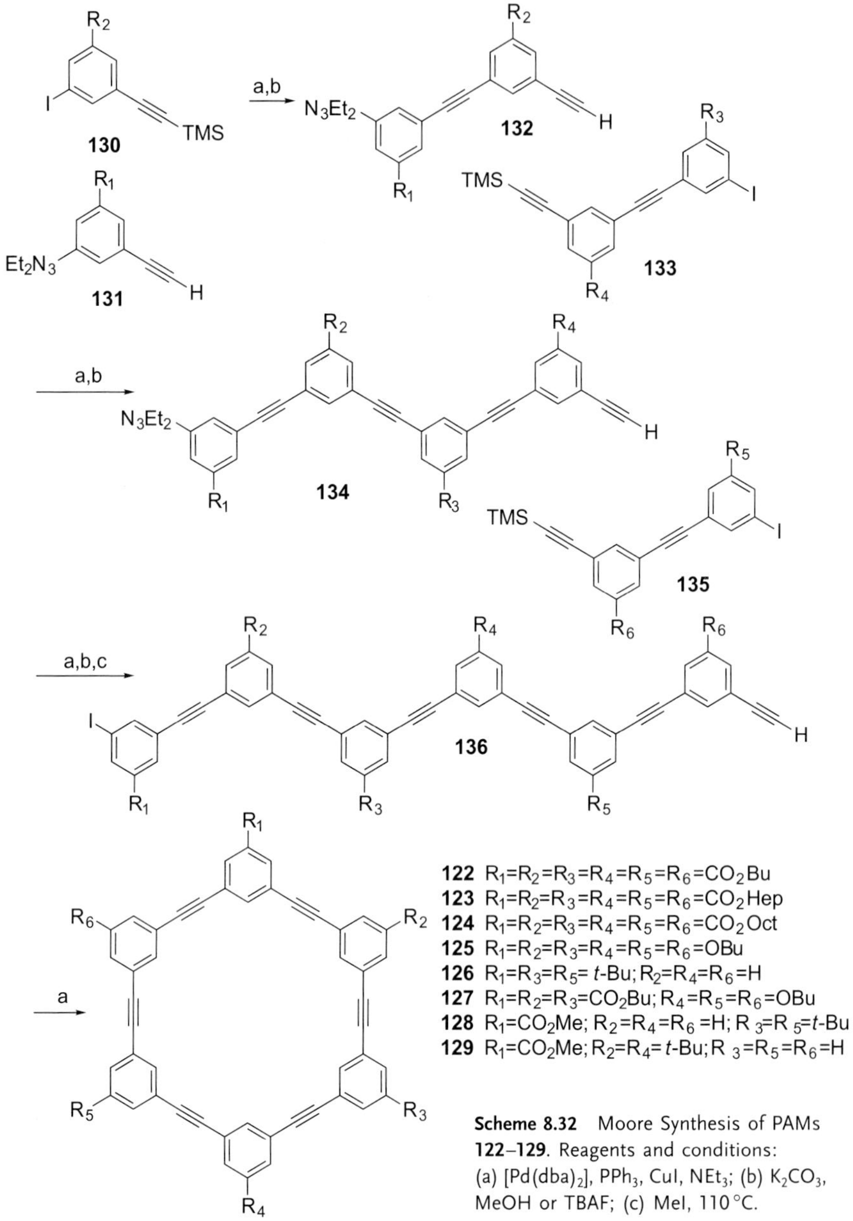

122 $R_1=R_2=R_3=R_4=R_5=R_6=CO_2Bu$
123 $R_1=R_2=R_3=R_4=R_5=R_6=CO_2Hep$
124 $R_1=R_2=R_3=R_4=R_5=R_6=CO_2Oct$
125 $R_1=R_2=R_3=R_4=R_5=R_6=OBu$
126 $R_1=R_3=R_5=$ *t*-Bu; $R_2=R_4=R_6=H$
127 $R_1=R_2=R_3=CO_2Bu$; $R_4=R_5=R_6=OBu$
128 $R_1=CO_2Me$; $R_2=R_4=R_6=H$; $R_3=R_5=$*t*-Bu
129 $R_1=CO_2Me$; $R_2=R_4=$*t*-Bu; $R_3=R_5=R_6=H$

Scheme 8.32 Moore Synthesis of PAMs **122–129**. Reagents and conditions: (a) [Pd(dba)$_2$], PPh_3, CuI, NEt_3; (b) K_2CO_3, MeOH or TBAF; (c) MeI, 110 °C.

The hexameric PAMs have attracted much interest thanks to a slew of novel properties. In 1992, Zhang and Moore reported that the chemical shifts in the proton NMR spectrum of ester **122** were concentration-dependant, which indicated the compound was spontaneously aggregating into ordered assemblies in solution due to π stacking [47]. A variety of differently functionalized phenylacetylene macrocycles were prepared for subsequent investigation. The results showed that those molecules containing ester groups with linear alkyl side chains (e. g., **123**) aggregated in solution, while aggregation was reduced significantly, or not observed at all, for macrocycles not possessing electron-withdrawing substituents. No self-association was observed, for example, for hexaether derivative **125** [47, 48].

If suitably functionalized, phenylacetylene macrocycles such as **137** and **138** display liquid crystalline behavior [49]. These compounds were shown to exhibit both ordered isotropic and fluid phases and demonstrated self-organization in the discotic nematic phase upon cooling from the isotropic melt. More recently, PAMs have been shown to exhibit a discotic liquid crystalline phase [50].

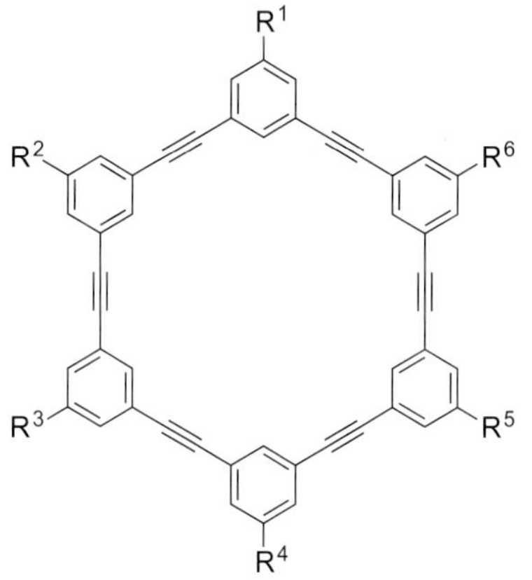

137 R^1=R^2=R^3=R^4=R^5=R^6= OHep
138 R^1=R^2=R^3=R^4=R^5=R^6=OCOHep
139 R^1=R^2=R^3=R^4=*t*-Bu;R^5=R^6=$COO^{-+}NBu_4$

One of the rich variety of properties displayed by phenylacetylene macrocycles is the ability of some to form ordered monolayers [46c]. For example, when the tetrabutylammonium carboxylate salt **139** is transferred onto various substrates, it adopts a stable two-dimensional organization displaying a high degree of polar and conformational order.

To date, only one example of a *meta*-hexaethynyl PAM with only internal ring substituents has been reported [51]: the macrocycle **140** was prepared in two steps from stilbene **141** (Scheme 8.33). McMurry coupling of **141** in DME/toluene afforded hexaene **142**, along with the tetraene **143**, in 14 and 23 % yields, respectively. Bromination/dehydrobromination of **142** furnished **140** in 52 % yield as a poorly soluble solid.

Examples of *meta*-hexaethynyl PAMs containing both internal and external ring substituents are known. Oda and co-workers prepared macrocycle **144** by bromination/dehydrobromination of substrate **145** (Scheme 8.34) [51]. In turn, compound **145** was prepared by McMurray coupling of the dialdehyde **146**, with the reaction also affording the trimeric macrocycle **147**. An X-ray structure was obtained for

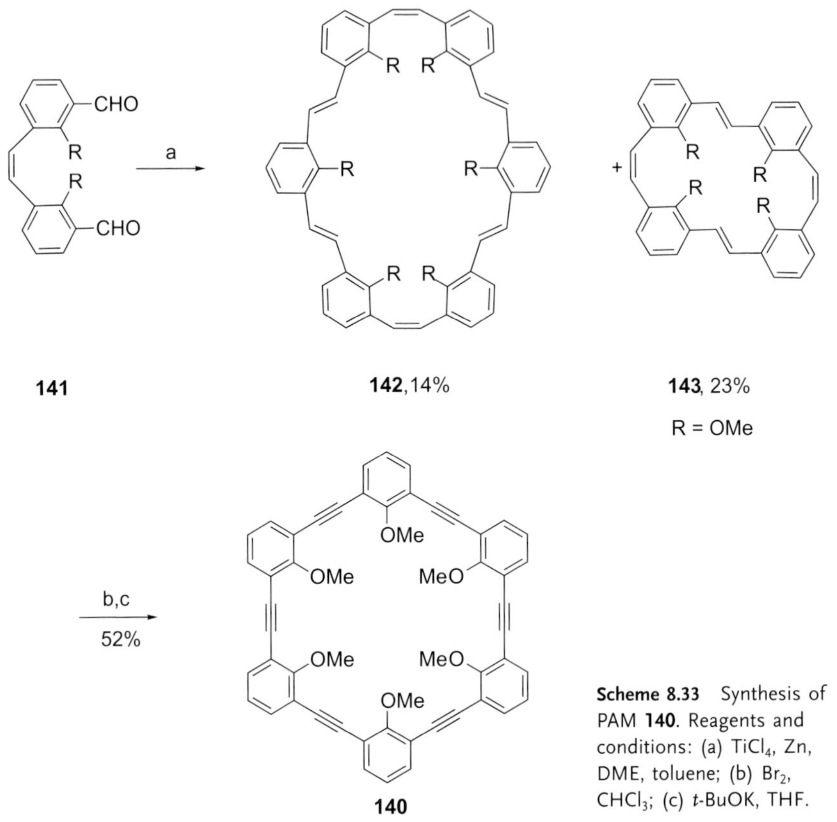

Scheme 8.33 Synthesis of PAM **140**. Reagents and conditions: (a) $TiCl_4$, Zn, DME, toluene; (b) Br_2, $CHCl_3$; (c) *t*-BuOK, THF.

macrocycle **144** [51]: the compound is essentially planar, with the methoxy groups alternating *syn* and *anti*. The molecule has a cavity of ~5Å and exhibits good ionophoric selectivity for the ammonium ion.

PAM **148** was prepared, along with the heptameric analogue, by Bunz et al., by alkyne metathesis of the dipropynylated benzene derivative **149**, albeit only in 0.5 % yield (Scheme 8.35) [45]. The formation of oligomeric and polymeric by-products presumably accounts for the low yield in which **148** was formed.

Cho and co-workers very recently prepared *meta*-PAM **150**, containing two different types of donor groups: OMe and NBu_2 (Scheme 8.36) [52]. The terminal alkyne **151** was cross-coupled with the requisite diiodobenzene to give **152**. Treatment of this with TMSA and then base afforded **153**, which, when cross-coupled with **152**, gave the target macrocycle **150** in 15 % yield. A detailed investigation of the photophysical properties of **150** revealed that the molecule has a Stokes shift of 3026–4152 cm^{-1}; the value increases in more polar solvents, which indicates a significant change in the charge-transfer state. The two-photon absorption cross-section of **150** is comparable to that of rhodamine, and so the compound may find application as a two-photon absorption chromophore.

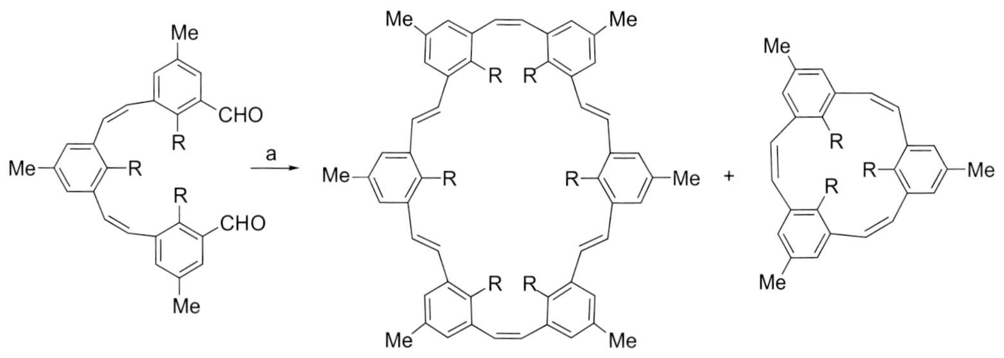

146 **145**, 15% **147**, 4%

R = OMe

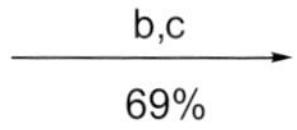

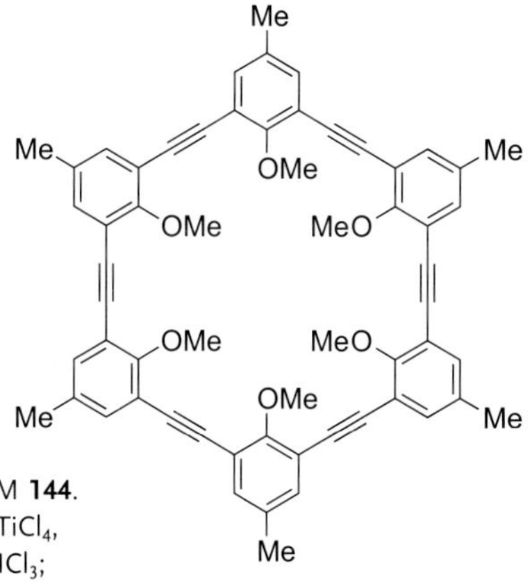

144

Scheme 8.34 Synthesis of PAM **144**. Reagents and conditions: (a) $TiCl_4$, Zn, DME, toluene; (b) Br_2, $CHCl_3$; (c) *t*-BuOK, THF.

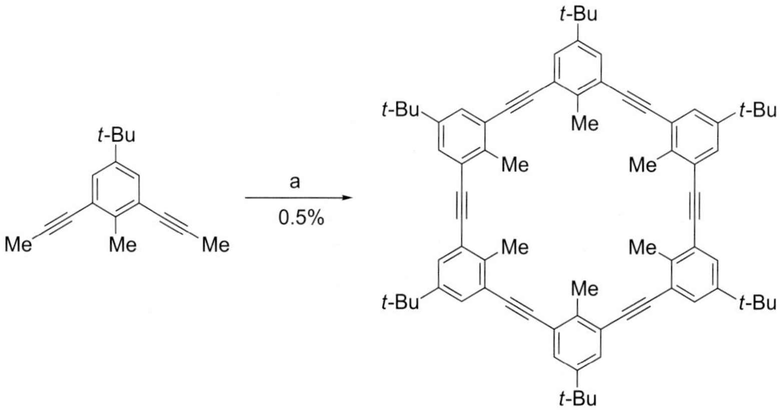

149 **148**

Scheme 8.35 Synthesis of PAM **148**. Reagents and conditions: (a) $Mo(CO)_6$, 4-chlorophenol, 150 °C.

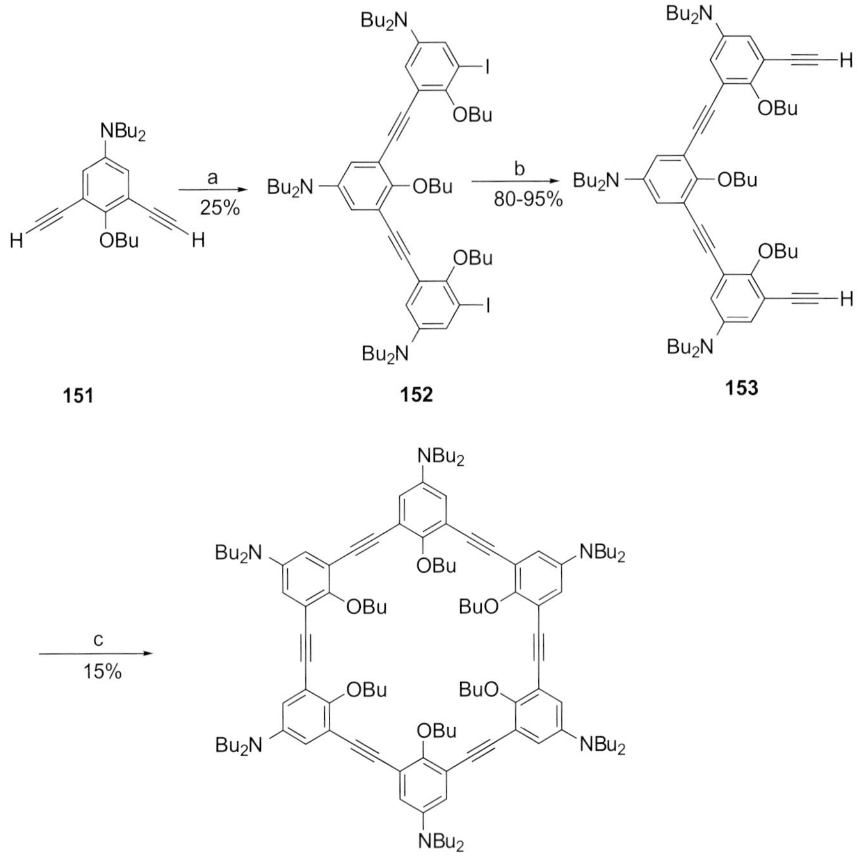

151 152 153

150

Scheme 8.36 Synthesis of PAM **150**. Reagents and conditions: (a) 4-butoxy-*N,N*-dibutyl-3,5-diiodoaniline (2 equiv.), [$PdCl_2(PPh_3)_2$], CuI, Et_3N, THF; (b) i] TMSA, [$PdCl_2(PPh_3)_2$], CuI, Et_3N, THF, ii] 1 M KOH, MeOH, THF; (c) **152**, [$PdCl_2(PPh_3)_2$], CuI, Et_3N.

An example of a hepta-*meta*-PAM, compound **154**, was prepared by the iterative route shown in Scheme 8.32 [43]. Although no X-ray structure of **154** was obtained, molecular modeling suggests the molecule has a flexible, non-planar geometry. Consequently, π stacking interactions between the aromatic rings of the compound are disfavored.

Some of the largest *meta*-PAMs to have been prepared are the branched macrobicyclic arrays **155** and **156** [53] and the macrotricycle **157** [54]. The last compound is a freely hinged system with a sizeable (and collapsible) 36 × 12 × 12 Å molecular cavity.

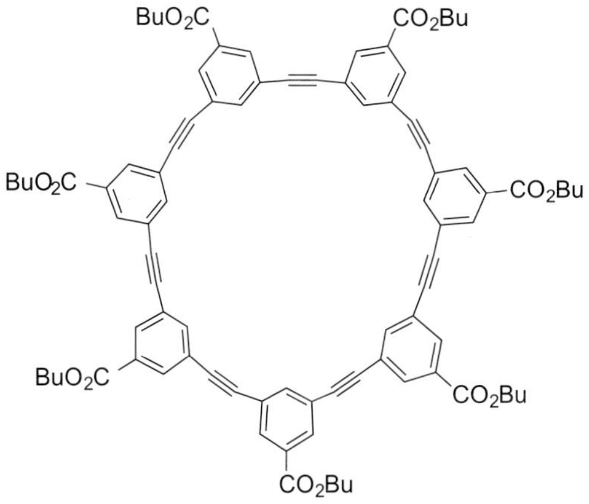

154

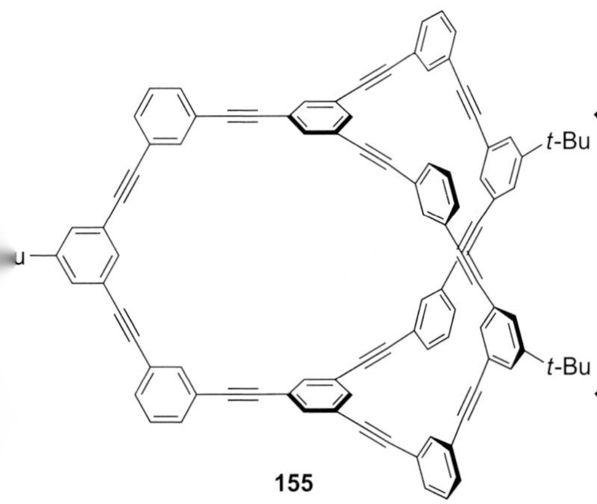

155

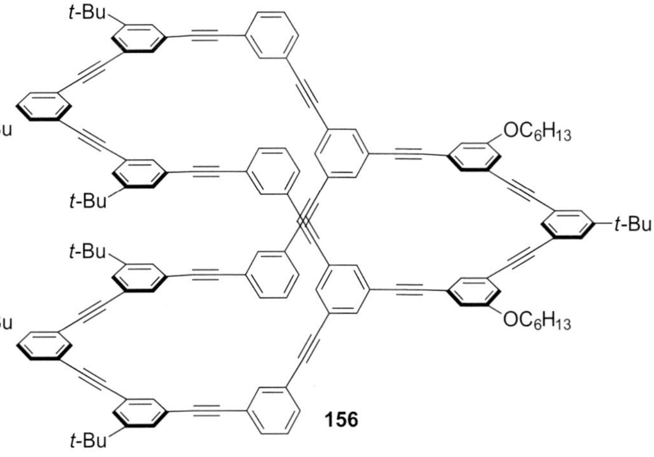

156

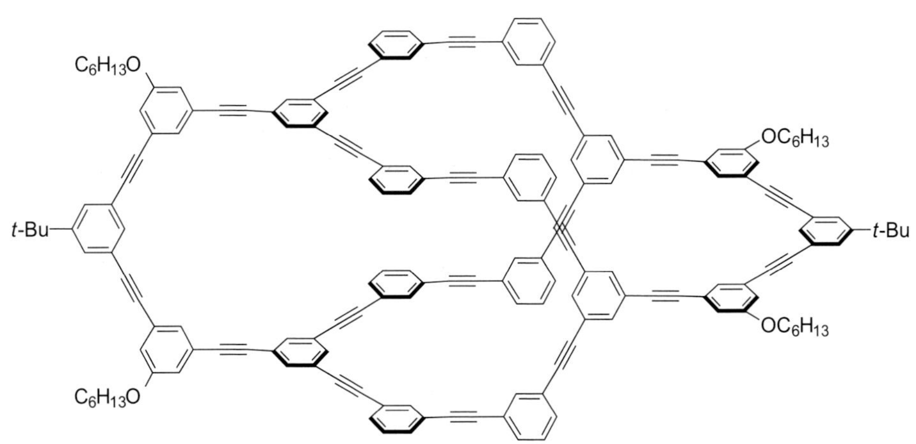

157

8.3.3
Para-PAMs

Relatively few examples of all-*para*-linked phenylacetylene macrocycles are known. The belt shaped PAMs **158–161** were prepared from the requisite macrocyclic synthons **162–165**, which were in turn obtained from the McMurray coupling of the appropriate stilbene (Scheme 8.37) [55]. Bromination/dehydrobromination of an inseparable mixture of **162** and **163** furnished a 4:1 ratio of the belt-shaped, air-sensitive macrocycles **158** and **159** in a combined 85 % yield; the compounds were separable by gel permeation chromatography. More recently, Oda et al. prepared relatively clean samples of synthons **162** and **163**, which afford the appropriate macrocycles in significantly higher yields and in almost pure forms [56]. The belt-shaped macrocycles **160** and **161** were prepared, by the same technique, in 20 and 4 % yields, respectively, from contaminated samples of synthons **164** and **165** [56]. Attempts to prepare the strained *para*-tetra(phenylacetylene) macrocycle by bromination/dehydrobromination of **166** were unsuccessful [55].

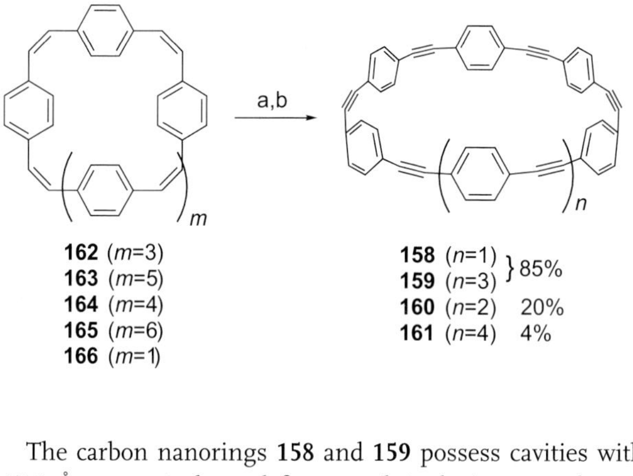

162 (*m*=3)
163 (*m*=5)
164 (*m*=4)
165 (*m*=6)
166 (*m*=1)

158 (*n*=1) } 85%
159 (*n*=3)
160 (*n*=2) 20%
161 (*n*=4) 4%

Scheme 8.37 Synthesis of PAMs **158–161**. Reagents and conditions: (a) Br_2, $CHCl_3$; (b) *t*-BuOK, THF.

The carbon nanorings **158** and **159** possess cavities with diameters of 13.2 and 17.3 Å, respectively, and form weak inclusion complexes with toluene and hexamethylbenzene [57]. Oda and his group reported that **158** forms stable 1:1 crystalline complexes both with C_{60} and with bis(ethoxycarbonyl)methanofullerene [58]. An X-ray structure of the methanofullerene · **158** complex shows that each molecule of the structure is associated with two solvent molecules. The C_{60} cage is not embedded deeply in **158** and adopts a bowl-shaped conformation, and the average distance between the host and guest is approximately 3.4 Å. Attempts to form host-guest complexes from the larger phenylacetylene macrocycle **159** and the aforementioned fullerenes were unsuccessful. Interestingly, nanorings **158** and **161** associate with one another to give "onion-type" complexes [59].

Oda and his group very recently prepared the acid- and oxygen-sensitive PAMs **167** and **168**, which can be stored in dilute solution at 0 °C over a month [60]. These macrocycles were shown to associate with both C_{60} and C_{70} and to form extremely stable complexes. PAM **167** also associates with other nanorings in an

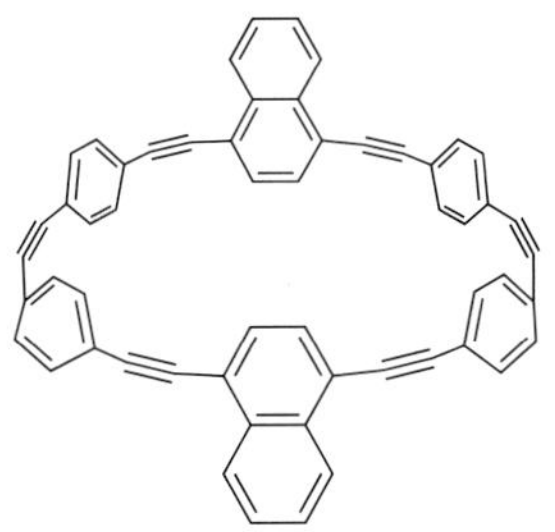

167

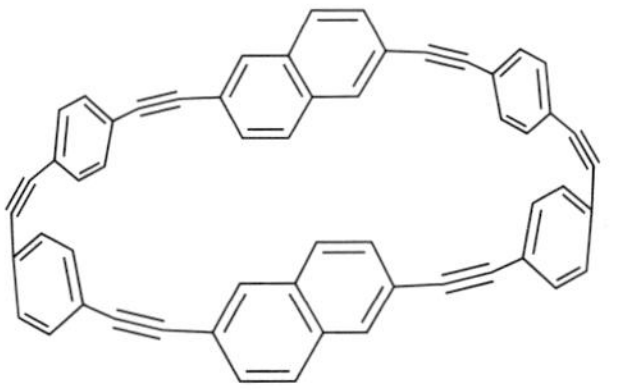

168

"onion-type" complex, which in turn includes C_{60} in the presence of excess fullerene [59]. Unfortunately, X-ray structures of these complexes have yet to be obtained.

8.3.4
Mixed PAMs

Vollhardt and his group prepared PAM **169**, containing both *ortho* and *meta* linkages, in 19 % yield by alkyne metathesis of a 1:1 mixture of the dipropynylated arenes **13** and **170** with the tungsten reagent [(*t*-BuO)$_3$W≡C*t*-Bu] (Scheme 8.38) [16]. An X-ray structure of **169** shows the molecule to be nearly planar with dihedral angles of 7.1° between the planes of the *ortho*- and *meta*-fused rings.

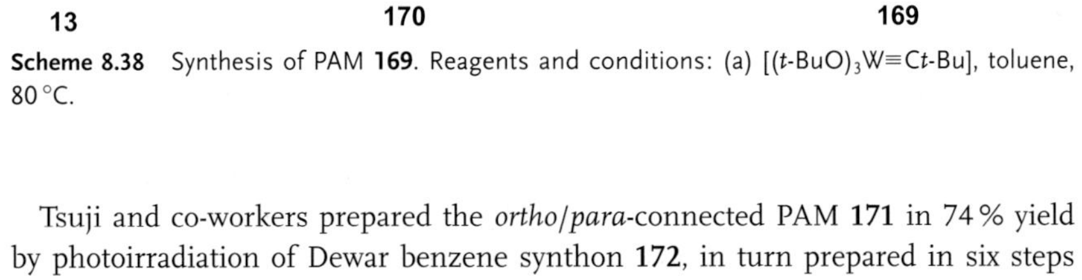

13 170 169

Scheme 8.38 Synthesis of PAM **169**. Reagents and conditions: (a) [(*t*-BuO)$_3$W≡C*t*-Bu], toluene, 80 °C.

Tsuji and co-workers prepared the *ortho*/*para*-connected PAM **171** in 74 % yield by photoirradiation of Dewar benzene synthon **172**, in turn prepared in six steps from 1,2-diiodobenzene (Scheme 8.39) [61]. Conversion of this last compound into **173** was accomplished by standard transformations. Protiodesilylation of **173** and subsequent Pd-catalyzed cross-coupling of the terminal alkyne with **174** afforded compound **175**, which upon deprotection and homocoupling under phase-transfer conditions gave **176**. Acidic hydrolysis of **176** and subsequent treatment with TIPSOTf furnished PAM **171** in 72 % yield. An X-ray structural analysis of macrocycle **171** shows the two *p*-substituted benzene rings to be nearly planar, with an interplanar distance of 3.48 Å, and tilted by 62.5 and 64.2° with respect to the plane of the macrocycle. As expected, the acetylene moieties are distorted, deviating from the normal linear arrangement by 8.6–12°.

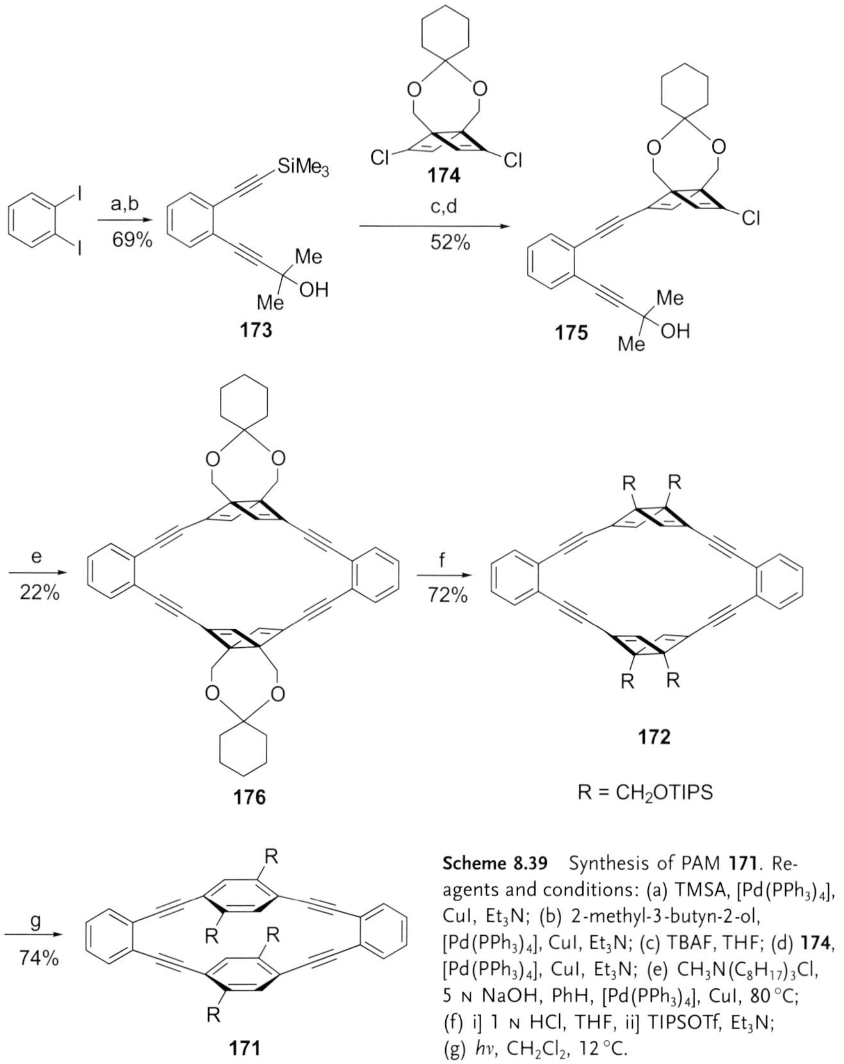

Scheme 8.39 Synthesis of PAM **171**. Reagents and conditions: (a) TMSA, $[Pd(PPh_3)_4]$, CuI, Et_3N; (b) 2-methyl-3-butyn-2-ol, $[Pd(PPh_3)_4]$, CuI, Et_3N; (c) TBAF, THF; (d) **174**, $[Pd(PPh_3)_4]$, CuI, Et_3N; (e) $CH_3N(C_8H_{17})_3Cl$, 5 N NaOH, PhH, $[Pd(PPh_3)_4]$, CuI, 80 °C; (f) i] 1 N HCl, THF, ii] TIPSOTf, Et_3N; (g) *hν*, CH_2Cl_2, 12 °C.

Using combinations of *ortho-*, *meta-*, and *para*-connected phenylacetylene monomers, Moore and co-workers have prepared a number of macrocyclic geometries based on a trigonal lattice [5c, 46a, 62, 63]. PAMs **177–179** represent a few examples of the many macrocycles prepared to date. Molecular modeling of compound **177**, which contains 12 arene rings, shows that the inner diameter (hydrogen to hydrogen) is slightly greater than 22 Å [63].

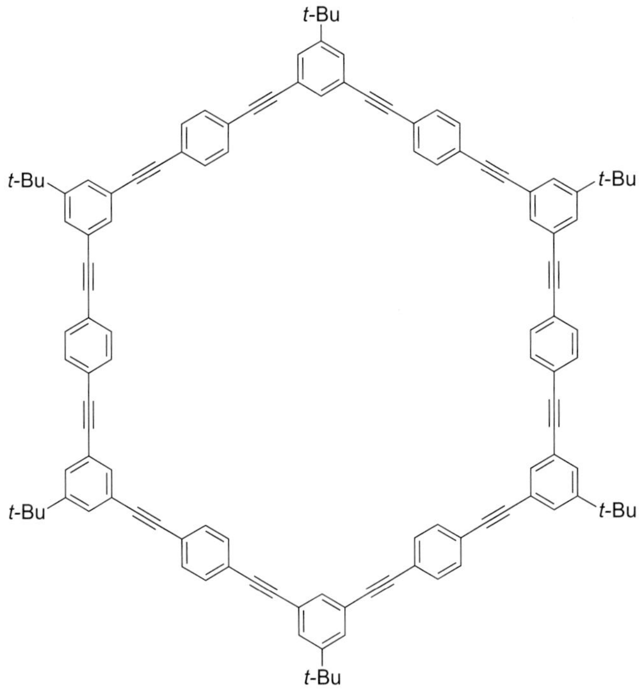

177

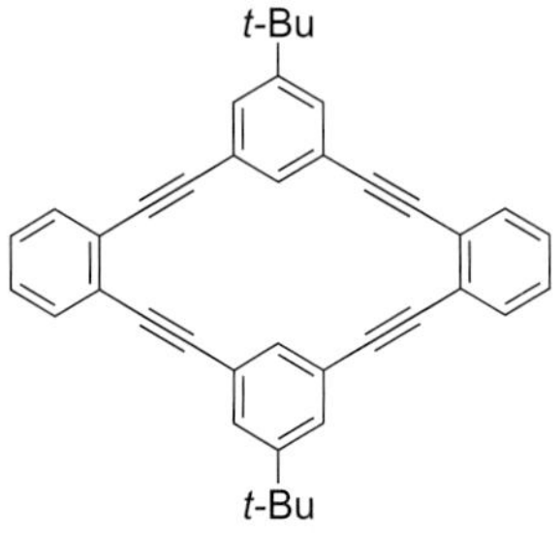

178

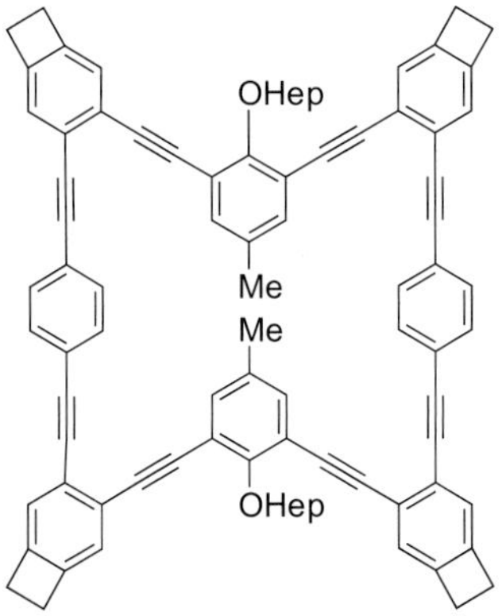

179

Fascinating examples of PAMs containing mixed linkages are the "molecular turnstiles" (**180–182**) prepared by Moore and Bedard [64]. Models of macrocycle **180** show that the inner ring can rotate freely. Indeed, variable-temperature NMR experiments suggest the spindle of the turnstile exhibits dynamic behavior. Similarly, the turnstile of **181** rotates freely at ambient temperature and has an estimated barrier to spindle rotation of ~13.4 kcal mol^{-1}. In contrast, the more sterically congested spindle of **182** is conformationally locked and no evidence of dynamic behavior was obtained.

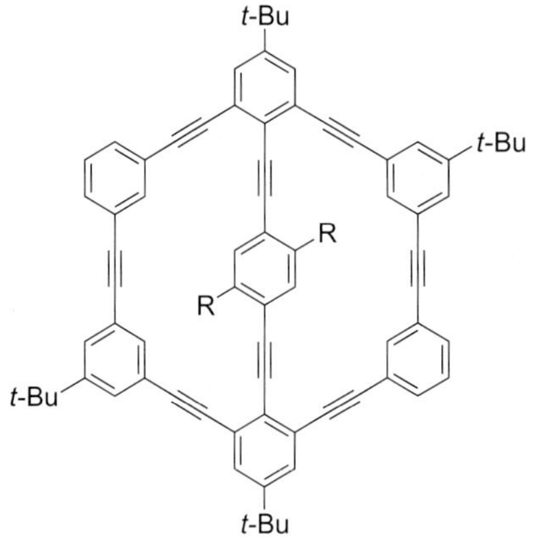

180 R=H
181 R=CH_2OCH_3
182 R=CH_2O

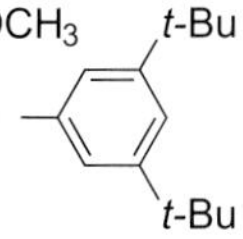

8.4
Phenyldiacetylene Macrocycles

In the thirty years following Eglinton's groundbreaking work on **3**, no reports on the preparation of related phenyldiacetylene macrocyclic systems were published. Since the 1990s, however, the chemistry of PDMs has witnessed an explosive resurgence concomitant with the advent of novel synthetic methodology tailored towards the construction of carbon-carbon bonds through the use of organotransition metal complexes. Although PDMs are typically less thermodynamically stable and more reactive than the homologous PAM derivatives, they exhibit a slew of novel electronic and optical properties. As such, they serve as potential synthons for technologically important compounds such as highly conjugated polymers, nonlinear optical materials, and new carbon allotropes.

In the 1970s, diacetylene monomers were found to undergo topochemical polymerizations to generate highly ordered crystalline polymers possessing conjugated backbones and displaying interesting conductive and optical properties [65]. If this reactivity could be extended to include PDM derivatives such as **3**, a potential route to highly conjugated networks of sp and sp^2 carbon atoms, with potentially interesting properties, could be developed. With this goal in mind, several research groups over the last decade have prepared a variety of derivatized PDMs for topochemical investigation.

8.4.1
Ortho-PDMs

Although the molecular structure of **3** (Scheme 8.1), the first phenyldiacetylene macrocycle to be synthesized and characterized [8], was obtained in 1959 [9], it took another 40 years before the packing behavior of this highly strained and reactive molecule was established [66]. The flat molecules stack parallel and form columns, with these columns staggered to form a "brick wall" motif. The orientation of the molecules within the column, however, precludes a topochemical polymerization. Diffraction experiments with "polymerized" samples showed that the material had lost all single-crystalline order. PDM **3** could also be co-crystallized with hexafluorobenzene and with TCNQ.

In 1994, with the goal of polymerization in mind, Swager et al. prepared annulene **3** and various derivatives for topochemical investigation [10]. In addition to diyne **2**, the Swager group studied the Cu-mediated cyclooligomerizations of 4,5-disubstituted-1,2-diethynylbenzenes (**183**). The requisite dialkylbenzenes were prepared in two steps from *o*-dichlorobenzene (ODCB; Scheme 8.40). The appropriate alkyl Grignard salt was cross-coupled with ODCB in the presence of catalytic [$NiCl_2$(dppp)] to afford the disubstituted benzene derivative, which was doubly iodinated with $I_2/NaIO_3$ in glacial AcOH. Cross-coupling of this arene with TMSA under Sonogashira conditions, followed by protiodesilylation with KOH, furnished alkylated derivatives of **183** (R = Bu, Hex, Dec, Dod). The 1,2-diiodo-4,5-didecoxybenzene was assembled in an analogous manner; however, the cross-coupling reaction required 2 molar equivalents of CuI to furnish **183** (R = ODec) in a reasonable yield. Although the exact origin of this effect is unclear, it is possible that the lone pairs on the heteroatom act as a Lewis base and form a weakly bound aryl-copper complex that prevents the Cu-acetylide coupling partner from forming.

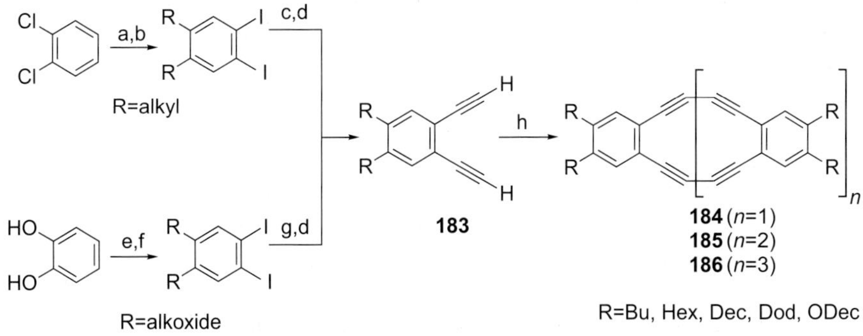

Scheme 8.40 Synthesis of PDMs **184–186**. Reagents and conditions: (a) $H(CH_2)_nMgBr$, [$NiCl_2$(dppp)], Et_2O; (b) I_2, $NaIO_3$, H_2SO_4, AcOH; (c) TMSA, [$PdCl_2(PPh_3)_2$], CuI, *i*-Pr_2NH; (d) aq. KOH, THF, MeOH; (e) $C_{10}H_{21}Br$, K_2CO_3, acetone; (f) $Hg(OAc)_2$, I_2, CH_2Cl_2, (g) TMSA, [$PdCl_2(PPh_3)_2$], xs CuI, *i*-Pr_2NH, 70 °C; (h) CuCl, TMEDA, O_2, ODCB.

Cyclooligomerization of arenes **183** under Hay conditions (CuCl, TMEDA, O_2) furnished mixtures of dimeric (**184**), trimeric (**185**), and tetrameric (**186**) PDMs along with some polymeric material [10]. Purification of the different products from the mixture proved laborious and was accomplished only through repetitive chromatography and fractional crystallization. The yields of the PDMs were found to vary widely and depended upon the length of the hydrocarbon tail, the scale of the reaction, and the exact experimental conditions. Thus, while derivatives of **183** possessing short alkyl and alkoxy groups tended to produce higher yields of larger macrocycles and polymeric material, the oligomerization of substrates with larger alkyl tails favored dimer formation. Moreover, fast, small-scale reactions were found to give the optimal overall yields of macrocyclic products. Overall yields and the product ratios of PDMs were also dependent on the order of reagent addition. As a whole, the above problems illustrate the weakness of the intermolecular cyclooligomerization reaction.

Because of the presumed unusual reactivity of the strained dimers **184**, several attempts to maximize their formation were made [10]. However, variation of reaction times and reduction of substrate concentrations in the cyclooligomerization reactions failed to improve significantly upon dimer formation. Accordingly, the tetrahexyl-substituted dimer **5** (Scheme 8.7; **184**, R = Hex) was synthesized by a six-step stepwise approach via synthons **24** and **25**. Ring-closure under Hay conditions furnished dimer **5** and tetramer **7** (**186**, R = Hex) in 13 % and 45 % yields, respectively. This product distribution is surprising since intramolecular reactions are typically more facile than intermolecular reactions.

An X-ray structural analysis of dimer **184** (R = Bu) reveals similarities to **3**, in that the unsaturated portion of the molecule is planar, with the angles between adjacent acetylenic bonds deviating from linearity by 13–15°. Since the connection of the alkyne moieties to the aromatic rings is only slightly compressed (2–3°), distortion of the acetylenic linkages appears as the major source of instability in these macrocycles. Unlike **3**, the molecules no longer stack in columns but are offset in the crystal lattice because of the hexyl substituents.

Differential scanning calorimetry (DSC) experiments performed on the various dimers **184** show that the molecules thermally polymerize abruptly between 100–125 °C. While the narrow temperature range for this exothermic process is suggestive of a chain reaction, the absence of acetylenic functionalities in the IR spectra of the polymerized materials, as well as broad peaks in the X-ray powder diffraction and solid-state ^{13}C NMR data, indicate that highly ordered topochemical diacetylene polymers are not formed. DSC analysis of the various trimeric PDMs **185** is even more disappointing. Although more thermally robust, these macrocycles polymerize with very broad exotherms at sufficiently high temperatures (ca. 200 °C) yielding only intractable tars.

To improve understanding of the reactivity of PDMs **184**, their chemistry in solution was investigated [10]. Addition of I_2 to an Ar-saturated benzene solution of the individual dimeric PDMs afforded compounds **187**, containing tetraiodinated 6-5-6-5-6 fused ring systems, in 50–67 % yields (Scheme 8.41). Eglinton reported formation of the same pentacyclic architecture on treatment of the parent **3** with

Na/NH_3 [8]. Although this reaction presumably occurs through radical intermediates, attempts to trap such species produced only uncharacterizable material. Exposure of **187** to oxygen or to iodine in the presence of oxygen gave dione **188**. Labeling experiments showed that the carbonyl oxygen was incorporated directly by oxidation of **187** with molecular oxygen rather than as a result of hydrolysis.

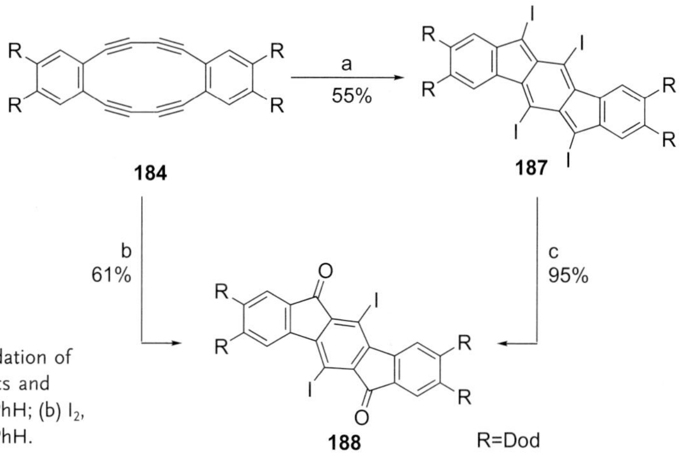

Scheme 8.41 Oxidation of PDM **184**. Reagents and conditions: (a) I_2, PhH; (b) I_2, O_2, PhH; (c) O_2, PhH.

Whereas the Hay conditions had failed to yield the parent tetramer **23** in Swager's study, Youngs et al. found that the use of highly dilute Glaser conditions with bubbling of air through the pyridine solvent afforded **23** in 20% yield along with a high yield of dimeric PDM **3** (Scheme 8.7) [22]. As is often the case in Cu-mediated acetylenic homocoupling, the exact reactions conditions play an important role and can be the crucial difference between high yield or no yield of product.

The dimeric and trimeric perfluorinated analogues of **3** and **1**, PDMs **189** and **190**, respectively, were recently reported by the Komatsu group [67]. Pd-catalyzed cross-coupling of TMSA with 1,2-diiodoperfluorobenzene, followed by protiodesilylation with KOH, gave 1,2-diethynylperfluorobenzene (Scheme 8.42). Cyclooligomerization under Hay conditions afforded the electron-deficient π-systems **189** and **190** in 20–45 % and 10–20 % yields, respectively, along with minor amounts of tetramer.

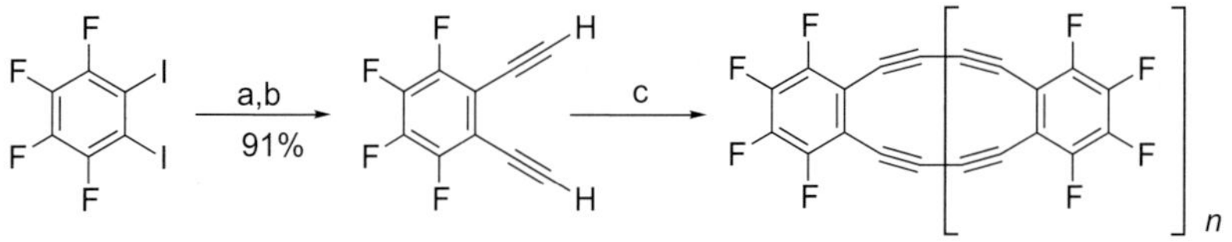

189 (n = 1, 20-40%)
190 (n = 2, 10-20%)

Scheme 8.42 Synthesis of PDMs **189** and **190**. Reagents and conditions: (a) TMSA, $[PdCl_2(PPh_3)_2]$, CuI, toluene, *i*-Pr_2NH; (b) KOH, H_2O, MeOH; (c) CuCl, TMEDA, O_2.

For *trans*-topochemical polymerization of 1,3-diynes to occur most efficiently, the diyne monomers should pack in an offset manner with an intermolecular distance $d \approx 5$ Å and a stacking angle $\Theta \approx 45°$ [65], while for *cis* polymerization $d \approx 3.5$ Å and $\Theta \approx 90°$ are required [68]. The X-ray data of the crystal derived from **189** in chloroform shows the PDM packs in a parallel and slanted arrangement with d = 4.99 Å and Θ = 39°, while that obtained from benzene solution adopts a similar arrangement but the stacks are more slanted (d = 6.35 Å and Θ = 26°); additionally, **189** crystallizes with a solvent molecule in a 1:1 ratio. The co-crystal derived from mixing of equimolar amounts of **189** and **3** packs in an unexpected 2:1 ratio in which the PDMs alternate in a face-to-face stacking array (d = 3.69–3.87 Å and Θ = 75–78°). While these results show the macrocycles stack in the proper orientation for topochemical polymerizations, attempts to induce polymerization upon photoirradiation were unsuccessful. DSC analyses of the PDMs show strong exotherms, but analysis of the thermoproducts suggests that they are not topological polymers but instead carbon-rich materials such as amorphous carbon and graphite, which have been shown to result from the thermal reactions of other dehydrobenzoannulenes [69, 70]. To date, ordered polymerization of the Eglinton–Galbraith dimer has yet to be observed. Attempts to grow X-ray-quality crystals either of **190** or of a co-crystal of **190** and **2** have been unsuccessful.

Very recently, Faust and Ott reported the synthesis of the Ru(bpy)$_2$-coordinated PDM **191** by treatment of the tetra-Boc-protected **192** with [(bpy)$_2$Ru(phenanthroline-5,6-dione)]$_2^+$(PF$_6^-$)$_2$ under acidic conditions (Scheme 8.43) [71a]. PDM **192** was generated through protiodesilylation/homocoupling of tetrayne **193**, which was assembled in five steps from 1,2-diiodo-4,5-dinitrobenzene by conventional methodology. Not surprisingly, the outcome of the final Cu-mediated homocoupling reaction was concentration-dependant. A 0.4 mM solution of deprotected **193** furnished dimer **192** exclusively in 72% yield, while use of a 1.5 mM solution of the terminal acetylene gave an inseparable mixture of **192** and tetramer **194** in a 2:1 ratio [71b]. The redox properties of **191** reveal two reduction waves at values comparable to that required for the first reduction of C$_{60}$. Furthermore, photoirradiation of **191** induces luminescence; preliminary measurements indicate that the lifetime of the excited state is 28 nm.

The π–frameworks of PDMs **192** and **194** are extended further upon treatment of a suspension of the annulene mixture and benzil with TFA (Scheme 8.44) [71b]. The resultant macrocycles **195** and **196** are obtained as sparingly soluble solids. In contrast, repetition of this reaction with 1,6-bis(TIPS)-hexa-1,5-diyne-3,4-dione affords PDMs **197** and **198**, which are readily separable by gel permeation chromatography. Unlike in **195**/**196**, the effect of the additional alkyne units in **197**/**198** show in the absorption spectra, with a bathochromic shift of ca. 20 nm relative to **195**/**196**.

Anthony and Gallagher have synthesized a series of linearly fused oligomers (**199**–**201**) based on the Eglinton–Galbraith dimer **3** (Scheme 8.45) [72]. Starting with tetrayne **202**, partial desilylation with catalytic AgNO$_3$ and 3 equiv. NBS gave a 1:1 mixture of mono- and dibromoacetylides. Exposure of the mixture to Sonogashira conditions with **203**, followed by removal of the remaining TMS groups

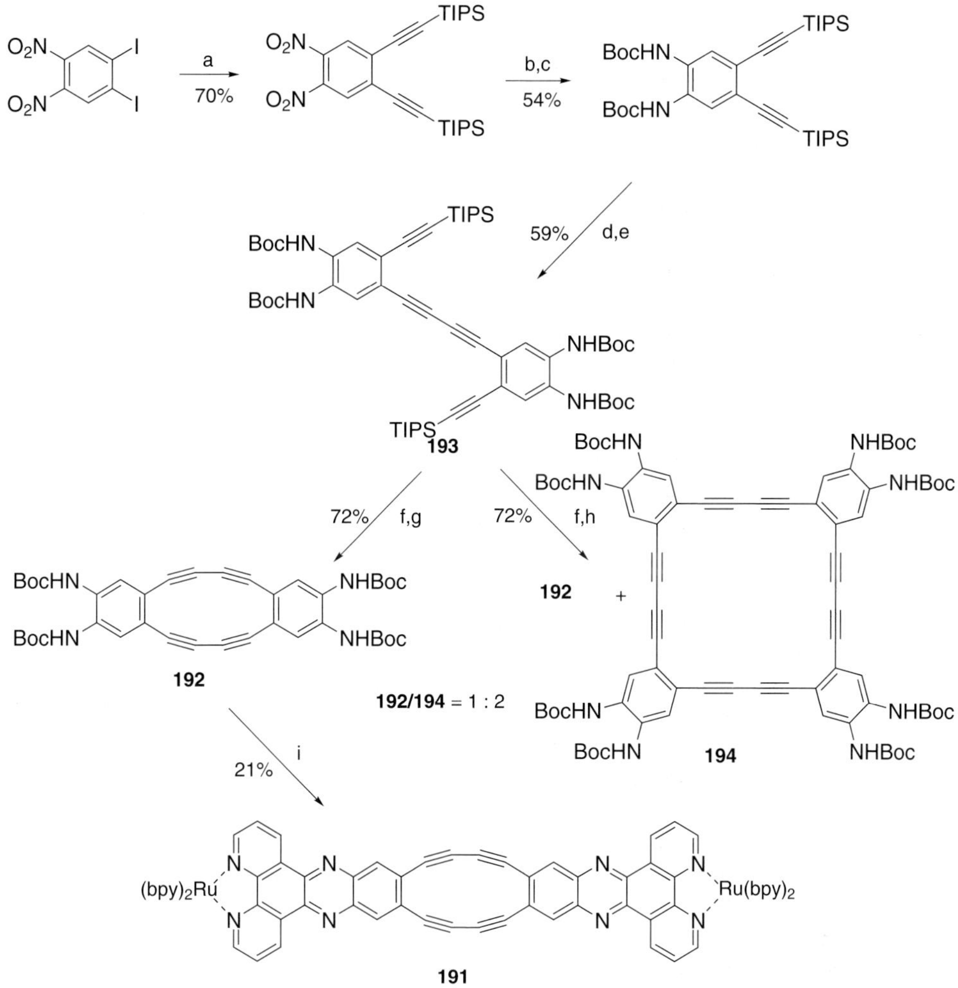

Scheme 8.43 Synthesis of PDMs **192** and **194**. Reagents and conditions: (a) TIPSA, $[PdCl_2(PPh_3)_2]$, CuI, NEt_3, 80 °C; (b) Sn, HCl, EtOH, 70 °C; (c) Boc_2O, *i*-Pr_2NH, THF, Δ; (d) TBAF, THF, 0 °C; (e) $Cu(OAc)_2$, CuCl, py, MeOH, 60 °C; (f) TBAF, THF, rt.; (g) $Cu(OAc)_2$, CuCl, py, MeOH, 60 °C, 0.4 mM; (h) $Cu(OAc)_2$, CuCl, py, MeOH, 60 °C, 1.5 mM; (i) $[(bpy)_2Ru(phenanthroline\text{-}5,6\text{-}dione)]_2^{+}(PF_6^{-})_2$, TFA, MeCN, 70 °C.

and oxidative alkyne coupling, gave precursors **204–206** as a mixture. Removal of the TIPS groups and alkyne homocoupling under Hay conditions furnished the brightly colored cyclic oligomers, which were separated by flash chromatography. Whereas **199** and **200** are freely soluble and can be stored for weeks, PDM **201** is only sparingly soluble and decomposes slowly in solution. The electronic absorption spectra of **199–201** show a progression of the lower-energy absorptions with increased oligomer length. An estimated cut-off of the infinite polymer derived

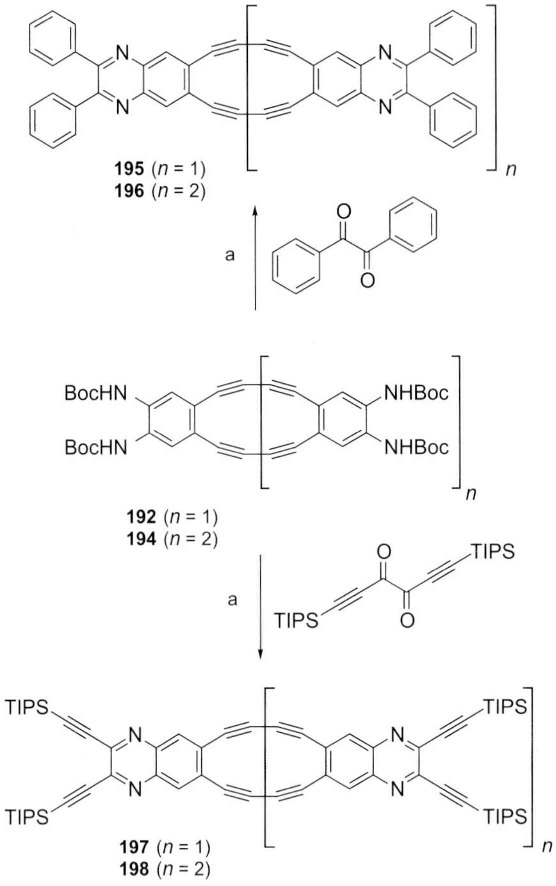

Scheme 8.44 Synthesis of PDMs **195–198**. Reagents and conditions: (a) AcOH, TFA, 60 °C.

from this series is 625 nm (1.98 eV), comparable to other linear acetylene-based polymers.

As shown in Scheme 8.8, Vollhardt and Youngs prepared hybrid PAM/PDM **26** through an intramolecular cyclization [23]. Like **3**, the molecule contains highly strained acetylene bonds that impart interesting solid-state behavior. As shown by X-ray crystallography, the monoacetylene units in **26** bow inward toward the center of the macrocycle by 3.9–11.5° while the diacetylene moiety bows outward by 8.6–11.2°. The molecules are stacked (d = 6.3 Å, Θ = 35.5°) in the same slanted and parallel fashion as **3**, but upon photoirradiation or application of pressure, **26** polymerizes to form an insoluble violet colored material with a metallic luster. Complete polymerization can be achieved through thermal annealing (150 °C by DSC) or high pressure (20,000 psi, 1 h). Solid-state ^{13}C NMR spectroscopy of the polymer displays two new peaks at 145.1 and 150.0 ppm, which are assigned to the alkene carbons of the polydiacetylene chain. Additional evidence is provided by TOF-MS, which detects oligomers containing up to nine macrocyclic units.

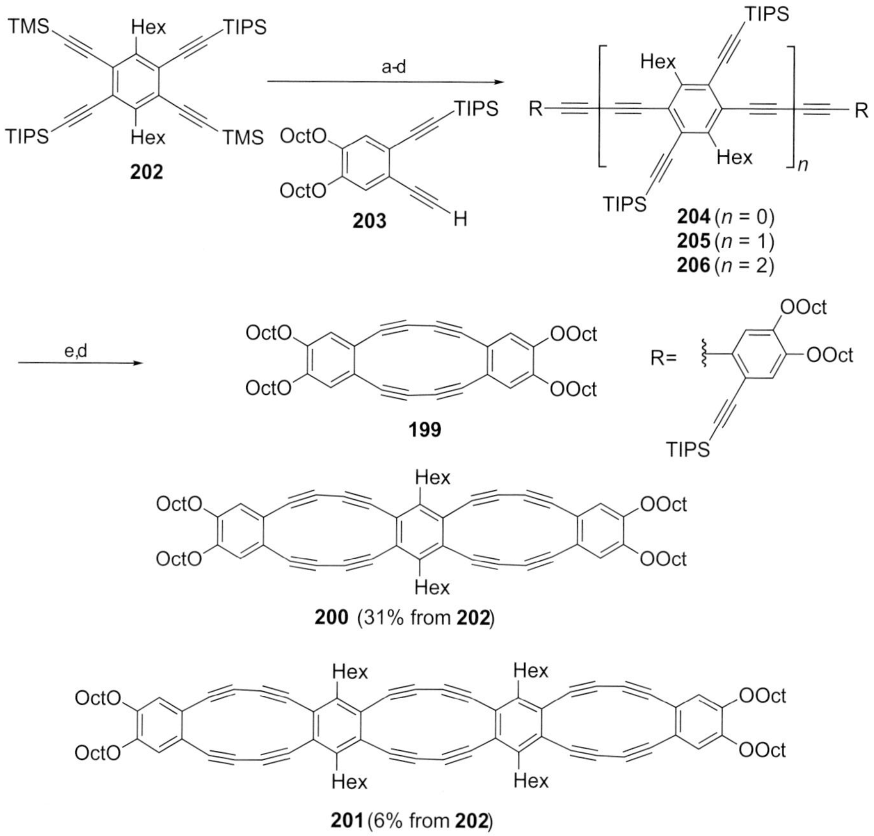

Scheme 8.45 Synthesis of PDMs **199–201**. Reagents and conditions: (a) NBS, $AgNO_3$, acetone, rt.; (b) **203**, $[PdCl_2(PPh_3)_2]$, CuI, *i*-Pr_2NH, 40 °C; (c) K_2CO_3, MeOH; (d) CuCl, TMEDA, acetone; (e) TBAF, THF, 0 °C.

Despite a number of related studies, the confirmed topochemical polymerization of **26** remains unique among PDMs.

An alternate decomposition pathway for dehydrobenzoannulenes and related PDMs was reported by the Vollhardt group in 1997 [69]. Treatment of the terminal acetylene **36** with $Cu(OAc)_2$ affords the non-planar dimer **207** in 13 % yield along with small amounts of trimer **208** (Scheme 8.46). The former PDM decomposes explosively at ca. 250 °C to produce a nearly pure carbon residue. Whereas the black solid did not contain any soluble carbonaceous material (e. g., fullerenes), TEM analysis of the residue showed that "bucky onions" and "bucky tubes" had been formed, in addition to amorphous carbon and graphite. Although the amount of "bucky" material composed only 1–2 % of the carbonaceous product, subsequent studies have shown that inclusion of metal complexes in the phenylacetylene starting materials results in considerably higher conversion to "bucky" compounds [73].

These results clearly validate the argument that PAMs and PDMs represent a potential route to such novel carbon-rich materials.

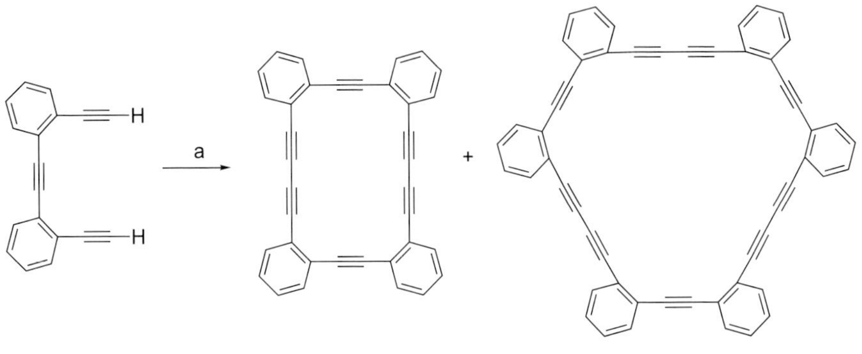

36 **207** (13%) **208** (5%)

Scheme 8.46 Synthesis of PDMs **207** and **208**. Reagents and conditions: (a) $Cu(OAc)_2 \cdot H_2O$, py, MeOH, Et_2O.

As discussed previously, the Haley group recently synthesized bisPDMs **30** and **31** from the same precursor (**29**, Scheme 8.9), with the structure of the final molecule dependent upon the method of alkyne homocoupling [24]. Scheme 8.47 illustrates the selectivity for ring size between the Cu- and Pd-mediated reactions. Both the 14- and the 15-membered ring precursors (**209** and **210**, respectively) were prepared by Sonogashira cross-coupling of **28** with the appropriate dibromodiiodobenzene. Protiodesilylation and alkyne homocoupling gave either **211** or **212**. While Cu-mediated ring-closure of **210** afforded the 15-membered ring (**212**) in very good yield (76–80 %), the analogous reaction with substrate **209** afforded the 14-membered ring (**211**) in much lower yield (24–35 %). Conversely, Pd-mediated ring-closure of **209** furnished PDM **211** in very good (67–76 %) yield, while analogous treatment of synthon **210** furnished the PDM **212** in a low (12–24 %) yield. The authors attribute this difference to the strain in the metal-containing intermediate prior to reductive elimination (alkyne homocoupling). Whereas a *cis* arrangement of the alkynes in the presumed Pd-bis(σ-acetylide) complex would favor formation of **211**, a similar *cis* arrangement to afford **212** would be highly strained. On the other hand, the Cu-acetylide intermediate leading to **212** should prefer a less strained, *trans*-like arrangement [4].

Confirmation that bisPDM **30** was indeed the product of the Cu-mediated cyclization of **29** was provided by independent synthesis of each 15-membered ring [24]. Pd-catalyzed cross-coupling of **212** with **28** furnished intermediate **213** in 43 % yield (Scheme 8.48). Removal of the TIPS groups with TBAF followed by oxidative cyclization with $Cu(OAc)_2$ afforded (57 % yield) target molecule **30**, the spectral data of which are an identical match with those obtained for the compound generated by the route shown in Scheme 8.9.

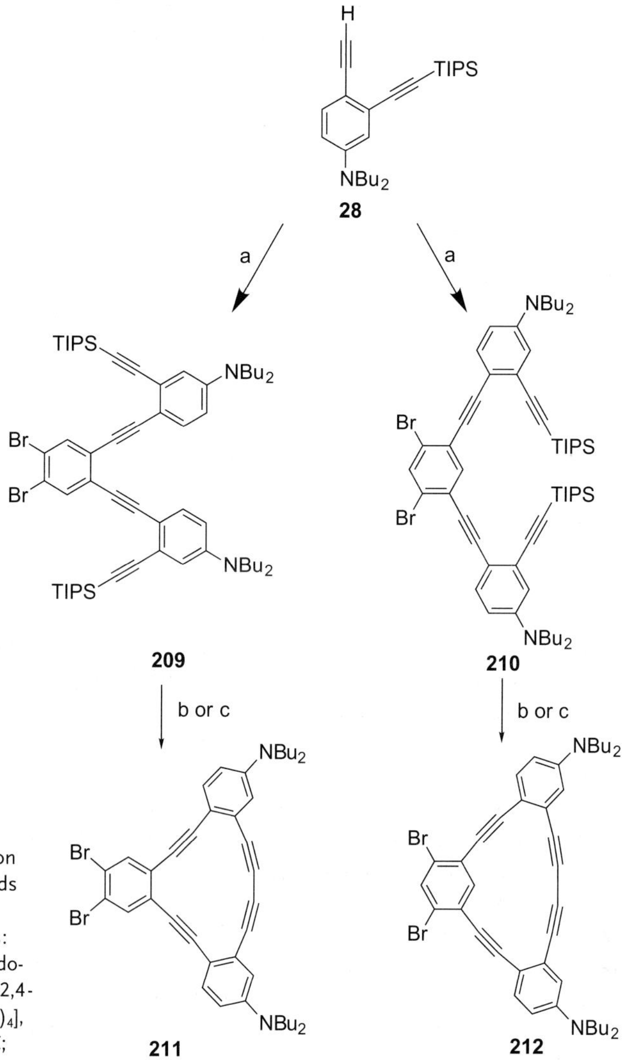

28

209 210

211 212

211		212
24%	$Cu(OAc)_2$	80%
35%	CuCl	76%
67%	$[PdCl_2(PPh_3)_2]$	24%
76%	$[PdCl_2(dppe)]$	12%

Scheme 8.47 Comparison of homocoupling methods for PDMs **211** and **212**. Reagents and conditions: (a) 1,2-dibromo-4,5-diiodobenzene or 1,5-dibromo-2,4-diiodobenzene, $[Pd(PPh_3)_4]$, CuI, *i*-Pr_2NH, THF, 45 °C; (b) i] TBAF, MeOH, THF, ii] 'Cu', py, 60 °C; (c) i] TBAF, MeOH, THF, ii] 'Pd', CuI, I_2, *i*-Pr_2NH, THF, 50 °C.

The Rubin group synthesized PDMs **214–216** from 1,2,3,4/5,6-differentially terminated hexaethynylbenzene **217** (Scheme 8.49) [74]. Preparation of these perethynylated derivatives requires more intricate synthetic maneuvering prior to cyclooligomerization. Arene **217** was prepared though a Diels–Alder reaction between the suitably derivatized tetraalkynylcyclopentadienone **218** and OHCC≡CCH(OEt)$_2$, followed by expulsion of CO to afford arene **219**. This compound

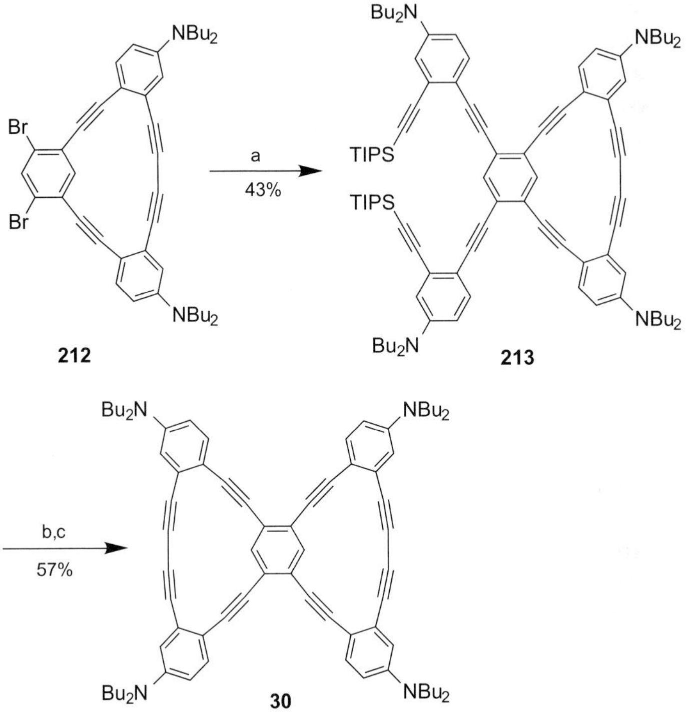

Scheme 8.48 Stepwise synthesis of BisPDM **30**. Reagents and conditions: (a) **28**, $[Pd(PPh_3)_4]$, CuI, *i*-Pr_2NH, THF, 40 °C; (b) TBAF, MeOH, THF; (c) $Cu(OAc)_2$, py, 60 °C.

was then converted into **217** via **220** by successive Corey–Fuchs alkynylation reactions. Attempts to introduce the alkynes directly by cycloaddition of **218** with bis-silylated hexatriynes were unsuccessful. Calculations suggest that the central triple bond in the triynes is sterically inhibited in the transition state by the bulky dienophile with the large capping groups on the dienone. Cyclooligomerization of **217** under Hay conditions furnished **214** in an isolated yield of 25 % along with an inseparable mixture of higher homologues **215** and **216** in a combined 13 % yield.

Compound **215** is a substructure of and potential precursor to the theoretical all-carbon-containing network graphdiyne (**221**), a homologue of graphyne (**78**) in which the acetylene linkages are replaced with butadiyne units. This and other diacetylenic all-carbon networks have attracted considerable attention due to predictions of technological importance, but have to date remained elusive because of synthetic difficulties [75]. PDM **215** can be viewed as a molecular fragment of graphdiyne featuring a fringe of *tert*-butyl-capped acetylene moieties around the periphery. This feature not only increases the potential for nonlinear optical activity by extending electron conjugation, but it also serves to stabilize the characteristically reactive dimer **214**. The difficulty of preparing the differentially terminated hexaethynylbenzenes limits this route to the above examples.

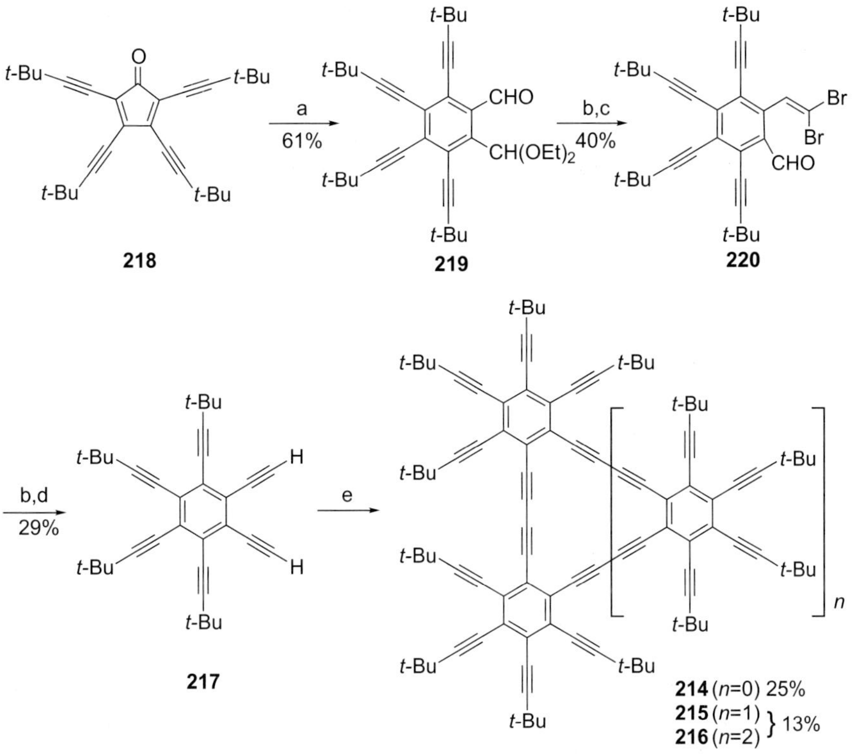

Scheme 8.49 Synthesis of PDMs **214–216**. Reagents and conditions: (a) $OHCC{\equiv}CCH(OEt)_2$, PhH; (b) CBr_4, PPh_3, CH_2Cl_2; (c) SiO_2; (d) i] LDA, THF, ii] aq. NH_4Cl; (e) CuCl, TMEDA, O_2, acetone.

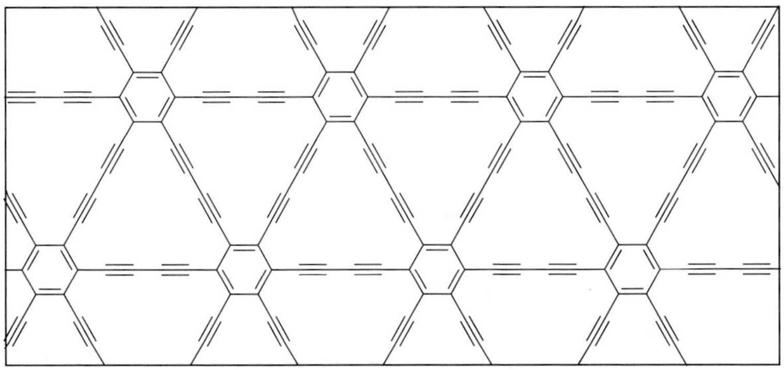

221

Surprisingly, the synthesis of the simplest graphdiyne subunit, PDM **1**, was not reported until 1997 [76]. The Haley group prepared **1** by an intramolecular route, since cyclooligomerization of *o*-diethynylbenzene had purportedly failed to afford the desired macrocycle. Conversion of 1-bromo-2-iodobenzene into triyne **222** was accomplished by standard cross-coupling procedures (Scheme 8.50), but attempts to monodeprotect **222** and to couple it with 1,2-diiodobenzene provided intractable gums. This problem was circumvented by generation of the free butadiyne in situ under Sonogashira conditions, which then provided the bis-coupled product **223** in 71 % yield. Desilylation and use of high-dilution conditions in the oxidative homocoupling reaction gave **1** in 35 % yield as a poorly soluble, cream-colored powder. The poor isolated yield of PDM **1** is presumably accounted for by its minimal solubility, but this provided an important clue as to why it had not yet been reported. To manipulate **1** best, it is necessary to work up the homocoupling reactions with CH_2Cl_2 instead of the more traditional Et_2O. Armed with this fact, repetition of Scheme 8.1 with this subtle modification now afforded a 58 % yield of **3** and an inseparable 3:2 mixture of PDMs **1** and **23** in 20 % yield [77].

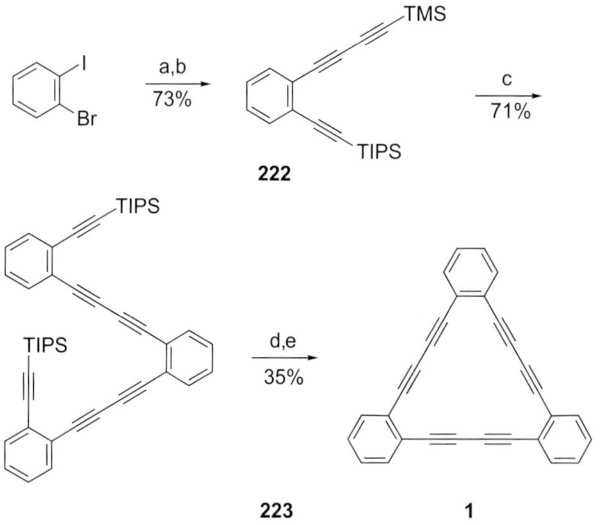

Scheme 8.50 Stepwise synthesis of PDM **1**. Reagents and conditions: (a) TMSC≡CC≡CH, $[PdCl_2(PPh_3)_2]$, CuI, Et_3N; (b) TIPSA, $[PdCl_2(PPh_3)_2]$, CuI, Et_3N; (c) *o*-diiodobenzene, $[Pd(PPh_3)_4]$, CuI, Et_3N, THF, aq. KOH; (d) TBAF, THF, EtOH; (e) $Cu(OAc)_2 \cdot H_2O$, py, MeOH.

A number of larger graphdiyne substructures (**224–228**) have been prepared from appropriately derivatized substrates by the intramolecular cyclization technique [77, 78]. The need for solubilizing substituents was recognized from the outset, and Scheme 8.51 illustrates the preparation of **225**. Triazene **229**, prepared from the corresponding aniline by the chemistry described in Scheme 8.23, could be transformed into triyne **230** in 60–75 % yield. Fourfold in situ desilylation/alkynylation of **230** gave dodecayne **231** in ca. 60–70 % yield. Subsequent removal of the four TIPS groups with TBAF followed by double intramolecular homocoupling furnished **225**. Whereas the use of *tert*-butyl groups afforded a solid virtually insoluble in common solvents, inclusion of decyl substituents solubilized **225** effectively, in that the compound was isolated in 63 % yield. The decyl versions of **224** and **226–228** were prepared in an analogous fashion.

The electronic absorption data for **1** and **224–228** each exhibit a characteristic pattern of four absorption bands, which are assigned primarily to $\pi \rightarrow \pi^*$ transi-

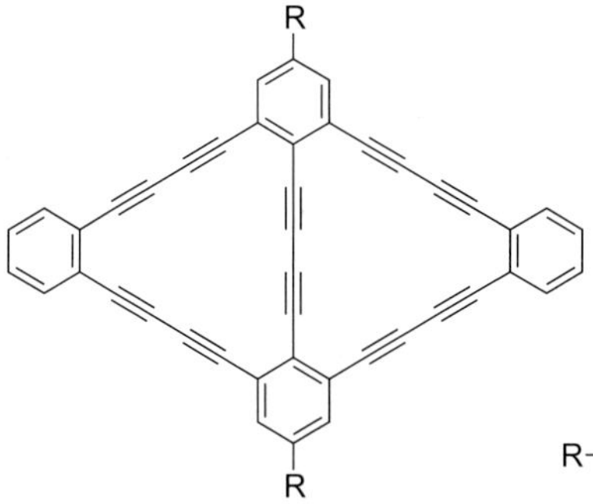

R=Dec
224

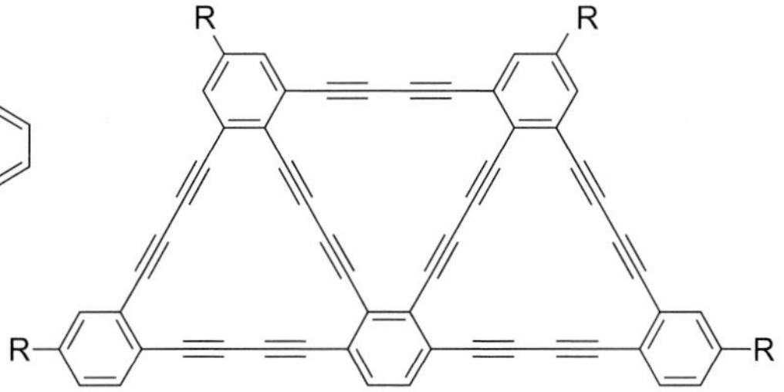

R=Dec
227

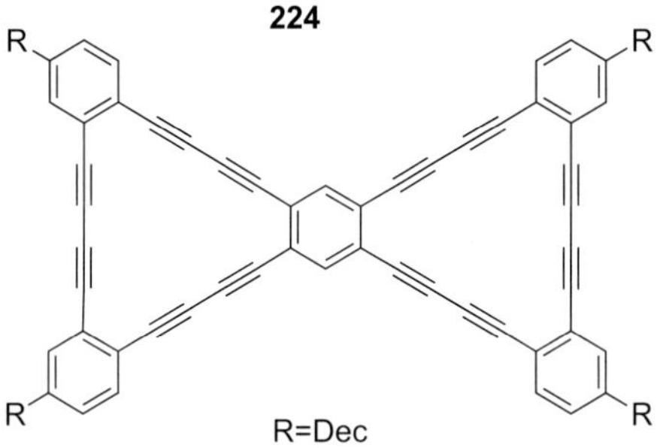

R=Dec
225

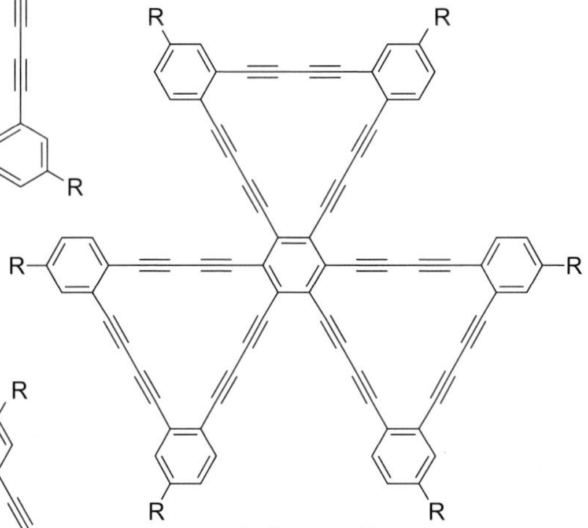

R=Dec or *t*-Bu
228

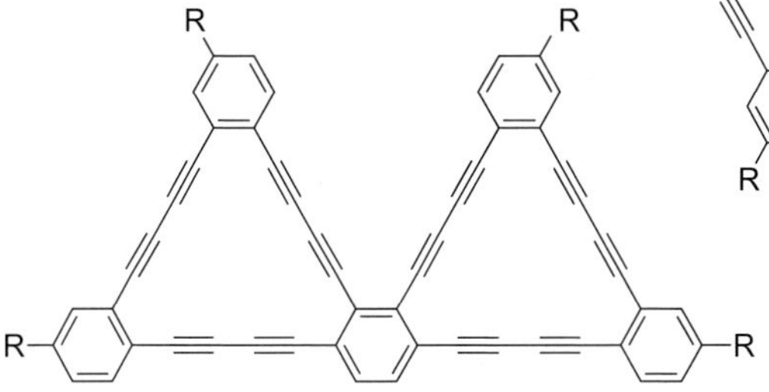

R=Dec
226

tions in the [18]annulene skeleton. These peaks shift to longer wavelengths and possess greater intensities as the number of longer 1,4-bis(phenylbutadiynyl)benzene chromophores is increased. The peaks of trisPDM **228** are shifted nearly 100 nm toward lower energy than those of **1**, reflecting the increased linear π-conjugation. Interestingly, the λ_{max} values of **1** and **224–228** all show the effect of locking the π-electron-rich backbone into planarity. This effect is most pronounced in **228**, in which λ_{max} is red-shifted by nearly 60 nm relative to its acyclic precursor.

Although the size of graphdiyne subunits is limited by solubility constraints, major advances towards the synthesis of "supersize" substructures such as **232**

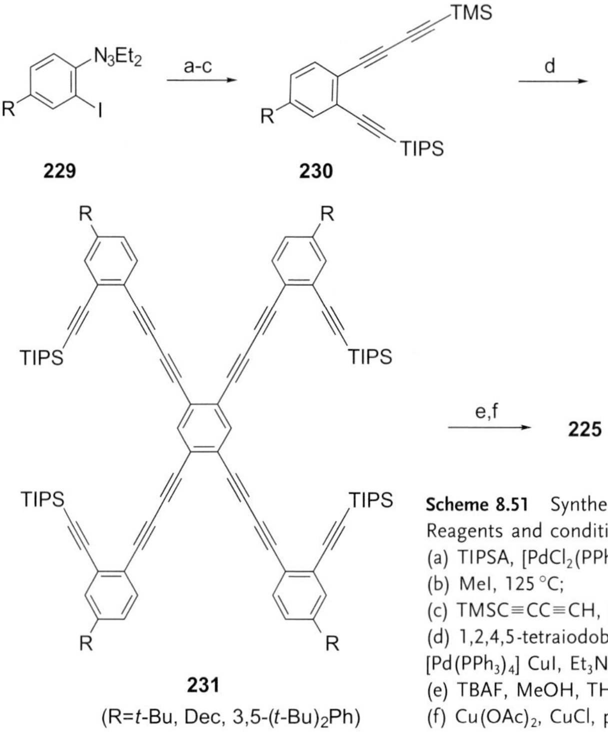

Scheme 8.51 Synthesis of PDM **225**. Reagents and conditions: (a) TIPSA, $[PdCl_2(PPh_3)_2]$, CuI, Et_3N; (b) MeI, 125 °C; (c) TMSC≡CC≡CH, $[PdCl_2(PPh_3)_2]$, CuI, Et_3N; (d) 1,2,4,5-tetraiodobenzene, aq. KOH, $[Pd(PPh_3)_4]$ CuI, Et_3N, THF; (e) TBAF, MeOH, THF; (f) $Cu(OAc)_2$, CuCl, py.

and **233** have been made in the Haley laboratories [79]. Triazene **234**, which permits the introduction of three different alkynylated groups, is converted into the asymmetric coupling unit **235** by standard techniques (Scheme 8.52). Protiodesilylation and Pd-mediated cross-coupling of the last compound with appropriately substituted benzene derivatives affords the corresponding macrocyclic precursors. Attempts to deprotect and cyclize those substrates containing decyl functionalization were unsuccessful, presumably because of the extreme insolubility of the resulting macrocycles. Whereas tetraether functionalization imparted greater solubility to the precursors, this route was abandoned because of the need for tedious and repeated chromatographic separations. Gratifyingly, the 3,5-di(*tert*-butyl)phenyl unit imparts excellent solubilizing characteristics; repetition of the chemistry shown in Scheme 8.51, for example, yielded the 3,5-di(*tert*-butyl)phenyl version of **225**, which is 10–15 times more soluble in halocarbon solvents than the corresponding decylated PDM. Use of a combination of decyl tails and 3,5-(*tert*-butyl) phenyl substituents furnished tricyclic PAM **232** (R^1 = 3,5-*t*-Bu_2Ph, R^2 = Dec) in approximately 8% yield from **235**. Preparation of the 3,5-di(*tert*-butyl)phenyl-functionalized hexaPDM **233** is currently underway.

Through the use of synthetic techniques similar to those described in Schemes 8.50 and 8.51, a wide variety of *ortho*-PDMs can be assembled quickly from common intermediates [70]. The key to these successes is efficient construc-

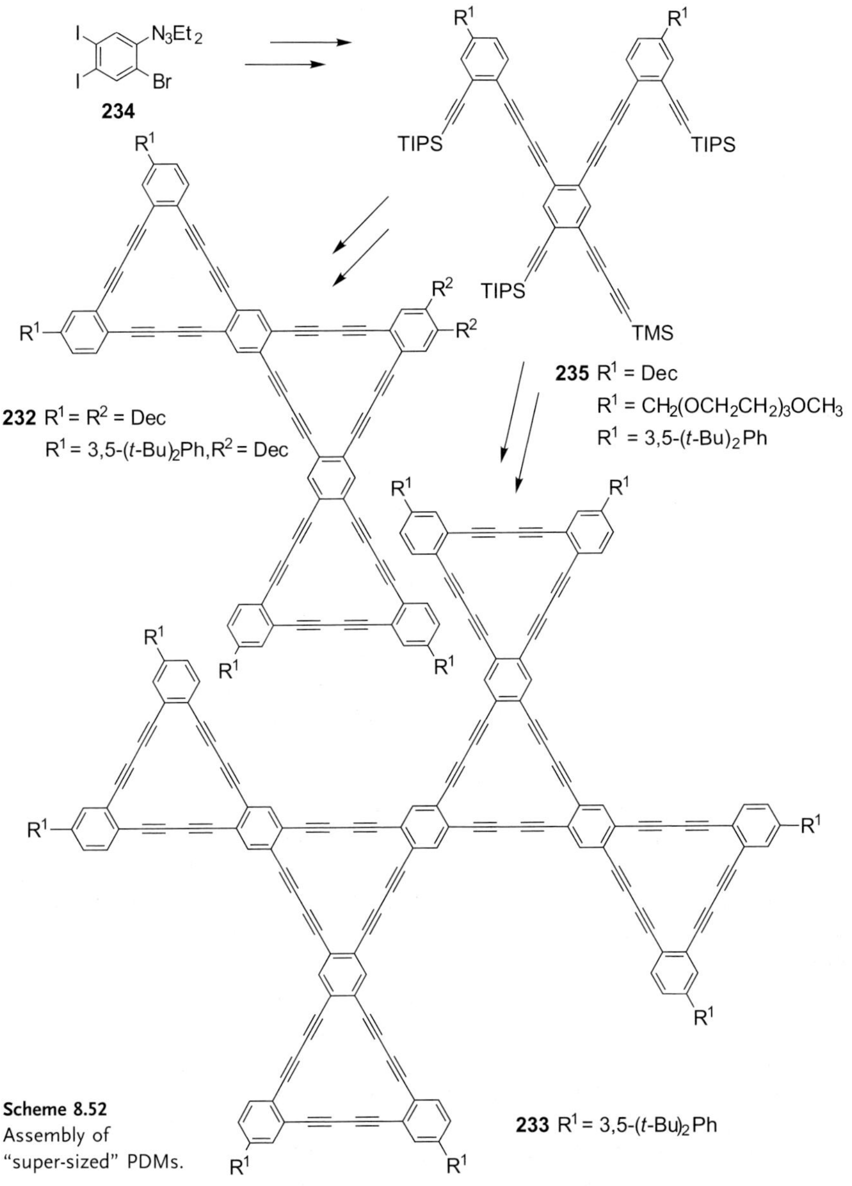

Scheme 8.52 Assembly of "super-sized" PDMs.

tion of the PDM carbon skeleton by the use of phenylbutadiynes generated in situ. The resultant polyyne can then be homocoupled under pseudo-high-dilution conditions to afford a single product. Among the structures created are PDMs either unavailable by traditional routes (**236–238**) or obtainable only in low yield (**7**, **207**). Thanks to the nonplanarity and/or lower symmetry of the PDMs, inclusion of solubilizing groups is not needed.

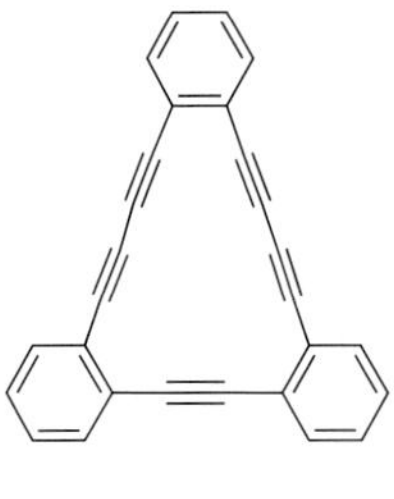

236

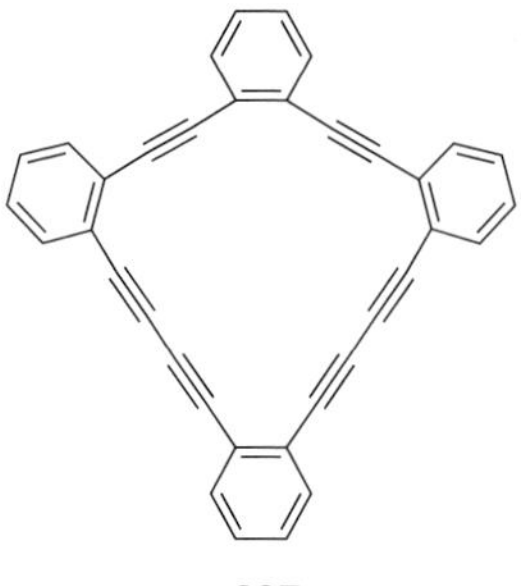

237

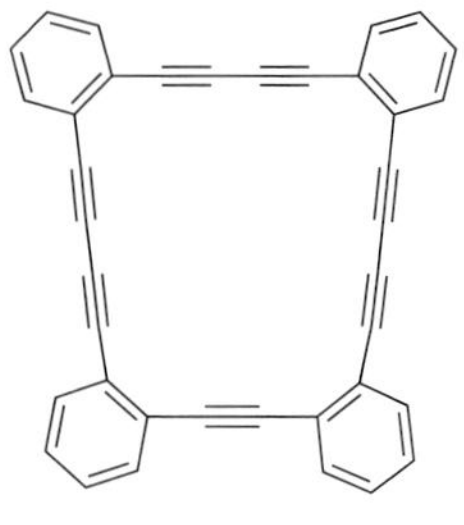

238

The advent of rational synthetic techniques for PDM assembly (vide supra) now makes site-specific placement of functional groups on the arene rings possible and thus allows access to PDM topologies that would otherwise be difficult to prepare by cyclooligomerization of functionalized 1,2-diethynylbenzenes. Structures **239**–**248** illustrate some of the various symmetries prepared and substituents utilized to date [80–82]. A representative synthesis, that of the C_s-symmetric PDM **248**, is shown in Scheme 8.53. Electron-donor and -acceptor groups can be introduced at a number of positions during the initial stages of PDM assembly, either as triyne "legs" (e. g., **249**–**250**) or as diiodoarene "heads" (e. g., **251**). In the case of **251**, selective cross-coupling reactions provided the desired hexayne precursor **253**. Desilylation and homocoupling furnished **248** in a remarkable 52 % overall yield for the

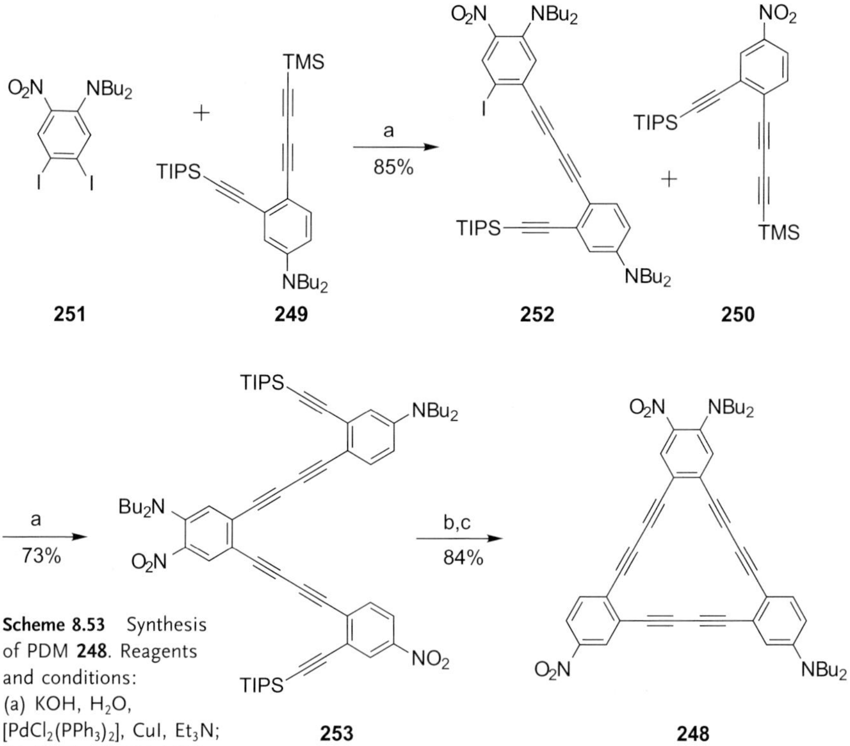

Scheme 8.53 Synthesis of PDM **248**. Reagents and conditions: (a) KOH, H_2O, $[PdCl_2(PPh_3)_2]$, CuI, Et_3N; (b) TBAF, MeOH, THF; (c) $Cu(OAc)_2$, CuCl, py.

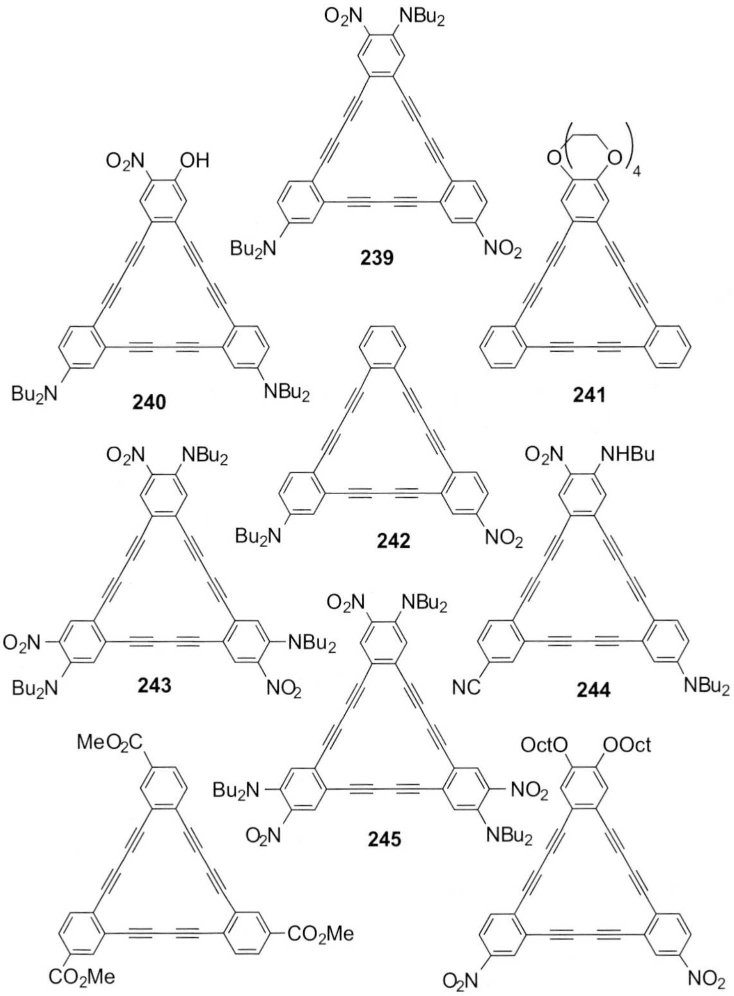

239 **240** **241** **242** **243** **244** **245** **246** **247**

four steps. More importantly, the molecule was the sole product of the cyclization reaction, making isolation/purification of the PDM trivial.

The extreme polarizability of the above donor/acceptor PDM derivatives results in novel UV/Vis absorption properties. The electronic spectra display the four characteristic absorbances typical for [18]annulenes, but these are weakened, and the low-energy bands are extended and broadened greatly. These shifts are maximized when opposing groups are positioned at each end of the macrocyclic framework, such as in **248**, which can be attributed to the increased polarization of the conjugated backbone [80]. The high polarization makes the donor/acceptor PDMs excellent candidates for nonlinear optical materials. By strategic placing of functional groups on the periphery of the macrocycle the physical properties of the material can be tuned. Furthermore, since these systems are also locked into planarity, the NLO efficiency is increased further, due to enhancement of π-conjugation and communication between chromophores. The β values for PDMs **1**, **239**, **242**, **243**, **245**, and **248** have been measured by the hyper Raleigh scattering technique and were found to be two to three times greater than that of the standard 4-dimethylamino-4′-nitrostilbene reference chromophore [81].

8.4.2
Meta-PDMs

The first examples of PDMs based on *meta* linkages were reported by Tobe and co-workers in 1996 [83]. The diacetylenic homologues of Moore's *meta*-PAM derivatives, the PDMs were also expected to self-associate in solution in a manner analogous to Moore's *meta*-PAMs [47], and so their solution-phase self-association properties were studied. The attractive force at work, π-π stacking of aromatic moieties, plays an important role in supramolecular chemistry. As examples, the existence of stable DNA double helices is dependant upon the weak associative forces between vertically stacked adjacent base pairs, while the packing orientation of aromatic molecules in the solid state is also determined by π-stacking. Consequently, understanding of the structural characteristics that contribute to the ability of a molecule to engage in this type of behavior has important real-world implications.

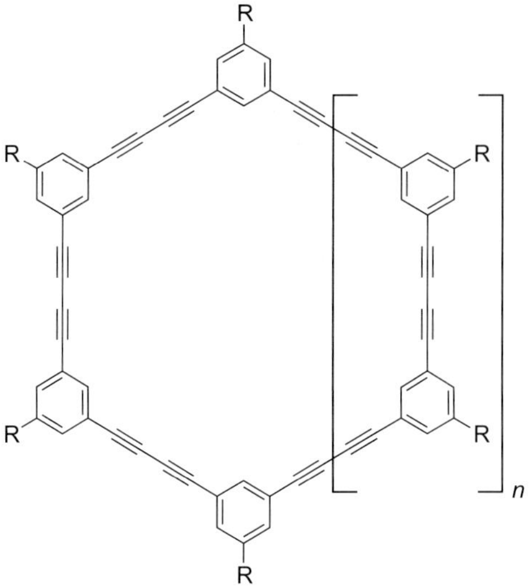

254 $n = 0$ a: $R = CO_2C_8H_{17}$
255 $n = 1$ b: $R = CO_2C_{16}H_{33}$
256 $n = 2$ c: $R = CO_2(CH_2CH_2O)_3CH_3$

With this as a goal, the Tobe group prepared *meta*-PDMs **254–256** with three different ester linkages (Scheme 8.54) [83]. Sequential Sonogashira reactions with the benzoate ester **257** afforded the differentially protected monomer **258**. Treatment of **258** with LiOH removed the TMS group and hydrolyzed the ester. The resultant acid was re-esterified with DCC and the appropriate alcohol, yielding **259**, which was then oxidatively dimerized to furnish **260**. Initial attempts at PDM formation focused on intermolecular oxidative coupling of the desilylated dimer **261b**, but cyclization with $Cu(OAc)_2$ provided very low yields of **254b** (10 %) and **256b** (2 %) and none of **255b** [83a]. It was thus necessary to prepare **254–256** by stepwise

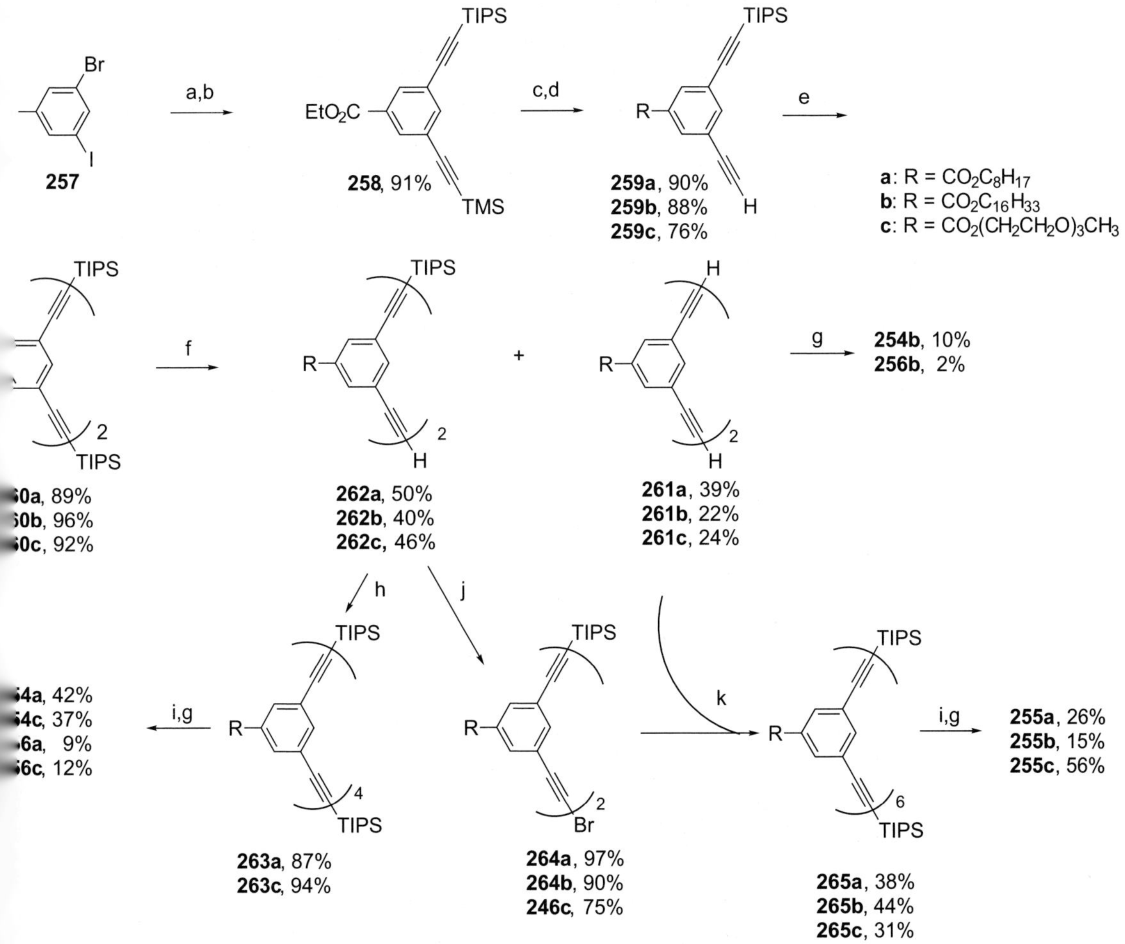

Scheme 8.54 Synthesis of PDMs **254–256**. Reagents and conditions: (a) TIPSA, $[Pd_2(dba)_3 \cdot CHCl_3]$, PPh_3, CuI, NEt_3; (b) TMSA, $[Pd_2(dba)_3 \cdot CHCl_3]$, PPh_3, CuI, NEt_3; (c) KOH, THF, H_2O; (d) R'OH, DCC, DMAP, DCE; (e) $Cu(OAc)_2$, py; (f) TBAF, THF, H_2O; (g) $Cu(OAc)_2$, py, PhH; (h) CuCl, TMEDA, acetone, O_2; (i) TBAF, THF; (j) NBS, $AgNO_3$, acetone; (k) $[Pd_2(dba)_3 \cdot CHCl_3]$, CuI, *i*-$Pr_2NH$, PhH.

assembly of the open-chain oligomers and subsequent intramolecular cyclization. Partial deprotection of **260** gave **262** as the major product, along with **261** and unreacted starting material. Homocoupling of **262** furnished the tetramer **263**, which was then deprotected and cyclized to give **254a/c** and **256a/c** in combined yields considerably higher (ca. 50%) than had been obtained in the intermolecular method (12%). Conversion of **262** to bromoalkyne **264** and cross-coupling to give **261** provided the open-chain hexamer **265**. Deprotection and cyclization as before gave **255a–c**.

The self-aggregation behavior of the PDMs was investigated by ^{1}H NMR spectroscopy and by vapor pressure osmometry. In $CDCl_3$ solution at 303 K the binding constants, K_{assoc}, for **254a–c**, **255c**, and **256a** were found to range from approximately 20–43 M^{-1}, while the K_{assoc} values of the hexamers **255a** and **255b** were calculated as 173 and 150 M^{-1}, respectively. In more polar solvents, association of *meta*-PDMs is enhanced through solvophobic effects. In fact, evidence suggests that large, nanotubular aggregates of PDMs **254c** and **255c** are formed. Surprisingly, the *meta*-PDMs aggregate more strongly in aromatic solvents than in $CDCl_3$, though the reason for this is not fully understood. Overall, the study concludes that PDMs promote π–π stacking interactions effectively due to the esters' electron-withdrawing nature. Furthermore, self-association properties are controlled by the size of the PDM, and the cyclic hexamers, with their rigid, planar backbones, associate to the greatest extent. The PDMs in fact aggregate more readily than the corresponding PAMs, which is attributed to withdrawal of electron density from the aromatic rings by the butadiyne linkages, thus facilitating π-π stacking interactions.

PDMs **266** and **267**, in which *tert*-butyl groups have replaced the ester moieties, have also been prepared by Tobe's group (Scheme 8.55) [83a, 84]. The two macrocycles were synthesized analogously to those in Scheme 8.54. As before, intermolecular cyclization of the dimer **268** afforded the PDMs in much lower combined yield than intramolecular cyclization of tetramer **269**. Both the Hay and Breslow cyclization conditions were also attempted, but PDM yields were lower with the former method and no PDMs were isolated by use of the latter procedure. Unlike the ester-substituted systems, PDMs **266** and **267** exhibit no self-association behavior in solution.

In 1998 Tobe et al. reported the synthesis of **270**, containing interior binding cyano groups [85]. PDM **270** was constructed by the same stepwise methodology as depicted in Scheme 8.54, the sole change being the inclusion of a cyano group on the benzoate ester. Unlike **255b**, **270** does not aggregate at all. The lack of self-association of this compound is most probably due to the electrostatic repulsion of the cyano groups. Surprisingly, upon mixing of **255b** and **270** in $CDCl_3$, aggregation does indeed occur but is considerably more complex, as **255b** interacts with **270** to form a heterodimer and also higher oligomeric aggregates. Thus, while the cyano groups of **270** inhibit self-association, the electron-withdrawing groups enhance attractive π-π stacking interactions with **255b**. In addition to this unusual aggregation behavior, PDM **270** forms 1:1 and 2:1 host/guest complexes with both tropylium and guanidinium cations.

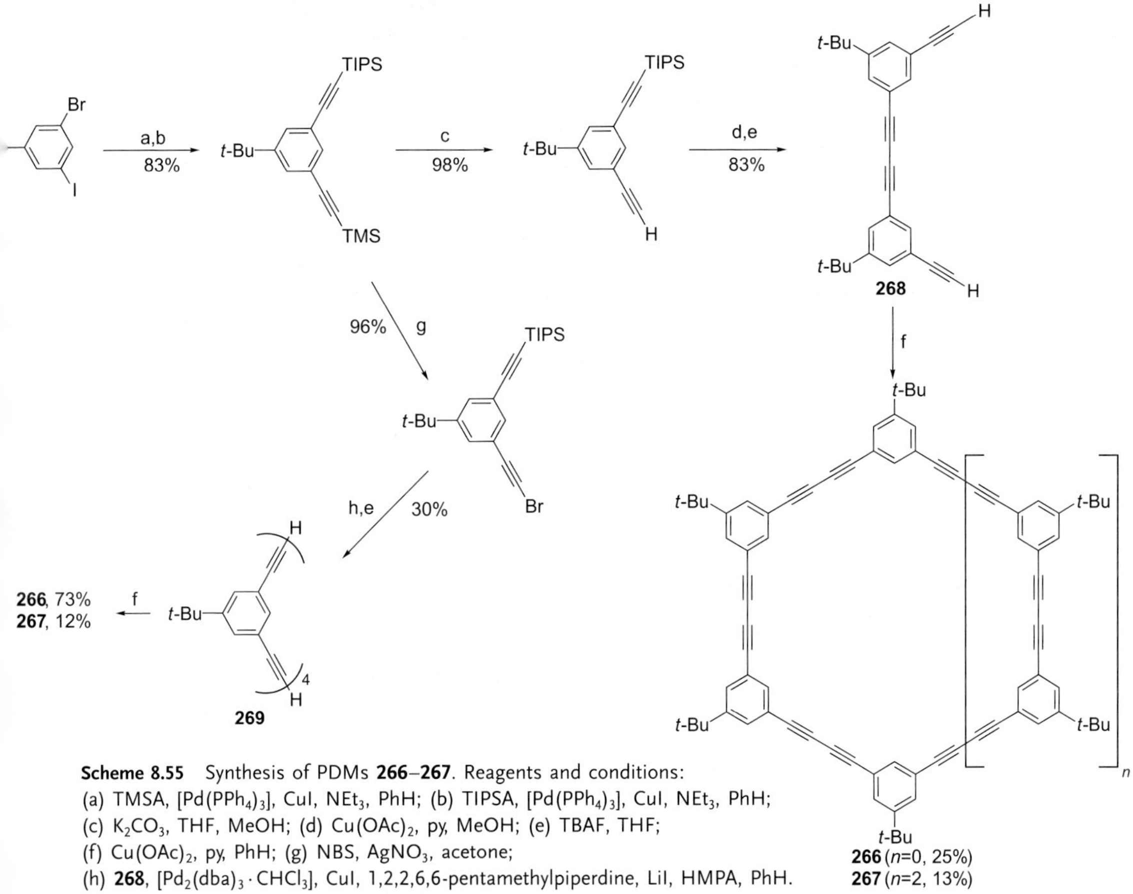

Scheme 8.55 Synthesis of PDMs **266**–**267**. Reagents and conditions: (a) TMSA, $[Pd(PPh_4)_3]$, CuI, NEt_3, PhH; (b) TIPSA, $[Pd(PPh_4)_3]$, CuI, NEt_3, PhH; (c) K_2CO_3, THF, MeOH; (d) $Cu(OAc)_2$, py, MeOH; (e) TBAF, THF; (f) $Cu(OAc)_2$, py, PhH; (g) NBS, $AgNO_3$, acetone; (h) **268**, $[Pd_2(dba)_3 \cdot CHCl_3]$, CuI, 1,2,2,6,6-pentamethylpiperdine, LiI, HMPA, PhH.

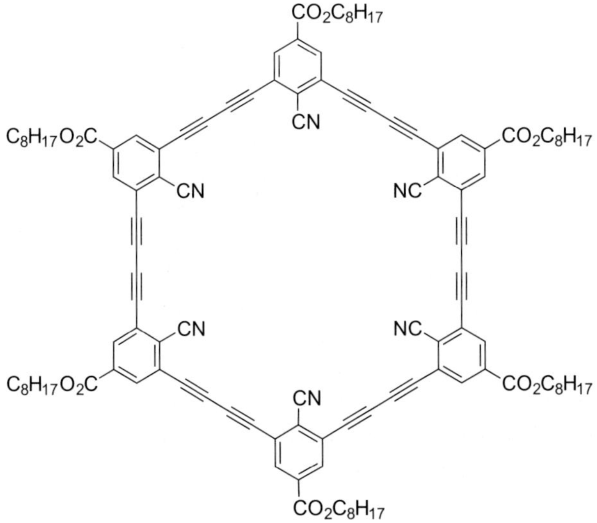

270

Mayor and Lehn created a series of phenylthio-linked tetrameric (**271**), hexameric (**272**), octameric (**273**), and decameric (**274**) *meta*-PDMs in order to observe the electrochemical properties of the PDM-based "molecular batteries" (Scheme 8.56) [86]. Reduction of the cyano moieties of isophthalonitrile **275** and subsequent fourfold nucleophilic aromatic substitution with *p-tert*-butylthiophenyl anion furnished sulfide **276**. Corey–Fuchs alkynylation of **276** and anion quenching with TBDMSCl gave monomer **277**. Partial deprotection with TBAF, followed by oxidative homocoupling, provided dimer **278**, which upon full deprotection and Cu-mediated cyclization afforded the PDMs **271–274** in 41% combined yield. Somewhat surprisingly, the intramolecular cyclization of desilylated **279**, again prepared by partial deprotection followed by oxidative homocoupling, gave the tetrameric PDM **271** in a low 14% yield.

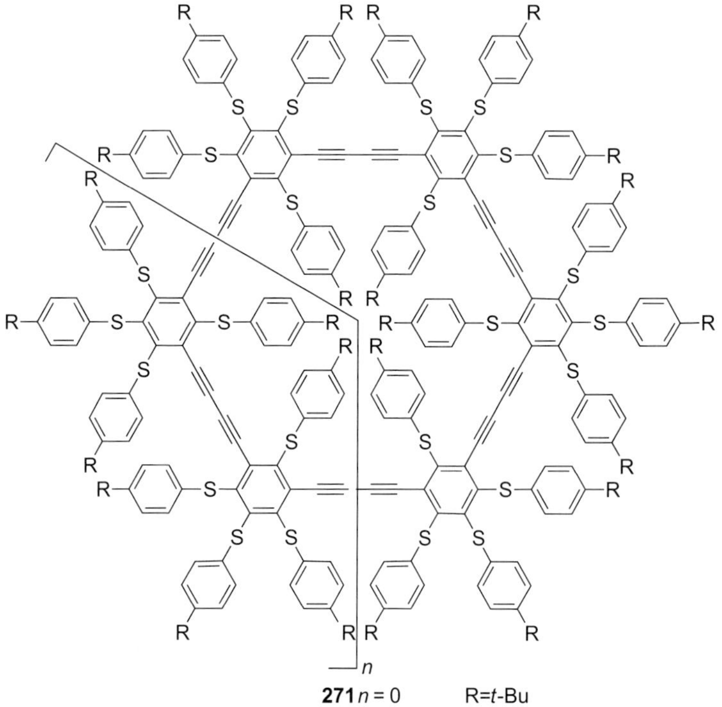

271 $n = 0$ R=*t*-Bu
272 $n = 1$
273 $n = 2$
274 $n = 3$

Cyclic voltammetry measurements of the sulfur-containing PDMs show no oxidation peaks up to –1.3 V, but several reversible reduction waves are observed for **271** (–1.07, –1.19, –1.44 V) and **272** (–1.09, –1.29, –1.40 V), while **273** displays one broad peak (–1.09 V) and a large shoulder toward more negative values, presumably arising from unresolved reduction waves. The multiple reversible reductions of these compounds point to the potential of these macrocycles to act as "electron reservoirs", which makes them potential candidates for "molecular batteries".

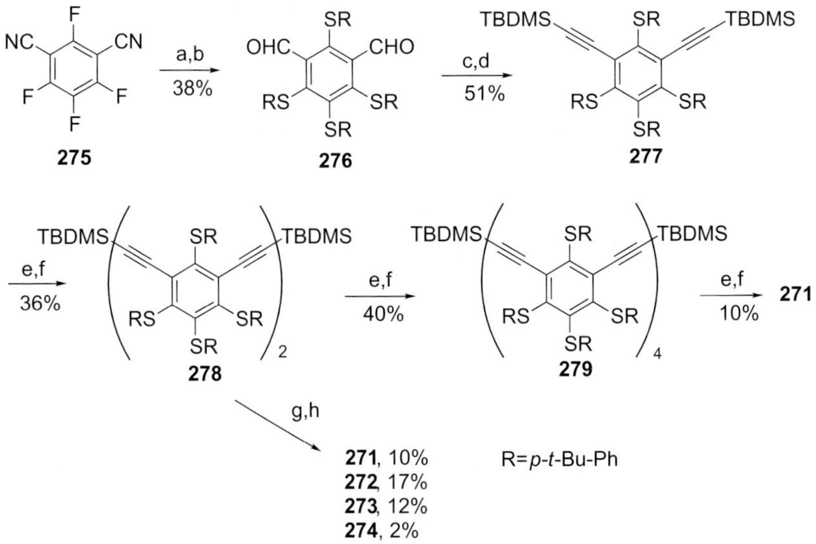

Scheme 8.56 Synthesis of PDMs **271–274**. Reagents and conditions: (a) DIBAL, toluene; (b) DMI, NaSR; (c) CBr_4, PPh_3, CH_2Cl_2; (d) i] LDA, THF, −78 °C, ii] TBDMSCl; (e) THF, AcOH, TBAF; (f) $Cu(OAc)_2 \cdot H_2O$, $CHCl_3$, 60 °C; (g) THF, TBAF; (h) 2×10^{-4} M in $CHCl_3$, 5.5 equiv. $Cu(OAc)_2 \cdot H_2O$, $CHCl_3$, 60 °C.

8.4.3
Para-PDMs

The sole *para*-linked PDM reported in the literature to date is the belt-shaped macrocycle **280**, available in six steps from substrate **174** (Scheme 8.57) [87]. Conversion of **174** into the differently protected diacetylene **281** was accomplished by standard Pd-catalyzed cross-coupling techniques. In situ removal of the TMS moiety and successive homocoupling of the free alkyne afforded tetrayne **282** in 85 % yield. Treatment of the tetrayne with TBAF and then $Cu(OAc)_2$/CuCl afforded the hexamer **283**, which, upon acidic hydrolysis/silylation and subsequent photoirradiation, furnished **280** as an air-sensitive solid that decomposed gradually over several days under ambient conditions. Molecular modeling studies suggest that **280** adopts a cylindrical conformation and that the acetylene moieties deviate from linearity by 8.8–9.2°. Moreover, rotation of the phenyl rings at 25 °C is rapid, as evidenced by the proton and carbon NMR spectra, which display only four and nine lines, respectively.

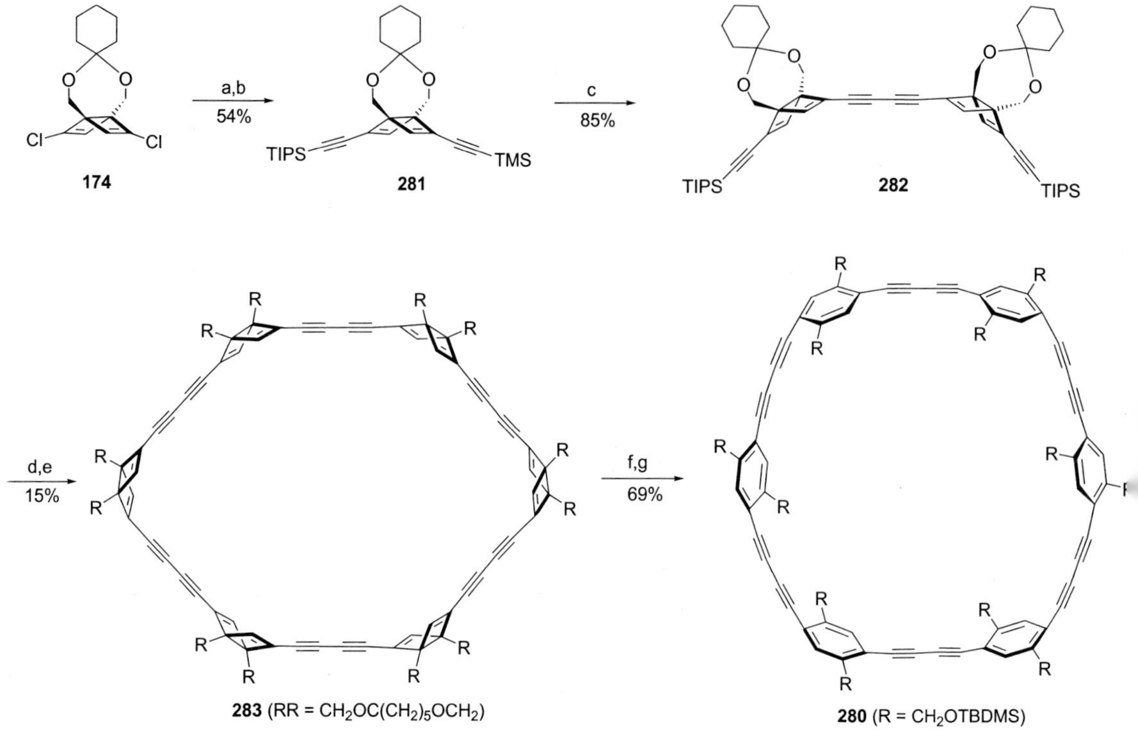

Scheme 8.57 Synthesis of PDM **280**. Reagents and conditions:
(a) TMSA, $[Pd(PPh_3)_4]$, CuI, Et_3N; (b) TIPSA, $[Pd(PPh_3)_4]$, CuI, Et_3N; (c) $Cu(OAc)_2$, CuCl, py, rt.; (d) TBAF, THF; (e) $Cu(OAc)_2$, CuCl, py, 60 °C; (f) i] 1 N HCl, THF, ii] TBDMSOTf, Et_3N; (g) *hν*, PhH, 12 °C.

8.4.4
Mixed PDMs

The shape-persistent, structurally well-defined natures of macrocycles based on phenyl and acetylene building blocks makes them ideal candidates for binding guests within their cavities. Although this review illustrates a few salient examples of large, shape-persistent macrocycles and their host-guest chemistry, the reader is referred to Chapter 10 by Höger for a more detailed discussion of this topic [88].

In 1995, Höger and Enkelmann constructed the *meta*/*para*-PAM/PDM hybrid **284** with hydrophobic/hydrophilic functionality specifically for use in host-guest chemistry (Scheme 8.58) [89]. The macrocycle was assembled in a straightforward manner from arene **285** by standard alkyne cross-coupling/deprotection chemistry. Cyclodimerization of **286** and hydrolysis of the THP ethers gave **284** in 45 % yield for the last two steps.

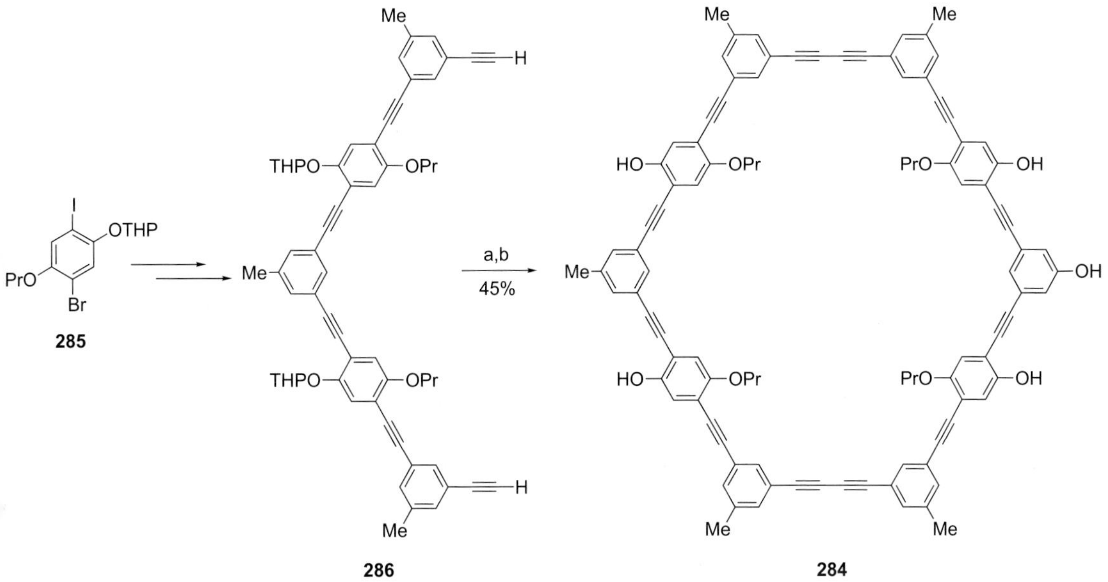

Scheme 8.58 Synthesis of PDM **284**. Reagents and conditions: (a) CuCl, $CuCl_2$, py; (b) CH_2Cl_2, MeOH, H^+.

X-ray structural analysis of **284** shows the macrocycle to be nonplanar, with a torsion angle of 6.7° about the diyne units. The interior cavity is approximately 2.0 × 2.4 nm and is occupied by the lipophilic propoxy groups. Furthermore, the macrocycle co-crystallizes with four molecules of pyridine, which form hydrogen bonds to the phenolic hydroxy groups on the exterior of the ring. Solution studies of **284** (^{1}H NMR) suggest that the compound exhibits dynamic behavior and rapidly interconverts between several different conformations, although the authors were unable to ascertain whether the hydroxy groups were inside or outside the macrocycle. The use of guest molecules of a suitable size, such as tetraamine **287**, results in reversal of the binding topology in **284** [90]. In fact, guest **287** fits extremely well in the cavity of the PDM, so that hydrogen bonding now occurs exclusively in the interior of the macrocycle (K_{assoc}= 160 M^{-1}).

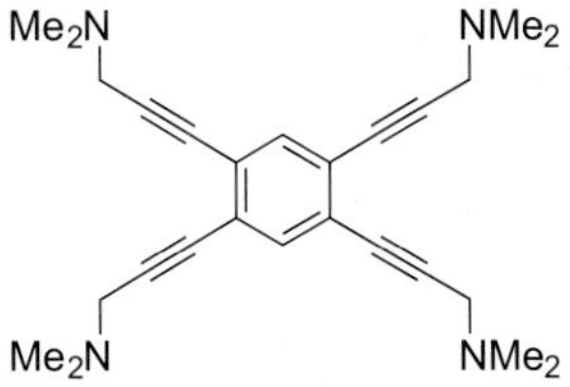

287

The cyclization of **286** was found to be the lowest-yielding step in the synthesis of the macrocyclic amphiphile **284**, and Höger and co-workers investigated two strategies to overcome this problem. The simpler of the two was a study of the effect of temperature and structure on the cyclooligomerization reaction [91]. Temperatures of ca. 60 °C were found to be optimum for product formation when $CuCl/CuCl_2$ in pyridine was used to promote the cyclization. The structures of the cyclization precursors were found to be equally important. The presence of groups that could possibly be templated by the Cu salts markedly increased yields.

A more ingenious method for cyclooligomerization utilized covalently bound templates to direct the cyclization reaction [92]. Condensation of **288** with 1,3,5-benzenetricarboxylic acid gave the templated ester **289** (Scheme 8.58). With the interior ester functionality binding the tetrayne moieties together, the terminal alkynes were then homocoupled, and the template was subsequently hydrolyzed to furnish PDM **290** in an excellent 89 % yield. This is a significant improvement over the non-templated pathway, which afforded **290** in ca. 20 % yield along with a number of inseparable cyclooligomeric byproducts. Use of a longer hydroxyundecyl tether in place of the 3-hydroxypropyl group furnished the target macrocycle in 84 % isolated yield, which suggests that, as long as the geometry is preorganized, proximal spatial constraint is not necessary for the high-yielding cyclization to occur [92a].

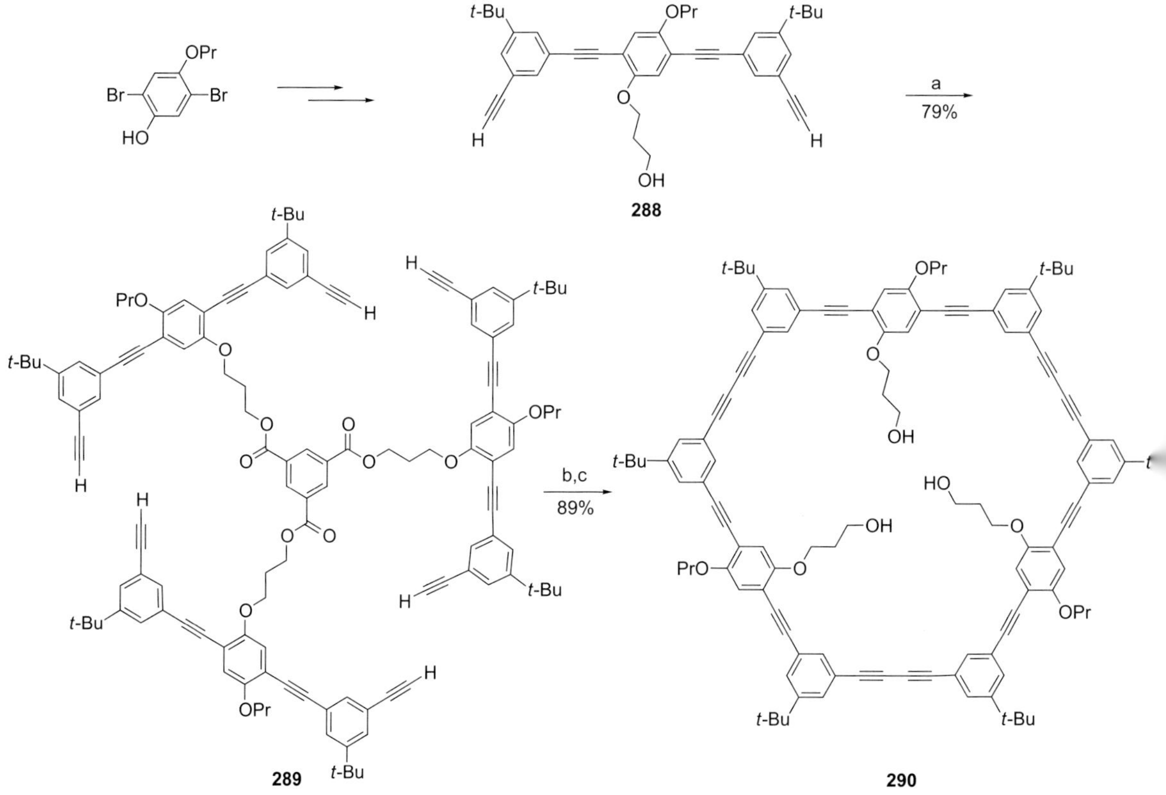

Scheme 8.59 Synthesis of PDM **290**. Reagents and conditions: (a) 1,3,5-benzenetricarboxylic acid, DEAD, PPh_3, THF; (b) CuCl, $CuCl_2$, py; (c) CH_2Cl_2, MeOH, H^+.

The use of "outside" templates affords PDMs in comparable impressive yields [92b]. Cu-mediated ring-closure of substrate **291**, for example, furnished PDM **292** in an impressive 86 % yield (Scheme 8.60). Subsequent experiments showed that tether length played an important role in the product yield in the "outside" case. If it is assumed that cyclization is a stepwise process, use of a short tether should promote the first dimerization but sterically prohibit the second. Indeed, shortening of the tether of the acyclic precursor by 10 carbon atoms reduced the yield of PDM by 15 %. Overall, the utilization of templates has proven to be a very powerful method in PAM/PDM synthesis and is only limited by the template introduction/removal. For further discussion of larger, shape-persistent, mixed PDMs, please refer to Chapter 10 by Höger [88].

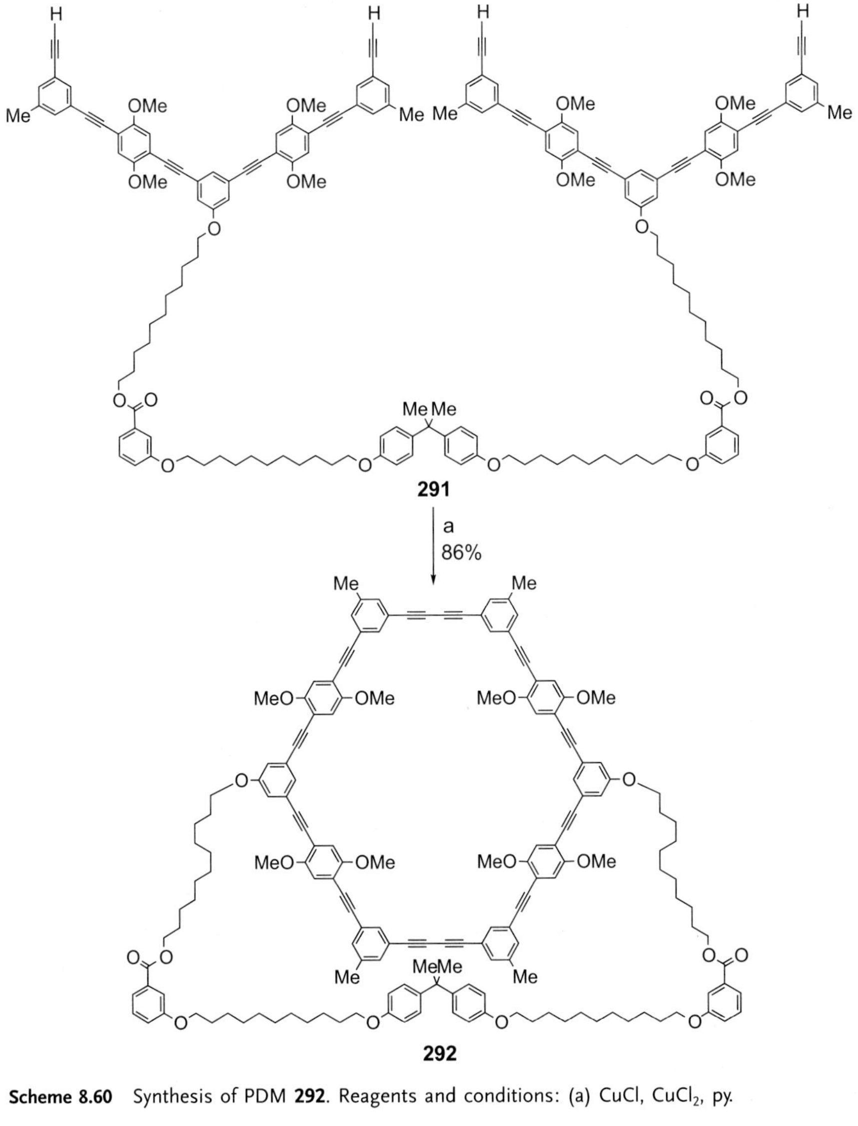

Scheme 8.60 Synthesis of PDM **292**. Reagents and conditions: (a) CuCl, $CuCl_2$, py.

A number of considerably simple mixed PDM derivatives have been prepared by Fallis and co-workers. Although originally seeking collapsible helical structures, the Canadian team made some interesting discoveries during their PDM studies. As an example, it was hoped that cyclization of **294**, prepared from **293** by sequential Suzuki coupling with 1,4-dibromobenzene and desilylation, might generate the intermolecular dimer **295**, but it instead furnished the intramolecular product **296** in 90 % yield (Scheme 8.61) [93]. If the Eglinton coupling reaction was performed at room temperature for 3 days instead of at 90 °C for 3 hours, **295** could be isolated in ca. 20–25 % yield along with ca. 45–55 % of **296**. An X-ray crystal structure of a derivative of **296** shows that the carbon backbone is highly strained, with the triple bonds distorted from linearity by about 16°. The terphenyl unit is also perturbed, deviating from planarity by ca. 18°. An X-ray structure of **295** shows that the molecule twists and collapses to give a saddle-like structure, as similarly observed in other PAMs and PDMs with four *ortho*-benzo-fused vertices (e. g., **10**, **23**, **207**, **237**).

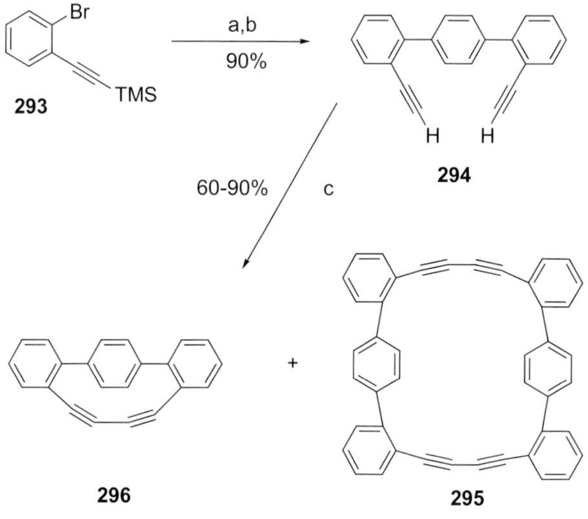

Scheme 8.61 Synthesis of PDMs **295** and **296**. Reagents and conditions: (a) i] BuLi, THF, –78°C, ii] $ZnBr_2$, THF, 0 °C, iii] 1,4-dibromobenzene, [$Pd(PPh_3)_4$], reflux; (b) K_2CO_3, THF, MeOH, H_2O; (c) $Cu(OAc)_2$, py, Et_2O.

Insertion of two acetylene units into the terphenyl moiety of **294** results in a different outcome of the cyclodimerization reaction (Scheme 8.62) [94]. Unlike those in **294**, the terminal acetylenes in **297** are too far apart to undergo intramolecular homocoupling, so dimer **298** is isolated as the sole product. Like **295**, PDM **298** adopts a twisted helical geometry.

It was hoped that modification of the PDM cores of **295** and **298** with long-chain groups might produce liquid crystalline materials [94], so compounds **299**–**301** were prepared by chemistry analogous to that shown in Scheme 8.61 and 8.62. As before, **299** and **300** were both produced by Cu-mediated homocoupling, whereas **301** was the sole cyclization product. The formation of **300** and **301** was more complex as both compounds were produced as racemic mixtures of two dif-

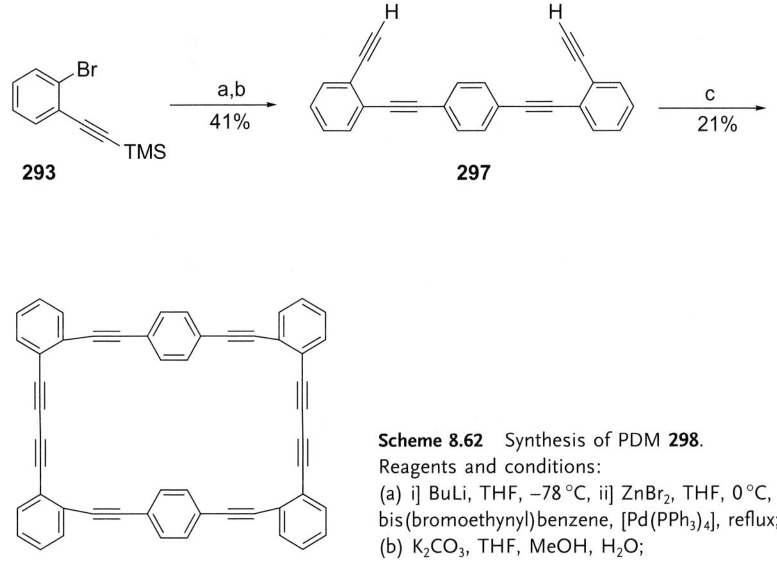

293 297

298

Scheme 8.62 Synthesis of PDM **298**.
Reagents and conditions:
(a) i] BuLi, THF, –78 °C, ii] $ZnBr_2$, THF, 0 °C, iii] 1,4-bis(bromoethynyl)benzene, $[Pd(PPh_3)_4]$, reflux;
(b) K_2CO_3, THF, MeOH, H_2O;
(c) $Cu(OAc)_2$, py, Et_2O.

ferent regioisomers. Of the whole series of PDMs, only one regioisomer of **301b** exhibited evidence of potential liquid crystalline behavior. Definitive proof will require the synthesis of an enantiomerically pure product, which is not possible by the cyclooligomerization route.

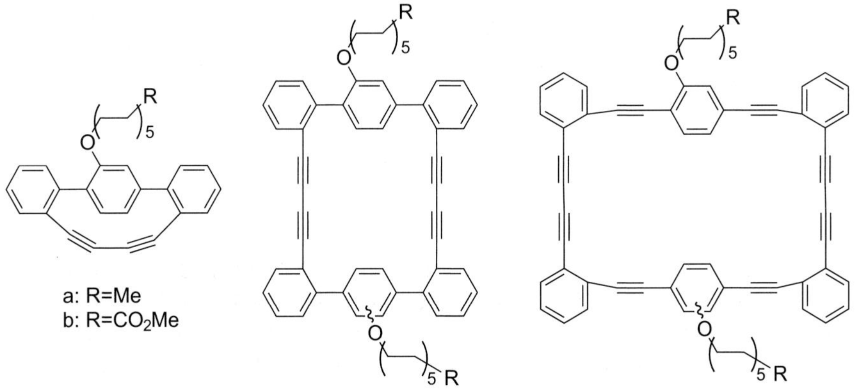

299 **300** **301**

Expanding on their previous work, the Fallis group constructed PDM **302**, containing *ortho* and *meta* linkages in the 60-carbon core [95]. Compound **302** was prepared in 15 % yield through in situ desilylation/dimerization of **303**, which was in turn obtained though a modified Cadiot–Chodkiewicz reaction between 1,3-bis-

(bromoethynyl)benzene and terminal acetylene **28** (Scheme 8.63). The flat cyclic representation of PDM **302** does not accurately depict its actual conformation. Computer modeling indicates that two different conformers are possible: the "bow-tie" macrocycle **302a** and the "butterfly" **302b**. The two isomers of **302** were produced by the cyclodimerization reaction in a 3:1 ratio and were separable by semipreparative HPLC. The major isomer, which according to its NMR data has higher symmetry than the minor isomer, is consistent with the "bow-tie" **302a**. Attempts to interconvert the two atropisomers at 100 °C were unsuccessful, which suggests the compounds are conformationally and configurationally stable and must have a high barrier to interconversion.

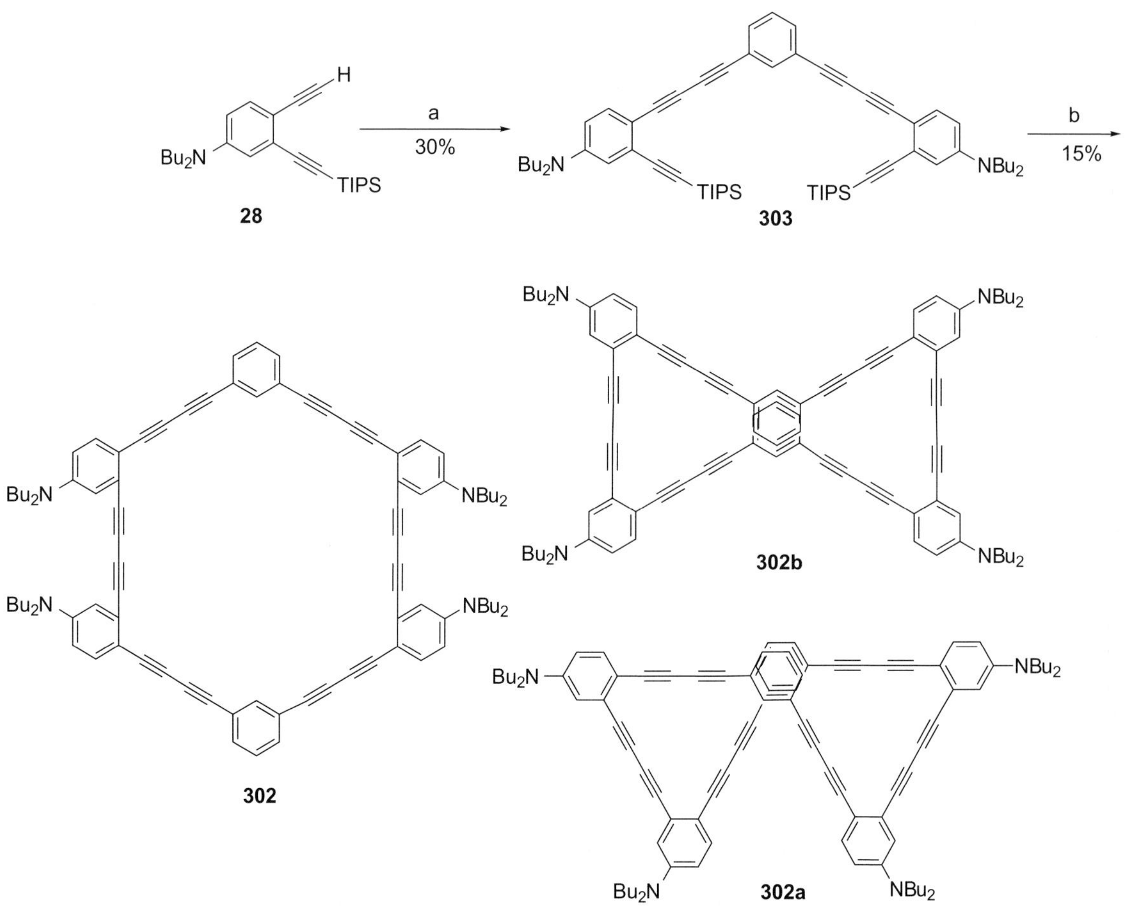

Scheme 8.63 Synthesis of PDM **302**. Reagents and conditions: (a) 1,4-bis(bromoethynyl)benzene, $[Pd_2(dba)_3]$, CuI, 1,2,2,6,6-pentamethylpiperidine, THF; (b) TBAF, $Cu(OAc)_2$, py, Et_2O.

Tsuji and co-workers have reported the synthesis of the *meta/para*-linked PDM **304** by photoirradiation of the Dewar benzene **305** [61] (Scheme 8.64). Compound **305** was prepared in five steps from the readily available diyne **306**. An X-ray structure of **304** shows the molecule to be nearly planar and two of the silane moieties residing within the sizeable molecular cavity (7.3 × 11.5 Å). The acetylene moieties deviate from linearity by 0.1–11.3°. The proton NMR spectrum reveals that the methylene protons are enantiotropic, which suggests that, in solution, the phenyl rings must rotate rapidly between rotamers **304a** and **304b**.

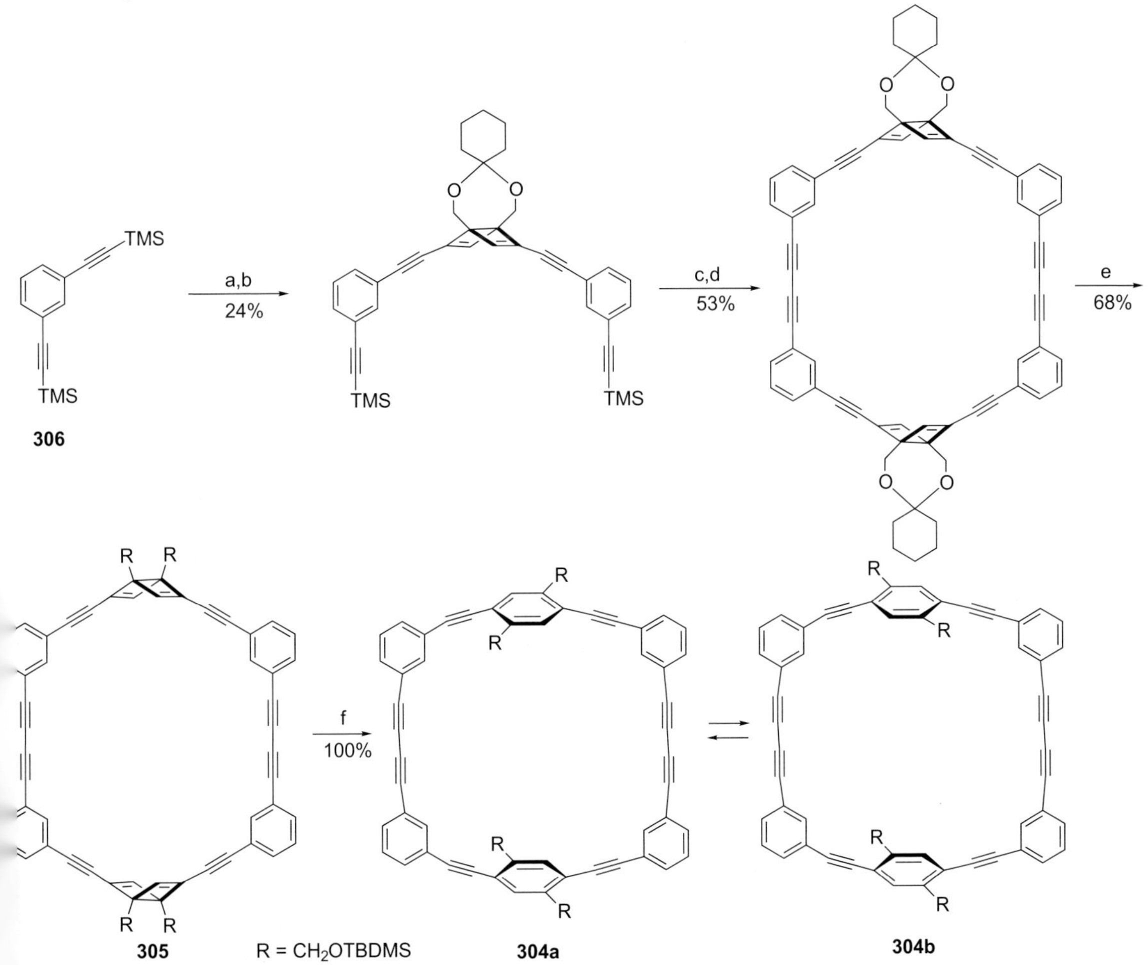

Scheme 8.64 Synthesis of PDM **304**. Reagents and conditions: (a) 1 N KOH, MeOH; (b) **174**, [Pd(PPh_3)$_4$], CuI, NEt_3; (c) TBAF, THF; (d) Cu(OAc)$_2$, CuCl, py, 60 °C, (e) i] 1 N HCl, THF, ii] TBDMSOTf, NEt_3; (f) *hν*, CH_2Cl_2, 12 °C.

The assembly of two unusual PDM-fused PDMs is depicted in Scheme 8.65 [96]. Both **307** and **308** were prepared from iodoarene **309**, synthesized in turn from the corresponding aniline in four steps by the chemistry shown in Scheme 8.23. Installation of the terminal alkyne by Stille reaction, followed by either cross-coupling or homocoupling, gave polyynes **310** and **311**, respectively. Desilylation of **310** and Pd-catalyzed homocoupling furnished PDM **307**, the first member of a previously unattainable class of π-extended fenestranes. The molecule exhibits remarkable stability and solubility, the latter due to a slight bowl shape (as predicted by calculations). Alternatively, desilylation of **311** and Cu-mediated homocoupling furnished PDM **308**. Interestingly, intramolecular homocoupling orients the molecule in such a way that intermolecular homocoupling now occurs. Whereas the protons on the outer phenyls of **307** show a slight shielding (δ 7.06, 7.03) due to the bent alkynes, the analogous protons on **308** exhibit a much more significant upfield shift (δ 6.94, 6.74) due to the bent alkynes together with the antiaromatic ring current of the two fused [12]annulenes.

Attempts to prepare three-dimensional PDM structures have met with varied success. As illustrated in Schemes 8.61–8.63, the distance between the terminal acetylenes in the homocoupling reaction can be used as a qualitative measure as to whether the reaction yields intra- or intermolecularly coupled products, or both. For example, the distance between the terminal acetylenes in **312** is less than 4.3 Å, and homocoupling thus yields highly strained **313** instead of **314** (Scheme 8.66) [97].

Expansion of the *meta*-terphenyl core with acetylene linkages, while increasing the spacing between the terminal acetylenes to 5.1 Å, is still insufficient for intermolecular reactivity. Tobe et al. discovered that tetrayne **315**, a simplified model for a 3D C_{48} precursor, furnishes intramolecular PDM **316**, analogously to reactivity shown by the C_{48} precursor (Scheme 8.67) [98].

The next higher homologue, in which diyne linkages connect the benzenes, was reported by Rubin in 1997 [99]. Once desilylated, the terminal acetylene carbons in **317** are over 10 Å apart, and so only the intermolecular PDM **318** was formed in the homocoupling reaction (Scheme 8.68). It was hoped that the molecule would "zip-up" to afford fullerene derivatives, but neither **318** nor its ethylenic analogue (benzene replaced by ethene) [100] possessed significant propensity to lose hydrogen and, as such, fullerenes were not obtained.

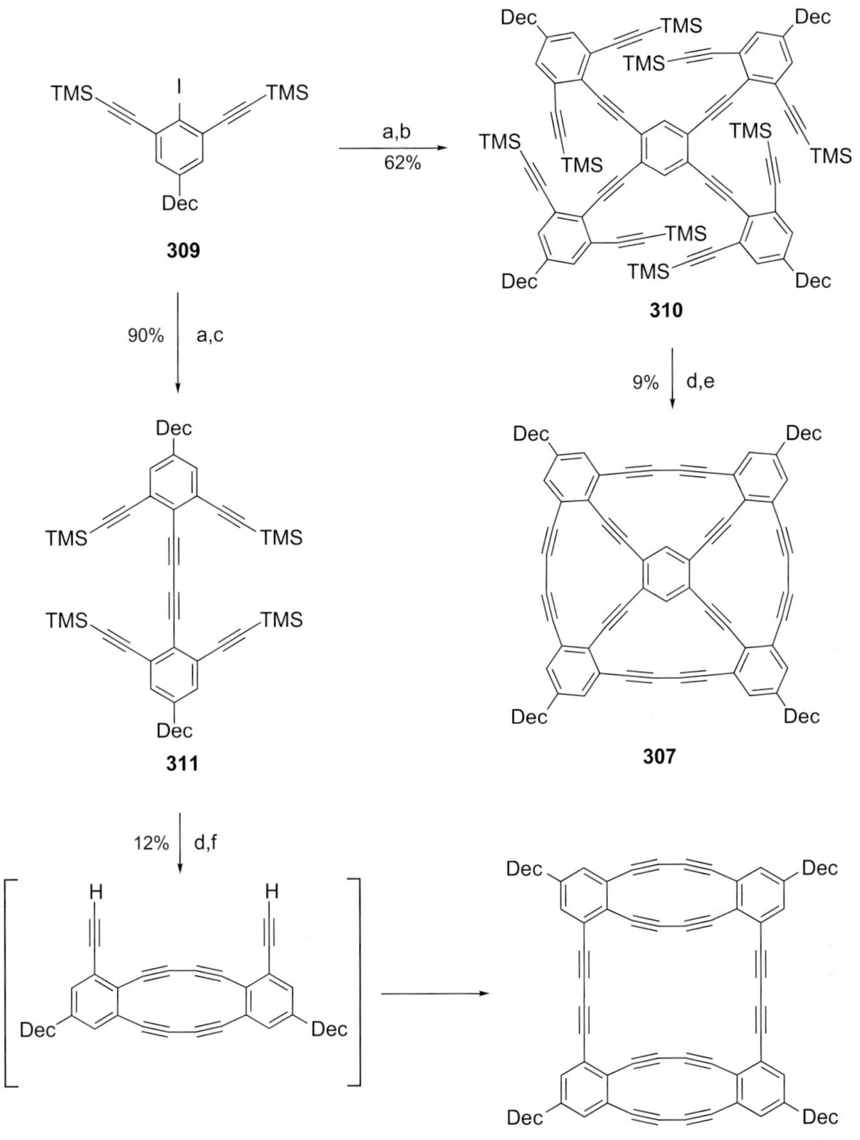

Scheme 8.65 Synthesis of PDMs **307** and **308**. Reagents and conditions: (a) $Bu_3SnC{\equiv}CH$, $[Pd(PPh_3)_4]$, THF, reflux; (b) 1,2,4,5-tetraiodobenzene, $[Pd(PPh_3)_4]$, CuI, *i*-Pr_2NH, THF, reflux; (c) $[PdCl_2(PPh_3)_2]$, CuI, I_2, *i*-Pr_2NH, THF; (d) TBAF, THF; (e) $[PdCl_2(dppe)]$, $[PdCl_2(PPh_3)_2]$, CuI, I_2, *i*-Pr_2NH, THF, 50 °C; (f) $Cu(OAc)_2$, CuCl, py, 50 °C.

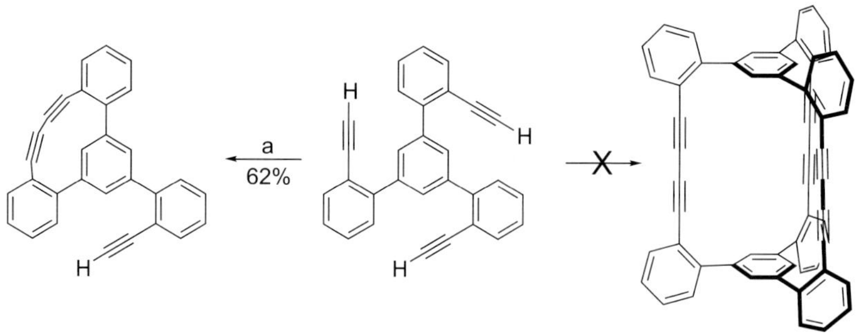

313 312 314

Scheme 8.66 Synthesis of PDM **313**. Reagents and conditions: (a) $Cu(OAc)_2$, py, Et_2O.

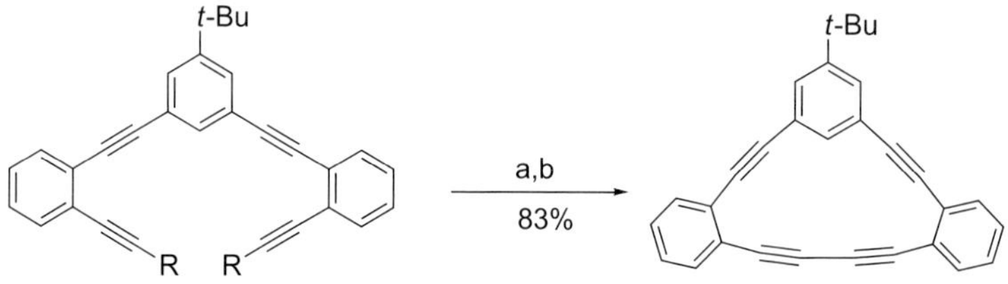

315
R = TMS

316

Scheme 8.67 Synthesis of PDM **316**. Reagents and conditions: (a) K_2CO_3, THF, MeOH; (b) $Cu(OAc)_2$, py, toluene.

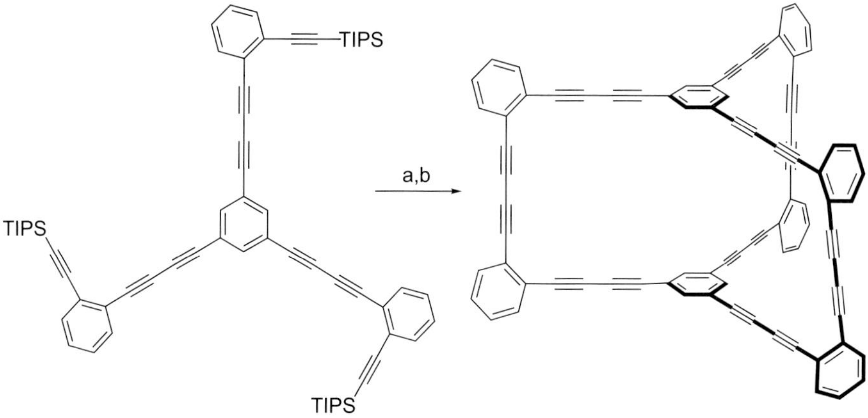

317 **318**

Scheme 8.68 Synthesis of PDM **318**. Reagents and conditions: (a) TBAF, THF; (b) CuCl, TMEDA, O_2, ODCB.

8.5 Phenyltriacetylene Macrocycles

Very recently, the Tobe group synthesized the simplest phenyltriacetylene macrocycles (PTrMs): **319** and **320** (Scheme 8.69) [101]. The two macrocycles were formed by loss of indan upon irradiation of precursors **321** and **322**, respectively, at 254 nm. The precursors were in turn generated from dichloride **323** by standard coupling/cyclization procedures. Highly reactive PTrM **319** could be generated and trapped in solution by Diels–Alder addition with furan in 20% yield. Alternatively, **319** could be generated and characterized in an Ar matrix at 20 K by FTIR and UV/Vis spectroscopy. PTrM **320** proved to be more stable and was characterized by ^{1}H NMR spectroscopy as a mixture with indan and **322** in deuterated THF.

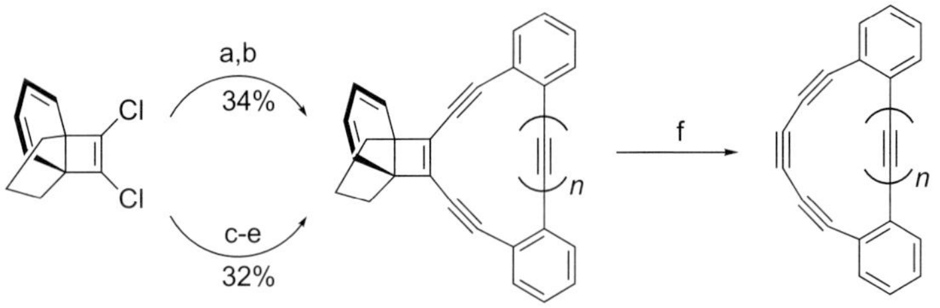

323 **321** (n=1) **322** (n=2) **319** (n=1) **320** (n=2)

Scheme 8.69 Synthesis of PTrMs **319** and **320**. Reagents and conditions: (a) 2-methyl-3-butyn-2-ol, [Pd(PPh$_3$)$_4$], CuI, BuNH$_2$; (b) **37**, [Pd(PPh$_3$)$_4$], CuI, PPh$_3$, NaOH, H$_2$O, CH$_3$N(C$_8$H$_{17}$)$_3$Cl, PhH, 80 °C; (c) 1-ethynyl-2-(triisopropylsilyl)ethynylbenzene, [Pd$_2$(dba)$_3$ · CHCl$_3$], CuI, piperidine, THF; (d) TBAF, EtOAc, THF; (e) Cu(OAc)$_2$, py; (f) *hν*.

PTrMs were initially reported by the Haley group in 1998 [70b, 102]. Because the assembly of such macrocycles through cyclooligomerization chemistry was not feasible, **324** was prepared by intramolecular acetylenic homocoupling (Scheme 8.70). The requisite building block **325** was prepared from **326** by selective desilylation and a subsequent modified Cadiot–Chodkiewicz coupling with TMSC≡CC≡CBr. In situ generation of the highly unstable phenylhexatriyne under Sonogashira conditions, followed by deprotection and Cu-mediated cyclization, gave **324**. PTrM **327** was also constructed by similar synthetic methods.

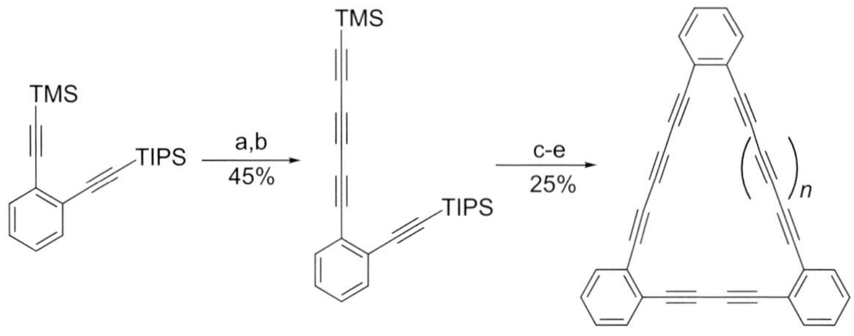

326 **325** **324** (n=1)
327 (n=0)

Scheme 8.70 Synthesis of PTrM **324**. Reagents and conditions: (a) K_2CO_3, MeOH; (b) i] BuLi, ii] CuBr, iii] TMSC≡CC≡CBr, py; (c) *o*-diiodobenzene, aq. KOH, $[Pd(PPh_3)_4]$, $[PdCl_2(PPh_3)_2]$, CuI, Et_3N, THF; (d) TBAF, EtOH; (e) $Cu(OAc)_2$, CuCl, py.

8.6
Phenyltetraacetylene Macrocycles

The first tetrayne-linked PTeMs (**328**/**329**) were prepared in 1997 (Scheme 8.71) [103]. Key to this success was the generation of the phenylbutadiynes in situ under Eglinton homocoupling conditions, yielding **330** and **331**. Subsequent removal of the TIPS groups and cyclization under pseudo-high-dilution conditions furnished PTeMs **328** and **329** in moderate overall yields. No intramolecular homocoupling was observed, as the terminal acetylenes are nearly 11 Å apart.

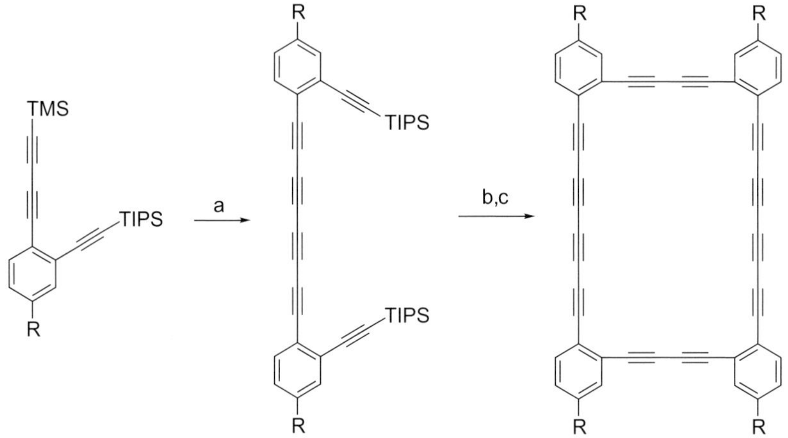

222 R=H
230 R=*t*-Bu

330 R=H (61%)
331 R=*t*-Bu (70%)

328 R=H (51%)
329 R=*t*-Bu (64%)

Scheme 8.71 Synthesis of PTeMs **328** and **329**. Reagents and conditions: (a) $Cu(OAc)_2 \cdot H_2O$, K_2CO_3, py, MeOH; (b) TBAF, EtOH, THF; (c) $Cu(OAc)_2$, CuCl, py.

The Fallis group prepared the dimeric PTeM **332** and the strained tetrayne **333** (Scheme 8.72) [104]. Sonogashira cross-coupling of **334** with 1,4-diethynylbenzene gave hexayne **335**. Once again, the distance between the terminal acetylenes, ca. 7.2 Å in the case of **335**, was an important factor. Generation of the phenylbutadiyne in situ with TBAF under Eglinton homocoupling conditions afforded only **333** (46 %) at low concentrations, whereas higher concentrations gave **332** (5 %) and **333** (26 %).

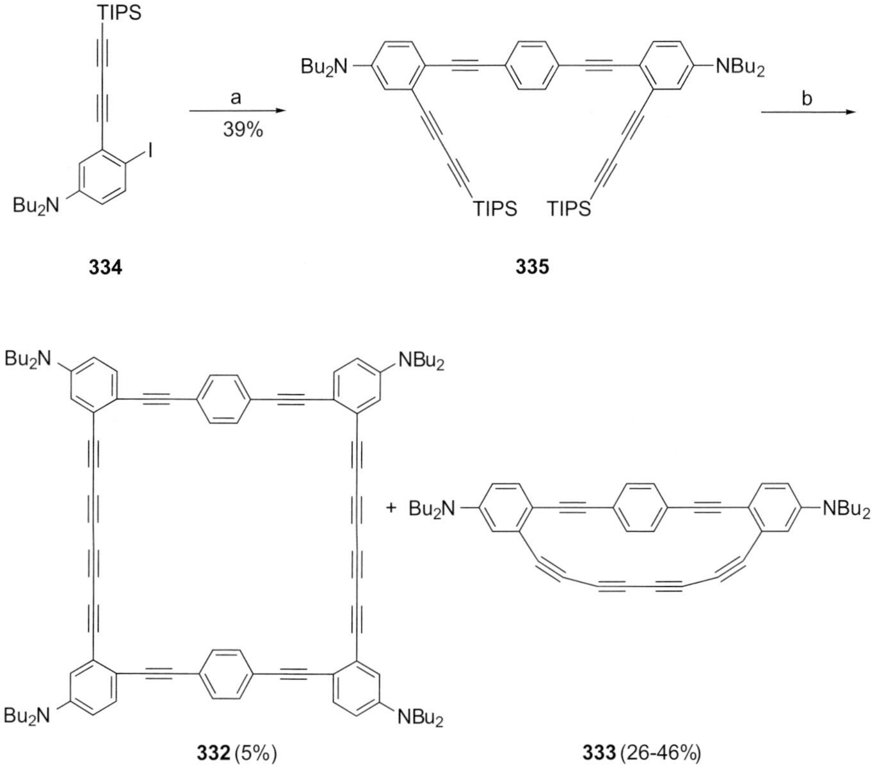

Scheme 8.72 Synthesis of PTeMs **332** and **333**. Reagents and conditions: (a) 1,4-diethynylbenzene, $[PdCl_2(PPh_3)_2]$, CuI, Et_3N, THF, Δ; (b) $Cu(OAc)_2$, TBAF, py, Et_2O.

A different route to tetra-donor PTeMs is shown in Scheme 8.73 [95]. In situ desilylation/homocoupling of triyne **249**, followed by removal of the TIPS groups, afforded hexayne **337**. Cross-coupling of **337** with 1-bromo-4-iodobenzene, subsequent addition of $[PdCl_2(PhCN)_2]$ and $P(t\text{-}Bu)_3$, and then dropwise addition of a second equivalent of **337** into the above mixture gave PTeM **336** in low (4 %) yield. The proton NMR spectrum of **336** reveals that the molecule adopts a highly symmetrical conformation in solution. Molecular modeling along with analysis of the NMR data suggests that **336** is C_2-symmetric.

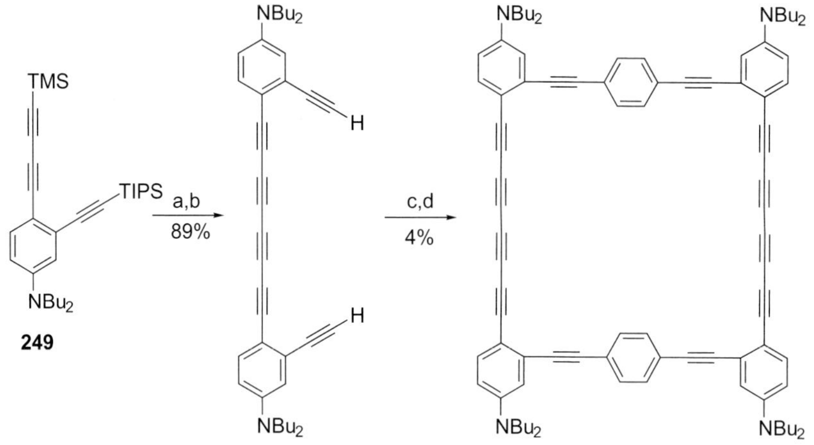

Scheme 8.73 Synthesis of PTeM **336**. Reagents and conditions: (a) K_2CO_3, $Cu(OAc)_2$, MeOH, py; (b) TBAF, THF; (c) 1-bromo-4-iodobenzene, $[PdCl_2(PPh_3)_2]$, CuI, Et_3N, THF, Δ; (d) **337**, $[PdCl_2(PhCN)_2]$, $P(t\text{-Bu})_3$.

An alternate approach to tetrayne-linked systems is the use of organometallic fragments to stabilize a strained annulene with the subsequent liberation of the hydrocarbon. With *sp* bond angles calculated to be around 162°, PTeM **338** would be highly strained and was therefore expected to be quite reactive [105]. The octacobalt complex **339**, on the other hand, should be readily isolable. Masked PTeM **339** was prepared easily from **340** in five steps, and was isolated as stable, deep maroon crystals (Scheme 8.74). All spectroscopic data supported formation of the strain-free dimeric structure. Unfortunately, all attempts to liberate **338** from the cobalt units led only to insoluble materials. Given the high reactivity of **319** and several other strained polyynes, it is likely that **338** is too unstable to be successfully isolated and characterized.

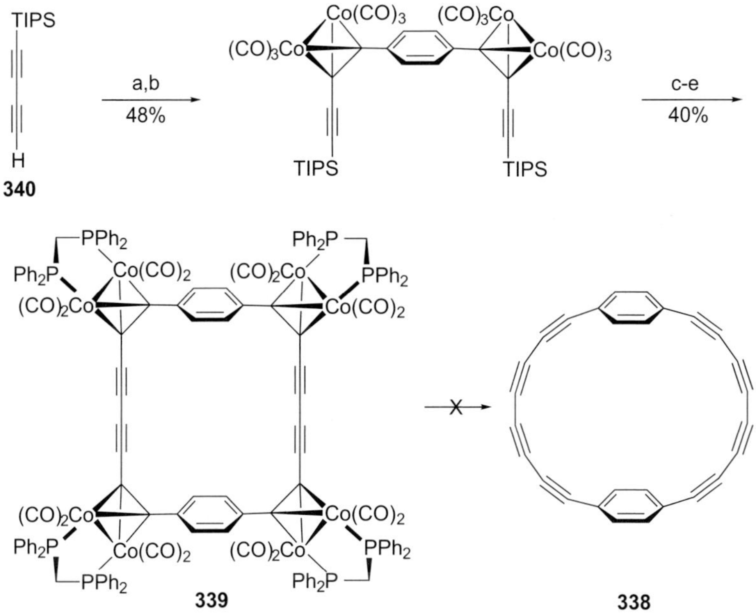

Scheme 8.74 Attempted synthesis of PTrM **338**. Reagents and conditions: (a) 1,4-diiodobenzene, $[PdCl_2(PPh_3)_2]$, CuI, Et_3N; (b) $Co_2(CO)_8$, Et_2O; (c) dppm, toluene; (d) TBAF, THF, EtOH; (e) $Cu(OAc)_2 \cdot H_2O$, py.

8.7
Phenyloligoacetylene Macrocycles

Some of the most exciting examples of macrocycles based on phenyl and acetylene units are the supercyclophanes **341–344**, which have been reported by both the Tobe [106] and the Rubin [107] groups. In 1998, both groups generated **343** in the mass spectrometer by laser desorption of the appropriate polyyne precursors and showed that these coalesce to C_{60}. Subsequent work by the Japanese team has shown that halogenated cyclophanes **342** and **344** are superior precursors to fullerenes, as the chlorine atoms are much easier to lose. Unfortunately, all attempts to generate fullerenes outside the mass spectrometer via these type of cyclophanes have been unsuccessful. The reader is referred to the excellent chapter by Tobe (Chapter 9) for a detailed discussion of this topic [108].

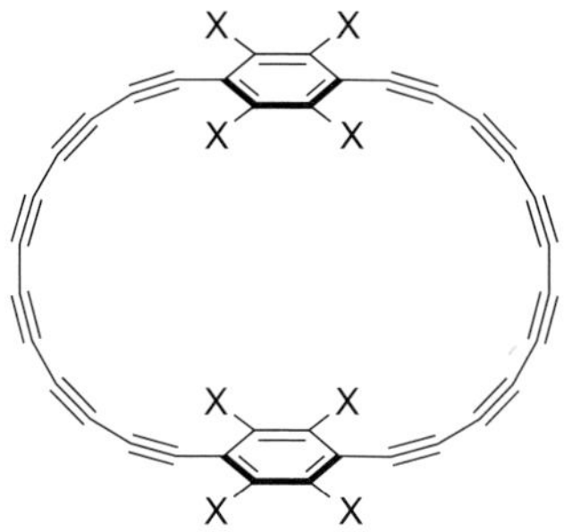

341 X=H
342 X=Cl

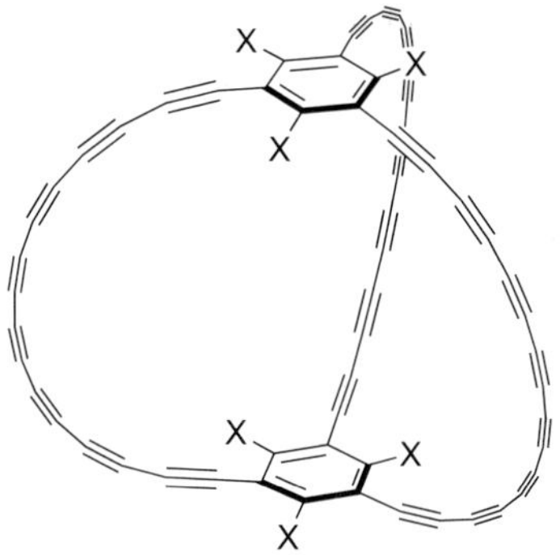

343 X=H
344 X=Cl

8.8 Conclusions

Although the groundwork for this field was laid over 45 years ago by Eglinton's pioneering studies, it has only been within the last 12–15 years that research on macrocycles comprised of phenyl and acetylene units has firmly established itself and grown at an amazing rate. Facile carbon-carbon bond formation through transition metal catalysis, in conjunction with the widespread availability of silyl-protected acetylenes, have made the assembly of complex macrocyclic structures a straightforward process. These synthetic developments now make it possible to tune and tailor the physiochemical properties of macrocycles, a definite first step if technological applications are to be achieved. Since the early 1990s a new generation of alkyne chemists have recognized that phenylacetylene structures could be designed with technologically significant applications in mind. Numerous phenylacetylene macrocycles have been assembled by utilization of many of the synthetic advances described in the previous pages, and have proven useful for diverse applications including liquid crystal displays (LCDs), supramolecular chemistry, self-assembly of nanostructures, nonlinear optical (NLO) devices, and all-carbon molecules and networks. Given the tremendous growth in this area in recent years, the future of phenylacetylene macrocycles appears to be rich with possibilities.

8.9 Experimental

8.9.1 Preparation of 8 From [(*t*-BuO)$_3$W≡C*t*-Bu)] Catalysis of 13

A 25 mL Schlenk flask was charged, under an atmosphere of N_2, with the propynylated arene **13** (0.25–0.35 mmol), [(*t*-BuO)$_3$W≡C*t*-Bu)] (20 mol%) and toluene (20 mL). The solution was stirred at 80 °C for 8 h. The solvent was removed in vacuo, and the residue was subjected to flash chromatography on silica (hexane/EtOAc elution) to give **8** (54%) [16].

8.9.2
Synthesis of 8 and 10 from Copper (2-Iodophenyl)acetylide

A 1 L flask was charged with copper(I) (2-iodophenyl)acetylide **9** (17.9 g, 0.0616 mol) and placed under an active vacuum to remove volatiles and to effect constant weight. Pyridine (500 mL) was added and the mixture was heated at reflux for 6.5 h. The pyridine was removed in vacuo, the crude product was extracted with Et_2O or acetone (1 L), and the insolubles were removed by filtration, after which the solvent was removed on the rotavapor. PAM **8** was sublimed at 160 °C on a diffusion-pump vacuum line (10^{-6} Torr) for approximately 48 h (yield 2.9 g, 48% based on the copper acetylide). Extraction of the residue left in the sublimator with acetone, followed by column chromatography (CH_2Cl_2/hexanes, 1:4 elution), gave **10** (0.5 g, 8%), together with a trace amount of **11** [13].

8.9.3
Preparation of 31 by Pd-catalyzed Cyclization of 29

TBAF (1 M in THF, 2.3 mL, 2.3 mmol) was added to a solution of octayne **29** (200 mg, 0.117 mmol) in THF (30 mL), and the system was stirred for 15 min. The solution was then diluted with Et_2O, washed with brine (4 × 50 mL), and dried with $MgSO_4$. The solution was then concentrated, rediluted to ca. 50 mL with THF, and added by syringe pump over 24 h to a solution of [$PdCl_2$(dppe)] (6.7 mg, 0.012 mmol), CuI (3.1 mg, 0.017 mmol), and I_2 (14.7 mg, 0.058 mmol) in THF (100 mL) and *i*-Pr_2NH (100 mL) at 55 °C. The reaction mixture was stirred for another 4 h at 55 °C after injection was complete. The mixture was then concentrated in vacuo, diluted with CH_2Cl_2/hexanes (1:1, 200 mL), and filtered through a small cake of silica gel. The crude material was triturated with hexanes and filtered to give **31** (105 mg, 84%) as an orange powder [24].

8.9.4
Preparation of 122 by Pd-mediated Cyclization of 136

A Schlenk flask charged with PPh_3 (0.09 g, 0.34 mmol), [$Pd(dba)_2$] (0.03 g, 0.052 mmol), and CuI (0.015 g, 0.076) was evacuated and back-filled three times with N_2. Dry NEt_3 (25 mL) was added to the flask. While the mixture was stirred under N_2 at 75 °C, a solution of **136** (0.281 g, 0.226 mmol) in dry NEt_3 (25 mL) and dry benzene (25 mL) was added to the flask by syringe pump at a rate of ~2.5 mL h^{-1}. After the addition was complete, the solvent was removed with a rotary evaporator. The reaction mixture was then subjected to flash chromatography (first eluted with a 1:3 mixture, then a 1:2 mixture of CH_2Cl_2/hexane) followed by precipitation with MeOH to give **122** (0.191 g, 76%) [43].

8.9.5
Synthesis of 148 from 149 and $Mo(CO)_6$

Diyne **149** (2 g, 9.52 mmol), $Mo(CO)_6$ (126 mg, 0.477 mmol), and 4-chlorophenol (1.23 g, 9.57 mmol) were dissolved in ODCB (50 mL) and heated for 16 h at 150 °C under a light stream of N_2. The reaction mixture was washed with water, dilute HCl, and dilute NaOH, followed by precipitation into MeOH. Thick layer chromatography (CH_2Cl_2/hexanes, 1:6) gave a mixture of **148** and the heptamer (2.3 mg, 0.5 %) [45].

8.9.6
Preparation of 189 and 190 from 1,2-Diiodotetrafluorobenzene under Hay Conditions

A solution of CuCl (309 mg, 3.11 mmol) and TMEDA (19 mL) in ODCB (200 mL) was saturated with O_2 by bubbling the gas though the solution. A solution of 1,2-diethynyl-3,4,5,6-tetrafluorobenzene (490 mg, 2.47 mmol) in ODCB (200 mL) was then added dropwise over a period of 1 h. The mixture was stirred for 5 h with O_2 bubbling and overnight under air. The reaction was quenched with HCl (0.5 N, 200 mL), and the organic layer was extracted with $CHCl_3$, washed with aqueous $NaHCO_3$, and dried over $MgSO_4$. The combined organic layers were evaporated under reduced pressure (~ 20 mmHg) to remove the $CHCl_3$ and then at 0.11 mmHg to remove the ODCB. The crude mixture was separated with preparative GPC to give crude **189** (233 mg, 0.59 mmol, 48 %) and **190** (78 mg, 0.13 mmol, 16 %). The products were recrystallized from $CHCl_3$ for **189** and from benzene for **190** to give pure **189** and **190** as yellow crystals [67].

8.9.7
Preparation of 1 by Deprotection and Cyclization of 223

EtOH (10–20 drops) and TBAF (1 M in THF, 1.21 mL, 1.21 mmol) were added with stirring at rt. to the silyl-protected substrate **223** (400 mg, 0.58 mmol) dissolved in THF (15 mL). The reaction was monitored by TLC. Upon completion, the mixture was diluted with Et_2O, washed three times with water and twice with brine, and dried over $MgSO_4$. After filtration though a short pad of silica gel and removal of the solvent, the resulting product was dissolved in a small volume of pyridine and was used immediately in the next step. The deprotected acetylene was added by syringe pump to a suspension of CuCl (175 mg, 1.76 mmol) and Cu $(OAc)_2 \cdot H_2O$ (320 mg, 1.6 mmol) in pyridine (145 mL) over 16–20 h at 60 °C. Upon completion, the reaction mixture was concentrate in vacuo and extracted with CH_2Cl_2. The organic layer was subsequently washed with dilute HCl solution and several times with water. The organic layer was dried ($MgSO_4$), filtered, and evaporated. Purification by chromatography on silica gel (CH_2Cl_2), followed by recrystallization in hot CH_2Cl_2, gave **1** as a yellow solid (76 mg, 35 %) [77].

8.9.8
Synthesis of 304 by Photolysis of Dewar Benzene 305

A solution of **305** (11 mg, 0.009 mmol) in CH_2Cl_2 (25 mL) was placed in a Pyrex test tube (18 mm × 18 cm), bubbled with Ar for 15 min at 0 °C, and irradiated with a high-pressure Hg lamp at 12 °C. The reaction was monitored by HPLC, which showed the formation of a single product. After 30 min (100 % conversion), the irradiation was terminated, and the mixture was evaporated to give **304** (11 mg, 100 %) as a colorless powder [61].

8.9.9
Preparation of 332 and 333 by Deprotection/Cyclization of 335 in situ

TBAF (1 M, 1.19 mL) was added over 2 h by syringe pump to a stirred solution of compound **335** (508 mg, 0.54 mmol) and $Cu(OAc)_2$ (590 mg, 3.23 mmol) in pyridine/Et_2O (3:1, 100 mL). Once the addition was complete, the solution was poured into Et_2O and HCl (1 M). The organic phase was washed thoroughly with HCl (1 M) until all pyridine had been removed, and the organic phase was dried and concentrated to yield a crude solid. Column chromatography (petroleum ether/CH_2Cl_2, 2:1) yielded **332** (16 mg, 5 %) as an orange solid and **333** (155 mg, 46 %) as a red solid [104].

Acknowledgments

We thank the National Science Foundation, the Petroleum Research Fund (administered by the American Chemical Society), and the Camille and Henry Dreyfus Foundation (Camille Dreyfus Teacher-Scholar Award to MMH) for past and present support of our research in phenylacetylene chemistry. We are grateful to the numerous contributions of our co-workers whose names appear in the Haley group references. We also thank Prof. A. Fallis, Prof. M. Iyoda, and Prof. Y. Tobe for sharing relevant results prior to publication.

References

[1] *Modern Acetylene Chemistry*, P. J. Stang, F. Diederich (eds.), VCH, Weinheim, **1995**; (b) *Acetylene Chemistry – Chemistry, Biology, and Materials Science*, F. Diederich, R. R. Tykwinski, and P. J. Stang (eds.), Wiley-VCH, Weinheim, **2005**.

[2] *Metal-Catalyzed Cross-Coupling Reactions*, F. Diederich, P. J. Stang (eds.), Wiley-VCH, Weinheim, **1998**; (b) *Transition Metal Catalyzed Reactions-IUPAC Monographs Chemistry for the 21st Century*, S. G. Davies, S. Murahashi (eds.), Blackwell Science, Oxford, **1998**;

(c) *Handbook of Organopalladium Chemistry for Organic Synthesis* (Ed.: E. Negishi), Wiley, New York, **2002**; (d) *Metal-Catalyzed Cross Coupling Reactions*, 2nd Ed. A. de Meijere, F. Diederich (eds.), Wiley-VCH, Weinheim, **2004**.

[3] J. A. Marsden, M. M. Haley, in reference 2d; pp 317–394; (b) K. Sonogashira, *J. Organomet. Chem.* **2002**, *653*, 46–49; (c) E. Negishi, L. Anastasia, *Chem. Rev.* **2003**, *103*, 1979–2017

[4] P. Siemsen, R. C. Livingston, F. Diederich, *Angew. Chem.* **2000**, *112*, 2740–2767; *Angew. Chem. Int. Ed.* **2000**, *39*, 2632–2657.

[5] M. M. Haley, J. J. Pak, S. C. Brand, *Top. Curr. Chem.* **1999**, *201*, 81–129; (b) U. H. F. Bunz, Y. Rubin, Y. Tobe, *Chem. Soc. Rev.* **1999**, *28*, 107–119; (c) D. Zhao, J. S. Moore, *Chem. Commun.* **2003**, 807–818; (d) J. A. Marsden, G. J. Palmer, M. M. Haley, *Eur. J. Org. Chem.* **2003**, 2355–2369.

[6] C. Grave, A. D. Schlüter, *Eur. J. Org. Chem.* **2002**, 3075–3098; (b) Y. Yamaguchi, Z. Yoshida, *Chem. Eur. J.* **2003**, *9*, 5430–5440; (c) U. H. F. Bunz, *J. Organomet. Chem.* **2003**, *683*, 269–287.

[7] G. Eglinton, A. R. Galbraith, *Proc. Chem. Soc.* **1957**, 350–351.

[8] O. M. Behr, G. Eglinton, R. A. Raphael, *Chem. Ind.* **1959**, 699–700; (b) O. M. Behr, G. E. Eglinton, A. R. Galbraith, R. A. Raphael, *J. Chem. Soc.* **1960**, 3614–3625.

[9] W. K. Grant, J. C. Speakman, *Proc. Chem. Soc.* **1959**, 231.

[10] Q. Zhou, P. C. Carroll, T. M. Swager, *J. Org. Chem.* **1994**, *59*, 1294–1301.

[11] D. Campbell, G. Eglinton, W. Henderson, R. A Raphael, *J. Chem. Soc., Chem. Commun.* **1966**, 87–89.

[12] C. E. Castro, R. D. Stephens, *J. Org. Chem.* **1963**, *28*, 2163; (b) A. M. Sladkov, I. R. Golding, *Russ. Chem. Rev.* **1979**, *48*, 868–896.

[13] D. Solooki, J. D. Ferrara, D. Malaba, J. D. Bradshaw, C. A Tessier, W. J. Youngs, *Inorg. Synth.* **1997**, *31*, 122–128.

[14] C. Huynh, G. Linstrumelle, *Tetrahedron* **1988**, *44*, 6337–6344.

[15] S. Pham, M. M. Haley, unpublished observations.

[16] O. S. Miljanic, K. P. C. Vollhardt, G. D. Whitener, *Synlett* **2003**, 29–34.

[17] H. A. Staab, F. Graf, *Tetrahedron Lett.* **1966**, 751–757.

[18] M. Iyoda, A. Vorasingha, Y. Kuwatani, M. Yoshida, *Tetrahedron Lett.* **1998**, *39*, 4701–4704.

[19] K. P. Baldwin, J. D. Bradshaw, C. A. Tessier, W. J. Youngs, *Synlett* **1993**, 853–855; (b) K. P. Baldwin, R. S. Simons, J. Rose, P. Zimmerman, D. M. Hercules, C. A. Tessier, W. J. Youngs, *Chem. Commun.* **1994**, 1257–1258.

[20] J. M. Kehoe, J. H. Kiley, J. J. English, C. A. Johnson, R. C. Petersen, M. M. Haley, *Org. Lett.* **2000**, *2*, 969–972.

[21] O. M. Behr, G. Eglinton, I. A. Lardy, R. A. Raphael, *J. Chem. Soc.* **1964**, 1151–1154.

[22] L. Guo, J. D. Bradshaw, C. A. Tessier, W. J. Youngs, *Chem. Commun.* **1994**, 243–244.

[23] K. P. Baldwin, A. J. Matzger, D. A. Scheiman, C. A. Tessier, K. P. C. Vollhardt, W. J. Youngs, *Synlett* **1995**, 1215–1218.

[24] J. A. Marsden, J. J. Miller, M. M. Haley, *Angew. Chem.* **2004**, *116*, 1726–1729; *Angew. Chem. Int. Ed.* **2004**, *43*, 1694–1697.

[25] H. N. C. Wong, P. J. Garratt, F. Sondheimer, *J. Am. Chem. Soc.* **1974**, *96*, 5604–5605.

[26] S. Chaffins, M. Brettreich, F. Wudl, *Synthesis* **2002**, 1191–1194; (b) A. Orita, D. Hasegawa, T. Nakano, J. Otera, *Chem. Eur. J.* **2002**, *8*, 2000–2004.

[27] M. Iyoda, S. Sirinintasak, Y. Nishiyama, A. Vorasingha, F. Sultana, K. Nakao, Y. Kuwatani, H. Matsuyama, M. Yoshida, Y. Miyake, *Synthesis* **2004**, 1527–1531. (b) D. Zhang, C. A. Tessier, W. J. Youngs, *Chem. Mater.* **1999**, *11*, 3050–3057.

[28] C. Eickmeier, H. Junga, A. J. Matzger, F. Scherhag, M. Shim, K. P. C. Vollhardt, *Angew. Chem.* **1997**, *109*, 2194–2199; *Angew. Chem. Int. Ed.*

Engl. **1997**, *36*, 2103–2108;
(b) A. J. Matzger, M. Shim, K. P. C. Vollhardt, *Chem. Commun.* **1999**, 1871–1872.

[29] J. D. Kinder, C. A. Tessier, W. J. Youngs, *Synlett* **1993**, 149–150.

[30] Djebli, J. D. Ferrara, C. Tessier-Youngs, W. J. Youngs, *J. Chem. Soc., Chem. Commun.* **1988**, 548–549.

[31] J. D. Ferrara, C. Tessier-Youngs, W. J. Youngs, *J. Am. Chem. Soc.* **1985**, *107*, 6719–6721;
(b) J. D. Ferrara, A. A. Tanaka, C. Fierro, C. Tessier-Youngs, W. J. Youngs, *Organometallics* **1989**, *8*, 2089–2098;
(c) W. J. Youngs, J. D. Kinder, J. D. Bradshaw, C. A. Tessier, *Organometallics* **1993**, *12*, 2406–2407.

[32] J. D. Ferrara, C. Tessier-Youngs, W. J. Youngs, *Organometallics* **1987**, *6*, 676–678;
(b) J. D. Ferrara, C. Tessier-Youngs, W. J. Youngs, *Inorg. Chem.* **1988**, *27*, 2201–2202.

[33] J. D. Ferrara, A. Djebli, C. Tessier-Youngs, W. J. Youngs, *J. Am. Chem. Soc.* **1988**, *110*, 647–649.

[34] M. Ohkoshi, T. Horino, M. Yoshida, M. Iyoda, *Chem. Commun.* **2003**, 2586–2587.

[35] W. J. Youngs, A. Djebli, C. A. Tessier, *Organometallics* **1991**, *10*, 2089–2090;
(b) D. Malaba, A. Djebli, L. Chen, E. A. Zarate, C. A. Tessier, W. J. Youngs, *Organometallics* **1993**, *12*, 1266–1276.

[36] R. H. Baughman, H. Eckhardt, M. Kertesz, *J. Chem. Phys.* **1987**, *87*, 6687–6699;
(b) N. Narita, S. Nagai, S. Suzuki, K. Nakao, *Phys. Rev. B* **1998**, *58*, 11009–11014.

[37] Y. Zhou, S. Feng, *Solid State Commun.* **2002**, *122*, 307–310.

[38] M. Sonoda, Y. Sakai, T. Yoshimura, Y. Tobe, K. Kamada, *Chem. Lett.* **2004**, *33*, 972–973.

[39] J. D. Bradshaw, D. Solooki, C. A. Tessier, W. J. Youngs, *J. Am. Chem. Soc.* **1994**, *116*, 3177–3179;
(b) D. Solooki, J. D. Bradshaw, C. A. Tessier, W. J. Youngs, R. F. See, M. Churchill, J. D. Ferrara, *J. Organomet. Chem.* **1994**, *470*, 231–236.

[40] T. Kawase, N. Ueda, M. Oda, *Tetrahedron Lett.* **1997**, *38*, 6681–6684.

[41] T. Kawase, N. Ueda, H. R. Darabi, M. Oda, *Angew. Chem.* **1996**, *108*, 1658–1660; *Angew. Chem. Int. Ed. Engl.* **1996**, *35*, 1556–1558.

[42] T. Kawase, Y. Hosokawa, H. Kurata, M. Oda, *Chem. Lett.* **1999**, 745–746.

[43] J. Zhang, D. J. Pesak, J. L Ludwick, J. S. Moore, *J. Am Chem. Soc.*, **1994**, *116*, 4227–4239.

[44] H. A. Staab, K. Neunhoeffer, *Synthesis* **1974**, 424.

[45] P.-H. Ge, W. Fu, W. A. Herrmann, E. Herdtweck, C. Campana, R. D. Adams, U. H. F. Bunz, *Angew. Chem.* **2000**, *112*, 3753–3756; *Angew. Chem. Int. Ed.* **2000**, *39*, 3607–3610.

[46] J. S. Moore, *Acc. Chem. Res.* **1997**, *30*, 402–413;
(b) J. Zhang, J. S. Moore, Z. Xu, R. Aguirre, *J. Am. Chem. Soc.* **1992**, *114*, 2273–2274;
(c) A. S. Shetty, P. R. Fischer, K. F. Stork, P. W. Bohn, J. S. Moore, *J. Am. Chem. Soc.* **1996**, *118*, 9409–9414;
(d) S. Lahiri, J. L. Thompson, J. S. Moore, *J. Am. Chem. Soc.* **2000**, *122*, 11 315–11319.

[47] S. Shetty, J. Zhang, J. S. Moore, *J. Am. Chem. Soc.* **1996**, *118*, 1019–1027.

[48] J. Zhang, J. S. Moore, *J. Am. Chem. Soc.* **1992**, *114*, 9701–9702.

[49] J. Zhang, J. S. Moore, *J. Am. Chem. Soc.* **1994**, *116*, 2655–2656.

[50] S. Höger, V. Enkelmann, K. Borad, C. Tschierske, *Angew. Chem.* **2000**, *112*, 2355–2358; *Angew. Chem. Int. Ed.* **2000**, *39*, 2267–2270.

[51] Y. Hosokawa, T. Kawase, M. Oda, *Chem. Commun.* **2001**, 1948–1949.

[52] O. S. Pyun, W. Yang, M.-J. Jeong, S. H. Lee, K. M. Kang, S.-J. Jeon, B. R. Cho, *Tetrahedron Lett.* **2003**, *44*, 5179–5182.

[53] Z. Wu, S. Lee, J. S. Moore, *J. Am. Chem. Soc.* **1992**, *114*, 8730–8732.

[54] Z. Wu, J. S. Moore, *Angew. Chem.* **1996**, *108*, 320–322; *Angew. Chem. Int. Ed. Engl.* **1996**, *35*, 297–299.

[55] T. Kawase, H. R. Darabi, M. Oda, *Angew. Chem.* **1996**, *108*, 2803–2805; *Angew. Chem. Int. Ed. Engl.* **1996**, *35*, 2664–2666.

[56] T. Kawase, N. Ueda, K. Tanaka, Y. Seirai, M. Oda, *Tetrahedron Lett.* **2001**, *42*, 5509–5511.

[57] T. Kawase, Y. Seirai, H. R. Darabi, M. Oda, Y. Sarakai, K. Tashiro, *Angew. Chem.* **2003**, *115*, 1659–1662; *Angew. Chem. Int. Ed.* **2003**, *42*, 1621–1624.

[58] T. Kawase. K. Tanaka, N. Fujiwara, H. R. Darabi, M. Oda, *Angew. Chem.* **2003**, *115*, 1662–1666; *Angew. Chem. Int. Ed.* **2003**, *42*, 1624–1628.

[59] T. Kawase, K. Tanaka, N. Shiono, Y. Seirai, M. Oda, *Angew. Chem.* **2004**, *116*, 1754–1756; *Angew. Chem. Int. Ed.* **2004**, *43*, 1722–1724.

[60] T. Kawase, K. Tanaka, Y. Seirai, N. Shiono, M. Oda, *Angew. Chem.* **2003**, *115*, 5755–5758; *Angew. Chem. Int. Ed.* **2003**, *42*, 5597–5560.

[61] M. Ohkita, K. Ando, T. Suzuki, T. Tsuji, *J. Org. Chem.* **2000**, *65*, 4385–4390.

[62] J. K. Young, J. S Moore, in reference [1a], p 415.

[63] J. S. Moore, J. Zhang, *Angew. Chem.* **1992**, *104*, 873–875; *Angew. Chem., Int. Ed. Engl.* **1992**, *31*, 922–924.

[64] T. C. Bedard, J. S. Moore, *J. Am. Chem. Soc.* **1995**, *117*, 10 662–10671.

[65] G. Wegner, *Z. Naturforsch. B* **1969**, *24*, 824–832;
(b) V. Enkelmann, *Adv. Polym. Sci.* **1984**, *63*, 91–136.

[66] U. H. F. Bunz, V. Enkelmann, *Chem. Eur. J.* **1999**, *5*, 263–266.

[67] T. Nishinaga, N. Nodera, Y. Miyata, K. Komatsu, *J. Org. Chem.* **2002**, *67*, 6091–6096.

[68] G. W. Coates, A. R. Dunn, L. M. Henling, D. A. Dougherty, R. H. Grubbs, *Angew. Chem.* **1997**, *109*, 290–293; *Angew. Chem., Int. Ed. Engl.* **1997**, *36*, 248–251.

[69] R. Boese, A. J. Matzger, K. P. C. Vollhardt, *J. Am. Chem. Soc.* **1997**, *119*, 2052–2053.

[70] M. M. Haley, M. L. Bell, J. J. English, C. A. Johnson, T. J. R. Weakley, *J. Am. Chem. Soc.* **1997**, *119*, 2956–2957;
(b) M. L. Bell, R. C. Chiechi, C. A. Johnson, D. B. Kimball, A. J. Matzger, W. B. Wan, T. J. R. Weakley, M. M. Haley *Tetrahedron* **2001**, *57*, 3507–3520.

[71] S. Ott, R. Faust, *Chem. Commun.* **2004**, 388–389;
(b) S. Ott, R. Faust, *Synlett* **2004**, 1509–1512.

[72] M. E. Gallagher, J. E. Anthony, *Tetrahedron Lett.* **2001**, *42*, 7533–7536.

[73] P. I. Dosa, C. Erben, V. S. Iyer, K. P. C. Vollhardt, I. M. Wasser, *J. Am. Chem. Soc.* **1999**, *121*, 10430–10431;
(b) M. Laskoski, W. Steffen, J. G. M. Morton, M. D. Smith, U. H. F. Bunz, *J. Am. Chem. Soc.* **2002**, *124*, 13814–13818;
(c) V. S. Iyer, K. P. C. Vollhardt, R. Wilhelm, *Angew. Chem.* **2003**, *115*, 4515–4519; *Angew. Chem. Int. Ed.* **2003**, *42*, 4379–4383.

[74] J. D. Tovar, N. Jux, T. Jarrosson, S. I. Khan, Y. Rubin, *J. Org. Chem.* **1997**, *62*, 3432–3433;
(b) corrections: *J. Org. Chem.* **1997**, *62*, 5656 and **1998**, *63*, 4856.

[75] W. B. Wan, M. M. Haley, in *Advances in Strained and Interesting Molecules*, B. Halton (ed.), JAI Press, Stamford, CT, **2000**, pp 1–41.

[76] M. M. Haley, S. C. Brand, J. J. Pak, *Angew. Chem.* **1997**, *109*, 863–866; *Angew. Chem. Int. Ed. Engl.* **1997**, *36*, 835–838.

[77] W. B. Wan, S. C. Brand, J. J. Pak, M. M. Haley, *Chem. Eur. J.* **2000**, *6*, 2044–2052.

[78] W. B. Wan, M. M. Haley, *J. Org. Chem.* **2001**, *66*, 3893–3901.

[79] J. A. Marsden, M. M. Haley, unpublished results.

[80] J. J. Pak, T. J. R. Weakley, M. M. Haley, *J. Am. Chem. Soc.* **1999**, *121*, 8182–8192.

[81] Sarkar, J. J. Pak, G. W. Rayfield, M. M. Haley, *J. Mater. Chem.* **2001**, *11*, 2943–2945.

[82] D. B. Kimball, A. J. Boydston, M. M. Haley, unpublished results.

[83] Y. Tobe, N. Utsumi, K. Kawabata, K. Naemura, *Tetrahedron Lett.* **1996**, *37*, 9325–9328;
(b) Y. Tobe, N. Utsumi, K. Kawabata, A. Nagano, K. Adachi, S. Araki, M. Sonoda, K. Hirose, K. Naemura, *J. Am. Chem. Soc.* **2002**, *124*, 5350–5364.

[84] Y. Tobe, N. Utsumi, A. Nagano, M. Sonoda, K. Naemura, *Tetrahedron* **2001**, *57*, 8075–8083.

[85] Y. Tobe, N. Utsumi, A. Nagano, K. Naemura, *Angew. Chem.* **1998**, *110*, 1347–1349; *Angew. Chem. Int. Ed.* **1998**, *37*, 1285–1287.

[86] M. Mayor, J.-M. Lehn, *J. Am. Chem. Soc.* **1999**, *121*, 11231–11232.

[87] M. Ohkita, K. Ando, T. Tsuji, *Chem. Commun.* **2001**, 2570–2571.

[88] S. Höger, in *Acetylene Chemistry – Chemistry, Biology, and Materials Science*, F. Diederich, R. R. Tykwinski, and P. J. Stang (eds.), Wiley-VCH, Weinheim, **2005**.

[89] S. Höger, V. Enkelmann, *Angew. Chem.* **1995**, *107*, 2917–2919; *Angew. Chem., Int. Ed. Engl.* **1995**, *34*, 2713–2716;
(b) S. Höger, A.-D. Meckenstock, S. Müller, *Chem. Eur. J.* **1998**, *4*, 2423–2434.

[90] D. L. Morrison, S. Höger, *Chem. Commun.* **1996**, 2313–2314.

[91] S. Höger, K. Bonrad, L. Karcher, A.-D. Meckenstock, *J. Org. Chem.* **2000**, *65*, 1588–1589.

[92] S. Höger, A.-D. Meckenstock, H. Pellen *J. Org. Chem.* **1997**, *62*, 4556–4557;
(b) S. Höger, A.-D. Meckenstock, *Tetrahedron Lett.* **1998**, *39*, 1735–1736;
(c) S. Höger, A.-D. Meckenstock, *Chem. Eur. J.* **1999**, *5*, 1686–1691.

[93] S. K. Collins, G. P. A. Yap, A. G. Fallis, *Angew. Chem.* **2000**, *112*, 393–396; *Angew. Chem. Int. Ed.* **2000**, *39*, 385–388.

[94] S. K. Collins, G. P. A. Yap, A. G. Fallis, *Org. Lett.* **2000**, *2*, 3189–3192.

[95] M. A. Heuft, S. K. Collins, A. G. Fallis, *Org. Lett.* **2003**, *5*, 1911–1914.

[96] J. A. Marsden, M. J. O'Connor, M. M. Haley, *Org. Lett.* **2004**, *6*, 2385–2388.

[97] S. K. Collins, G. P. A. Yap, A. G. Fallis, *Org. Lett.* **2002**, *4*, 11–14.

[98] Y. Tobe, J. Kishi, I. Ohki, M. Sonoda, *J. Org. Chem.* **2003**, *68*, 3330–3332.

[99] T. C. Parker, S. I. Khan, C. L. Holliman, S. W. McElvany, Y. Rubin, unpublished results; cited in: Y. Rubin, *Chem. Eur. J.* **1997**, *3*, 1009–1016.

[100] Y. Rubin, T. C. Parker, S. I. Khan, C. L. Holliman, S. W. McElvany, *J. Am. Chem. Soc.* **1996**, *118*, 5308–5309.

[101] Y. Tobe, I. Ohki, M. Sonoda, H. Niino, T. Sato, T. Wakabayashi, *J. Am. Chem. Soc.* **2003**, *125*, 5614–5615;
(b) I. Hisaki, T. Eda, M. Sonoda, Y. Tobe, *Chem. Lett.* **2004**, *33*, 620–621.

[102] W. B. Wan, D. B. Kimball, M. M. Haley, *Tetrahedron Lett.* **1998**, *39*, 6795–6798.

[103] M. M. Haley, M. L. Bell, S. C. Brand, D. B. Kimball, J. J. Pak, W. B. Wan, *Tetrahedron Lett.* **1997**, *38*, 7483–7486.

[104] M. A. Heuft, S. K. Collins, G. P. A. Yap, A. G. Fallis, *Org. Lett.* **2001**, *3*, 2883–2886.

[105] M. M. Haley, B. L. Langsdorf, *Chem. Commun.* **1997**, 1121–1222.

[106] Y. Tobe, N. Nakagawa, K. Naemura, T. Wakabayashi, T. Shida, Y. Achiba, *J. Am. Chem. Soc.* **1998**, *120*, 4544–4545;
(b) Y. Tobe, N. Nakagawa, J. Kishi, M. Sonoda, K. Naemura, T. Wakabayashi, T. Shida, Y. Achiba, *Tetrahedron* **2001**, *57*, 3629–3636;
(c) Y. Tobe, R. Furukawa, M. Sonoda, T. Wakabayashi, *Angew. Chem.* **2001**, *113*, 4196–4198; *Angew. Chem. Int. Ed.* **2001**, *40*, 4072–4074.

[107] Y. Rubin, T. C. Parker, S. J. Pastor, S. Jalisatgi, C. Boulle, C. L. Wilkins, *Angew. Chem.* **1998**, *110*, 1353–1356; *Angew. Chem. Int. Ed.* **1998**, *37*, 1226–1229.

[108] Y. Tobe, in *Acetylene Chemistry – Chemistry, Biology, and Materials Science*, F. Diederich, R. R. Tykwinski, and P. J. Stang (eds.), Wiley-VCH, Weinheim, **2005**.

9
Carbon-Rich Compounds: Acetylene-Based Carbon Allotropes

Yoshito Tobe and Tomonari Wakabayashi

9.1
Introduction

There has been longstanding interest in relatively small carbon clusters in the fields of spectroscopy and theoretical chemistry because of their structure and bonding and of their possible astrophysical significance. From the extensive theoretical and spectroscopic studies it has become apparent that the natures of the most abundant structures, whether they are linear or cyclic, are principally a function of the size of the clusters. With the discovery of the fullerenes as unusually stable clusters and their production in macroscopic quantities, increasing interest has been focused on larger carbon clusters in connection with the mechanism of fullerene formation. Like linear carbon clusters, linear polyynes $R(C{\equiv}C)_nR$ end-capped with a variety of atoms and groups – including hydrogen, halogen, alkyl, aryl, alkylsilyl, and transition metal groups – have been attracting considerable interest with regard to their structures, spectroscopy, and isolation limits. Moreover, thanks to their rod-like shapes and cylindrically surrounding π electrons around the axis, there has been renewed interest in polyynes in connection with molecular wires. In view of these ever-growing activities in sp carbon-based, all-carbon, and carbon-rich compounds, it seems appropriate to summarize the state of the art in this area. This chapter commences with a review of spectroscopic studies of small carbon clusters mainly of the linear form, followed by a survey of research into the hitherto unknown carbon allotrope carbyne, an infinite linear chain of sp carbon atoms. The next section deals with linear polyynes $R(C{\equiv}C)_nR$ end-capped with various atoms including transition metal elements. Finally, in the context of relatively large carbon clusters, monocyclic and multicyclic forms of carbon clusters are reviewed with emphasis on their relevance to their transformation into fullerene structures.

Acetylene Chemistry. Chemistry, Biology and Material Science. Edited by F. Diederich, P. J. Stang, R. R. Tykwinski

ISBN 3-527-30781-8

9.2 Linear Carbon Clusters

Linear carbon clusters C_n ($n \leq \sim30$) are chemical species that pose challenges in various experimental contexts, including production, detection, and characterization. Since they are not particularly fluorescent (except for C_2 and C_3) and are so reactive as to coagulate easily, forming soot particles, the detection of such transient species is naturally a challenge for chemists. Identification of novel bonding structures consisting of highly reactive sp carbon moieties can be also a motivation for research. So far, researchers have developed matrix isolation, laser vaporization, and sputtering techniques for production, and mass-, laser-, and electron-spectroscopy for detection and characterization.

In 1942, Herzberg reproduced the cometary emission band system at 405 nm under laboratory conditions [1a]. Almost a decade later, the carrier was identified as the linear C_3 [1b]. Thereafter, larger clusters C_n ($n \geq 4$) were detected as linear species both in the gas phase and in solid rare gas matrices. In the early 1960s, Weltner and co-workers pioneered the matrix isolation spectroscopy of trapped carbon vapor molecules in rare gas solids such as solid Ar and Ne and found various electronic and vibrational transitions [2]. In the 1980s, guided by the infrared frequency obtained by matrix isolation spectroscopy, rotationally resolved spectra were measured in the gas phase by diode laser spectroscopy, from which the size n of linear C_n could be deduced in order to provide unequivocal assignment of the vibrational transitions.

Another research stream was opened by the pulsed-laser vaporization of graphite in a supersonic gas expansion source that enabled production of C_n, C_n^-, and C_n^+ species of various sizes. From these results, in combination with those obtained from time-of-flight mass spectrometry (TOF-MS) and ultraviolet photoelectron spectroscopy (UPS), it was suggested that the stable structure changes from linear ($n \geq 3$), to monocyclic ($n \geq 10$), to cage forms ($n \geq \sim30$) depending on their sizes of up to $n = 84$ [3]. In 1989, Weltner and Van Zee reviewed both the experimental and the theoretical works [4a]. In 1998, Van Orden and Saykally updated the review, covering progress in mass-selective matrix isolation spectroscopy and high-resolution infrared absorption spectroscopy [4b].

In the 1990s, progress towards the assignment of electronic transitions of linear carbon clusters was made by mass-selective matrix isolation spectroscopy. Maier and his colleagues developed a continuous anion-cluster source in which cesium vapor atoms transfer electrons to neutral C_n clusters, efficiently producing anionic C_n^- [5a]. These anions are mass-selected by a Q-pole mass filter and deposited on a cold surface with an excess of neon gas to form solid Ne matrices. For study of neutral species, the trapped anions are photobreached to detach the electron, forming the corresponding neutral clusters. They observed a series of the dipole-allowed $^1\Sigma_u^+ \leftarrow X^1\Sigma_g^+$ transitions of C_{2n+1} ($n = 2$–7) in the UV regions [5c] and $^3\Sigma_u^- \leftarrow X^3\Sigma_g^-$ transitions of C_{2n} ($n = 3$–5) in the visible regions [5a, 5b]. The $^3\Sigma_u^- \leftarrow X^3\Sigma_g^-$ transition of linear C_4 at $\sim$380 nm was reported separately [5d]. As would be expected for the delocalized π-electron systems, the transition wavelength shifts almost linearly as a function of the cluster size, showing, for example, a uniform increment of about 40–43 nm between the bands of C_{2n+1} and C_{2n+3} [5c]. By extrapolation of the systematic relationship, the assignment of the $^1\Sigma_u^+ \leftarrow X^1\Sigma_g^+$ electronic transition was extended to linear C_{21} [6]. Some infrared absorption lines were further identified for the mass-selected clusters in Ne matrices [7].

For the mass-selected anions trapped in Ne matrices, the electronic absorption spectra of C_{2n}^- ($n = 2$–7) [8a, 8b] and of C_{2n+1}^- ($n = 2$–5) [8c], and also the emission of C_4^- [9] were reported. Surface plasmon polariton-enhanced Raman spectra (SERS) were obtained for size-selected C_{2n} ($n = 7$–10) clusters in nitrogen matrices and compared with the calculated vibrational frequencies for the ring and linear chain isomers [10].

New spectroscopic information in the gas phase has also been accumulated. Diode laser (DL) infrared absorption spectroscopy provides rotationally resolved spectra, from which the momentum of inertia of the molecule is deduced to give unequivocal assignment for the size n of linear C_n. Since the observable range with a diode is not especially wide, the absorption frequency from matrix isolation spectroscopy can be of help. By 1998 [4b], the IR spectra of antisymmetric stretching vibrations of linear C_n of $n = 3$–7, 9, and 13 [11] had been reported. Recently, the IR spectra of linear C_8 [12a] and C_{10} [12b] were observed.

Cavity ring-down spectroscopy (CRD) is a highly sensitive intra-cavity absorption spectroscopy of transient molecules, radicals, and intermediates in the gas phase at the resolution of the pulsed laser used. The decay time (or ring-down time) of a light pulse reflecting back and force in a cavity of high reflectivity (> 99.99%) is recorded as a function of the wavelength that gives an absorption spectrum of the molecules in the cavity. Linnartz et al. detected electronic transitions of linear carbon clusters and end-capped polyynic species by CRD. Among them, rotationally resolved spectra of linear carbon clusters were measured for the $^3\Sigma_u^- \leftarrow X\,^3\Sigma_g^-$ transition of C_4 at 379 nm [13a] and the $^1\Pi_u \leftarrow X^1\Sigma_g^+$ transition of C_5 at 511 nm [13b]. CRD spectroscopy in the infrared region (IR-CRD) was very recently demonstrated through the development of a new pulsed-IR light source by the stimulated Raman effect, for which the optical dye-laser output is down-converted in a tandem long-pass H_2 gas cell [14]. The rotationally resolved IR absorption spectrum of linear C_9 was revisited with IR-CRD [14].

Ultraviolet photoelectron spectroscopy (UPS) of anionic clusters C_n^- goes beyond the cluster size reach accessible by the other spectroscopic methods mentioned above [3b, 15]. UPS provides information on electron affinity (EA) and the electronic and vibrational structures of neutral C_n. In 1988, Yang et al. found two distinct EA series, one for clusters of C_n ($n \leq 9,11$) and the other for larger C_n ($10 \leq n \leq 29$), and suggested linear and monocyclic structures for the corresponding series of anions [3a]. Arnold et al. measured at higher resolution to identify vibrational structures of linear C_n up to $n = 11$ [16a]. Xu et al. refined the spectroscopic data for C_6 [16b], while Kohno et al. found a series of transitions associated with the lowest triplet states of linear C_{2n+1} ($n = 3$–9) [17]. In the above experiments [3a–b, 15, 16a–b, 17], cluster anions C_n^- are produced by laser vaporization of graphite in a supersonic helium expansion source. Recently, UPS has been applied to laser-desorbed [18] or discharged [19] organic molecules in order to distinguish the isomers of the fragments, which may retain skeletal carbon framework structures in the precursory molecules.

Anionic clusters C_n^- were studied by their electronic spectra in the gas phase. Resonance two-color photodetachment (R2CPD) spectroscopy is a high-sensitivity method applicable to the anionic species. Tulej et al. observed electronic transitions of C_4^-, C_5^-, C_7^-, and C_9^- [20], finding coincidence of the C_7^- lines with some DIB (diffuse interstellar bands) features [21], although this was eventually questioned by new astronomical observations [22]. Zhao et al. investigated the spectra and electron-detachment dynamics of C_4^-, C_6^-, and C_8^- [23]. Using resonance-enhanced multi-photon electron detachment (REMPED: the one-color version of R2CPD), Ohara et al. reported spectra of linear cluster anions of C_5^- [24a], C_7^-, C_9^-, and C_{11}^- [24b], C_{2n}^- (n = 6–8), and C_{2n+1}^- (n = 6–10) [24c]. For C_{10}^-, the authors found two distinctly different spectra depending on the cluster source conditions, one with sharp lines attributable to the linear isomer and the other with broad detachment features, for which the authors suggested a large deformation of the monocyclic ring isomer in its anionic state [25].

Matrix isolation spectroscopy has played a major role in carbon cluster science [2a–c, 4a]. Trapped in an inert, solid, rare-gas matrix, the reactive clusters can be studied under conditions where further reactions and aggregations are prevented. Recently, solid parahydrogen was used as a matrix for high-resolution spectroscopy, and this showed that the linear carbon clusters C_n (n = 3, 5, 6, 7, and 9) are stably trapped in the molecular matrix media [26a–c]. Figure 9.1 shows typical UV/Vis absorption spectra of carbon vapor molecules produced from laser ablation of graphite in vacuum and then trapped in solid Ne at 5 K. The difficulty in the assignment of spectral features arises from the fact that the sample is a mixture of various clusters C_n. The assignment of bands should thus be correlated by analogy with the results of mass-selective matrix isolation spectroscopy [5a–d, 27] and the recent vacuum UV (VUV) spectroscopy of the $^1\Sigma_u^+ \leftarrow X^1\Sigma_g^+$ transition of C_3 [28].

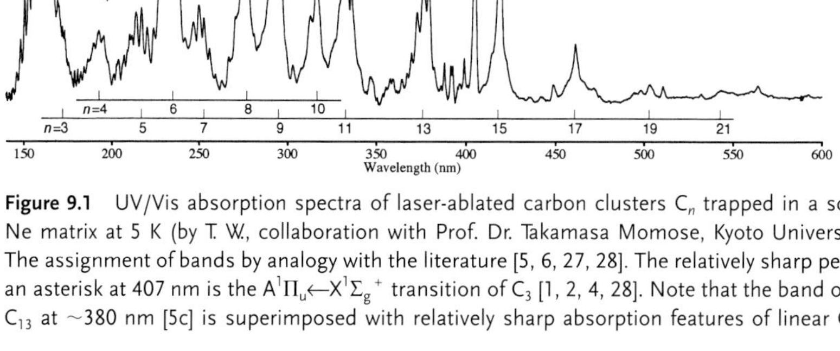

Figure 9.1 UV/Vis absorption spectra of laser-ablated carbon clusters C_n trapped in a solid Ne matrix at 5 K (by T. W., collaboration with Prof. Dr. Takamasa Momose, Kyoto University). The assignment of bands by analogy with the literature [5, 6, 27, 28]. The relatively sharp peak with an asterisk at 407 nm is the $A^1\Pi_u \leftarrow X^1\Sigma_g^+$ transition of C_3 [1, 2, 4, 28]. Note that the band of linear C_{13} at ~380 nm [5c] is superimposed with relatively sharp absorption features of linear C_4 [6].

One of the advantages of the mixed-cluster matrix sample over the mass-selective approach is the larger number (orders of magnitude) of accumulated clusters. Szczepanski et al. warmed up a carbon-containing Ar matrix in order to enhance the growth of larger carbon clusters, and observed electronic transitions in the UV/Vis regions [29]. They assigned the bands to long carbon chain clusters C_n ($n = 5–29$) and anions C_n^- ($n = 6–36$) on the basis of the linear dependence of the transition wavelengths as a function of the cluster size [29]. Szczepanski et al. obtained resonance Raman spectra for C_6^- [27], C_{17}, C_{21}, and C_{23} by band-selective photoexcitation with a tunable light source [30], to find a systematic decrease in the frequency of a specific vibrational mode. The relatively intense, major electronic transitions of neutral C_n clusters in Ne and Ar matrices are summarized in Table 9.1.

Table 9.1 Major electronic transitions of neutral C_n clusters.

Carbon cluster	*Electronic transition (nm)*		
	$^1\Sigma_u^+ \leftarrow X^1\Sigma_g^+$ Ne[a,c,d] Ar[a,b]	$^3\Sigma_u^- \leftarrow X^3\Sigma_g^-$ Ne[e] ($^1E_{1u} \leftarrow X^1A_{1g}$)	$^3\Sigma_u^- \leftarrow X^3\Sigma_g^-$ Ne[f,g,h]
C_3	140–180[a] 145–190[a]		
C_4		–	379.9[f]
C_5	– 226.1[b]		
C_6		237.5[e]	511.3[g,h]
C_7	252.8[c] –		
C_8		277.2[e]	639.8[h]
C_9	295.0[c] 307.9[b]		
C_{10}		(316.1)[e]	735.5[h]
C_{11}	336.3[c] 354.0[b]		
C_{12}		(332.1)[e]	860.2[h]
C_{13}	379.6[c] 400.5[b]		
C_{14}		(346.8)[e]	980.9[h]
C_{15}	419.6[c] 447.0[b]		
C_{17}	460.3[d] 493.0[b]		
C_{19}	503.2[d] 539.2[b]		
C_{21}	543.7[d] 586.4[b]		

a) Ref. 28.
b) Ref. 29.
c) Ref. 5c.
d) Ref. 6.
e) Ref. 27.
f) Ref. 5d.
g) Ref. 5a.
h) Ref. 5b.

The IR absorption lines of carbon clusters in rare gas matrices are relatively sharp (~0.1–0.5 cm^{-1} in FWHM (full width at half maximum)) because the rotational structures are smeared out. Figure 9.2 shows the typical IR absorption spectra of carbon vapor molecules trapped in solid Ar. The assignment of peaks is taken from the literature [4b, 7, 31], and the numbers indicate the size (n) of linear C_n.

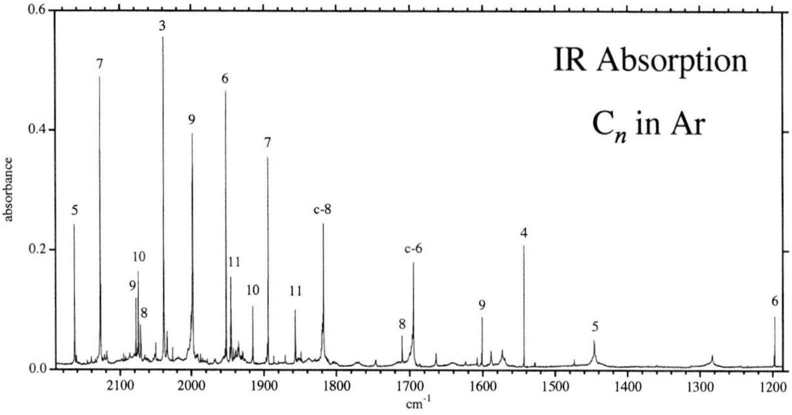

Figure 9.2 IR absorption spectra of carbon vapor molecules C_n trapped in a solid Ar matrix at 6 K (by T. W., collaboration with Prof. Dr. Wolfgang Krätschmer, Max-Plank-Institut für Kernphysik, Heidelberg). The assignment of peaks is by analogy with the literature [4b, 7, 31]. The numbers indicate the cluster size n of linear C_n. The two peaks indicated as c-6 and c-8 presumably belong to cyclic species [37] (see text [38] for details).

In recent years, FT-IR data have been accumulated for linear C_n [31a, 31b], their anions C_n^- [8, 32a, 32b], sulfides C_nS [33a], and water complexes $C_n(H_2O)_m$ [33b]. ^{13}C isotope enrichment gives more information on the fundamental and combination-band frequencies for the isotopomers, and the aid of calculations by density functional theory (DFT) provides firmer bases for vibrational assignment [31a, 31b]. The observed vibrational frequencies of linear C_n clusters in solid Ne and Ar and in the gas phase are summarized in Table 9.2.

Closing the section on linear carbon clusters, we briefly address the observation of, and controversy relating to, cyclic carbon clusters. Cyclic carbon clusters have been conjectured since the observation of the UPS spectra in 1988 [3a]. Anion photoelectron spectroscopy clearly showed a distinct series of spectra for C_{10}^- and anions of larger size, reasonably attributable to the monocyclic ring isomers [3a]. Ion chromatography (IC) experiments indicated that the cyclic isomers exist for C_7^+ and larger sizes of cations [34a, 34b] and for C_{10}^- and larger sizes of anions [34c], consistently with observations in anion photoelectron spectroscopy [3a]. For neutrals, circumstantial evidence of the stability and abundance of cyclic C_{10} has been reported by one-photon ionization mass spectroscopy, which may avoid secondary fragmentation upon ionization [35, 36]. No evidence of the highly strained smaller cyclic species of $n = 6$ has appeared, at least in the gas phase.

Table 9.2 Observed vibrational frequencies of linear C_n.

Carbon cluster	*Frequency (cm⁻¹)*			*Mode assignment*
	Ne	*Ar*	*Gas*	
C_3	2036.4[a]	2038.9[a]	2040.02[a]	ν_3 IR-str.
	1226[a]	1214.0[a]	1224.5[a]	ν_1
	75[a]	83[a]	63.42[a]	ν_2 IR
C_4			2032(50)[a]	ν_1
	1547.2[a]	1543.4[a]	1548.94[a]	ν_3 IR-str.
			352(15)[a]	ν_4 IR
		172.4[a]	160(4)[a]	ν_5
C_5	2166.4[a]	2164.3[a]	2169.44[a]	ν_3 IR-str.
	1444.3[a]	1446.6[a]		ν_4 IR-str.
		775.8[c]	779(10)[a]	ν_2
			535(10)[a]	ν_6 IR
			218(13)[a]	ν_5
			118(3)[a]	ν_7 IR
C_6		2050.3[c]	2089 (50)[e]	ν_1
	1958.7[a]	1952.5[a]	1959.86[a]	ν_4 IR-str.
		1665.8[c]	1694(50)[e]	ν_2
	1199.4[a]	1197.3[a]		ν_5 IR-str.
		627.3[c]	637(50)[e]	ν_3
C_7	2134.6[a]	2127.8[a]	2138.32[a]	ν_4 IR-str.
	1897.5[a]	1894.3[a]	1898.38[a]	ν_5 IR-str.
		581.7[c]	548(90)[a]	ν_3
			496(110)[a]	ν_7
C_8	2067.8[a,b]	2071.5[c]		ν_5 IR-str.
	1707.4[a,b]	1710.5		ν_6 IR-str.
			~565(8)[a]	ν_4
C_9	2081.1[a,b]	2078.1[c]	2079.67[a]	ν_5 IR-str.
	2010.0[a]	1998.0[a]	2014.28[a]	ν_6 IR-str.
		1601.0[a]		ν_7 IR-str.
			1258(50)[a]	ν_3
		447.6[c]	484(48)[a]	ν_4
			30(20)[a]	ν_{15}
C_{10}	2074.5[a,b]			IR-str.
	1915.4[a,b]			IR-str.
C_{11}	1938.6[a,b]	1945.7[d]		ν_7 IR-str.
	1853.4[a,b]	1856.6[d]		ν_8 IR-str.
			~440[a]	ν_5
C_{12}	2003.9[a,b]			
C_{13}			1808.96[f]	IR-str.

a) Duplicated from Table 1 in Ref. 7.
b) Original data from Ref. 7.
c) Ref. 31c.
d) Ref. 31b
e) Ref. 16b.
f) Ref. 11.

In Ar matrices, on the other hand, two independent groups assigned an IR absorption line at 1695 cm^{-1} in solid Ar to cyclic C_6 with D_{3h} symmetry [37a, 38a], although for cyclic C_8 the lines assigned to the same mode of vibration disagreed [37b, 38b, 38c]. Two lines in Figure 9.2, at 1695 cm^{-1} and 1818 cm^{-1} and indicated as c-6 and c-8, respectively, are the lines assigned to cyclic C_6 and cyclic C_8 from isotopomeric studies by Presilla-Márquez et al. [37]. These two lines show a similar tendency to increase upon annealing of a carbon-containing matrix sample [39]. So far, no UV/Vis electronic counterpart to the IR line has yet been identified. Note that the intensity correlation between the 1695 cm^{-1} IR line and the 586 nm Vis band in solid Ar was reported earlier [40].

For mass-selected C_{10}^- trapped in a solid Ne matrix, Grutter et al. suggested that the UV band at 316 nm (see Figure 9.1) is attributable to cyclic C_{10} rather than to linear C_{10} [27]. The assignment is based on extrapolation of the calculated excitation energy for cyclic C_6 to the energy for cyclic C_{10} at the MRD-CI (multi-reference single and double excitation configuration interaction) level of theory [24]. No IR line for cyclic C_{10} has yet been identified.

9.3 Carbyne

Carbyne had long been thought to exist as the third allotropic form of carbon (i. e., together with graphite (sp^2) and diamond (sp^3)) [41]. The infinitely long sp-hybridized linear carbon chains are ideal constituents that may stack into a new crystalline form. However, their fragility and reactivity have thwarted attempts to isolate such a solid form. After the 1990 macroscopic production of fullerene C_{60}, a sort of modification of the graphitic (sp^2) structure [42], that molecule or its solid form seemed to gain the honor of the third position. In the hope that the carbyne story might experience a rapid rise similar to that enjoyed by C_{60} [42, 43], increasing interest has been focused on the macroscopic preparation of the hitherto never-isolated system of stacking sp-carbon chains. The goals include its full characterization and understanding of its electronic, magnetic, optical, mechanical, thermal, and chemical properties. Here we briefly address early investigations and recent attempted syntheses of carbyne-rich solid materials.

The early story of the linear carbon allotrope is related to its natural occurrence associated with meteorite impact or shock-induced phenomena [44]. The materials in nature were tiny crystals embedded in a graphitic or an amorphous carbon matrix and observed by electron microscopy or its associated methods such as electron diffraction (ED), energy dispersed X-ray emission (EDX), and electron energy loss spectroscopy (EELS) [44]. In 1968, while studying diamonds in meteorite ejecta, El Goresy and Donnay found shiny, crystalline portions in polished sections of shock-fused graphite gneisses from the Ries Crater in Bavaria [45] and coined the name Chaoite in honor of the geologist E. C. T. Chao, who had investigated this impact crater thoroughly. The X-ray diffraction powder patterns indicated large cell dimensions a = 8.948 Å, c = 14.078 Å with a primitive hexagonal lattice [45].

After the first report, researchers were encouraged to publish their observations of d-spacings similar to those found in Chaoite (e. g., 0.45 nm). The larger spacing than the inter-layer spacing in graphite (0.34 nm) was believed to be evidence of linear carbon chains suspended between the graphitic layers. Amongst the reports on carbynes and Chaoite-like materials, some fascinating hypothesis were raised concerning pre-solar grains and carriers of noble gases with isotope anomalies in meteorites [46]. Eventually, Smith and Buseck questioned the carbyne concept, arguing that the d-spacings of quartz and some layer silicates such as talc can also reproduce the X-ray diffraction data of Chaoite and carbynes [47]. This cast a shadow over the whole concept of carbyne and rather muted interest in the field.

Attempts to synthesize the linear carbon allotrope started even earlier, around 1961 [48]. The synthetic approaches reported so far have involved oxidative dehydropolycondensation of acetylene, polycondensation of halides, chemical or photo-induced dehydrohalogenation of polymers, dehydrogenation of polyacetylene at high static pressure, decomposition of hydrocarbons, and pyrolysis of organic polymers [41]. Condensation of carbon vapor has also been tested [41]. In addition to the experimental studies, theoretical considerations of the electronic and phonon properties have been reported, not only for the infinitely long carbon chains [49, 50] but also for two-dimensional mixed-valence systems possessing sp and sp^2 carbon atoms [51, 52]. In 1999, Heimman, Evsyukov, and Kavan edited a treatise on the state of the art in this field [41].

Electrochemical carbonization of polymers has become a reality for the bulk synthesis of carbyne-rich compounds [53–55]. Kavan and co-workers performed electrochemical reduction of polytetrafluoroethylene (PTFE), in which a thin foil of PTFE was dipped into alkali metal amalgam (Hg/M, M = Li, Na, and K) to eliminate fluorine as MF salts [53]. Characterization of the material in situ showed a rather broad but strong Raman signal around 2100 cm^{-1}, attributed to the stretching vibration of C≡C triple bonds [53]. The intensity of the "carbyne" band increases relative to the intensity of the "D + G" band ("disorder-induced" + "graphite" band, spanning between 1200 and 1650 cm^{-1} [56]), as the excitation wavelength changed from red to blue [54]. This is explained by the resonance Raman effect assisted by π-π^* absorption possibly located on nearby UV regions. As a mechanism for the stabilization of the linear-carbon system it was proposed that the embedded by-products (i. e., nano-crystalline MF salts) prevent collapse through cross-linking reactions between neighboring carbon chains [55].

Cluster-beam deposition under ultrahigh vacuum is a method preferable to atomic beams for cluster-assembled films. Milani and his colleagues deposited C_n clusters of a mean size of $n \sim 600$ to obtain a thin film of $\sim$200 nm thickness [57, 58]. Raman spectroscopy in situ showed high-frequency bands at 2100 cm^{-1} and 1980 cm^{-1}, to which the authors assigned the stretching vibrations of alternating triple bonds -$(C{\equiv}C)_n$- (polyynes) and conjugated double bonds $= (C = C)_n =$ (polycumulenes), respectively [57]. It was suggested that the carbyne structures are supported in-between the fullerene-like structures. Upon exposure to various gases – H_2, He, N_2, and dry air – the carbyne fraction decayed exponentially with time [58]. It turned out that the carbyne moieties are sensitive to oxygen and elevated temperatures, from room temperature to ~200 °C [58].

Matrix-isolated carbon clusters can be precursors to carbyne structures. So far, the growth of longer-chain clusters upon annealing of a solid rare-gas matrix containing small carbon clusters has been reported [59]. Recently, linear C_n clusters ($n \leq \sim 20$) embedded in such a cold matrix were coagulated upon sublimation of the matrix to form a brownish material on the surface [60]. Figure 9.3 shows the preliminary in situ Raman spectrum of the aggregates showing the "carbyne" or "C" band at 1900–2200 cm^{-1} in addition to the "D + G" band at 1250–1650 cm^{-1}.

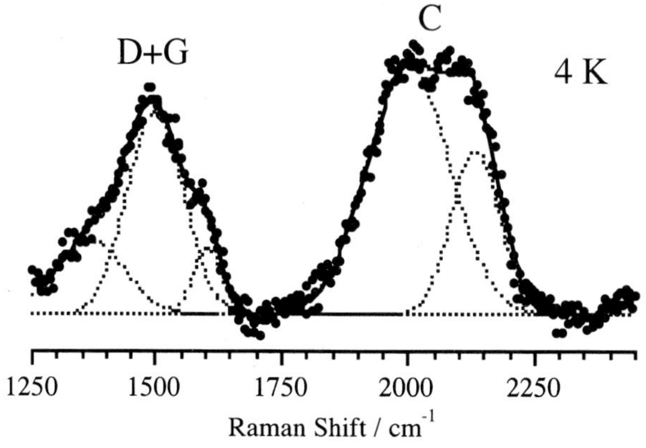

Figure 9.3 The in situ Raman spectrum of the carbon-cluster aggregates that form upon sublimation of a solid Ne matrix containing small carbon clusters [60]. The sample was excited with cw laser light at 532 nm and the scattered light was detected by use of a monochromator at a resolution of ~12 cm^{-1}. The "C" or "carbyne" band at 1900–2200 cm^{-1} is attributed to the stretching vibration of linear carbon moieties of sp hybridization [53–55, 57, 58]. The "D + G" band at 1250–1650 cm^{-1} is due to highly disordered structures of reduced crystallinity composed of sp^2 and sp^3 carbon atoms [56].

The spectral features in Figure 9.3 resemble those obtained for the electrochemically formed carbyne [53, 54] and the carbyne-rich films generated by the cluster-beam deposition [57, 58]. The cluster aggregates on the cold surface (< 50 K) readily undergo a highly exothermic reaction that heats up the product particles to ~2500 K [60, 61]. The recent experiments described above were performed in vacuum and mostly based on in situ Raman spectroscopy. Further progress for the production of stoichiometric material with a well defined structure is still awaited.

9.4 Linear Polyynes

Preparative work on linear polyynes has been conducted in order to assess the isolation limit of sp carbon chains and in connection with natural products containing polyyne units [62]. With respect to the increasing interest in carbyne (Section 9.3), linear polyynes are regarded as models of this unknown allotrope. Conjugated polyyne chains are also regarded as molecular wires, which may communicate electronic interaction or transport electrons through π bonds between the two ends. When the termini are substituted by redox-active transition metals, a mixed valence state may occur. In view of these points of interest, synthetic and spectroscopic research on linear polyynes has been vigorous. Methods of synthesis, remarks on stability, and characterization status of relatively long known linear polyynes $H(C{\equiv}C)_nH$ and their substituted derivatives $R(C{\equiv}C)_nR$, where R is an alkyl, aryl, halogen, or cyano group, are listed in Table 9.3, while those properties for polyynes end-capped by transition metals $L_mM(C{\equiv}C)_nML_m$ are summarized in Table 9.4.

Table 9.3 Synthesis, stability, and characterization of linear polyynes $R(C{\equiv}C)_nR$.

R	*Number of C≡C unit n*	*(a) Method of preparation, (b) stability and other data for those with large n, and (c) characterization status*	*Reference*
H	2–5	(a) desilylation of TMS derivatives (n=4, 5) (c) UV: n=2–3 (both in solution and in gas phase), n=4, 5 (in solution)	63
H	4–12	(a) deprotection of the Et_3Si derivatives (c) UV/Vis of an eluate from chromatography	64
H	4–8	(a) laser ablation of graphite suspended in a hydrocarbon solvent (c) detected by HPLC and identified by UV/Vis	65
H	4–9	(a) arc discharge of graphite rods in an organic solvent such as CH_3CN (c) detected by HPLC and identified by UV/Vis	66
H	6–13	(a) arc discharge of butadiyne in Ar (c) UV/Vis in Ne matrix at 6 K	67
Me	5, 6	(a) n=5: dehydrochlorination, n=6: Hay coupling (b) n=5: cream-colored needles decomposing easily under light and at moderate temp., n=6: yellow crystals turning brown rapidly (c) IR, UV/Vis	68a
t-Bu	3–7	(a) dehydrochlorination (b) n=7: yellow needles decomposing >150 °C (c) UV/Vis	69
t-Bu	8, 10	(a) Eglinton coupling (b) n=8: yellow-brown needles decomposing >130 °C, n=10: orange needles decomposing ca. 100 °C (c) UV/Vis	68c

Table 9.3 Continue

R	*Number of C≡C unit n*	*(a) Method of preparation, (b) stability and other data for those with large n, and (c) characterization status*	*Reference*
t-Bu	12	(a) Eglinton coupling (b) Red-brown needles turning black at 20 °C during 8 min, also decomposing ca. 50 °C (c) UV/Vis	70
Et_3Si	4–12, 16	(a) Hay coupling (b) *n*=6: dark solid that blackened rapidly upon warming to 20 °C: larger homologues were not isolated. (c) UV/Vis (with *n*=8–12 and 16 only for chromatography eluate)	64
t-$BuMe_2Si$	6	(a) flash vacuum pyrolysis (650 °C) of a cyclobutene-dione type precursor (c) IR, ^{1}H and ^{13}C NMR, MS	71
Ph	6, 8	(a) Hay coupling (b) *n*=6: golden-yellow needles decomposing >100 °C, *n*=8: extremely unstable copper-colored needles (c) UV/Vis	68b
Ph	10	(a) Hay coupling (b) dark red needles that blackened and conflagrated violently during filtration under nitrogen (c) IR, UV/Vis (for chromatography eluate)	70
Ph	5	(a) flash vacuum pyrolysis (650 °C) of a cyclobutene-dione-type precursor (c) X-ray analysis	71
Ar[a]	4–6	(a) dehydrochlorination (c) UV/Vis	73
Mes[b]	6, 8	(a) Hay coupling (b) *n*=6: yellow needles decomposing >170 °C, *n*=8: dark red solid decomposing >150 °C (c) IR, UV/Vis	70
3,5-disiloxyPh[c]	6, 8	(a) Hay coupling (b) slowly decomposing orange solid (*n*=8) (c) IR, UV/Vis, ^{1}H and ^{13}C NMR, MS	72
Dendron[d]	6, 8, 10	(a) Hay coupling (b) yellow, orange, and red solids, respectively (c) Raman, UV/Vis, ^{1}H and ^{13}C NMR, MS	72
Porphyrin[e]	2, 4, 6, 8	(a) Eglinton coupling (b) deeply colored stable solids under ambient conditions (c) IR, UV/Vis, ^{1}H NMR (for *n*=8), MS (for *n*=8)	74
Br	4	(a) information available from the author on request (b) highly explosive (c) IR, ^{13}C NMR, MS	75
I	3–4	(a) iodination of the corresponding Me_3Si derivatives with NIS (b) solids that explode at 100 °C (*n*=3) or at 85 °C (*n*=4) (c) ^{13}C NMR (*n*=3, 4), X-ray analysis for a 1:1 co-crystal with Ph_3PO (*n*=3)	76

Table 9.3 Continue

R	*Number of C≡C unit n*	*(a) Method of preparation, (b) stability and other data for those with large n, and (c) characterization status*	*Reference*
CN	3–8	(a) resistive heating of a graphite rod in the presence of dicyanogen, $(CN)_2$, under He followed by extraction (b) colorless crystals only stable at rt. in the dark: concentrated solutions of the longest one (*n*=6) slowly decompose. (c) IR, UV/Vis, MS for *n*=7 and 8, IR, UV/Vis, MS, and ^{13}C NMR for *n*=3–6	77

a: 4-biphenyl, 1- and 2-naphthyl, 2-fluorenyl, 1- and 9-anthryl, 2-, 3-, and 9-phenanthryl, 1- and 2-pyrenyl, 6-chrysenyl.
b: 1,3,5-trimethylphenyl (mesityl).
c: 3,5-bis(*tert*-butyldimethylsiloxy)phenyl derivatives **1a**–**b**: an inseparable mixture of polyynes of *n*=8, 9, and 10 was prepared by the same method as a red solid.
d: third-generation aliphatic polyether dendrimers **3a**–**c**.
e: Ni(II) complexes **4a**–**d** of *meso*-diarylporphyrin.

It is known that parent polyynes $H(C{\equiv}C)_nH$ become drastically less stable with increasing number of triple bonds [62]. Although butadiyne, for example, can be distilled, it polymerizes rapidly above 0 °C, and hexatriyne is often reported to explode violently. Accordingly, polyynes were prepared and characterized either in the gas phase, in dilute solutions, or in rare gas matrices at low temperature to prevent their (explosive) decomposition and polymerization. UV spectroscopic studies for polyynes $H(C{\equiv}C)_nH$ of n = 3–5, prepared by deprotection of the corresponding trimethylsilyl (TMS) derivatives, in solution (also in the gas phase for n = 3 and 4) were undertaken by Kloster-Jensen in the early 1970s [63]. Similarly, polyynes with n = 4–12, prepared by deprotection of the corresponding triethylsilyl (TES) derivatives, were characterized by UV/Vis spectroscopy in the chromatography eluates of mixtures containing these polyynes [64]. Recently, Tsuji and Cataldo independently reported new methods for the formation of linear polyynes by application of graphite ablation techniques, typically used for the formation of fullerenes in the gas phase, in organic solvents. Namely, laser ablation of graphite suspended in a hydrocarbon solvent at room temperature allowed a mixture of linear polyynes with n = 4–8 to be produced, and these were identified by GC-MS analysis of the mixture and the UV/Vis spectra of the HPLC eluate [65]. Similarly, arc discharge of graphite rods in an organic solvent gave a mixture of linear polyynes with n = 4–9, also identified by the UV/Vis spectra of the HPLC eluate [66]. Although polyynes with n up to 9 were produced by the arc discharge method at −40 °C, the relative amount of polyyne drops considerably with increasing number of carbon atoms. Maier et al. conducted electronic spectroscopic studies of polyynes with n = 6–13 isolated in neon matrices at 6 K [67]. Polyynes were generated by electron impact on a butadiyne-argon mixture, mass selection of the polyyne anions, codeposition of an anion with neon into a matrix, and electron detachment by UV irradiation. A linear relationship between the number of the carbon atoms (n) and the electronic absorption maximum was observed, although it deviated from linearity with increasing n. For the measurement in the gas phase by resonant two-color two-photon ionization spectroscopy, the mass-selected polyyne anions, generated by the same method as above, were neutralized by discharge in the gas phase. The C≡C stretching bands were observed between 1945 and 1826 cm^{-1}, indicating that the compounds possess localized polyyne structures with large bond alternation. It can be concluded that the polyyne $H(C{\equiv}C)_nH$ with 24 carbon atoms (n = 12) is the current limit of characterization in solution at ambient temperature, while that with 26 carbon atoms (n = 13) is the longest one generated and characterized in a matrix at low temperature.

Attempts to isolate stable derivatives of even longer linear polyynes were initiated during the 1950s by Jones et al. and by Bohlman [68, 69]. For alkyl-substituted derivatives, the bulkiness of the endgroups has a remarkable effect on their stabilities. In the series of polyynes end-capped with methyl groups ($Me(C{\equiv}C)_nMe$), for example, the derivatives with n = 5–6 were the longest members isolated and characterized by their IR and UV/Vis spectra [68a]. In contrast, *tert*-butyl derivatives are much more stable; $^tBu(C{\equiv}C)_n{}^tBu$ with n = 7 was isolated by Bohlman [69] and those with up to 10 were characterized by Jones et al.[68c]. The longest homologue, with n = 12, was isolated later by Johnson and Walton [70], but decomposed rapidly even at ambient temperature.

With regard to steric protection, Walton et al. employed the bulky TES (triethylsilyl) group, which is also useful in terms of its ease of removal, making further elongation of the polyyne chain feasible [64]. The longest isolated polyyne of this series, $TES(C{\equiv}C)_6TES$, a dark solid that blackened rapidly at 20 °C, however, is rather short in relation to the *tert*-butyl derivatives. While the compounds were not isolated, solutions containing $TES(C{\equiv}C)_nTES$ with n = 8–12 and n = up to 16 were prepared by repetition of partial deprotection and Hay coupling sequences, and their UV/Vis spectra were recorded. This allowed ready access to the otherwise difficult to prepare parent polyynes $H(C{\equiv}C)_nH$ in solution as described above. The longest fully characterized silyl derivative is the *tert*-butyldimethylsilyl derivative of dodecahexayne prepared by Diederich et al. by flash vacuum pyrolysis of its cyclobutenedione-type precursor [71]. The polyyne chain of this compound is nearly linear and is composed of alternating single and triple bonds.

Aryl groups are also used for kinetic protection. $Ph(C{\equiv}C)_nPh$ with n = 8 was isolated as an extremely unstable solid by Jones et al. in the 1950s [68b]. Since the polyyne with n = 10 later prepared by Walton decomposed violently during filtration under nitrogen, it was only characterized in a chromatography eluate [70]. The longest diphenylpolyyne characterized by X-ray crystallographic analysis is $Ph(C{\equiv}C)_5Ph$ [71]. Mesityl derivatives seem to be more stable than phenyl derivatives; polyynes with n = 6 and 8 were isolated as yellow and dark red solids, respectively, and were characterized by their IR and UV/Vis spectra [70]. Recently, the 3,5-bis(*tert*-butyldimethylsiloxy)phenyl derivatives **1a** (n = 6) and **1b** (n = 8) were prepared, isolated, and fully spectroscopically characterized by Hirsch et al. [72]. Moreover, these researchers also utilized sterically demanding dendrimers for steric protection [72]. Polyynes **2a–c** with *n* up to 6 bearing first- to third-generation dendrimers of Fréchet's benzyl ether type were prepared and fully spectroscopically characterized. While dilute solutions of hexaynes **2a–c** were stable, however, the solid materials decomposed rather quickly. On the other hand, polyynes **3a** (n = 6), **3b** (n = 8), and **3c** (n = 10), possessing third-generation aliphatic dendrimers, turned out to be more stable; **3a–c** were isolated and fully characterized. The icosadecayne **3c** (n = 10) represents one of the longest known linear polyynes to have been isolated. As the extinctions of the most intense electronic absorption band increase with increasing number of the triple bonds, the value for the longest polyyne **3c** ($\epsilon = 6 \times 10^5$ M^{-1} cm^{-1}) represents one of the largest measured for this class of compounds.

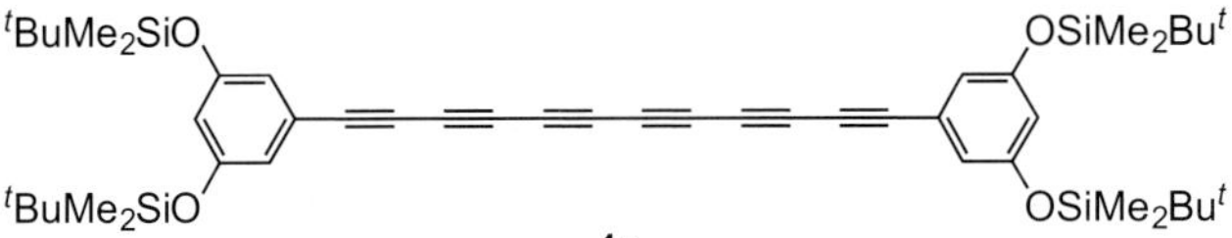

1a

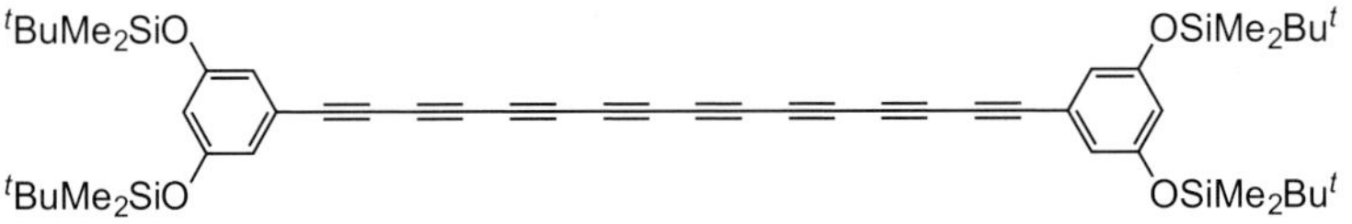

1b

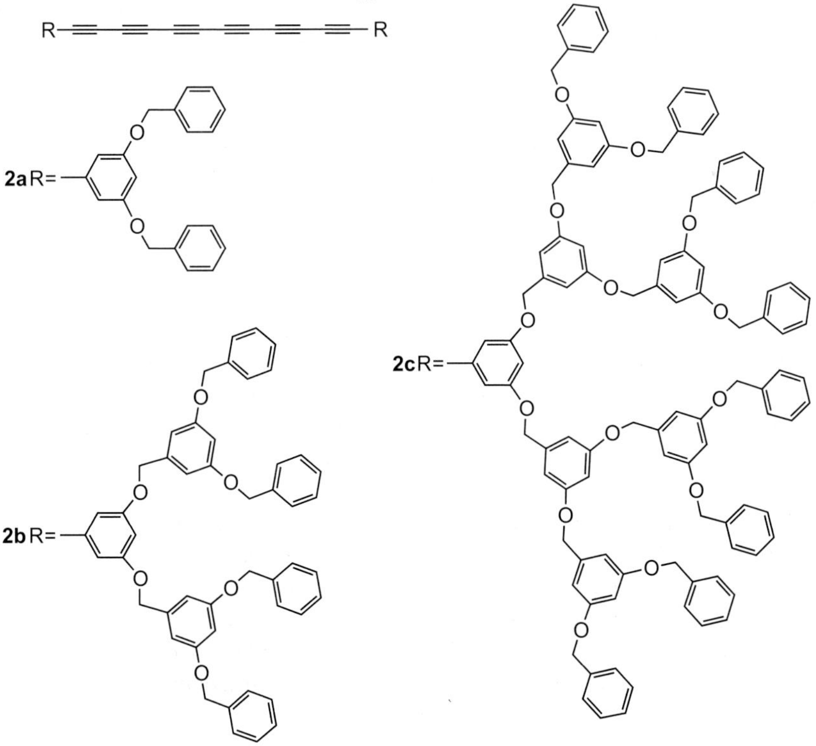

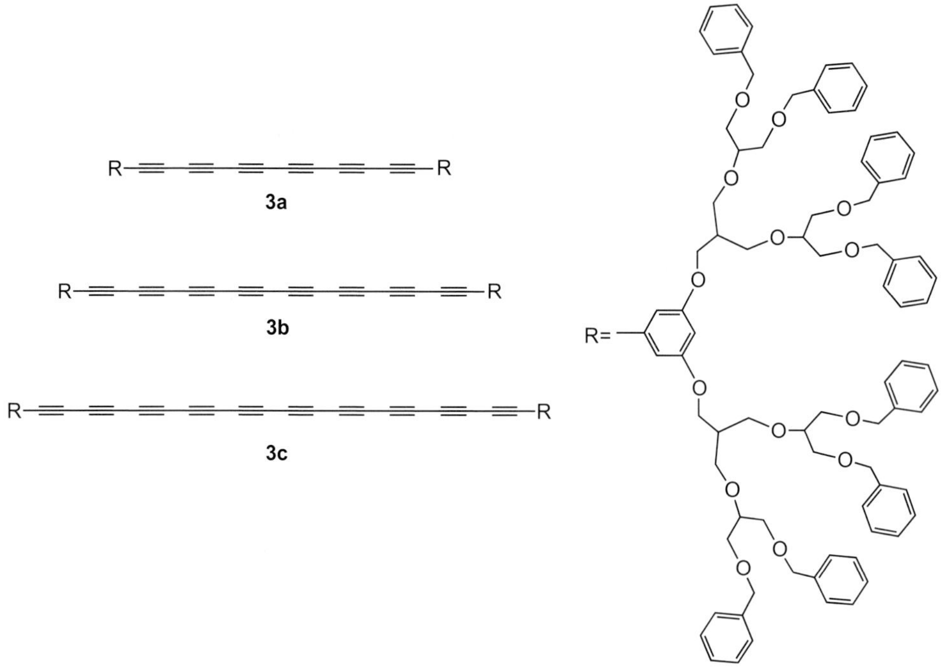

3a

3b

3c

Diaryl-substituted polyynes are also regarded as models for investigation of electronic interactions between two π systems connected by a polyyne chain. For this purpose diarylpolyynes Ar(C≡C)$_n$Ar with n = 4–6 and possessing a variety of aromatic rings were prepared and their electronic interactions investigated by UV/Vis spectroscopy by Nakagawa and Akiyama et al. [73]. More recently, porphyrin-substituted polyynes **4a–d**, with n = 2, 4, 6, and 8, respectively, were prepared for a similar but more intriguing reason: investigation of whether the polyyne chain may serve as a molecular wire that can transmit electrons in through-bond fashion [74]. The splitting (1380 cm^{-1}) of the Soret band of **4a** (n = 2) indicated the presence of exciton interactions between the porphyrin rings. However, the splitting becomes smaller with increasing chain length, suggesting a decrease in the electronic interactions between the porphyrins with increasing extension of the polyyne unit.

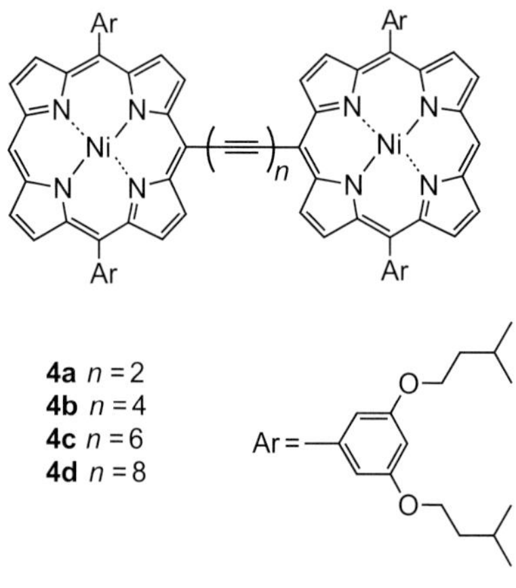

4a n = 2
4b n = 4
4c n = 6
4d n = 8

Halogen-substituted polyynes are useful building blocks for the preparation of longer polyynes. The longest known dibromopolyyne, $Br(C{\equiv}C)_4Br$, was prepared and characterized successfully by IR and ^{13}C NMR spectroscopy and mass spectrometry [75]. Because of its highly explosive properties, special care should be taken; details of the synthesis are available only upon personal request to the author. The diiodo compounds, $I(C{\equiv}C)_nI$, with n = 3–4, were prepared by Gao and Goroff by treatment of the TMS derivatives with *N*-iodosuccinimide [76]. Although they explode when heated, they were successfully characterized by conventional spectroscopic methods. The chemical shifts of the terminal carbon atoms in the ^{13}C NMR spectra are considerably solvent-dependent, indicating that the iodine atoms interact strongly with the negative poles of dipolar solvents. Moreover, $I(C{\equiv}C)_3I$ formed a stable 1:1 co-crystal with triphenylphosphine oxide, its structure was characterized by X-ray analysis.

Hirsch et al. prepared a series of cyano-substituted polyynes, $NC(C{\equiv}C)_nCN$ with n = 3–8, by resistive heating of a graphite rod (the method employed for the first preparative-scale formation of fullerenes by Krätschmer and Huffman) in the presence of dicyanogen and helium buffer gas [77]. The same reaction conducted in the presence of chlorine, however, did not give the corresponding chloropolyynes but chlorinated aromatic hydrocarbons. The dicyanopolyynes were separated by chromatography and were reasonably characterized by spectroscopic methods. While dilute solutions of the dicycanopolyynes kept at low temperature were stable, concentrated solutions and solid materials gradually decomposed. On the basis of the observed stability of the dicyanopolyynes, the authors questioned whether the material obtained by Lagow et al. under similar conditions was the sp carbon allotrope as proposed [78].

In general, linear polyynes display a series of fine-structured UV/Vis absorption bands. The wavelengths for each set of polyynes have been correlated with the number of triple bonds (n) according to the Lewis–Calvin equation [79]: $\lambda^2 = kn$, where λ is the wavelength of the longest-wavelength absorption band (*A* band) or that of the most intense absorption band (*L* band). The slope (k) of the plot depends largely on which band is taken for the plot. The slope for the longest-wavelength band (*A* band) of the phenyl-substituted derivatives, for example, is $28 \times 10^3\,nm^2$ per triple bond, while that for the highest-intensity band (*L* band) is $15 \times 10^3\,nm^2$ per triple bond [70]. Similarly, the slope k for the *A* band of the methyl-substituted polyynes is $30 \times 10^3\,nm^2$ per triple bond [68a], while those for the *L* bands of the parent polyynes [64], the *tert*-butyl derivatives [70], and cyano derivatives [77b] are all $12 \times 10^3\,nm^2$ per triple bond. On the other hand, the slopes for the *L* bands of the two 3,5-bis(*tert*-butyldimethylsiloxy)phenyl derivatives **1a** and **1b** and aliphatic dendrimers such as **3a–c** are $20 \times 10^3\,nm^2$ per triple bond, larger than those observed for the other sets of polyynes [72]. The absorption wavelengths can also be treated with the empirical equation $E_{exp} = a + b/n$, which is more appropriate for long-chain polyenes and polyynes, since for larger n the plots of E_{exp} versus $1/n$ become linear and approach a limiting value of a as $n \rightarrow \infty$ [80]. From the plots for cyanopolyynes $NC(C{\equiv}C)_nCN$ [77b] and those for both 3,5-bis (*tert*-butyldimethylsiloxy)phenyl derivatives and aliphatic dendrimers [72], the limiting values as $n \rightarrow \infty$ were estimated as 550 and

569 nm, respectively. Since the limiting values can be regarded as the absorption maximum of the sp carbon allotrope carbyne, these observations suggest that carbyne should be a colored insulating material.

Transition metal endgroups stabilize polyynes through the bulkiness of the ligands on the metals, which protect the reactive sp carbons from intermolecular contacts. In this field, a number of relatively long-chain (longer than $n = 8$) polyyne complexes, $L_mM(C≡C)_nML_m$, end-capped by transition metal groups, have been synthesized. The longest polyyne chain ($n = 12$) of an isolated metal complex is longer than that of the longest C-terminated polyyne ($n = 10$). Moreover, polyyne-bridged metal complexes exhibit a variety of redox properties relating to molecular wires, in which an electron, charge, or exciton communicates through the π bonds of the rod-like molecules. In principle, acetylenic complex **I** can be transformed into the dicationic species **III**, with a cumulenic bridge, by two one-electron oxidations through the intermediate radical cation **II** with a mixed valence state, as shown in Scheme 9.1. In addition, when the metals are bridged by an odd number of sp carbon atoms, the carbyne-like form **V** may also occur in addition to the cumulenic form **IV** (Scheme 9.2). For these reasons, organometallic polyynes have been studied intensively by the groups of Gladysz, Lapinte, Akita, Bruce, and others, as summarized in Table 9.4 [81].

I $L_nM—C≡C{-}(C≡C){-}C≡C—ML_n$

$+e^- \updownarrow -e^-$

II $[L_nM—C≡C{-}(C≡C){-}C≡C—ML_n]^{·+}$

$+e^- \updownarrow -e^-$

III $[L_nM=C=C{-}(C=C){-}C=C=ML_n]^{2+}$

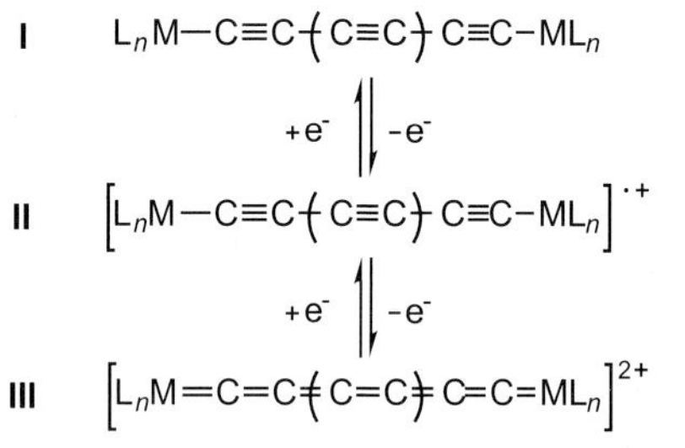

Scheme 9.1 Redox cycle of polyynes with transition metal endgroups and even number of sp carbons.

$L'_nM'=C{-}(C=C){-}C=C=M'L'_n \longleftrightarrow L'_nM'≡C{-}(C≡C){-}C≡C—M'L'_n$

            IV   **V**

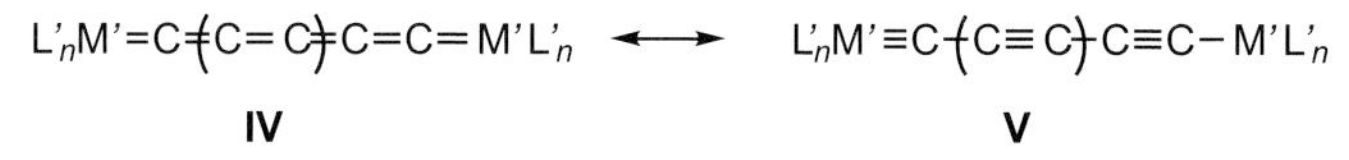

Scheme 9.2 Resonance structures of polyynes with transition metal endgroups and odd number of sp carbons.

Tetrayne-bridged diiron complex **5**, with (η^5-C_5Me_5)Fe(dppe) endgroups (dppe = 1,2-bis(diphenylphosphino)ethane), was synthesized by Coat and Lapinte [82]. Cyclic voltammetry (CV) of **5** in dichloromethane showed two reversible one-electron processes with a large wave separation, $\Delta E = |E_1° - E_2°| = 0.43$ V, indicating strong interaction between the metal centers through the carbon chain separated as long as 13 Å. Moreover, the thermally stable mixed-valence complex [Fe^{II}–C_8–Fe^{III}](PF_6^-) was prepared by oxidation of **5** with a ferrocenium salt. These results indicate that the tetrayne chain serves as a molecular wire connecting two electron-rich metal centers.

Table 9.4 Synthesis and characterization of linear polyynes $L_mM(C{\equiv}C)_nML_m$ end-capped by transition metal groups.

ML_n	*Number of C≡C unit n*	*(a) Method of preparation and (b) characterization*	*Reference*
$Fe(C_5Me_5)(dppe)^{a,b}$	4	(a) Eglinton coupling (b) IR, ^1H, ^{13}C, and ^{31}P NMR, CV (reversible two one-electron oxidations)	82
$Fe(C_5Me_5)(CO)_2{}^{b}$	1–4, 6	(a) Hay coupling (b) IR, ^1H and ^{13}C NMR, X-ray (*n*=2, 6)	83
$Ru_2(dpf)_4{}^{c}$	4	(a) Hay coupling (b) IR, ^1H and ^{13}C NMR, X-ray, CV (irreversible oxidation)	84
$Ru(C_5H_5)(PPh_3)_2{}^{d}$	3, 4	(a) substitution at the metal by acetylenes generated *in situ* (b) IR, ^1H NMR, MS	85
$W(C_5H_5)(CO)_3{}^{d}$	2, 4	(a) Hay coupling (b) IR, ^1H and ^{13}C NMR, MS	86
$Au(PCy_3)^{e}$	1–4	(a) substitution at the metal by acetylenes generated *in situ* (b) IR, UV/Vis, ^1H, ^{13}C, and ^{31}P NMR, MS, X-ray (*n*=3)	87
Re ($^tBu_2bpy$) $(CO)_3{}^{f}$	4, 6	(a) Eglinton coupling (b) IR, UV/Vis, ^1H NMR, MS	88
$Re(C_5Me_5)(NO)(PAr_3)^{b,g}$	4–6, 8, 10	(a) Eglinton coupling (b) IR, Raman, UV/Vis, ^1H, ^{13}C, and ^{31}P NMR, MS, X-ray (*n*=4), CV (reversible two, one-electron oxidations for *n*=4–6, 8)	75, 89, 90
$Re(C_5Me_5)(NO)(PPh_3)^{b}$ $Fe(C_5H_5)(dppe)^{a,b,d}$	2	(a) substitution at Fe(II) by acetylenes generated *in situ* (b) IR, UV/Vis, ^1H, ^{13}C, and ^{31}P NMR, MS, X-ray, CV (reversible, two one-electron oxidations)	91
$Re(C_5Me_5)(NO)$ $(PPh_2R)^{a,h}$	4	(a) Eglinton coupling, followed by olefin metathesis or intramolecular Eglinton coupling (b) IR, ^1H, ^{13}C, and ^{31}P NMR, MS	92
$Pt(Ar)(PAr'_3)_2{}^{i}$	4, 6, 8, (10, 12)	(a) Hay coupling (b) IR, UV/Vis, ^1H, ^{13}C, and ^{31}P NMR, MS, X-ray (*n*=4, 6, 8), CV (reversible two one-electron oxidations except for *n*=8), only UV/Vis for *n*=10 and 12	93
$Pt(C_6F_5)(PAr_2R)_2{}^{j}$	4, 6	(a) Hay coupling, followed by olefin metathesis or phosphine ligand exchange (b) ^1H, ^{13}C, and ^{31}P NMR, X-ray	94

a) dppe = diphenylphosphinoethane
b) C_5Me_5 = η^5-pentamethylcyclopentadienyl
c) dpf = *N,N'*-diphenylformamidine
d) C_5H_5 = η^5-cyclopentadienyl
e) PCy_3 = tricyclohexylphosphine
f) $^tBu_2bpy$ = di-*tert*-butylbipyridyl
g) Ar = phenyl, *p*-tolyl, 4-*tert*-butylphenyl, 4-biphenyl, cyclohexyl
h) R = methylene chain
i) Ar = *p*-tolyl, pentafluorophenyl; Ar' = phenyl, *p*-tolyl
j) R = methylene chain; Ar = phenyl, *p*-tolyl, 4-*tert*-butylphenyl

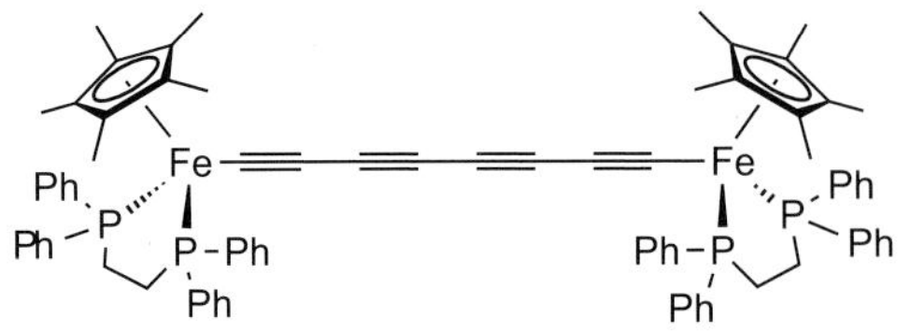

5

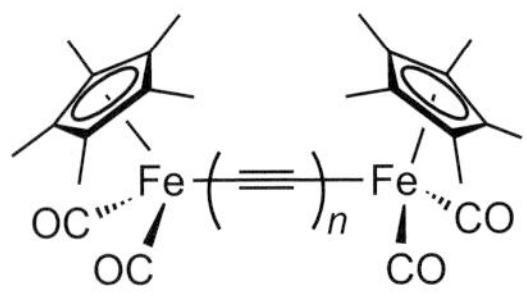

6a $n = 2$
6b $n = 4$
6c $n = 6$

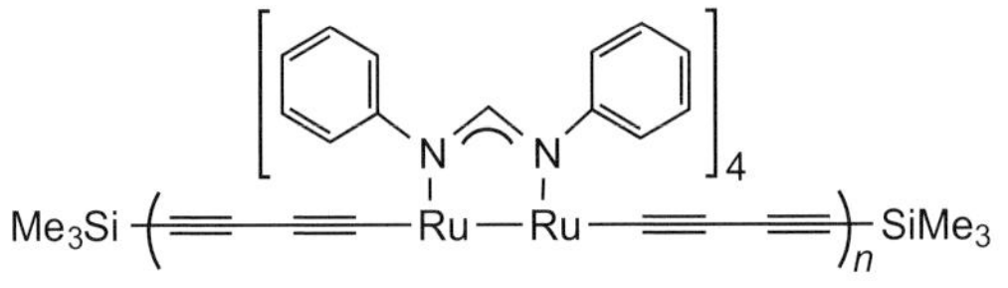

7a $n = 2$
7b $n = 3$

A series of diiron complexes **6a–c** with $(\eta^5\text{-}C_5Me_5)Fe(CO)_2$ endgroups and $(C{\equiv}C)_n$ chains with $n = 2, 4, 6$ were prepared by Akita et al. [83]. The bond lengths determined by X-ray crystallographic analyses of **6a** and **6c** suggested small contributions by the cumulenylidene forms. The C_{12} chain of **6c** deviates considerably from linearity, with the largest deviation of 8° at the C_2 position.

Conjugated linear diruthenium complexes **7a** and **7b** bridged by tetrayne units were prepared as models of molecular wires, though **7b** was not fully characterized [84]. X-ray analysis of **7a** revealed significant deviation from linearity, with a Ru–Ru–C bond angle of 163°. The CV of **7a** exhibits two irreversible one-electron oxidations with ΔE of 0.31 V, indicating that the first radical cation is efficiently delocalized over the system.

Ruthenium complex **8** with $(\eta^5\text{-}C_5H_5)Ru(PPh_3)_2$ endgroups [85], tungsten complex **9** with $(\eta^5\text{-}C_5H_5)W(CO)_3$ endgroups [86], gold complex **10** with $Au(PCy_3)$ endgroups (Cy = cyclohexyl) [87], all with a $(C{\equiv}C)_n$ chain up to $n = 4$, and rhenium complexes **11a–c** with $Re(^tBu_2bpy)(CO)_3$ endgroups (bpy = bipyridine) and polyyne bridges of $n = 2, 4, 6$ have been isolated and characterized spectroscopically [88].

Gladysz et al. synthesized a series of redox-active rhenium complexes **12a–g** with $(\eta^5\text{-}C_5Me_5)Re(NO)(PPh_3)$ endgroups and polyyne chains $(C{\equiv}C)_n$ with $n = 2, 3, 4, 5, 6, 8$, and 10, and investigated their spectral, structural, and redox properties [75, 89]. Because of the existence of *meso* and *dl* isomers due to the chirality of the rhenium center, however, the longest carbon chain belonging to a structurally fully characterizable polyyne – **12c** ($n = 4$) – in this series is rather short. In general, the spectral features exhibit constant shifts with increasing chain length. The IR stretching frequencies of the NO group, for example, increase from 1630 cm^{-1} for **12a** to the limit of 1653–1654 cm^{-1} for **12e–g** ($n = 6, 8, 10$) due to the increasing inductive effect of the polyyne group. Not only the wavelengths of the absorption maxima but also the molar extinction coefficients increase with increasing chain length; ϵ of the longest polyyne **12g** approaches $2 \times 10^5\ \text{M}^{-1}\ cm^{-1}$. The absorption

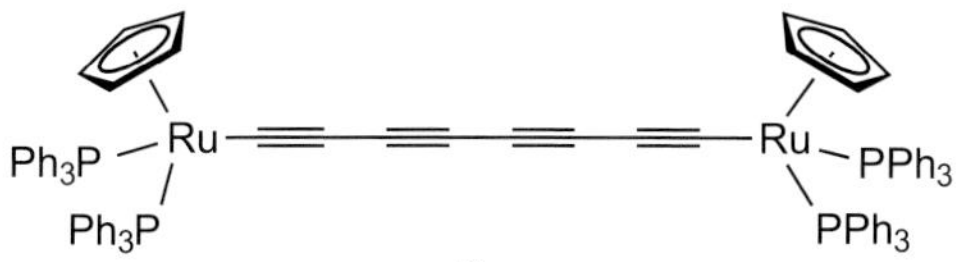

8

9

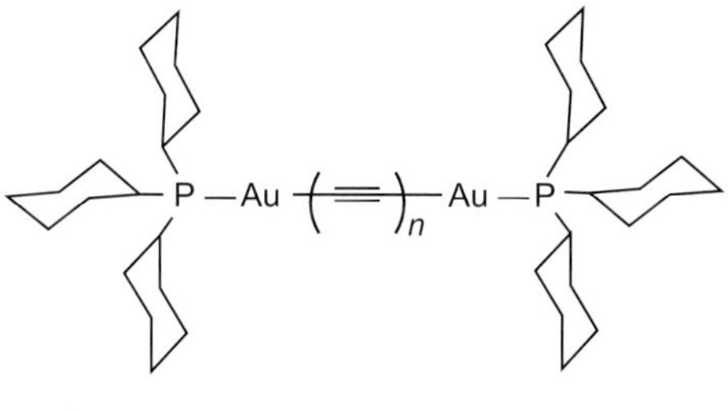

10 $n = 1\ 4$

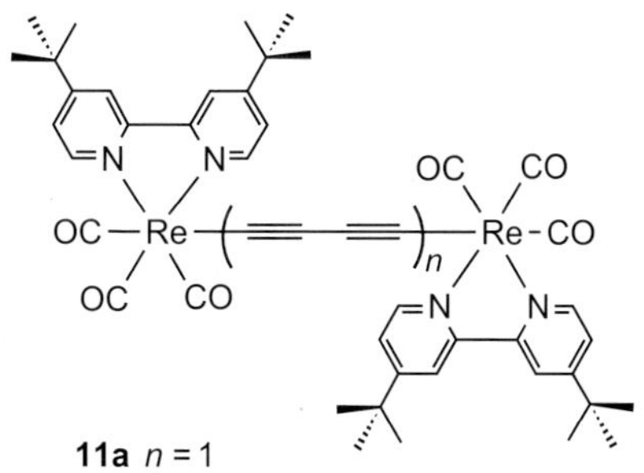

11a $n = 1$
11b $n = 2$
11c $n = 3$

wavelengths plotted to the empirical equation $E_{exp} = a + b/n$ gave a good linear relationship, yielding a limiting value a of 565 nm for $n \rightarrow \infty$, similar to those obtained for the linear polyynes with other endgroups. The CV of **12a** exhibits reversible two one-electron oxidations with ΔE of 0.53 V, indicating strong communication between the two metal centers. The redox reversibility and the ΔE values decrease with increasing chain length, implying that the two metal centers become more independent. Although it is to be expected that the oxidation potential should decrease with increasing conjugation length, the opposite trend was observed in a result ascribed to predominance of the polyyne chain's inductive effect over the resonance effect. While in the case of **12a** the corresponding radical cation salt **12a**$^{+}$ PF_6^- and dication salt **12a**$^{2+}$ $2PF_6^-$ were isolated, attempted chemical oxidation of **12b** ($n = 3$) and **12c** ($n = 4$) failed, even on changing the ligand to more sterically demanding and/or electron-rich phosphines [90]. However, in analogy with the successful isolation of radical cation/dication salts of octatetrayne-bridged diiron complex **5** and dirhenium complex **12a**, chemical oxidation of the mixed iron/rhenium complex **13** afforded the corresponding radical cation salt **13**$^{+}$ PF_6^- and dication salt **13**$^{2+}$ $2PF_6^-$, which constitute the first "conjugal" consanguineous family of the polyyne complexes [91]. An odd number of electrons in **13**$^{+}$ PF_6^- is delocalized between the two metal centers. Owing to the antiferromagnetic coupling between unpaired spins on the rhenium and iron endgroups in dication $\mathbf{13}^{2+}$ $2PF_6^-$, the singlet cumulenic state is stabilized over the triplet butadiynyl form. In order to prepare stable radical cation and dication salts of the rhenium complex with a $(C{\equiv}C)_4$ linkage, complexes **14** and **15** possessing shielding alkyl chains were prepared by intramolecular Grubbs coupling (ring-closure metathesis) or by intramolecular oxidative coupling of the butadiyne units as the last step of the synthesis [92]. However, the corresponding oxidation products have yet to be prepared.

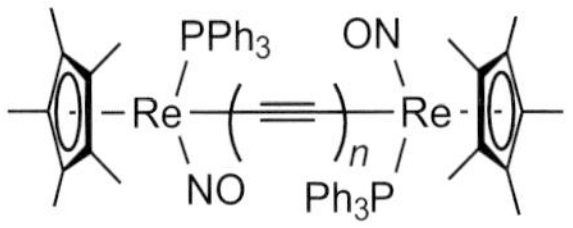

12a *n* = 2
12b *n* = 3
12c *n* = 4
12d *n* = 5
12e *n* = 6
12f *n* = 8
12g *n* = 10

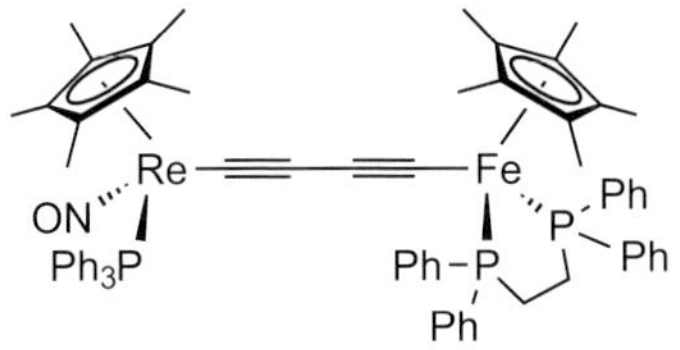

13

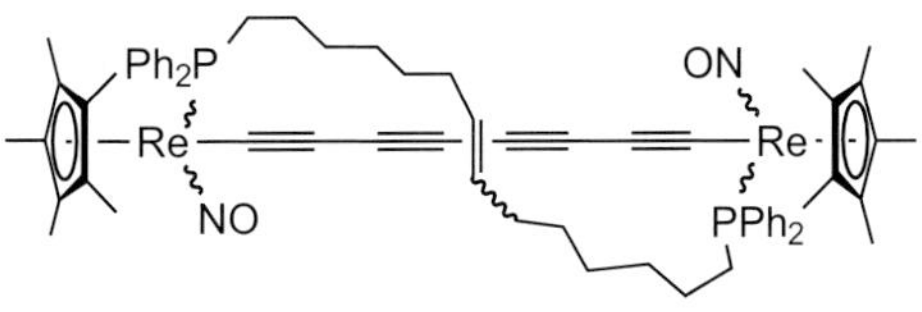

14

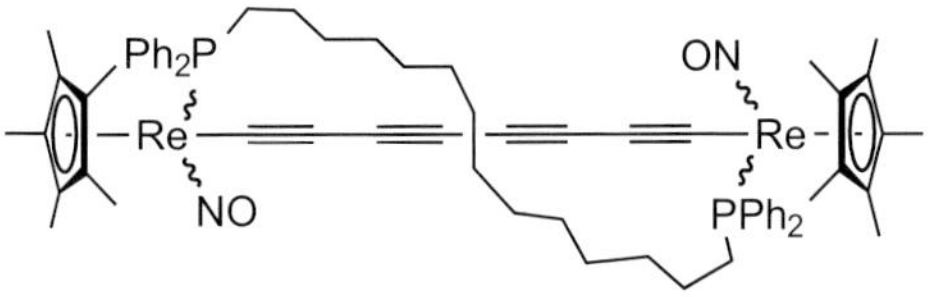

15

Platinum complexes **16a**, **16b** and **17a–f** with long polyyne bridges were also prepared by Gladysz et al. [93]. Polyyne complexes **16a** and **16b** with (*p*-tol)Pt[P(*p*-tol)$_3$]$_2$ endgroups (tol = tolyl) and $(C{\equiv}C)_n$ linkages with n = 4 or 6 were prepared and their structures were characterized by X-ray analysis. Attempts to prepare longer homologues were hampered by the lability of the C_8 terminal alkyne complex. This problem was solved by the use of more powerfully electron-withdrawing aryl group on the metal. Thus, complexes **17a–d**, with (C_6F_5)Pt[P(*p*-tol)$_3$]$_2$ endgroups and $(C{\equiv}C)_n$ linkages of n = 2, 4, 6, 8, were characterized both spectroscopically and by X-ray structural analyses. While the homologues **17e** and **17f**, with longer polyyne chains

(n = 10 and 12, respectively), were also isolated, only UV/Vis spectra were recorded, owing to the small quantities obtained. The most striking feature of the X-ray analysis of the polyyne-bridged platinum complexes is the remarkable bending of the polyyne chain observed in the co-crystal of **17c** (n = 6: with four benzene molecules and one ethanol molecule) with average Pt−C−C and C−C−C angles of 174.6°. In contrast, the sp carbon chains of **16a** and **17b** (n = 4), **16b** and a second crystal of **17c** (without solvates, n = 6), and **17d** (n = 8) are nearly straight. The slightly S-shaped structure of **17d** represents the longest structurally characterized polyyne. Only one-electron oxidation processes were observed in the CV of **17a–d**, with decreasing reversibility with increasing chain length. The oxidation potential increases with increasing number of sp carbons, indicating that the HOMO levels decrease with chain length in spite of the increasing conjugation. On the other hand, the UV/Vis spectra of **17a–f** exhibit monotonic bathochromic shifts with increasing chain length. This indicates that the LUMO levels decrease more dramatically than the HOMO levels with chain length, making the HOMO/LUMO gaps smaller. The absorption spectra also show progressively more intense bands with increasing chain length; the ϵ value of **17d** (6×10^5 M^{-1} cm^{-1}) represents one of the most intense polyyne absorptions.

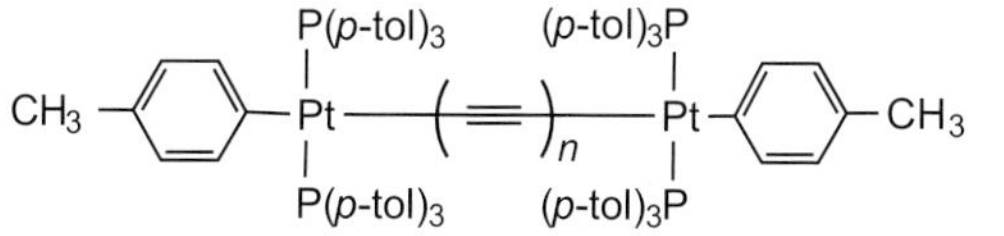

16a n = 4
16b n = 6

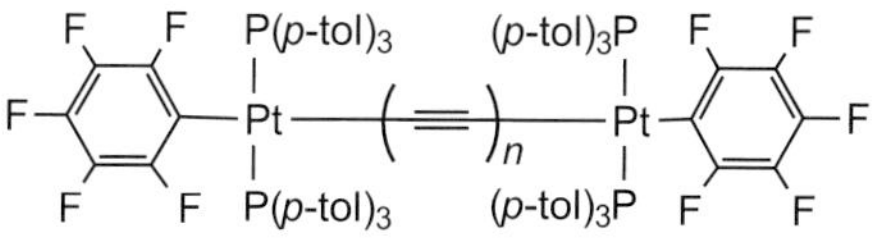

17a n = 2
17b n = 4
17c n = 6
17d n = 8
17e n = 10
17f n = 12

In order to stabilize charged and/or radical cation species originating from oxidation of polyyne-bridged metal complexes by steric shielding, Gladysz et al. synthesized the novel platinum end-capped polyynes **18a–f**, some of which have double helical methylene chains reminiscent of insulating molecular wires [94]. The methylene bridges were introduced either by Grubbs metathesis of the alkenyl chain on the phosphine ligand followed by hydrogenation (for **18a** (n = 4)) or by ligand-exchange reactions of **17b** (n = 4) or **17c** (n = 6) with long-chain diphosphines (for **18a–f**). The structures of the methylene bridges are dominated by matching between the lengths of the sp carbon chains and those of the bridges. Thus, complexes **18a–c** with $(C{\equiv}C)_4$ linkages and $(CH_2)_{14}$ bridges feature double helical conformations of the methylene bridges, as revealed by X-ray structure analysis. On the other hand, the methylene chains in the two polymorphs of **18d** with shorter bridges ($(CH_2)_{10}$) are not helical: one has a straight bridge laterally shielding the sp carbons and the other has the bridge warped in the same direction exposing the bent polyyne moiety. Two independent molecules are present in the

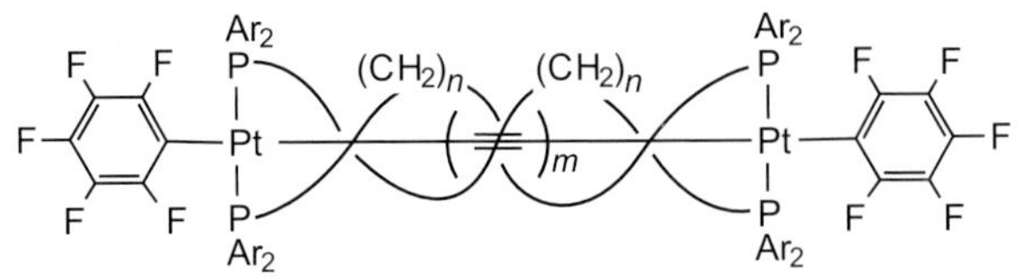

18a $m = 4$, $n = 14$, Ar = Ph
18b $m = 4$, $n = 14$, Ar = *p*-tol
18c $m = 4$, $n = 14$, Ar = *p*-$C_6H_4Bu^t$
18d $m = 4$, $n = 10$, Ar = Ph
18e $m = 4$, $n = 11$, Ar = Ph
18f $m = 6$, $n = 18$, Ar = Ph

crystal of **18f** with a $(C{\equiv}C)_6$ linkage and $(CH_2)_{18}$ bridges: one with helical and the other with nonhelical conformations. The oxidation in the CV of **18a** was much more reversible than that recorded for **17b** without shielding, indicating the formation of kinetically more stable radical cation species.

9.5 Monocyclic Carbon Clusters: Cyclo[*n*]carbons

Among the various forms of carbon clusters, monocyclic clusters, known as cyclo[*n*]carbons, have been of considerable interest from many points of view and for several reasons, ranging from spectroscopy and theoretical chemistry to synthetic organic chemistry [95]. Firstly, in carbon cluster chemistry, little has been done towards the spectroscopic determination of the structures of monocyclic carbon rings, compared to the extensive studies for the linear forms as described in Section 9.2. Secondly, it is likely that monocyclic carbon clusters play a key role during the early stages of the mechanism of fullerene formation [96], although the mechanism has not been clarified. Namely, ample experimental [97] and theoretical [98] evidence that has been accumulating suggests that carbon nucleation occurs through coalescence of medium-sized carbon clusters, rather than by sequential addition of small pieces such as C_2 and C_3, followed by annealing to form the cage structure. Thirdly, cyclo[*n*]carbons can be regarded as completely dehydrogenated derivatives of annulenes. Partially dehydrogenated annulenes had been extensively studied in association with the Hückel rule. In addition, many derivatives have recently been prepared for purposes involving potential opto-electronic applications [99]. It is quite natural to ask whether the Hückel rule is also valid for this class of novel π systems [100], and a number of theoretical calculations have been undertaken. However, there still remains controversy regarding the most stable geometries of cyclo[*n*]carbons with $4n + 2$ π electrons, particularly for cyclo[18]carbon [101]. In addition, in view of the expected high reactivity of such molecules, it might be possible to prepare the hitherto unknown, two-dimensional carbon network known as graphyne [52], made up of sp and sp^2 carbon atoms, by controlled [2 + 2 + 2] cyclotrimerization of the triple bonds of *cyclo*-C_{18} [95a, 102]. Finally, cyclo[*n*]carbons have been attracting interest as possible interstellar materials. Sarre et al., for example, reported, from analysis of ultra-high resolution absorption spectra of the diffuse interstellar bands and molecular rotational contour calculations, that large carbon ring molecules with planar oblate symmetric structures (i. e., cyclo[*n*]carbons) are good candidate sources of these bands [103].

In this context, a number of investigations to produce cyclo[*n*]carbons have been undertaken, both from graphite, as described in Section 9.2, and from organic precursors in order to determine their molecular structures by spectroscopic methods. A pioneering project to the synthesis of cyclo[*n*]carbons from organic precursors was initiated by Diederich and Rubin, who first prepared precursor **19**, possessing three dibenzobicyclo[2.2.2]octatriene units, from which expulsion of three molecules of anthracene in a retro-Diels–Alder sense was expected to generate cyclo[18]carbon [104]. The formation of C_{18} was indeed observed in the laser-desorption time-of-flight (LD TOF) mass spectra of **19**. Attempts to prepare macroscopic quantities of C_{18} were unsuccessful, however. Next, stable dinuclear cobalt complexes **20a** and **20b**, which can be regarded as transition metal complexes of C_{18} and C_{24}, respectively, were prepared [104b, 105]. Attempted decomplexations of **20a** and **20b** have also been unsuccessful. The third type of precursor are the dehydroannulenes **21a–c**, incorporating cyclobutenedione units, which would produce cyclo[*n*]carbons by multi-step decarbonylation [97d, 104b, 106]. The formation of C_{18}, C_{24}, and C_{30} ions, produced by the stepwise loss of carbon monoxide, was indeed observed in both positive and negative modes of the laser-desorption (LD) Fourier transform (FT) mass spectra of the respective precursors **21a–c**. Interestingly enough, C_{60}^+, which proved to possess the fullerene structure, was formed by coalescence between C_{30}^+ and C_{30} derived from the precursor **21c**. Irradiation of **21a** dispersed in a low-temperature glass matrix resulted in the formation of ketene intermediates and subsequent loss of carbon monoxide. However, definite spectroscopic evidence for the formation of cyclo[18]carbon has yet to be obtained [104b].

As stable precursors of cyclocarbons, Tobe and Wakabayashi utilized [4.3.2]propella-1,3,11-triene units, which might generate carbon-carbon triple bonds in a retro [2 + 2] sense ([2 + 2] cycloreversion). To this end, dehydroannulenes **22a–c** and **23a** and **23b**, annelated with propellatriene units, were prepared as the precursors of cyclo[*n*]carbons with $n = 12, 16, 20, 18$, and 24, respectively [107].

In the negative mode LD TOF mass spectra of **22a–c** and of **23a** and **23b**, the formation of cyclo[*n*]carbon anions ($n = 12, 16, 18, 20, 24$) resulting from the successive expulsion of the indane fragments from the precursors was indeed observed. The structures of the generated carbon cluster anions (C_{12}^-, C_{16}^-, C_{18}^-, C_{20}^-, and C_{24}^-) were investigated by ultraviolet photoelectron spectroscopy (UPS), the spectra being compared with those of the anions generated by laser vaporization of graphite followed by mass selection [22]. From the splitting frequencies due to vibrational excitation of the neutral cyclocarbons produced by detachment of single electrons from the corresponding anions, it was deduced that cyclo[*n*]carbons with 4*m*

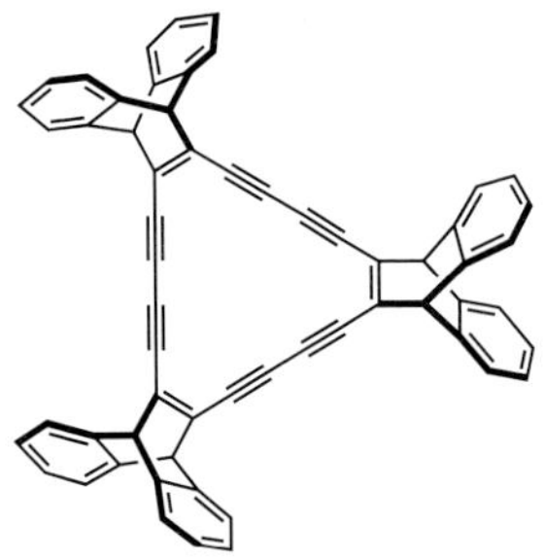

19

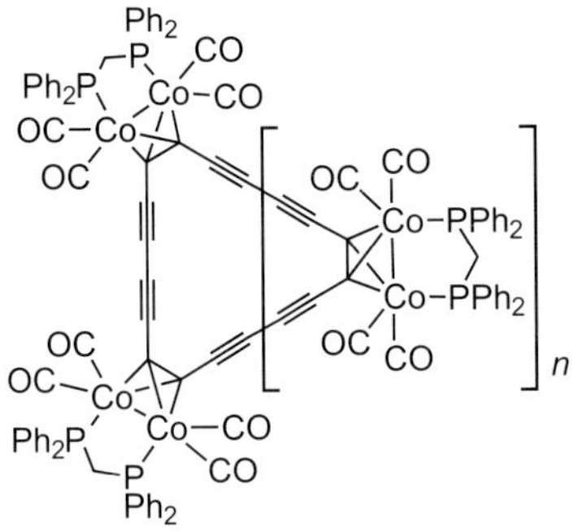

20a $n = 1$
20b $n = 2$

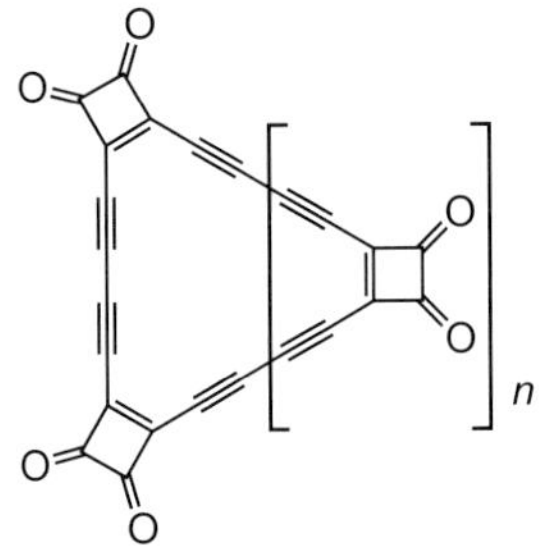

21a $n = 1$
21b $n = 2$
21c $n = 3$

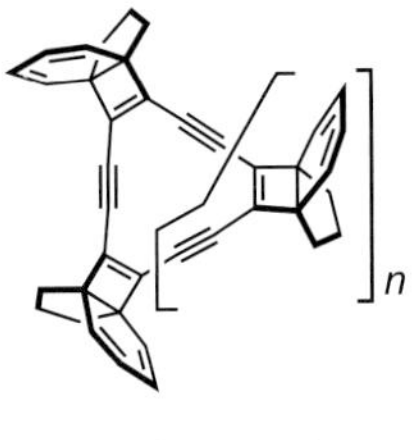

22a $n = 1$
22b $n = 2$
22c $n = 3$

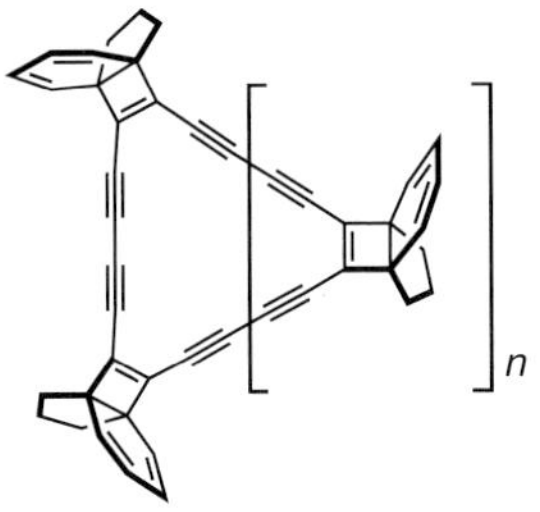

23a $n = 1$
23b $n = 2$

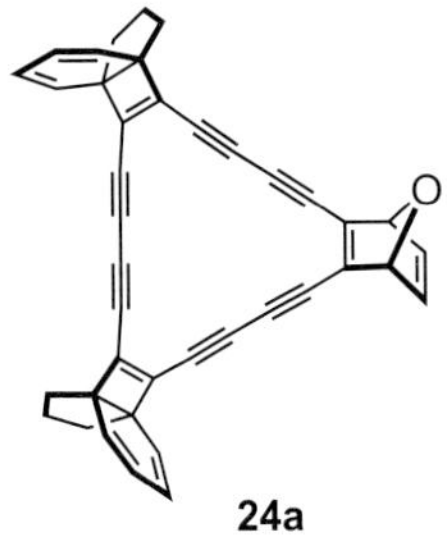

24a

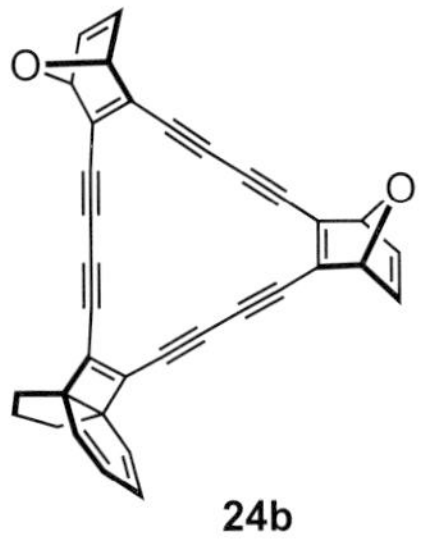

24b

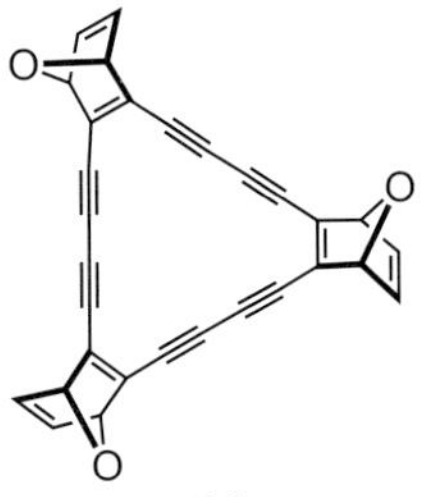

24c

carbon atoms (i. e., C_{16}, C_{20}, and C_{24}) should possess polyyne structures with alternating single and triple bonds, while cyclo[18]carbon, with $4m + 2$ carbon atoms, would have to have a cumulenic structure.

While UV irradiation of a solution of **23a** in THF-d_8 at 0 °C resulted in the formation of polymeric materials, indicating that the intermediates are too reactive for characterization by NMR, irradiation in furan afforded three oxanorbornadiene derivatives **24a–c**, in which one, two, or all three indane units are replaced by furan moieties. These results suggest that the [2 + 2] cycloreversion did take place at least in stepwise fashion [107b, 107d] However, the limited solubility in conjunction with low volatility precluded the isolation of the precursors either in rigid glass or in rare gas matrices.

In contrast, photochemical [2 + 2] cycloreversion was successfully employed in the generation of dibenzooctadehydro[12]annulene (**26a**), which can be regarded as a tetrahydro derivative of cyclo[12]carbon (Scheme 9.3). Annulene **26a** was produced by extrusion of an indane unit from its precursor **25a** by UV irradiation and was characterized by UV/Vis spectroscopy in a rigid glass matrix at 77 K and by FT-IR spectroscopy in an argon matrix at 20 K [108]. Although **26a** was too reactive for observation in solution, its higher homologue dibenzodecadehydro[14]annulene (**26b**), similarly produced by photolysis of its precursor **25b**, was stable enough for characterization in solution by NMR spectroscopy [109].

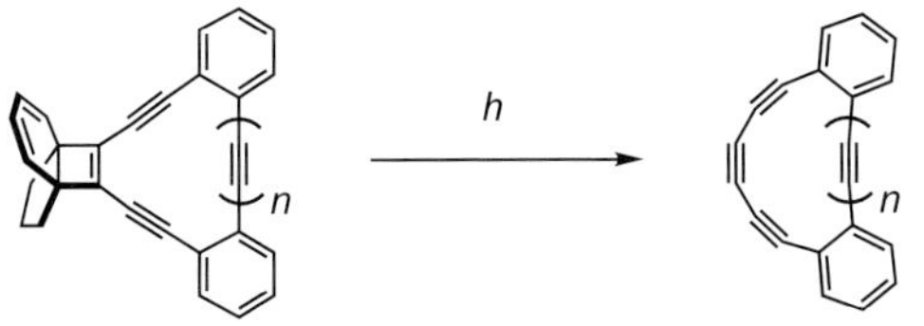

25a $n = 1$
25b $n = 2$

26a $n = 1$
26b $n = 2$

Scheme 9.3 Photochemical [2 + 2] cycloreversion of precursors **25a** and **25b** to form highly strained dehydrobenzoannulenes **26a** and **26b**.

In addition to the above precursors containing four-membered rings, those with three-membered rings also act as potential precursors of cyclocarbons. Thus, [2 + 1] cheletropic fragmentation of "exploded" [*n*]rotanes such as the cyclic dehydrooligomers of diethynyl[4.4.1]propellatetraenes **27a** and **27b** with appended fullerene moieties have been prepared, and their fragmentation was investigated by mass spectrometry [110]. The MALDI TOF mass spectra of **27a** and **27b** exhibited peaks due to the stepwise loss of the fullerene fragments, resulting in the formation of mono-fullerene adducts of C_{15} and C_{20}. Observation of cyclocarbons was not achieved, however, because of the extensive fragmentation under strong laser power. Dehydro-oligomers of 1,1-diethynylcyclopropane such as **28a** and **28b**, from which extrusion of tetramethylethylene or ethylene could form cyclo[5*n*]carbons, have been prepared. However, it remains to be seen whether these form C_{30} and C_{60}, respectively, by fragmentation in the gas phase [111].

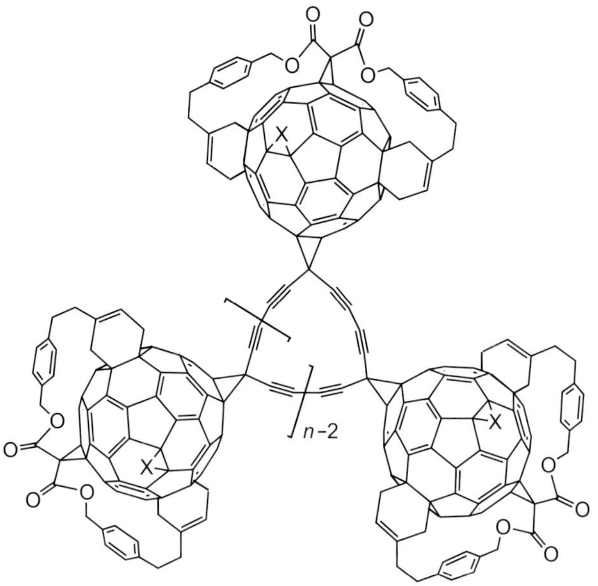

27a $n = 3$
27b $n = 4$

$X = (EtO_2CCH_2O_2C)_2C$

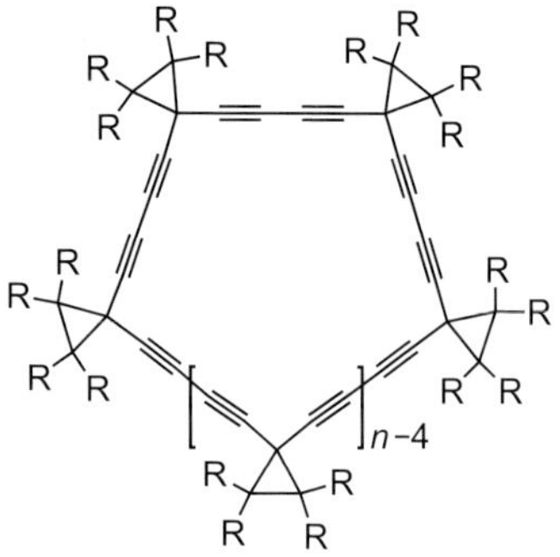

28a $R = H, n = 12$
28b $R = Me, n = 6$

In the case of the [4.3.2]propellatriene-type precursors **23a** and **23b**, because of the relatively small angle (ca. 90°) between the triple bonds of the diethynyl[4.3.2] propellatriene unit, only the cyclic dehydrotrimer **23a** and the dehydrotetramer **23b** were obtained; the larger cyclic dehydro-oligomers such as pentamer and hexamer could not be prepared. As new precursors for large-ring cyclocarbons, radialenes **29a–d**, containing bicyclo[4.3.1]deca-1,3,5-triene units, were designed, since the C(sp)–C(sp^2)–C(sp) bond angle of the *exo* methylene carbon of the diethynylated precursor was expected to be nearly 120° [112].

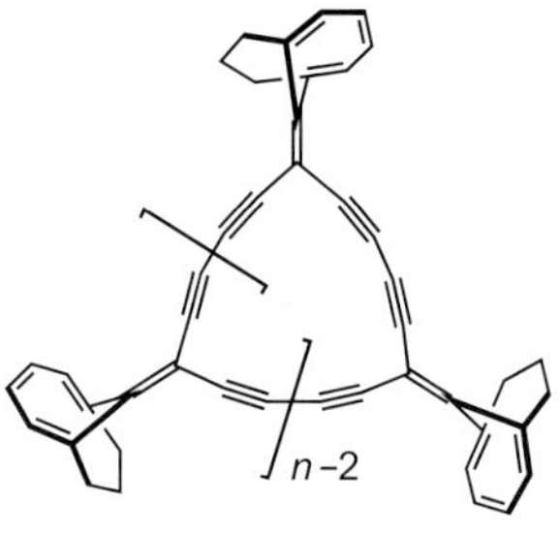

29a $n = 3$
29b $n = 4$
29c $n = 5$
29d $n = 6$

It was speculated that extrusion of an aromatic fragment (indane) from **29a–d** should produce the corresponding vinylidenes, which would then isomerize to form polyynes as illustrated in Scheme 9.4 for C_{36}. Large cyclic dehydro-oligomers, such as the pentamer **29c** and the hexamer **29d** were indeed obtained, together with smaller homologues **29a** and **29b**. The negative mode LD TOF mass spectra of **29a–d** exhibited peaks due to the corresponding carbon anions (C_{18}^-, C_{24}^-, C_{30}^-, and C_{36}^-) formed by stepwise loss of the aromatic indane fragments. Although no spectroscopic evidence as to the structures of the carbon cluster anions produced was provided, it was assumed that these should possess cyclocarbon structures based on the relative stability considerations and in view of the results of photolysis of acyclic model compounds. In the case of the only weakly observed C_{36}^-, however, the formation of carbon cluster anions of different structures seems likely since the monocyclic forms of carbon clusters are most stable within the $C_{10}-C_{30}$ regime only.

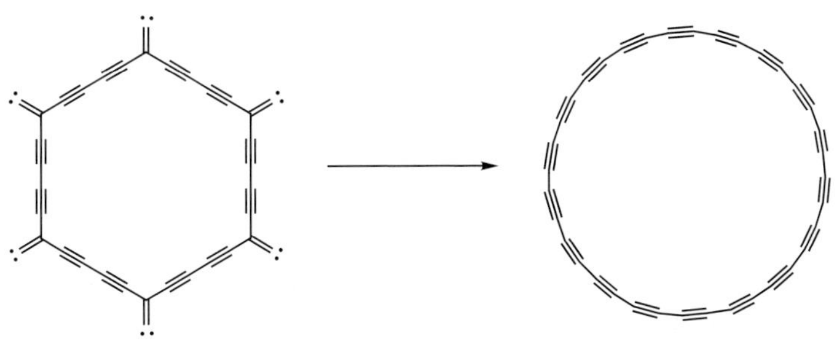

Scheme 9.4 Vinylidene-to-acetylene rearrangement to form cyclo[36]carbon.

9.6 Three-Dimensional Multicyclic Polyynes

Unlike the monocyclic forms of carbon clusters, none of the structures of multicyclic polyyne-based carbon clusters has been characterized since many structural isomers exist, most of them thermodynamically more stable than the polyyne-based isomers. Collapse of the polyynes takes place to the more stable forms, presumably with fullerene-like cage structures. C_{60}^+, for example, was formed by ion-molecule reaction of monocyclic C_{30}, cyclo[30]carbon, and it has been shown to be identical with the [60]fullerene ion [97d, 106b]. An inspiring mechanism for the formation of [60]fullerene from a [2 + 2] dimer of C_{30} was proposed, consisting of sequential ring-formation/-cleavage steps, followed by the spontaneous cyclization of the polyyne chains to form the five- and six-membered rings of the [60]fullerene cage [97e]. Multicyclic polyynes have attracted considerable interest in this respect, in view of their size-selective, and – more intriguingly – structure-selective, transformation into the fullerene structures.

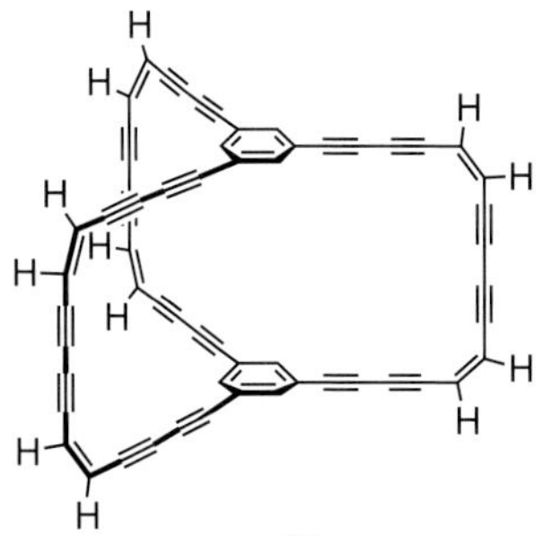

30

Inspired by the polyyne cyclization mechanism, Rubin's and Tobe's groups independently investigated the synthesis of cage-shaped polyynes and their transformation into [60]fullerene. As an initial step, Rubin et al. prepared $C_{60}H_{18}$ (**30**), made up of two benzene rings and three C_{16} enyne bridges [113]. In the MALDI and APCI (atmospheric pressure chemical ionization) mass spectra (negative mode) of **30**, partial dehydrogenation down to $C_{60}H_{14}^-$ was observed, suggesting the possibility of complete dehydrogenation from more highly unsaturated precursors to C_{60}. The desired precursor $C_{60}H_6$ (**31a**), with three C_{16} polyyne chains (Scheme 9.5), was deemed too reactive for isolation in view of the previous studies on the synthesis of linear polyynes as described in Section 9.4. Rubin et al. therefore utilized the decarbonylation of cyclophane **32**, possessing cyclobutenedione units, to generate **31a** [114]. As expected, in the negative mode ICR (ion cyclotron resonance) LD mass spectrum of **32**, not only $C_{60}H_6^-$ but also a strong peak due to C_{60}^- were observed, suggesting that cyclization of the polyyne chain had taken place efficiently, accompanied by dehydrogenation to form the fullerene cage.

Tobe and Wakabayashi et al., on the other hand, used the [2 + 2] cycloreversion strategy to produce polyyne $C_{60}H_6$ (**31a**) [115]. The positive mode LD TOF mass spectrum of the [4.3.2]propellatriene-containing precursor **33a** exhibited a peak due to C_{60}^+ formed by the loss of six indane fragments and six hydrogen atoms. Although C_2 loss down to C_{50}^+ (an observation frequently encountered for carbon cluster cations generated by the laser desorption method) was observed, it has not yet been confirmed whether or not the structure of the C_{60} cation retains the I_h symmetry of fullerene. Moreover, hexachloro derivative **33b** exhibited a very strong peak for C_{60}^+ in its positive mode LD TOF mass spectrum, together with weak peaks due to C_2 loss and others up to C_{120}^+, which might be formed by an ion-molecule reaction of C_{60} and the subsequent fragmentation of the dimeric cluster ion. The formation of C_{60}^+ is facilitated by electron capture by the chlorine atoms, as supported by the observation of a strong Cl^- peak in the negative mode spectrum. The negative mode LD TOF mass spectrum of **33a**, in contrast with the positive mode spectrum, showed the peak due to $C_{60}H_6^-$, formed by the loss of all indane units. The mass spectrum of **33a** also showed a small peak due to C_{60} anion formation by spontaneous dehydrogenative removal of all the hydrogen atoms in $C_{60}H_6^-$, as judged from the isotope distribution. The negative mode LD TOF mass

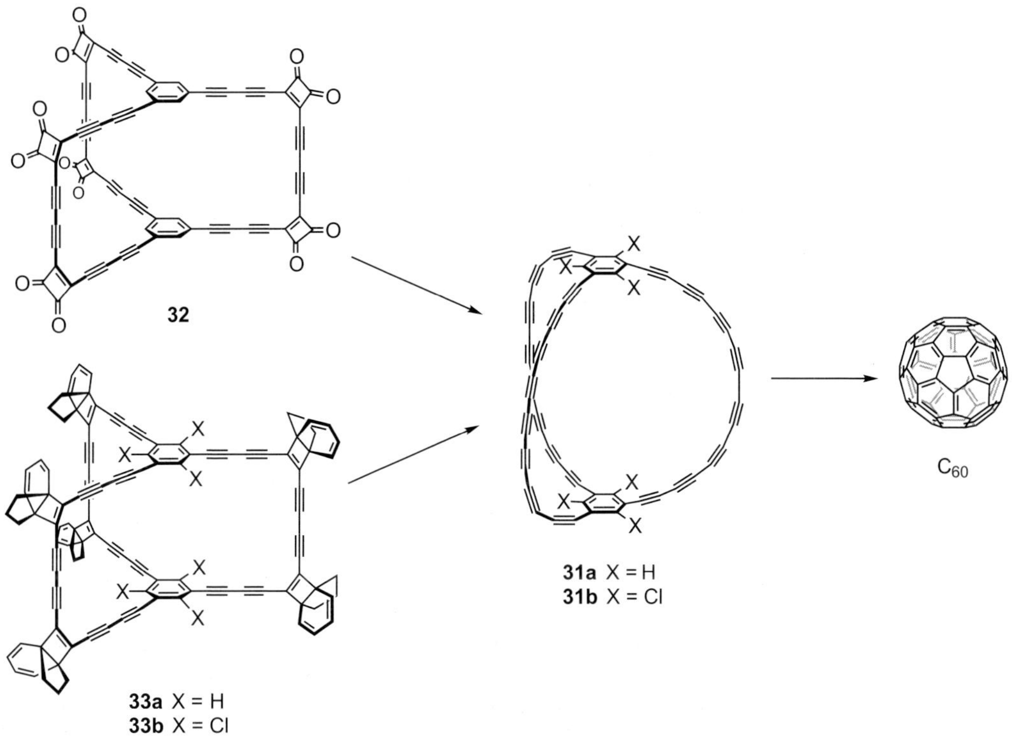

Scheme 9.5 Generation of highly reactive three-dimensional polyynes **31a** and **31b** from precursors **32**, **33a**, and **33b** and their transformation into C_{60} fullerene.

spectrum of **33b** also exhibited a peak due to $C_{60}Cl_6^-$, from which stepwise loss of chlorine atoms was observed, leading ultimately to C_{60}^-.

Although C_{60} and some higher fullerenes are produced on a commercial basis by arc-vaporization of graphite or combustion of hydrocarbons [116], it is still important to synthesize fullerenes in a rational manner based on the procedures of organic synthesis, as it should then be possible to construct the carbon cages in size-selective and structure-defined manner. It should also be possible to incorporate heteroatoms in the cage or to encapsulate transition metals in it, giving rise to the as yet undiscovered endohedral transition-metallofullerenes. Intensive studies in this area directed at the designed synthesis of C_{60} have been carried out [117]. Recently, Scott et al. succeeded in the synthesis of C_{60} by high-temperature pyrolysis of a polycyclic aromatic chlorohydrocarbon, during which dehydrogenation/dechlorination took place to form the closed cage [118]. The cyclization of reactive three-dimensional polyynes **31a** and **31b** as described above is a viable alternative route to C_{60}, even though [60]fullerene has not yet been obtained in a preparative scale by this method. It would be useful to extend the method for the size-selective formation of small and large fullerenes, because the required polyyne precursors should be easily preparable.

In view of the recently reported interception of C_{50} fullerene as a decachloro derivative $C_{50}Cl_{10}$ [119], many small fullerenes are likely to exist, albeit only in the gas phase. Some carbon clusters with magic numbers, such as C_{20} and C_{36}, have been shown to possess fullerene structures when they are produced from specific precursors or under certain conditions. C_{20} fullerene cation, for example, was produced selectively by field evaporation of carbon nanotubes [120]. Moreover, both positive and negative ions of [20]fullerene were formed by debromination of polybrominated dodecahedrane under mass spectrometric conditions, the cage structure being characterized by UPS [23]. Even though controversy remains regarding the structure of C_{36} [121], no attempt to produce C_{36} from an appropriate organic precursor has been made so far. Just as in the case of the proposed transformation of multicyclic polyyne $C_{60}H_6$ (**31a**) into [60]fullerene under mass spectrometric conditions, it was assumed that a similar cyclization of polyyne-bridged paracyclophanes **34a** and **34b** would give C_{36} of D_{6h} symmetry. Polyynes **34a** and **34b** are larger homologues of the polyyne-bridged [8.8]paracyclophane that has eluded synthesis [122], and are generated from the corresponding propellane derivatives **35a** and **35b** by [2 + 2] cycloreversion [123]. In the LD TOF mass spectrum of **35a** (negative mode), the cycloreversion took place cleanly to form $C_{36}H_8^-$ (**34a**$^-$). Unlike in the case of $C_{60}H_6^-$ (**31a**$^-$), however, subsequent dehydrogenation to afford C_{36}^- was not observed, which can be ascribed both to thermodynamic (C_{36} not as stable as C_{60}) and to kinetic (the polyyne chains of **34a** are located in remote positions) causes. In contrast, the mass spectrum of the octachloro derivative **35b** exhibits not only the $C_{36}Cl_8^-$ (**34b**$^-$) peak, but also those due to the stepwise loss of chlorine atoms down to C_{36}^-, analogously to the degradation of $C_{60}Cl_6^-$ (**31b**$^-$). Since the intensity of the C_{36}^- signal is considerably larger than those of $C_{36}Cl^-$, $C_{36}Cl_2^-$, and $C_{36}Cl_3^-$, it was assumed that the dechlorination was accompanied by drastic skeletal changes, possibly affording C_{36}^- with a cage structure, though this remains to be confirmed.

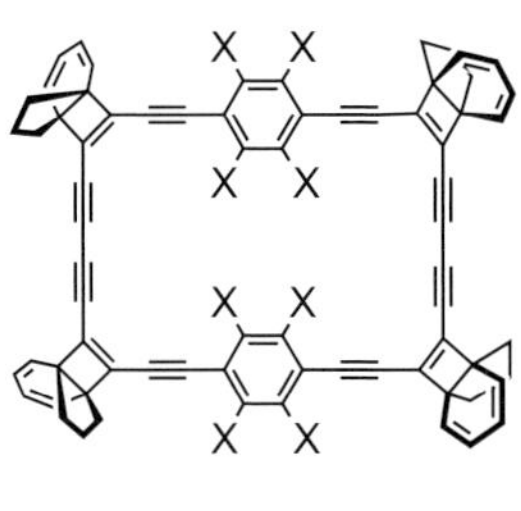

34a X = H
34b X = Cl

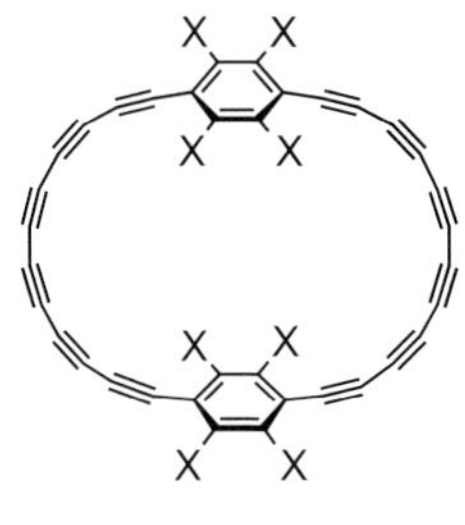

35a X = H
35b X = Cl

The structurally novel precursor **36** possessing an expanded cubane structure was prepared by Diederich et al. [124]. While the MALDI TOF mass spectrum (negative mode) of **36** exhibited only the peak due to $[M\text{–}OCH_3]^-$ without further fragmentation peaks, high-resolution FT-ICR MALDI mass spectra exhibited the peaks due to the stepwise loss of methoxy groups down to the weakly visible peak for C_{56}^-, corresponding to the cubane framework of **36**. In addition, strong peaks due to C_2 loss from C_{56}^- down to C_{50}^- were observed, suggesting that the initially formed C_{56}^- ion rearranged into a fullerene-like structure. In the positive mode, the loss of methoxy groups took place more readily. Namely, although C_{56}^- was not detected, the fragments due to C_2 loss (C_{54}^- to C_{50}^-) and the ions formed by ion-molecule reactions of these ions (C_{100}^- to C_{106}^-) as well as those derived from C_2 fragmentation (C_{94}^- to C_{98}^-) were observed.

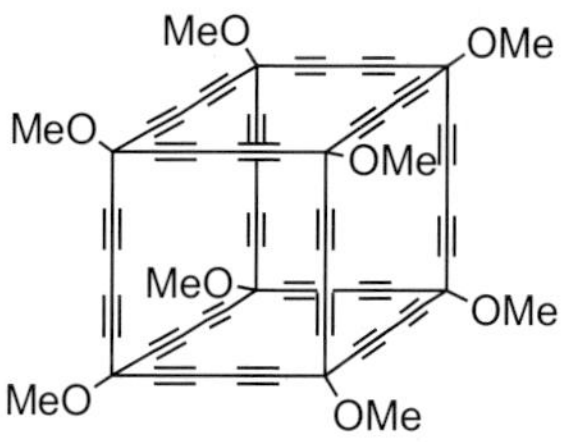

36

Regarding large fullerenes, it is well known that the number of possible geometrical isomers increases dramatically with increasing number of constituent carbon atoms [125]. It is therefore extremely interesting to see if geometry-selective synthesis is possible under kinetically controlled conditions based on organic transformations. For C_{78}, for example, there are five isomers that satisfy the isolated pentagon rule [126], and three of them – with C_{2v}, D_3, and $C_{2v'}$ symmetry – have been isolated [127]. The difference between the distributions of the isomers reported by three different groups was interpreted in terms of the foreign gas pressure, reaction temperature, and carbon atom density, suggesting the kinetically controlled formation of the isomers rather than their equilibration [128]. In order to produce [78]fullerene in size-selective manner, three-dimensional cyclophanes **37a** and **37b** containing [4.3.2]propellane units were prepared as precursors of polyynes **38a** and **38b**, respectively [129]. In the negative mode LD TOF mass spectrum of **37a**, an intense peak due to $C_{78}H_{18}^-$ (**38a**$^-$) formed by expulsion of all indane units was observed. Only partial dehydrogenation down to $C_{78}H_{14}^-$ took place, however, the result being reminiscent of the case of $C_{60}H_{18}$ (**30**). In the case of the hexachloro derivative **37b**, on the other hand, fragmentation of $C_{78}H_{12}Cl_6^-$ (**38b**$^-$) formed by the loss of all indane fragments from **37b** took place to release hydrogen and chlorine atoms simultaneously, resulting in the formation mainly of C_{78}^- and $C_{78}H_2^-$ in an approximate 1:1 peak ratio. Although the structure of the C_{78} ion remains to be established, the observed C_2 loss *only* from C_{78}^- (not from $C_{78}H_2^-$) down to C_{70}^- and resultant C_2 addition *both* to C_{78}^- and to $C_{78}H_2^-$ are strong indications of a fullerene structure for the C_{78}^- ion.

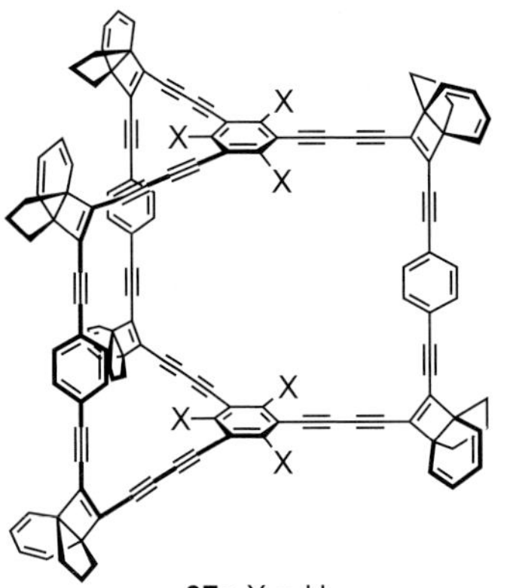

37a X = H
37b X = Cl

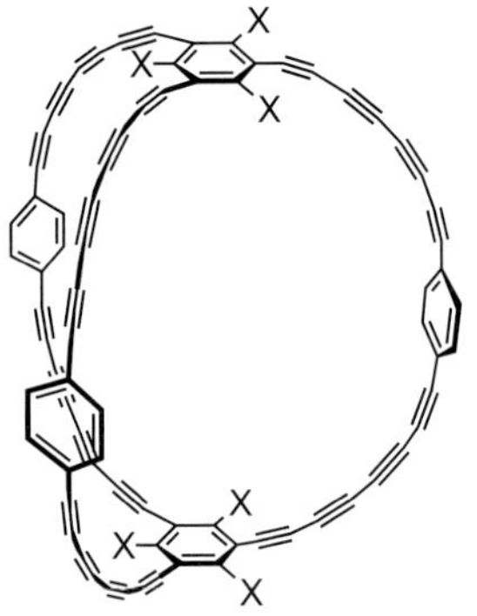

38a X = H
38b X = Cl

9.7 Conclusion

The field of acetylene-based all-carbon and carbon-rich molecules has attracted increasing interest, owing to the development of novel technologies for their production and spectroscopic investigation, which expand the previous limits in every respect, and also to novel perspectives and challenges that this research opens, particularly for materials chemistry.

Acknowledgements

This work was supported by CREST, Japan Science and Technology Agency (JST).

References

[1] (a) G. Herzberg, *Astrophys. J.* **1942**, *96*, 314–315.
(b) A. E. Douglas, *Astrophys. J.* **1951**, *114*, 466–468.

[2] (a) W. Weltner, Jr., P. N. Walsh, C. L. Angell, *J. Chem. Phys.* **1964**, *40*, 1299–1305.
(b) W. Weltner, Jr., D. McLeod, Jr., *J. Chem. Phys.* **1964**, *40*, 1305–1316.
(c) W. Weltner, Jr., D. McLeod, Jr., *J. Chem. Phys.* **1966**, *45*, 3096–3105.

[3] (a) S. Yang, K. J. Taylor, M. J. Craycraft, J. Conceicao, C. L. Pettiette, O. Cheshnovski, R. E. Smalley, *Chem. Phys. Lett.* **1988**, *144*, 431–436.
(b) S. H. Yang, C. L. Pettiette, J. Conceicao, O. Cheshnovski, R. E. Smalley, *Chem. Phys. Lett.* **1987**, *139*, 233–238.

[4] For reviews, (a) W. Weltner, Jr., R. J. Van Zee, *Chem. Rev.* **1989**, *89*, 1713–1747.

(b) A. Van Orden, R. J. Saykally, *Chem. Rev.* **1998**, *98*, 2313–2357.

[5] (a) D. Forney, J. Fulara, P. Freivogel, M. Jakobi, D. Lessen, J. P. Maier, *J. Chem. Phys.* **1995**, *103*, 48–53.
(b) P. Freivogel, J. Fulara, M. Jakobi, D. Forney, J. P. Maier, *J. Chem. Phys.* **1995**, *103*, 54–59.
(c) D. Forney, P. Freivogel, M. Grutter, J. P. Maier, *J. Chem. Phys.* **1996**, *104*, 4954–4960.
(d) P. Freivogel, M. Grutter, D. Forney, J. P. Maier, *Chem. Phys. Lett.* **1996**, *249*, 191–194.

[6] M. Wyss, M. Grutter, J. P. Maier, *Chem. Phys. Lett.* **1999**, *304*, 35–38.

[7] P. Freivogel, M. Grutter, D. Forney, J. P. Maier, *Chem. Phys.* **1997**, *216*, 401–406.

[8] (a) P. Freivogel, M. Grutter, D. Forney, J. P. Maier, *J. Chem. Phys.* **1997**, *107*, 22–27.
(b) P. Freivogel, M. Grutter, D. Forney, J. P. Maier, *J. Chem. Phys.* **1997**, *107*, 4468–4472.
(c) D. Forney, M. Grutter, P. Freivogel, J. P. Maier, *J. Phys. Chem. A* **1997**, *101*, 5292–5295.

[9] M. Schäfer, M. Grutter, J. Fulara, D. Forney, P. Freivogel, J. P. Maier, *Chem. Phys. Lett.* **1996**, *260*, 406–408.

[10] (a) G. A. Rechtsteiner, C. Felix, A. K. Ott, O. Hampe, R. P. Van Duyne, M. F. Jarrold, K. Raghavachari, *J. Phys. Chem. A* **2001**, *105*, 3029–3033.
(b) A. K. Ott, G. A. Rechtsteiner, C. Felix, O. Hampe, M. F. Jarrold, R. P. Van Duyne, *J. Chem. Phys.* **1998**, *109*, 9652–9655.

[11] T. F. Giesen, A. Van Orden, H. J. Hwang, R. S. Fellers, R. A. Provencal, R. J. Saykally, *Science* **1994**, *265*, 756–759.

[12] (a) P. Neubauer-Guenther, T. F. Giesen, U. Berndt, G. Fucks, G. Winnewisser, *Spectrochimica Acta Part A* **2003**, *59*, 431–441.
(b) T. F. Giesen, U. Berndt, K. M. T. Yamada, G. Fucks, R. Schieder, G. Winnewisser, R. A. Provencal, F. N. Keutsch, A. Van Orden, R. J. Saykally, *Chem. Phys. Chem.* **2001**, 242–247.

[13] (a) H. Linnartz, O. Vaizert, T. Motylewski, J. P. Maier, *J. Chem. Phys.* **2000**, *112*, 9777–9779.
(b) T. Motylewski, O. Vaizert, T. F. Giesen, H. Linnartz, J. P. Maier, *J. Chem. Phys.* **1999**, *111*, 6161–6163.

[14] R. Casaes, R. Provencal, J. Paul, R. J. Saykally, *J. Chem. Phys.* **2002**, *116*, 6640–6647.

[15] H. Handschuh, G. Ganteför, B. Kessler, P. S. Bechthold, W. Eberhardt, *Phys. Rev. Lett.* **1995**, *74*, 1095–1098.

[16] (a) D. W. Arnold, S. E. Bradforth, T. N. Kitsopoulos, D. M. Neumark, *J. Chem. Phys.* **1991**, *95*, 8753–8764.
(b) C. Xu, G. R. Burton, T. R. Taylor, D. M. Neumark, *J. Chem. Phys.* **1997**, *107*, 3428–3436.

[17] M. Kohno, S. Suzuki, H. Shiromaru, Y. Achiba, *J. Chem. Phys.* **1999**, *110*, 3781–3784.

[18] T. Wakabayashi, M. Kohno, Y. Achiba, H. Shiromaru, T. Momose, T. Shida, K. Naemura, Y. Tobe, *J. Chem. Phys.* **1997**, *107*, 4783–4787.

[19] H. Prinzbach, A. Weiler, P. Landenberger, F. Wahl, J. Wörth, L. T. Scott, M. Gelmont, D. Olevano, B. v. Lessendorff, *Nature* **2000**, *407*, 60–63.

[20] M. Tulej, D. A. Kirkwood, G. Maccaferri, O. Dopfer, J. P. Maier, *Chem. Phys.* **1998**, *228*, 293–299.

[21] D. A. Kirkwood, H. Linnartz, M. Grutter, O. Dopfer, T. Motylewski, M. Pachkov, M. Tulej, M. Wyss, J. P. Maier, *Faraday Discuss.* **1998**, *109*, 109–119.

[22] B. J. McCall, J. Thorburn, L. M. Hobbs, T. Oka, D. G. York, *Astrophys. J.* **2001**, *559*, L49–L53.

[23] Y. Zhao, E. de Beer, C. Xu, T. Taylor, D. M. Neumark, *J. Chem. Phys.* **1996**, *105*, 4905–4919.

[24] (a) M. Ohara, H. Shiromaru, Y. Achiba, K. Aoki, K. Hashimoto, S. Ikuta, *J. Chem. Phys.* **1995**, *103*, 10 393–10394.
(b) M. Ohara, H. Shiromaru, Y. Achiba, *J. Chem. Phys.* **1997**, *106*, 9992–9995.
(c) M. Ohara, D. Kasuya, H. Shiromaru, Y. Achiba, *J. Phys. Chem. A.* **2000**, *104*, 8622–8626.

[25] M. Ohara, M. Suwa, T. Ishigaki, H. Shiromaru, Y. Achiba, W. Krätschmer, *J. Chem. Phys.* **1998**, *109*, 1329–1333.

[26] (a) M. Miki, T. Wakabayashi, T. Momose, T. Shida, *J. Phys. Chem.* **1996**, *100*, 12 135–12137.
(b) S. Tam, M. Macler, M. E. Fajardo, *J. Chem. Phys.* **1997**, *106*, 8955–8963.
(c) H. Hoshina, Y. Kato, Y. Morisawa, T. Wakabayashi, T. Momose, *Chem. Phys.* **2004**, *300*, 69–77.

[27] M. Grutter, M. Wyss, E. Riaplov, J. P. Maier, S. D. Peyerimhoff, M. Hanrath, *J. Chem. Phys.* **1999**, *111*, 7397–7401.

[28] G. Monninger, M. Förderer, P. Gürtler, S. Kalhofer, S. Petersen, L. Nemes, P. G. Szalay, W. Krätschmer, *J. Phys. Chem. A* **2002**, *106*, 5779–5788.

[29] J. Szczepanski, J. Fuller, S. Ekern, M. Vala, *Spectrochimica Acta Part A* **2001**, *57*, 775–786.

[30] J. Szczepanski, E. Auerbach, M. Vala, *J. Phys. Chem. A* **1997**, *101*, 9296–9301.

[31] (a) J. F. Fuller, J. Szczepanski, M. Vala, *Chem. Phys. Lett.* **2000**, *323*, 86–92.
(b) L. Lapinski, M. Vala, *Chem. Phys. Lett.* **1999**, *300*, 195–201.
(c) J. Szczepanski, S. Ekern, C. Chapo, M. Vala, *Chem. Phys.* **1996**, *211*, 359–366.

[32] (a) J. Szczepanski, R. Hodyss, M. Vala, *J. Phys. Chem. A* **1998**, *102*, 8300–8304.
(b) J. Szczepanski, S. Ekern, M. Vala, *J. Phys. Chem. A* **1997**, *101*, 1841–1847.

[33] (a) J. Szczepanski, R. Hodyss, J. Fuller, M. Vala, *J. Phys. Chem. A* **1999**, *103*, 2975–2981.
(b) M. Dibben, J. Szczepanski, C. Wehlburg, M. Vala, *J. Phys. Chem. A* **2000**, *104*, 3584–3592.

[34] (a) G. von Helden, M.-T. Hsu, N. Gotts, M. T. Bowers, *J. Phys. Chem.* **1993**, *97*, 8182–8192.
(b) G. von Helden, W. E. Palke, M. T. Bowers, *Chem. Phys. Lett.* **1993**, *212*, 247–252.
(c) N. G. Gotts, G. von Helden, M. T. Bowers, *Int. J. Mass Spectrom. Ion Processes* **1995**, *149/150*, 217–229.

[35] K. Kaizu, K. Kohno, S. Suzuki, H. Shiromaru, T. Moriwaki, Y. Achiba, *J. Chem. Phys.* **1997**, *106*, 9954–9956.

[36] (a) T. Wakabayashi, T. Momose, T. Shida, H. Shiromaru, M. Ohara, Y. Achiba, *J. Chem. Phys.* **1997**, *107*, 1152–1155.
(b) T. Wakabayashi, T. Momose, T. Shida, *J. Chem. Phys.* **1999**, *111*, 6260–6263.
(c) Y. Kato, T. Wakabayashi, T. Momose, *J. Chem. Phys.* **2003**, *118*, 5390–5394.
(d) Y. Kato, T. Wakabayashi, T. Momose, *Chem. Phys. Lett.* **2004**, *386*, 279–285.

[37] (a) J. D. Presilla-Márquez, J. A. Sheehy, J. D. Mills, P. G. Carrick, C. W. Larson, *Chem. Phys. Lett.* **1997**, *274*, 439–444.
(b) J. D. Presilla-Márquez, J. Harper, J. A. Sheehy, P. G. Carrick, C. W. Larson, *Chem. Phys. Lett.* **1999**, *300*, 719–726.

[38] (a) S. L. Wang, C. M. L. Rittby, W. R. M. Graham, *J. Chem. Phys.* **1997**, *107*, 6032–6037.
(b) S. L. Wang, C. M. L. Rittby, W. R. M. Graham, *J. Chem. Phys.* **1997**, *107*, 7025–7033.
(c) S. L. Wang, C. M. L. Rittby, W. R. M. Graham, *J. Chem. Phys.* **2000**, *112*, 1457–1461.

[39] See, for example, Figure 2 in T. Wakabayashi, A.-L. Ong, D. Strelnikov, W. Krätschmer, *J. Phys. Chem. B* **2004**, *108*, 3686–3690.

[40] J. Szczepanski, M. Vala, *J. Phys. Chem.* **1991**, *95*, 2792–2798.

[41] For a review, R. B. Heimann, S. E. Evsyukov, L. Kavan, *Carbyne and Carbynoid Structures*; Kluwer Academic Publishers: Dordrecht, 1999.

[42] W. Krätschmer, L. D. Lamb, K. Fostiropoulos, D. R. Huffman, *Nature* **1990**, *347*, 354–358.

[43] H. W. Kroto, J. R. Heath, S. C. O'Brien, R. F. Curl, R. E. Smalley, *Nature* **1985**, *318*, 162–163.

[44] K. W. R. Gilkes, C. T. Pillinger, in *Carbyne and Carbynoid Structures*; R. B. Heimann, S. E. Evsyukov, L. Kavan, Eds., Kluwer Academic Publishers: Dordrecht, 1999; pp. 17–30.

[45] A. El Goresy, G. Donnay, *Science* **1968**, *161*, 363–364.

[46] A. G. Whittaker, E. J. Watts, R. S. Lewis, E. Anders, *Science* **1980**, *209*, 1512–1514.

[47] (a) A. G. Whittaker, *Science* **1985**, *229*, 485–486.
(b) P. P. K. Smith, P. R. Buseck, *Science* **1985**, *229*, 486–487.

[48] Y. P. Kudryavtsev, in *Carbyne and Carbynoid Structures*; R. B. Heimann, S. E. Evsyukov, L. Kavan, Eds., Kluwer Academic Publishers: Dordrecht, 1999; pp. 39–45.

[49] M. J. Rice, S. R. Phillpot, A. R. Bishop, D. K. Campbell, *Phys. Rev. B* **1986**, *34*, 4139–4149.

[50] J. Kürti, C. Magyar, A. Balázs, P. Rajczy, *Synth. Metals* **1995**, *71*, 1865–1866.

[51] R. B. Heimann, J. Kleiman, N. M. Salanski, *Nature* **1983**, *306*, 164–167.

[52] R. H. Baughman, H. Eckhardt, M. Kertesz, *J. Chem. Phys.* **1987**, *87*, 6687–6699.

[53] L. Kavan, J. Hlavaty, J. Kastner, H. Kuzmany, *Carbon* **1995**, *33*, 1321–1329.

[54] J. Kastner, H. Kuzmany, L. Kavan, F. P. Dousek, J. Kürti, *Macromolecules* **1995**, *28*, 344–353.

[55] L. Kavan, *Chem. Rev.* **1997**, *97*, 3061–3082.

[56] J. R. Dennison, M. Holtz, G. Swain, *Spectroscopy* **1996**, *11*, 38–46.

[57] L. Ravagnan, F. Siviero, C. Lenardi, P. Piseri, E. Barborini, P. Milani, C. S. Casari, A. Li Bassi, C. E. Bottani, *Phys. Rev. Lett.* **2002**, *89*, 285506/1–285506/4.

[58] C. S. Casari, A. Li Bassi, L. Ravagnan, F. Siviero, C. Lenardi, P. Piseri, G. Bongiorno, C. E. Bottani, P. Milani, *Phys. Rev. B* **2004**, *69*, 075422/1–075422/7.

[59] I. Cermak, G. Monninger, W. Krätschmer, *Advances in Molecular Structure Research* **1997**, *3*, 117–146, and references therein.

[60] T. Wakabayashi, A.-L. Ong, D. Strelnikov. W. Krätschmer, *J. Phys. Chem. B* **2004**, *108*, 3686–3690.

[61] Y. Yamaguchi, T. Wakabayashi, *Chem. Phys. Lett.* **2004**, *388*, 436–440.

[62] For a review, W. D. Huntsman, in *The Chemistry of the Carbon-Carbon Triple Bond*; Part 2, S. Patai, Ed., John Wiley & Sons: Chichester, 1978; pp. 553–620.

[63] (a) E. Kloster-Jensen, *Angew. Chem.* **1972**, *84*, 483–485; *Angew. Chem. Int. Ed.* **1972**, *11*, 438–439.
(b) E. Kloster-Jensen. H.-J. Haink, H. Christen, *Helv. Chim. Acta* **1974**, *57*, 1731–1744.

[64] R. Eastmond, T. R. Johnson, D. R. M. Walton, *Tetrahedron* **1972**, *28*, 4601–4616.

[65] (a) M. Tsuji, T. Tsuji, S. Kuboyama, S.-H. Yoon, Y. Korai, T. Tsujimoto, K. Kubo, A. Mori, I. Mochida, *Chem. Phys. Lett.* **2002**, *355*, 101–108.
(b) M. Tsuji, S. Kuboyama, T. Matsuzaki, T. Tsuji, *Carbon* **2003**, *41*, 2141–2148.

[66] (a) F. Cataldo, *Carbon* **2003**, *41*, 2671–2674.
(b) F. Cataldo, *Tetrahedron Lett.* **2004**, *45*, 141–145.
(c) F. Cataldo, *Carbon* **2004**, *42*, 129–142.

[67] (a) M. Grutter, M. Wyss, J. Fulara, J. P. Maier, *J. Phys. Chem. A* **1998**, *102*, 9785–9790.
(b) T. Pino, H. Ding, F. Güthe, J. P. Maier, *J. Chem. Phys.* **2001**, *114*, 2208–2212.

[68] (a) C. L. Cook, E. R. H. Jones, M. C. Whiting, *J. Chem. Soc.* **1952**, 2883–2891.
(b) J. B. Armitage, N. Entwistle, E. R. H. Jones, M. C. Whiting, *J. Chem. Soc.* **1954**, 147–154.
(c) E. R. H. Jones, H. H. Lee, M. C. Whiting, *J. Chem. Soc.* **1960**, 3483–3489.

[69] F. Bohlman, *Chem. Ber.* **1953**, *86*, 657–667.

[70] T. R. Johnson, D. R. M. Walton, *Tetrahedron* **1972**, *28*, 5221–5236.

[71] Y. Rubin, S. S. Lin, C. B. Knobler, J. Anthony, A. M. Boldi, F. Diederich, *J. Am. Chem. Soc.* **1991**, *113*, 6943–6949. Note added in proof: For a new recard, A. D. Slepkov, F. A. Hegmann, S. E. Eisler, E. Elliot, R. R. Tykwinski, *J. Chem. Phys.* **2004**, *120*, 6807–6810.

[72] T. Gibtner, F. Hampel, J.-P. Gisselbrecht, A. Hirsch, *Chem. Eur. J.* **2002**, *8*, 408–432.

[73] (a) S. Akiyama, M. Nakagawa, *Tetrahedron Lett.* **1964**, 719–725.
(b) S. Akiyama, K. Nakasuji, K. Akashi, M. Nakagawa, *Tetrahedron Lett.* **1968**, 1121–1126.
(c) K. Nakasuji, S. Akiyama, K. Akashi, M. Nakagawa, *Bull. Chem. Soc. Jpn.* **1970**, *43*, 3567–3576.
(d) S. Akiyama, M. Nakagawa, *Bull. Chem. Soc. Jpn.* **1971**, *44*, 2237–2248.
(e) M. Nakagawa, S. Akiyama, K. Nakasuji, K. Nishimoto, *Tetrahedron* **1971**, *27*, 5401–5418.

[74] K. Nakamura, T. Fujimoto, S. Takara, K.-i. Sugiura, H. Miyasaka, T. Ishii, M. Yakmashita, Y. Sakata, *Chem. Lett.* **2003**, *32*, 694–695.

[75] R. Dembinski, T. Bartik, B. Bartik, M. Jaeger, J. A. Gladysz, *J. Am. Chem. Soc.* **2000**, *122*, 810–822.

[76] K. Gao, N. S. Goroff, *J. Am. Chem. Soc.* **2000**, *122*, 9320–9321.

[77] (a) T. Grösser, A. Hirsch, *Angew. Chem.* **1993**, *105*, 1390–1392; *Angew. Chem. Int. Ed.* **1993**, *32*, 1340–1342.
(b) G. Schermann, T. Grösser, F. Hampel, A. Hirsch, *Chem. Eur. J.* **1997**, *3*, 1105–1112.

[78] R. J. Lagow, J. J. Kampa, H.-C. Wei, S. L. Battle, J. W. Genge, D. A. Laude, C. J. Harper, R. Bau, R. C. Stevens, J. F. Haw, E. Munson, *Science* **1995**, *267*, 362–367.

[79] G. N. Lewis, M. Calvin, *Chem. Rev.* **1939**, *25*, 273–328.

[80] E. J. Ginsburg, C. B. Gorman, R. H. Grubbs, in *Modern Acetylene Chemistry;* P. J. Stang, F. Diederich, Eds., VCH: Weinheim, 1995; pp. 353–383.

[81] For reviews, (a) M. I. Bruce, *Coordination Chem. Rev.* **1997**, *166*, 91–119.
(b) F. Paul, C. Lapinte, *Coordination Chem. Rev.* **1998**, *178–180*, 431–509.
(c) H. Lang, *Angew. Chem.* **1994**, *106*, 569–572; *Angew. Chem. Int. Ed.* **1994**, *33*, 547–550.
(d) U. H. F. Bunz, *Angew. Chem.* **1996**, *108*, 1047–1049; *Angew. Chem. Int. Ed.* **1996**, *35*, 969–971.
(e) S. Szafert, J. A. Gladysz, *Chem. Rev.* **2003**, *103*, 4175–4205.

[82] F. Coat, C. Lapinte, *Organometallics* **1996**, *15*, 477–479.

[83] (a) M. Akita, M.-C. Chung, A. Sakurai, S. Sugimoto, M. Terada, M. Tanaka, Y. Moro-oka, *Organometallics* **1997**, *16*, 4882–4888.
(b) A. Sakurai, M. Akita, Y. Moro-oka, *Organometallics* **1999**, *18*, 3241–3244.

[84] K.-T. Wong, J.-M. Lehn, S.-M. Peng, G.-H. Lee, *Chem. Commun.* **2000**, 2259–2260.

[85] M. I. Bruce, B. D. Kelly, B. W. Skelton, A. H. White, *J. Organomet. Chem.* **2000**, *604*, 150–156.

[86] M. I. Bruce, M. Ke, P. J. Low, *Chem. Commun.* **1996**, 2405–2406.

[87] W. Lu, H.-F. Xiang, N. Zhu, C.-M. Che, *Organometallics* **2002**, *21*, 2343–2346.

[88] V. W.-W. Yam, *Chem. Commun.* **2001**, 789–796.

[89] (a) M. Brady, W. Weng, J. A. Gladysz, *J. Chem. Soc., Chem. Commun.* **1994**, 2655–2656.
(b) T. Bartik, B. Bartik, M. Brady, R. Dembinski, J. A. Gladysz, *Angew. Chem.* **1996**, *108*, 467–469; *Angew. Chem. Int. Ed.* **1996**, *35*, 414–417.
(c) M. Brady, W. Weng, Y. Zhou, J. W. Seyler, A. J. Amoroso, A. M. Arif, M. Böhme, G. Frenking, J. A. Gladysz, *J. Am. Chem. Soc.* **1997**, *119*, 775–788.

[90] W. E. Meyer, A. J. Amoroso, C. R. Horn, M. Jaeger, J. A. Gladysz, *Organometallics* **2001**, *20*, 1115–1127.

[91] F. Paul, W. E. Meyer, L. Toupet, H. Jiao, J. A. Gladysz, C. Lapinte, *J. Am. Chem. Soc.* **2000**, *122*, 9405–9414.

[92] (a) C. R. Horn, J. M. Martín-Alvarez, J. A. Gladysz, *Organometallics* **2002**, *21*, 5386–5393.
(b) C. R. Horn, J. A. Gladysz, *Eur. J. Inorg. Chem.* **2003**, 2211–2218.

[93] (a) T. B. Peters, J. C. Bohling, A. M. Arif, J. A. Gladysz, *Organometallics* **1999**, *18*, 3261–3263.
(b) W. Mohr, J. Stahl, F. Hampel, J. A. Gladysz, *Inorg. Chem.* **2001**, *40*, 3263–3264.
(c) W. Mohr, J. Stahl, F. Hampel, J. A. Gladysz, *Chem. Eur. J.* **2003**, *9*, 3324–3340.

[94] J. Stahl, J. C. Bohling, E. B. Bauer, T. B. Peters, W. Mohr, J. M. Martín-Alvarez, F. Hampel, J. A. Gladysz, *Angew. Chem.* **2002**, *114*, 1951–1957; *Angew. Chem. Int. Ed.* **2002**, *41*, 1871–1876.

[95] For reviews, (a) F. Diederich, Y. Rubin, *Angew. Chem.* **1992**, *104*, 1123–1146; *Angew. Chem., Int. Ed.* **1992**, *31*, 1101–1123.
(b) F. Diederich, *Nature* **1994**, *369*, 199–207.
(c) F. Diederich, in *Modern Acetylene Chemistry*, P. J. Stang, F. Diederich, Eds., VCH: Weinheim, 1995; pp. 443–471.
(d) F. Diederich, L. Gobbi, *Top. Curr. Chem.* **1999**, *201*, 43–79.
(e) Y. Tobe, in *Advances in Strained and Interesting Organic Molecules;* B. Halton, Ed., JAI Press: Stamford, 1999; Vol. 7, pp. 153–184.

[96] For reviews, (a) H. Schwarz, *Angew. Chem.* **1993**, *105*, 1475–1478; *Angew. Chem., Int. Ed.* **1993**, *32*, 1412–1415.
(b) N. S. Goroff, *Acc. Chem. Res.* **1996**, *29*, 77–83.

[97] (a) J. Hunter, J. Fye, M. F. Jarrold, *Science* **1993**, *260*, 784–786.
(b) G. von Helden, N. G. Gotts, M. T. Bowers, *Nature* **1993**, *363*, 60–63.
(c) G. von Helden, N. G. Gotts, M. T. Bowers, *J. Am. Chem. Soc.* **1993**, *113*, 4363–4364.
(d) S. W. McElvany, M. M. Ross, N. S. Goroff, F. Diederich, *Science* **1993**, *259*, 1594–1596.
(e) J. M. Hunter, J. L. Fye, E. J. Roskamp, M. F. Jarrold, *J. Phys. Chem.* **1994**, *98*, 1810–1818.

[98] D. L. Strout, G. E. Scuseria, *J. Phys. Chem.* **1996**, *100*, 6492–6498.

[99] For recent reviews, (a) M. M. Haley, *Synlett* **1998**, 557–565.
(b) M. M. Haley, J. J. Pak, S. C. Brand, *Top. Curr. Chem.* **1999**, *201*, 81–130.
(c) M. M. Haley, W. B. Wan, in *Advances in Strained and Interesting Organic Molecules*; B. Halton, Ed., JAI Press: Stamford, 2000; Vol. 8, pp. 1–41.
(d) J. A. Marsden, G. J. Palmer, M. M. Haley, *Eur. J. Org. Chem.* **2003**, 2355–2369.

[100] R. Hoffmann, *Tetrahedron* **1966**, *22*, 521–538.

[101] For a review, D. A. Plattner, Y. Li, K. N. Houk, in *Modern Acetylene Chemistry*, P. J. Stang, F. Diederich, Eds., VCH: Weinheim, 1995; pp. 1–32.

[102] Y. Rubin, F. Diederich, in *Stimulating Concepts in Chemistry*, F. Vögtle, J. F. Stoddart, M. Shibasaki, Eds., Wiley-VCH: Weinheim, 2000; pp. 163–186.

[103] T. H. Kerr, R. E. Hibbins, J. R. Miles, S. J. Fossey, W. B. Somerville, P. J. Sarre, *Mon. Not. R. Astron. Soc.* **1996**, *283*, L105–L109.

[104] (a) F. Diederich, Y. Rubin, C. B. Knobler, R. L. Whetten, K. E. Schriver, K. N. Houk, Y. Li, *Science* **1989**, *245*, 1088–1090.
(b) F. Diederich, Y. Rubin, O. L. Chapman, N. S. Goroff, *Helv. Chim. Acta* **1994**, *77*, 1441–1457.

[105] Y. Rubin, C. B. Knobler, F. Diederich, *J. Am. Chem. Soc.* **1990**, *112*, 4966–4968.

[106] (a) Y. Rubin, C. B. Knobler, F. Diederich, *J. Am. Chem. Soc.* **1990**, *112*, 1607–1617.
(b) Y. Rubin, M. Kahr, C. B. Knobler, F. Diederich, C. L. Wilkins, *J. Am. Chem. Soc.* **1991**, *113*, 495–500.

[107] (a) Y. Tobe, T. Fujii, K. Naemura, *J. Org. Chem.* **1994**, *59*, 1236–1237.
(b) Y. Tobe, T. Fujii, H. Matsumoto, K. Naemura, Y. Achiba, T. Wakabayashi, *J. Am. Chem. Soc.* **1996**, *118*, 2758–2759.
(c) Y. Tobe, H. Matsumoto, K. Naemura, Y. Achiba, T. Wakabayashi, *Angew. Chem.* **1996**, *108*, 1924–1926; *Angew. Chem., Int. Ed.* **1996**, *35*, 1800–1802.
(d) Y. Tobe, T. Fujii, H. Matsumoto, K. Tsumuraya, D. Noguchi, N. Nakagawa, M. Sonoda, K. Naemura, Y. Achiba, T. Wakabayashi, *J. Am. Chem. Soc.* **2000**, *122*, 1762–1775.

[108] Y. Tobe, I. Ohki, M. Sonoda, H. Niino, T. Sato, T. Wakabayashi, *J. Am. Chem. Soc.* **2003**, *125*, 5614–5615.

[109] I. Hisaki, T. Eda, M. Sonoda, Y. Tobe, *Chem. Lett.* **2004**, *33*, 620–621.

[110] (a) L. Isaacs, P. Seiler, F. Diederich, *Angew. Chem.* **1995**, *107*, 1636–1639; *Angew. Chem., Int. Ed.* **1995**, *34*,

1466–1469.
(b) L. Isaacs, F. Diederich, R. F. Haldiman, *Helv. Chim. Acta* **1997**, *80*, 317–342.

[111] (a) A. de Meijere, S. I. Kozhushkov, *Top. Curr. Chem.* **1999**, *201*, 1–42.
(b) A. de Meijere, S. I. Kozhushkov, *Chem. Eur. J.* **2002**, *8*, 3195–3202.

[112] (a) Y. Tobe, N. Iwasa, R. Umeda, M. Sonoda, *Tetrahedron Lett.* **2001**, *42*, 5485–5488.
(b) Y. Tobe, R. Umeda, N. Iwasa, M. Sonoda, *Chem. Eur. J.* **2003**, *9*, 5549–5559.

[113] Y. Rubin, T. C. Parker, S. I. Khan, C. L. Holliman, S. W. McElvany, *J. Am. Chem. Soc.* **1996**, *118*, 5308–5309.

[114] Y. Rubin, T. C. Parker, S. J. Pastor, S. Jalisatgi, C. Boulle, C. L. Wilkins, *Angew. Chem.* **1998**, *110*, 1353–1356; *Angew. Chem. Int. Ed.* **1998**, *37*, 1226–1229.

[115] (a) Y. Tobe, N. Nakagawa, K. Naemura, T. Wakabayashi, T. Shida, Y. Achiba, *J. Am. Chem. Soc.* **1998**, *120*, 4544–4545.
(b) Y. Tobe, N. Nakagawa, J.-y. Kishi, M. Sonoda, K. Naemura, T. Wakabayashi, T. Shida, Y. Achiba, *Tetrahedron* **2001**, *57*, 3629–3636.

[116] For reviews on the production of fullerenes, (a) M. S. Dresselhaus, G. Dresselhaus, P. C. Eklund, *Science of Fullerenes and Carbon Nanotubes*; Academic Press: San Diego, 1995; pp. 110–142.
(b) J. C. Withers, R. O. Loutfy, T. P. Lowe, *Fullerene Sci. Technol.* **1997**, *5*, 1–31.
(c) M. Ozawa, P. Deota, E. Osawa, *Fullerene Sci. Technol.* **1999**, *7*, 387–409.
(d) H. Murayama, S. Tomonoh, J. M. Alford, M. E. Karpuk, *Fullerenes Nanotubes Carbon Nanostructures* **2004**, *12*, 1–9.

[117] For a review, L. T. Scott, *Angew. Chem.* **2004** *116*, 5102–5116; *Angew. Chem. Int. Ed.* **2004**, *43*, 4994–5007.

[118] L. T. Scott, M. M. Boorum, B. J. McMahon, S. Hagen, J. Mack, J. Blank, H. Wegner, A. de Meijere, *Science* **2002**, *295*, 1500–1503.

[119] S.-Y. Xie, F. Gao, X. Lu, R.-B. Huang, C.-R. Wang, X. Zhang, M.-L. Liu, S.-L. Deng, L.-S. Zheng, *Science* **2004**, *304*, 699.

[120] K. Hata, M. Ariff, K. Tohji, Y. Saito, *Chem. Phys. Lett.* **1999**, *308*, 343–346.

[121] (a) C. Piskoti, J. Yarger, A. Zettl, *Nature* **1998**, *393*, 771–774.
(b) P. G. Collins, J. C. Grossman, M. Côté, M. Ishigami, C. Piskoti, S. G. Louie, M. L. Cohen, A. Zettl, *Phys. Rev. Lett.* **1999**, *82*, 165–168.
(c) A. Koshio, M. Inakuma, T. Sugai, H. Shinohara, *J. Am. Chem. Soc.* **2000**, *122*, 398–399.

[122] M. M. Haley, B. L. Langsdorf, *Chem. Commun.* **1997**, 1121–1122.

[123] Y. Tobe, R. Furukawa, M. Sonoda, T. Wakabayashi, *Angew. Chem.* **2001**, *113*, 4196–4198; *Angew. Chem. Int. Ed.* **2001**, *40*, 4072–4074.

[124] P. Manini, W. Amrein, V. Gramlich, F. Diederich, *Angew. Chem.* **2002**, *114*, 4515–4519; *Angew. Chem. Int. Ed.* **2002**, *41*, 4339–4343.

[125] P. W. Fowler, D. E. Manolopoulos, *An Atlas of Fullerenes*; Clarendon Press: Oxford, 1995.

[126] (a) H. W. Kroto, *Nature* **1987**, *329*, 529–531.
(b) T. G. Schmalz, W. A. Seitz, D. J. Klein, G. E. Hite, *J. Am. Chem. Soc.* **1988**, *110*, 1113–1127.

[127] (a) F. Diederich, R. L. Whetten, C. Thilgen, R. Ettl, I. Chao, M. M. Alvarez, *Science* **1991**, *254*, 1768–1770.
(b) K. Kikuchi, N. Nakahara, T. Wakabayashi, S. Suzuki, H. Shiromaru, Y. Miyake, K. Saito, I. Ikemoto, M. Kainosho, Y. Achiba, *Nature* **1992**, *357*, 142–145.
(c) R. Taylor, G. J. Langley, T. J. S. Dennis, H. W. Kroto, D. R. M. Walton, *J. Chem. Soc., Chem. Commun.* **1992**, 1043–1046.

[128] T. Wakabayashi, K. Kikuchi, S. Suzuki, H. Shiromaru, Y. Achiba, *J. Phys. Chem.* **1994**, *98*, 3090–3091.

[129] Y. Tobe, R. Umeda, M. Sonoda, T. Wakabayashi, unpublished results.

10
Shape-Persistent Acetylenic Macrocycles for Ordered Systems

Sigurd Höger

10.1
Introduction

In 1894, Emil Fischer's *"lock and key"* principle laid out the basis for what is known today as *molecular recognition*, one of the major areas of *supramolecular chemistry* [1]. Supramolecular chemistry, "chemistry beyond the molecule" according to Lehn, can be divided into two broad, partially overlapping areas: the field of supermolecules, well defined oligomolecular species formed by the association of only a few components (e. g., host-guest complexes), and the field of supramolecular assemblies, polymolecular entities formed by the spontaneous assembly of a rather large and often ill-defined number of objects into a specific phase (e. g., micelles) [2].

Macrocyclic compounds play a special role in supramolecular chemistry and especially in the field of supermolecules. Pedersen's work on crown ethers has shown that these macrocycles have properties that cannot be found in their open-chain analogues [3]. Besides these rather flexible macrocycles, rigid macrocycles have also been well investigated in terms of their synthesis and binding properties. Remarkable are Cram's spherands, which combine the molecular stiffness of Staab's phenylene rings and Vögtle's concept of intraannular groups, resulting in highly selective and strong binders for metal ions [4–6]. However, the size of the rings did not reach the nanometer regime for some time. At the beginning of the 1990s, progress in synthetic (metal)organic chemistry and the broad success of supramolecular chemistry inspired the investigation of shape-persistent macrocycles with larger interiors based on the phenylene-ethynylene backbone [7, 8].

This article concentrates on shape-persistent phenylene-ethynylene macrocycles with interiors in the nanometer regime, excluding both small rigid cycles and large, but rather flexible compounds. In addition, this article focuses on the super- and supramolecular chemistry of these compounds, as the synthesis of phenylene-ethynylene macrocycles is presented in Chapter 8.

The term "shape-persistent macrocycle" means that the building blocks of the ring are reasonably rigid and their connections are made in such a way that the final structure cannot collapse. A more precise definition can be made through

Acetylene Chemistry. Chemistry, Biology and Material Science. Edited by F. Diederich, P. J. Stang, R. R. Tykwinski

ISBN 3-527-30781-8

analogy with shape-persistent linear oligomers and polymers, in which the end-to-end distance corresponds, on average, to the contour length of the molecular backbone of the molecule [8b, 9]. Accordingly, shape-persistent macrocycles have an interior (lumen) d that is, on average, equal to the contour length l of their molecular backbone divided by π (Figure 10.1).

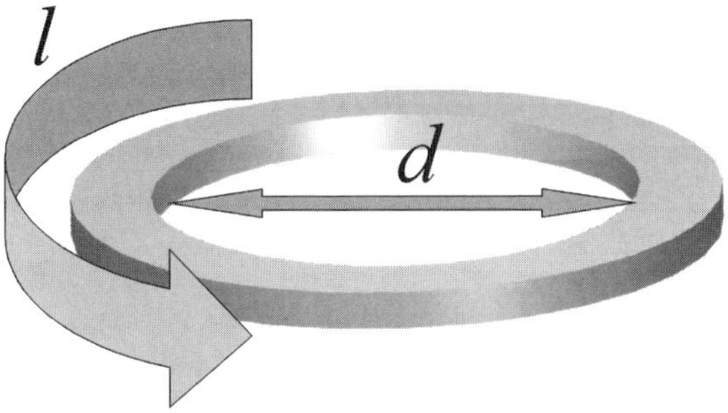

Figure 10.1 For shape-persistent macrocycles the following relationship applies: $\langle d \rangle = l/\pi$ [reproduced from ref. 8g, with permission].

In order to keep these rigid structures tractable, side groups have to be attached to the backbone. Their orientation has a strong influence on the compound's properties. Macrocycles with an orthogonal arrangement of polar functional groups (orthogonal substituents, Figure 10.2, left) are able to form tubular superstructures in the solid state [10]. If the polar functional groups point towards the outside (extraannular substituents, Figure 10.2, second from left) this may result in the formation of a two-dimensional network in the solid state [11]. Shape-persistent macrocycles with polar groups pointing towards the inside (intraannular substituents, Figure 10.2, second from right) can act as host molecules for the recognition of appropriate guest molecules by the *lock and key principle* [12], and shape-persistent macrocycles with an adaptable arrangement of the polar groups can change their conformation according to an external parameter (adaptable substituents, Figure 10.2, right) [13]. If the conformational change of the macrocycle is influenced by the guest molecule, the binding process resembles the *induced fit mechanism* [14]. The adaptable functional group orientation requires a low rotational barrier for parts of the molecule, and this, in combination with overall structural rigidity, is a feature fulfilled by the phenylene-ethynylene moiety [15].

Experimental evidence of the adaptable nature of *para*-phenylene moieties in shape-persistent phenylene-ethynylene macrocycles was provided by the single-crystal X-ray analysis of the amphiphilic macrocycle **1a** (Scheme 10.1). Depending on the polarity of the solvent from which the crystals were grown, different conformational states could be observed. From THF solution, all non-polar hexyloxy side groups point outwards and the phenol OH groups point inside the cavity of the macrocycle (these are stacked on top of each other, thus forming large channels). In crystals grown from the more polar pyridine, however, the macrocycle adopts a conformation in which only two of the polar phenol OH groups point inwards, the remaining two being directed outwards [16].

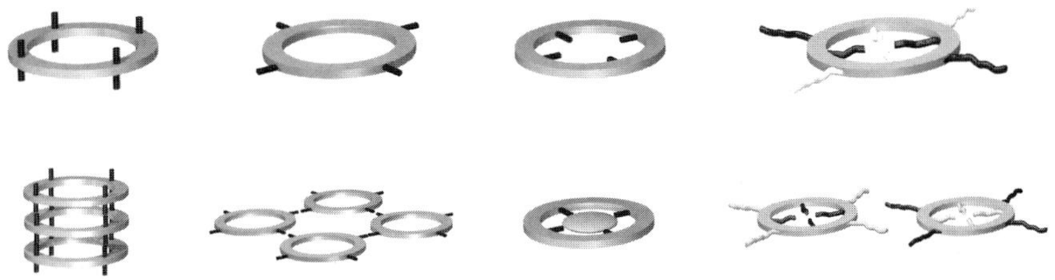

Figure 10.2 Orientations of polar (dark gray) and non polar (light gray) side groups in shape-persistent macrocycles and their respective properties or functions.

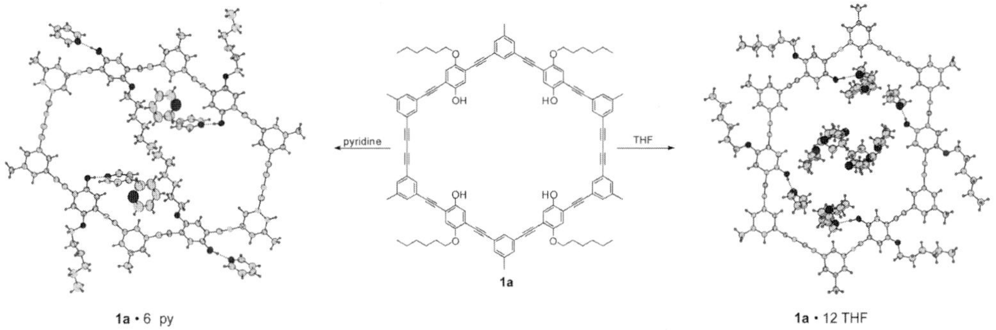

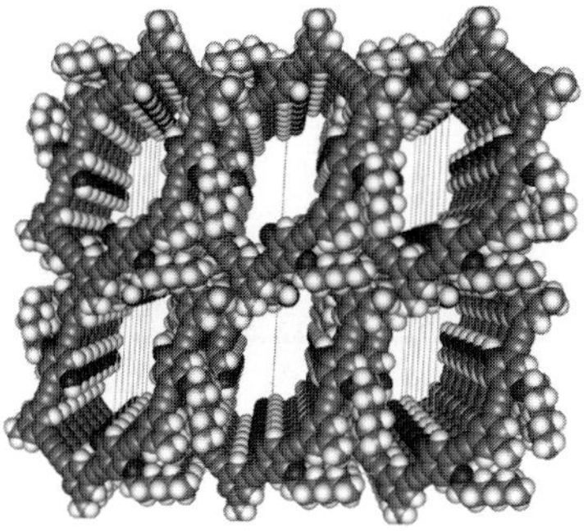

Scheme 10.1 Crystal structures (ORTEP drawings) of **1a** · 6py and of **1a** · 12THF (top). View down the crystallographic *b* axis of **1a** · 12THF showing large channels within the crystal (solvent molecules are removed) (bottom) [reprinted from ref. 16, with permission; copyright (2002) American Chemical Society].

10.2 Ordered Systems

10.2.1 Host-Guest Complexes

The stiffness of the rigid core in combination with appropriate intraannular binding sites makes shape-persistent macrocycles ideal candidates for investigation of host-guest complexes.

Sanders et al. investigated macrocycles with porphyrin moieties (e. g., **2**) in great detail [17]. In contrast with Cram's spherands, not only metal ions but also clusters and organic molecules can be bound [18]. The size of the interior also allows the binding of more than one substrate, and the ability of the macrocycles to catalyze Diels–Alder reactions and acyl-transfer reactions (the latter without product inhibition) has been demonstrated [19]. Recently, even larger porphyrin phenylene-ethynylene macrocycles were described by Lindsey et al. (**3**) and Gossauer et al. (**4**) (Figure 10.3) [20, 21]. The porphyrin metal can bind guest molecules with donor

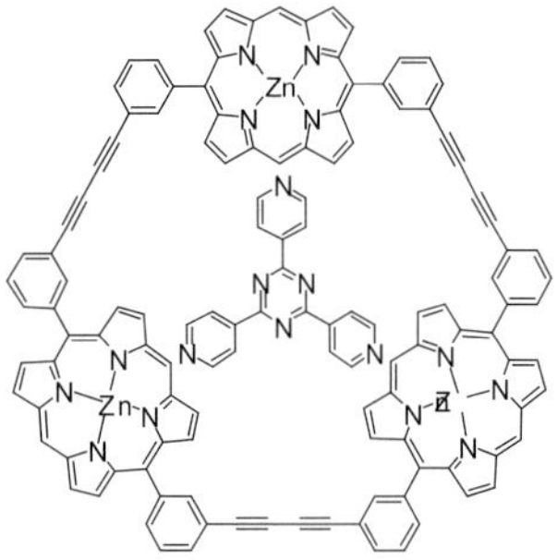

2

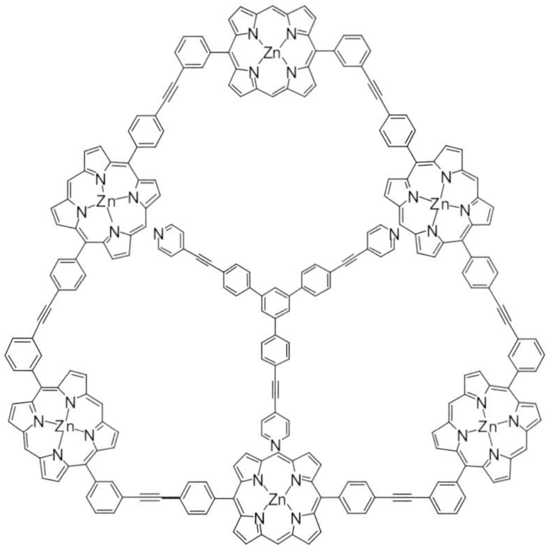

3

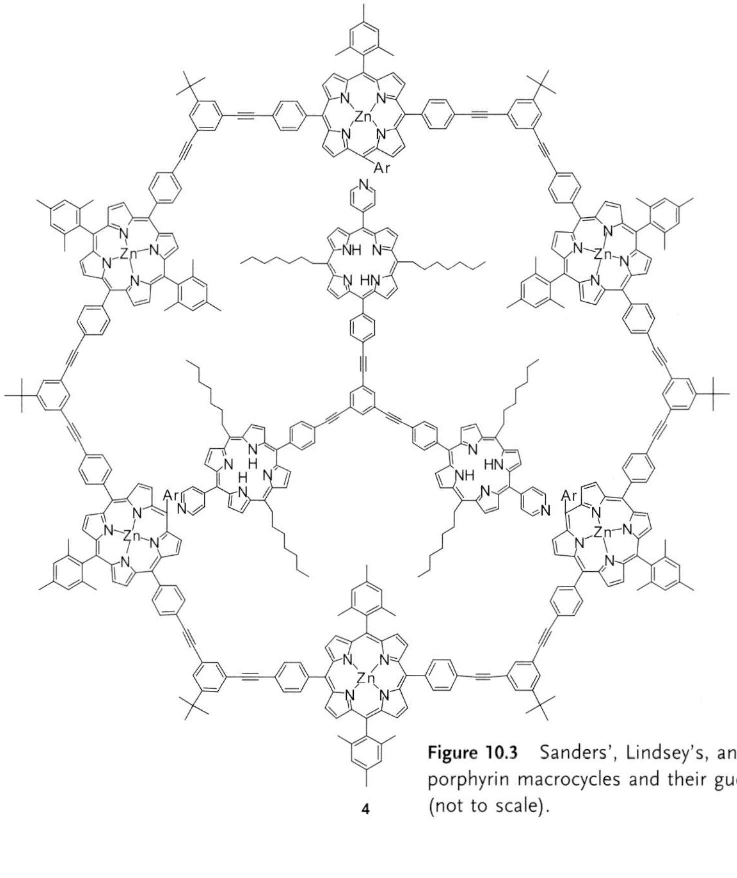

4

Figure 10.3 Sanders', Lindsey's, and Gossauer's porphyrin macrocycles and their guest molecules (not to scale).

atoms such as pyridine derivatives, and multiple N-Zn interactions between the rigid hosts and guests result in remarkably high association constants in some cases ($K_{assoc} \sim 10^{10}\ \text{M}^{-1}$). These self-assembled porphyrin arrays display interesting light-harvesting properties and can serve as model compounds for the light-harvesting complexes involved in photosynthesis.

Instead of having the metal center in the macrocyclic backbone and binding a nitrogen donor, the macrocycle can contain the donor group and then act as macrocyclic metal ligand (Figure 10.4). Actually, investigation of pyridine-, bipyridine-, terpyridine-, and phenanthroline-containing shape-persistent phenylene-ethynylene macrocycles is a rapidly growing research field [8f, 22–25].

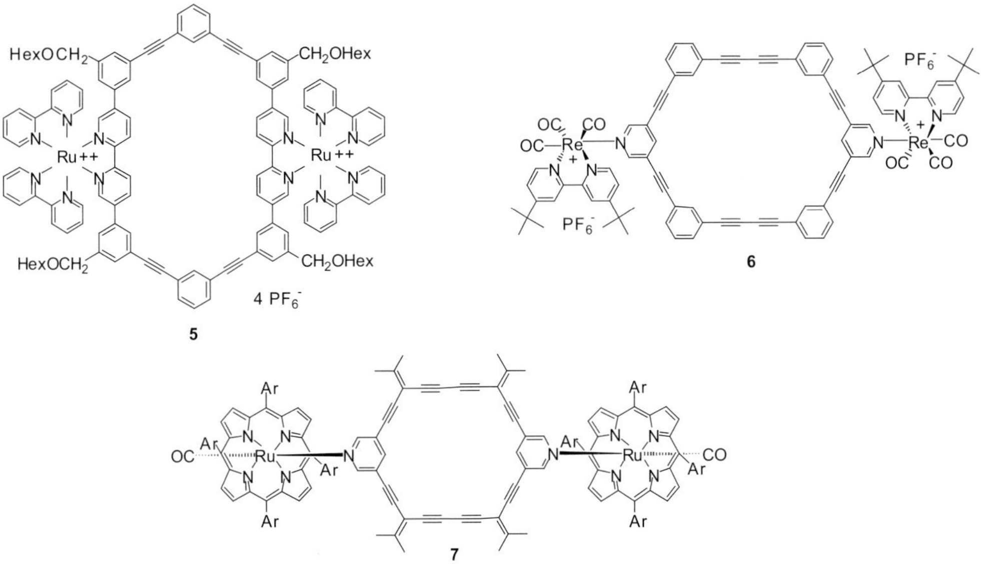

5 **6** **7**

Figure 10.4 *N*-heterocyclic macrocycles **5**–**7** and their respective metal complexes.

Several metal complexes of these large ligands have been described, including Schlüter's Ru- (**5**), Lees' Re-(**6**), or Tykwinski's Ru-complexes (**7**) [26–28]. In all metal-macrocycle complexes the donor atoms of the ring are exocyclically oriented. In **5**, for steric reasons, and in **6** and **7** they have fixed exocyclic orientations. These macrocycles might therefore be viewed as expanded 4,4′-bipyridines with the potential to form porous solid-state superstructures.

Macrocycles with intraannular polar groups have attained considerable stature in the area of host-guest chemistry. In most cases the binding sites of the ring are highly preorganized, resulting in guest recognition according to the lock-and-key principle. Diederich's chiral binaphthol macrocycles **8**, for example, are able to recognize 1-*O*-octylglycopyranosides and can even discriminate, albeit only slightly, between different anomers [29]. Tobe's *"expanded cyanospherand"* **9** binds cations such as tropylium or guanidinium, most probably through ion-dipole interactions, whereas the non-cyclic hexameric analogue does not bind such guests [30]. An *"expanded spherand"* **10** described by Oda et al. binds ammonium ions but not cesium ions, a notable difference in comparison with other host molecules such as 18-crown-6 that bind both metal cations and ammonium ions of similar size [31]. Höger's adaptable macrocycles **1** bind large tetraamines inside their nanometer-size cavities. In contrast with the shape-persistent macrocycles described above, though, the orientations of the binding sides are not determined by the ring structure but are influenced by the guest molecules, thus recalling the induced fit mechanism [32].

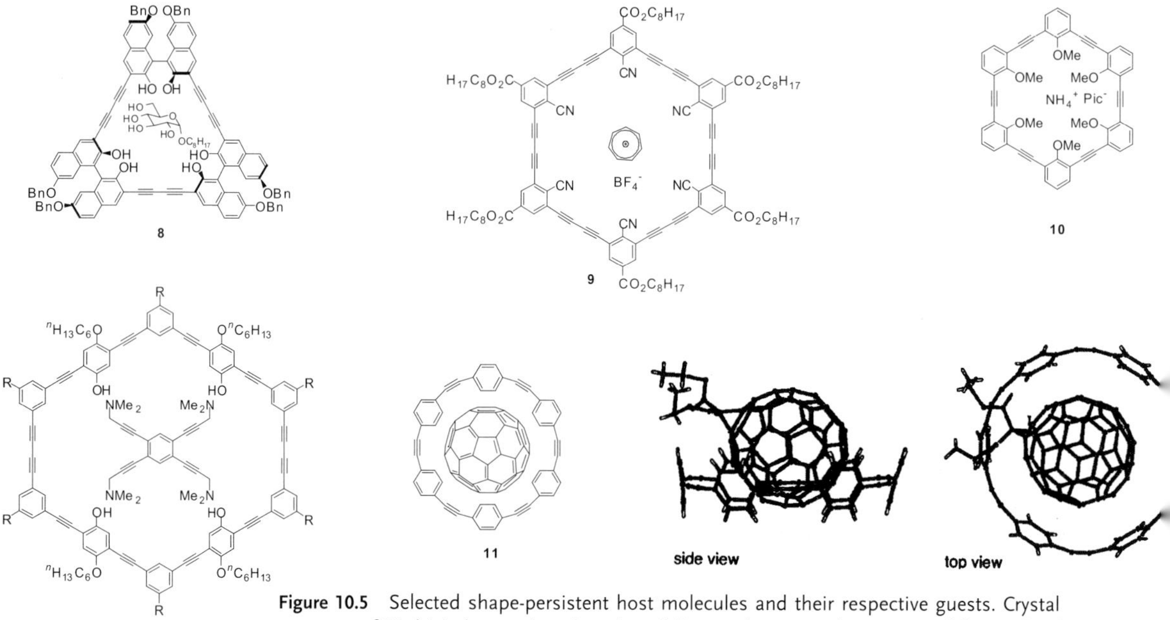

Figure 10.5 Selected shape-persistent host molecules and their respective guests. Crystal structure of **11** · bis(ethoxycarbonyl)methanofullerene (bottom right) [reprinted from ref. 33b, with permission].

Oda's cyclic *para*-phenylene-ethynylene rings have attracted special attention [33]. These rings have p-orbitals in the macrocycle plane, whereas the p-orbitals of the rings described above are oriented perpendicular to the ring plane. The size of the cavity in [6]paraphenyleneacetylene **11** is suitable for the inclusion of C_{60}, and Kawase and Oda et al. have estimated that the association constant for **11** · C_{60} is very large ($K_{assoc} \sim 10^{10}\ \text{M}^{-1}$) at −100 °C in CD_2Cl_2. In addition, a single-crystal X-ray analysis of a stable complex of **11** with bis(ethoxycarbonyl)methanofullerene has been reported, and this showed that the benzene rings of **11** were facing the fullerene. Recent binding experiments between a carbon nanoring containing two 2,6-naphthylene units at diametrically opposed positions in place of the 1,4-phenylene units of **11** have shown that these new rings have a selectivity for C_{70} over C_{60}. An approach towards tailor-made hosts for the separation of higher fullerenes and carbon nanotubes is thus now potentially available.

10.2.2
Tubular Superstructures in Solution

As briefly discussed in Section 10.1, macrocycles with orthogonal polar groups can form tubular superstructures through functional group interaction. An alternative approach towards hollow cylindrical superstructures is the aggregation of shape-persistent macrocycles through purely dissipative forces. The solvophobically favored stacking of large phenylene-ethynylene macrocycles may offer the potential not only to create one-dimensional superstructures, but also to form *functionalized* one-dimensional superstructures in which the polar functionality remains free. In other words, the functional groups will not be used for the superstructure organization, but the superstructure formation will be used for the organization of the functional groups. In the ideal case, the functional groups will be only marginally influenced by the superstructure formation and will be further usable for the recognition of appropriate guest molecules.

Moore et al. first reported the concentration-dependence of the NMR chemical shifts of a series of phenylene-ethynylene macrocycles **12** in chloroform [34]. The observed NMR data were analyzed in terms of a monomer-dimer model, it being assumed that this process was predominant in solution and that no higher aggregates were formed. Electron-withdrawing substituents (ester groups, **12a**) favored the aggregation, which was only slightly affected by the length of the aliphatic side chain. Vapor pressure osmometry (VPO) studies of **12a** in chloroform confirmed that higher aggregates were not formed to any significant degree, thus providing good agreement between the NMR and the VPO studies. Bulky side groups, although electron-withdrawing (*tert*-Bu esters, **12b**), inhibit aggregation, most probably for steric reasons. Electron-donating groups also disfavor the aggregation: **12c** does not show any concentration-dependence in its NMR signals. As one might expect, no aggregation was observed for **12a** in aromatic solvents such as benzene.

Since the aggregation depends strongly on the nature of the solvent, more polar solvents should, given sufficient macrocycle solubility, favor aggregation through increased solvophobicity of the aromatic macrocycle backbone. Compound **12d** fulfills this requirement, and an aggregation constant of $K_{assoc} \sim 1.5 \cdot 10^4\ \text{M}^{-1}$ (equal K-model for indefinite self association) could be observed in acetone, probably in conjunction with the formation of higher aggregates [35]. Unfavorable electronic properties have a dramatic effect, however, so **12e** has a lower association tendency ($1.4 \cdot 10^2\ \text{M}^{-1}$) in the same solvent. Surprisingly, **12d** also aggregates in aromatic solvents, and its association tendency here exceeds its association tendency in chloroform by more than a factor of 20.

Expanded analogues of the phenylene-ethynylene macrocycles **12** are the phenylene-butadiynelene macrocycles **13** described by Tobe et al. [36]. These rigid macrocycles also tend to aggregate, and **13a** has an association constant in chloroform four times larger than **12a**. Since electron-withdrawing aromatic systems produce stronger aggregation tendencies, the behavior of **13a** can be attributed to the strongly electron-withdrawing nature of the butadiyne group. Electrostatic potential calculations are in agreement with this hypothesis, and studies of the temperature

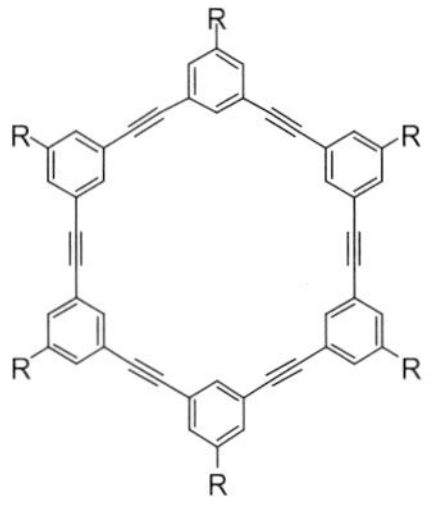

12a: R = $CO_2{}^nBu$
12b: R = $CO_2{}^tBu$
12c: R = O^nBu
12d: R = $CO_2(CH_2CH_2O)_3CH_3$
12e: R = $O(CH_2CH_2O)_3CH_3$

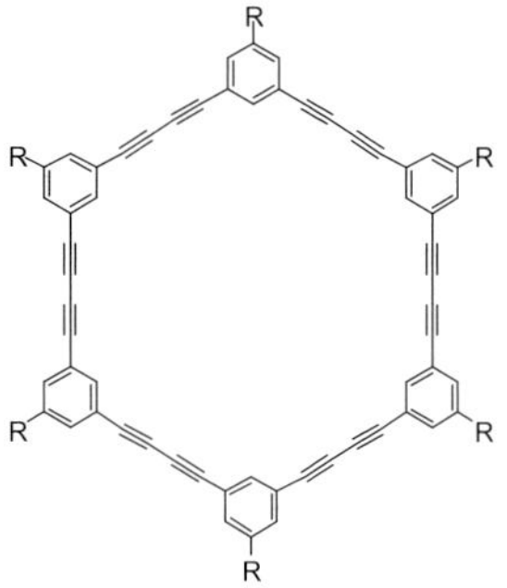

13a: R = $CO_2C_8H_{17}$
13b: R = ^tBu
13c: R = $CO_2(CH_2CH_2O)_3CH_3$

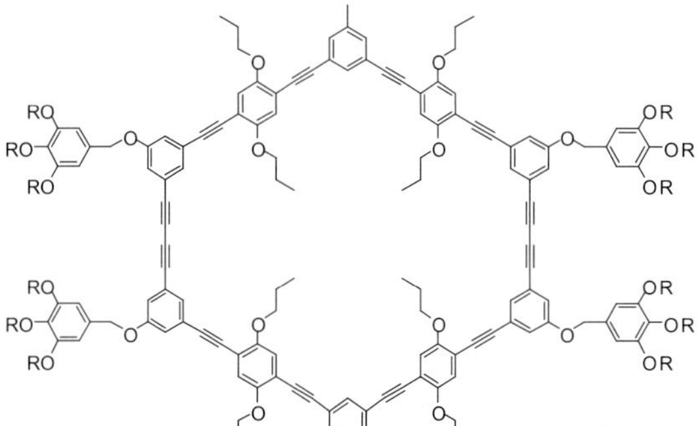

14a: R = $^nC_6H_{13}$
14b: R = $^nC_{12}H_{25}$

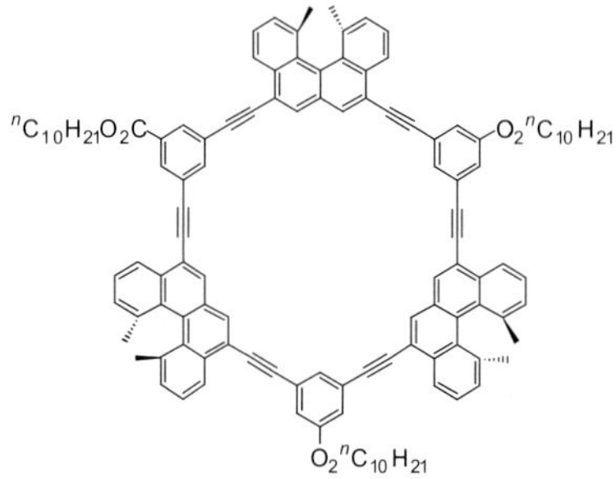

M,M,M-**15**

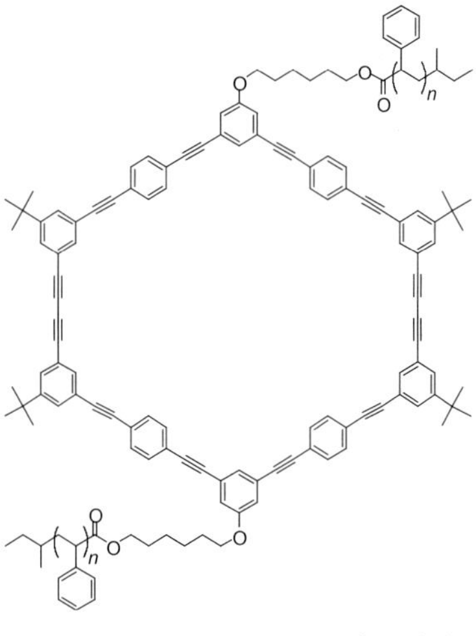

16a M_w (PS) = 1000 g mol^{-1}
16b M_w (PS) = 1500 g mol^{-1}
16c M_w (PS) = 2500 g mol^{-1}
16d M_w (PS) = 3500 g mol^{-1}
16e M_w (PS) = 5000 g mol^{-1}

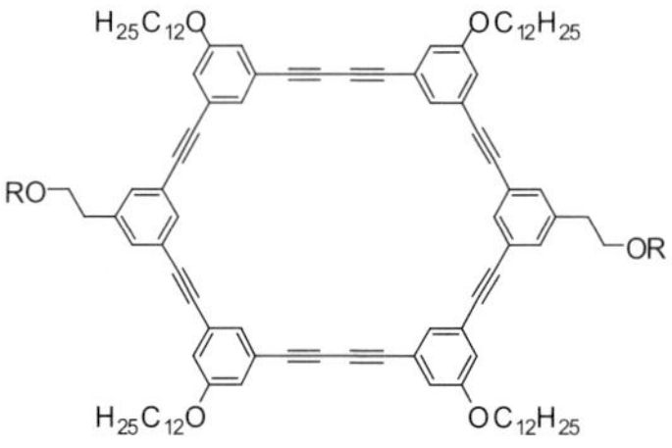

17a: R = $Si(CH_3)_2{}^tBu$
17b: R = H

Figure 10.6 Selected shape-persistent macrocycles capable of forming aggregates in solution.

dependence of the association also show that a larger enthalpy change is responsible for the observed behavior. Again, if the macrocycles are decorated with sterically demanding side groups such as *tert*-Bu (**13b**), no aggregation is observed in chloroform. Polar solvents also favor aggregation in this case, and the extrapolated association constant for **13c** in methanol is of the order of $10^6\ \text{M}^{-1}$, so the formation of nanotubular structures can reasonably be anticipated. Again, contrary to expectations, aromatic solvents do not prevent aggregation: macrocycles **13a** and **13c** both self-associate in toluene and *o*-xylene with association constants ($K_{assoc} \sim 2.1 \cdot 10^5\ \text{M}^{-1}$ for **13a**, $K_{assoc} \sim 3 \cdot 10^4\ \text{M}^{-1}$ for **13c**, both in toluene) higher than those found in chloroform. For the aggregation of **13c** in acetone and in toluene, analysis of the association data obtained by VPO studies suggests that the formation of higher aggregates is more favorable than the formation of dimers, suggesting a nucleation mechanism.

From the finding that electron-withdrawing substituents strongly enhance the aggregation, one might expect a high association tendency for **9**, because it is even more electron-deficient than **13a**. However, the NMR signals of **9** are concentration-independent, which can be explained in terms of a non-planar macrocycle conformation as a result of the electrostatic repulsion of the intraannular CN groups. On the other hand, mixtures of **9** and **13a** show concentration dependence of the NMR signals of both compounds, indicating the formation of heteroaggregates [30].

Solvophobic association of large macrocycles decorated with oligoalkyl side groups on their exteriors was reported by Höger and co-workers [37]. In THF and in aromatic and halogenated solvents, no concentration dependence of the NMR signals was found for macrocycle **14b**. In a mixed solvent system containing dichloromethane and hexane, the latter a poor solvent for the rigid core of the compound, concentration dependence of the NMR shifts for the macrocycle could, however, be observed. Application of a monomer-dimer model indicates that the tendency of the macrocycles to aggregate increases with increasing hexane content ($K_{assoc} \sim 130\ \text{M}^{-1}$ in dichloromethane/hexane 1:3 and $\sim 790\ \text{M}^{-1}$ in dichloromethane/hexane 1:6), but restricted compound solubility in apolar solvents did not permit the observation of higher dimerization constants.

An interesting form of macrocycle aggregation was described by Yamaguchi [38]. Cyclic phenylene-ethynylene trimers containing helicene corner pieces aggregate in chloroform and benzene, and the aggregation tendency depends on the chirality of the helicenes in the macrocyclic backbone. The cycloalkyne (*M*,*M*,*M*)-**15** displays a higher dimerization constant than the stereoisomeric (*M*,*P*,*M*)-**15**, and the tendency to aggregate is even lower in racemic mixtures of the macrocycles. Even more interesting is the observation that only dimers and no higher aggregates are formed. This can be explained in terms of the structure of the dimer, in which the two macrocycles are slightly bent toward each other for better stacking, while the surfaces above and below the dimer prevent good π-overlap with additional macrocycles. These observations opened the door to new aspects of macrocycle organization [39]. It has been shown that the aggregation of chemically linked macrocycles is influenced by the linker moiety. Flexible linkers give rise to very strong intramolecular aggregation ("*castagnet structure*"), while rigid linkers produce intermolecular dimer formation.

Since aggregation of a solvophobic compound is often not observable due to limited macrocycle solubility, Höger et al. prepared coil-ring-coil block copolymers **16**, each containing a rigid macrocyclic core and a periphery of two narrowly dispersed styrene oligomers [40–42]. All the block copolymers are readily soluble in THF, toluene, and halogenated solvents at room temperature, but the solubility of the block copolymers in aliphatic solvents depends strongly on the size of the coiled block. While **16a** only forms a suspension even in warm cyclohexane, **16b–e** are quite soluble in cyclohexane at elevated temperatures. Upon cooling, compound **16b** forms a gel at concentrations above 0.5 wt%. Under the same conditions, **16c** rapidly forms a highly viscous solution, as does **16d** after several days. All block copolymer solutions are strongly birefringent, with the single exception of solutions of **16e**, which exhibit neither unusual viscosity nor birefringence. The birifringence and the viscosities of the solutions are indications of the existence of extended surpamolecular structures. Their formation can be explained by the different solubilities of the rigid and the flexible parts of the molecule, keeping in mind that cyclohexane is a Θ-solvent for PS but, unlike THF or toluene, a poor solvent for the rigid core. Cyclohexane solutions of **16c** were investigated in more detail. Dynamic light scattering (DLS) was performed on solutions of **16c** in toluene and cyclohexane (Figure 10.7, left), revealing that only one species, with a hydrodynamic radius of approximately 2 nm, corresponding to the size of a simple block copolymer molecule **16c**, is present in toluene (Figure 10.8, left). In contrast, the light scattering data in cyclohexane indicated the presence of species with a broad distribution of hydrodynamic radii around 60 nm, corresponding to total lengths of the objects between 250 and 1300 nm and a high "virtual" persistent length of over 100 nm. Additional X-ray scattering measurements indicated that the coil-ring-coil block copolymers aggregate in solution into *hollow* cylinder-shaped polymer brushes (Figure 10.8, right).

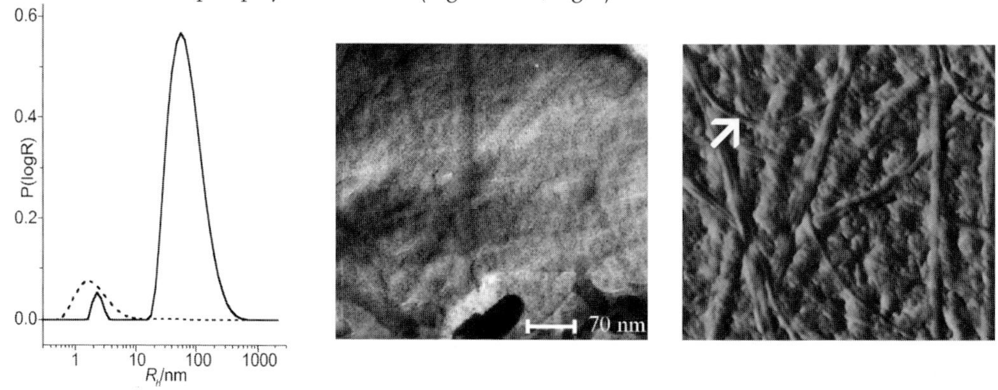

Figure 10.7 CONTIN-fit: rate distribution of **16c** in toluene (---) and in cyclohexane (—) at a concentration of 0.11 wt% (left); TEM: C/Pt shadowed film obtained by freeze drying of a 0.15 wt% cyclohexane solution of **16c** (middle); AFM (amplitude picture, 1.5 × 1.5 μm^2): film obtained by dipping mica into a 0.15 wt% cyclohexane solution of **16c** (right) [reprinted from ref. 40a, with permission].

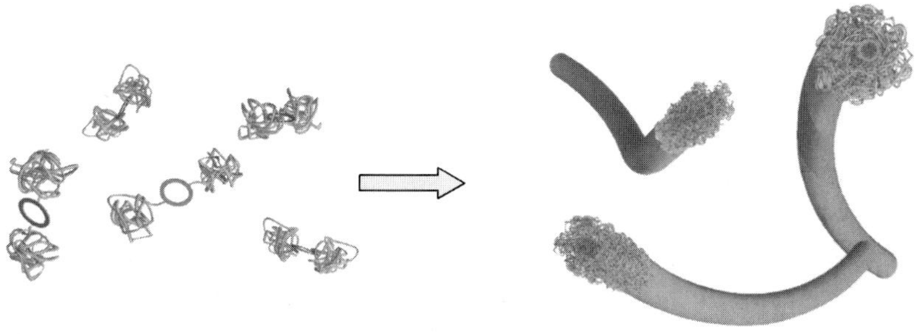

Figure 10.8 Schematic representation of the aggregation of the coil-ring-coil block copolymers (left) into hollow cylindrical brushes (right) [reprinted from ref. 40a, with permission].

Further support for the proposed aggregation is provided by an investigation of solid samples of **16c** prepared under "non-equilibrium conditions" (i. e., by fast solvent evaporation). The transmission electron micrograph (TEM) of a sample obtained by freeze-drying of a cyclohexane solution (Pt/C shadowed film) shows ribbons of different widths at the sample surface, the narrowest in the 15 nm range (Figure 10.7, middle). Atomic force microscopy (AFM) images of a polymer film on mica show long bundles of two or three cylindrical aggregates, together with individual aggregates (Figure 10.7, right), the latter having diameters of approximately 10–15 nm. The dimensions obtained by TEM and AFM correspond well with the dimensions obtained by scattering methods and are also in accordance with the dimensions of the molecular building blocks. Moreover, surveying of the most curved cylindrical object found in the AFM (Figure 10.7, right, white arrow) and treating it as a worm-like chain results in a persistence length of approximately 350 nm for the cylindrical aggregates. It is interesting to note that the aggregate formation in these copolymers is hindered neither by the bulky *tert*-Bu side groups nor by the electron-donating ether substituents on the ring. The light scattering data also indicate a nucleation mechanism, since only monomeric and high molecular weight aggregates of the block copolymers could be observed. It can be assumed that the oligomeric side groups and their interaction with the solvent (cyclohexane is a θ-solvent for polystyrene) play a more complex role than simply improving the compound solubility.

Side group-assisted aggregation was also reported by Tour et al. [43]. As expected, **17a** does not show any concentration dependence of its NMR signals since it is decorated with electron-donating substituents. The NMR spectrum of **17b** is concentration-dependent, however, and although the association constant is rather small (3.7 M^{-1} in $CDCl_3$), the π-stacking of the macrocycles is hydrogen-bond-assisted. In addition, sliding motions might be reduced, resulting in a larger and better defined inner diameter of the stacks.

In this context it is worth noting that aggregation of macrocycles can also be observed in the gas phase. Höger et al. described the MALDI-TOF (matrix-assisted laser desorption/ionization time-of-flight) mass spectrum of the pure macrocycle **1a** (M = 1598 Da). In addition to the molecular ion signal, peaks for dimers, trimers, and so forth are also present, each accompanied by peaks representing rings that have lost one or more hexyl side chains (Figure 10.9, left; the relative peak intensities depend on the laser power) [44]. In order to confirm that traces of an impurity were not responsible for the extra signals in the spectrum, a mixture of the two macrocycles **1a** and **1b** (M = 1851 Da) was investigated (Figure 10.9, right). In addition to the signals for the two macrocycles and their clusters with the matrix, the signals for the homo-dimers and -trimers as well as the mixed dimers and trimers could be observed, undoubtedly a result of aggregate formation in the matrix or during the evaporation process. These results show that information obtained by MALDI-TOF spectrometry should be verified by an alternative analytical method.

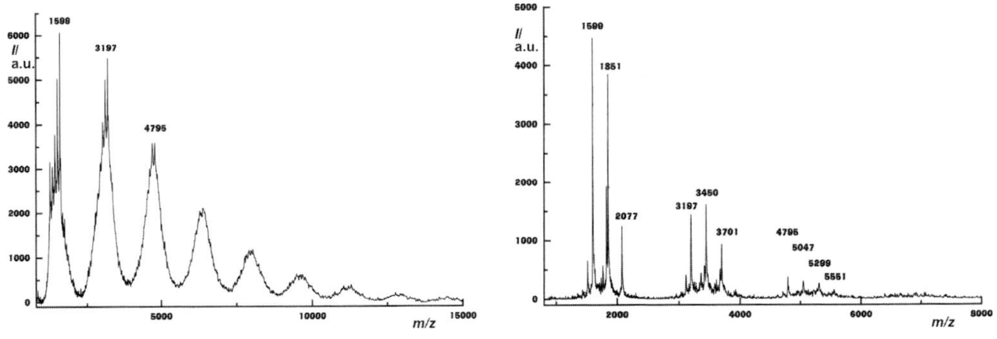

Figure 10.9 Mass spectrum of **1a**, showing aggregates of the ring up to the octamer (left). Mass spectrum of a mixture of **1a** and **1b** showing the rings, their clusters with the matrix (1,8,9 trihydroxyanthracene, M = 226 Da), the dimers of **1a** and **1b**, and the mixed dimers. Small amounts of homo- and mixed trimers are also detectable (right) [reprinted from ref. 44, with permission; copyright (1997) American Chemical Society].

10.2.3 Thermotropic Liquid Crystals

Liquid crystals (LCs) based on disk-like or cyclic molecules were first described by Chandrasekhar et al. over 25 years ago [45]. Macrocyclic compounds are of special interest in this sense because they could form supramolecular channel structures were they to adopt a columnar superstructure and the molecular backbone not collapse [46]. The typical design principle of discotic liquid crystals is a rigid core structure surrounded by a flexible periphery (Figure 10.10, top left). Therefore, shape-persistent macrocycles decorated with a flexible periphery are ideal candidates for this approach towards nanotubular superstructures. Indeed, Moore et al. were able to show that cyclic phenylene-ethynylene hexamers such as **18** exhibit liquid crystalline behavior [47]. Compounds with ester substituents (**18a**) showed evidence of ordered fluid phases with no clearing point up to 300 °C, most probably due to strong intermolecular attraction between the rings (behavior similar to what is observed in solution). Macrocycles **18b** and **18c** exhibit discotic nematic and isotropic phases, and **18d**, upon cooling from the isotropic melt (I), exhibits a discotic nematic (D_N) and discotic columnar phase (D) before it crystallizes (C) at 110 °C (I 200 D_N 144 D 110 C). A detailed study of the columnar phase found that the X-ray scattering data matched an oblique net with the lattice parameters a = 27.65, b = 51.25 Å and φ = 113.56°, with two molecules per mesh (Z = 2) [48]. Subsequent experiments showed that **18d** could be doped with small quantities (up to 2%) of silver triflate without changing the peak positions in the diffraction pattern, indicating that the metal salt was located inside the nanotubes.

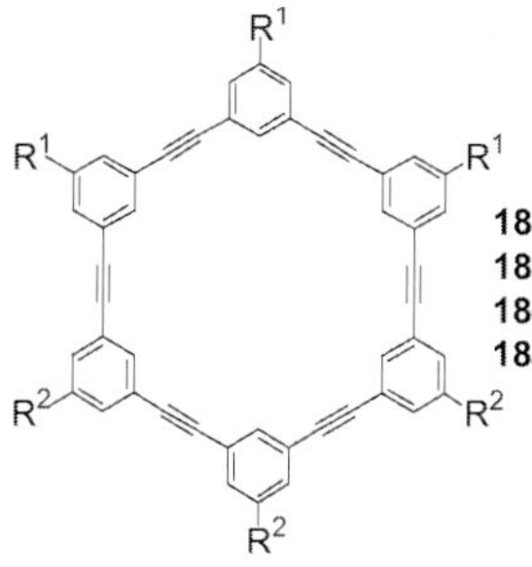

18a: $R^1 = R^2 = CO_2{}^nC_8H_{17}$
18b: $R^1 = R^2 = O^nC_7H_{15}$
18c: $R^1 = R^2 = OCO^nC_7H_{15}$
18a: $R^1 = CO_2{}^nC_6H_{13}$ $R^2 = O^nC_6H_{13}$

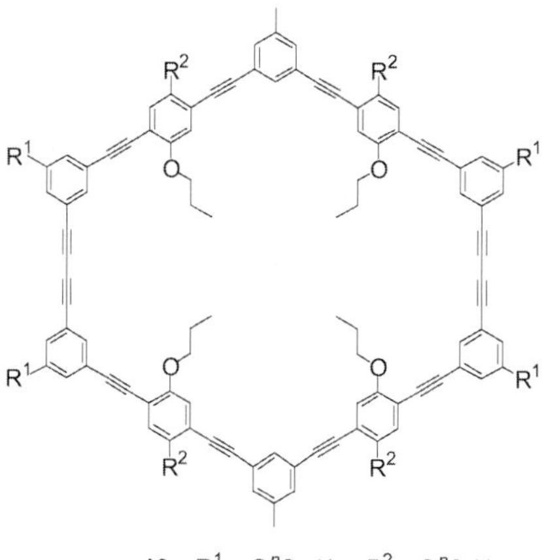

19a: $R^1 = O^nC_{18}H_{37}$ $R^2 = O^nC_3H_7$
19b: $R^1 = CH_3$ $R^2 = O^nC_{18}H_{37}$

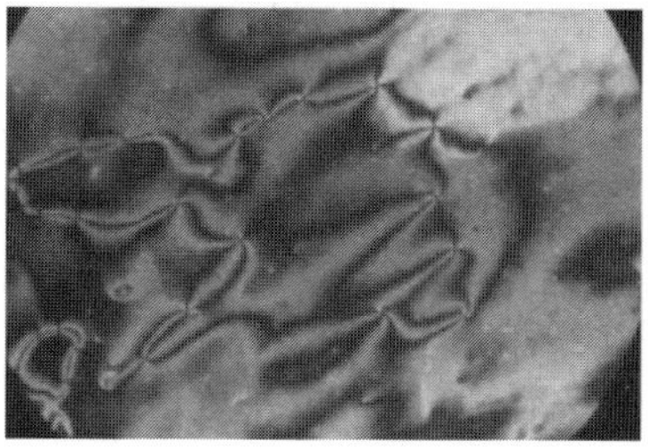

Figure 10.10 Design principle of cyclic molecules capable of forming thermotropic mesophases (top left); texture of **19b** at 200 °C between crossed polarizers (bottom right) [reprinted from ref. 49, with permission].

An interesting structural motive for discotic liquid crystals based on shape-persistent phenylene-ethynylene macrocycles was discovered by Höger et al. [49]. In order to obtain tubular superstructures with larger internal voids, macrocycles **19a** and **19b** were prepared. Their thermal behavior was investigated, and it was found that only the latter compound exhibits a stable thermotropic mesophase (C 185 N_D 207 I) (Figure 10.10, bottom right). At first glance, both compounds correlate with the design principle of discotic liquid crystals as described above (namely, a rigid core surrounded by a flexible periphery), and they differ only in the position in which the long, flexible alkyl chains are attached to the ring: at the corner elements in **19a** and in the middle of the sides of the ring in **19b**. In order to explain the different thermal behavior of **19a** and **19b** it was speculated that the large internal void might destabilize an ordered fluid phase through the frustration between the molecular anisotropy and the empty space inside the rings. Since back-folding of the alkyl periphery of **19a** is rather improbable, the macrocycle interior would have to be filled by the alkyl chains of adjacent rings, resulting in competition between orientational correlation and optimal space filling that could finally prevent the formation of a stable LC phase. Although **19b** should, according to Figure 10.10, suffer from the same handicap, single-crystal X-ray analysis found that the macrocycle can fill the interior with its own alkyl chains and so eliminate intermolecular entanglements (Figure 10.11, left).

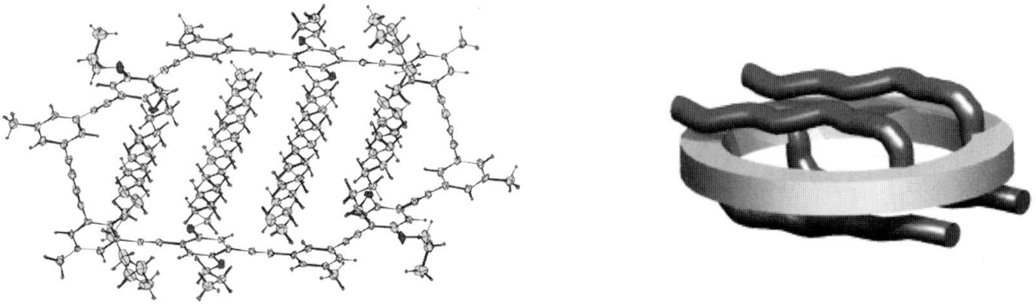

Figure 10.11 Crystal structure (ORTEP drawing) of **19b** (left). Schematic presentation of the arrangement of the alkyl groups in **19b** [reprinted from ref. 49, with permission].

This is possible because the long alkyl chains in **19b** are located at the adaptable positions of the ring, and so can point inwards without adopting entropically or energetically unfavorable conformations. Even more interesting is the finding that the topology of **19b** is inverted with regard to all discotic liquid crystals described so far: a flexible core is surrounded by a rigid periphery (Figure 10.11, right). In order to explain the stable LC phase of **19b**, the conformation in the LC phase was derived from the X-ray data, as it was not known whether the adaptable alkyl chains also point inside the macrocycle in the LC phase. Since the side group orientation of **19b** in the LC phase could not be determined experimentally, compound **20** was prepared. In this case, the long intraannular alkyl chains are attached at the corner elements of the ring and cannot rotate outwards. Intraannular orientation of the alkyl groups is also observed in the solid-state structure as determined by single-crystal X-ray analysis (Figure 10.12, middle). Investigation of the thermal behavior exhibited a stable nematic LC phase (C 134 N_D 159 I), and so the question as to whether liquid crystalline macrocycles with inverted topology really exist was answered positively [50].

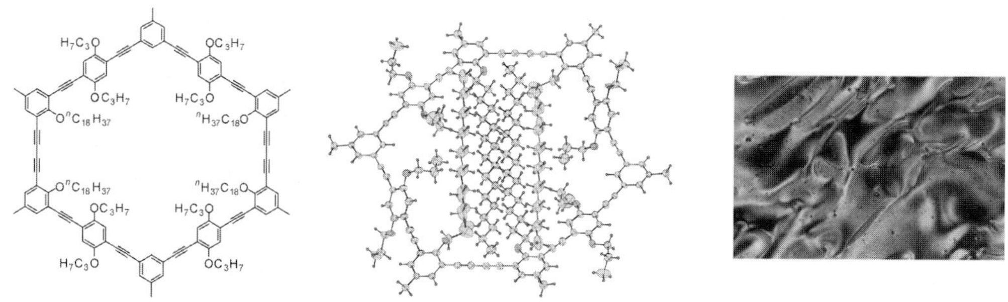

20

Figure 10.12 Structure of **20** (left), crystal structure (ORTEP drawing) of **20** (middle) and texture of **20** at 150 °C viewed between crossed polarizers (right).

In the search for liquid crystalline shape-persistent macrocycles with very large and open cavities, derivatives of **19a** with increased numbers of flexible segments have been investigated. Although **14a** and **14b** have lower melting points (**14a**: mp.: 202–204 °C; **14b**: mp.: 122–126 °C), no stable mesophase could be observed [37]. While this can be explained by the hypothesis described above, one might alternatively assume that the balance between rigid and flexible parts of the macrocycles is not correct. Filling the compounds' interiors answered these questions, and investigation of the thermotropic behavior of **21a** and **21b** revealed that both exhibit stable mesophases with beautiful fan-shaped textures under a polarizing microscope, typical for a columnar order of the molecules (**21a**: C 93 D 195 I; **21b**: C 82 D 145 I) (Figure 10.13) [51]. These results show that the balance between the core dimensions and the flexible periphery in **14** and **21** was correct and supports the hypothesis that the internal void in **14** can destabilize a mesophase. The finding that the melting point of **21a** is about 30 °C below the melting point of **14b**, even though the former compound contains additional intraannular polar side groups, supports the hypothesis that a large cavity inside the macrocycles can act as a physical cross-link [49].

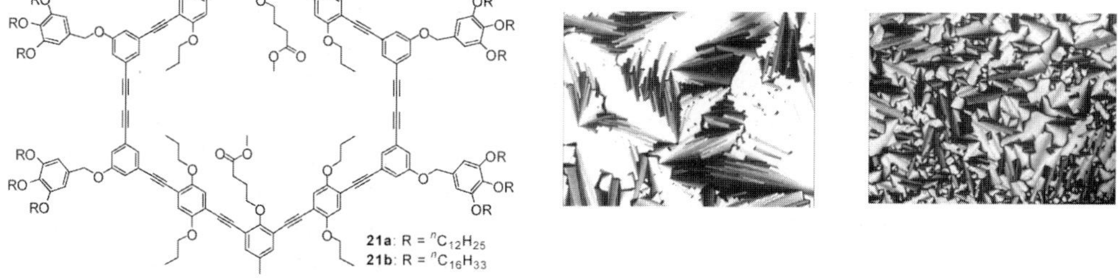

Figure 10.13 Structure of **21** (left); texture of **21a** (middle, 150 °C) and **21b** (right, 110 °C) between crossed polarizers [reprinted from ref. 51, with permission; copyright (2004) American Chemical Society].

The observed LC phase X-ray powder data for **21a** and **21b** fit with an oblique net with the basal periodicities $a = 67.6$ and $b = 42.5$ Å and angles between the main directions $\gamma = 109.6°$ for **21a** and $a = 72.6$, $b = 45.6$ Å, and $\gamma = 98.5°$ for **21b**. In both compounds, two stacks of the macrocycles run through one mesh of this oblique net ($Z = 2$), a feature pointing most probably to restricted rotation of the macrocycles within a stack [52]. Advanced solid-state ^1H and ^{13}C NMR investigations of **21b** provided further insight into the molecular dynamics of this macrocycle in the liquid crystalline phase. Individual degrees of mobility could be determined for different segments in the macrocycle by solid-state NMR, showing the macrocyclic core to be rather rigid in the LC phase, not rotating at kHz or higher frequencies within the column, consistently with the X-ray observations. The substituents, in contrast, show significant mobility above the kHz range, and the mobile alkyl chains fill the space around the core.

10.2.4
Two-Dimensional Organization

As indicated in Figure 10.2, macrocycles with extraannular polar groups can form two-dimensional superstructures in the solid state. As in the case of one-dimensional superstructures, the formation of well defined two-dimensional superstructures through the use only of non-specific forces is an attractive goal since these structures might act as platforms for the construction of ordered supramolecular three-dimensional structures (Figure 10.14) [53]. The driving force for the formation of the two-dimensional pattern is non-specific interaction with an appropriate substrate, so functionalized macrocycles should form functionalized two-dimensional superstructures capable of acting as templates for the creation of three-dimensional structures by recognition of appropriate guest structures.

Figure 10.14 Nanostructured (left), nanofunctionalized (middle), and nanofunctionalized surface with appropriate guest molecules (right) (schematic).

A purely dissipative platform for the formation of regular two-dimensional structures is the surface of highly oriented pyrolytic graphite (HOPG) [54]. One common technique for investigation of the self-assembled monolayers (SAMs) at the interface is scanning tunneling microscopy (STM), which can be performed under liquid [55]. STM images for several shape-persistent phenylene-ethynylene macrocycles have been reported, including macrocycles such as **23** and **24** (Figure 10.15), containing heterocyclic building blocks [8d, 25, 37, 51, 56].

The brightness of the STM images is related to the level of the detected tunneling current. Aromatic and conjugated moieties of molecules are known to display

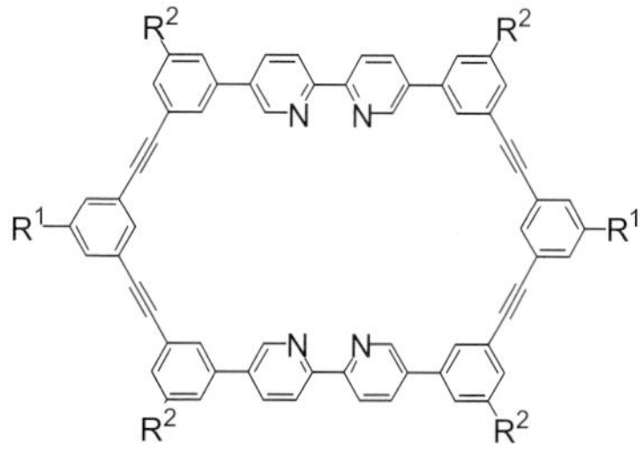

23: $R^1 = CH_2O^nC_6H_{13}$ $R^2 = OCH_2OMOM$

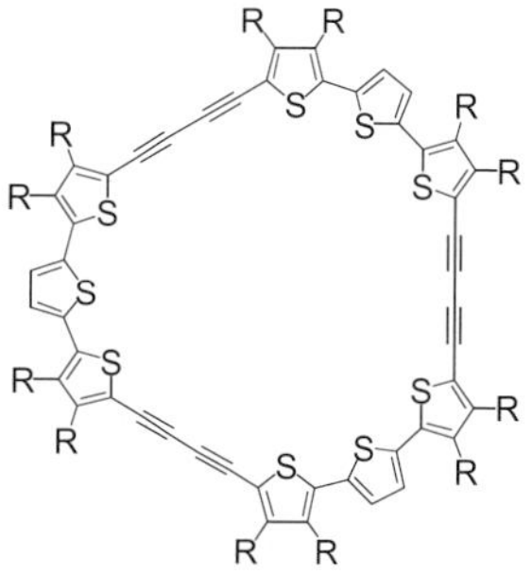

24: $R = {}^nC_4H_9$

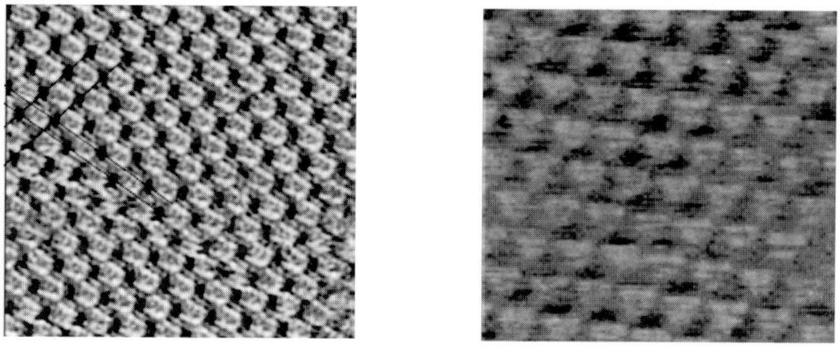

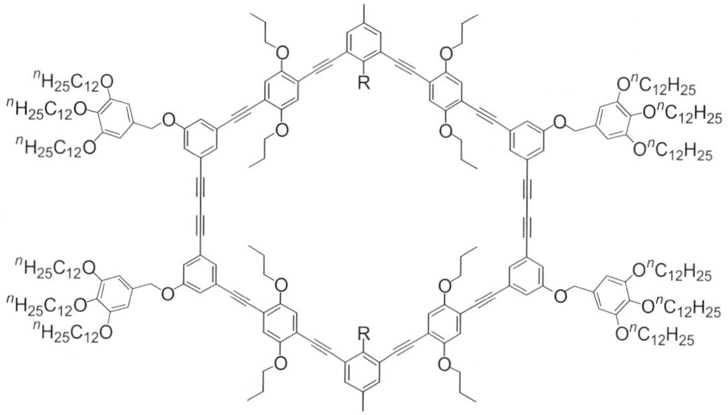

14b: R = H
21a: $R = O(CH_2)_3COOCH_3$

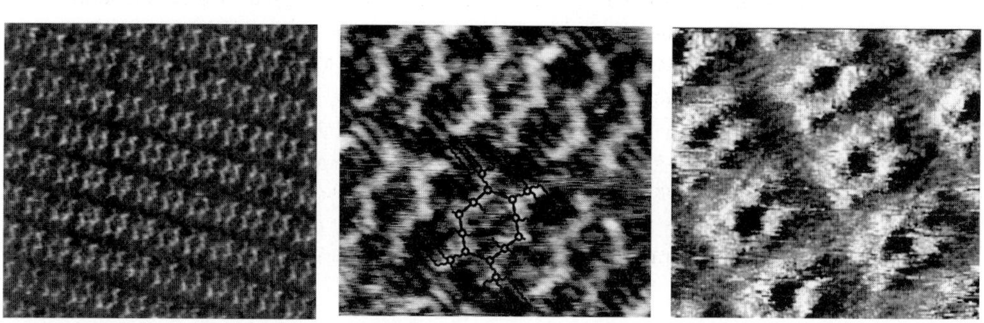

Figure 10.15 STM images of the phenylene-ethynylene macrocycles **23** and **24** (top), and of the macrocycles **14b** (bottom left and middle) and **21a** (bottom right) at the HOPG surface [reprinted from refs. 8d, 37, 51, and 56, with permission].

higher tunneling efficiency than aliphatic alkyl segments, so the bright parts of an image can be attributed to the shape-persistent backbones of the compounds. Contrast within a macrocycle often correlates with the density of the highest occupied molecular orbital (HOMO) or of the lowest unoccupied molecular orbital (LUMO) of the macrocycle, and may depend on the bias voltage [37, 57]. The molecular dimensions observed for the macrocycles **14** and **21** on the graphite surface correspond well with those obtained by single-crystal X-ray analysis of similar macrocycles [13, 16, 49].

Computer simulations (performed on the molecular backbone of **14**) suggest that the macrocycle has (in vacuum) two energetically similar nonplanar geometries: a "boat" and a "chair" conformation. The latter is also often found in the solid-state structures of shape-persistent macrocycles. The flat conformation is less stable than the nonplanar conformations by about 30 kJ mol^{-1}. When the macrocycle is brought into contact with the graphite surface, however, the situation changes dramatically. In the nonplanar conformations the macrocycle can not approach the surface closely, while the macrocycles in the flat conformation can reach a minimum distance of 0.4 nm. The calculated difference in the adsorption enthalpy is much larger than the differences between the three conformations. The calculations thus predict a transition from the nonplanar to the planar conformation during the adsorption process [37].

The degree of ordering on the surface varies from compound to compound and is also a function of experimental parameters. A comparison between macrocycles **14b** and **21a** shows that the extent of ordering depends on the solvent used for the investigations. While **14b** forms rather large monodomains on the graphite surface (Figure 10.15, bottom left) when adsorbed from either 1,2,4-trichlorobenzene or 1-phenyloctane, the situation for **21a** is more complex. When physisorbed from a 1,2,4-trichlorobenzene solution, the degree of ordering is often more pronounced (although far from ideal) than in images obtained for monolayers physisorbed from 1-phenyloctane, although the image resolution is worse. It is assumed that the main reason for the lack of long-range order in this case is the balance or competition between kinetically and thermodynamically controlled monolayer formation. The interplay between solute-solvent interaction and solute-substrate interaction will be an important factor. Each molecule contains twelve long alkyl chains, which should result in considerable interaction with the graphite substrate upon adsorption (in addition to the interaction between the aromatic cores and the substrate). Even if the molecule is not packed in an ideal fashion, the molecules/substrate interaction should be large enough to immobilize the molecules on the substrate to some degree, and as such, they could be kinetically trapped. Given that 1,2,4-trichlorobenzene is a better solvent for the macrocycles than 1-phenyloctane, the extent of kinetic trapping should be less in 1,2,4-trichlorobenzene, resulting in, on the timescale of the experiment, more long-range ordered systems.

While these macrocycles have the ability to form well organized monolayers on HOPG, the capability to form organized monolayers on other substrates, such as gold, is an attractive goal for several applications (sensing, metal deposition, etc.) [58]. Attempts to form SAMs of **14b** on gold substrates were not successful. One reason for that might be the lack of functionalization needed to bind the macrocycle to Au(111). Since alkanethiols and also alkyl sulfides have been shown to produce steady SAMs on Au(111), macrocycle **25** (Scheme 10.2) was investigated. The synthesis of **25** is based on the finding that the yield of the desired cyclization product can be dramatically enhanced if the bisacetylenes are covalently connected to an appropriate template (4,4′-bis(hydroxymethyl)biphenyl) prior to the oxidative coupling reaction [59]. The template simultaneously acts as a protecting group for the carboxylic acid functionality in this case.

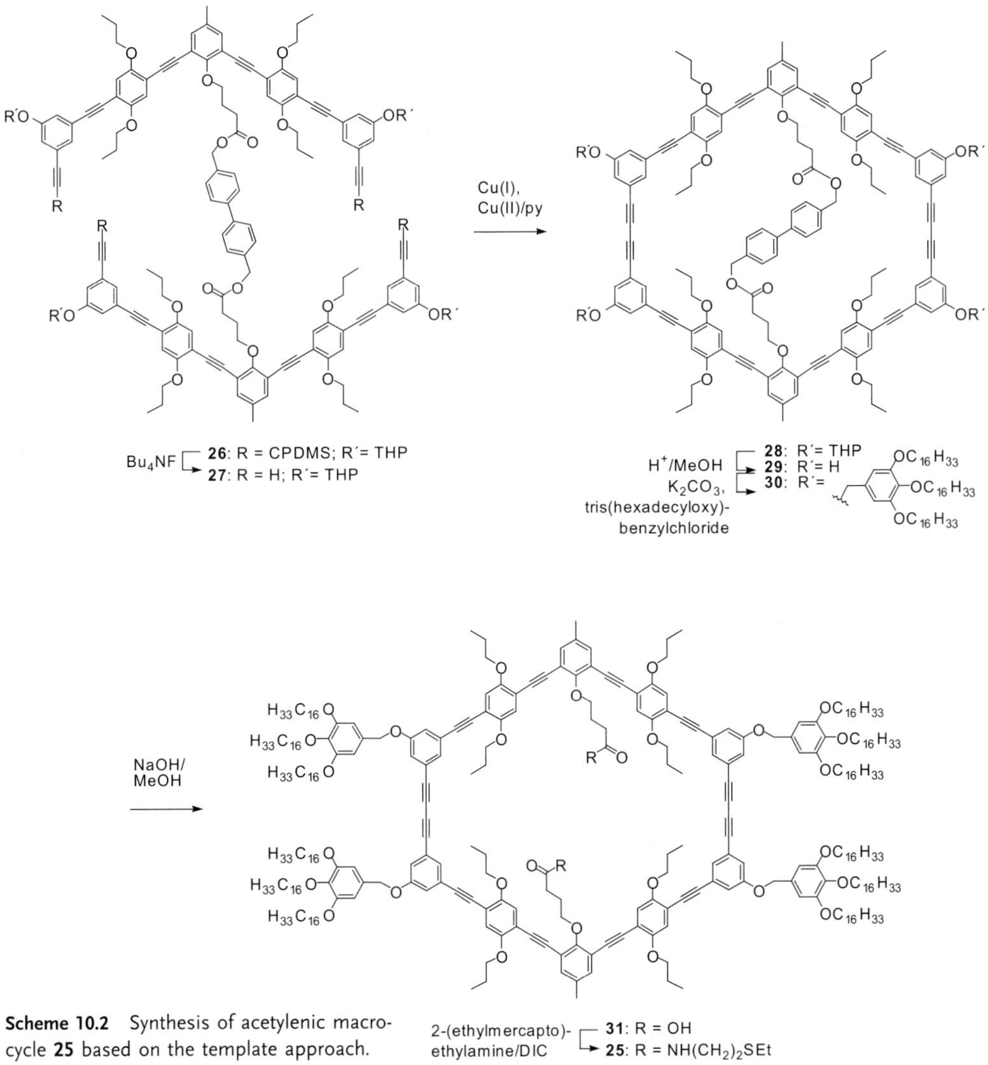

Scheme 10.2 Synthesis of acetylenic macrocycle **25** based on the template approach.

Macrocycle **25** contains intraannular functional groups (amides) and additional thioether groups for binding to the Au substrate. The effect of the sulfur functionalization on the adsorption properties of the macrocycle is remarkable, and STM images showing an ordered assembly of **25** on single-crystal Au(111) were obtained (Figure 10.16) [60]. The molecules are organized into rows with a width of approximately 5 nm. X-ray photoelectron spectroscopy and ultraviolet photoelectron spectroscopy measurements have been carried out and indicate that some "thiolate-like" sulfur is formed on the Au surface, producing an enhanced adhesion strength of the monolayer onto the Au.

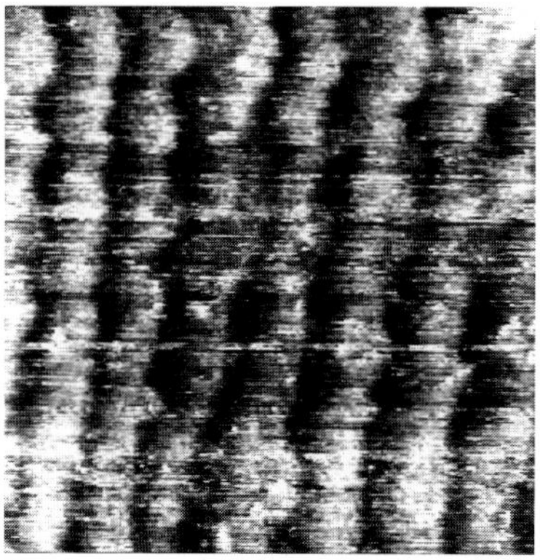

Figure 10.16 STM image of macrocycle **25** adsorbed on Au(111) [reprinted from ref. 60, with permission; copyright (2004) American Chemical Society].

These investigations have shown that shape-persistent macrocycles are an excellent tool for the extension of the concept of *nano-patterning* towards *nano-functionalization* of the HOPG and Au(111) surfaces in the multi-nanometer regime.

10.3 Conclusions

The continuously growing interest in well defined nanometer-scale organic materials during the last two decades has resulted in shape-persistent macrocycles being perceived as an interesting class of compounds containing interiors and exteriors that can be addressed separately. The presence of non-functionalized side groups at the periphery can provide thermotropic liquid crystalline compounds or materials that can aggregate in solution to form either dimers or extended tubular aggregates. If the side groups direct inwards, new liquid crystalline topologies not available from disc-like molecules can be attained.

Macrocycles with functionalized side groups at the periphery allow for the creation of supramolecular structures through specific interactions with appropriate connectors, and macrocycles with intraannular functional groups are host molecules for guest molecules of complementary size, polarity, and functionality. Advanced synthetic methods also allow the combination of different (separately addressable) side groups within one macrocycle. The investigation of these compounds is only just beginning, but the potential to create complex ordered structures based on functionalized shape-persistent phenylene-ethynylene macrocycles is clearly recognizable.

10.4 Experimental Procedures

10.4.1 Deprotection of a CPDMS-Protected Acetylene (26 → 27) [61]

Bu_4NF (1 M in THF, 3 mL) was added to a solution of **26** (0.37 g, 0.13 mmol) in THF (15 mL). After three hours stirring at room temperature, the mixture was poured into Et_2O and water, and the organic phase was separated, extracted with water and brine, and dried over $MgSO_4$. The solvent was evaporated, and the remaining residue was purified by repeated column chromatography over silica gel with Et_2O/PE (1:1) (R_f = 0.28) as eluent to give **27** (0.23 g, 76%) as a yellow solid. ^1H NMR (CD_2Cl_2): δ 7.50 (d, J = 8.1 Hz, 4 H), 7.35–7.12 (m, 20 H), 7.05 (s, 4 H), 7.01 (s, 4 H), 5.40 (t, J = 3.1 Hz, 4 H), 5.00 (s, 4 H), 4.41 (t, J = 6.1 Hz, 4 H), 3.98 (t, J = 6.5 Hz, 8 H), 3.97 (t, J = 6.3 Hz, 8 H), 3.90–3.78 (m, 4 H), 3.65–3.53 (m, 4 H), 3.15 (s, 4 H), 2.75 (t, J = 7.5 Hz, 4 H), 2.31 (s, 6 H), 2.25–2.10 (m, 4 H), 2.05–1.50 (m, 40 H), 1.07 (t, J = 7.3 Hz, 12 H), 1.06 (t, J = 7.4 Hz, 12 H); MS (MALDI-TOF): m/z 2351 $[M + Na]^+$.

10.4.2 Template-based Oxidative Cyclodimerization of a Rigid Bisacetylene (27 → 28)

A solution of **27** (187 mg, 0.08 mmol) in dry pyridine (20 mL) was added over 96 h at room temperature to a suspension of CuCl (1.17 g, 118 mmol) and $CuCl_2$ (235 mg, 17 mmol) in pyridine (170 mL). After completion of the addition, the mixture was allowed to stir for an additional day and was then poured into CH_2Cl_2 and water. The organic phase was extracted with 25% NH_3 solution, water, 10% acetic acid, water, 10% aqueous sodium hydroxide, and brine, and dried over $MgSO_4$. After evaporation of the solvent to a small amount (about 15 mL), the product was precipitated by the addition of methanol and collected by filtration. Purification was performed by column chromatography over silica gel with CH_2Cl_2 as eluent (R_f = 0.30). Compound **28** (151 mg, 81%) was obtained as a slightly yellow solid. ^1H NMR (CD_2Cl_2): δ 7.62 (d, J = 8.2 Hz, 4 H), 7.43 (d, J = 8.2 Hz, 4 H), 7.36–7.33 (m, 4 H), 7.30 (s, 4 H), 7.27–7.24 (m, 4 H), 7.21–7.16 (m, 4 H), 7.09 (s, 4 H), 7.06 (s, 4 H), 5.44 (t, J = 3.1 Hz, 4 H), 5.08 (s, 4 H), 4.36 (t, J = 5.6 Hz, 4 H), 4.03 (t, J = 6.5 Hz, 8 H), 4.01 (t, J = 6.6 Hz, 8 H), 3.94–3.80 (m, 4 H), 3.68–3.57 (m, 4 H), 2.75 (t, J = 7.6 Hz, 4 H), 2.33 (s, 6 H), 2.23–2.08 (m, 4 H), 2.08–1.50 (m, 40 H), 1.09 (t, J = 7.4 Hz, 12 H), 1.07 (t, J = 7.4 Hz, 12 H); MS (MALDI-TOF): m/z 2348 $[M + Na]^+$.

10.4.3 Deprotection of a Macrocyclic THP-Protected Tetraphenol (28 → 29)

p-Toluenesulfonic acid (5 mg) was added to a solution of **28** (122 mg, 0.05 mmol) in $CHCl_3$ (70 mL) and methanol (5 mL). The mixture was stirred for three days with exclusion of light. After evaporation of the solvent to a small volume, the residue was dissolved in THF, and the product was precipitated by the addition of methanol. After filtration and vacuum drying in the dark, **29** was obtained as a slightly yellow solid (100 mg, quantitative) and used as received. ^{1}H NMR (THF d-8): δ 7.64 (d, J = 8.2 Hz, 4 H), 7.46 (d, J = 8.5 Hz, 4 H), 7.28 (s, 4 H), 7.20–7.16 (m, 4 H), 7.14 (s, 4 H), 7.10 (s, 4 H), 6.96–6.93 (m, 4 H), 6.92–6.88 (m, 4 H), 5.11 (s, 4 H), 4.38 (t, J = 5.7 Hz, 4 H), 4.05 (t, J = 6.3 Hz, 8 H), 4.02 (t, J = 6.3 Hz, 8 H), 2.82–2.68 (m, 4 H), 2.30 (s, 6 H), 2.23–2.10 (m, 4 H), 1.92–1.76 (m, 16 H), 1.10 (t, J = 7.6 Hz, 12 H), 1.06 (t, J = 7.4 Hz, 12 H); MS (FD): m/z 1988 $[M]^+$.

10.4.4 Alkylation of a Macrocyclic Tetraphenol (29 → 30)

K_2CO_3 (2.00 g, 14.5 mmol) was added to a solution of **29** (100 mg, 0.05 mmol) and 3,4,5-tris(hexadecyloxy)benzyl chloride (255 mg, 0.30 mmol) in dry DMF (50 mL), and the mixture was stirred at 60 °C in the dark for three days. The mixture was poured into CH_2Cl_2 and water, and the organic phase was separated and washed with water and brine, and dried over $MgSO_4$. Purification by repeated column chromatography over silica gel with CH_2Cl_2/PE (3:1) as eluent (R_f = 0.90) gave **30** as a slightly yellow solid (188 mg, 72 %). ^{1}H NMR (CD_2Cl_2): δ 7.62 (d, J = 8.2 Hz, 4 H), 7.43 (d, J = 7.9 Hz, 4 H), 7.39–7.28 (m, 8 H), 7.22–7.08 (m, 8 H), 7.08 (s, 4 H), 7.05 (s, 4 H), 6.63 (s, 8 H), 5.07 (s, 4 H), 4.98 (s, 8 H), 4.36 (t, J = 5.4 Hz, 4 H), 4.09–3.88 (m, 40 H), 2.74 (t, J = 7.6 Hz, 4 H), 2.32 (s, 6 H), 2.24–2.08 (m, 4 H), 1.94–1.65 (m, 40 H), 1.52–1.16 (m, 312 H), 1.14–1.00 (m, 24 H), 0.94–0.81 (m, 36 H); MS (MALDI-TOF): m/z 5341 $[M + Ag]^+$.

10.4.5 Hydrolysis of a Macrocycle with Two Intraannular Ester Groups (30 → 31)

Aqueous potassium hydroxide (10 %, 2 mL) was added to a solution of **30** (80 mg, 15 μmol) in THF (20 mL). The mixture was heated at reflux for three days. After the mixture had cooled to room temperature, hydrochloric acid (10 %, 2 mL) was added, and the solvent was removed to a small amount. The residue was dissolved in THF and precipitated with methanol. The precipitate was filtered and washed with ethanol to give **31** as a slightly yellow solid that was used as received (76 mg, quantitative). ^{1}H NMR (THF d-8): δ 7.35–7.32 (m, 4 H), 7.29 (s, 4 H), 7.21–7.18 (m, 4 H), 7.17–7.15 (m, 4 H), 7.14 (s, 4 H), 7.10 (s, 4 H), 6.71 (s, 8 H), 5.02 (s, 8 H), 4.39 (t, J = 5.9 Hz, 4 H), 4.10–3.90 (m, 40 H), 2.67 (t, J = 7.6 Hz, 4 H), 2.30 (s, 6 H), 2.23–2.11 (m, 4 H), 1.96–1.68 (m, 40 H), 1.60–1.20 (m, 312 H), 1.14 (t, J = 8.0 Hz, 12 H), 1.11 (t, J = 7.8 Hz, 12 H), 0.93–0.83 (m, 36 H); MS (MALDI-TOF): m/z 5162 $[M + Ag]^+$.

10.4.6
Formation of a Macrocycle with Two Intraannular Thioether Groups (31 → 25)

2-(Ethylmercapto)ethylamine hydrochloride (11 mg, 80 μmol), *N*-ethylmorpholine (18 mg, 160 μmol), and 1*H*-hydroxybenzotriazole (10 mg, 80 μmol) were added to a solution of **31** (25 mg, 5 μmol) in THF (7 mL). After the mixture had been cooled to 0 °C, diisopropylcarbodiimide (16 mg, 160 μmol) was added dropwise, and the solution was stirred for an additional hour at 0 °C and overnight at room temperature. The solvent was evaporated, the remaining residue was dissolved in a small amount of CH_2Cl_2, and **25** was precipitated with methanol. Filtration and washing with ethanol yielded **25** (22 mg, 84 %) as a yellow solid. ^{1}H NMR (CD_2Cl_2): δ 7.30–7.26 (m, 4 H), 7.22 (s, 4 H), 7.09–7.06 (m, 4 H), 7.03–6.99 (m, 4 H), 6.98 (s, 4 H), 6.97 (s, 4 H), 6.96 (br. s, 2 H), 6.54 (s, 8 H), 4.89 (s, 8 H), 4.34 (t, J = 5.5 Hz, 4 H), 3.99–3.78 (m, 40 H), 3.20–3.09 (m, 4 H), 2.48–2.31 (m, 12 H), 2.23 (s, 6 H), 2.11–1.96 (m, 4 H), 1.87–1.05 (m, 382 H), 0.84–0.72 (m, 36 H); MS (MALDI-TOF): m/z 5339 $[M + Ag]^+$.

Acknowledgements

We acknowledge financial support from the Deutsche Forschungsgemeinschaft, the Fonds der Chemischen Industrie, the Volkswagenstiftung, and the GDCh (Dr. Hermann Schnell-Stiftung).

References

[1] E. Fischer, *Ber. Dtsch. Chem. Ges.* **1894**, *27*, 2985–2993.

[2] J.-M. Lehn, *Angew. Chem.* **1988**, *100*, 91–116; *Angew. Chem. Int. Ed. Engl.* **1988**, *27*, 89–112.

[3] a) C. J. Pedersen, *J. Am. Chem. Soc.* **1967**, *89*, 2495–2496; b) C. J. Pedersen, *J. Am. Chem. Soc.* **1967**, *89*, 7017–7036.

[4] a) D. J. Cram, T. Kaneda, R. C. Helgeson, G. M. Lein, *J. Am. Chem. Soc.* **1979**, *101*, 6752–6754; b) D. J. Cram, *Angew. Chem.* **1986**, *98*, 1041–1060; *Angew. Chem. Int. Ed. Engl.* **1986**, *25*, 1039–1057.

[5] H. A. Staab, F. Binnig, *Chem. Ber.* **1967**, *100*, 293–305.

[6] a) F. Vögtle, P. Neumann, *Tetrahedron* **1970**, *26*, 5299–5318; b) E. Weber, F. Vögtle, *Chem. Ber.* **1976**, *109*, 1803–1831.

[7] For other pioneering work on shape-persistent macrocycles see, e. g.: a) H. A. Staab, F. Binnig, *Tetrahedron Lett.* **1964**, 319–321; b) H. A. Staab, K. Neunhoeffer, *Synthesis* **1974**, 424; c) G. R. Newkome, H.-W. Lee, *J. Am. Chem. Soc.* **1983**, *105*, 5956–5957; d) Y. Fujioka, *Bull. Chem. Soc. Jpn.* **1984**, *57*, 3494–3506.

[8] For recent reviews on shape-persistent macrocycles, mainly phenylene-ethynylene based, see: a) J. S. Moore, *Acc. Chem. Res.* **1997**, *30*, 402–413; b) S. Höger, *J. Polym. Sci. Part A: Polym. Chem.* **1999**, *37*, 2685–2698;

c) M. M. Haley, J. J. Pak, S. C. Brand, *Top. Curr. Chem.* **1999**, *201*, 81–130;
d) C. Grave, A. D. Schlüter, *Eur. J. Org. Chem.* **2002**, 3075–3098;
e) D. Zhao, J. S. Moore, *Chem. Commun.* **2003**, 807–818;
f) Y. Yamaguchi, Z. Yoshida, *Chem. Eur. J.* **2003**, *9*, 5430–5440;
g) S. Höger, *Chem. Eur. J.* **2004**, *10*, 1320–1329.

[9] A. Y. Grosberg, A. R. Khokhlov, *Giant Molecules*, Academic Press, San Diego, **1997**, p. 73.

[10] L. Tomasic, G. P. Lorenzi, *Helv. Chim. Acta* **1987**, *70*, 1012–1016;
b) M. R. Ghadiri, J. R. Granja, R. A. Milligan, D. E. McRee, N. Khazanovich, *Nature* **1993**, *366*, 324–327;
c) G. Gattuso, S. Menzer, S. A. Nepogodiev, J. F. Stoddart, D. J. Williams, *Angew. Chem.* **1997**, *109*, 1615–1617; *Angew. Chem. Int. Ed. Engl.* **1997**, *36*, 1451–1454;
d) D. T. Bong, T. D. Clark, J. R. Granja, M. R. Ghadiri, *Angew. Chem.* **2001**, *113*, 1016–1041; *Angew. Chem. Int. Ed.* **2001**, *40*, 988–1011.

[11] D. Venkataraman, S. Lee, J. Zhang, J. S. Moore, *Nature* **1994**, *371*, 591–593.

[12] A. D. Hamilton, *Bioorg. Chem. Front.* **1991**, *2*, 115–174.

[13] a) S. Höger, V. Enkelmann, *Angew. Chem.* **1995**, *107*, 2917–2919; *Angew. Chem. Int. Ed. Engl.* **1995**, *34*, 2713–2716;
b) S. Höger, A.-D. Meckenstock, S. Müller, *Chem. Eur. J.* **1998**, *4*, 2423–2434.

[14] D. E. Koshland Jr., *Angew. Chem.* **1994**, *106*, 2468–2472; *Angew. Chem. Int. Ed. Engl.* **1994**, *33*, 2375–2378.

[15] J. K. Young, J. S. Moore in *Modern Acetylene Chemistry* (Eds.: P. J. Stang, F. Diederich), VCH, Weinheim, **1995**, pp. 415–442.

[16] S. Höger, D. L. Morrison, V. Enkelmann, *J. Am. Chem. Soc.* **2002**, *124*, 6734–6736.

[17] a) H. L. Anderson, J. K. M. Sanders, *J. Chem. Soc., Chem. Commun.* **1989**, 1714–1715;
b) H. L. Anderson, A. Bashall, K. Henrick, M. McPartlin, J. K. M. Sanders, *Angew. Chem.* **1994**, *106*, 445–447; *Angew. Chem. Int. Ed. Engl.* **1994**, *33*, 429–431;
c) S. Anderson, H. L. Anderson, A. Bashall, M. McPartlin, J. K. M. Sanders, *Angew. Chem.* **1995**, *107*, 1196–1200; *Angew. Chem. Int. Ed. Engl.* **1995**, *34*, 1096–1099.

[18] H. L. Anderson, J. K. M. Sanders, *J. Chem. Soc., Chem. Commun.* **1992**, 946–947.

[19] a) C. J. Walter, H. L. Anderson, J. K. M. Sanders, *J. Chem. Soc., Chem. Commun.* **1993**, 458–460;
b) L. G. Mackay, R. S. Wylie, J. K. M. Sanders, *J. Am. Chem. Soc.* **1994**, *116*, 3141–3142.

[20] a) A. Ambroise, J. Li, L. Yu, J. S. Lindsey, *Org. Lett.* **2000**, *2*, 2563–2566;
b) K. Tomizaki, L. Yu, L. Wei, D. F. Bocian, J. S. Lindsey, *J. Org. Chem.* **2003**, *68*, 8199–8207.

[21] S. Rucareanu, O. Mongin, A. Schuwey, N. Hoyler, A. Gossauer, *J. Org. Chem.* **2001**, *66*, 4973–4988.

[22] M. Schmittel, H. Ammon, *Synlett* **1999**, 750–752.

[23] Y. Tobe, A. Nagano, K. Kawabata, M. Sonoda, K. Naemura, *Org. Lett.* **2000**, *2*, 3265–3268.

[24] P. N. W. Baxter, *Chem. Eur. J.* **2003**, *9*, 5011–5022.

[25] C. Grave, D. Lentz, A. Schäfer, P. Samori, J. P. Rabe, P. Franke, A. D. Schlüter, *J. Am. Chem. Soc.* **2003**, *125*, 6907–6918.

[26] O. Henze, D. Lentz, A. Schäfer, P. Franke, A. D. Schlüter, *Chem. Eur. J.* **2002**, *8*, 357–365.

[27] S.-S. Sun, A. J. Lees, *Organometallics* **2001**, *20*, 2353–2358.

[28] a) K. Campbell, R. McDonald, N. R. Branda, R. R. Tykwinski, *Org. Lett.* **2001**, *3*, 1045–1048;
b) K. Campbell, R. McDonald, R. R. Tykwinski, *J. Org. Chem.* **2002**, *67*, 1133–1140;
c) K. Campbell, C. J. Kuehl, M. J. Ferguson, P. J. Stang, R. R. Tykwinski, *J. Am. Chem. Soc.* **2002**, *124*, 7266–7267.

[29] S. Anderson, U. Neidlein, V. Gramlich, F. Diederich, *Angew. Chem.* **1995**,

107, 1722–1725; *Angew. Chem. Int. Ed. Engl.* **1995**, *34*, 1596–1600.

[30] Y. Tobe, N. Utsumi, A. Nagano, K. Naemura, *Angew. Chem.* **1998**, *110*, 1347–1349; *Angew. Chem. Int. Ed.* **1998**, *37*, 1285–1287.

[31] Y. Hosokawa, T. Kawase, M. Oda, *Chem. Commun.* **2001**, 1948–1949.

[32] D. L. Morrison, S. Höger, *Chem. Commun.* **1996**, 2313–2314.

[33] a) T. Kawase, H. R. Darabi, M. Oda, *Angew. Chem.* **1996**, *108*, 2803–2805; *Angew. Chem. Int. Ed. Engl.* **1996**, *35*, 2664–2666;
b) T. Kawase, K. Tanaka, N. Fujiwara, H. R. Darabi, M. Oda, *Angew. Chem.* **2003**, *115*, 1662–1666; *Angew. Chem. Int. Ed.* **2003**, *42*, 1624–1628;
c) T. Kawase, K. Tanaka, Y. Seirai, N. Shiono, M. Oda, *Angew. Chem.* **2003**, *115*, 5755–5758; *Angew. Chem. Int. Ed.* **2003**, *42*, 5597–5600;
d) T. Kawase, K. Tanaka, N. Shiono, Y. Seirai, N. Shiono, M. Oda, *Angew. Chem.* **2004**, *116*, 1754–1756; *Angew. Chem. Int. Ed.* **2004**, *43*, 1722–1724.

[34] a) J. Zhang, J. S. Moore, *J. Am. Chem. Soc.* **1992**, *114*, 9701–9702;
b) A. S. Shetty, J. Zhang, J. S. Moore, *J. Am. Chem. Soc.* **1996**, *118*, 1019–1027.

[35] S. Lahiri, J. L. Thompson, J. S. Moore, *J. Am. Chem. Soc.* **2000**, *122*, 11315–11319.

[36] Y. Tobe, N. Utsumi, K. Kawabata, A. Nagano, K. Adachi, S. Araki, M. Sonoda, K. Hirose, K. Naemura, *J. Am. Chem. Soc.* **2002**, *124*, 5350–5364.

[37] S. Höger, K. Bonrad, A. Mourran, U. Beginn, M. Möller, *J. Am. Chem. Soc.* **2001**, *123*, 5651–5659.

[38] K. Nakamura, H. Okubo, M. Yamaguchi, *Org. Lett.* **2001**, *3*, 1097–1099.

[39] a)Y. Saiki, H. Sugiura, K. Nakamura, M. Yamaguchi, T. Hoshi, J. Anzai, *J. Am. Chem. Soc.* **2003**, *125*, 9268–9269;
b) Y. Saiki, K. Nakamura, Y. Nigorikawa, M. Yamaguchi, *Angew. Chem.* **2003**, *115*, 5348–5350; *Angew. Chem. Int. Ed.* **2003**, *42*, 5190–5192.

[40] a) S. Rosselli, A.-D. Ramminger, T. Wagner, B. Silier, S. Wiegand, W. Häußler, G. Lieser, V. Scheumann, S. Höger, *Angew. Chem.* **2001**, *113*, 3234–3237; *Angew. Chem. Int. Ed.* **2001**, *40*, 3138–3141;
b) S. Höger, K. Bonrad, S. Rosselli, A.-D. Ramminger, T. Wagner, B. Silier, S. Wiegand, W. Häußler, G. Lieser, V. Scheumann, *Macromol. Symp.* **2002**, *177*, 185–191;
c) S. Rosselli, A.-D. Ramminger, T. Wagner, G. Lieser, S. Höger, *Chem. Eur. J.* **2003**, *9*, 3481–3491.

[41] For reviews about rod-coil block copolymers see, e. g. a) M. Lee, B.-K. Cho, W.-C. Zin, *Chem. Rev.* **2001**, *101*, 3869–3892;
b) H.-A. Klok, S. Lecommandoux, *Adv. Mater.* **2001**, *13*, 1217–1229;
c) G. Mao, C. K. Ober, *Acta Polymer.* **1997**, *48*, 405–422.

[42] Unlike previously described macrocycles the block copolymers **16** contain a variety of different molecular species, although PD $<$ 1.05.

[43] C.-H. Lin, J. Tour, *J. Org Chem.* **2002**, *67*, 7761–7768.

[44] S. Höger, J. Spickermann, D. L. Morrison, P. Dziezok, H. J. Räder, *Macromolecules* **1997**, *30*, 3110–3111.

[45] a) S. Chandrasekhar, B. K. Sadashiva, K. A. Suresh, *Pramana* **1977**, *9*, 471–480;
b) S. Chandrasekhar, in *Handbook of Liquid Crystals, Vol. 2B* (Eds.: D. Demus, J. Goodby, G. W. Gray, H.-W. Spiess, V. Vill), Wiley-VCH, Weinheim, **1998**, pp. 749–780;
c) R. J. Bushby, O. R. Lozman, *Curr. Opin. Colloid & Inter. Sci.* **2002**, *7*, 343–354.

[46] a) J.-M. Lehn, J. Malthête, A.-M. Levelut, *J. Chem. Soc., Chem. Commun.* **1985**, 1794–1796;
b) G. Lattermann, *Mol. Cryst. Liq. Cryst.* **1990**, *182B*, 299–311.

[47] J. Zhang, J. S. Moore, *J. Am. Chem. Soc.* **1994**, *116*, 2655–2656.

[48] O. Y. Mindyuk, M. R. Stetzer, P. A. Heiney, J. C. Nelson, J. S. Moore, *Adv. Mater.* **1998**, *10*, 1363–1366.

[49] S. Höger, V. Enkelmann, K. Bonrad, C. Tschierske, *Angew. Chem.* **2000**, *112*, 2356–2358; *Angew. Chem. Int. Ed.* **2000**, 39, 2268–2270.

[50] S. Höger, A.-D. Ramminger, X. H. Cheng, V. Enkelmann, *in preparation.*

[51] M. Fischer, G. Lieser, A. Rapp, I. Schnell, W. Mamdouh, S. De Feyter, F. C. De Schryver, S. Höger, *J. Am. Chem. Soc.* **2004**, *126*, 214–222.

[52] a) F. C. Frank, S. Chandrasekhar, *J. Phys. (Paris)* **1980**, *41*, 1285–1288; b) A. M. Levelut, *J. Chim. Phys. Phys. Chim. Biol.* **1983**, *80*, 149–161.

[53] (a) C. Zeng, B. Wang, B. Li, H. Wang, J. G. Hou, *Appl. Phys. Lett.* **2001**, *79*, 1685–1687. (b) M. M. S. Abdel-Mottaleb, N. Schuurmans, S. De Feyter, J. Van Esch, B. L. Feringa, F. C. De Schryver, *Chem. Commun.* **2002**, 1894–1895; (c) S. Hoeppener, J. Wonnemann, L. Chi, G. Erker, H. Fuchs, *Chem. Phys. Chem.* **2003**, *4*, 490–494; (d) S. Hoeppener, L. Chi, H. Fuchs, *Chem. Phys. Chem.*, **2003**, *4*, 494–498.

[54] Phenylacetylene macrocycles studied on a Langmuir–Blodgett trough adopt an edge-on orientation at the interface, while the face-on orientation is not stable: A. S. Shetty, P. R. Fischer, K. F. Stork, P. W. Bohn, J. S. Moore, *J. Am. Chem. Soc.* **1996**, *118*, 9409–9414.

[55] J. Rabe, S. Buchholz, *Science* **1991**, *253*, 424–427.

[56] E. Mena-Osteritz, P. Bäuerle, *Adv. Mater.* **2001**, *13*, 243–246.

[57] a) R. Strohmaier, J. Petersen, B. Gompf, W. Eisenmenger, *Surf. Sci.* **1998**, *418*, 91–104; b) A. Miura, Z. Chen, H. Uji-i, S. De Feyter, M. Zdanowska, P. Jonkheijm, A. P. H. J. Schenning, E. W. Meijer, F. Würthner, F. C. De Schryver, *J. Am. Chem. Soc.* **2003**, *125*, 14968–14969.

[58] For Calixarenes on Au(111) see: G.-B. Pan, J.-M. Liu, H.-M. Zhang, L.-J. Wan, Q.-Y. Zheng, C.-L. Bai, *Angew. Chem.* **2003**, *115*, 2853–2857; *Angew. Chem. Int. Ed.* **2003**, 42, 2747–2751.

[59] For the synthesis of shape-persistent butadiyne macrocycles by the covalent template approach see, e. g.: a) S. Höger, A.-D. Meckenstock, H. Pellen, *J. Org. Chem.* **1997**, *62*, 4556–4557; b) S. Höger, A.-D. Meckenstock, *Chem. Eur. J.* **1999**, *5*, 1686–1691; c) S. Höger, *Macromol. Symp.* **1999**, *142*, 185–191; d) S. Höger, K. Bonrad, L. Karcher, A.-D. Meckenstock, *J. Org. Chem.*, **2000**, *65*, 1588–1589; e) M. Fischer, S. Höger, *Eur. J. Org. Chem.* **2003**, 441–446.

[60] D. Borissov, A. Ziegler, S. Höger, W. Freyland, *Langmuir* **2004**, *20*, 2781–2784.

[61] S. Höger, K. Bonrad, *J. Org. Chem.* **2000**, *65*, 2243–2245.

11
Chiral Acetylenic Macromolecules

Lin Pu

11.1
Introduction

The structural rigidity and electronic conjugation of aryleneethynylenes have made them very useful building blocks in the synthesis of both polymers and dendrimers. In addition, the mild conditions used for the coupling of terminal alkynes with aryl halides have provided a very convenient synthetic route for the preparation of the aryleneethynylene-based materials. Interesting electrical and optical properties have been discovered for the polymeric and dendritic aryleneethynylenes [1–8]. Dendrimers such as **1**, for example, have shown efficient light-harvesting properties [3]. Energy migration from the phenyleneethynylene units to the more conjugated pyrene core results in greatly enhanced fluorescence intensity. Linear conjugated polyphenyleneethynylenes such as **2** have shown high efficiency in fluorescent sensing [8].

In the past few years, our laboratory has used aryleneethynylenes as structural units for the construction of optically active dendrimers and polymers. Introduction of chirality into these materials makes them potentially useful for chiral sensing, asymmetric catalysis, asymmetric electrosynthesis, nonlinear optics, and polarized light emission. The chirality in all of the macromolecular aryleneethynylenes prepared in our laboratory is derived from 1,1′-binaphthyl moieties. Enantiomerically pure 1,1′-binaphthyl groups are incorporated into these materials either as the core of the dendrimers or as the main chain chiral units of the polymers. The synthesis and characterization of these materials are summarized here [9].

Acetylene Chemistry. Chemistry, Biology and Material Science. Edited by F. Diederich, P. J. Stang, R. R. Tykwinski

ISBN 3-527-30781-8

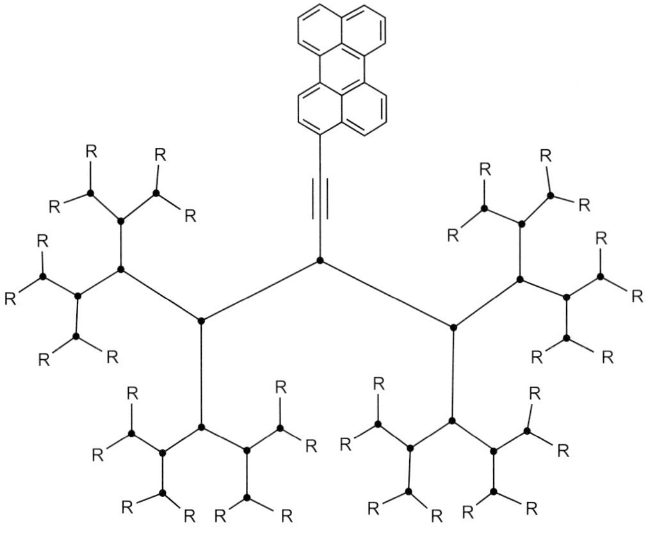

1

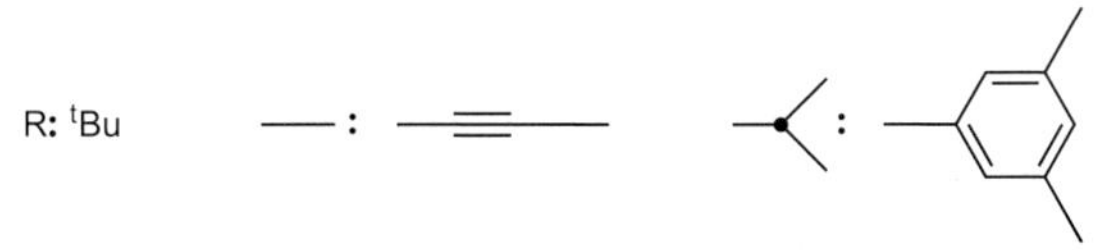

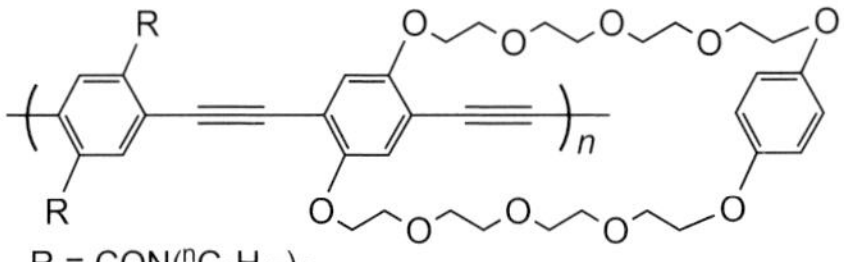

R = CON(^{n}C$_8$H$_{17}$)$_2$

2

11.2 Chiral Acetylenic Dendrimers

As shown in the study of dendrimer **1** [3], the light-harvesting properties of the phenyleneethynylenes in these dendrimers generated greatly enhanced fluorescence over their small-molecular counterparts. When a molecular receptor is incorporated into the core of these dendrimers, these materials should serve as much more sensitive sensors for fluorescent quenchers than the corresponding small molecules. Incorporation of chirality into the molecular receptor in the dendritic core could further make these sensors enantioselective. Enantioselective fluorescent sensors may potentially provide a real-time analytical tool for high-throughput chiral assay. We thus synthesized dendritic phenyleneethynylenes containing enantiomerically pure 1,1′-binaphthyl cores for enantioselective fluorescent sensing.

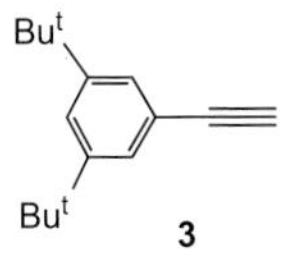

3

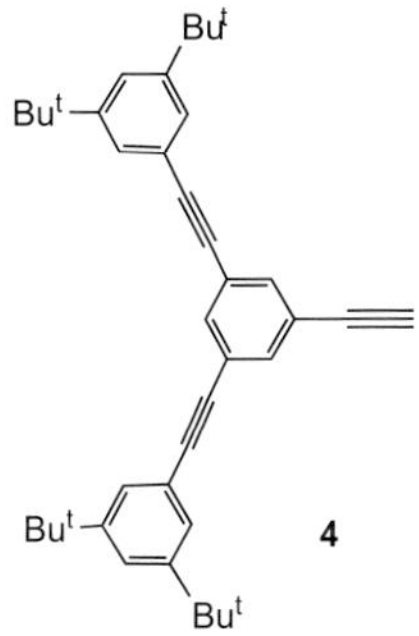

4

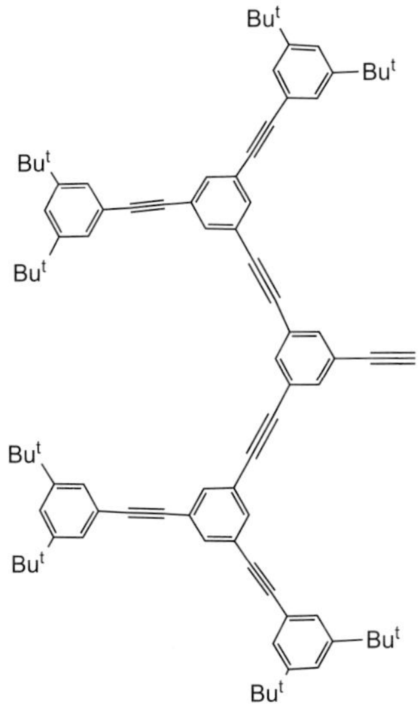

5

Dendrons **3**–**5** were prepared by Moore's procedure [3], while the 4,4′,6,6′-tetrabromo-1,1′-binaphthyl compounds (*S*)-**6** and (*S*)-**7** were prepared from the enantiomerically pure (*S*)-BINOL (Scheme 11.1) [10]. Since the coupling of the aryl bromides of (*S*)-**6** with the higher-generation dendrons was found to give incomplete reactions, the hexyl groups of (*S*)-**6** were converted into the less electron-donating acyl groups to make (*S*)-**7**. The aryl bromides in (*S*)-**7** were more reactive than those in (*S*)-**6**, and coupling of (*S*)-**7** with the dendrons **3**–**5** in the presence of $Pd(PPh_3)_4$/CuI and subsequent hydrolysis produced the three generations of chiral dendrimers (*S*)-**8** to (*S*)-**10** (Scheme 11.1).

Compounds (*S*)-**8**, (*S*)-**9**, and (*S*)-**10** were the generation 0 (G0), generation 1 (G1), and generation 2 (G2) dendrimers, respectively. The molecular weights of these compounds were determined by FAB and MALDI mass spectroscopic analyses. Table 11.1 summarizes the characterization data of these dendrimers. The specific optical rotations of the dendrimers decreased as the generation increased, and the molar optical rotation increase was also very small as the molecular weight increased by a factor of over four on going from the G0 dendrimer to the G2. This indicated that the phenyleneethynylene dendrons in these materials did not achieve a chiral order to amplify the chiral effect of the binaphthyl core. Consistently with this, all three generations of dendrimers gave very similar CD signals (Table 11.2). Two bands were observed in the UV spectra of the dendrimers: a stronger band at 250–330 nm and a weaker band at 340–400 nm (Table 11.2). From the G0 dendrimer to the G2 dendrimer, while the absorption at the short wavelength increased about nine fold, little change was observed at the long wavelength, so the *meta*-phenyleneethynylene dendrons in these materials were not increasing the conjugation length of the dendrimers. The ^{13}C NMR spectra of these dendrimers showed four alkynyl carbon signals for (*S*)-**8**, eight for (*S*)-**9**, and 12 for (*S*)-**10**, consistently with C_2 symmetry of these materials in solution.

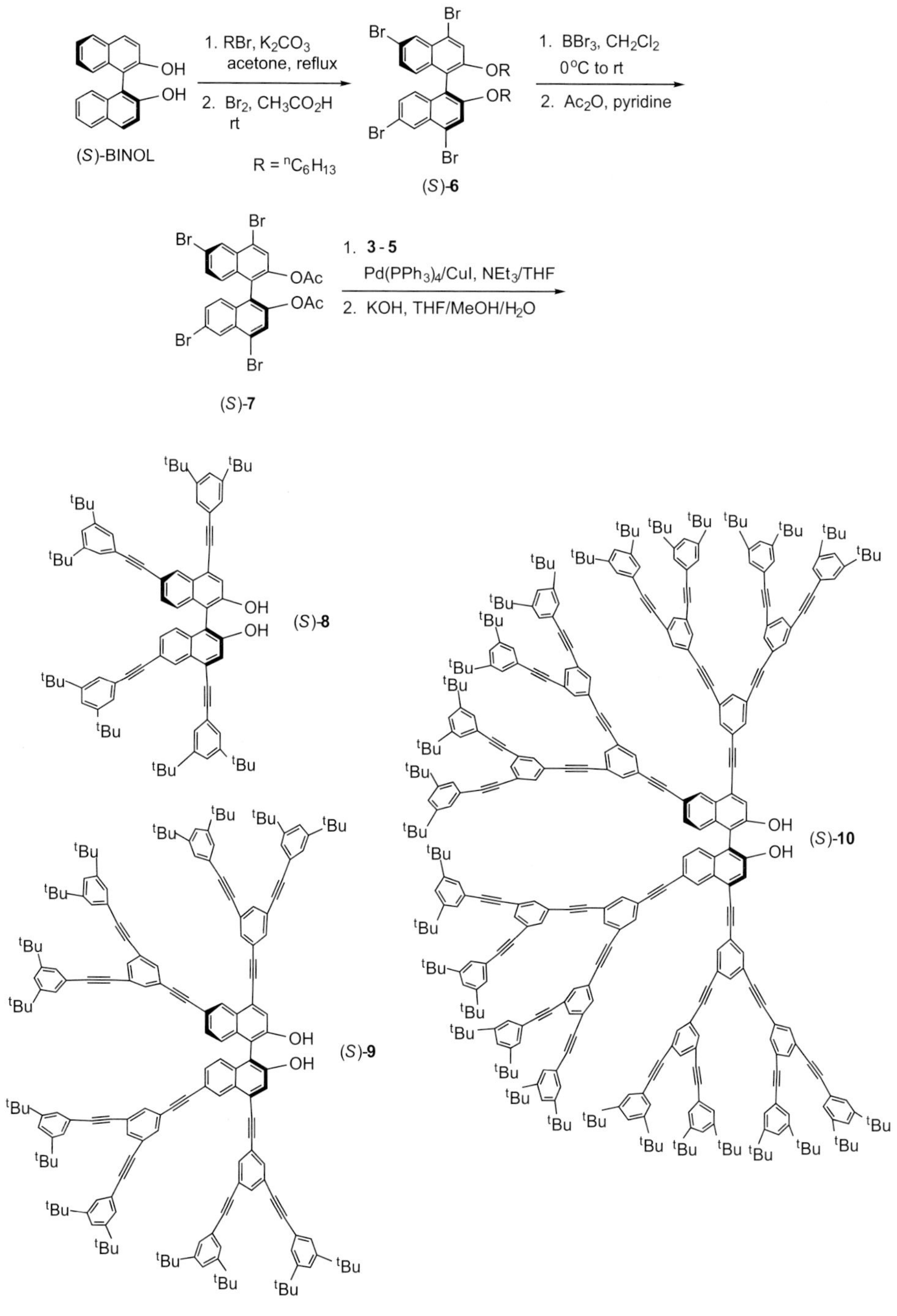

Scheme 11.1 Synthesis of the chiral dendrimers (*S*)-**8** to (*S*)-**10**.

Table 11.1 Characterization data for the chiral dendrimers.

Dendrimer	*Molecular weight by mass analysis (calculated)*	$[\alpha]_D$ (c = 1, CH_2Cl_2)	*Molar optical rotation [M]*
(*S*)-**8** (G0)	FAB, 1134.7190 (M, 1134.7254)	114.4	1299
(*S*)-**9** (G1)	FAB, 2385.5 (M+H, 2384.5)	67.5	1609
(*S*)-**10** (G2)	MALDI, 4882.5 (M+H, 4882.0)	36.2	1769
(*R*)-**10** (G2)	MALDI, 4883.0 (M+H, 4882.0)	−36.0	−1759

Table 11.2 CD and UV spectral data for the chiral dendrimers.

Compounds	(*S*)-**8**	(*S*)-**9**	(*S*)-**10**	**4**	**5**
CD [θ]	-1.6×10^5 (271)	-1.5×10^5 (255)	-1.4×10^5 (258)		
(λ nm)	-1.6×10^5 (279)	-1.0×10^5 (281)	-1.2×10^4 (282)		
(CH_2Cl_2)	4.5×10^5 (305)	4.5×10^5 (309)	4.2×10^5 (312)		
	-8.1×10^4 (363)	-9.2×10^4 (363)	-7.7×10^4 (363)		
λ_{max} (nm)	292 (8.02×10^4)	292 (3.13×10^5)	294 (8.20×10^5)	290 (6.79×10^4)	292 (1.85×10^5)
(ε, M^{-1} cm^{-1})	304 (1.04×10^5)	308 (3.52×10^5)	310 (9.16×10^5)	308 (6.25×10^4)	310 (2.03×10^5)
	374 (2.62×10^4)	378 (4.12×10^4)	380 (4.44×10^4)		

The fluorescence spectra of all the dendrimers (4.0×10^{-8} M in CH_2Cl_2) each gave a maximum at 421–422 nm and a shoulder at 441 nm when excited at 310 nm. No emission from the monodendrons **4** and **5** [**4**: λ_{emi} = 348 and 360 nm; **5**: λ_{emi} = 359 and 370 (sh) nm] was observed in the dendrimers. The G2 dendrimer (*S*)-**10** emitted four times more strongly than the G1 (*S*)-**9**, which in turn emitted three times more strongly than the G0 (*S*)-**8**. The excitation spectra of the dendrimers showed that excitation at 310 nm, where the diphenylethynylene units absorbed, gave the strongest emissions for all of the dendrimers. These observations supported an efficient intramolecular energy transfer from the phenyleneethynylene dendrons to the more conjugated core. The fluorescence quantum yields of the dendrimers (*S*)-**8**, (*S*)-**9**, and (*S*)-**10** were 0.30, 0.32, and 0.40, respectively, as estimated with quinine sulfate in 1 N sulfuric acid as the reference. The increase in the fluorescence quantum yield as the dendritic generation increased might be due to the more restricted rotation around the 1,1′-binaphthyl bond as the size of the dendritic phenyleneethynylene substituents increased.

The G2 dendrimer (*S*)-**10** was used to catalyze the reaction between benzaldehyde and diethylzinc to generate 1-phenylpropanol (**11**) (Scheme 11.2) [11]. This nanoscale material exhibited catalytic properties very different from those of the small BINOL molecule. In toluene solution at room temperature, while both dendrimer (*S*)-**10** and (*S*)-BINOL gave low enantioselectivity for this reaction, (*S*)-**10** was a much more active catalyst than (*S*)-BINOL. In 24 h, (*S*)-**10** (5 mol%) gave 98.6% conversion while BINOL (5 mol%) gave only 37% conversion. This experiment demonstrated two features of the dendrimer: (1) that the 1,1′-binaphthyl core of this macromolecule was accessible by small organic substrates for catalysis, and (2) that the longer dendritic arms of (*S*)-**10** might have prevented intermolecular aggregation by the zinc complex generated by reaction between the dendrimer core hydroxy groups and Et_2Zn. Aggregation of the BINOL-zinc complex, formed by reaction between BINOL and Et_2Zn, through intermolecular O-Zn-O bonds might have reduced the Lewis acidity of the BINOL-zinc complex, resulting in its lower catalytic activity in relation to the macromolecule. When the Et_2Zn addition was carried out in the presence of $Ti(O^iPr)_4$, both (*S*)-**10** and BINOL showed high enantioselectivity. Compound (*S*)-**10** (20 mol%) in combination with $Ti(O^iPr)_4$ (1.4 equiv.) catalyzed the addition of Et_2Zn to benzaldehyde with 89% *ee* and 100% conversion in 5 h. It also catalyzed the Et_2Zn addition to 1-naphthaldehyde with 90% *ee*, catalytic properties very similar to those of BINOL [12]. This indicated that the catalytically active species of the BINOL-$Ti(O^iPr)_4$ complex might be monomeric in nature rather than the aggregated BINOL-Ti(IV) complex, since the dendrimer should not be able to form such an aggregated structure through intermolecular interaction of the core. In this catalytic process, dendrimer (*S*)-**10** was easily recovered after the reaction by precipitation with methanol.

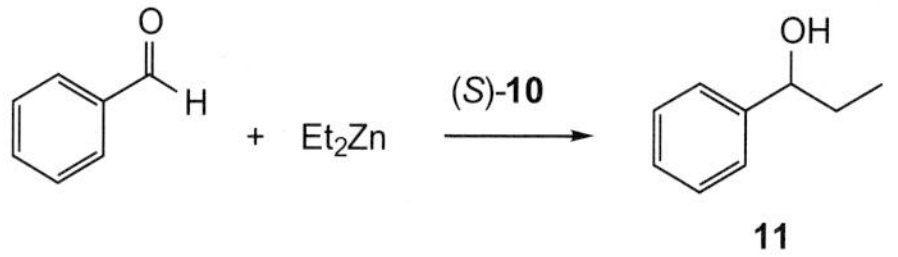

11

Scheme 11.2 Asymmetric diethylzinc addition to benzaldehyde catalyzed by (*S*)-**10**.

The fluorescence spectrum of (*S*)-**10** was compared with that of BINOL [13]. While a solution of (*S*)-**10** [4.0×10^{-8} M in benzene/hexane (20:80)] showed very intense fluorescence signals, BINOL under the same conditions gave only very weak emissions at the base line. The greatly enhanced fluorescence signal of the dendrimer made it more suitable than BINOL as a fluorescent sensor for interaction with a molecular quencher.

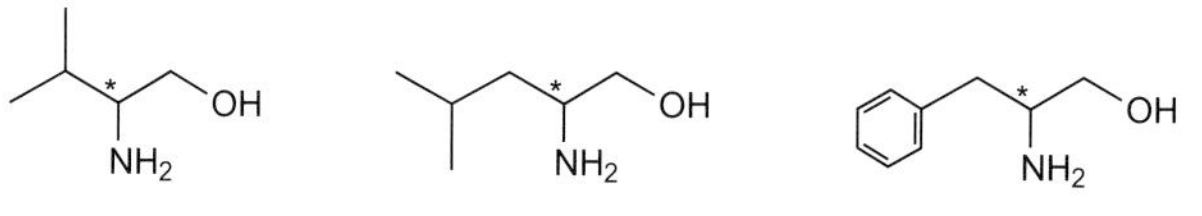

12 **13** **14**

The interactions of the dendrimers (*S*)-**8** to (*S*)-**10** with the enantiomers of the chiral amino alcohols **12**–**14** were also studied [13, 14]. These amino alcohols quenched the fluorescence of the dendrimers according to the Stern–Volmer equation: $I_0/I = 1 + K_{SV}[Q]$. In the equation, I_0 is the fluorescence intensity of the fluorophore, I is the fluorescence intensity in the presence of the quencher, K_{SV} is the Stern–Volmer constant, and [Q] is the quencher concentration. Table 11.3 summarizes the Stern–Volmer constants of these dendrimers in the presence of the *R* and *S* enantiomers of the amino alcohols. As can be seen from the data in Table 11.3, the *S* enantiomers of the amino alcohols generally quenched the fluorescence of the dendrimers more efficiently than the *R* enantiomers, the enantioselectivities represented by $K_{SV}(S)/K_{SV}(R)$ being in the 1.04–1.27 range.

Table 11.3 Stern–Volmer constants (M^{-1}) of the dendrimers in the presence of the enantiomers of the amino alcohols **12–14**.

Dendrimer	*(R)- and (S)-12*[a]		*K_{SV} (S) /K_{SV} (R)*	*(R)- and (S)-13*[a]		*K_{SV} (S) /K_{SV} (R)*	*(R)- and (S)-14*[b]		*K_{SV} (S) /K_{SV} (R)*
	K_{SV} (S)	K_{SV} (R)		K_{SV} (S)	K_{SV} (R)		K_{SV} (S)	K_{SV} (R)	
(*S*)-**8**	47	43	1.09	61	54	1.13	305	253	1.21
(*S*)-**9**	63	57	1.11	92	73	1.26	445	351	1.27
(*S*)-**10**	83	80	1.04	90	82	1.10	520	426	1.22
(*S*)-BINOL[c]							111	109	1.02

a In methylene chloride. The dendrimer concentrations were 4.0×10^{-8} M.
b In benzene/hexane (20:80). The dendrimer concentrations were 1.0×10^{-6} M.
c Concentration = 1.0×10^{-4} M.

The enantioselectivity of the dendrimers was higher than that of BINOL, and the enantioselective recognition of the amino alcohols by the dendrimers was also confirmed through the use of the enantiomers of the dendrimers to interact with the amino alcohols. Dendrimers (*S*)- and (*R*)-**10**, for example, were treated with (*S*)- and (*R*)-**13** in methylene chloride. As shown by the fluorescence quenching constants given in Table 11.4, there was a mirror image relationship, indicating chiral recognition, between dendrimers (*S*)- and (*R*)-**10** in the fluorescence responses to the amino alcohol enantiomers.

Table 11.4 Stern–Volmer Constants (M^{-1}) of the enantiomeric dendrimers (*S*)- and (*R*)-**10** in the presence of the amino alcohols (*R*)- and (*S*)-**13** in methylene chloride.

Dendrimer	*(R)- and (S)-13*[a]	
	K_{SV} (S)	K_{SV} (R)
(*S*)-**10**	90	82
(*R*)-**10**	80	93

a In methylene chloride. The dendrimer concentrations were 4.0×10^{-8} M.

Dendrimer (*S*)-**10** had a fluorescence lifetime of 1.6±0.2 nanoseconds, which did not change in the presence of various concentrations of (*R*)- and (*S*)-**14**, demonstrating that the fluorescence quenching of the dendrimer by the amino alcohol was due to static quenching. Formation of non-fluorescent ground-state hydrogen-bonded complexes between the dendrimer core and the amino alcohol was put forward as responsible for the static quenching. When the hydroxy groups of dendrimer (*S*)-**10** were replaced with methoxy groups, no significant fluorescence quenching by (*S*)-**14** was observed. The Stern–Volmer constants were found to be much larger in the less polar benzene/hexane (20:80) than in methylene chloride, which was consistent with a stronger tendency to form the hydrogen-bonded complexes in the less polar solvents. The Stern–Volmer constants also increased significantly as the dendrimer generation increased. Thus, the higher generation dendrimers were more sensitive to the quenchers than those of the lower generation.

11.3
Chiral Acetylenic Polymers

We have incorporated 1,1′-binaphthyl moieties into polyaryleneethynylenes to make main-chain chiral conjugated polymers. These materials would be expected to have more stable main-chain chiral configurations than those containing only side-chain chiral substituents [15]. The stable chiral configurations of these materials may provide advantageous properties in practical applications. The following sections discuss our work on the use of *p*-phenyleneethynylenes, *o*-phenyleneethynylenes, *m*-phenyleneethynylenes, and other aryleneethynylenes to couple with enantiomerically pure 1,1′-binaphthyl monomers.

11.3.1
Chiral Polymers Containing Main-Chain *para*-Phenyleneethynylenes

Two bromine atoms were specifically introduced into the 6,6′-positions of (*R*)-BINOL to give (*R*)-**15** (Scheme 11.3). Treatment of (*R*)-**15** with various alkyl halides resulted in the formation of monomers (*R*)-**16a–c**. Sonogashira coupling of monomers (*R*)-**16a–c** with 1,4-diethynylbenzene (**17**) gave polymers (*R*)-**18a–c** [16]. It was found that when the shorter alkyl-substituted monomers (*R*)-**16a** or (*R*)-**16b** were used, the resulting polymers (*R*)-**18a** and (*R*)-**18b** were insoluble in common organic solvents. Polymerization of the monomer (*R*)-**16c**, containing much longer octadecyl chains with **17**, gave the polymer (*R*)-**18c**. As a result of the introduction of the long-chain alkyl groups, this polymer was soluble in solvents such as chloroform, methylene chloride, and THF. Analysis by gel permeation chromatography (GPC) relative to polystyrene standards showed the molecular weight of polymer (*R*)-**18c** as Mw = 7,400 and Mn = 3,800 (PDI = 2.0).

The specific optical rotation of monomer (*R*)-**16c** was $[\alpha]_D$ = 20.5 (c = 1.0, CH_2Cl_2), and that of polymer (*R*)-**18c** was –154.9 (c = 1.0, CH_2Cl_2). The large change in optical rotation on going from the monomer to the polymer indicated very different electronic and steric structures. Coupling of the racemic monomer *rac*-**16c** with **17** gave the optically inactive polymer *rac*-**18c**. The molecular weight of *rac*-**18c** was Mw = 8,500 and Mn = 3,200 (PDI = 2.7).

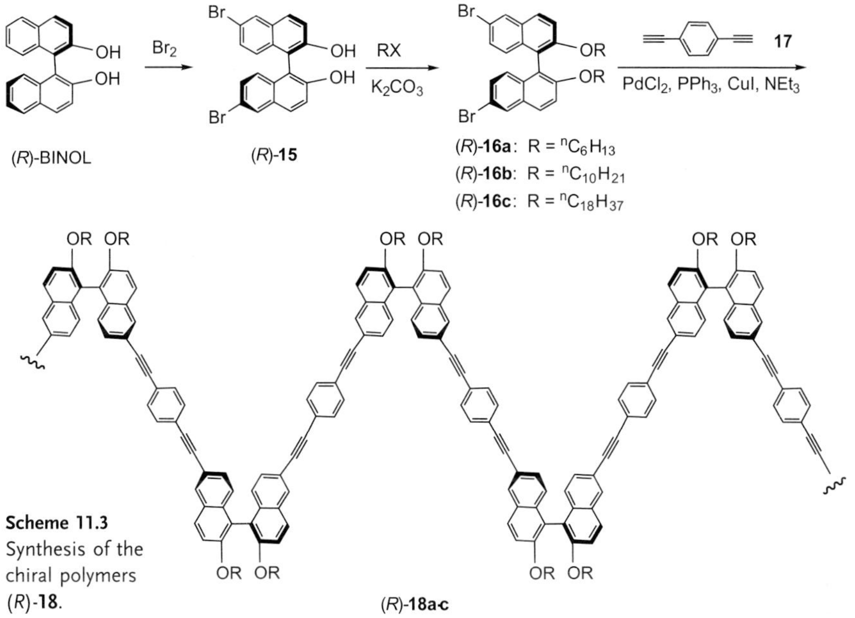

Scheme 11.3 Synthesis of the chiral polymers (*R*)-**18**.

Polymerization of (*R*)-**16c** with 4,4′-diethynylbiphenyl (**19**) in the presence of $PdCl_2$/CuI resulted in the formation of polymer (*R*)-**20a** (Scheme 11.4). This polymer was also soluble in common organic solvents. The specific optical rotation of (*R*)-**20a** was –158.9 (c = 1.0, CH_2Cl_2). GPC analysis of the polymer gave Mw = 9,300 and Mn = 4,800 (PDI = 1.9). The racemic monomer *rac*-**16c** was also coupled with **19** to give *rac*-**20a**. Its molecular weight was Mw = 6,600 and Mn = 3,900 (PDI = 1.7).

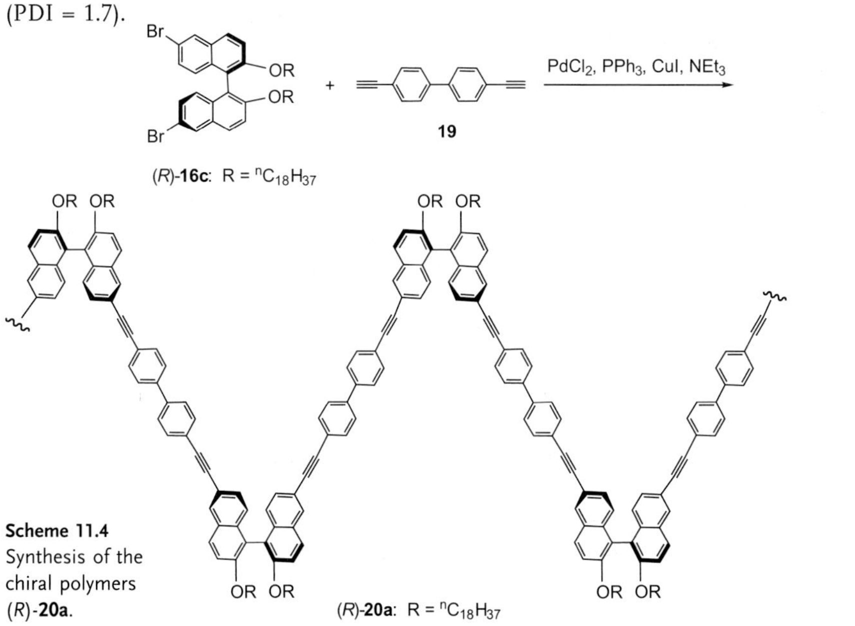

Scheme 11.4 Synthesis of the chiral polymers (*R*)-**20a**.

Another enantiomerically pure 1,1′-binaphthyl monomer (*R*)-**23** was also prepared (Scheme 11.5). Treatment of (*R*)-**16c** with 2-methyl-3-butyn-2-ol (**21**) gave (*R*)-**22**, which underwent base-promoted deprotection to form the 6,6′-diethynyl binaphthyl monomer (*R*)-**23**. Polymerization of (*R*)-**23** with 1,4-diiodobenzene in the presence of $PdCl_2$/CuI gave polymer (*R*)-**18d**, which should have had the same structure as (*R*)-**18c**. GPC, however, showed its molecular weight as Mw = 21,000 and Mn = 15,000 (PDI = 1.4), much higher than that of (*R*)-**18c**. This high molecular weight polymer was still soluble in common organic solvents, while its specific optical rotation was $[\alpha]_D = -278.2$ (c = 0.5, CH_2Cl_2), significantly higher than that of (*R*)-**18c**. We also used $Pd(PPh_3)_4$ to replace $PdCl_2$ for the polymerization of (*R*)-**23** with 1,4-diiodobenzene, which gave polymer (*R*)-**18e**. The molecular weight of (*R*)-**18e** was Mw = 29,000 and Mn = 12,000 (PDI = 2.4). When *rac*-**23** was polymerized with 1,4-diiodobenzene, the resulting *rac*-**18e** had an even higher molecular weight of Mw = 47,000 and Mn = 25,000 (PDI = 1.9) and was still soluble in common organic solvents. This demonstrated that the coupling of an aryl iodide with the diethynyl monomer in the presence of $Pd(PPh_3)_4$/CuI was the most efficient polymerization method and gave the polymer of highest molecular weight. The specific optical rotation of (*R*)-**18e** was $[\alpha]_D = -272.2$ (c = 0.5, CH_2Cl_2). A large increase in optical rotation was observed on going from the low molecular weight polymer (*R*)-**18c** to the high molecular weight polymers (*R*)-**18d** and (*R*)-**18e**, but the optical rotations of the high molecular polymers were very similar.

Monomer (*R*)-**23** was also polymerized with 4,4′-diiodobiphenyl in the presence of $PdCl_2$/CuI or $Pd(PPh_3)_4$/CuI to give polymers (*R*)-**20b** and (*R*)-**20c**, respectively (Scheme 11.5). The molecular weight of (*R*)-**20b** was Mw = 29,000 and Mn = 15,000 (PDI = 1.9), while that of (*R*)-**20c** was Mw = 37,000 and Mn = 18,000 (PDI = 2.1) as determined by GPC. Both polymers were soluble in common organic solvents. The specific optical rotation of (*R*)-**20b** was −250.6 (c = 0.5, CH_2Cl_2) and that of (*R*)-**20c** was −281.7 (c = 0.18, CH_2Cl_2). These values were significantly larger than that for the lower molecular weight polymer (*R*)-**20a**. The molecular weights of *rac*-**20b** and *rac*-**20c** were Mw = 20,000 (PDI = 1.7) and Mw = 54,000 (PDI = 2.2), respectively.

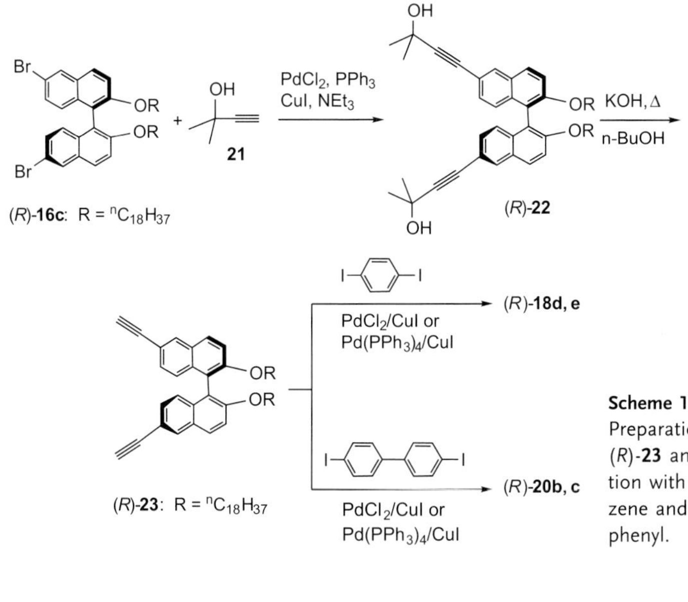

Scheme 11.5 Preparation of monomer (*R*)-**23** and its polymerization with 1,4-diiodobenzene and 4,4′-diiodobiphenyl.

All the polymers **18** and **20** were characterized by ^1H and ^{13}C NMR spectroscopy. In the ^{13}C NMR spectrum of (*R*)-**18e** in $CDCl_3$, the two alkyne carbon signals were observed at δ 91.9 and 88.9. The alkyne carbon signals of polymer (*R*)-**20c** were at δ 91.1 and 88.9. Thermogravimetric analyses (TGA) of polymers (*R*)-**18e** and (*R*)-**20c** under nitrogen showed that both polymers were very stable. Polymer (*R*)-**18e** started to lose weight at 396 °C. By 486 °C, it had lost all of its alkyl groups and became very stable up to 800 °C. Similarly, (*R*)-**20c** started to lose mass at 395 °C. By 540 °C it had lost all of its alkyl groups and showed no further mass loss up to 800 °C.

A laser light scattering (LLS) study of the polymers (*R*)-**20a**, (*R*)-**20b**, and *rac*-**20c** was undertaken. It was found that, in general, the GPC data for the molecular weights of these polymers relative to the polystyrene standards were significantly lower (by factors of 1.4 to 2.5) than those obtained by LLS (Table 11.5).

Table 11.5 Comparison of the GPC data for the polymers with the LLS data.

Polymer	*Mw (GPC)*	*Mw (g mol^{-1}, LLS data)*
(*R*)-**20a**	9,300	23,000
(*R*)-**20b**	20,000	32,000
rac-**20c**	54,000	73,000

From coupling of the naphthyl bromide **24** with **17** and **19** we obtained the repeating units **25** and **26**, respectively, of polymers (*R*)-**18** and (*R*)-**20** (Scheme 11.6). The bulky neopentyl groups in these molecules were introduced to make them soluble in organic solvents, since the linear alkyl-substituted analogues were insoluble. This was probably because the neopentyl groups could reduce the intermolecular packing of these rigid and planar compounds in the solid state.

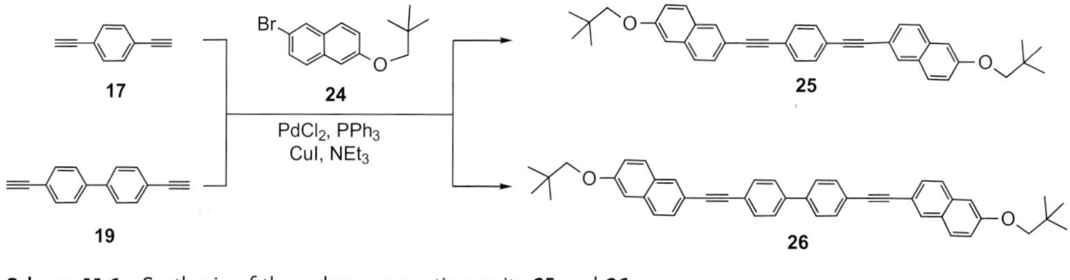

Scheme 11.6 Synthesis of the polymer repeating units **25** and **26**.

Table 11.6 summarizes the optical spectral data of the repeating units **25** and **26** and their corresponding polymers. The electronic absorptions of the polymers were very close to those of the corresponding repeating units, indicating that there was almost no extended conjugation between the repeating units in the polymers across the 1,1′-bonds of the binaphthyls. The conjugation of these chiral polymers

Table 11.6 Spectral data for the polymers and their repeating units.

	(R)-18e	*(R)-20c*	25	26
UV absorption	238	234	236	236
λ_{max} (nm)	300	306 (sh)	290	296 (sh)
(CH_2Cl_2)	354	352	350	348
	384 (sh)		374 (sh)	
Fluorescence	401 (sh)	407 (sh)		
	419	420		
	461 (sh)	460 (sh)		

was determined by the conjugation of their repeating units and was independent of the polymer chain length. UV and fluorescence spectroscopic studies showed that the optically active polymers gave almost the same absorption and emission maxima as those obtained from the racemic monomers. The UV absorptions of (*R*)-**18e** and **25** were also somewhat red-shifted from those of (*R*)-**20c** and **26**, so the additional phenyl ring in the repeating unit of polymer (*R*)-**20c** might have disrupted the planarity of the conjugated units in the repeating unit of polymer (*R*)-**18e**, giving shorter-wavelength absorption.

The circular dichroism (CD) spectra of polymers (*R*)-**18e** and (*R*)-**20c** in methylene chloride were obtained. They showed signals for (*R*)-**18e** at [θ] (λ, nm) = 5.8×10^4 (235), -1.8×10^4 (262), -1.4×10^4 (295 sh), and -1.5×10^4 (380), and for (*R*)-**20c** at [θ] (λ, nm) = 2.7×10^4 (240), -1.3×10^4 (262), and -0.6×10^4 (372).

A binaphthyl-based polyphenyleneethynylene with much more extended conjugation in the repeating units was synthesized [17]. As shown in Scheme 11.7, coupling of (*R*)-**16c** with compound **27** gave (*R*)-**28**. Base-promoted deprotection of (*R*)-**28** gave (*R*)-**29**, and polymerization of (*R*)-**29** with 1,4-diiodobenzene in the presence of $Pd(PPh_3)_4$/CuI gave polymer (*R*)-**30** containing 54 conjugated π electrons in its repeating unit. The molecular weight of this polymer was Mw = 38,200 and Mn = 29,000 (PDI = 1.3) as determined by GPC. Four distinctive ^{13}C NMR signals, at δ 92.4, 91.2, 90.8 and 88.8, were observed for the alkyne carbons of (*R*)-**30**, indicating a well defined structure. The specific optical rotation of (*R*)-**30** was $[\alpha]_D$ = −288.6 (c = 0.5, CH_2Cl_2), very close to those of polymers (*R*)-**18d**, (*R*)- **18e**, and (*R*)-**20c**, and suggesting that the polymer chain structure might contribute little to the optical rotation and the individual 6,6′-phenyleneethynylene-substituted 1,1′-binaphthyl unit was the most important factor in determining the optical rotation. This prompted a further assumption that there should not be a propagating helical chain structure in these polymers, or much bigger difference in the optical rotation contributed by the main chain helical structure should be observed.

The UV spectrum of (*R*)-**30** in methylene chloride showed λ_{max} at 238 and 366 nm. Its fluorescence spectrum gave λ_{emi} at 420, 464 and 500 (sh) nm. The polymer *rac*-**20** [GPC: Mw = 42,000 (PDI = 2.1)], made from the racemic monomer *rac*-**29**, gave the same spectral data. Comparison of (*R*)-**30** with (*R*)-**18e** and (*R*)-**20** indicated red shifts in the absorption and emission maximum peaks, arising from

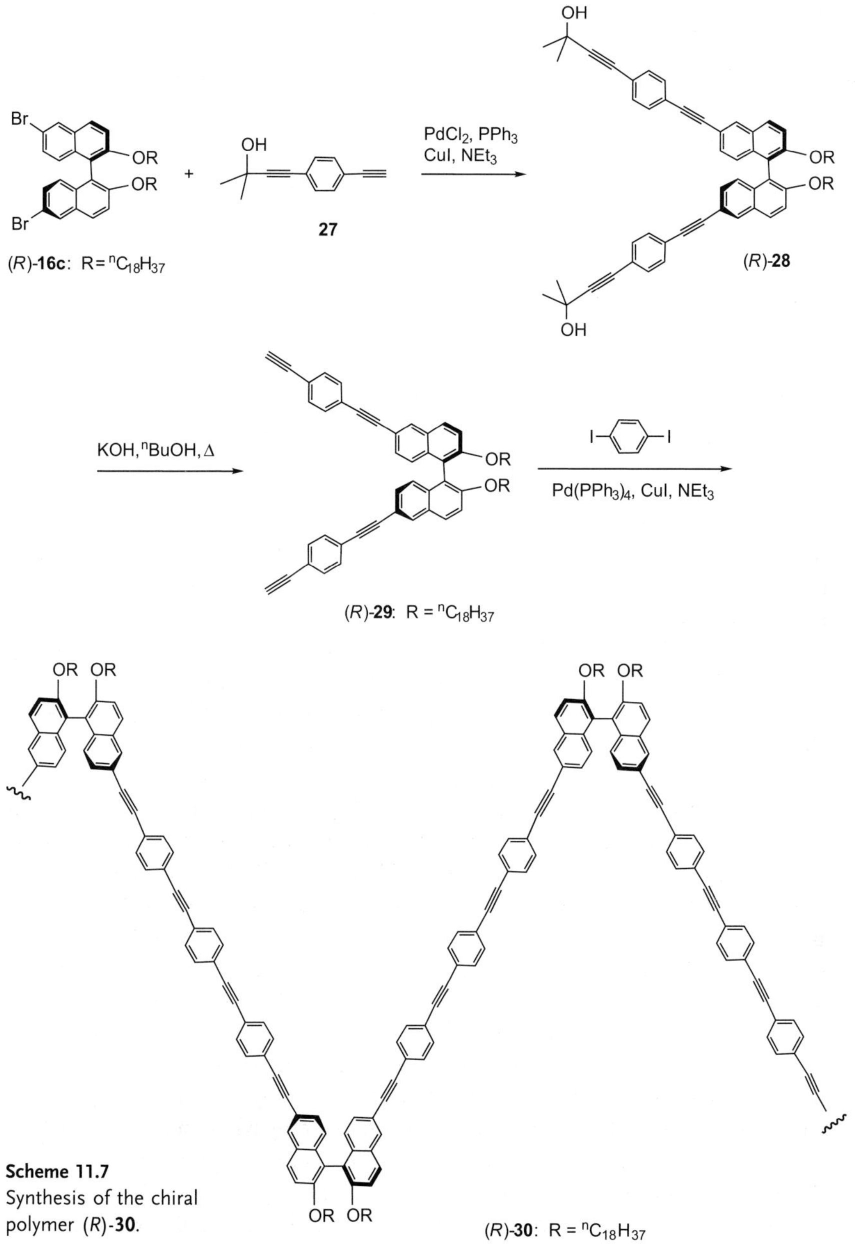

Scheme 11.7 Synthesis of the chiral polymer (*R*)-**30**.

the increased conjugation in the repeating unit of (*R*)-**30**. The CD spectrum of (*R*)-**30** in methylene chloride displayed signals at [θ] (λ, nm) = +4.5 × 10^5 (235), −7.7 × 10^4 (262), −1.2 × 10^5 (284), +3.9 × 10^4 (338), and −2.0 × 10^5 (385). The long-wavelength CD signals of (*R*)-**30** were much stronger than those of (*R*)-**18e** and (*R*)-**20**.

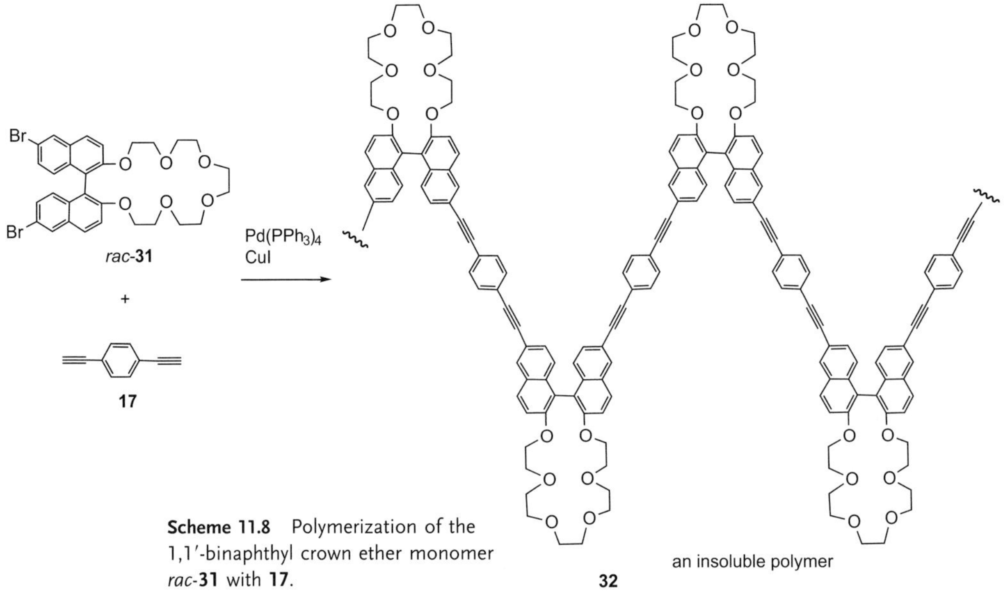

Scheme 11.8 Polymerization of the 1,1′-binaphthyl crown ether monomer *rac*-**31** with **17**.

Crown ether groups were also introduced into the chiral polyaryleneethynylenes [18]. We first attempted the polymerization of **17** with the racemic crown ether monomer *rac*-**31** (Scheme 11.8), but the resulting polymer **32** was found to be insoluble in common organic solvents. We then synthesized the acetylenic monomer **33**, containing two flexible long-chain alkoxy groups, in order to improve the solubility of the polymer. Polymerization of **33** with the enantiomerically pure monomer (*S*)-**31** in the presence of $Pd(PPh_3)_4$/CuI afforded the crown ether polymer (*S*)-**34** (Scheme 11.9). This polymer was soluble in common organic solvents such as THF, chloroform and methylene chloride. GPC analysis gave Mw = 10,500 and Mn = 5,300 (PDI = 2.0). The ^{13}C NMR spectrum of (*S*)-**34** gave two multiplets at δ 86.1 and 95.5 for the alkyne carbons. The $[\alpha]_D$ value for (*S*)-**34** was 120.9 (c = 0.95, THF) with the opposite sign to the monomer (*S*)-**31** {$[\alpha]_D = -8.3$ (c = 0.95, THF)}. The UV spectrum of (*S*)-**34** in methylene chloride gave absorptions at λ_{max} = 240, 296, 334, and 378 nm. This polymer was also strongly fluorescent, with emission maxima at λ_{emi} = 420, 445, and 455 nm when excited at 378 nm in methylene chloride. The CD spectrum of (*S*)-**34** gave signals at [θ] (λ, nm) = -3.72×10^5 (233), 1.88×10^5 (251), 5.21×10^4 (285), -8.86×10^3 (349), and 3.46×10^4 (398).

Coupling of the racemic monomer *rac*-**31** with another racemic monomer *rac*-**29** gave polymer **35** (Scheme 11.10), with a higher molecular weight than (*S*)-**34**. GPC analysis showed its molecular weight as Mw = 27,900 and Mn = 11,500 (PDI = 2.4). The R groups of monomer *rac*-**29** made the polymer soluble in organic solvents, and the UV spectrum of **35** gave absorption maxima at λ_{max} = 246, 300, and 354 nm. The UV spectrum of (*S*)-**34** had significantly longer-wavelength absorptions than **35** because of the two electron-donating alkoxy groups on the phenylene

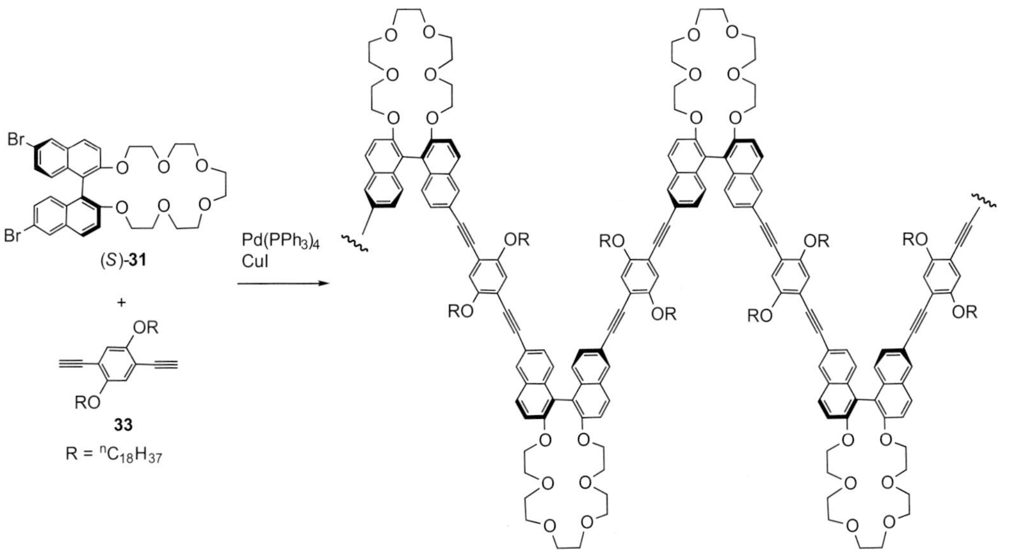

Scheme 11.9 Synthesis of the crown ether polymer (*S*)-**34**. (*S*)-**34**: R = $^{n}C_{18}H_{37}$

linkers. These crown ether polymers were prepared for potential applications in molecular recognition and chiral sensing.

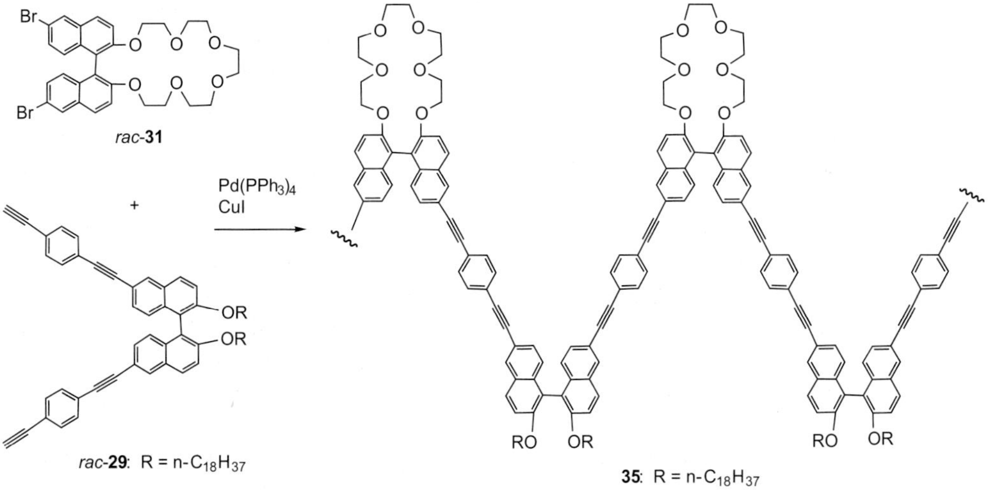

35: R = n-$C_{18}H_{37}$

Scheme 11.10 Coupling of the crown ether monomer *rac*-**31** with the binaphthyl monomer *rac*-**29**

.

11.3.2 Chiral Polymers Containing Main-Chain *ortho*-Phenyleneethynylenes

11.3.2.1. Non-Dipolar Polymers Based on BINOL

We synthesized the *ortho*-phenyleneethynylene-based polybinaphthyls for potential nonlinear optical applications [19]. As can be seen in Figure 11.1, the repeating unit (**37**) in the *ortho*-phenyleneethynylene-based polymer **36** contained electron donors (D) and electron acceptors (A) across a conjugated π system. If the aromatic rings of the repeating units were assumed to be coplanar, the adjacent repeating units linked through the enantiomerically pure 1,1′-binaphthyl bonds should progressively rotate along the polymer axis to form a propeller-like helical structure. At infinite polymer chain length, all the dipole moments of the repeating units would be cancelled. Such non-dipolar multipole chiral polymers may have interesting second-order nonlinear optical properties.

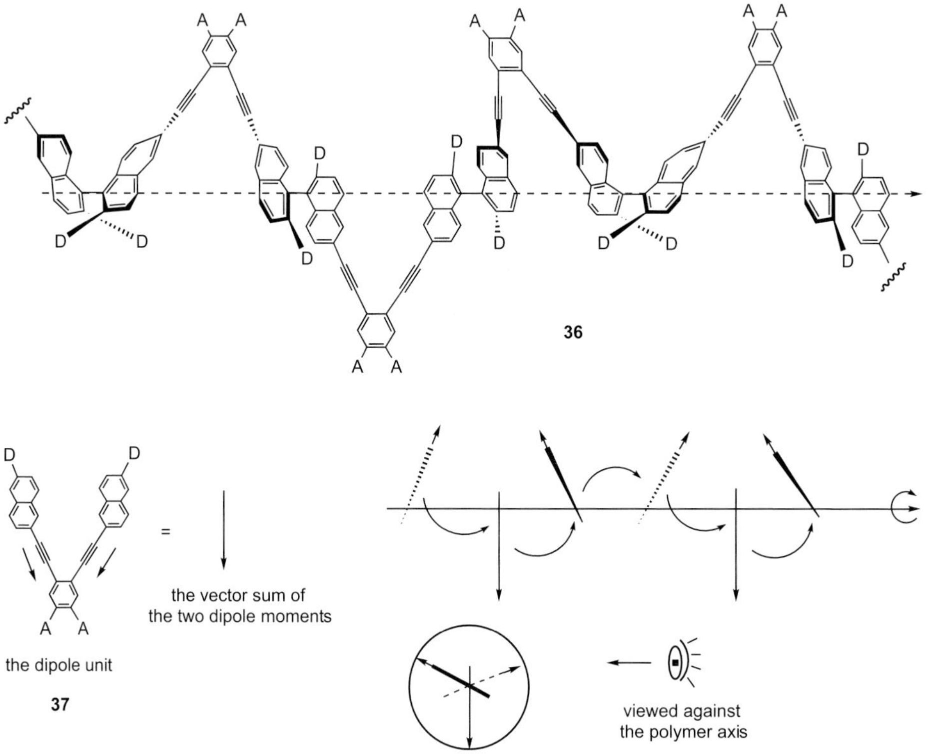

Figure 11.1 An *ortho*-phenyleneethynylene-based chiral polybinaphthyl.

We first conducted the cross-coupling of enantiomerically pure (*R*)-**23** with 1,2-dibromo-4-nitrobenzene (**38**) in the presence of $Pd(PPh_3)_4/CuI$ in refluxing toluene to generate polymer (*R*)-**39** (Scheme 11.11). In (*R*)-**39**, because the mononitro group on the repeating unit was randomly distributed at positions 1 or 2, the net dipole moment would be cancelled as indicated by **36**. Polymer (*R*)-**39** was soluble in common organic solvents. GPC gave its molecular weight as Mw = 7200 and Mn = 5100 (PDI = 1.4), while its specific optical rotation was $[\alpha]_D = -160$ (c = 0.15, CH_2Cl_2). The racemic monomer *rac*-**23** was also polymerized with **38** to give *rac*-**39** with a molecular weight of Mw = 7400 and Mn = 5200 (PDI = 1.4).

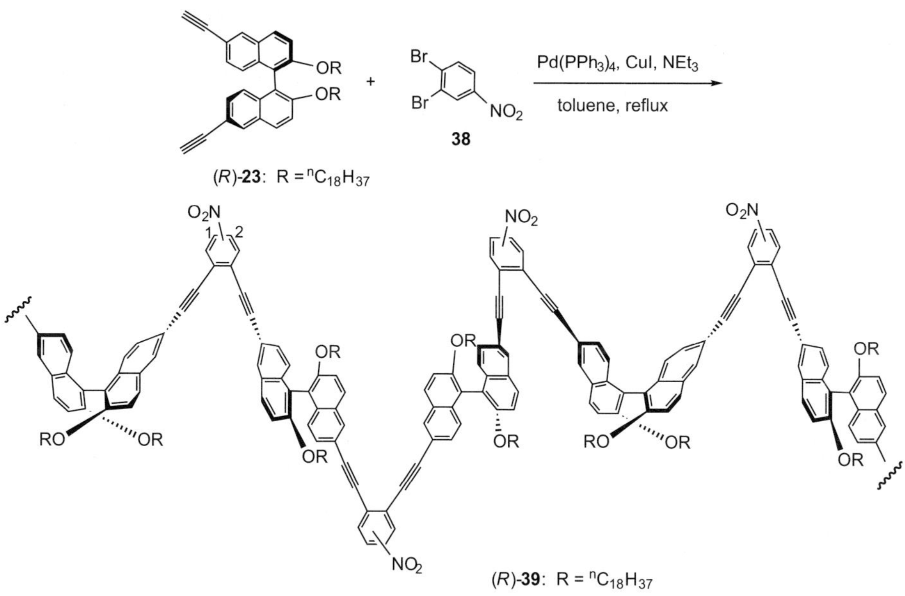

Scheme 11.11 Polymerization of (*R*)-**23** with **38** to give polymer (*R*)-**39**.

When the coupling of (*R*)-**40**, a close analogue of (*R*)-**23**, with **38** was conducted at room temperature, (*R*)-**41** was isolated in high yield (Scheme 11.12). This demonstrated that the two bromides, *para* or *meta* to the nitro groups of **38**, had very different reactivities, the *para*-bromide being much more active than the *meta* one. Compound (*R*)-**41** was treated with two equivalents of the monoprotected 6,6′-ethynyl binaphthyl compound (*R*)-**42**, which generated both the alkyne-aryl bromide coupling product and the alkyne-alkyne homocoupling product (*R*)-**43**. The significant yield of (*R*)-**43** (~50%) indicated that polymer (*R*)-**39** should contain a mixture of the aryl bromide-alkyne coupling fragments and the alkyne-alkyne coupling fragments in the polymer chain. This was consistent with the complicated NMR spectrum of (*R*)-**39**.

In order to avoid the above structural complications in the formation of this polymer, 1,2-dibromo-4,5-dinitrobenzene (**44**) was prepared as the *ortho*-phenylene monomer (Scheme 11.13). Because both the bromides in **44** were *para* to nitro groups, this compound was very reactive and could couple with (*R*)-**23** even at 0 °C in the presence of the catalysts. We conducted the polymerization of (*R*)-**23**

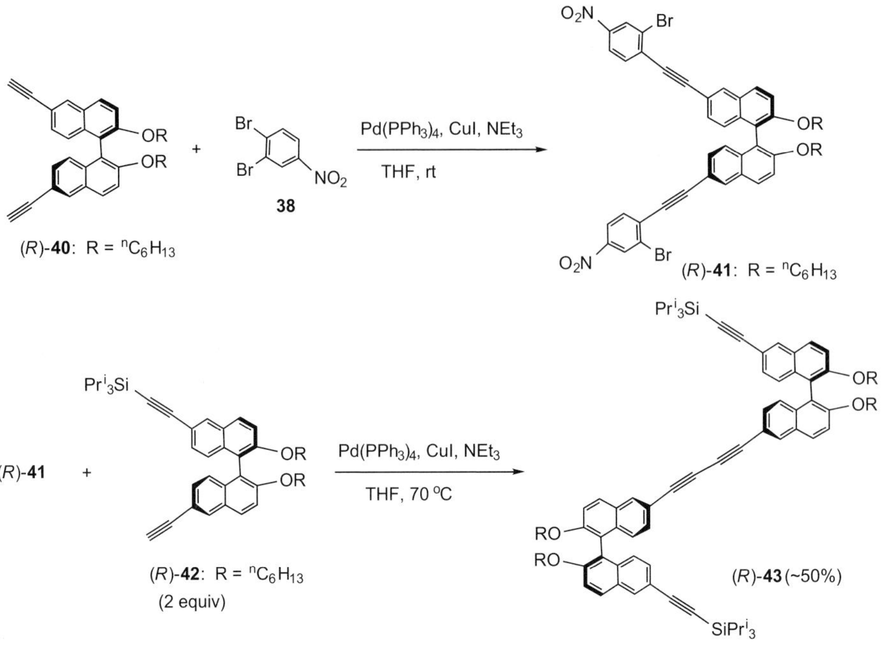

Scheme 11.12 Cross couplings of (*R*)-**40** and (*R*)-**41** with **38** and (*R*)-**42**, respectively.

with **44** in the presence of $Pd(PPh_3)_4$/CuI at room temperature, which gave polymer (*R*)-**45** with a molecular weight of Mw = 18,000 and Mn = 12000 (PDI = 1.5) as measured by GPC analysis. The molecular weight of (*R*)-**45** was significantly higher than that of (*R*)-**39** because the reactions of the more active aryl bromides in **44** had suppressed the homocoupling of (*R*)-**23**. Both the ^{1}H and the ^{13}C NMR spectra of (*R*)-**45** showed a polymer much more structurally defined than (*R*)-**39**. The specific optical rotation of (*R*)-**45** was $[\alpha]_D = -163.9$ (c = 0.16, CH_2Cl_2), and a powder X-ray diffraction study of (*R*)-**45** indicated that this polymer was completely amorphous. Polymer *rac*-**45** made from the racemic monomer had a molecular weight of Mw = 14,200 and Mn = 10,000 (PDI = 1.4).

The repeating unit of polymer (*R*)-**45**, compound **47**, was synthesized by coupling of **46** with **44** (Scheme 11.14). The high yield (>95 %) of **47** at room temperature from this reaction and the similarity of the NMR signal distribution of this model compound to that of the polymer further supported the structure of (*R*)-**45**; the two different alkyne carbons of **47** gave ^{13}C NMR signals at δ 102.0 and 85.6, while those of the polymer (*R*)-**45** were observed at δ 101.9 and 85.8. The UV spectrum of **47** showed λ_{max} values of 236, 300, 354, and 406 nm, very similar to those of (*R*)-**45** [λ_{max} 238, 277 (sh), 309 (sh), 358, and 418 nm]. This was consistent with what had been observed in the comparison of polymers (*R*)-**18e** and (*R*)-**20c** with their corresponding repeating units **25** and **26**; that is, there was little extended conjugation across the 1,1′-bond of the binaphthyl unit.

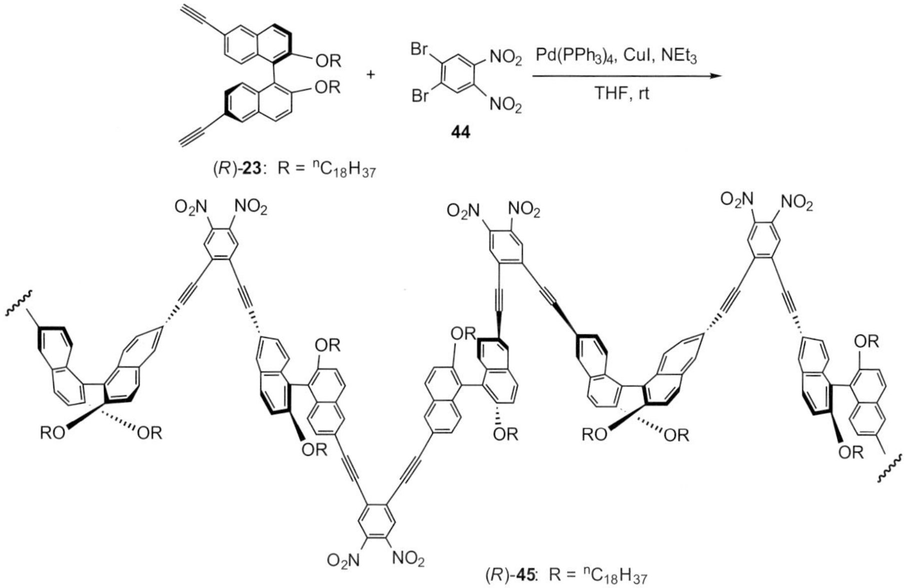

Scheme 11.13 Synthesis of the *ortho*-phenyleneethynylene-based chiral polymer (*R*)-**45**.

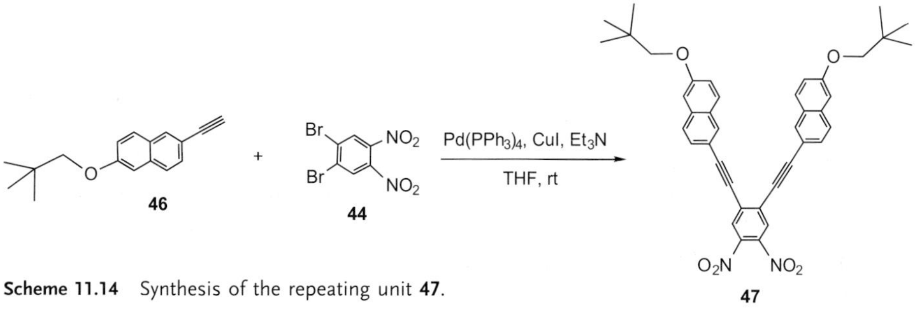

Scheme 11.14 Synthesis of the repeating unit **47**.

Several optically active and inactive analogues of polymer (*R*)-**45** were synthesized. Polymer (*R*)-**48**, made by coupling between (*R*)-**23** and 3,4,5,6-tetrafluoro-1,2-diiodobenzene (Scheme 11.15), had a molecular weight of Mw = 7200 and Mn = 5100 (PDI = 1.4). Its specific optical rotation was $[\alpha]_D = -103$ (c = 0.29, CH_2Cl_2), and its ^{13}C NMR spectrum gave two major alkyne carbon signals at δ 101.9 and 79.2. The ^{19}F NMR also gave two dominant signals at δ 121.2 and 141.2 and two other very small peaks. The racemic *rac*-**23** was polymerized with 3,4,5,6-tetrafluoro-1,2-diiodobenzene to give *rac*-**48** [Mw = 7600 and Mn = 5300 (PDI = 1.4)]. Polymerization of *rac*-**23** with 1,2-diiodobenzene gave the polymer *rac*-**49** [Mw = 8100 and Mn 5400 (PDI = 1.5)] which showed four alkyne signals in its ^{13}C NMR spectrum. This indicated that polymer *rac*-**49** contained a mixture of different structures, probably due to the competitive homocoupling of *rac*-**23** in the cross-coupling polymerization.

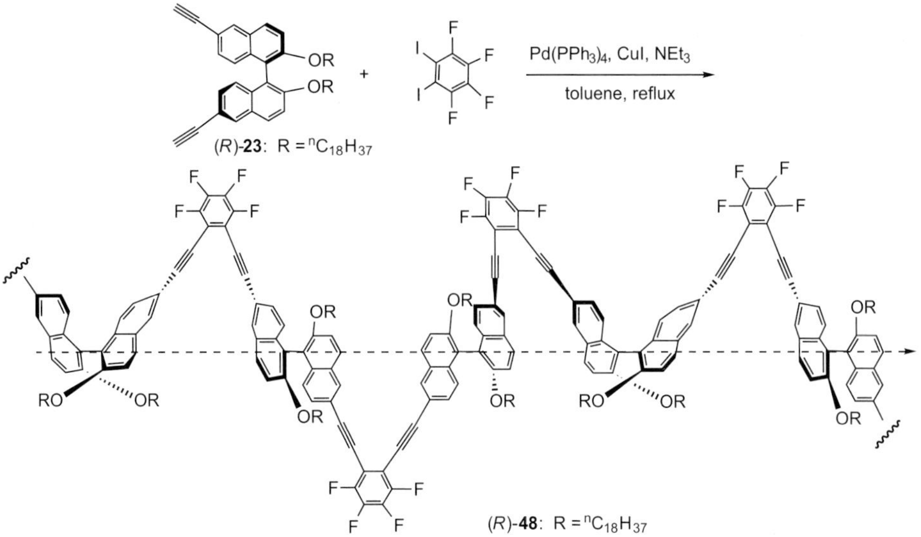

Scheme 11.15 Synthesis of the chiral polymer (*R*)-**48**.

Polymerization of (*R*)-**29** with **44** at room temperature gave polymer (*R*)-**50**, which had more extended conjugation than (*R*)-**45** (Scheme 11.16). Polymer (*R*)-**50** had a molecular weight of Mw = 10,700 and Mn = 6000 (PDI = 1.8), and its specific optical rotation was $[\alpha]_D = -228$ (c = 0.13, CH_2Cl_2). Compound (*R*)-**29** was also polymerized with the mononitro-substituted dibromo monomer **38** to give polymer (*R*)-**51**, with a molecular weight of Mw = 14,200 and Mn = 6900 (PDI = 2.1) and a specific optical rotation of $[\alpha]_D = -284$ (c = 0.14, CH_2Cl_2). The corresponding optically inactive polymer *rac*-**51** had a molecular weight of Mw = 10,500 and Mn = 6000 (PDI = 1.8). Polymerization of *rac*-**29** with *ortho*-diiodobenzene gave *rac*-**52** with a molecular weight of Mw = 11,600 and Mn = 6000 (PDI = 1.8).

Table 11.7 summarizes the UV and CD spectral data of the propeller-like polymers. Among the polymers, (*R*)-**45** gave the longest-wavelength absorption. Even though polymer (*R*)-**50** had a longer conjugation length in its repeating unit than (*R*)-**45**, its UV absorption maxima were significantly blue-shifted, indicating that the conjugation of the electron donor and acceptor groups in (*R*)-**50** was much weaker than that in (*R*)-**45**, which might be due to the much longer distances between the donors and acceptors in (*R*)-**50**. Polymer (*R*)-**39**, with only one nitro group in its repeating unit, gave significantly shorter-wavelength absorptions than polymer (*R*)-**45**, with two nitro groups. Polymer (*R*)-**51**, with one nitro group in its repeating unit but a longer conjugation length than (*R*)-**39**, also gave blue-shifted absorptions relative to those of (*R*)-**39**. Although the fluorine atoms of polymer (*R*)-**48** have high electronegativity, their strong π-donating effects make them poor electron acceptors, resulting in much shorter-wavelength absorptions for (*R*)-**48** than for (*R*)-**45**. Polymers *rac*-**49** and *rac*-**52** with no electron acceptors gave short-wavelength absorptions. All the optically active polymers gave exci-

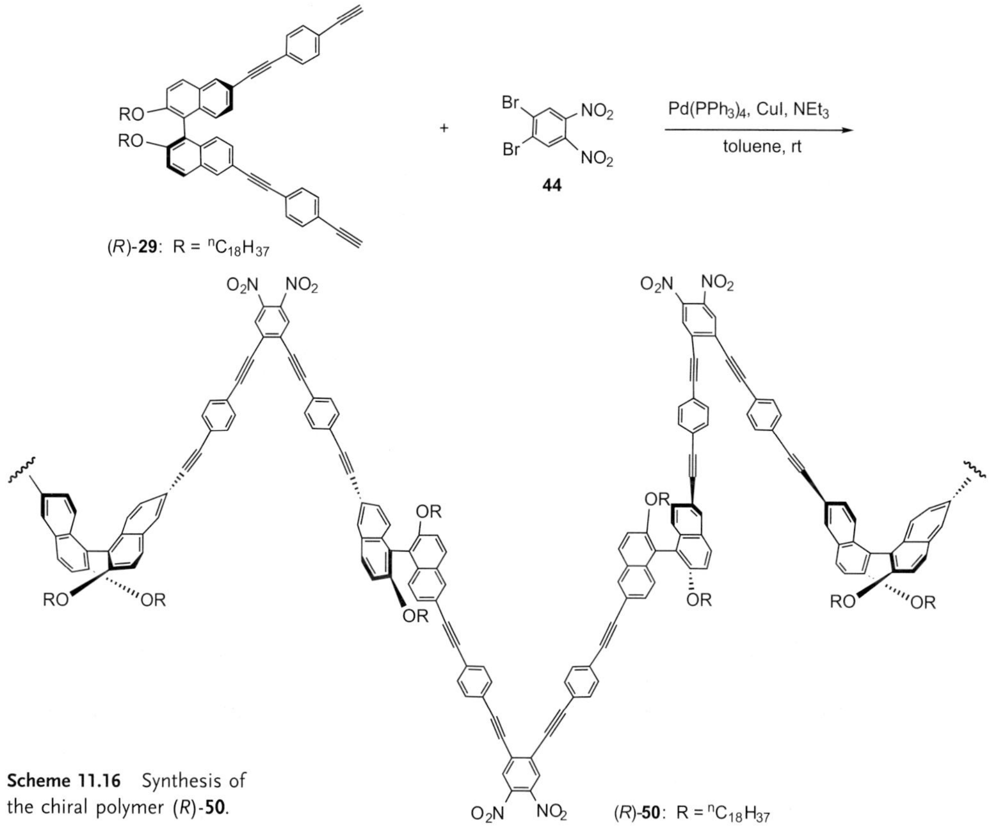

Scheme 11.16 Synthesis of the chiral polymer (*R*)-**50**.

ton coupling CD signals at 230–280 nm, corresponding to their (*R*)-1,1′-binaphthyl units. Their major differences were in the long-wavelength region because of the different repeating units.

Table 11.7 Spectral data for the propeller-like polymers.

	(R)-39	*(R)-45*	*46*	*(R)-48*	*rac-49*	*(R)-50*	*(R)-51*	*rac-52*
λ_{max}	234	238	236	262	236	234	232	234
(nm)	295 (sh)	277 (sh)	300	278	276	296	296	302
	346	309 (sh)	354	344	340 (sh)	346	362	342
	380	358	406					
		418						
[θ]	2.19×10^5 (235)	2.68×10^5 (234)		2.38×10^5 (237)		2.29×10^5 (234)		
(λ, nm)	-5.42×10^4 (264)	-1.40×10^5 (277)		-2.14×10^5 (284)		-4.63×10^4 (264)		
	4.82×10^3 (351)	9.14×10^3 (356)		-2.38×10^4 (381)		-6.56×10^4 (296)		
	-5.05×10^3 (396)	-9.71×10^3 (419)				-2.78×10^4 (383)		

The fluorescence of the polymers was studied. The nitro-substituted polymers all showed either very weak or no fluorescence signals, but polymers *rac*-**49** and (*R*)-/*rac*-**48** gave very strong fluorescence signals in the 420–460 nm range. Electron transfer between the donors and acceptors in the nitro-containing polymers might account for the quenched fluorescence.

The molecular weights of the polymers were studied by laser light scattering (LLS) and compared with those determined from the GPC data. It was found that the LLS data were in general ~20% higher than those obtained by GPC, although the LLS molecular weight of polymer (*R*)-**39** was as much as three times larger than its GPC molecular weight. This indicated that the polystyrene standards used in the GPC analysis had three-dimensional structures very different from that of (*R*)-**39**.

Differential scanning calorimetry (DSC) analysis of the polymers showed that polymers (*R*)-/*rac*-**39**, (*R*)-/*rac*-**45**, (*R*)-**50**, (*R*)-/*rac*-**51** and the repeating unit **46** each had a highly exothermic decomposition point at 176–202 °C ($\Delta H = -58$ to -113 J g^{-1}). To account for the decomposition of these polymers, an enediyne cyclization (the Bergman cyclization) was proposed (Scheme 11.17) [20]; Bergman cyclization of *ortho*-diethynyl benzene (**53**) was observed at 191 °C in chlorobenzene in the presence of 1,4-cyclohexadiene [20c]. Polymers (*R*)-/*rac*-**48** and *rac*-**52** showed much smaller heat release, with $\Delta H = -16$ to 36 J g^{-1}. This suggested that these polymers had much smaller amounts of the enediyne fragments in their polymer chain because of the homocoupling of the binaphthyl monomers as shown in the formation of (*R*)-**43**. In an extreme case, polymer *rac*-**49** did not decompose up to 300 °C.

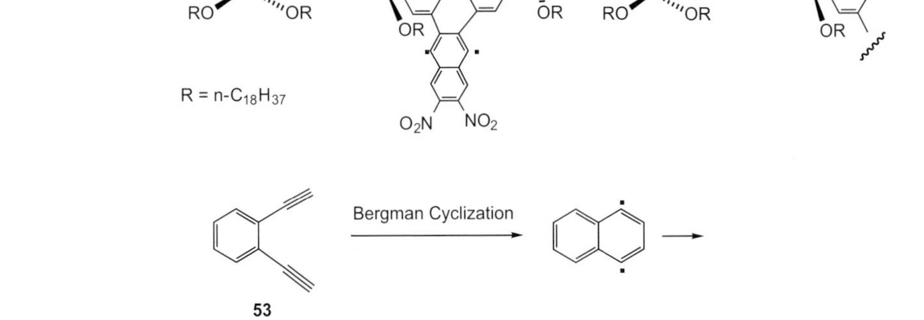

Scheme 11.17 Proposed thermal decomposition of (*R*)-**45** and the Bergman cyclization.

TGA analyses of the polymers showed that all of these materials started to lose 5 % of their masses at 310–380 °C. The polymers with nitro groups decomposed at lower temperatures than those without nitro groups. After these polymers had lost their alkyl groups, they were very stable from 500 °C to 800 °C.

11.3.2.2 **Dipole-Oriented Polymers Based on BINOL**

In polymer (*R*)-**39**, because of the randomly distributed NO_2 groups at positions 1 and 2 of the repeating units, dipole moments should cancel at infinite polymer chain length. If the NO_2 in polymer (*R*)-**39** were incorporated specifically at only one of the positions 1 or 2 over the entire polymer chain. as shown in (*R*)-**54**, the dipole moment of the repeating unit would tilt toward only one direction of the polymer chain, resulting in a large collective dipole moment (Figure 11.2).

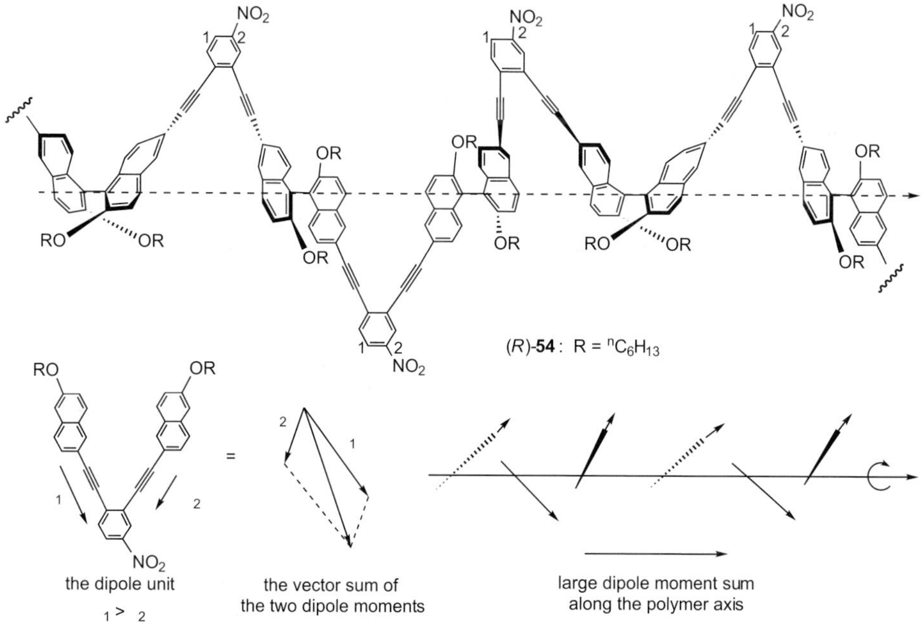

Figure 11.2 Structure of the dipole-oriented polymer (*R*)-**54**.

In order to synthesize polymer (*R*)-**54**, an AB monomer was prepared as shown in Scheme 11.18 [21]. Compound (*R*)-**55**, containing two different alkyne protecting groups, triisopropylsilyl and trimethylsilyl, was prepared. Removal of the trimethylsilyl group of (*R*)-**55** with base gave (*R*)-**56**, and treatment of (*R*)-**56** with 3,4-diiodo-1-nitrobenzene at −10 to 0 °C in the presence of $Pd(PPh_3)_4/CuI$, followed by deprotection with Bu_4NF, gave a mixture of the desired AB monomer (*R*)-**58** (53 %) and the bisbinaphthyl compound (*R*)-**59** (21 %).

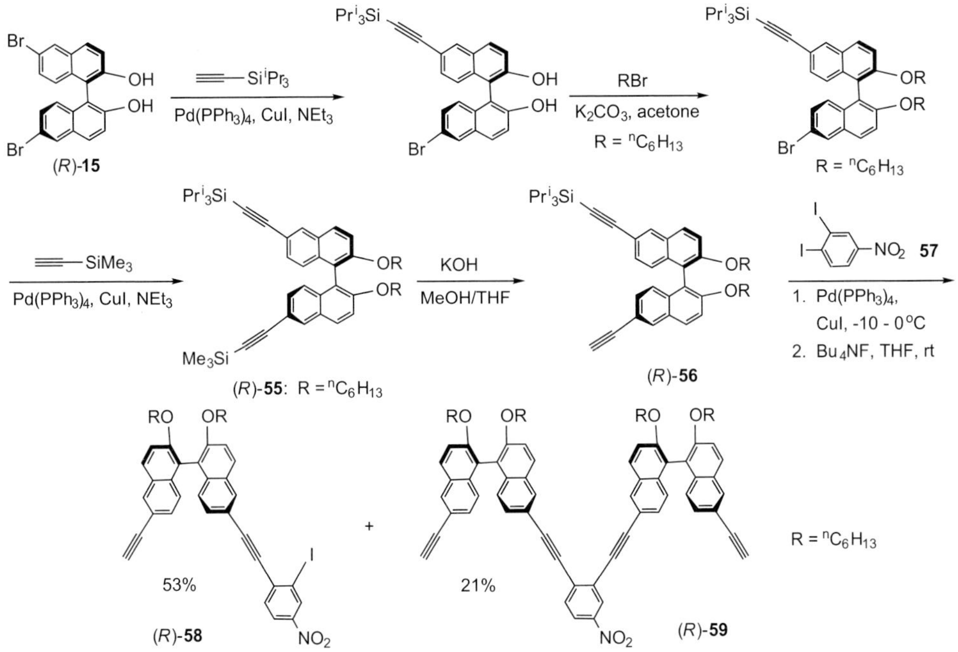

Scheme 11.18 Synthesis of the AB monomer (*R*)-**58**.

Monomer (*R*)-**58** (0.24 M) was polymerized at 60 °C in the presence of Pd(PPh_3)$_4$/CuI in Et_3N and THF (Scheme 11.19). Polymer (*R*)-**54** was obtained with a molecular weight of Mw = 34,000 and Mn = 17,000 (PDI = 1.97).

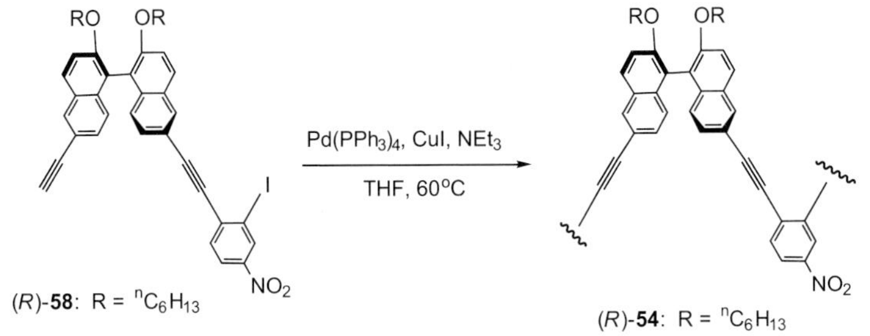

Scheme 11.19 Homopolymerization of the AB monomer (*R*)-**58**.

The ^{13}C NMR spectrum of (*R*)-**54** gave four alkyne carbon signals at δ 100.2, 97.0, 87.4, and 86.6, the specific optical rotation of the polymer was $[\alpha]_D = -197.6$, and the UV spectrum of (*R*)-**54** was found to be very similar to that of monomer (*R*)-**58**, with λ_{max} = 250, 300, 354, and 390 nm. This demonstrated that the conjugation of the polymer was determined by the nitro-alkoxy push-pull conjugated system in the repeating unit. The CD spectrum of the polymer gave strong Cotton effects at [θ] (λ, nm) = 2.7×10^5 (232), -1.6×10^5 (271 sh), -1.8×10^5 (282), 1.9×10^4 (350), and -1.8×10^4 (387). DSC analysis of (*R*)-**54** showed an irreversible exothermic peak at 209 °C ($\Delta H = -90.2\ J\ g^{-1}$), which might be due to the Bergman cyclization of the ene-diyne fragment in the polymer chain as shown in Scheme 11.17. TGA analysis showed that (*R*)-**54** lost its hexyl groups from 300 °C to 474 °C, and the decomposition then became very slow up to 800 °C.

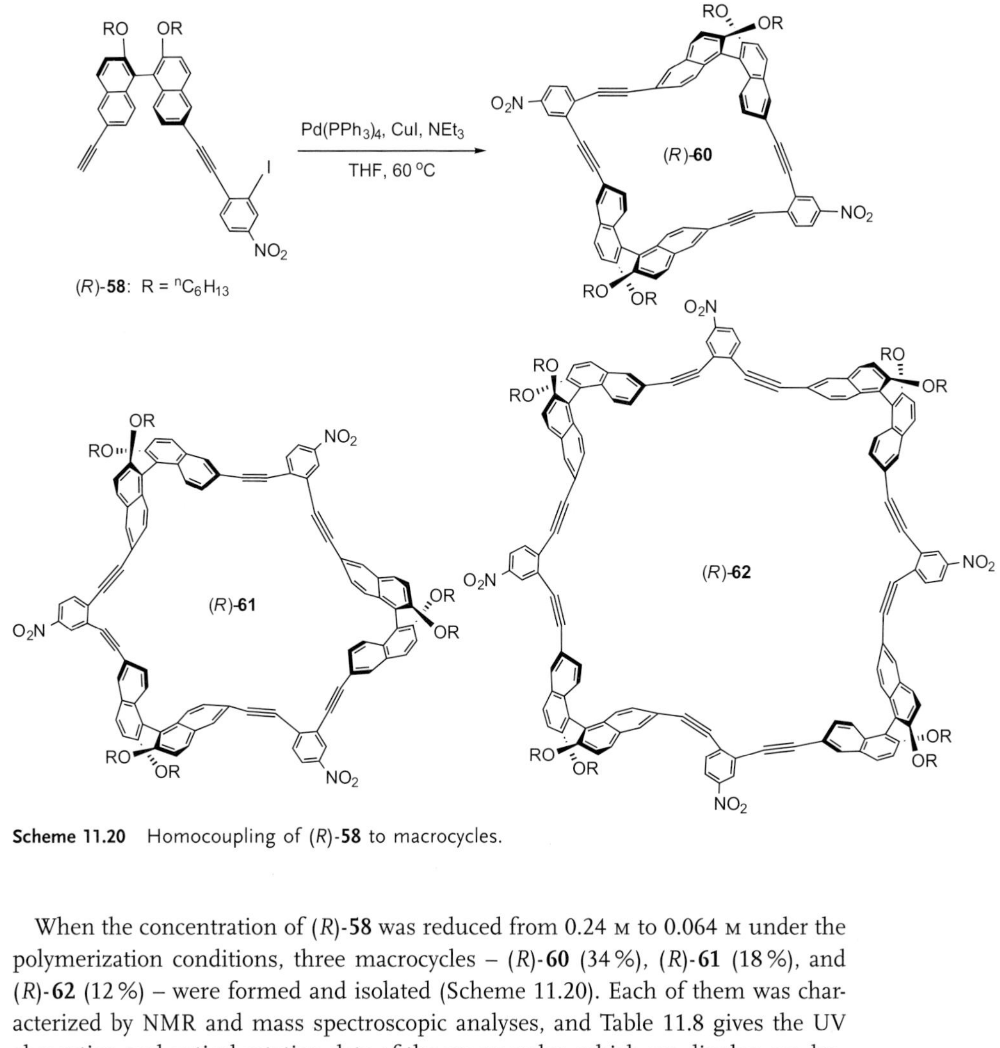

Scheme 11.20 Homocoupling of (*R*)-**58** to macrocycles.

When the concentration of (*R*)-**58** was reduced from 0.24 M to 0.064 M under the polymerization conditions, three macrocycles – (*R*)-**60** (34 %), (*R*)-**61** (18 %), and (*R*)-**62** (12 %) – were formed and isolated (Scheme 11.20). Each of them was characterized by NMR and mass spectroscopic analyses, and Table 11.8 gives the UV absorption and optical rotation data of the macrocycles, which are dipolar, quadrupolar, and octupolar materials. All the macrocycles had similar UV absorptions and optical rotations, but although the UV absorptions of the macrocycles were also very close to those of polymer (*R*)-**54**, the specific optical rotations of the macrocycles were much larger.

Table 11.8 UV absorptions and specific optical rotations of macrocycles (*R*)-**60** to (*R*)-**62**.

	(R)-60	*(R)-61*	*(R)-62*
λ_{max} (nm)	254	236	234
	286	286	286
	342	346	344
	390	388	384
$[\alpha]_D$ (CH_2Cl_2)	−301.3	−282.3	−297.0

11.3.2.3 **1,1′-Binaphthyl-2,2′-diamine-Derived Polymers**

We also used 1,1′-binaphthyl-2,2′-diamine ((*R*)-**63**) to prepare propeller-like polymers, as amine groups are stronger electron donors than alkoxy groups [22]. Scheme 11.21 shows the conversion of (*R*)-**63** into the desired monomer (*R*)-**66**. Acylation of (*R*)-**63** followed by reduction gave (*R*)-**64**. The 6- and 6′-positions of (*R*)-**64** were specifically brominated by treatment with $^nBu_4NBr_3$ to give (*R*)-**65**, and coupling of (*R*)-**65** with triisopropylsilylacetylene followed by methylation and desilylation gave monomer (*R*)-**66**. The two alkyl groups were introduced onto the nitrogen atoms of (*R*)-**66** in order to make the subsequent polymer more soluble and stable.

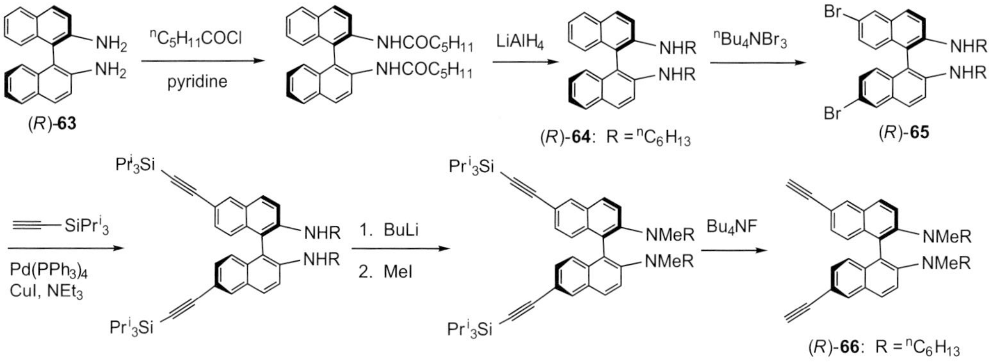

Scheme 11.21 Synthesis of the binaphthyldiamine-based monomer (*R*)-**66**.

Polymerization of (*R*)-**66** with **44** in the presence of $Pd(PPh_3)_4$ and CuI gave polymer (*R*)-**67** (Scheme 11.22). GPC analysis of (*R*)-**67** gave Mw = 23,500 and Mn = 12,400 (PDI = 1.9), while the alkyne carbon signals in the ^{13}C NMR spectrum of (*R*)-**67** were multiplets at δ 102.5 and 86.1.

Heck coupling between (*R*)-**68** and **69** in the presence of $Pd(OAc)_2$ and CuI resulted in the formation of polymer (*R*)-**70**, containing double bonds rather than triple bonds (Scheme 11.23). The molecular weight of this polymer was Mw = 2800 and 1900 (PDI = 1.45), much smaller than that of (*R*)-**67**.

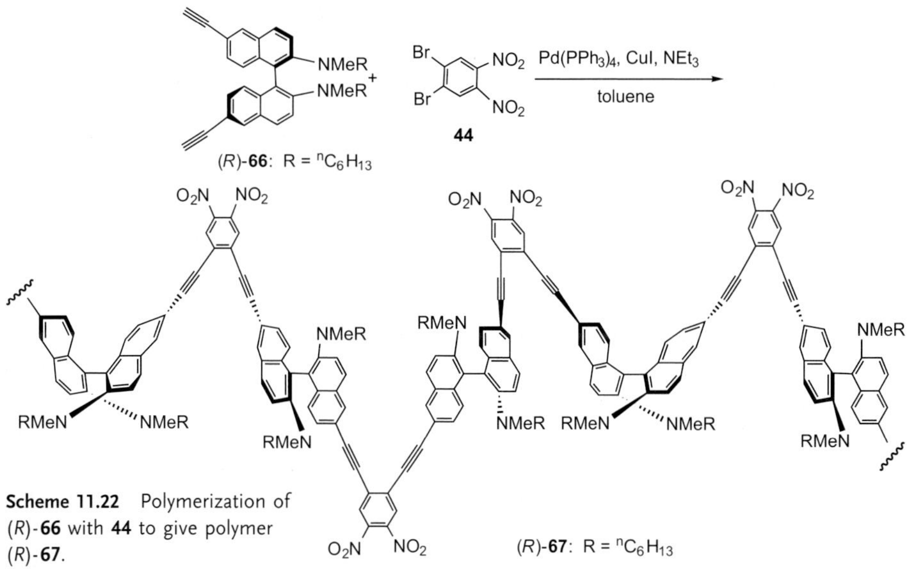

Scheme 11.22 Polymerization of (*R*)-**66** with **44** to give polymer (*R*)-**67**.

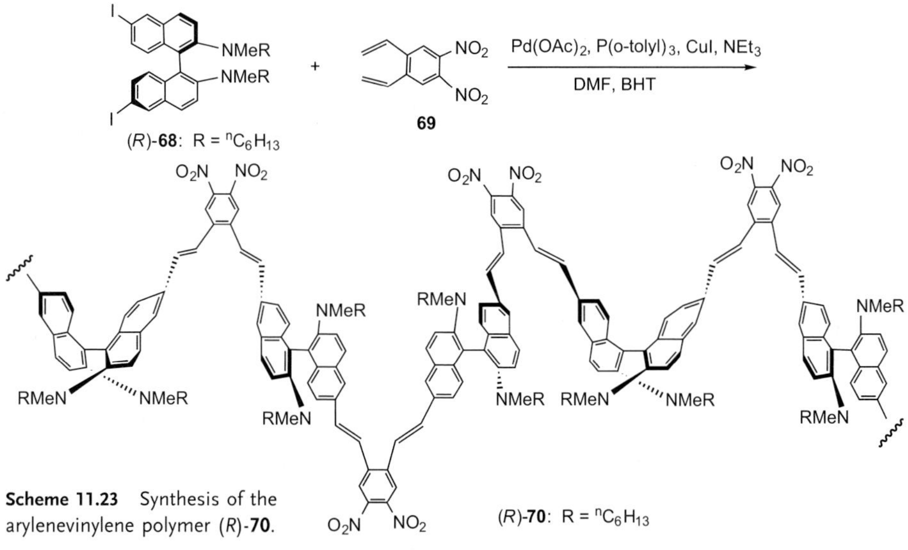

Scheme 11.23 Synthesis of the arylenevinylene polymer (*R*)-**70**.

The UV and CD spectral data of the binaphthylamine-derived polymers (*R*)-**67** and (*R*)-**70** were compared with those of the BINOL-derived polymer (*R*)-**45**, as shown in Table 11.9. The amine groups in polymer (*R*)-**67** were much stronger electron donors than the alkoxy groups in (*R*)-**45**, and so the push-pull conjugation between the donors and acceptors in the repeating units of (*R*)-**67** should be stronger than that in (*R*)-**45**, resulting in the large red shifts of the UV absorptions from (*R*)-**45** to (*R*)-**67**. Polymer (*R*)-**70** had further red-shifted absorptions over those of

Table 11.9 Comparison of the spectral data for polymers (*R*)-**67** and (*R*)-**70** with those for polymer (*R*)-**45**.

	(R)-67	**(R)-70**	**(R)-45**
λ_{max} (nm)	272	254	238
	324	308	277 (sh)
	383	400	309 (sh)
	470	486	358
			418
[θ] (λ_{max} nm)	2.30×10^5(220, sh)	1.86×10^5 (223)	2.68×10^5 (234)
	-1.18×10^5 (286)	-5.68×10^4 (278)	-1.40×10^5 (277)
	1.50×10^3 (377)	-1.49×10^4 (356)	9.14×10^3 (356)
	-1.70×10^4 (409, sh)	4.74×10^3 (415)	-9.71×10^3 (419)
	-1.40×10^4 (469, sh)	-3.34×10^4 (503)	

(*R*)-**67** because the π-electrons of the double bonds in (*R*)-**70** were more polarizable than those of the triple bonds in (*R*)-**67**. The CD signals of (*R*)-**67** were very similar to those of (*R*)-**45**, indicating similar stereostructures, while the long-wavelength CD signals of (*R*)-**70** were very different from those of (*R*)-**67**, the vinylene-based polymer being structurally very different from the ethynylene-based polymer.

11.3.2.4 **Nonlinear Optical Study of the Polymers**

In order to study the nonlinear optical properties of the chiral polymers, Langmuir–Blodgett (LB) films of the optically active polymers (*R*)-**45**, (*R*)-**67**, (*R*)-**70**, and (*R*)-**71** were prepared (Figure 11.3) [23]. Dilute chloroform solutions of these polymers were spread on water, and after evaporation of the solvent, the monolayer was compressed to form LB films with thicknesses of between one and eight monomolecular layers. The second harmonic response (SHG) of the films were measured with a Q-switched Nd:YAG laser (10 ns, 50 Hz, 1064 nm). The films of polymers (*R*)-**45**, (*R*)-**71**, and (*R*)-**70** were of very good optical quality, but those of polymer (*R*)-**67** were poor, since large holes were visible. The long-chain alkyls of (*R*)-**45** and the double bonds of (*R*)-**71** and (*R*)-**70** were therefore very good for generating high quality films because they introduced more structural flexibility.

It was found that the maximum SHG intensities from the two polymers (*R*)-**45** and (*R*)-**71** were very weak. In addition, the SHG signals of polymer (*R*)-**71** decreased to 75 % of their initial values within minutes because of the instability of this polymer upon irradiation by the infrared laser light. Polymer (*R*)-**67** showed a maximum SHG intensity a factor of 50 higher than those of polymers (*R*)-**45** and (*R*)-**71** even though the LB films of (*R*)-**67** were of poor optical quality. This was attributed to the more strongly electron-donating amine groups in (*R*)-**67**. Replacement of the triple bonds of (*R*)-**67** with double bonds resulted in high quality films of (*R*)-**70**. This polymer showed SHG signals four times as efficient as (*R*)-**67** and 200 times as efficient than (*R*)-**45** and (*R*)-**71**. Through analysis of the polarization dependence

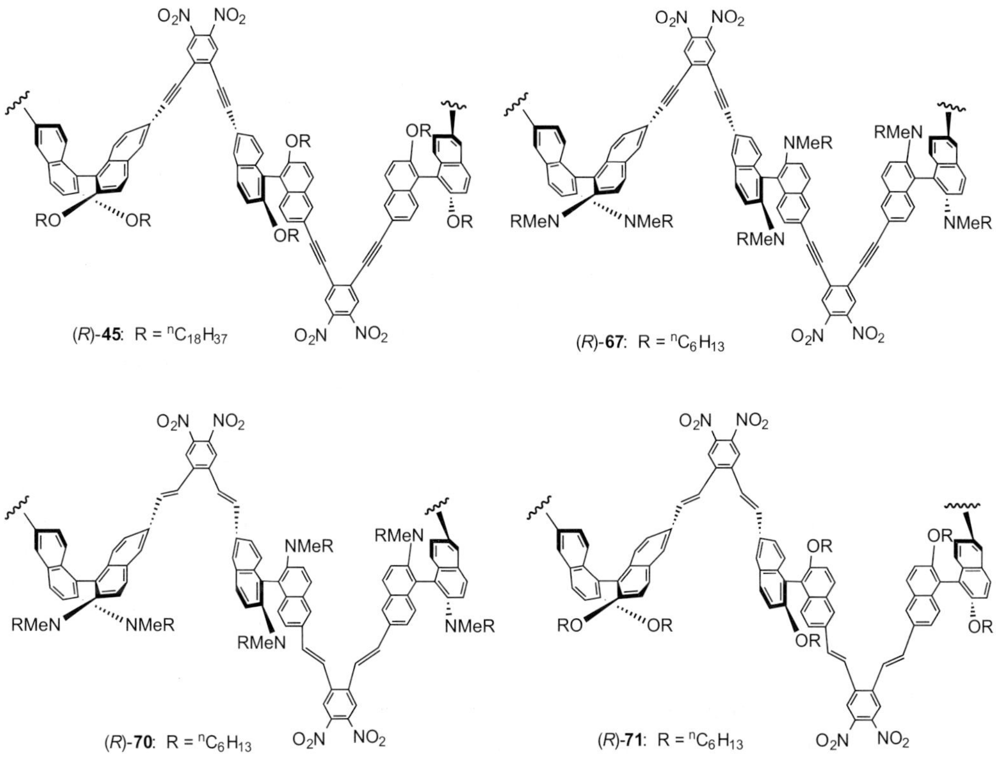

Figure 11.3 Polymers used in nonlinear optical studies.

of the SHG, the tensor components of the nonlinear susceptibility $\chi_{ijk}^{(2)}$ were determined for polymer (*R*)-**70** (Table 11.10). Although the second order susceptibility values in Table 11.10 were resonantly enhanced, they demonstrated that this optically active polybinaphthyl was a new class of promising NLO material. The chirality of the LB films was reflected by the xyz component of 2 pm V^{-1}.

Table 11.10 The second order susceptibility [$\chi_{ijk}^{(2)}$] of polymer (*R*)-**70**.

Second order susceptibility component	*Absolute value (pm V^{-1})*	*Chirality classification*
xyz	2	chiral
xxz	7	achiral
zxx	5	achiral
zzz	35	achiral

These polymers also displayed different second harmonic efficiency when they were irradiated with left- and right-handed circularly polarized light [23b]. The experiments were conducted on LB films of (*R*)-**45** by spreading the chloroform solution of the polymer on water. The thickness of the films was 10 X-type layers with an absorption maximum at 430 nm. The fundamental beam of a Q-switched and injection-seeded Na:YAG laser (1064 nm) was used to pump the chiral LB films with a 45° angle of incidence with respect to the sample. The fundamental beam was initially p-polarized with respect to the films and the polarization was varied by a quarter-wave plate. The signals of the s- and p-components of the transmitted second harmonic fields were detected and showed significant circular difference responses. These circular difference (CD) effects can be represented by the following equation: $\Delta I/I = 2(I_l - I_r)/(I_l + I_r)$, where I_l and I_r are the intensities of the second harmonic light generated when the fundamental beam is left- or right-handedly circularly polarized. The second harmonic CD effect for (*R*)-**45** was 65%, orders of magnitude higher than the linear circular dichroism effect ($\Delta\epsilon/\epsilon$) in solution. Second harmonic CD effects between 25 and 60% were also observed for other chiral polymers of this class.

11.3.3
Chiral Polymers Containing Main-Chain *meta*-Phenyleneethynylenes

The *meta*-dibromo phenylene monomer **72** was polymerized with (*R*)-**23** in the presence of $Pd(PPh_3)_4/CuI$ at room temperature to afford polymer (*R*)-**73** (Scheme 11.24) [19]. The molecular weight of (*R*)-**73** was Mw = 30,500 and Mn = 16,300 (PDI = 1.9), and its ^{13}C NMR spectrum gave two singlets at δ 83.9 and 77.2 for the alkyne carbons. The polymer *rac*-**73** was prepared by coupling between *rac*-**23** and **72** and had a molecular weight of Mw = 20,000 and Mn = 12000 (PDI = 1.6).

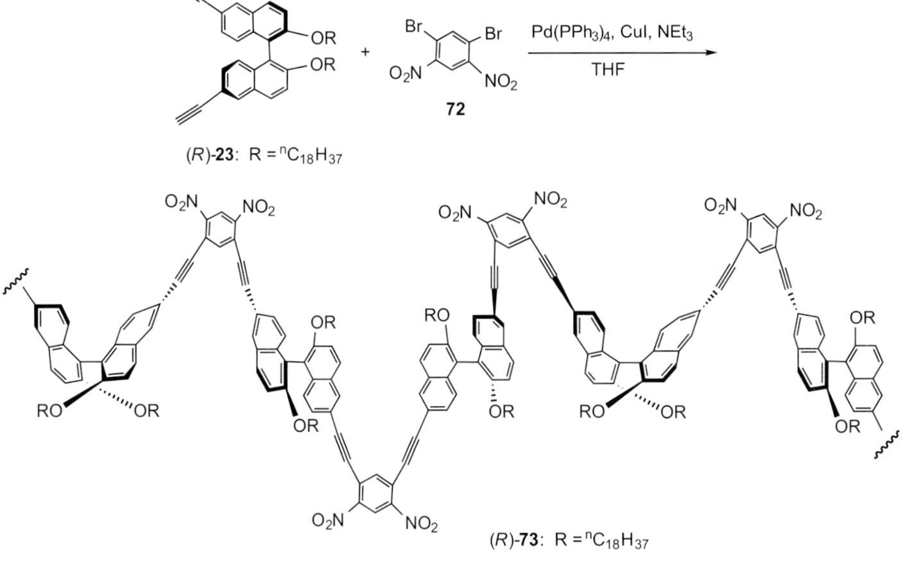

Scheme 11.24 Polymerization of (*R*)-**73** with the *meta*-phenylene linker **72**.

Compound **74** (Scheme 11.24) was synthesized as the repeating unit of the polymer by treatment of **46** (Scheme 11.14) with **72**. The UV spectra of both polymers gave λ_{max} at 256, 284 (sh), 359 (sh), and 442 nm. The model compound **74** showed λ_{max} at 244, 282, 354, and 430 nm, similarly to the polymers. These UV absorption wavelengths were significantly longer than those seen in polymer (*R*)-**45**, with *ortho*-phenylene linkers. This is because each of the alkoxy donors of (*R*)-**73** is in conjugation with both the *para*- and the *ortho*-nitro groups in the polymer repeating unit. Conversely, each alkoxy donor in (*R*)-**45** can conjugate with only one of the two nitro groups. The stronger push-pull conjugation in (*R*)-**73** resulted in much longer-wavelength absorptions. The CD spectrum of (*R*)-**73** in methylene chloride showed [θ] (λ, nm) = 2.53×10^5 (238), -1.16×10^5 (270), and -3.11×10^4 (470).

A DSC plot of (*R*)-**73** gave a highly exothermic chemical decomposition at 154 °C with $\Delta H = -359.6$ (J g^{-1}). For *rac*-**73**, the decomposition was also observed at 154 °C with $\Delta H = -447.9$ (J g^{-1}). For the repeating unit **74**, decomposition was observed at the much higher temperature of 226 °C, with $\Delta H = -466.4$ (J g^{-1}). The heat released from the decomposition of (*R*)-**73** was ca. four times that from (*R*)-**45**, indicating a very different decomposition reaction. An intramolecular double cyclization was proposed to account for the decomposition of (*R*)-/*rac*-**73** and **74** (Scheme 11.25).

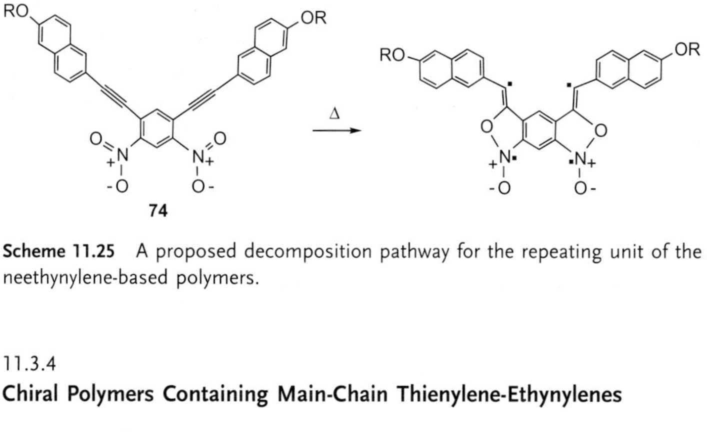

74

Scheme 11.25 A proposed decomposition pathway for the repeating unit of the *meta*-phenyleneethynylene-based polymers.

11.3.4
Chiral Polymers Containing Main-Chain Thienylene-Ethynylenes

We have also incorporated thiophenes into 1,1′-binaphthyl-based arylene-ethynylene polymers [24]. Scheme 11.26 shows the syntheses of the thienylene-ethynylene monomers **75** and **76**.

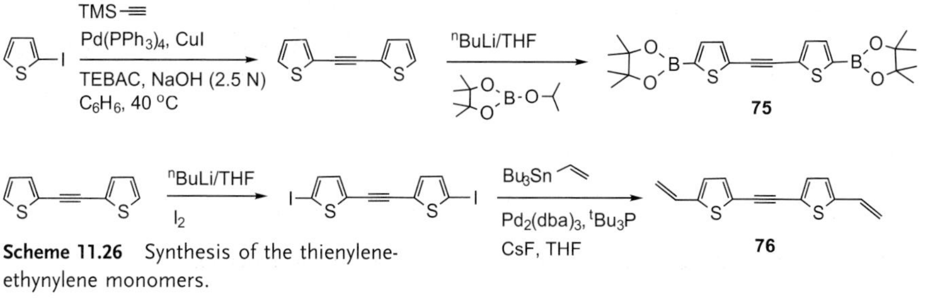

Scheme 11.26 Synthesis of the thienylene-ethynylene monomers.

Compound **76** was not very stable in air over time and so was prepared immediately prior to its subsequent polymerization. Suzuki coupling between **75** and the binaphthyl monomer (*R*)-**77** gave polymer (*R*)-**78** (Scheme 11.27). GPC showed its molecular weight as Mw = 7500 and Mn = 3000 (PDI = 2.5).

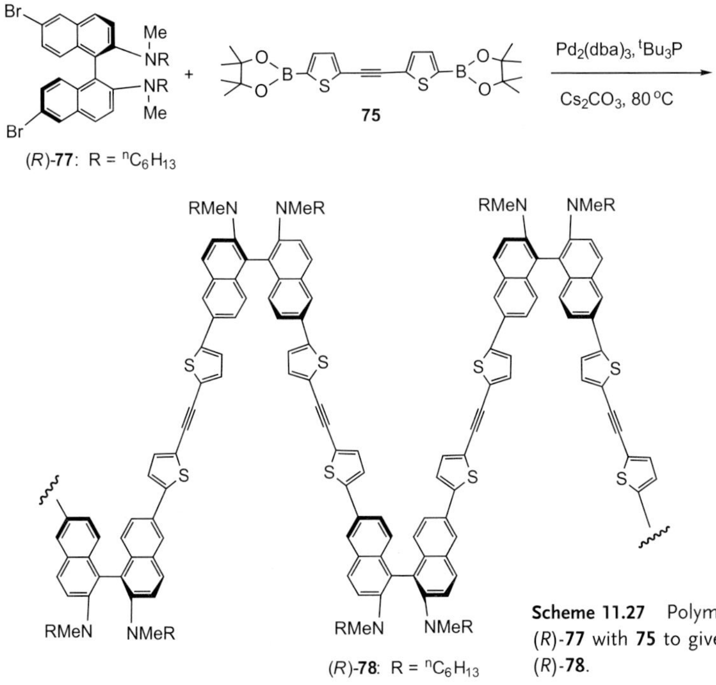

Scheme 11.27 Polymerization of (*R*)-**77** with **75** to give polymer (*R*)-**78**.

Heck couplings of (*R*)-**77** and (*R*)-**79** were also conducted, giving polymers (*R*)-**80** and (*R*)-**81**, respectively (Scheme 11.28). BHT was added to inhibit the radical polymerization of **76** in this reaction.

The molecular weight of (*R*)-**80** was Mw = 12,000 (PDI = 2.4) and that of (*R*)-**81** was Mw = 11,800 (PDI = 2.5). The presence of different halogen atoms and alkyl chains between monomers (*R*)-**77** and (*R*)-**79** did not produce any significant effect on the molecular weights of the polymers. UV spectroscopic study of these polymers showed that polymers (*R*)-**80** and (*R*)-**81** had longer-wavelength absorption than (*R*)-**78**, because of the longer conjugation in the repeating units of (*R*)-**80** and (*R*)-**81**. Table 11.11 gives electronic absorption and CD data for the polymers. The fluorescence spectra of (*R*)-**81** in solvents including methylene chloride, benzene, and hexane were studied; the polymer showed the most intense fluorescence signal in hexane, with λ_{emi} at 491 nm.

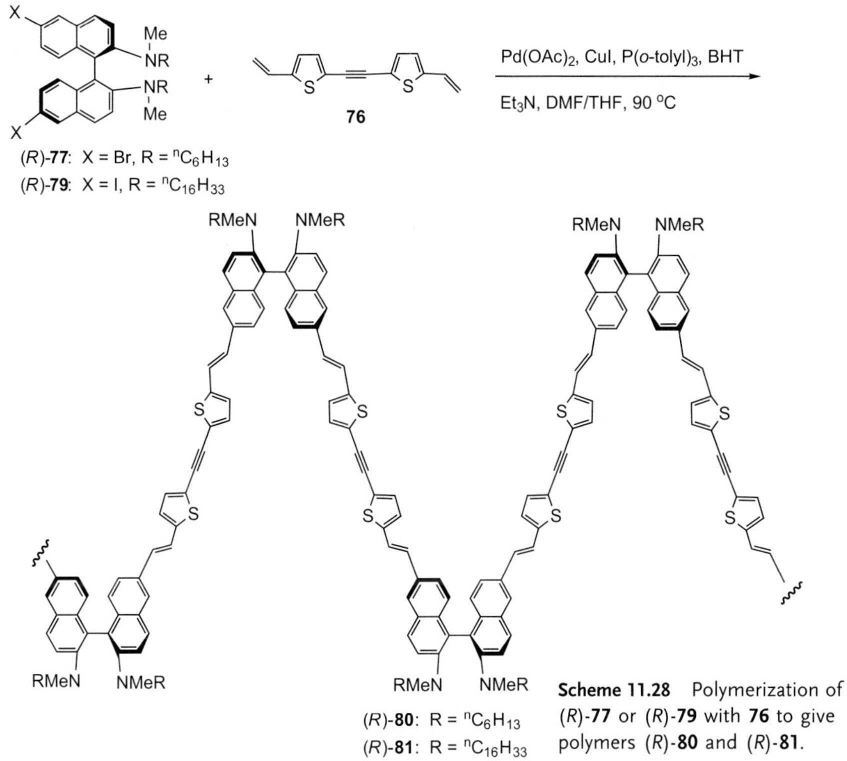

Scheme 11.28 Polymerization of (*R*)-**77** or (*R*)-**79** with **76** to give polymers (*R*)-**80** and (*R*)-**81**.

Table 11.11 Spectral data for polymers (*R*)-**78**, (*R*)-**80** and (*R*)-**81** in chloroform.

	(R)-78	***(R)-80***	***(R)-81***
λ_{max} (ϵ) nm	263 (4.73×10^4) 376 (2.88×10^4)	263 (4.25×10^4) 397 (2.83×10^4)	260 (6.34×10^4) 401 (5.25×10^4)
[θ] (λ, nm)	-6.78×10^4 (275)	-1.48×10^5 (275)	-1.28×10^5 (277)

11.3.5
Chiral Polymers Containing Side-Chain Phenyleneethynylenes

1,1′-Binaphthyl compounds contain two chiral grooves, major and minor, as indicated in the case of the structure of (*R*)-**82** (Scheme 11.29). All the polymers discussed in the previous sections were obtained by polymerization at the major grooves of the binaphthyl monomers. Suzuki coupling between (*R*)-**82** and **83**, however, allowed polymerization at the minor groove of the binaphthyl to give polymer (*R*)-**84** (Scheme 11.29) [25].

This polymer contained electron-rich side-chain phenyleneethynylene units, incorporated in order to conduct an intramolecular cyclization to generate a polycyclic aromatic system as seen in Swager's work (Scheme 11.30) [26], thanks to the

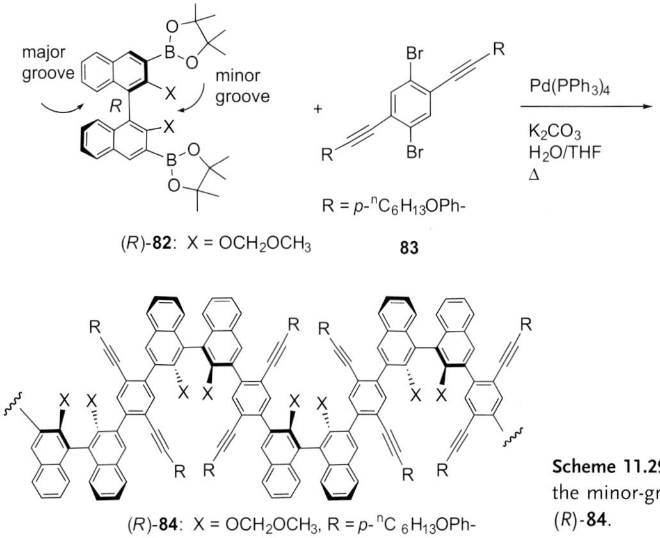

Scheme 11.29 Synthesis of the minor-groove polybinaphthyl (*R*)-**84**.

mild conditions and high yield of this reaction. The molecular weight of (*R*)-**84** was Mw = 11,200 and Mn = 7300 (PDI = 1.54), its specific optical rotation was 36.8 (c = 0.1, CH_2Cl_2), and its ^{13}C NMR spectrum displayed two alkyne carbon signals at δ 94.7 and 98.5. The UV spectrum of (*R*)-**84** gave absorptions at λ_{max} (ϵ) = 341 (4.86×10^4) and 366 (sh, 3.58×10^4) nm and had a strong emission at 398 nm. The CD spectrum of (*R*)-**84** gave Cotton effects at [θ] (λ, nm) = 1.90×10^5 (258), -1.17×10^5 (292), and 5.86×10^4 (365).

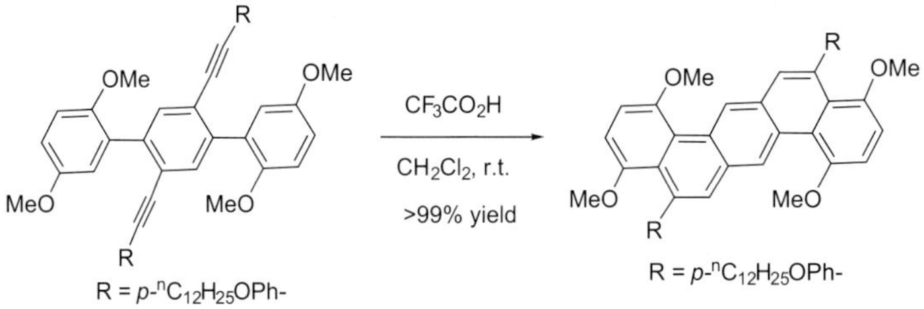

Scheme 11.30 Swager's acid-catalyzed cyclization of arylalkynes to provide polyaromatic compounds.

Polymer (*R*)-**84** was treated with CF_3CO_2H in methylene chloride by Swager's method to generate polymer (*R*)-**85**, through intramolecular cyclization of the electron-rich side-chain alkynes (Scheme 11.31).

NMR spectroscopic examination of polymer (*R*)-**85** showed that the CH_3OCH_2 protecting groups of (*R*)-**84** had been removed under the acidic conditions of this step, and the alkyne carbon signals of (*R*)-**84** had also disappeared in (*R*)-**85**. Polymer (*R*)-**85** was still soluble in common organic solvents, and GPC gave its molecular weight as Mw = 17,900 and Mn = 10,500 (PDI = 1.7). The difference in molecular weight between (*R*)-**84** and (*R*)-**85** is ascribed to their different struc-

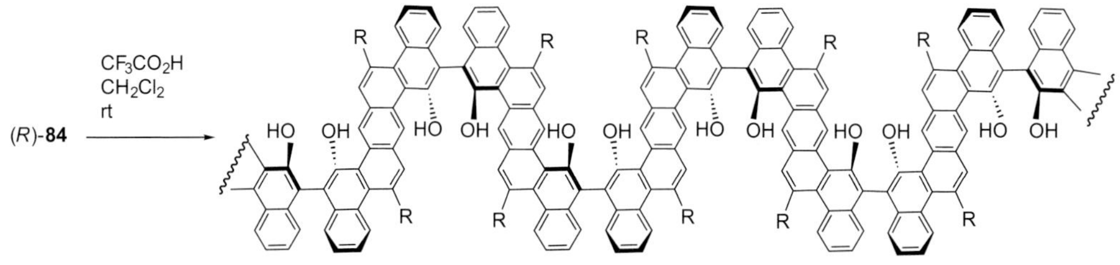

(*R*)-**85**: R = *p*-$^nC_6H_{13}OPh$-

Scheme 11.31 Synthesis of the minor-groove polybinaphthyl (*R*)-**84**.

tures relative to the polystyrene standards. Because of the dark brown color of this polymer, its optical rotation fluctuated at around −530 (c = 0.1, CH_2Cl_2). The UV spectrum of (*R*)-**85** showed large red shifts in absorptions relative to those of (*R*)-**84**, with λ_{max} (ϵ) = 344 (7.33×10^4) and 453 (1.26×10^4) nm. The fluorescence of (*R*)-**85** was observed at 521 nm, which was 123 nm red-shifted from that of (*R*)-**84**. The CD spectrum of (*R*)-**85** gave inverted signals at long wavelengths relative to those of (*R*)-**84**, with [θ] (λ, nm) = 1.23×10^5 (268), 7.49×10^4 (330), and -6.98×10^4 (360). All these data supported the formation of planar fused polyaromatic repeating units in (*R*)-**85**. The enantiomerically pure binaphthyl units of (*R*)-**85** and its fused aromatic rings would force the polymer chain to adopt a propagating helical chain structure as shown in Figure 11.4.

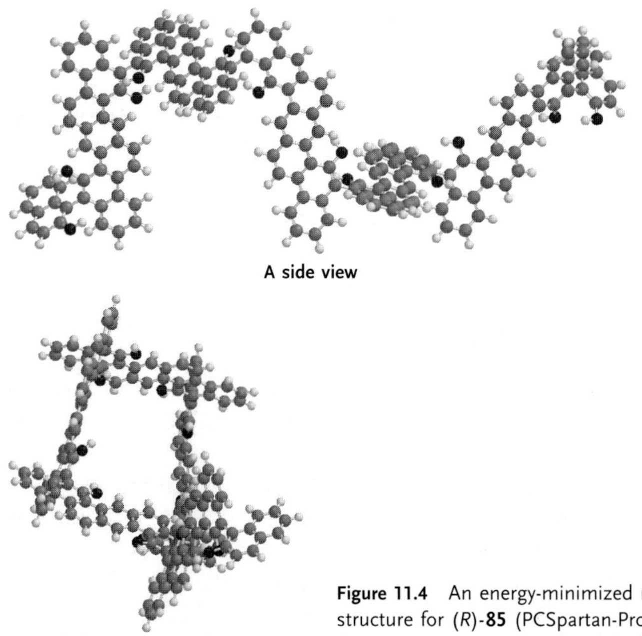

Figure 11.4 An energy-minimized molecular modeling structure for (*R*)-**85** (PCSpartan-Pro semiempirical AM1, the R groups are replaced with hydrogens).

The fluorescence of (*R*)-**85** was efficiently quenched in the presence of the amino alcohol (1*R*,2*S*)-(–)-*N*-methylephedrine (**86**). When 1.0 × 10^{-5} M solutions of (*R*)-**85** and (*R*)-BINOL in methylene chloride were treated with 3.0 × 10^{-2} M **86**, the fluorescence quenching I_0/I was 3.4 for (*R*)-**85** and 1.3 for (*R*)-BINOL, so the amino alcohol quenched the fluorescence of polymer (*R*)-**85** much more efficiently than that of (*R*)-BINOL. This could be ascribed to energy migration in the helical polymer (*R*)-**85**, since quenching at one binding site could be felt by other fluorescent units along the polymer chain. The lower π-π* band gap in (*R*)-**85** versus BINOL could also give rise to more efficient photoinduced electron transfer quenching by the amine electrons of **86**. The fluorescence quenching of (*R*)-**85** was enantioselective, with an observed *ef* (*ef*: enantiomeric fluorescence difference ratio = $|I_0-I_{R,S}|/|I_0-I_{S,R}|$) of 1.12. No enantioselectivity was observed for the fluorescence quenching of (*R*)-BINOL by **86**.

The polymer (*R*)-**88** (Scheme 11.32) was obtained by Suzuki coupling between (*R*)-**87** and **83** [25]. The molecular weight of (*R*)-**88** was Mw = 14,300 and Mn = 8700 (PDI = 1.64). Its specific optical rotation was $[\alpha]_D$ = –115.7 (c = 0.1, CH_2Cl_2. The UV spectrum of (*R*)-**88** in methylene chloride gave λ_{max} (ε) = 281 (5.53 × 10^4) and 320 (6.28 × 10^4) nm, and this polymer emitted at 408 nm. The CD spectrum of (*R*)-**88** gave strong signals at [θ] (λ, nm) = –1.64 × 10^5 (284), 8.07 × 10^3 (341), and –4.06 × 10^3 (359).

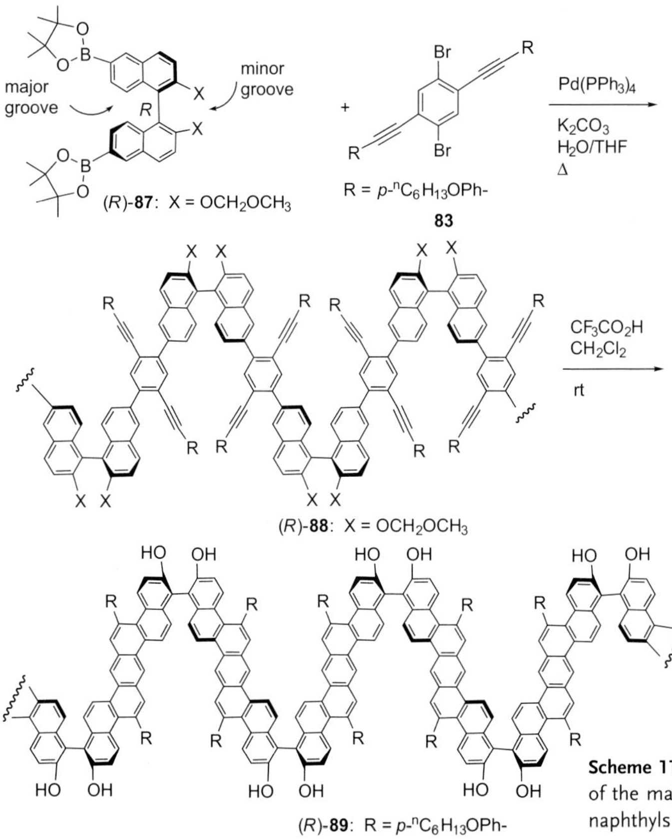

Scheme 11.32 Syntheses of the major-groove polybinaphthyls (*R*)-**88** and (*R*)-**89**.

Treatment of (*R*)-**88** with CF_3CO_2H in methylene chloride at room temperature gave polymer (*R*)-**89** (Scheme 11.32). It was speculated that the cyclization in the repeating units of (*R*)-**88** occurred at the α-positions of the naphthalene rings rather than at the β-positions because of the more reactive nature of the α-positions toward electrophilic substitutions. In the ^{13}C NMR spectra of (*R*)-**89**, both the alkyne and CH_3OCH_2 signals had disappeared. The molecular weight of (*R*)-**89** was Mw = 17,400 and Mn = 10,200 (PDI = 1.71). The dark brown color of (*R*)-**89** also made its optical rotation $[\alpha]_D$ fluctuate around –710 (c = 0.1, CH_2Cl_2). The UV spectrum of (*R*)-**89** showed large red shifts from the absorptions of (*R*)-**88**, with λ_{max} = 339 (6.06×10^4) and 411 (1.03×10^4) nm, and its emission maxima were observed at 450 and 476 (sh) nm, shifted greatly to the red in comparison with (*R*)-**88**. The CD spectrum of (*R*)-**89** was also very different from that of (*R*)-**88**, with [θ] (λ, nm) = -3.54×10^3 (273), 1.62×10^5 (320), and -2.86×10^5 (351). The long-wavelength Cotton effects were two orders of magnitudes greater than those of (*R*)-**88**, suggesting a more restricted conformation for the more conjugated repeating units of polymer (*R*)-**89**.

Monomer (*R*)-**90**, containing a bridging methylene group, was polymerized with **83** to give polymer (*R*)-**91**, which had a molecular weight of Mw = 38,800 and Mn = 12,100 (PDI = 3.22) (Scheme 11.33). Its specific optical rotation was $[\alpha]_D$ = −466.7 (c = 0.1, CH_2Cl_2), much greater than that of (*R*)-**88** and attributed to the bridging methylene groups in the repeating units of (*R*)-**91**, restricting the rotation of the binaphthyl units around the 1,1′-bond and greatly reducing the conformational flexibility.

Treatment of (*R*)-**91** with CF_3CO_2H gave polymer (*R*)-**92** (Scheme 11.33). The NMR spectra of (*R*)-**92** showed that the methylene groups of the polymer were intact after the acid treatment, unlike in the conversion of (*R*)-**88** to (*R*)-**89** in which deprotection had taken place. GPC of (*R*)-**92** gave Mw = 29,800 and Mn = 11,400 (PDI = 2.61).

The UV, fluorescence, and CD spectroscopic data of polymers (*R*)-**84**, (*R*)-**85**, (*R*)-**88**, (*R*)-**89**, (*R*)-**91**, and (*R*)-**92** are summarized in Table 11.12. Large red shifts in the absorption and emission values were observed after the transformation from (*R*)-**91** into (*R*)-**92**. The CD spectrum of (*R*)-**92** also gave inverted signals in relation to those of (*R*)-**91** at long wavelengths, similar to those observed in the conversion of other side-chain phenyleneethynylene polymers into the helical polybinaphthyls.

Table 11.12 Spectral data for the polymers (*R*)-**84**, (*R*)-**85**, (*R*)-**88**, (*R*)-**89**, (*R*)-**91**, and (*R*)-**92** in methylene chloride solution.

Polymer	*(R)-84*	*(R)-85*	*(R)-88*	*(R)-89*	*(R)-91*	*(R)-92*
UV λ_{max} (ε) nm	341 (4.86×10^4) 366 (sh, 3.58×10^4)	344 (7.33×10^4) 453 (1.26×10^4)	281 (5.53×10^4) 320 (6.28×10^4)	339 (6.06×10^4) 411 (1.03×10^4)	268 (5.28×10^4) 315 (7.23×10^4) 372 (sh, 3.47×10^4)	330 (7.14×10^4) 410 (1.28×10^4)
Fluorescence λ_{emi} (nm)	398	521	408	450 476 (sh)	408	455 484 (sh)
CD [θ] (λ, nm, 1.0×10^{-5} M)	1.90×10^5 (258) -1.17×10^5 (292) 5.86×10^4 (365)	1.23×10^5 (268) 7.49×10^4 (330) -6.98×10^4 (360)	-1.64×10^5 (284) 8.07×10^3 (341) -4.06×10^3 (359)	-3.54×10^3 (273) 1.62×10^5 (320) -2.86×10^5 (351)	3.90×10^4 (252) -1.75×10^5 (289) 2.80×10^4 (337)	1.56×10^5 (241) -8.19×10^4 (277) -4.76×10^4 (303) 6.05×10^4 (324) -2.49×10^5 (350)

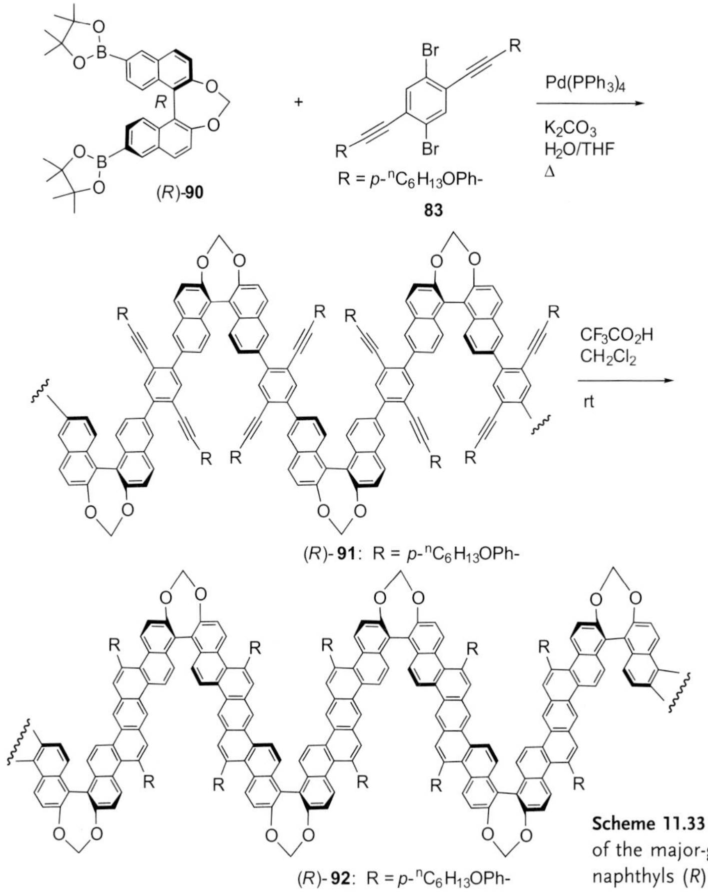

Scheme 11.33 Syntheses of the major-groove polybinaphthyls (*R*)-**91** and (*R*)-**92**.

11.4 Summary

A variety of chiral acetylenic dendrimers and polymers have been synthesized by utilization of the chirality of 1,1′-binaphthyls, the main synthetic method used to construct these chiral acetylenic macromolecules being Sonogashira coupling between terminal alkynes and aryl bromides or iodides. Various sets of reaction conditions for the polymer synthesis have been explored, and chiral macrocycles can also be obtained. Since these materials can be made soluble in organic solvents through the introduction of various alkyl substituents, a number of spectroscopic methods have been used for their characterization. Preliminary explorations of applications of these dendrimers and polymers as asymmetric catalysts, enantioselective fluorescent sensors, and nonlinear optical materials have been conducted. The light-harvesting properties of the dendrimers have produced large signal amplification in enantioselective fluorescent sensing.

Acknowledgements

Support for our work had been provided by the National Science Foundation, the Air Force Office of Scientific Research, the Office of Naval Research, the National Institute of Health, and the Petroleum Research Fund Administered by the American Chemical Society. I am most grateful for my post-doctoral and graduate students, whose names are cited in the references, for their excellent contributions.

11.5 Experimental Procedures

11.5.1 Preparation of the Chiral Dendrimers – A Typical Procedure

THF (10 mL) and Et_3N (15 mL) were added to a mixture of (*S*)- or (*R*)-**7** (0.3 mmol) and the alkyne dendron **3**, **4**, or **5** (1.56 mmol, 1.3 equiv. per bromo unit). After this solution had been degassed by bubbling with N_2 for 30 min, $Pd(PPh_3)_4$ (50 mg) and CuI (15 mg) were added. The mixture was heated at 70 °C for 24 h and then allowed to cool to room temperature. Hexanes were used for extraction, and the organic layer was washed with HCl (1 N), brine, saturated $NaHCO_3$, and brine. Evaporation of the solvent gave a yellow solid, which was subjected directly to hydrolysis with a mixture of KOH (1.12 g), THF (50 mL), MeOH (20 mL), and H_2O (10 mL) at room temperature for 12 h. Hexanes were added, and the mixture was neutralized with HCl (1 N). The organic layer was then washed with brine, saturated $NaHCO_3$, and brine. Evaporation of the solvent gave a yellow solid. Flash chromatography on silica gel (eluent: hexanes/EtOAc = 100:1.25 to 100:5) gave the desired dendrimers in 82–90 % yields as pale yellow solids.

11.5.2 Preparation of the Chiral Polymer (*R*)-18e

A mixture of (*R*)-**23** (210 mg, 0.25 mmol), 1,4-diiodobenzene (83 mg, 0.25 mmol), triethylamine (2 mL), and toluene (8 mL) in a 50 mL flame-dried Schlenk flask was degassed by bubbling with N_2 for 30 min. The flask was then placed in a dry-box, and tetrakis(triphenylphosphine)palladium(0) (14.4 mg, 0.013 mmol) and cuprous iodide (2.4 mg, 0.013 mmol) were added. After the reaction mixture had been heated under reflux under nitrogen for 48 h, it was filtered at room temperature to remove triethylammonium bromide precipitate. The salt was rinsed with diethyl ether, and the combined filtrates were evaporated to dryness. The residue was dissolved in a minimum amount of methylene chloride, which was then poured into methanol (75 mL) to precipitate out the polymer. This process was repeated once more. The resulting polymer was separated by centrifugation and dried under vacuum to give (*R*)-**18e** as a yellow solid in 96 % yield.

11.5.3
Preparation of the Chiral Polymer (*R*)-45

A 50 mL flame-dried Schlenk flask was loaded under nitrogen with (*R*)-**23** (210 mg, 0.25 mmol), **44** (82 mg, 0.25 mmol), triethylamine (1 mL), and THF (4 mL). The resulting solution was degassed for by bubbling with N_2 30 min, and tetrakis-(triphenylphosphine)palladium(0) (14.4 mg, 0.013 mmol) and cuprous iodide (2.4 mg, 0.013 mmol) were then added in a dry-box. After this reaction mixture had been stirred at room temperature for 2 d, it was filtered to remove the insoluble triethylamine hydrobromide salt. The salt was rinsed with diethyl ether until the filtrate was clear. The combined filtrate was evaporated to dryness to give a brown residue, which was dissolved in a minimum amount of CH_2Cl_2 and precipitated twice with methanol (75 mL). After filtration and drying under vacuum, (*R*)-**45** was obtained as an orange solid in 83 % yield (208 mg).

11.5.4
Preparation of the Helical Polymer (*R*)-85

Preparation of polymer (*R*)-**84**. In a dry-box, THF (6.0 mL) and degassed aqueous K_2CO_3 (2.0 mL, 1.0 M, 2.0 mmol) were added to a mixture of (*R*)-**82** (188 mg, 0.3 mmol), **83** (191 mg, 0.30 mmol), and $Pd(PPh_3)_4$ (21 mg, 0.018 mmol). The reaction mixture was heated at reflux for under N_2 90 h. The solvent was then removed under vacuum, and the residue was redissolved in CH_2Cl_2 (30 mL). The solution was washed with H_2O (3 × 30 mL) and brine, and dried over Na_2SO_4. After filtration and removal of solvent, the sticky residue was dissolved in a minimum amount of CH_2Cl_2. The polymer was precipitated out with the addition of hexane. This procedure was repeated twice more to afford polymer (*R*)-**84** as an orange solid in 91 % yield (230 mg).

Conversion of (*R*)-**84** to (*R*)-**85**. Trifluoroacetic acid (0.69 mL, 8.9 mmol) was added under nitrogen at room temperature to a solution of polymer (*R*)-**84** (150 mg, 0.18 mmol) in CH_2Cl_2 (20 mL). The color of the solution changed from yellow to dark green within 1 h. After having been stirred overnight, the solution was washed with 10 % $NaHCO_3$ (3 × 20 mL) and H_2O (2 × 20 mL), and dried over Na_2SO_4. After filtration and removal of solvent, the sticky residue was dissolved in a minimum amount of CH_2Cl_2. The polymer was precipitated out with the addition of hexane. This procedure was repeated twice more to afford polymer (*R*)-**85** as a dark brown solid in 95 % yield.

References

[1] R. Giesa, *Rev. Macromol. Chem. Phys.* **1996**, *C36*, 631–670.

[2] U. H. F. Bunz, *Chem. Rev.* **2000**, *100*, 1605–1644.

[3] C. Devadoss, P. Bharathi, J. S. Moore, *J. Am. Chem. Soc.* **1996**, *118*, 9635–9644.

[4] Z. Peng, Y. Pan, B. Xu, J. Zhang, *J. Am. Chem. Soc.* **2000**, *122*, 6619–6623.

[5] (a) J. S. Schumm, D. L. Pearson, J. M. Tour, *Angew. Chem., Int. Ed. Engl.* **1994**, *33*, 1360–1363; *Angew. Chem.* **1994**, *106*, 1445–1448;
(b) S. L. Huang, J. M. Tour, *J. Am. Chem. Soc.* **1999**, *121*, 4908–4909.

[6] (a) K. Sanechika, T. Yamamoto, A. Yamamoto, *Bull. Chem. Soc. Jpn.* **1984**, *57*, 752–755;
(b) M. Takagi, K. Kizu, Y. Miyazaki, T. Maruyama, K. Kubota, T. Yamamoto, *Chem. Lett.* **1993**, 913–916;
(c) C. Weder, M. S. Wrighton, *Macromolecules* **1996**, *29*, 5157–5165;
(d) T. Mangel, A. Eberhardt, U. Scherf, U. H. F. Bunz, K. Müllen, *Macromol. Rapid Commun.* **1995**, *16*, 571–580;
(e) M. Moroni, J. Le Moigne, S. Luzzati, *Macromolecules* **1994**, *27*, 562–571.

[7] (a) R. H. Grubbs, D. Kratz, *Chem. Ber.* **1993**, *126*, 149–157;
(b) R. Giesa, R. C. Schulz, *Makromol. Chem.* **1990**, *191*, 857–867.

[8] Q. Zhou, T. M. Swager, *J. Am. Chem. Soc.* **1995**, *117*, 12593–12602.

[9] Other researchers have also synthesized and studied chiral acetylenic polymers: (a) J. C. Nelson, J. G. Saven, J. S. Moore, P. G. Wolynes, *Science* **1997**, *277*, 1793–1796;
(b) R. B. Prince, T. Okada, J. S. Moore, *Angew. Chem., Int. Ed. Engl.* **1999**, *38*, 233–236; *Angew. Chem.* **1999**, *111*, 245–249;
(c) M. S. Gin, T. Yokozawa, R. B. Prince, J. S. Moore, *J. Am. Chem. Soc.* **1999**, *121*, 2643–2644;
(d) M. S. Gin, J. S. Moore, *Org. Lett.* **2000**, *2*, 135–138;
(e) C.-J. Li, W. T. Slaven, V. T. John, S. Banerjee, *J. Chem. Soc., Chem. Commun.* **1997**, 1569–1570;
(f) C.-J. Li, D. Wang, W. T. Slaven, *Tetrahedron Lett.* **1996**, *37*, 4459–4462.

[10] Q.-S. Hu, V. Pugh, M. Sabat, L. Pu, *J. Org. Chem.* **1999**, *64*, 7528–7536.

[11] L. Pu, H.-B. Yu, *Chem. Rev.* **2001**, *101*, 757–824.

[12] (a) M. Mori, T. Nakai, *Tetrahedron Lett.* **1997**, *38*, 6233–6236;
(b) F.-Y. Zhang, C.-W. Yip, R. Cao, A. S. C. Chan, *Tetrahedron: Asymmetry* **1997**, *8*, 585–589;
(c) F.-Y. Zhang, A. S. C. Chan, *Tetrahedron: Asymmetry* **1997**, *8*, 3651–3655.

[13] V. J. Pugh, Q.-S. Hu, L. Pu, *Angew. Chem. Int. Ed.* **2000**, *39*, 3638–3641; *Angew. Chem.* **2000**, *112*, 3784–3787

[14] V. J. Pugh, Q.-S. Hu, X. Zuo, F. D. Lewis, L. Pu, *J. Org. Chem.* **2001**, *66*, 6136–6140.

[15] L. Pu, *Acta Polym.* **1997**, *48*, 116–141.

[16] L. Ma, Q.-S. Hu, K. Y. Musick, D. Vitharana, C. Wu, C. M. S. Kwan, L. Pu, *Macromolecules* **1996**, *29*, 5083–5090.

[17] L. Ma, Q.-S. Hu, L. Pu, *Tetrahedron: Asymmetry* **1996**, *7*, 3103–3106.

[18] H. Cheng, L. Ma, Q.-S. Hu, X.-F. Zheng, J. Anderson, L. Pu, *Tetrahedron: Asymmetry* **1996**, *7*, 3083–3086.

[19] (a) L. Ma, Q.-S. Hu, D. Vitharana, C. Wu, C. M. S. Kwan, L. Pu, *Macromolecules* **1997**, *30*, 204–218; (b) L. Ma, Q.-S. Hu, D. Vitharana, L. Pu, *Polym. Prepr.* **1996**, *37(2)*, 462–463.

[20] Bergman cyclization references: (a) R. G. Bergman, *Acc. Chem. Res.* **1973**, *6*, 25–31; (b) K. N. Bharucha, R. M. Marsh, R. E. Minto, R. G. Bergman, *J. Am. Chem. Soc.* **1992**, *114*, 3120–3121; (c) J. W. Grissom, T. L. Calkins, *J. Org. Chem.* **1993**, *58*, 5422–5427.

[21] L. Ma, L. Pu, *Macromol. Chem. Phys.* **1998**, *199*, 2395–2401.

[22] H. Cheng, L. Pu, *Macromol. Chem. Phys.* **1999**, *200*, 1274–1283.

[23] (a) S. V. Elshocht, T. Verbiest, M. Kauranen, L. Ma, H. Cheng, K. Y. Musick, L. Pu, A. Persoons, *Chem. Phy. Lett.* **1999**, *309*, 315–320; (b) A. Persoons, M. Kauranen, S. Van Elshocht, T. Verbiest, L. Ma, L. Pu, B. M. W. Langeveld-Voss, E. W. Meijer, *Mol. Cryst. Liq. Cryst.* **1998**, *315*, 93–98.

[24] H.-C. Zhang, L. Pu, *Tetrahedron* **2003**, *59*, 1703–1709.

[25] H.-C. Zhang, L. Pu, *Macromolecules* **2004**, *37*, 2695–2702.

[26] M. B. Goldfinger, K. B. Crawford, T. M. Swager, *J. Am. Chem. Soc.* **1997**, *119*, 4578–4593.

Index

Acetylene Chemistry. Chemistry, Biology and Material Science. Edited by F. Diederich, P. J. Stang, R. R. Tykwinski

ISBN 3-527-30781-8

D

E

M

N